MAGNETIC CIRCUITS AND TRANSFORMERS

PRINCIPLES OF
ELECTRICAL ENGINEERING
SERIES

Magnetic Circuits and Transformers

A FIRST COURSE
FOR POWER AND COMMUNICATION
ENGINEERS

By Members of the Staff of the
Department of Electrical Engineering
Massachusetts Institute of Technology

THE M.I.T. PRESS
MASSACHUSETTS INSTITUTE OF TECHNOLOGY
CAMBRIDGE, MASSACHUSETTS

ISBN: 0-262-63063-X (Paperback)

Foreword

The staff of the Department of Electrical Engineering at the Massachusetts Institute of Technology has for some years been engaged in an extensive program of revising as a unit substantially its entire presentation of the basic technological principles of electrical engineering. This volume is one of a series covering this revised presentation.

The decision to undertake so comprehensive a plan rather than to add here and patch there came from the belief that the Department's large staff, with its varied interests in teaching and related research, could effect a new synthesis of educational material in the field of electrical engineering and evolve a set of textbooks with a breadth of view not easily approached by an author working individually.

Such a comprehensive series, it was felt, should be free from the duplications, repetitions, and unbalances so often present in unintegrated series of textbooks. It should possess a unity and a breadth arising from the organization of a subject as a whole. It should appeal to the student of ordinary preparation and also provide a depth and rigor challenging to the exceptional student and acceptable to the advanced scholar. It should comprise a basic course adequate for all students of electrical engineering regardless of their ultimate specialty. Restricted to material which is of fundamental importance to all branches of electrical engineering, the course should lead naturally into any one branch.

This book and the reorganized program of teaching out of which it has grown are thus products of a major research project to improve educational methods. The fact that this book and its companions are the result of collaborative endeavor involving substantially the entire staff of the Institute's Department of Electrical Engineering and that this collaboration has been carried out with sustained enthusiasm is striking testimony to each contributor's devotion to this broad objective of improved instruction. To those of us who have observed the progress of the work from the outside, this demonstration of unity and *esprit de corps* on the part of a large staff is perhaps the most impressive aspect of the entire effort.

KARL T. COMPTON

Preface

This book is the second volume in the Principles of Electrical Engineering series described in the Foreword, and extends into another field the circuit theory begun in the first volume. Computation of magnetic-circuit performance, principles associated with the concepts of interlinked electric and magnetic circuits, and their application to the analysis of transformers are the essential subject matter.

Like the other volumes in this series, this book is intended for a first basic course. Emphasis, hence, again is placed on fundamental principles important to students of electrical engineering regardless of what their special fields ultimately may be; both power and communication problems are considered. Rigor of thought and analysis rather than extensiveness of scope is likewise the intended feature of this book.

The treatment assumes that the reader has knowledge of electric-circuit theory as given in the first volume of the series or in other textbooks on a similar level. Mathematics through differential equations is freely used; not, however, as a substitute for the explanation of physical phenomena but rather as a means of describing quantitatively their consequences. Parallel to the mathematical analysis, explanations are given in terms of physical interpretations.

The text is divided into two parts. Part I, " Magnetic Circuits," starts with a discussion of the current theory of ferromagnetism, derives the magnetic-circuit concept, and continues with development of the fundamental principles for computation of the behavior of magnetic circuits. A chapter is devoted to discussion of iron-core reactors by means of model theory — a method of analysis of great power whenever nonlinear phenomena are involved. A brief summary of thermal-circuit problems is also included. Part II, " Transformers," begins with a short discussion pointing out that the theories from which the characteristics of all electric apparatus can be computed are developed through the combination of the fundamental concepts of the electric, dielectric, magnetic, thermal, and mechanical circuits contained in them, and that costs and other practical matters are of paramount importance in their influence on the development of electric apparatus. Part II then continues with a study of the applications of these general principles to transformers. The theory of transformers is developed both from the resolution of their magnetic fields into leakage and resultant mutual components and from the classical theory of coupled circuits, the interrelations between the two methods of analysis being emphasized.

Although the text is by no means exhaustive in any one field, nor is it intended to be a design book, an attempt has been made to discuss a variety of representative practical problems in both the heavy-current power and the light-current control, measurement, and communication applications of magnetic materials and transformers. As in the other volumes of the series, more material than is usually covered in a first course is presented, to provide for additional study by particularly apt or advanced students, and to increase the usefulness of the text as a reference book.

CAMBRIDGE, MASSACHUSETTS
January, 1943

Contents

PART I. MAGNETIC CIRCUITS

CHAPTER I

CHAPTER II

CHAPTER III

CHAPTER IV

PERMANENT MAGNETS AND ENERGY IN THE MAGNETIC FIELD 98

CHAPTER V

LOSSES IN MAGNETIC CORES CONTAINING TIME-VARYING FLUXES 124

CHAPTER VI

ALTERNATING-CURRENT EXCITATION CHARACTERISTICS OF IRON-CORE REAC-
TORS AND TRANSFORMERS 156

CHAPTER X

CHAPTER XI

CHAPTER XII

CHAPTER XIII

CHAPTER XXI

CHAPTER XXII

CHAPTER XXIII

CHAPTER XXIV

Table of Symbols

For space vectors, a bold-face italic or script letter is used to represent the vector, and an ordinary italic or script letter is used to represent merely the magnitude of the vector, for example: $\mathcal{B}$, $\mathcal{B}$, or $\mathcal{E}$, $\mathcal{E}$. For complex quantities, a Roman letter is used to represent the complex quantity, and an italic letter is used to represent merely the magnitude of the complex quantity, for example, E, E. Ordinary italic or script letters represent ordinary scalars, for example: t, $\mathcal{R}$. For voltage, current, charge, and time, capital letters usually represent fixed quantities, and lower-case letters usually represent variable quantities, for example: E, e. To some extent this system of capital and lower-case letters is carried out for other quantities; thus, lower-case letters are used in most instances to represent nonlinear parameters, for example: g_c, b_m. It should be recognized that the number of symbols required in a treatment of this sort is so large that it is neither feasible to be entirely consistent, nor possible to avoid duplications.

The symbols used for quantities are very often given special meanings by the addition of subscripts or superscripts. In such cases, the discussion in the text defines the meaning intended, and no attempt is made to include all these special designations in the list of symbols. However, certain subscripts appear in many places with the same meaning, as for example, I_φ; such symbols are listed in this table. Usually numerical subscripts are used to designate the winding of a transformer to which the symbol applies, for example: I_1, V_2. In the chapters dealing with polyphase connections of transformers, capital letter subscripts are used to designate the primary phases and lower-case letter subscripts the secondary phases, for example: V_{AB}, V_{ab}. Superscript primes and double primes are used for two purposes. In the treatment of a single transformer primes and double primes are used for indication of the winding to which a quantity is referred; for example, I'_L is the vector load current referred to the primary. In the treatment of groups of transformers operating in parallel or connected for phase transformation, however, the primes and double primes indicate the transformer of the group to which the symbol applies; for example, I''_1 is the vector current in the primary of the double-prime transformer of a parallel group of single-phase transformers.

Abbreviations used in the text are, in general, in accordance with the *American Tentative Standard Abbreviations for Scientific and Engineering Terms*, approved by the American Standards Association in 1932, wherein such abbreviations are listed.

English Letter Symbols

	Symbol		*Description*	*Defined or First Used Page*
A			Complex multiplying factor	457
	A		Area	
	A		Real part of a complex number	346
	a		Ratio of primary to secondary turns	339
	a		Ratio of turns in two similar reactors	216
	a		Number of current paths through armature	89
	a		Dimension of air gap	69
	a		Coefficient in Fourier series	173
	a		Core-loss component of unit cost	424
	B		Susceptance	
	B		Imaginary part of complex number	346
	$\mathcal{B}$	$\mathcal{B}$	Magnetic flux density	
	$\mathcal{B}$		Effective value of flux density	215
	$\mathcal{B}_r$		Residual flux density	21
	b		Dimension of air gap	69
	b		Coefficient in Fourier series	173
	b		Copper-loss component of unit cost	424
	b_m		Magnetizing susceptance	195
	b_{oc}		Open-circuit susceptance	352
	$\mathscr{b}$		Instantaneous flux density	136
	C		Capacitance	
	C		Carter's coefficient	91
	C_s		Cost of standard transformer	427
	c		Coefficient in Fourier series	174
	c		Cost of transformer for maximum economy	427
D	D		Determinant	
	D		Width of duct	93
	D		Diameter	
	D		Cost of demand per kva	423
	d		Tooth pitch	90
	d		Armature diameter	93
	d		Half-thickness of lamination	135
	E		Constant electromotive force	
E	E		Effective value of electromotive force	
	E		Cost of energy per kwhr	423
	$\mathcal{E}$	$\mathcal{E}$	Electric field intensity	
	e		Instantaneous electromotive force	
	e		Thermal emissivity coefficient	246
	F		Magnetomotive force	
	F		Effective value of magnetomotive force	215

Symbol		Description	Defined or First Used Page
	T	Time constant	213
	t	Time	
	t	Tooth width	90
	U	Difference of magnetic potential	50
	V	Constant voltage; difference of potential	
V	V	Effective value of alternating voltage	
	$\mathcal{V}$	Volume	
	v	Instantaneous voltage	
v	v	Velocity	5
	W	Energy	
	w	Energy per unit volume	116
	X	Reactance	
	X_{eq}	Equivalent reactance	349
	X_ℓ	Leakage reactance	333
	X_m	Mutual reactance	437
	X_{oc}	Open-circuit reactance	444
	X_{sc}	Short-circuit reactance	354
	x	Distance	
Y	Y	Admittance	
Y_{jj}	Y_{jj}	Self-admittance, node j	628
Y_{jk}	Y_{jk}	Mutual admittance, nodes j and k	628
$Y_{\overline{jk}}$	$Y_{\overline{jk}}$	Admittance of link connecting terminals j and k . .	634
Y_{oc}	Y_{oc}	Open-circuit admittance	365
Y_φ	Y_φ	Exciting admittance	215
	y	Distance	
y_{jj}	y_{jj}	Short-circuit driving-point admittance	627
y_{jk}	y_{jk}	Short-circuit transfer admittance	627
Z	Z	Impedance	
	Z	Number of armature conductors	88
Z_{eq}	Z_{eq}	Equivalent impedance	349
Z_{jj}	Z_{jj}	Self-impedance of winding j	625
Z_{jk}	Z_{jk}	Mutual impedance between windings j and k . . .	625
$Z_{(jk)}$	$Z_{(jk)}$	Short-circuit impedance of windings j and k acting as a two-circuit transformer	640
Z_{oc}	Z_{oc}	Open-circuit impedance	365
Z_{sc}	Z_{sc}	Short-circuit impedance	354
Z_φ	Z_φ	Exciting impedance	215
Z_{11}	Z_{11}	Self-impedance, loop 1	437
Z_{12}	Z_{12}	Mutual impedance, loops 1 and 2	437

GREEK LETTER SYMBOLS

α Alpha   Angular displacement

Symbol	Description	Defined or First Used Page
β Beta	Intrinsic flux density	21
Γ Gamma	Reciprocal inductance	
Δ Delta	Temperature difference	247
δ	Air-gap length	69
δ	Fractional change in inductance	230
ε Epsilon	Naperian base of logarithms (2.71828...)	
η Eta	Hysteresis coefficient	129
θ Theta	Power-factor angle	
θ	Angle between two vectors	
θ	Temperature	244
λ Lambda	Instantaneous flux linkages	159
λ	Effective value of flux linkages	215
λ	Annual rms load ratio	414
μ Mu	Permeability	
μ	Amplification factor of a vacuum tube	272
μ_{ac}	Incremental permeability	198
μ_i	Initial permeability	28
μ_0	Permeability of free space	
π Pi	Ratio of circumference to diameter of circle (3.14159...)	
ρ Rho	Volume resistivity	
σ Sigma	Leakage coefficient	439
σ	Factor used in determination of equivalent air-gap length	90
τ Tau	Lamination thickness	133
φ Phi	Magnetic flux	
φ	Effective value of flux	215
Φ	Vector representing alternating flux	
φ	Instantaneous flux	
φ_l	Instantaneous leakage flux	314
ψ Psi	Angular displacement	
ω Omega	Angular frequency	

OTHER SYMBOLS

$\bar{A}$	Conjugate of complex number A
~	Cycles per second
∠	At an angle of
⦠	The angle measured from A to B
≡	Identically equal to; equal by definition
≈	Approximately equal to
∝	Varies as
▶◀	Marks used for emphasis

Part I

MAGNETIC CIRCUITS

CHAPTER I

Properties of Ferromagnetic Materials

Sometimes the best way to realize the value of any object is to try to imagine what would happen if it were taken away. If the iron and steel in dynamo-electric machinery, power and audio-frequency transformers, telephone receivers, relays, loudspeakers, and hundreds of other electromagnetic devices should suddenly lose their magnetic properties, these devices could no longer function properly. It is hard to imagine how without these materials new designs could be made within a reasonable range of space and cost. Obviously, any substance which permits a large flux density for a specified magnetizing force or which makes possible the constraint of flux to definite desired paths is bound to be of inestimable value to the designer. These properties are found in certain forms of iron and its alloys with cobalt, tungsten, nickel, aluminum, and other metals, which are called the *ferromagnetic materials*. Easy to magnetize, such substances when used for the cores of apparatus make possible flux densities which are hundreds or even thousands of times greater than could be conveniently established with a practical coil without the ferromagnetic core.

1. IMPORTANCE OF FERROMAGNETIC MATERIALS

The commercial importance of ferromagnetic materials is indicated by the thousands of tons produced annually, comprising a wide range of physical and magnetic properties. The materials are available commercially in numerous forms such as sheets a few thousandths up to about one-fourth inch thick, as wires a few thousandths up to about one-fourth inch in diameter, as bars of various cross-sectional shapes, and as castings weighing a few ounces up to several hundred tons for use in machines or other devices.

The fields of application of these materials, which are discussed briefly in Art. 6, are so broad and the requirements of each application are so different that engineers, physicists, and metallurgists have had to develop a wide range of ferromagnetic alloys, each having particular qualities valuable in specific applications. The value of these accomplishments can hardly be overestimated, and many of the recent, perhaps startling, developments in the design-technique of electrical machinery have arisen because better materials have been made available by metallurgical research and because the engineer has utilized physical and magnetic properties to better advantage. To emphasize this point, it should be noted that the basic electromagnetic phenomena in a modern

3

165,000-kva generator are the same as in a 5-kw Edison bipolar generator of a little more than half a century ago. The increase in capacity and the efficient performance achieved in the large machines are due in no small measure to the improvements in the magnetic and other physical properties of steels, and to a better understanding of these properties by engineers.

Engineering utilization of electrical devices containing ferromagnetic materials necessitates quantitative description of the circuit parameters representing the electric-circuit behavior of these devices. In the presence of ferromagnetic materials, an electric-circuit element has a resistance parameter which depends not only upon the magnitude of the current in the circuit but also upon the way in which it varies. The inductance parameter is not only nonlinear but is not even a single-valued function of the current. If in addition it varies with time, the difficulty of describing it exactly can be readily appreciated.

Before any quantitative studies can be made, quantitative data concerning the properties of the ferromagnetic materials must be available, and the manner of using these data toward the desired end must be clearly understood. The major portion of this chapter is therefore devoted to a discussion of the properties of the ferromagnetic materials of chief interest to the electrical engineer. Numerous quantitative data are given, which form a basis for the solution of magnetic-circuit problems to be treated in later chapters. As an aid to a better visualization of the phenomena involved, a theory of magnetism is described in the next article.

2. THEORY OF MAGNETISM[1]

The effort to explain the phenomenon of magnetism and to increase the understanding of many of the nonlinear properties observed for magnetic materials has produced many attempts at a theory of magnetism. One of the earliest was made by Ampère, who suggested about 100 years ago that the magnetization of a substance results from the orientation of molecules containing circulating currents. While Ampère's idea was too vague to be useful except as a philosophical concept, it came surprisingly close to being an introduction to the present theory, which during the past few years has been developed to the stage where it can explain numerous observed effects qualitatively, and some even quantitatively. In spite of these advances, however, developments in the theory still lag far behind the recent improvements in the magnetic properties of the materials used in industry. These improvements have

---

[1] The subject matter of this article is taken largely from an excellent summarizing article by R. M. Bozorth, " Present Status of Ferromagnetic Theory," *A.I.E.E. Trans.*, *54* (1935), 1251–1261.

been achieved almost entirely as a result of empirical investigation, and only in comparatively recent times has the theory been developed to the stage where it can serve as a reliable guide to experimentation. Only the briefest qualitative statement of the present theory, which in its details includes many aspects of modern atomic and quantum theory, is given here.

Visualizing the magnetic behavior of actual samples of material in accordance with the present concepts of magnetism involves considering several different subdivisions of matter. The smallest particles concerned are the components of the atom; namely, the nucleus and its associated electrons. Next in size are the atoms. Larger than the atom are the domains, which are subcrystalline particles of varying shapes and sizes having a volume of about 10^{-9} cubic centimeter and containing about 10^{15} atoms. Widely varying numbers of domains are included in a crystal, and all but very special samples of magnetic materials contain many crystals.

2a. *Electrons and Nuclei.* — Magnetism is believed to be basically electrical. When an electric charge q moves with a velocity $\mathcal{v}$, a magnetic field is set up as indicated in Fig. 1 in which $\mathcal{v}$ shows the direction of motion and P indicates the point at which $\mathcal{H}$ is to be determined. If the charge is positive, and if the velocity vector and the point P are in the plane of

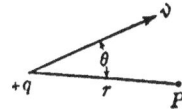

Fig. 1. Moving charge $+q$ produces a magnetic field in the surrounding space.

the paper, the direction of $\mathcal{H}$ is away from the reader perpendicular to the plane of the paper and its magnitude is

$$\mathcal{H} = \frac{q\mathcal{v}}{r^2} \sin \theta \qquad [1]$$

In the atom, electrons move in orbits about the heavier nucleus and at the same time each electron, as well as the nucleus, spins[2] about an axis of its own. A spinning electron or a spinning nucleus has a definite moment of momentum (angular momentum) and a definite magnetic moment. For the electron, these moments are in opposite directions and, for the nucleus, in the same direction. An electron in an atom has, in addition to its spin moments, a moment of momentum and a magnetic moment due to its motion in its orbit. The total magnetic moment of an atom is the vector sum of all its component magnetic moments.

2b. *The Atom.* — The latest atom model[3] as conceived by the physicists is pictured as a spinning nucleus composed of protons and neutrons surrounded by definite numbers of spinning electrons performing certain orbital motions. The simplest of the atoms is that of hydrogen having a

[2] K. K. Darrow, " Spinning Atoms and Spinning Electrons," *B.S.T.J.*, *16* (1937), 319–336.
[3] L. B. Loeb, *Atomic Structure* (New York: John Wiley & Sons, 1938).

proton for its nucleus and a single orbital electron. The atom as a whole has a magnetic moment composed of three parts — the magnetic moment of the positive charge spinning on its axis, that of the negative charge spinning on its axis, and that produced by the negative charge moving in its orbit. The magnetic moment associated with the orbital and spin motions of the electron are of an order of magnitude one thousand times the magnetic moment of the spinning proton. Hence the magnetic effect of the nucleus is difficult if not impossible to detect in experiments with hydrogen atoms unless the magnetic moments of the electron can be in some way neutralized. This neutralization actually takes place in the hydrogen *molecule* since, in the combination of two atoms, the electron spins oppose each other as do the orbital motions. As might be expected, the two proton spins also oppose each other in some of the molecules (parahydrogen), while in the others (orthohydrogen) the proton spins line up in the same direction. Under ordinary conditions, about three-fourths of the molecules are of the ortho type.

According to the above discussion, parahydrogen should not have any magnetic moment at all, but experiments disclose for it the presence of a magnetic moment of the same order of magnitude as that of ortho-hydrogen but somewhat smaller. The explanation is that the rotating charges precess about the direction of applied magnetic field. An analysis of this precession shows that, regardless of the direction of orbital motion or spin, the precessional motion of all charges is such as to set up a field in opposition to the applied field. In the parahydrogen molecule, then, the only resultant magnetic effect is this weakening of the field due to the precessional magnetic moments. This effect is known as *diamagnetism*. Orthohydrogen has, in addition to this effect, a nuclear spin moment of somewhat larger magnitude.

When a single electron has been removed from the hydrogen molecule, the result is a singly charged hydrogen ion. Here the neutralization of spin and orbital magnetic moments has been destroyed, and the ion tends to align its orbital and spin magnetic moments in the direction of the applied field. This strengthening of the field is known as *paramagnetism*. Both diamagnetic and paramagnetic substances lose their magnetic effects when the external field is removed: the diamagnetic because the precessions about a fixed direction cease when the field is removed; and the paramagnetic because, without an external field, the magnetic moments are so oriented, as specified by quantum considerations, that their sum is zero. The atoms of the noble gases are diamagnetic and so also are certain ions such as Na+, Mg+ +, and Cl−. The greater number of atoms and ions are paramagnetic.

The properties of the ferromagnetic substances are not determined entirely from their atoms, but the foregoing discussion indicates that the atom of chromium with six excess positive-spin electrons, the atom of iron with four, and the atom of cobalt with three should be good building blocks of which a highly magnetic solid might be built. The facts are, however, that when these atoms are assembled in crystalline structures, the spin orientations are such that chromium is nonferromagnetic, while iron, cobalt, nickel, and certain alloys are highly ferromagnetic. The determining feature[4] is the orientation of the atoms in the crystal lattice.

If this is such as to give a resultant magnetic moment, the material is ferromagnetic. If it is such that the vector sum of the magnetic moments is zero, the material is nonferromagnetic. The orientations in turn depend upon whether the lower potential energy of the crystalline assembly is achieved with the magnetic or the nonmagnetic orientation. Figure 2 shows a curve of the difference ΔW between the potential energy W_u of the unmagnetized arrangement and the potential energy W_m of the magnetized arrangement of atoms in the crystal versus the ratio of atomic separation r_a to the radius r_d of the orbit containing the excess positive-spin electrons. When ΔW is positive, the magnetized state is the stable one, because it is the state of lower potential energy. When ΔW is negative, the unmagnetized state is the

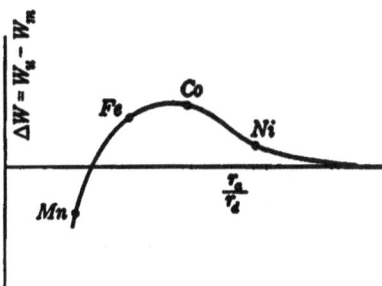

FIG. 2. For ferromagnetic substances, the potential energy of the crystal lattice is less when magnetized than when unmagnetized.

stable one. Thus iron, cobalt, and nickel are ferromagnetic but manganese is not.

2c. *The Domain.* — The parallel locking of atomic magnetic moments in the crystalline structure extends throughout a limited but somewhat indefinite volume of a ferromagnetic crystal. The reason for this limitation is not completely understood, but experimental evidences are cited in the next article to show that, even when an iron crystal as a whole is unmagnetized, tiny neighboring regions called *domains* are completely magnetized. The individual regions, however, have their magnetic moments in different directions and these moments add to zero over the whole crystal.

▶ Any substance which is made up of these spontaneously magnetized and saturated domains is said to possess *ferromagnetism.*◀

[4] W. Shockley, " The Quantum Physics of Solids — I," *B.S.T.J.*, *18* (1939), 645–723.

In the crystal structure within a domain, the atoms are arranged in definite orderly fashion. In the iron crystal,[5] for example, the atoms are at the corners of a cube with one at the center as shown in Fig. 3. This arrangement is called a *body-centered cubic lattice.* The grouping in a nickel crystal differs from this by having an atom in the center of each

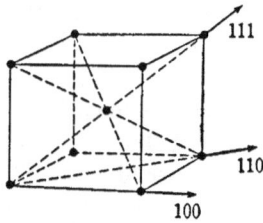

FIG. 3.   Body-centered crystal lattice of iron.

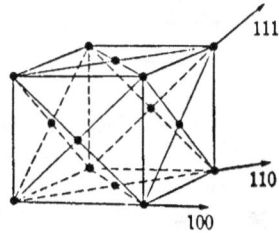

FIG. 4.   Face-centered crystal lattice of nickel.

face but none at the center of the cube as shown in Fig. 4. This is called a *face-centered cubic lattice.* A domain in an iron crystal, in the absence of an external magnetizing force, has its atomic magnetic moments all lined up in a single direction, the direction of one of the edges of the cubic lattice. This is called a 100 direction or axis. A diagonal of a face represents a 110 direction. A diagonal of the cube represents a 111 direction and is the one taken by the atomic magnetic moments of the nickel domain.

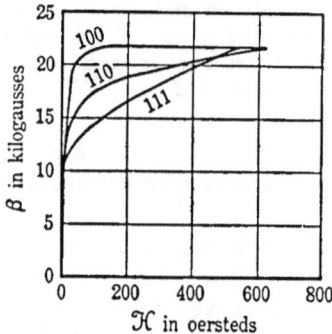

FIG. 5.   Magnetization curves for an iron crystal in the directions 100, 110, and 111.   (*From footnote reference 1.*)

2d. *The Crystal.* — For study of the behavior of a single crystal, specimens have been prepared which are large enough to be tested in the laboratory for their magnetic properties. When external fields are applied to an iron crystal along the 100 axis, the intrinsic* flux density β reaches a steady value at a small value of the external field $\mathcal{H}$, as shown in Fig. 5. This is explained by supposing that in the unmagnetized state, the domains have a random orientation in the six preferred directions in the crystal, as indicated in Fig. 6a. When the magnetizing force is applied along one of these directions, the domains having magnetic moments in the other five directions shift their moments into the direction of the

[5] R. M. Bozorth, " The Physical Basis of Ferromagnetism," *B.S.T.J.*, *19* (1940), 1–39.
* Intrinsic flux density is defined on p. 21.

applied field, as indicated in Fig. 6b. This 100 axis is called the direction of easy magnetization. If the field is applied along a face diagonal 110, four of the six magnetic moments first shift to the two directions having

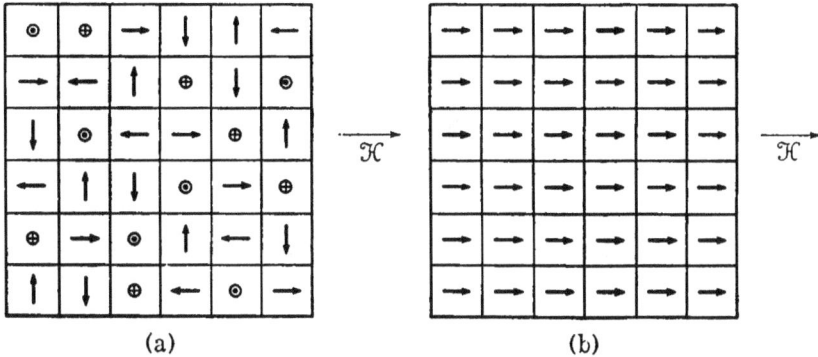

FIG. 6. (a) Domains have magnetic moments in all six 100 directions in an unmagnetized iron crystal; (b) in the same 100 direction when the crystal is fully magnetized in a 100 direction.

components along 110 and then, with increasing field, the moments of all domains are gradually shifted to the field direction. The 110 direction is for an iron crystal the direction of medium magnetization. Finally, if the field is applied along a cube diagonal 111, three of the magnetic moments shift to the three directions having components along 111 and then a further increase in the field brings all the moments into the 111 direction, known as the direction of hard magnetization. The corresponding curves for nickel are shown in Fig. 7.

The presence of domains is indicated in a striking way by the photographs produced by W. C. Elmore and shown in Fig. 8. The three parts represent three exposures of the same portion of the surface of a cobalt crystal under three different conditions of external field. In Fig. 8a, the crystal is magnetized by the application of an external field; in Fig. 8b, the crystal has been demagnetized and in Fig. 8c the direction of the field has been reversed. The patterns are believed to be produced by the action of the magnetized domains upon a colloidal suspension of iron oxide which is placed on the surface of the crystal and viewed under a microscope. Such arrange-

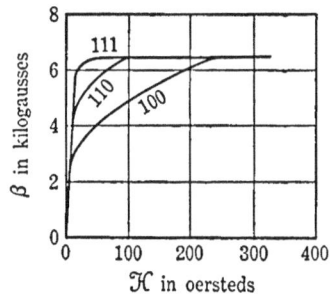

FIG. 7. Magnetization curves for a nickel crystal in the directions 100, 110, and 111. (*From footnote reference 1.*)

ments of the iron particles never take place on nonferromagnetic substances, but are produced by the spontaneously magnetized domains at the surface of an iron or a cobalt crystal even though the crystal as a whole is unmagnetized.

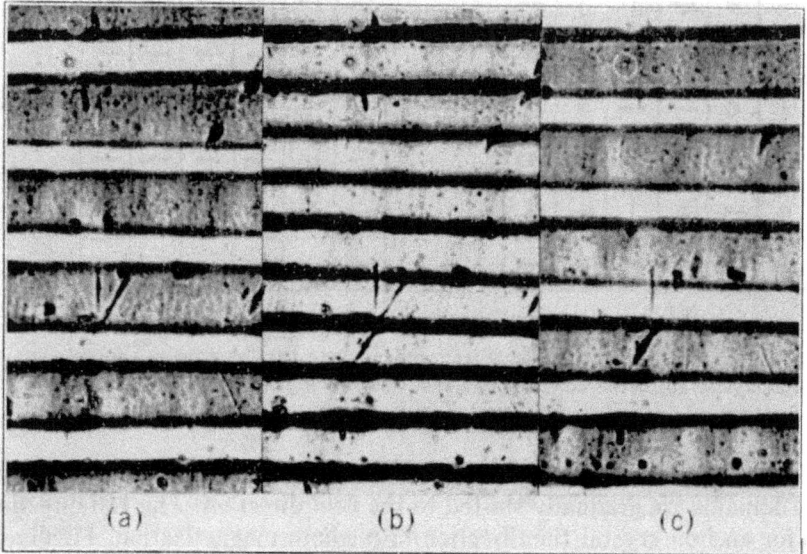

Fig. 8. Powder patterns on a plane surface parallel to the direction of easy magnetization of a cobalt crystal (a) magnetized toward reader, (b) demagnetized, (c) magnetized away from reader.

2e. *Polycrystalline Substances.* — A polycrystalline sample of iron or of a magnetic alloy is composed of many crystals packed rigidly together but with no definite directional alignment of their axes. When no external field is applied, each domain in such materials is spontaneously and completely magnetized in one of its 100 directions. The magnetic moments of the domains in each crystal are distributed equally in the six directions of easy magnetization, with the result that each crystal is in the unmagnetized state. The random alignment of crystals produces a random distribution of magnetic moments throughout the material.

Imposing an external magnetic field is believed to result first in enlarging all domains having magnetic moments in the general direction of the applied field by reducing the size of adjacent domains whose magnetic moments are in less favorable directions. The effect of the supposed enlargement of domains on the magnetic state of the material as observed externally is very small, and predominates only for extremely small applied fields.

As the applied field is increased beyond the extremely small values, a second effect becomes noticeable, in which the magnetic moments of individual domains are aligned in the direction of the preferred crystal axis nearest to that of the impressed field. This effect takes place, not by realignment of the domain as a rigid body, but by realignment of the axes of the spins of the individual electrons within the domain from one stable direction to another stable direction. Throughout any one domain, this realignment of spins takes place simultaneously. As the field con-

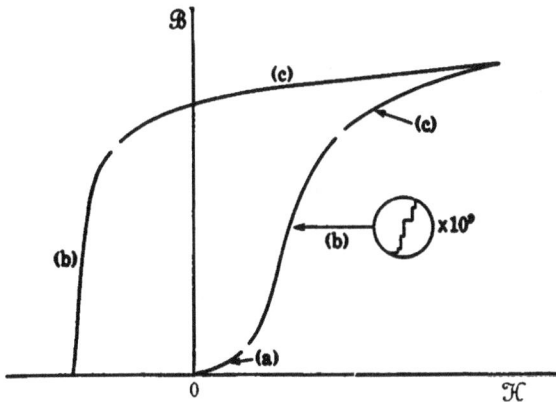

FIG. 9. The three regions of magnetization: (a) boundary displacement, (b) sudden change in orientation, and (c) slow change in orientation. (*From footnote reference 1.*)

tinues to be increased from the very small values, the flux density $\mathfrak{B}$ is observed experimentally to increase by finite jumps, each increase corresponding to the realignment of spins in one domain. This stepwise buildup, known as the *Barkhausen effect*, is illustrated by the magnified portion of the curve of Fig. 9. Of course, these steps do not occur in any curve actually plotted from experimental data, but the fact that the change in magnetization occurs in jumps can be observed by means of a loudspeaker connected to a coil surrounding the specimen, as shown in Fig. 10. Each reorientation of a domain suddenly changes the flux through the pickup coil and causes a click in the loudspeaker.

Because of the great number of crystal edges not parallel to the direction of the applied field, a third effect becomes apparent with externally applied magnetizing forces greater than those required to align the domain moments along the crystal edges most nearly parallel to the applied field vector. This last mode of magnetization consists of a smooth orientation of all domain moments from the position in line with the crystal edges to the position in line with the direction of the applied field. No sharp demarcation exists between regions (b) and (c) of the curve but the sudden jumps become less frequent as saturation is approached.

The above phenomena can be summarized briefly as follows: When a gradually increasing, but small, magnetizing force is applied to a mass of initially unmagnetized iron, the first mode of magnetization that predominates is caused by the slight growth in the size of domains magnetized in the direction most nearly parallel to the applied field, at the expense of the size of neighboring domains. A small amount of magnetization builds up, and the increase takes place smoothly. The portion of the magnetization curve for which this phenomenon occurs is marked (a) in Fig. 9. As the magnetizing force is increased further, a second mode of

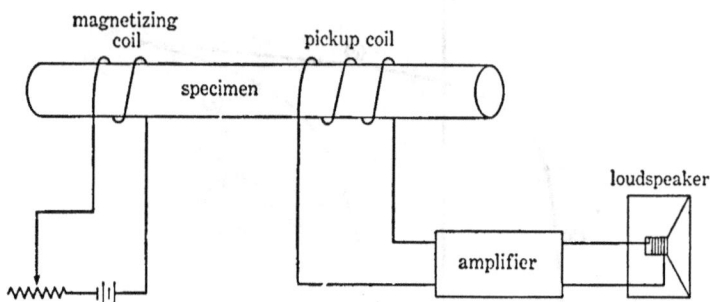

Fig. 10.   Circuit connections for demonstration of Barkhausen effect.

magnetization predominates, which is caused by the sudden alignments of groups of spins along the crystal axes most nearly parallel to the impressed field. The region for which this phenomenon principally occurs is marked (b) in Fig. 9. For values of magnetizing force larger than those for which a pronounced Barkhausen effect occurs, the final mode of magnetization predominates. This mode is described as a slow rotation or alignment of the domains into the direction of the field and away from the crystal axes nearest in line with the applied field. The region for which this mode principally occurs is marked (c) in Fig. 9. The first and last modes result in a much smaller change in flux density for a given change in magnetizing force than does the intermediate mode. During the last mode, the iron is becoming magnetically saturated, and, when all the domain moments are aligned in the direction of the applied field, the material is fully magnetized.

Ferromagnetic materials can be magnetized up to intrinsic flux densities ranging from 5,000 to 25,000 gausses, have relative permeabilities of the order of hundreds or thousands, exhibit a pronounced saturation effect, and retain some of their magnetism when the field is removed. Paramagnetic substances have very small positive relative permeabilities and lose all their magnetism when the external field is removed. Diamagnetic substances also respond weakly to magnetization, but their relative permeabilities are less than unity.

Just as the crystal of chromium is nonmagnetic* even though its atom has six uncompensated positive spins, the crystal lattice of iron alloyed with certain other elements may not have pronounced ferromagnetic properties. Alloys of this kind frequently encountered are certain manganese and stainless steels. Specifically, an alloy comprising about 0.3 per cent carbon, 10 per cent manganese, 6.5 per cent nickel, and the remainder iron is essentially nonmagnetic at room temperature. Curiously, however, certain alloys called Heusler alloys, formed from essentially nonmagnetic materials, are observed to exhibit strong magnetic properties. The most highly magnetic of these contains approximately 65 per cent copper, 20 per cent manganese, and 15 per cent aluminum. The best specimens can be magnetized about as easily as low-grade iron. Although they have not found any extensive commercial utilization at the present time, they have properties of marked interest in the development of theories of magnetism.

The preceding remarks concerning the phenomena of magnetization apply chiefly to single ferromagnetic crystals and isotropic assemblies of these crystals in an unstrained condition. A sample of iron comprising many crystals oriented wholly at random and free from strains has properties that are an average or composite of magnetization in all directions of a single crystal. Frequently, however, the polycrystalline ferromagnetic materials used in engineering are not strain free, nor are they isotropic; that is, their crystal axes are not always distributed equally in every direction. Rolled sheet steel is a good illustration of this condition and frequently exhibits widely different magnetic properties for different directions of magnetization.†

The subject of strains is closely connected with certain aspects of ferromagnetic theory. For instance, the magnetic forces in a crystal occurring as a result of the spins of the electrons are balanced by the electric forces occurring as a result of the electric charges on the electrons. When a rearrangement of the magnetic vectors occurs, the balance between the magnetic and electric forces is disturbed. Consequently, the physical dimensions of the material are observed to vary. This phenomenon is one of many effects which have to do with strains, mechanical or magnetic, and are referred to collectively as *magnetostriction*. The magnetostrictive properties of iron are opposite to those of nickel. The effect of mechanical tension on iron is to increase the magnetization for a specified magnetizing force, whereas the effect of mechanical compression is to decrease it. With nickel, the effect of tension is to decrease the magnetization for a specified magnetizing force, whereas the effect of compression is to increase it. With iron, the length of the material

* See p. 7.
† See footnote 1, Ch. III.

increases as the magnetization is increased (positive magnetostriction), whereas with nickel the length decreases (negative magnetostriction). The limiting values of magnetostriction occur simultaneously with the magnetic saturation of the material. The curve of Fig. 11 shows the saturation magnetostriction properties of the iron-nickel alloys commonly known as Permalloys.

Strains are produced in a material in several ways, such as by cold working, by the presence of impurities in the material which give rise to crystal lattice distortion (chemical strain), or by magnetostriction

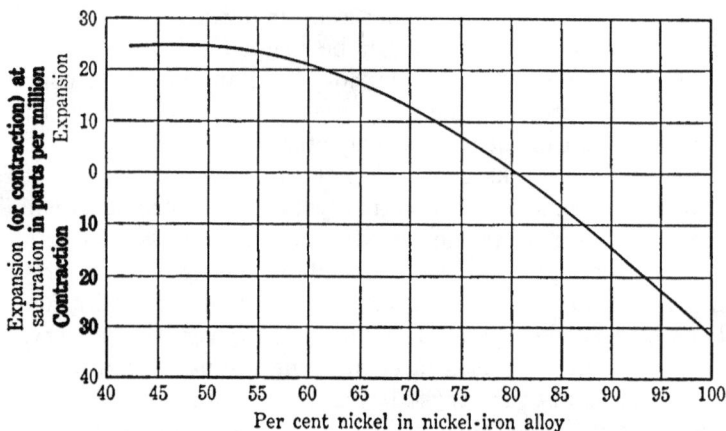

Fig. 11.   Saturation magnetostriction of the nickel iron alloys.   (*From footnote reference 1.*)

(latent or residual strain). Where permanence of magnetization is desired, high internal strain is beneficial, because the larger the strain the more stable is the direction of the magnetization of the domain. This condition is well borne out in permanent magnets, which frequently exhibit magnetic hardness simultaneously with the physical hardness that is typical of a metal with large internal strain.

On the other hand, when ease of magnetization is desired — that is, when a material is to have a large permeability — negligible internal strain or magnetostrictive effect is desirable. This condition is well illustrated by the data of Fig. 11, since the nickel-iron alloys which show small magnetostriction exhibit large permeabilities. Actually, the largest relative permeability thus far obtained, about 600,000, occurred when an alloy of 65 per cent nickel and 35 per cent iron was heated first in hydrogen for 18 hours at about 1,400 degrees centigrade to remove the nonmetallic impurities and to relieve the chemical strain and next at about 650 degrees for one hour, and then was cooled in hydrogen under the influence of a field intensity of about 16 oersteds. This treatment appears to have oriented the directions of the magnetostrictive strains (latent or residual strain) so as to give the easy axes of magnetization in the direction of the applied field.

3. General properties of magnetic materials

At present, the scope of the theory of magnetism is insufficient to allow prediction of the magnetic properties of materials on purely theoretical considerations, even when the composition of the material is accurately known. This insufficiency of the theory is not a very serious disadvantage, for even were such a prediction possible the procedure involved in the prediction would probably not be sufficiently simple to find extensive use. The customary procedure for determining the properties of magnetic alloys is to make measurements of each property on samples of each kind of material manufactured. Frequently, a few measurements are made at the mill on samples drawn from each batch as manufactured. The data so obtained are then used to prepare characteristic curves for the particular material. Finally when the material is fabricated into various forms for inclusion as constituent elements of a particular device, the performance of the device, or its parameters as a circuit element, can usually be computed from these characteristic curves with an accuracy sufficient for most engineering purposes.

Ferromagnetic materials are characterized by one or more of the following attributes:

(a) **They can be magnetized much more easily than other materials. This characteristic is indicated by a large relative permeability** μ/μ_0.

(b) They have a high maximum intrinsic flux density β_{max}.

(c) They are magnetized with widely different degrees of ease for different values of magnetizing force. This attribute leads to a nonlinear relation between flux density $\mathfrak{B}$ and magnetizing force $\mathfrak{K}$.

(d) An increase in magnetizing force produces in them a change in flux different from the change produced by an equal decrease in magnetizing force. This attribute indicates that the relationships expressing the flux density and the permeability μ as functions of magnetizing force are nonlinear and multivalued.

(e) They retain magnetization when the magnetizing force is removed.

(f) They tend to oppose a reversal of magnetization after once being magnetized.

The degree to which certain of these characteristics are important in particular applications of magnetic materials depends upon the circumstances encountered in each application.

Of the materials available, iron finds the most extensive use. Its permeability is large and its cost per pound is least of all the ferromagnetic materials available. In its commercially pure form, it is used frequently in the structures of numerous machines. It is used also as the base element for practically all the ferromagnetic alloys. Probably the alloy produced in the largest quantity is that composed of essentially pure iron and between 1 and 4 per cent of silicon, depending upon the purpose for which the material is required. When this alloy is given a particular heat treatment, a material is obtained which, compared with iron, has better magnetic properties at low values of magnetizing force and larger resistivity. As shown later, both these properties are desirable. This alloy is rolled into sheets and strip, principally in the thickness range 0.014 to 0.025 inch, and annealed; it is designated in the trade literature as silicon steel sheets and strip. The sheet form is convenient for punching into many shapes used in the construction of electromagnetic apparatus. Typical shapes of punchings are shown in Fig. 12.

The silicon sheet steels used in the electrical industry are known among steel manufacturers[6,7] and electrical designers by certain descriptive names.

Field grade sheets contain about one-fourth of 1 per cent silicon and have a resistivity of about 16 microhm-centimeters. This grade is used for small low-priced motors.

Armature grade sheets contain about one-half of 1 per cent silicon and have a resistivity of about 19 microhm-centimeters. This grade is relatively soft and for this reason easy to punch. It is used in small motor and generator field poles, armatures, and other devices in which high flux densities are required but core losses are not of great importance.

Electrical grade sheets contain about 1 per cent silicon and have a resistivity of about 26 microhm-centimeters. This grade is widely used in commercial motors and generators of small and moderate sizes and medium efficiencies, and in transformers, relays, and other devices designed for intermittent operation.

Motor grade sheets contain about 2.5 per cent silicon and have a resistivity of approximately 42 microhm-centimeters. This material is used in medium-size motors and generators of good efficiencies, in control apparatus, and in inexpensive radio transformers.

Dynamo grade sheets contain about 3.5 per cent silicon and have a resistivity of about 50 microhm-centimeters. This grade is used in high-

[6] Carnegie-Illinois Steel Corporation, Pittsburgh, Pa., *Electrical Steel Sheets, Technical Bulletin No. 2* (1941).

[7] Allegheny Steel Co., Brackenridge, Pa., *Magnetic Core Materials Practice* (1937).

efficiency motors and generators, small power-distribution transformers, and radio transformers.

Several *transformer grades* are available, the principal ones being designated 72 (radio C), 65 (radio B), 58 (radio A), and 52. The num-

Courtesy Carnegie-Illinois Steel Corp.

FIG. 12.   Typical lamination punchings.

bers are the core losses of 29-gauge sheets in hundredths of watts per pound at 60 cycles per second as found by the standard Epstein test described in Ch. V. The silicon content increases as the losses decrease. These sheets are used primarily for power and radio transformers, and for large high-efficiency alternators, motors, and synchronous condensers.

Because of the multitude of ferromagnetic alloys available at the present time, to mention the properties of more than a few of those frequently encountered would be impractical. Certain properties of a selected group of the more common alloys, and some of their fields of usefulness are discussed in Art. 6; but, before these matters are considered, the properties of the magnetization curves of typical ferromagnetic materials when subjected to a direct magnetizing force will be investigated more fully.

4. CHARACTERISTICS AND MAGNETIZATION CURVES OF MAGNETIC MATERIALS

The relation between the magnetizing force $\mathcal{H}$ and the flux density or magnetic induction $\mathcal{B}$ which it produces in a ferromagnetic material is of considerable importance in the engineering uses of the material. It is most conveniently expressed by means of characteristic curves.[8,9] For purposes of explanation, a ferromagnetic material is considered to be placed in a region where the magnetic field intensity can be varied. A possible arrangement[10] consists of a toroidal ring of the material on which is wound a coil of wire as in Fig. 13a. The magnetizing force is varied through changing the current in the coil. The material is originally demagnetized. If the flux density $\mathcal{B}$ is measured by an appropriate method, as a function of the magnetizing force $\mathcal{H}$ for values of $\mathcal{H}$ up to a maximum, say $+\mathcal{H}_{max}$, and the relation is plotted, a curve similar to *oab* of Fig. 13a is obtained. This curve is sometimes referred to as the *rising* magnetization curve. If $\mathcal{H}$ is now decreased, a different relation between $\mathcal{B}$ and $\mathcal{H}$ is found, as is typified by the curve *bc*, which lies above the rising curve. A flux density, given by *oc* on the plot, remains when $\mathcal{H}$ is made zero. This flux density is called the *remanence* or *remanent magnetism*. To reduce the flux density to zero, a magnetizing force *od* must be applied in the direction opposite to that of the force formerly applied. For certain conditions of magnetization, as discussed below, this magnetizing force is called the *coercive force*.

As $\mathcal{H}$ is made more negative until it reaches the value $-\mathcal{H}_{max}$, the relation between $\mathcal{B}$ and $\mathcal{H}$ follows the curve *db'*. Then if $\mathcal{H}$ is increased from $-\mathcal{H}_{max}$ through zero to the value $+\mathcal{H}_{max}$, the curve follows a path such as *b'c'd'e*. The point *e* differs from the point *b* by a small amount, and the path does not yet form a closed loop. If $\mathcal{H}$ is varied through

[8] F. Bitter, *Introduction to Ferromagnetism* (New York: McGraw-Hill Book Co., Inc., 1937).

[9] Carnegie-Illinois Steel Corporation, Pittsburgh, Pa., *Electrical Steel Sheets, Technical Bulletin No. 2* (1941).

[10] Thomas Spooner, *Properties and Testing of Magnetic Materials* (New York, McGraw-Hill Book Co., Inc., 1927), Chs. xv and xvii.

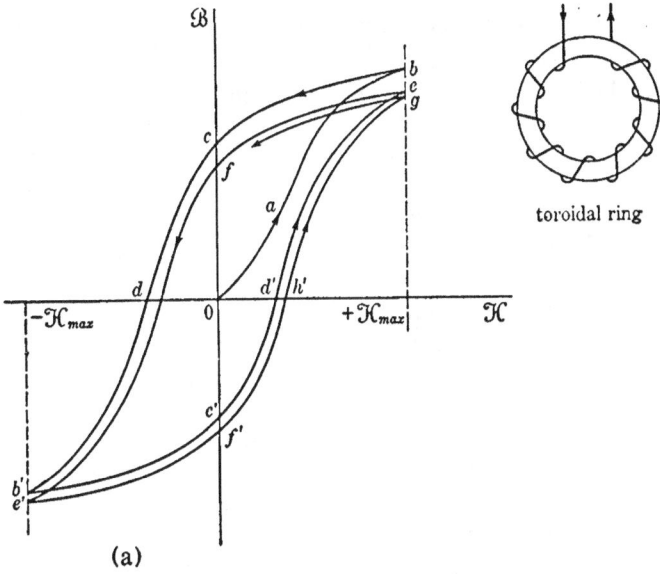

(a)

toroidal ring

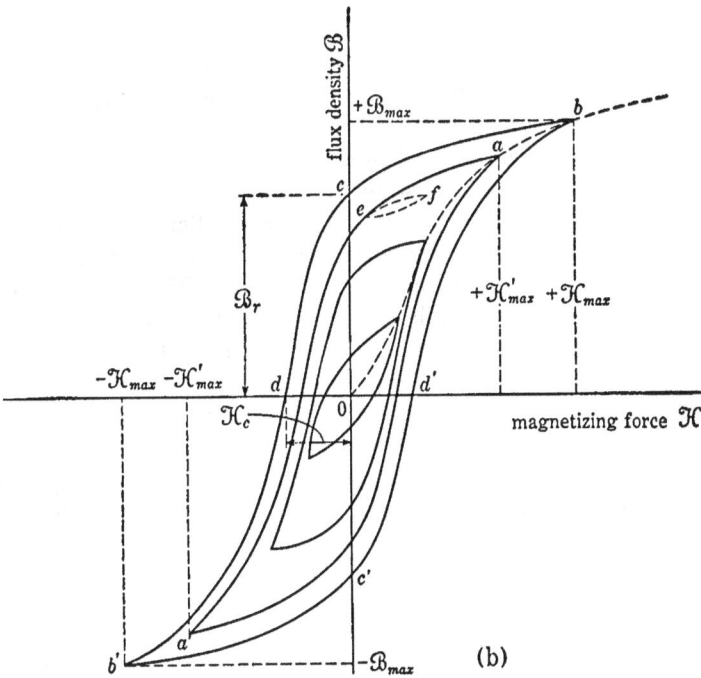

(b)

FIG. 13. (a) $\mathcal{B}(\mathcal{H})$ curves from initially unmagnetized sample. (b) Typical hysteresis loops.

another cycle between the same limits of $\mathcal{K}$, the relation between $\mathcal{B}$ and $\mathcal{K}$ follows the path $efe'f'h'$ to g. As the variation in $\mathcal{K}$ is carried through additional identical cycles, the path gradually approaches a fixed curve. Finally, after many cycles the path becomes a closed loop, as indicated by the loops of Fig. 13b. If the positive and the negative values of $\mathcal{K}_{max}$ are equal, the loop is symmetrical about the origin. The steel is then in its symmetrically cyclically magnetized condition, sometimes abbreviated to *cyclic condition*, for the particular numerical value $\mathcal{K}_{max}$.

Mention has already been made that the values of $\mathcal{B}$ on the falling curve are greater than those on the rising curve. Thus the material has the property of tending to oppose a change in the value of the flux density. This property is known as *hysteresis*, which is a term meaning *a lagging behind*. The closed loop obtained when the magnetizing force is taken through a complete cycle of values is known as a *hysteresis loop*. Although the word hysteresis implies a time lag, the hysteresis phenomenon does not depend on time but only on whether the magnetizing force is decreasing or increasing. If at any instant the magnetizing force is raised to a new value, and the magnetic material is not jarred, the flux density apparently never increases above or settles below its new initial value. The hysteresis phenomenon results in a dissipation of energy, called *hysteresis loss*, within the material when cyclic variations of magnetizing force are considered. The distinction between hysteresis loss and the above hysteresis phenomenon is explained in Ch. V.

If the magnetic material is subject to a cycle involving smaller values of $\mathcal{K}$ and $\mathcal{B}$, as from $+\mathcal{K}'_{max}$ to $-\mathcal{K}'_{max}$, a smaller hysteresis loop $aa'a$, Fig. 13b, is obtained. If the magnetizing force is not varied continuously in one direction between the maximum values of $\mathcal{K}$, small internal loops are introduced. If, for example, after descending from point a, Fig. 13b, to e at which $\mathcal{K}$ is equal to $\mathcal{K}_e$, $\mathcal{K}$ is increased to $\mathcal{K}_f$ and then brought back to $\mathcal{K}_e$, a loop ef is introduced into the curve.

When the properties of different magnetic materials are compared, some of the properties are denoted by special terms. Those of major interest in this part of the treatment are (a) remanence, (b) residual flux density or residual induction, (c) retentivity, (d) coercive force, and (e) coercivity. The usually accepted definitions[11] of these terms are as follows:

(a) *Remanence* is the flux density, or magnetic induction, which remains in a magnetic material after the removal of an applied magnetizing force.

[11] These definitions are in substantial agreement with the definitions of the American Society for Testing Materials standards A 34.

(b) *Residual flux density*, or *residual induction* $\mathcal{B}_r$, in a magnetic material is the value of the flux density for the condition of zero magnetizing force, when the material is being symmetrically cyclically magnetized. It is distinguished from *remanence* by the symmetrically cyclic requirement.

(c) *Retentivity* is the flux density that remains in the material after a magnetizing force sufficient to cause saturation flux density or saturation induction has been removed.

(d) *Coercive force* $\mathcal{H}_c$ for a magnetic material is the magnitude of the magnetizing force at which the flux density is zero when the material is being symmetrically cyclically magnetized.

(e) *Coercivity* is the coercive force required to reduce the flux density in the material to zero from a condition corresponding to saturation flux density or saturation induction.

The foregoing discussion shows that the flux density associated with a given magnetizing force is not single valued. It can have any value between certain limits, depending on the history of the material. In many magnetic problems, the history of the material is unknown. Existing magnetization may then have occurred by an increase of the magnetizing force from a small value up to the value under consideration, or by a decrease from a larger value. Many magnetic calculations are therefore performed using a magnetization curve, called *the normal magnetization curve*, which is obtained by drawing a single-valued curve through the tips of a series of increasingly larger symmetrical hysteresis loops. Such a curve is shown as *oab* in Fig. 13b.

Figure 14a shows a typical hysteresis loop for 24-gauge electrical-grade sheet steel. The co-ordinates are in aemu; that is, the magnetizing force is given in gilberts per centimeter, or oersteds, and the flux density is in flux lines per square centimeter, or gausses. The curve of Fig. 14a shows the relationship between $\mathcal{B}$ and $\mathcal{H}$ [herein abbreviated to $\mathcal{B}(\mathcal{H})$] for a maximum induction of 10,000 gausses, which is the flux density at which the hysteresis loops, residual induction $\mathcal{B}_r$, and coercive force $\mathcal{H}_c$, of electrical sheet steels are compared according to standards of the American Society for Testing Materials. From Fig. 14a it is seen that the value of $\mathcal{B}_{r10}$ for electrical-grade sheet steel is observed to be 8,100 gausses, and the value of $\mathcal{H}_{c10}$ is 0.7 oersted. In Fig. 14b is shown a typical normal magnetization curve of the same grade of electrical sheet steel as represented by the data of Fig. 14a. The co-ordinates of this curve are also expressed in aemu. This grade of material does not become saturated until $\mathcal{B}$ approaches 20,500 gausses as shown by the curve of *intrinsic flux density* β. The intrinsic flux density β is $\mathcal{B}$ minus $\mu_0\mathcal{H}$, and is a measure of that part of the flux density attributable to the ferromagnetic

characteristic of the material. When the material becomes saturated, it
can make no further contribution to the flux density; hence as $\mathcal{H}$ is

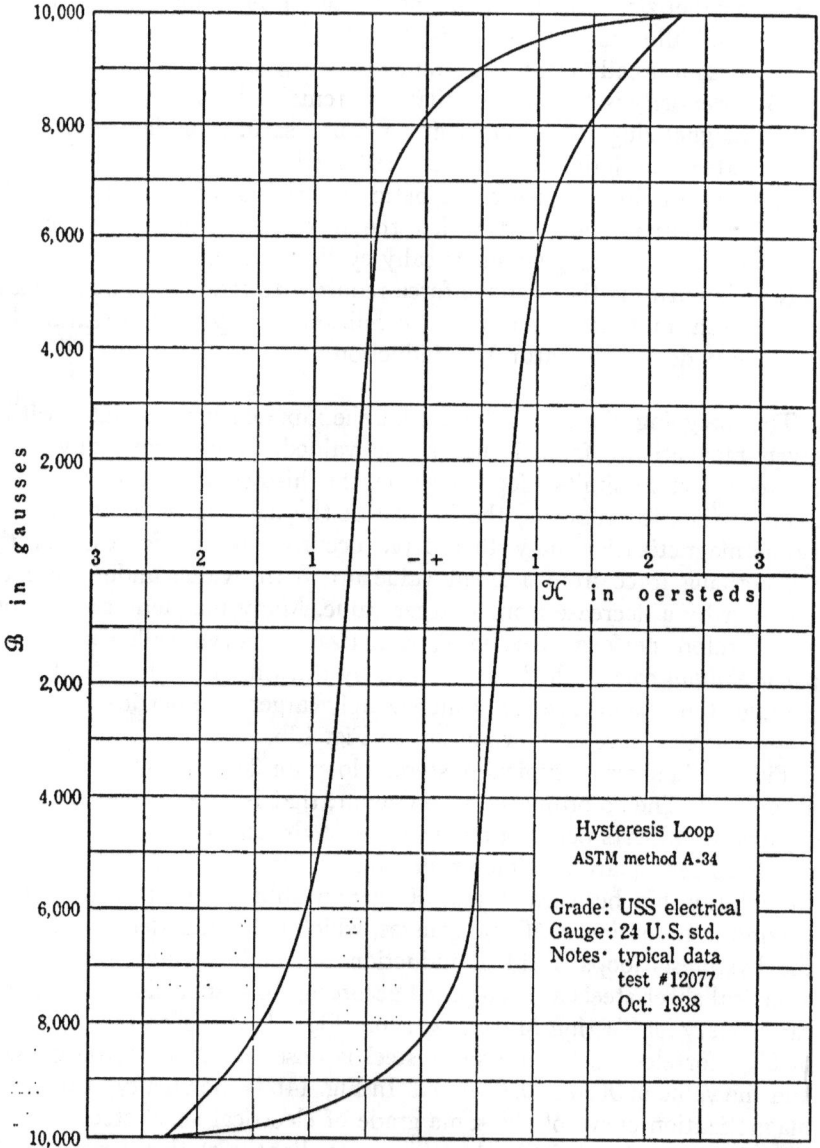

FIG. 14a.   Hysteresis loop.

increased beyond the values which saturate the material, the slope of
the $\mathcal{B}(\mathcal{H})$ curve becomes μ_0, the permeability of free space, and β be-

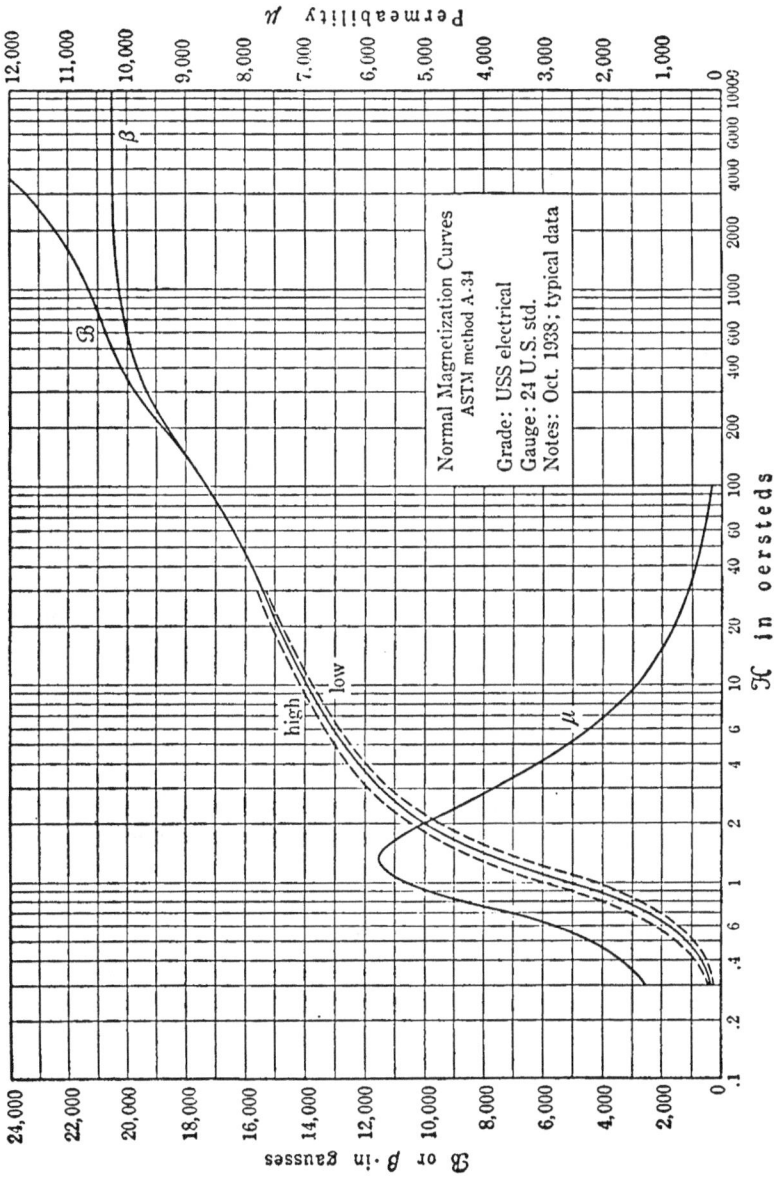

Normal Magnetization Curves
ASTM method A-34

Grade: USS electrical
Gauge: 24 U.S. std.
Notes: Oct. 1938; typical data

Courtesy Carnegie-Illinois Steel Corp.

FIG. 14b. Curves of flux density and permeability.

comes constant. With most ferromagnetic materials, the difference between *normal flux density* $\mathcal{B}$ and *intrinsic flux density* β is negligible until saturation conditions are approached.

The *static permeability* μ of the material is defined by the relation

$$\mu = \frac{\mathcal{B}}{\mathcal{H}} , \qquad\qquad [2]$$

in which $\mathcal{B}$ is the total flux density and $\mathcal{H}$ is the magnetizing force, $\mathcal{B}$ and $\mathcal{H}$ being related by the normal magnetization curve. In other words, the static permeability (as distinguished from the dynamic or incremental permeability defined in Ch. VI) at any point on the magnetization curve is the slope (gausses/oersteds) of a line drawn through the origin to the point on the normal magnetization curve. In Fig. 14b, a curve of static permeability is shown as a function of the magnetizing force $\mathcal{H}$. Static permeability and its reciprocal, called *reluctivity*, are sometimes useful quantities, but, for most practical engineering work involving calculations on magnetic circuits, the characteristic curves giving the relation between $\mathcal{B}$ and $\mathcal{H}$ contain the same information in a more useful form.

The general form of the hysteresis loop of nearly all the ferromagnetic materials is similar to that shown in Fig. 14a. The proportions of such a loop frequently vary considerably among different materials, and, for a given material, vary with the heat treatment and mechanical working to which the material is subjected. Normal intrinsic magnetization curves for a wide variety of magnetic materials are shown in Fig. 15a. Figure 15b shows hysteresis loops for some of the typical permanent-magnet materials. The chief purpose in presenting the curves of various materials on the same sheet and to the same scale is to facilitate comparison of the magnetic properties among these materials.

Representative properties for a selected group of magnetic materials are given in Table I. These figures represent an average of values for the various materials given by several sources.[12] In this table, the magnetically soft materials are arranged in order according to their maximum permeabilities, the material having highest permeability appearing at the top. This order is also essentially that of decreasing relative magnetic softness, a magnetically soft material having a relatively narrow hysteresis loop and hence a small coercivity, usually less than about 3 oersteds. Magnetically soft materials are used in devices where high permeabilities are desired, and where the materials are subjected to alternating fields. Magnetically hard materials are used for permanent magnets; their uses are discussed in Ch. IV.

[12] V. E. Legg, " Survey of Magnetic Materials and Applications in the Telephone System," *B.S.T.J.*, *18* (1939), 438–464.

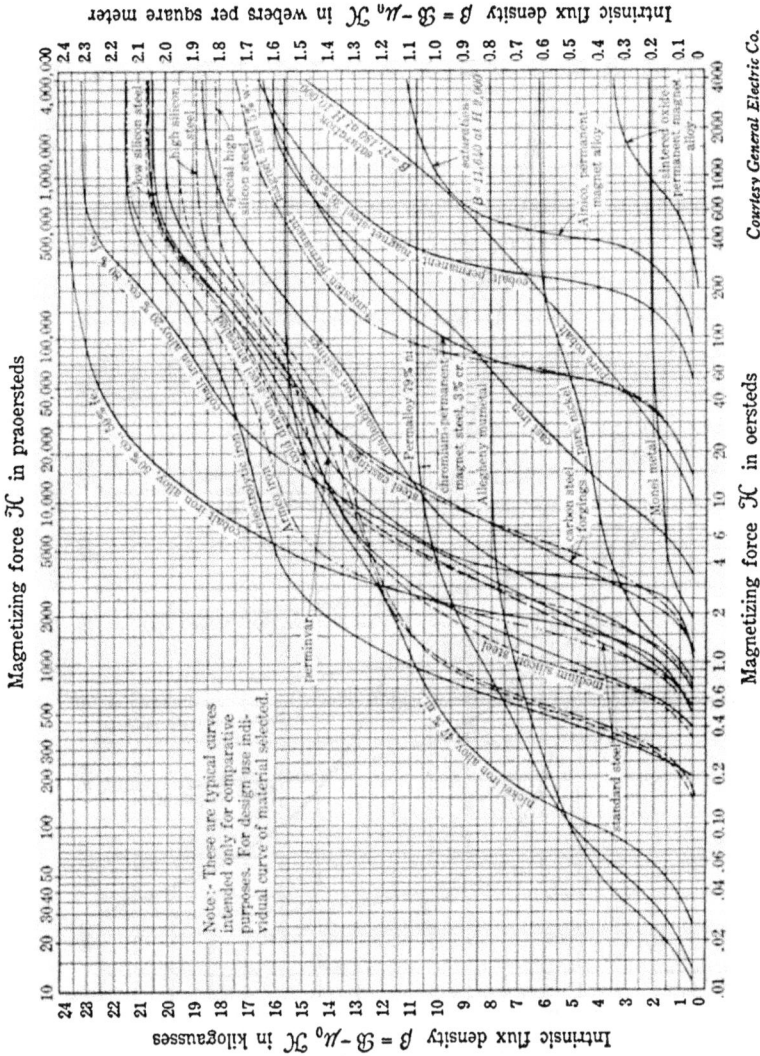

FIG. 15a.   Normal d-c magnetization curves for various magnetic materials.

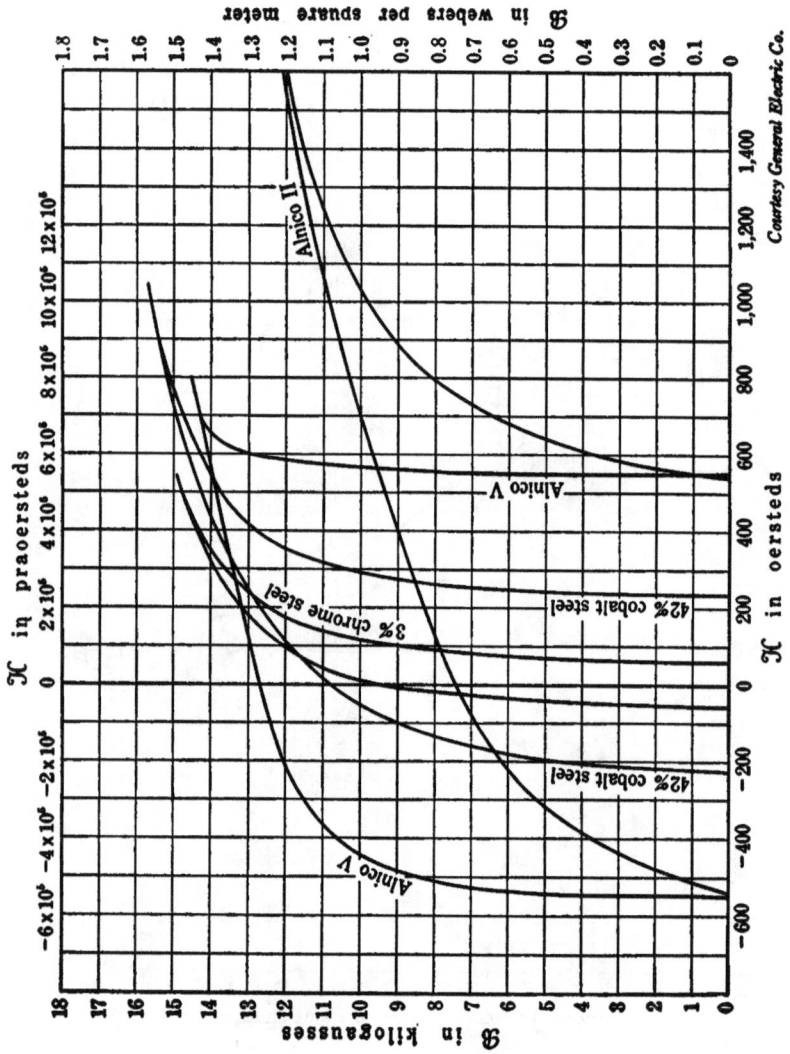

FIG. 15b. Typical hysteresis loops for permanent-magnet materials.

TABLE I — REPRESENTATIVE PROPERTIES OF FERROMAGNETIC MATERIALS‡

Magnetically Soft Materials	Elements	Composition approx. parts per 100	Intrinsic Saturation B_{max} kilogausses	Residual Flux Density B_r kilogausses	Coercive Force H_c oersteds	Typical Maximum Permeability μ_{max} kilogausses/oersted	Typical Initial Permeability μ_0 kilogausses/oersted	Cost cents/lb
Magnetic Iron (Purified in H)	Fe	99.98	21.5	13.6	0.05	275.	25.	7
Sendust	Fe, Si, Al	85, 9.5, 5.5	10.0	5.0	0.05	120.	30.	3
78.5 Permalloy	Ni, Fe, Mn	78.5, 20.9, 0.6	10.7	6.0	0.05	105.	9.	28
Hipernik	Fe, Ni	50, 50	15.0	7.5	0.06	90.	6.	18
Mumetal	Ni, Fe, Cu, Mn	74, 20, 5, 1	8.5	6.0	0.05	80.	7.	27
4-79 Mo-Permalloy	Ni, Fe, Mo, Mn	79, 16.4, 4, 0.6	8.5	5.0	0.05	72.	22.	32
High-Silicon Steel	Fe, Si	95.5, 4.5	19.0	5.0*	0.5*	8.3	0.750	8
Low-Silicon Steel	Fe, Si	99, 1	21.0	8.5*	0.7*	5.2	0.350	7
Permendur	Fe, Co	50, 50	24.5	14.0	2.0	5.0	0.800	69
7-70 Perminvar	Ni, Fe, Co, Mn	70, 22.4, 7, 0.6	12.5†	2.4†	0.6†	4.0†	0.850	35
45-25 Perminvar	Ni, Fe, Co, Mn	45, 29.4, 25, 0.6	15.5†	3.3†	1.4†	1.8†	0.365	50
Conpernik	Fe, Ni	50, 50	15.0	....		1.0	0.175	18
Cast Steel	Fe, C	....	20.0	....	5.0*	1.0	0.175	1
Cast Iron Annealed	Fe	....	16.0	5.5	11.0*	0.3	0.125	1
12.5-80 Mo-Permalloy	Ni, Mo, Fe	80, 12.5, 7.5 (Loses magnetic properties at 40 degrees C.)					9.	40

* B_r and H_c from B_{max} = 10 kilogausses instead of from saturation. † These are not the conditions under which Perminvar is useful. See p. 28.

Magnetically Hard Materials	Elements	Composition approx. parts per 100	Retentivity kilogausses	Coercivity oersteds	$(BH)_{max}$	Cost cents/lb
Alnico V Cast	Fe, Co, Ni, Al, Cu	51, 24, 14, 8, 3	12.5	550	4.5×10^6	56
Honda Metal	Fe, Co, Ni, Ti, Al	45, 27, 18, 6.7, 3.3	7.1	780	2.0	53
Alnico II Cast	Fe, Ni, Co, Al, Cu	54.5, 17, 12.5, 10, 6	7.2	540	1.6	26
Alnico I Cast	Fe, Ni, Al, Co	63, 20, 12, 5	7.3	430	1.4	17
Mishima Metal	Fe, Ni, Al	58, 29, 13	6.0	550	1.4	14
Remalloy	Fe, Mo, Co	71, 17, 12	10.5	250	1.1	33
Cobalt Steel	Fe, W, C	64, 35, 1	9.5	260	1.05	62
Tungsten Steel	Fe, W, C	93, 6, 1	10.0	80	0.34	9
Chrome Steel	Fe, Cr, C, Mn	96, 3, 1, 0.4	9.7	65	0.30	1

‡Data for Table I taken chiefly from reference 12, p. 24.

The characteristics tabulated for the magnetically soft materials are: the maximum intrinsic flux density β_{max}, the residual flux density $\mathfrak{B}_r$, the coercive force $\mathfrak{H}_c$ (also the retentivity and the coercivity respectively since the symmetrical cyclic magnetization was carried to saturation, except as noted), the maximum and initial permeabilities μ_{max} and μ_i, and the cost of the raw material. The characteristics tabulated for the magnetically hard materials are the retentivity, the coercivity, the maximum value of the product of $\mathfrak{B}$ and $\mathfrak{H}$ occurring between the points $\mathfrak{H} = 0$ and $\mathfrak{H} = -\mathfrak{H}_c$, and the cost. The product $(\mathfrak{B}\mathfrak{H})_{max}$ is a measure of the effectiveness of the material for use in permanent magnets; its significance is discussed in Ch. IV. The composition of the alloys in parts per hundred is given for both hard and soft materials. Inadvertent impurities are not indicated, although they frequently produce serious effects. Almost all the materials mentioned in the table are commercially available, but a few are in the stage of laboratory development.

The extent of the dependence of certain of the magnetic properties of all ferromagnetic materials on the heat treatment to which they are subjected cannot be overemphasized. A discussion of the heat-treatment procedure[13] is beyond the scope of this textbook, but the importance of the treatment may be illustrated by the fact that Hipernik and Conpernik, although having the same composition, display widely different magnetic properties. The differences are brought about wholly by a difference in the heat treatment to which each material is subjected. The Perminvars maintain a practically constant permeability μ_i if not magnetized above about 1,000 gausses. Magnetization above this value destroys the property of constant permeability and this can be restored only by further heat treatment. Of the characteristics tabulated, the maximum intrinsic flux density is the one subject to the least variation. It is dependent chiefly on the percentage of ferromagnetic material present in the alloy. The degree to which the other characteristics are subject to variation increases in the order, residual induction, coercive force, and permeability.

From the curves of Figs. 15a and 15b and the data of Table I, the material best suited from the magnetic point of view for any specific purpose can be chosen. Whether the material should be used depends, of course, on numerous other factors such as availability, cost, mechanical strength, and machinability. Characteristics of magnetic materials subjected to alternating fields are given in Ch. V. Additional magnetic data on commercial materials may be obtained from the manufacturers or from the literature.

[13] C. E. Webb, " Recent Developments in Magnetic Materials," *I.E.E.J.*, *82* (1938), 303–323.

5. MAGNETIC TESTING

The normal magnetization curve and a family of hysteresis loops for a magnetic material give the information necessary for the solution of problems involving the use of the material in direct-current applications. In order to get these curves experimentally, some kind of permeameter[14] is required in which a sample of the material can be tested.

Of the many types of permeameter, only those most commonly used will be described here. These are the *Fahy simplex* permeameter, which

FIG. 16a.   The Fahy simplex permeameter.

gives good precision up to magnetizing forces of 300 oersteds, and with an adapter up to 1,000 oersteds, and the *high-$\mathcal{H}$* permeameter, which permits a range up to 5,000 oersteds.

The Fahy simplex permeameter is pictured in Fig. 16a and is represented diagrammatically in relation to its associated equipment in Fig. 16b. The circuit shown is somewhat simpler than the actual laboratory layout, since the object of this treatment is to set forth the method of testing rather than to stress the technique of manipulation. The ballistic galvanometer is first calibrated by the adjustment of its resistances R_S, R_B, and R_H, so that a change of some convenient unit of flux density in coil B (called the $\mathcal{B}$-coil) or a change of some definite unit of field

[14] R. L. Sanford, " Magnetic Testing," *Circ. Nat. Bur. Stand.* C415 (1937).

intensity in coil H (called the $\mathcal{H}$-coil) produces a galvanometer deflection of one main scale division. This calibration requires the use of a standard mutual inductor not shown in the diagram.

The specimen S to be tested is clamped between the solid iron posts P and the ends of the laminated iron yoke Y, forming a magnetic circuit in which flux is set up by the magnetomotive force of coil M. The current in this magnetizing winding is supplied from a steady direct-current source and can be adjusted by the variable resistor R_M. The switch S_2

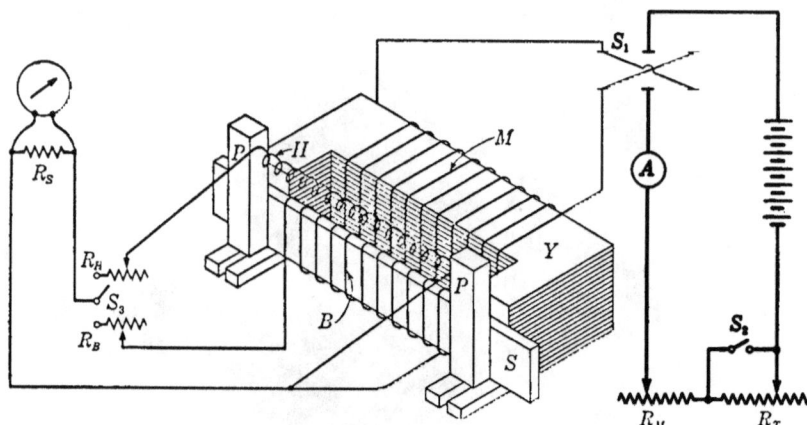

Fig. 16b.   Diagram of connections for Fahy permeameter.

is closed when the normal magnetization curve is being taken and the switch S_1 serves to reverse the direction of the current in the coil. The coil H is an air-core solenoid of many turns of fine wire that provides a means for measuring the difference in magnetic potential between the posts P. This difference in magnetic potential divided by the distance between posts is the magnetizing force $\mathcal{H}$ in the specimen.

Before data for the normal $\mathcal{B}(\mathcal{H})$ curve are taken, the sample is demagnetized, and, with switch S_2 closed, rheostat R_M is set for the lowest point on the curve. Switch S_1 is placed in either closed position and S_3 is left open. The current in coil M is reversed several times by means of switch S_1 in order to put the sample in cyclically magnetized condition. Switch S_3 is now closed in the lower position and S_1 is moved from one pair of contacts to the other. The deflection of the ballistic galvanometer then indicates the change in flux density $\mathcal{B}$ in the sample. A repetition of the process with S_3 in the upper position, using the same value of magnetizing current as before, gives the corresponding change in magnetizing force $\mathcal{H}$ in the sample. Other points on the curve can be obtained through increasing the current in coil M, putting the sample in cyclic condition by several reversals of S_1, and then taking data as for the first point. The normal magnetization curve is the locus of points plotted from these data as indicated by the broken line in Fig. 17.

When a hysteresis loop is to be determined, the cyclic condition is first established at the desired value of $\mathcal{B}_{max}$. The rheostat R_T is set to

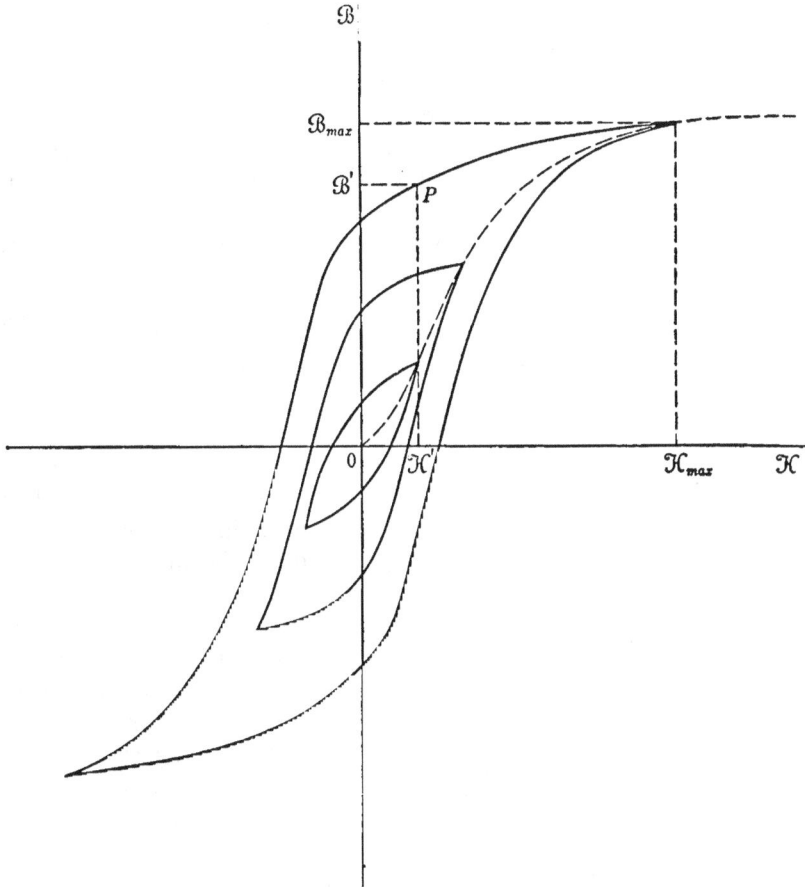

Fig. 17.   Points on hysteresis and normal magnetization curves obtained with permeameter

give a small decrease in the reading of ammeter A when S_2 is opened. With S_3 in the lower position, S_2 is opened and the galvanometer deflection is noted. The reading indicates the change in $\mathcal{B}$ as it is decreased from $\mathcal{B}_{max}$. This will give the value $\mathcal{B}_{max} - \mathcal{B}'$ as indicated in Fig. 17. The cyclic condition at the tip of the loop is now restored through reversing S_1, closing S_2, and then reversing S_1 again. If, with S_3 in the upper position, S_2 is again opened, the value $\mathcal{H}_{max} - \mathcal{H}'$ is obtained. Points farther down the loop can be obtained in the same way by the use of larger values of resistance R_T, each reading being made from a cyclical condition at the tip of the loop. This makes it possible to establish identical initial conditions for the $\mathscr{B}$ and $\mathscr{H}$ readings and also to avoid cumulative errors which would accrue in going from point to point.

When the points on the hysteresis loop are near the horizontal ($\mathscr{H}$) axis, small errors in the determination of $\mathscr{B}$ are not of much consequence, because the curve is nearly vertical, but errors in $\mathscr{H}$ are of importance. The method just described for the determination of $\mathscr{H}$ is likely to introduce a considerable error in this region, since the values of $\mathscr{H}$ are found through taking the difference between $\mathscr{H}_{max}$ and another large value $\mathscr{H}_{max} - \mathscr{H}$ determined from the test. Greater precision in these values of $\mathscr{H}$ is obtained by the use of an attachment which makes possible the measurement of $\mathscr{H}$ directly at a fixed magnetization of the specimen through quickly rotating the coil H 180 degrees. Not only is greater precision possible in a single reading by the use of this device, but several readings of $\mathscr{H}$ can be taken without changing the magnetic condition of the specimen.

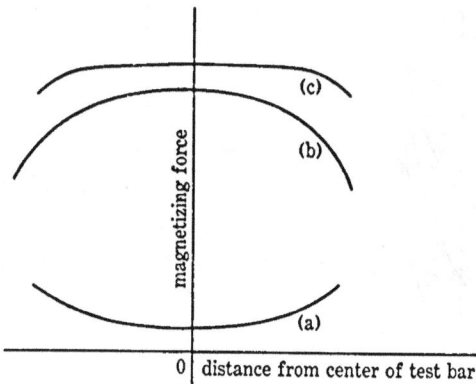

Fig. 18. Distribution of magnetizing force $\mathscr{H}$ along test bar for: (a) coil on yoke, (b) coil surrounding specimen, (c) combination of (a) and (b).

Since a considerable difference in magnetic potential exists between the ends of the test bar, a leakage of flux away from the bar through the surrounding air is bound to occur. For this reason, the flux density decreases along the bar from the ends toward the center with a corresponding decrease in magnetizing force as shown in curve (a) of Fig. 18. The greater the total applied magnetomotive force, the greater will be this leakage and resulting nonuniformity in $\mathscr{H}$ along the specimen. For magnetizing forces greater than 300 oersteds, this effect becomes great enough to interfere with the precision of the instrument. The recognition of this situation and the need for testing the new hard permanent-magnet alloys led to the development of the super-$\mathscr{H}$ adapter. This device is clamped against the yoke Y in place of the sample and auxiliary posts P and holds near its center a short length of test material and an accompanying short $\mathscr{H}$-coil. Magnetizing forces up to 2,500 oersteds can be applied to the sample without overheating the apparatus, but the precision of the measurements above about 1,000 oersteds is doubtful.

The curve (b) of Fig. 18 shows the variation of magnetizing force along a specimen when the magnetizing coil is a solenoid surrounding the specimen rather than a winding on the yoke. Here the value of $\mathcal{H}$ is greatest near the center and drops off toward the end. Curve (c) represents a combination of these two curves and indicates that, with a part of the magnetizing winding on the yoke and a part surrounding the specimen, a uniform field can be obtained over a considerable length about the center of the test bar. This principle is used in the Babbitt[15] permeameter, which gives good precision up to 1,000 oersteds.

The most recent development is the Sanford-Bennett[16,17] high-$\mathcal{H}$ permeameter of the U. S. Bureau of Standards. This instrument is claimed to measure both $\mathcal{B}$ and $\mathcal{H}$ in small samples (up to cross-sectional dimensions 0.75 in. $\times$ 1.5 in.) for magnetizing forces up to 5,000 oersteds with an error not exceeding 1 per cent. Values of $\mathcal{H}$ up to 9,000 oersteds can be used without overheating the instrument or the sample, but the precision for the higher values has not yet been determined. Figure 19 shows this permeameter. The two large coils in the center surround the pole pieces and

Courtesy National Bureau of Standards.

FIG. 19.   The Sanford-Bennett high-$\mathcal{H}$ permeameter.

comprise the main magnetizing winding. Each of these coils has 2,690 turns of 14 AWG silk-enamel wire. Each of the auxiliary or yoke coils has 1,600 turns. Several $\mathcal{B}$-coils are available in order that one may be chosen which fits snugly around the sample to be tested. Each of these coils has 25 turns of fine wire wound on a thin brass shell. This is clearly visible at the middle of the picture, as is also the $\mathcal{H}$-coil just beneath the $\mathcal{B}$-coil. The device at the front is a motor-driven rotator for the $\mathcal{H}$-coil, in order that $\mathcal{H}$ may be measured at a fixed magnetization

[15] B. J. Babbitt, " An Improved Permeameter for Testing Magnet Steel," *J.O.S.A.* and *R.S.I.*, *17* (1928), 47–58.

[16] R. L. Sanford and E. G. Bennett, " An Apparatus for Magnetic Testing at High Magnetizing Forces," *J. Res. Nat. Bur. Stand.*, *10* (1933), 567–573.

[17] R. L. Sanford and E. G. Bennett, " An Apparatus for Magnetic Testing at Magnetizing Forces up to 5,000 Oersteds," *J. Res. Nat. Bur. Stand.*, *23* (1939), 415–425.

of the specimen. The instrument connections are essentially the same as for the Fahy simplex permeameter.

6. ENGINEERING UTILIZATION OF FERROMAGNETIC MATERIALS

A detailed study of the data of Figs. 15a and 15b and Table I shows that the various alloys mentioned display the characteristics typical of ferromagnetic materials to markedly varying degrees. The predominance of certain characteristics determines the usefulness of a particular material for a specific purpose. For example, for an electromagnet, a material that has a relatively large permeability and a large saturation flux density is usually desirable in order that the magnetomotive force and the volume of material required shall be small. But these properties must not be achieved at too great a sacrifice of the desirable mechanical properties, or at too large a cost. A relatively pure iron or low-silicon-content sheet steel is frequently used for such purposes. However, when the flux is varying with time, as it does in transformers, reactors, and rotating machines, energy is dissipated in the iron in a manner explained in Ch. V. Because of heating, and for economic reasons, this dissipation of energy must be small. Electrical-sheet steel with between 2 and 4 per cent silicon is used frequently in such devices. Its cost per pound is relatively small and it is commercially available in many forms convenient for fabrication into desired shapes. Some of these forms are illustrated in Fig. 12. Because of its silicon content, it has relatively high resistivity and a hysteresis loop of relatively small area,* which reduce the energy loss in an alternating field. When more than about 4 per cent of silicon is used, the improvement in the magnetic properties is negligible, and the material becomes brittle and difficult to punch into desired shapes. Manufacturing difficulties are then increased.

In some communications equipment, a material having an extremely high permeability is desirable. For such applications, certain nickel-iron alloys, known as Permalloys[18,19] (*permeability alloy*) or Hipernik in the United States, or a nickel-iron-copper alloy, known as Mumetal[20] and developed in England, may be used. The use of these alloys involves an increased cost per pound of material but not of cost per unit of inductance.

---

* Representative values of core loss for symmetrical cyclic magnetization and representative values of resistivity for average samples of various materials are tabulated in Art. 3, Ch. V.

[18] G. W. Elmen, " Magnetic Alloys of Iron, Nickel, and Cobalt," *A.I.E.E. Trans.*, *54* (1935), 1292–1299.

[19] V. E. Legg and F. J. Given, " Compressed Powdered Molybdenum Permalloy for High-Quality Inductance Coils," *B.S.T.J.*, *19* (1940), 385–406.

[20] W. F. Randall, " Nickel-Iron Alloys of High Permeability with Special Reference to Mumetal," *I.E.E.J.*, *80* (1937), 647.

The large permeability of these alloys is chiefly the result of their having been subjected to a particular heat-treatment procedure, for which careful control is necessary. Also, their permeability is very susceptible to strains introduced by mechanical working; hence their use may increase manufacturing problems.

In Ch. V an analysis is presented showing that eddy-current loss can be reduced through building ferromagnetic cores of laminated sheets insulated from each other by their surface oxides or by an insulating lacquer. Another method of reducing eddy currents is to reduce the

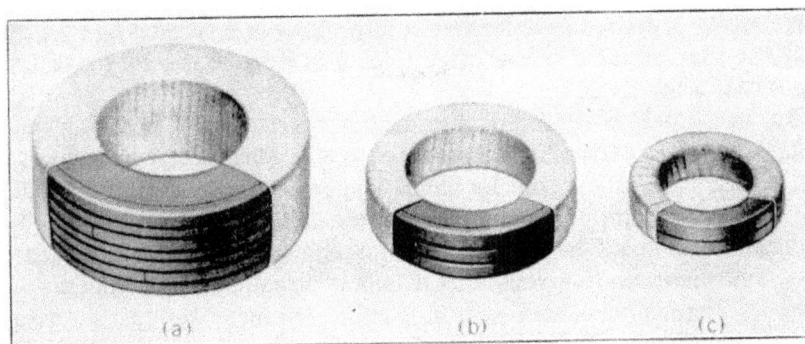

Courtesy Bell Telephone Laboratories.

FIG. 20.   Reduction in size of compressed-powder loading-coil cores in 20 years: (a) iron, (b) 80-Permalloy, and (c) 2–81 Mo-Permalloy.

material to a powder and then to insulate the particles from one another by an insulating substance tough enough to withstand the forces of compression when the powder is molded into the cores of various shapes. Figure 20 is indicative of the advance given to this art by the engineers and physicists of the Bell System during the past 20 years.

In certain other communications problems, extreme constancy of permeability throughout the working range is imperative. A nickel-iron-cobalt alloy known as Perminvar (*perm*eability *invar*iable) and a nickel-iron alloy known as Conpernik possess this property at relatively low flux densities. Unfortunately, the permeability of these alloys is low. They have the unique properties of negligible residual magnetism and of negligible coercive force throughout their useful range. Perminvar has a low resistivity, of the order of 20 microhm-centimeters, but by the addition of about 7 per cent of molybdenum its resistivity is increased by a factor of about four without appreciable sacrifice of its magnetic properties.

When a material is to be used in rotating machinery, one having properties different from the above is required. Mechanical strength becomes important, since in the rotors of turbo-alternators, centrifugal forces are often large and a material with large tensile strength is then essential. When the magnetizing forces are essentially constant, solid steel forgings of nickel steel or nickel-chromium steel are often used. These alloys possess the high values of saturation flux density desired in steels for these applications. In the parts of the machine where the flux is alternating, as it is in the neighborhood of the alternating-current windings, the resulting energy losses are of considerable concern. For these parts, a suitable silicon steel is used.

In rotating-armature machines, a special problem arises. The flux density in the teeth is large, and a material with a very large saturation flux density is desired in order that a large flux per pole shall be available without uneconomical values of excitation in spite of the constriction at the roots of the teeth.

An iron-cobalt alloy called Permendur (*permeability enduring*) may hold interesting possibilities for these applications, although it has not yet been used to any extent for them. The present high cost of this alloy, its brittleness, and its low resistivity are deterrents to wide use. The addition of about 1.7 per cent of vanadium, however, decreases the brittleness, and increases the resistivity to about 24 microhm-centimeters. At present, Permendur is used only in small electromagnets, relays, or telephone equipment. In certain applications such as in aeronautical or other equipment where space or weight is at a premium, the item of its cost should not be a serious one. The important factors are the relation between the cost per pound of the material, the cost per pound of the finished article, and the percentage by weight of the material in the finished article.

When ferromagnetic materials are used as permanent magnets, other properties become primary in importance. Permanent-magnet applications usually are in instruments, meters, and small motors or generators, where the cost per pound of the finished product is relatively large. The magnets are subject frequently to severe mechanical shock, to alternating fields, and to moderate changes in temperature. The mechanical strength of the magnets is usually considerably in excess of that necessary. The coercivity and retentivity of the material become, in a general way, the figures of merit. For many applications, the determining factor is the maximum value of the product $(\mathcal{B}\mathcal{H})$ along the demagnetization curve between $\mathcal{H}$ equals 0 and $\mathcal{H}$ equals $\mathcal{H}_c$ as is explained in Ch. IV. This product is tabulated as $(\mathcal{B}\mathcal{H})_{max}$ in Table I for those materials that are suitable for use as permanent magnets.

During the early days of the electrical industry, the carbon steels used for permanent magnets were seriously weakened magnetically by vibration, alternating fields, and temperature variations, and were therefore of little use where constancy of field was desired. The development of tungsten-steel and chromium-steel alloys provided improved permanent-magnet materials which still find wide application where the cost of the material is of major importance and space is not. These steels are subject to the same kind of defects as carbon steel but to a lesser degree.

The development of cobalt steel provided a material superior magnetically to either tungsten or chromium steel. Not only can stronger fields be produced by a cobalt-steel magnet, but alternating fields and mechanical vibration have relatively little effect on its permanency. Recently has come the development of other alloys containing chiefly iron, nickel, cobalt, and aluminum. These alloys, which are known by the trade name Alnico in the United States and Nipermag in England, are less expensive to produce and lighter in weight than cobalt steel, and yet have better magnetic properties. The alloy Alnico[21] is produced with various compositions, an average composition being about 20 per cent nickel, 12 per cent aluminum, 5 per cent cobalt, 0.4 per cent manganese plus silicon, and the remainder iron. It has a large $(\mathfrak{BH})_{max}$ and its properties are changed only slightly by vibration, alternating fields, or even a surprisingly wide range of temperature variation. The use of this alloy introduces a new problem, however, for it cannot be machined readily and frequently must be cast in the form in which it is to be used. Hence use of it may involve a complicated manufacturing problem. This problem has been partly solved by sintering,[22] a process in which the powdered constituents are diffused together at high temperature and under tremendous pressure. Some of the parts formed with considerable accuracy by this process are shown in Fig. 21. Larger elements, as well as the ones shown, are commercially available and can be combined with soft iron or steel in the construction of magnetic devices.

Certain other alloys, and also the oxides of iron and iron-cobalt, show desirable properties as permanent magnets. The oxide magnets must be molded and are relatively brittle. Future developments in this part of the art may yield useful products. In many devices, no single ferromagnetic material is adequate. For example, the modern telephone receiver shown in Fig. 22 has Remalloy permanent magnets, 45-Permalloy pole pieces, and a Permendur diaphragm.

In certain devices, such as relays, resistors, and transformers, a material having magnetic properties very sensitive to temperature can be used to advantage. This condition has only recently become of engineering importance, although the change in the magnetic property with temperature has been of theoretical significance since the era of early attempts to formulate a theory of ferromagnetism. By appropriate composition, an alloy is obtained having the temperature at which it ceases to be ferro-

---

[21] J. Q. Adams, " Alnico — Its Properties and Possibilities," *G.E. Rev.,* *41* (1938), 518–523.
[22] G. H. Howe, " Sintering of Alnico," *Iron Age,* *145* (1940), 27.

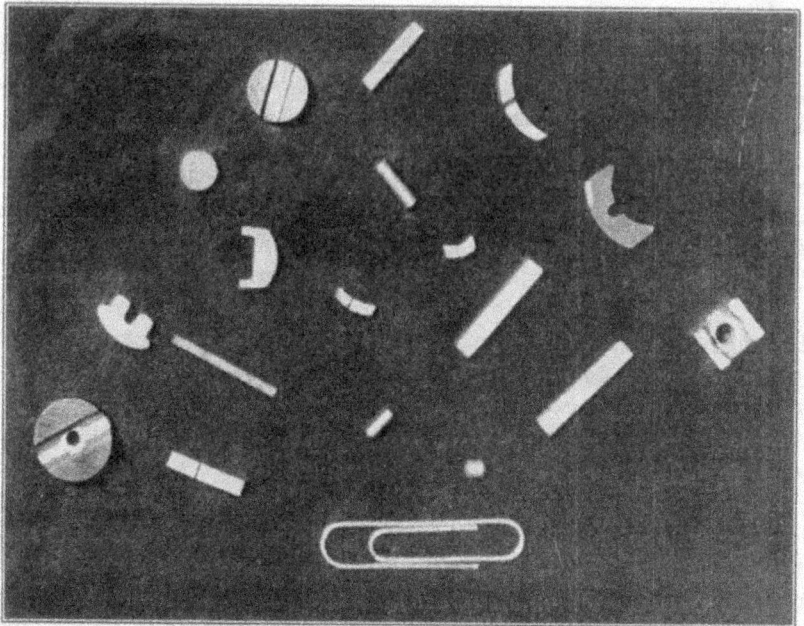

Courtesy General Electric Co.

FIG. 21.   Representative collection of small Alnico magnets.   Paper clip
included to indicate size.

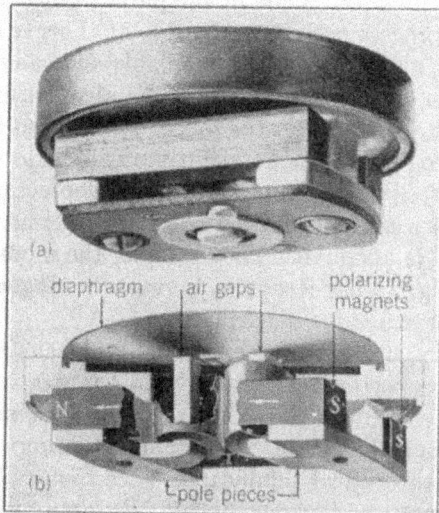

Courtesy Bell Telephone Laboratories.

FIG. 22.   Magnetic circuit of telephone receiver.

magnetic (that is, its Curie point) in a useful range for many engineering applications. Alloys having this property are used for the temperature-sensitive element in temperature-sensitive reactors, contactors, and transformers. The possibilities for useful application of alloys of this kind may be better appreciated when it is realized that an alloy[23] comprising approximately 35 per cent nickel, 5 per cent chromium, 60 per cent iron, and 0.3 per cent silicon, appropriately heat-treated, has a saturation flux density of about 6,000 gausses at a temperature of 60 degrees centigrade but is essentially nonmagnetic at 160 degrees. Another of these alloys, 12.5–80 Mo-Permalloy, becomes nonmagnetic at 40 degrees centigrade, or slightly above normal room temperature.

The magnetic properties of many ferromagnetic materials change gradually with time, a phenomenon known as aging. All the steels behave this way to some extent since their iron-carbon compounds are not stable, but change from one form to another with time. The changes may proceed rapidly at high temperatures, but appropriate sudden cooling as part of the heat treatment can so retard the rate that the whole effect is distributed over a period of years. However, the aging cannot be completely stopped. The silicon steels show only a slight aging effect because of the reaction occurring among the silicon, oxygen, and carbon during their production.

While the effects of aging on the magnetic properties of all steels are important, usually causing a decrease in the permeability and an increase in coercivity of magnetically soft materials, the effects are especially serious when they occur in magnetically hard steels used for magnets in instruments and meters. Here the strength of the magnet is decreased, and the accuracy of the device is impaired by even a very small change in the strength. The usefulness of the magnetically hard carbon steels can be increased by artificial aging. For example, if the steel is subjected to a temperature of about 100 degrees centigrade for several hours, a rapid aging and considerable change in strength occur. The subsequent rate of change of magnetic strength at normal room temperature is then greatly reduced.

Many of the newer alloys do not contain carbon in significant amounts and are not subject to these aging effects. Some of the compounds formed, notably Alnico, are stable even up to temperatures above 600 degrees centigrade. Welding is even possible with some of these alloys after they have been heat-treated, without appreciably impairing their desired magnetic properties. The advantages offered by such alloys for manufacturing processes may therefore be readily appreciated.

[23] L. R. Jackson and H. W. Russell, " Temperature-Sensitive Magnetic Alloys and Their Uses," *Instruments, 11* (1938), 280–282.

PROBLEMS

1. Describe briefly any new developments in the theory of magnetism which have appeared in the periodical literature in the fields of physics and electrical engineering since this textbook was published.

2. Give the physical properties including characteristic curves of any important ferromagnetic materials not included in this chapter.

3. A strip of specially treated stainless steel (18% chromium, 8% nickel), when first placed under a powerful permanent magnet, shows very little tendency to be lifted, but, after the lapse of about one and one-half minutes, is strongly attracted to the poles of the magnet. The special treatment consists in removing the nitrogen from the metal and quenching in water from a temperature of 1,100 F. The magnetic experiment is performed immediately after quenching. How can this change in magnetic properties be explained? Does your explanation indicate that the strip regains its nonferromagnetic properties when removed from the magnetic field?[24]

4. When a sinusoidal voltage of 2,300 v effective value is applied at the primary terminals of a transformer, the maximum flux density in the core is 1 weber per sq m. The data for the hysteresis loop of the steel in the transformer core are given in the following table:

$\mathcal{B}$ in webers/sq m	$\mathcal{H}$ in praoersteds
−1.00	−3,750
−0.92	−1,500
−0.77	0
−0.50	+1,400
−0.20	+2,100
0.00	+2,350
+0.45	+2,500
+0.75	+2,950
+1.00	+3,750

Plot, on the same horizontal axis, a complete cycle of sinusoidal voltage and the corresponding cycles of flux density and magnetizing force. Use a horizontal scale of 40 degrees = 1 in. Use vertical scales of 1,000 v = 1 in., 0.40 weber per sq m = 1 in., and 1,000 praoersteds = 1 in. The waveform of $\mathcal{H}$ is similar to that of the magnetizing current of the transformer described in Ch. VI.

[24] See "Magnetic Delay," *G.E. Rev.*, *45* (1942), 245, 246.

CHAPTER II

The Magnetic-Circuit Concept

When in engineering practice a field of magnetic flux is desired, specially shaped structures of ferromagnetic material and appropriately located current-carrying conductors, or appropriately located permanent magnets, usually are used. The ability to compute the magnetic field intensity $\mathcal{H}$ and the flux density $\mathcal{B}$ throughout such a structure is necessary for the solution of many problems. In general, $\mathcal{H}$ and $\mathcal{B}$ are functions of space and time, determined by the geometry of the conductors and the magnetic structure, by the properties and history of the magnetic material, and by the values and derivatives or frequencies of the currents or the strength of the magnets. A field problem involving materials which have nonlinear characteristics is thus encountered. Suitable simplifying assumptions make possible in many instances the reduction of the general field problem to a simpler one involving the circuit concept.

1. THE GENERAL PROBLEM

When the currents and fluxes vary with time as well as with the three co-ordinates of space, the problem is of the most general type and its solution may be extremely difficult, often requiring the step-by-step technique described in Ch. XIII of the volume on electric circuits. In order to simplify the treatment in this chapter the conditions will be limited to constant or slowly varying currents and fluxes; that is, the *quasi-stationary state.*

The general analytical problem for the quasi-stationary state can be given as follows: For a specified configuration of current-carrying conductors and ferromagnetic material, what are the direction and magnitude of the flux-density vector $\mathcal{B}$ at some or all points of the region as functions of the currents in the conductors or of the known properties of the permanent-magnet material? Or, conversely, what configuration of current-carrying conductors and ferromagnetic materials are required to establish desired directions and magnitudes of the flux-density vector $\mathcal{B}$ at some or all points in the region? The solution of this general problem within a specified region of space requires that the three following conditions be simultaneously satisfied:

(a) The field-intensity vector $\mathcal{H}$ is at every point equal to the flux-density vector $\mathcal{B}$ divided by the permeability μ which is in general a variable dependent upon $\mathcal{B}$.

(b) The surface integral of the normal component of $\mathfrak{B}$ over any closed surface in the region must be zero.

(c) The line integral of the tangential component of $\mathfrak{K}$ around any closed contour in the region must be equal to $4\pi I$ where I is the current linking the contour.

Except in its very simplest form, this general problem is wholly insolvable analytically with present-day knowledge of materials and methods. In fact, even if only nonmagnetic material is present, the work involved for relatively simple geometrical arrangements of conductors is considerable. Consequently, for a working engineering solution of the more difficult problems, rough assumptions regarding some details must be made. The extent to which the results are reliable depends in no small degree upon the quality of what is called, for the want of a better name,

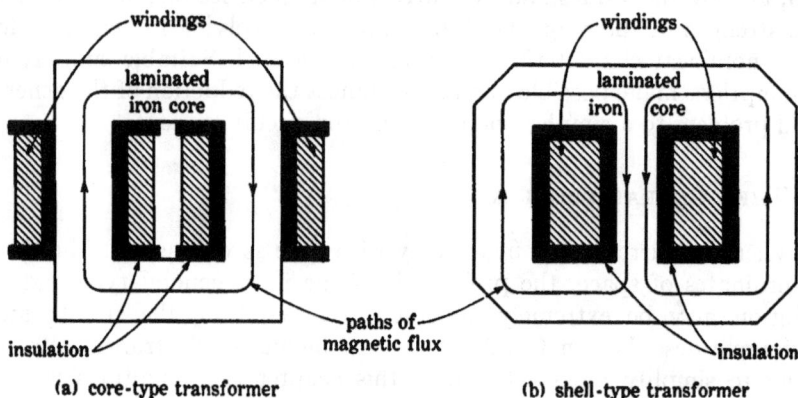

(a) core-type transformer   (b) shell-type transformer

FIG. 1.   Magnetic structures of typical transformers.

engineering judgment. However, when by a combination of analytical and experimental procedure a satisfactory design has been obtained for a specific magnetic structure, the design of similar magnetic structures can be determined by use of model theory, as outlined in Ch. VII.

A secondary problem, easily solved once the field is known, is that of obtaining the total *magnetic flux* ϕ crossing a surface. This flux is merely the component of $\mathfrak{B}$ normal to the surface integrated over the surface. The relation is expressed mathematically as

$$\phi = \int \mathfrak{B} \cos \theta \, ds, \qquad [1]$$

where θ is the angle between the flux-density vector $\mathfrak{B}$ and the normal to the element of surface ds; or in vector notation as

$$\phi = \int \mathfrak{B} \cdot n \, ds, \qquad [2]$$

where *n* is the outward normal unit vector associated with the element of surface *ds*. The converse of this secondary problem, usually simpler to solve, is the problem of determining the currents or permanent-magnet arrangements that produce a desired total flux across a surface.

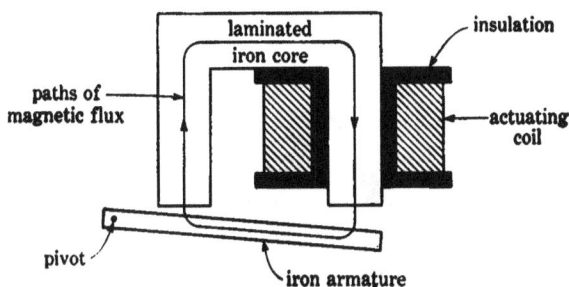

FIG. 2. Magnetic structure of a relay.

Fortunately, in numerous practical applications the general three-dimensional problem can be reduced with good approximation for analysis to a one-dimensional problem. This reduction results in a tre-

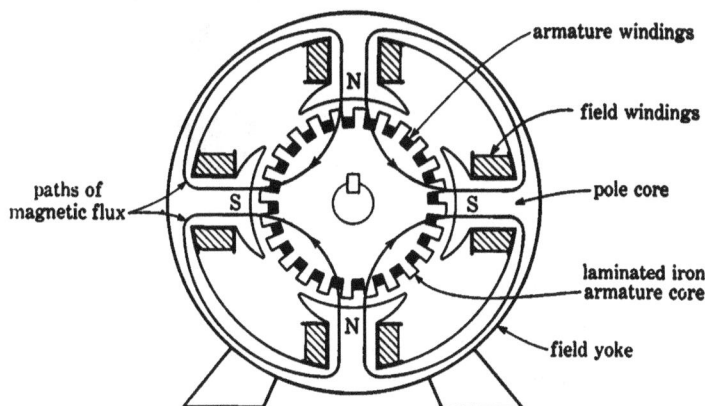

FIG. 3. Magnetic structure of a 4-pole d-c dynamo.

mendous simplification through which the three-dimensional *field* becomes a one-dimensional *circuit*, called a *magnetic circuit*. In general, a magnetic circuit consists throughout most of its length of high-permeability material with substantially uniform cross section in which the magnetic flux is largely confined.

Typical examples of magnetic structures that may be analyzed by use of the circuit concept are shown in Figs. 1, 2, 3, and 4. In Fig. 1, the flux paths are shown as being entirely in the iron cores. Actually a small amount of flux is also present in the space adjacent to the iron, for reasons discussed qualitatively in Art. 2, and quantitatively in Chs. III and IV. In Fig. 2, the iron portion of the flux circuit is interrupted at two places by air gaps; yet the flux path is largely in high-permeability material. The structure of Fig. 3 has a more complicated magnetic circuit of a series-

FIG. 4.   Rotor and stator for a 3-phase, 2,300-v 60-∿ synchronous generator.

parallel form. The magnetic structures in these figures are laminated to minimize core losses, as discussed in Ch. V. Figure 4 shows a photograph of a typical three-phase synchronous generator. The rotating member of a large machine may weigh several hundred tons, a substantial portion of which is ferromagnetic material of high quality.

2. ANALOGIES BETWEEN ELECTRIC AND MAGNETIC CIRCUITS

The concept of a magnetic circuit as mentioned in Art. 1 is founded on the idea that a constant or slowly varying flux tends to confine itself to the high-permeability paths of a ferromagnetic structure in a manner that resembles the tendency of a constant or slowly varying current to confine itself to high-conductivity paths in an electric circuit. This concept leads to the realization of a certain similarity between the behavior of a nonlinear resistive circuit with constant or slowly varying currents and the behavior of a ferromagnetic circuit, which is inherently non-linear, with constant or slowly varying flux.

In an important respect, however, the magnetic circuit differs markedly from a simple electric analogue. When used to carry current, an electrical conductor is normally separated from other conductors by insulating material. A moderately good insulator, such as rubber, has a conductivity of about 10^{-20} times that of copper, so that in most electric circuits the conduction current in the insulating material is negligibly small in comparison with that in the conductors. In contrast, no magnetic insulator of similar properties is known. The most diamagnetic substance known, namely bismuth, has a permeability about 0.9998 that of air, while in usual practice the relative permeability of the air, which ordinarily is the flux insulator, is seldom less than about 10^{-4} and often only about 10^{-2} times that of the ferromagnetic material. An added complication results from the fact that the desired magnetic path often crosses an air gap, as shown in Figs. 2 and 3, which is in parallel magnetically with another air path having only perhaps five or ten times the magnetic insulating effect of the air gap. This condition serves to by-pass a considerable percentage of the magnetic flux from the useful path into the parallel air path where it is unusable. This shunting action gives rise to a condition known as *flux leakage*.

The shunting action of air paths can be visualized in terms of a bare copper electric circuit immersed in an electrolyte having a conductivity of perhaps 10^{-3} times that of copper. If the copper conductor is short and continuous and has a large cross section, relatively little current traverses the electrolyte. The analogous condition in a magnetic circuit is illustrated by the structures of Fig. 1. If, on the other hand, the conductor has a break or gap in it such that the current is forced across even a short length of path in the electrolyte, the current spreads to a considerable extent into the entire surrounding region. This spreading is analogous to what occurs in the magnetic circuits of Figs. 2 and 3, where the useful flux crossing the air gap may be only 80 or 90 per cent, or for some designs less than 50 per cent of the flux traversing much of the path in the iron. The determination of the current paths in and the resistance of the electrolyte in the electric circuit, or the flux paths in and magnetic reluctance of the material in the magnetic circuit, is a problem not susceptible in general to exact calculation. The leakage effects must be borne in mind in all magnetic calculations.

Within the scope of certain simplifying assumptions, an understanding of the magnetic circuit concept is aided by quantitative consideration of a magnetic configuration and a closely analogous electric configuration. Two such analogous configurations are shown in Fig. 5. Owing to the symmetry of form of each configuration, explicit solutions can be obtained directly through applying the field equations to them. These solutions indicate that the simple magnetic arrangement of Fig. 5b can be analyzed in terms of a lumped-parameter magnetic circuit in a manner similar to that by which the arrangement of Fig. 5a can be analyzed in terms of a lumped-parameter electric circuit.

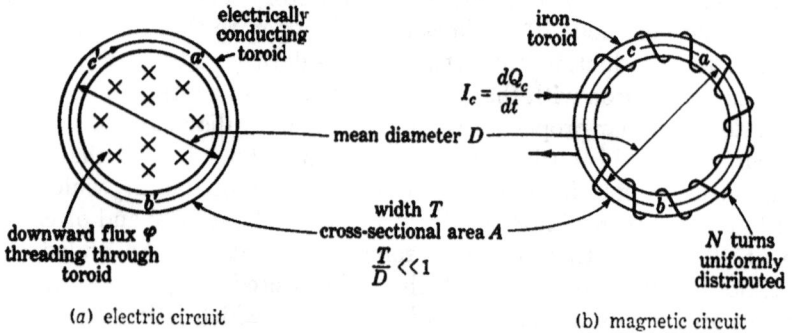

electrically conducting toroid

iron toroid

$$I_c = \frac{dQ_c}{dt}$$

mean diameter D

downward flux φ threading through toroid

width T

cross-sectional area A

$$\frac{T}{D} \ll 1$$

N turns uniformly distributed

(a) electric circuit   (b) magnetic circuit

FIG. 5.   Idealized actual analogous electric and magnetic circuits.

Figure 5a shows a homogeneous electrically conducting toroid through which a magnetic flux is directed downward into the paper. For any cap bounded by a line such as $a'b'c'a'$ within the toroid, the flux φ is, by Eq. 2,*

$$\varphi = \int \mathcal{B} \cdot n \, ds. \qquad [3]$$

This flux may be set up by any suitable means. If φ has a time derivative, the Faraday induction law (Eq. 1, Ch. I of the volume on electric circuits in this series) gives, for any closed path similar to $a'b'c'a'$ within the conductor,

$$E = \oint \mathcal{E} \cdot d\ell = -\frac{d}{dt} \int \mathcal{B} \cdot n \, ds \qquad [4]$$

or

$$E = -\frac{d\varphi}{dt}. \qquad [5]$$

The quantity E is called the *electromotive force* and is proportional only to the time derivative of the flux threading or linking the toroid.

In Fig. 5b, an analogous magnetic arrangement is shown. The toroid is made of iron or other homogeneous ferromagnetic material, and is wound with N turns uniformly distributed and carrying a current I_c.

* In this chapter, the equations are written in such a form that any consistent unrationalized system of units can be used. Unfortunately in engineering literature and practice, a mixture of units is encountered which requires the use of numerous conversion factors in the equations. This clumsiness is avoided in the literal equations of this chapter.

Along any closed path such as *abca* within the iron of the toroid, the line integral of the component of magnetic field intensity $\mathcal{H}$ in the direction of the element of path *dℓ*, for a steady or slowly varying current, is

$$F = \oint \mathcal{H} \cos \theta \, d\ell, \qquad [6]$$

or, in vector notation,

$$F = \oint \mathcal{H} \cdot d\ell. \qquad [7]$$

On the basis of Eq. 63, Ch. I of the volume on electric circuits,

$$F = 4\pi \int \mathcal{J}_c \cdot n \, ds = 4\pi N I_c, \qquad [8]$$

in which $\mathcal{J}_c$ is the current density in the winding, or

$$F = 4\pi N \frac{dQ_c}{dt} = 4\pi \frac{dQ}{dt}. \qquad [9]$$

In Eq. 8, the integration is over any cap bounded by line *abcd*. In Eq. 9, $N dQ_c/dt$ or dQ/dt is the time rate of total charge passing through the plane of the toroid.

By analogy to the electromotive force E of Eqs. 4 or 5, the quantity F of Eqs. 8 or 9 is called the *magnetomotive force*, often being abbreviated to mmf. The analogy between dQ/dt and $d\varphi/dt$ is essentially mathematical rather than physical, since the charge as visualized is actually moving down through the plane of the toroid in the wires at a rate dQ/dt; whereas the flux φ merely describes the state of a medium, the state changing with time at the rate $d\varphi/dt$. In the physical sense no idea of a velocity is associated with $d\varphi/dt$.

If the two toroids are considered as electric and magnetic circuits, respectively, the analogies can be extended further and the effects of E and F on the respective circuits can be compared. If the electric field intensity is uniform along the path $a'b'c'a'$ of Fig. 5a, and if $a'b'c'a'$ is located at the mean radius of the conducting toroid, the magnitude of the electric field intensity is

$$\mathcal{E} = \frac{E}{\ell} = \frac{E}{\pi D}, \qquad [10]$$

where *ℓ* is the length of the path and D is its mean diameter. The electric field gives rise to a current density $\mathcal{J}$ which is a function of the electric field intensity $\mathcal{E}$ and the resistivity ρ of the material of the toroid:

$$\mathcal{J} = \frac{\mathcal{E}}{\rho}. \qquad [11]$$

If the ratio of the thickness T of the toroid to the mean diameter D is so small that ℓ is substantially the same for all paths,

$$\mathcal{J} = \frac{E}{\ell\rho} \qquad [12]$$

and the total current in the toroid is

$$I = A\mathcal{J} = A\frac{E}{\ell\rho} = \frac{E}{R}, \qquad [13]$$

where

$$R \equiv \frac{\rho\ell}{A}, \qquad [14]$$

A is the cross-sectional area of the toroid, and R is its electrical resistance. Thus, in so far as the current in the conductor is concerned, the toroidal configuration linked by a flux having a time derivative is equivalent to a lumped source of electromotive force E, connected in series with a path of resistance R. The magnitude of the voltage of the source is $\frac{d\varphi}{dt}$, and in the portion of the circuit external to the source the drop in electric potential $V_{a'b'}$ from any point a' to another point b' is

$$V_{a'b'} = \int_{a'}^{b'} \mathcal{E} \cdot d\ell, \qquad [15]$$

or

$$V_{a'b'} = IR_{a'b'}, \qquad [16]$$

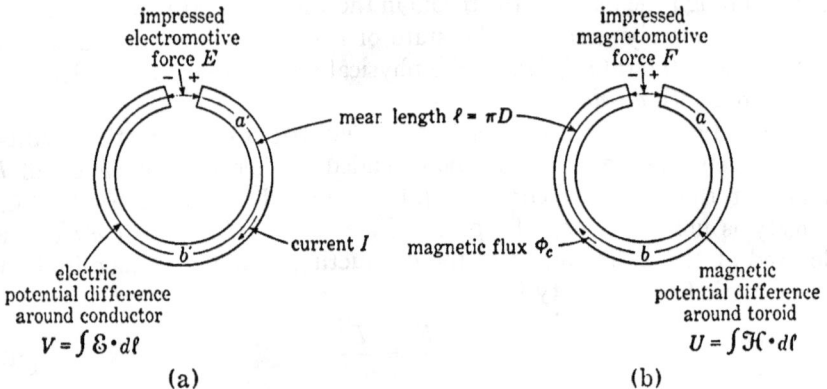

FIG. 6. Schematic diagrams showing motive forces lumped for idealized actual analogous electric and magnetic circuits.

where $R_{a'b'}$ is the resistance between a' and b'. These conditions are illustrated by Figs. 6a and 7a. When the resistivity of the material is not constant, the resistance is a nonlinear function of the current,

$$R = f(I). \qquad [17]$$

For the magnetic configuration of Fig. 5b, relations mathematically analogous to those expressed by Eqs. 10 to 17 can be written. The mag-

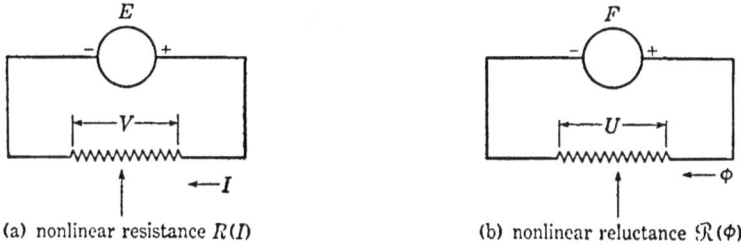

(a) nonlinear resistance $R(I)$ (b) nonlinear reluctance $\mathfrak{R}(\phi)$

FIG. 7.   Circuit representations for idealized actual analogous electric and magnetic circuits.

nitude of the magnetic field intensity $\mathfrak{H}$ at any point on a path *abca* of Fig. 5b, if *abca* is located at the mean radius, is

$$\mathfrak{H} = \frac{F}{\pi D} = \frac{F}{\ell} = \frac{4\pi N I_c}{\ell}, \qquad [18]$$

where I_c is the current in each of the N turns on the toroid.

This magnetic field intensity establishes a magnetic flux density $\mathfrak{B}_c$ which is a function of $\mathfrak{H}$ and the permeability μ of the ferromagnetic material:

$$\mathfrak{B}_c = \mu\mathfrak{H}. \qquad [19]$$

For ferromagnetic materials, this function can seldom be considered linear. It is usually expressed by an experimentally determined curve such as that shown in Fig. 13a of Ch. I. If the ratio of T to D is so small that $\mathfrak{B}_c$ is substantially the same for all paths,

$$\mathfrak{B}_c = \mu\frac{F}{\ell}, \qquad [20]$$

and the total flux in the core is

$$\phi_c = A\mathfrak{B}_c = A\mu\frac{F}{\ell} = \frac{F}{\mathfrak{R}}, \qquad [21]$$

where

$$\mathfrak{R} \equiv \frac{\ell}{\mu A} \qquad [22]$$

and is called the *reluctance* of the magnetic path. The reluctance of the iron toroid is analogous to the resistance of the electrical conducting toroid. Equations 21 and 22 show that in so far as flux in the iron is concerned, the iron toroid having a uniformly distributed winding carrying a current I_c is equivalent to a lumped source of magnetomotive force F connected in series with a magnetic path of reluctance $\mathcal{R}$. The magnitude of the source of magnetomotive force F is $4\pi \dfrac{dQ}{dt}$. In the portion of the circuit external to the source, the drop in magnetic potential U_{ab} from any point a to another point b is

$$U_{ab} = \int_a^b \mathcal{H} \cdot d\ell, \qquad\qquad [23]$$

or

$$U_{ab} = \phi_c \mathcal{R}_{ab}, \qquad\qquad [24]$$

where $\mathcal{R}_{ab}$ is the reluctance between a and b. These conditions are illustrated by Figs. 6b and 7b. When the permeability of the ferromagnetic material is not constant, the reluctance is a nonlinear function of the flux,

$$\mathcal{R} = f(\phi_c). \qquad\qquad [25]$$

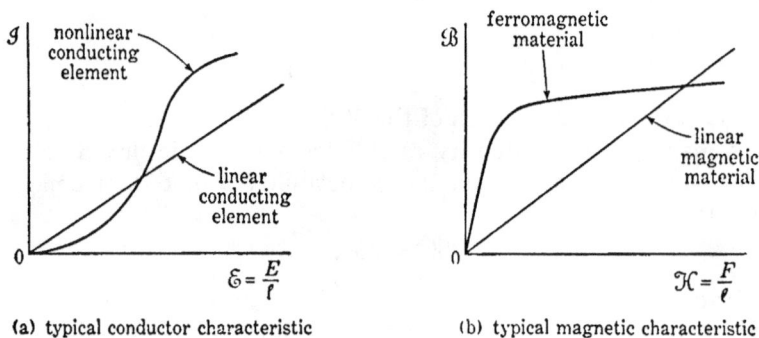

(a) typical conductor characteristic   (b) typical magnetic characteristic

FIG. 8. Characteristic curves of materials for idealized actual analogous electric and magnetic circuits.

The forms of the characteristic curves for the materials of the toroids for idealized actual analogous electric and magnetic circuits are shown in Fig. 8.

The use of reciprocals of R and $\mathcal{R}$ often is convenient:

$$G \equiv \frac{1}{R} \qquad\qquad [26]$$

and

$$\mathcal{P} \equiv \frac{1}{\mathcal{R}}, \qquad\qquad [27]$$

in which G is the *conductance* of the electrical conductor and $\mathcal{P}$ is the *permeance* of the magnetic path. In terms of these **reciprocals, Eqs. 16** and 24 become, respectively,

$$I = GV \qquad [28]$$

and

$$\phi_c = \mathcal{P}U. \qquad [29]$$

The ideas of permeability, reluctance, and permeance are useful as aids in reasoning. Only when the relationship between $\mathcal{B}$ and $\mathcal{H}$ can be assumed to be a linear function — that is, when μ is essentially a constant — are μ, $\mathcal{R}$, or $\mathcal{P}$ useful for computation.

In the foregoing magnetic relations, the ratio T/D is assumed to be small compared with unity. Otherwise the toroid must be divided into elementary concentric rings for each of which l is a constant, and $\mathcal{H}$ must be determined for each ring. The flux density $\mathcal{B}$ for each ring must then be determined from the curve of $\mathcal{B}$ as a function of $\mathcal{H}$ for the material, because of the nonlinear relationship involved. Then, to obtain the total flux ϕ, the flux density $\mathcal{B}$ must be integrated over the cross-sectional area of the toroid. The analogous argument applies to the calculation of I if T is appreciable compared with D, and all of the magnetic flux of Fig. 5a is enclosed by the inner ring of the toroid.

The concept of a magnetic circuit with lumped parameters is made clearer by a consideration of methods for computing ϕ_c in the toroid. In the analogous electrical circuit, the current can be and usually is computed as if the toroid were cut and opened as in Fig. 6a and the electromotive force E were applied between the faces thus formed. Similarly the magnetic toroid may be visualized as cut and opened as in Fig. 6b, and the entire magnetomotive force F as lumped and applied between the faces thus formed. The ferromagnetic material may now be visualized as external to the source, so that, whenever a flux ϕ_c is present in the material, a magnetic potential difference U must exist between its end faces. Since the toroid is assumed to be of uniform cross-sectional area and material, the potential difference is, from Eq. 24,

$$U = \phi_c \mathcal{R}, \qquad [30]$$

which, from Eq. 7, must equal the sum of the lumped sources of magnetomotive force. Hence,

$$F = U = \phi_c \mathcal{R}. \qquad [31]$$

The direction of the magnetomotive force of a magnetic source is determined by the right-hand screw rule: If the screw is turned in the direction of the current around the coil, the screw advances in the direction of increase of magnetic potential. Outside the source, the decrease of magnetic potential is in the direction of the flux. If a magnetic circuit is completely traced, the sum of the source rises must equal the sum of the reluctance drops.

Dividing Eq. 31 by l gives

$$\frac{F}{l} = \frac{U}{l} = \phi_c \frac{\mathcal{R}}{l} = \frac{\phi_c}{\mu A} = \frac{\mathcal{B}}{\mu} = \mathcal{H}. \qquad [32]$$

In Eq. 32, the quantity U/l is the magnetic potential difference per unit length of the toroid. However, for a circuit having nonuniform cross-sectional area, or nonhomogeneous construction, U is taken across only a homogeneous portion of path of uniform cross-sectional area and length l, to give $\mathcal{H}$ in that portion of path.

The foregoing concepts lead to the idea of a magnetic-circuit-connection diagram as is shown schematically in Fig. 7b, and serve to show a parallelism to the electric-circuit-connection diagram of Fig. 7a. The circuit-element concept is illustrated when the properties of the materials are represented in characteristic curves as in Figs. 8a and 8b. In the above discussion, however, E is assumed to be caused by a $d\varphi/dt$, and F caused by a dQ/dt. To maintain a constant E by means of a constant $d\varphi/dt$ is impractical because finally an infinite φ is required; but to maintain a constant F by a constant dQ/dt is practical, and is customarily done, because a constant current can be maintained. On the contrary, E can be maintained practically by a battery of negligible internal resistance connected between the faces of the cut in Fig. 6a, but F cannot be maintained by current in a coil having a core of negligible internal reluctance, connected between the faces of the cut in Fig. 6b, because materials of very small reluctivity are not available. When the magnetic problem is visualized in terms of a circuit with a magnetomotive force applied between the faces of a cut in Fig. 6b, one should recognize that the cut is hypothetical and does not introduce an air gap into the path of the magnetic flux. Actually, of course, the problem is fundamentally one involving a three-dimensional field; the fact should not be overlooked that only for special configurations can it be reduced to a circuit problem.

Since the permeability of a ferromagnetic material is a variable, the reluctance of the path through such material is a variable, and hence is not a particularly useful quantity. Neither the permeability nor the reluctance should be used except to provide a visualization of the magnetic circuit in terms of its more familiar electrical analogue as an aid in reasoning. Except for the use of incremental permeability, which can be treated as a constant, as illustrated in Art. 14, Ch. VI, numerical calculations on magnetic structures can best be carried out by use of the curves of flux density plotted as functions of magnetizing force as is done in Ch. III.

Magnetic reluctance elements connected in series or parallel can be shown to combine in the same manner as do resistance elements in an electric circuit of corresponding configuration.

Although the magnetic and electric circuits are mathematically analogous in many respects, they have certain important physical differences. For example, in the analogous nonlinear electric circuit of Fig. 7, except at extremely low temperature, power equal to I^2R is dissipated in the circuit whenever the current I is present. However, no power is dissipated or absorbed directly in the circuit portion of the magnetic toroid by the magnetic field so long as the flux density is constant everywhere within the circuit. The fact that permanent magnets, if properly cared for, retain their magnetism throughout extremely long periods is a good justification for this statement. When the magnetic field is established by an exciting winding on the structure, a power equal to $I_c^2R_c$ is dissipated in the coil, where R_c is the resistance of the coil and I_c is the current in it.

CHAPTER III

Principles of Magnetic-Circuit Computations

The analogies between the electric circuit and the magnetic circuit seem to indicate, upon first consideration, that the Ohm's-law relationship among magnetomotive force, flux, and reluctance ought to provide a straightforward method for solving magnetic-circuit problems. The direct application of this method is made difficult, however, by the relatively large flux leakage encountered in magnetic-circuit problems and by the dependence of the reluctance of a ferromagnetic material upon the flux density. For this reason, special ingenious techniques have been devised for the solution of magnetic-circuit problems. These techniques are the subject of this chapter and Ch. IV. The methods of solution are of course not dependent upon the kind of units employed, but the solutions are somewhat facilitated by the selection of a convenient set of units.

1. Units for Practical Magnetic Calculations

Unfortunately, usage of systems of magnetic units in English-speaking countries is not standardized. In scientific work, the cgs electromagnetic system is used, but in practical design computations the ampere-turn, the ampere-turn per inch, and the maxwell per square inch are the commonly used units. These of course fit into no recognized system but prove convenient when dimensions of apparatus are expressed in inches.

The adoption by the International Electrotechnical Commission of the mks practical system of units,[1] including as it does all the practical electrical units, increases the probability of general use of its magnetic units. Thus engineers need to be familiar with two, and possibly three, sets of magnetic units; namely, the cgs electromagnetic and the mks practical, the mixed English sets. In this treatment, the mixed English units are used for magnetic-circuit arithmetical calculations.

Since in magnetic-circuit computations only the units for F (or U), $\mathcal{H}$, and $\mathcal{B}$ are used frequently, less complication arises in the use of the mixed English set than may at first appear. In this set, the unit for F (or U) is the ampere-turn. That for $\mathcal{H}$ is the ampere-turn per inch. The unit for $\mathcal{B}$ is commonly expressed in lines per square inch, one line being the same as one maxwell, or in kilolines (thousands of lines) per square inch, since commonly used densities are many thousands of lines per square inch.

---

[1] Systems of units are discussed in this textbook series in Appendix C of the volume on electric circuits.

The conversion factors given in Table I are useful for dealing with the various systems of units.

TABLE I

Multiply	by	to obtain
F in ampere-turns	$4\pi = 12.57$	F in pragilberts
F in ampere-turns	$0.4\pi = 1.257$	F in gilberts
F in gilberts	10	F in pragilberts
$\mathcal{H}$ in ampere-turns/in.	$\dfrac{4\pi 100}{2.54} = 495$	$\mathcal{H}$ in praoersteds
$\mathcal{H}$ in ampere-turns/in.	$\dfrac{0.4\pi}{2.54} = 0.495$	$\mathcal{H}$ in oersteds
$\mathcal{H}$ in oersteds	1000	$\mathcal{H}$ in praoersteds
$\mathcal{B}$ in lines/sq in.	$\dfrac{10^{-4}}{6.45} = 1550 \times 10^{-8}$	$\mathcal{B}$ in webers/sq m
$\mathcal{B}$ in lines/sq in.	$\dfrac{1}{6.45} = 0.1550$	$\mathcal{B}$ in gausses
$\mathcal{B}$ in gausses	10^{-4}	$\mathcal{B}$ in webers/sq m

The permeability μ_0 of free space is:

$$\text{1 cgs electromagnetic unit}$$
$$\text{3.192 mixed English units}$$
$$10^{-7} \text{ mks practical unit}$$

In Figs. 1a and 1b are shown representative normal magnetization curves for specimens of several ferromagnetic materials commonly used in the construction of electromagnetic apparatus. The magnetizing force $\mathcal{H}$ in Fig. 1a is plotted on a logarithmic scale only to open up the curves for low values of $\mathcal{H}$. The use of the arithmetic scale for $\mathcal{H}$ in Fig. 1b gives the curves in a form particularly useful for certain problems treated in subsequent articles. Since the magnetic properties of the ferromagnetic materials are markedly affected by small changes in chemical composition, by heat treatment, and by mechanical working, the properties of different specimens of a ferromagnetic material may differ considerably from those represented by the curves of Figs. 1a and 1b.

2. CALCULATIONS FOR MAGNETIC PATHS WHOLLY IN IRON

As previously stated, ferromagnetic materials may be used for any of several reasons, such as to conduct the flux along some particular path, to increase the flux obtained by a fixed magnetomotive force, or to concentrate the flux at some particular place in a device. These uses require

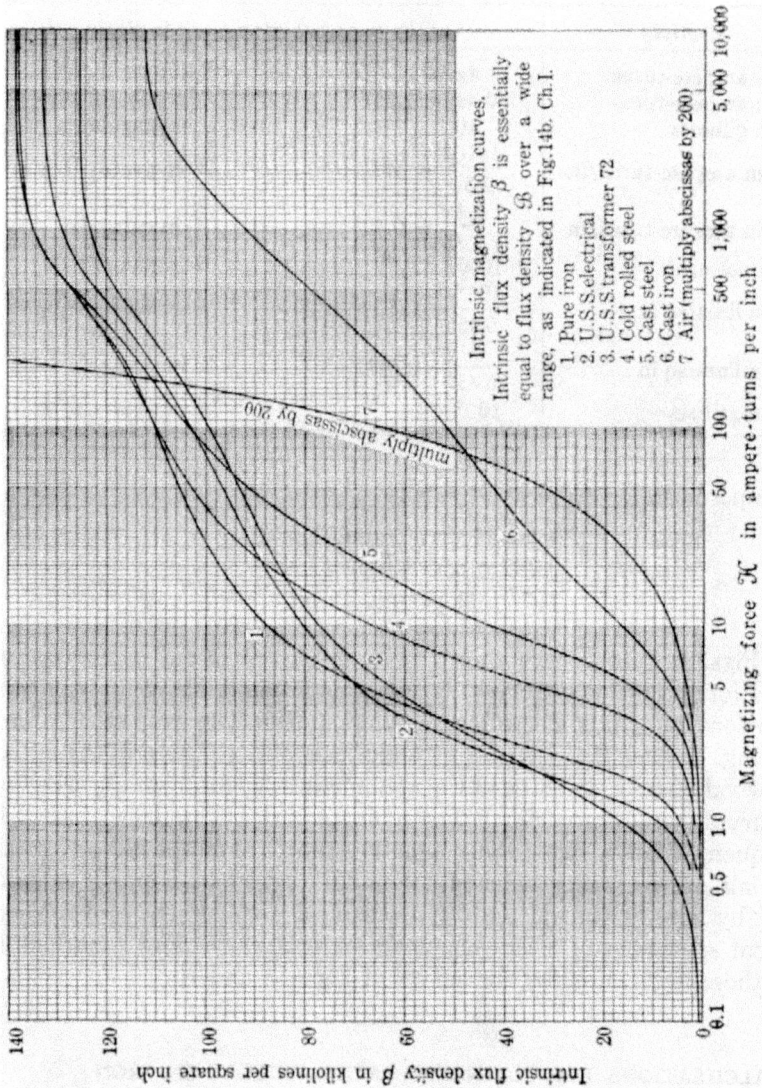

Intrinsic magnetization curves.
Intrinsic flux density β is essentially equal to flux density $\mathcal{B}$ over a wide range, as indicated in Fig. 14b. Ch. I.
1. Pure iron
2. U.S.S. electrical
3. U.S.S. transformer 72
4. Cold rolled steel
5. Cast steel
6. Cast iron
7. Air (multiply abscissas by 200)

Magnetizing force $\mathcal{H}$ in ampere-turns per inch

Intrinsic flux density β in kilolines per square inch

FIG. 1a.

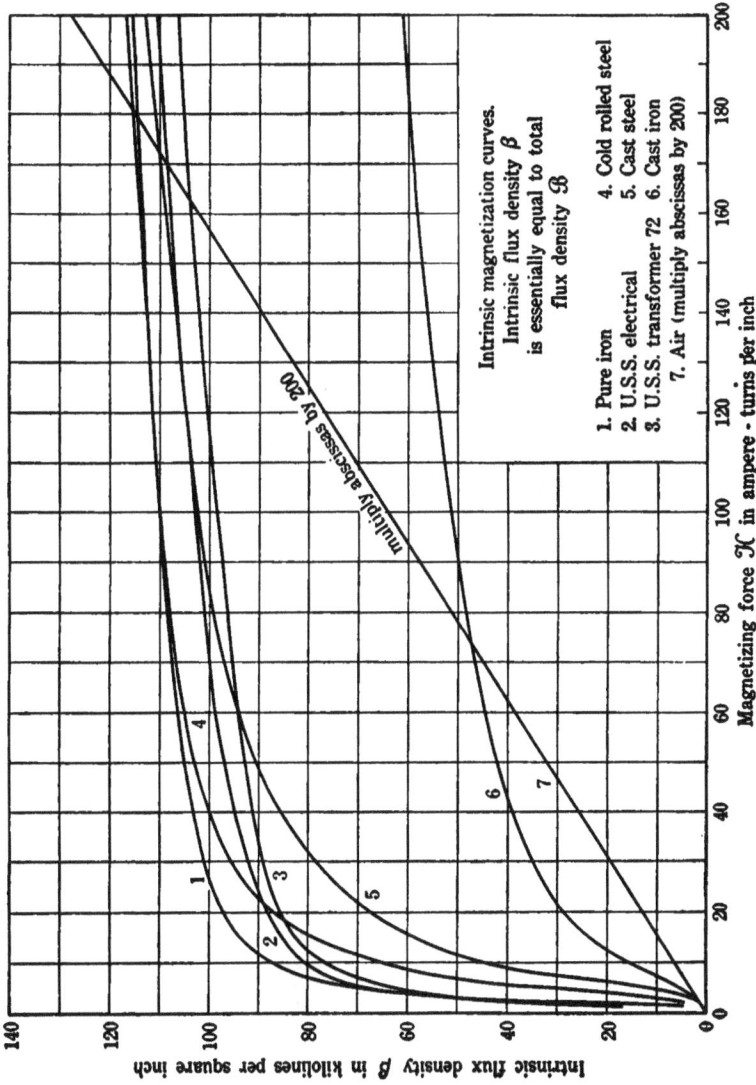

Intrinsic magnetization curves.
Intrinsic flux density β
is essentially equal to total
flux density $\mathcal{B}$

1. Pure iron   4. Cold rolled steel
2. U.S.S. electrical   5. Cast steel
3. U.S.S. transformer 72 6. Cast iron
          7. Air (multiply abscissas by 200)

Magnetizing force $\mathcal{H}$ in ampere - turns per inch

FIG. 1b.

Intrinsic flux density β in kilolines per square inch

two broad classes of magnetic-circuit calculations:

- (a) The determination of the magnetomotive force required to produce a desired flux or flux density in a specified part of the structure.
- (b) The determination of the flux or flux density produced at specified places in a magnetic structure by magnetomotive forces impressed at various places throughout the structure.

Strictly speaking, the magnetic-circuit method of analysis does not yield flux densities except as averages of total fluxes over cross-sectional areas of the circuit; the determination of flux distribution is a field problem.

When the problem is to determine the magnetomotive force required to produce a desired total flux or flux density, that is, the calculation as in (a), the procedure is direct, provided the leakage flux is neglected or estimated. In each portion of a series magnetic path having a cross-sectional area A, the average value of flux density $\mathcal{B}$ is equal to the ratio of the total flux ϕ to the area A. The value of magnetizing force $\mathcal{H}$ required to establish this value of $\mathcal{B}$ is determined from a curve of $\mathcal{B}$ plotted as a function of $\mathcal{H}$ for the particular material. This value of $\mathcal{H}$ is then multiplied by the length of that portion of the path for which $\mathcal{B}$ is assumed constant, to give the magnetic potential difference U between the ends of that portion of the path; that is,

$$U_{ab} = \mathcal{H}\ell_{ab}, \qquad [1]$$

where the distance a to b is the length of path of uniform material and cross-sectional area. If the path includes portions of different kinds of ferromagnetic material, the value of $\mathcal{H}$ for each material is multiplied only by the length of path in that material to give the magnetic potential difference for that portion of the path. The sum of the magnetic potential differences U_{ab}, U_{bc}, $\cdots$ for all such portions of paths a–b, b–c, c–d, $\cdots$ taken around the series circuit gives the total magnetomotive force required; that is,

$$F = U_{ab} + U_{bc} + U_{cd} + \cdots + U_{na}. \qquad [2]$$

If the construction of the circuit is such that the average flux density differs markedly from the extremes of flux density on the cross-sectional area, special procedure based on that outlined on p. 51 must be employed.

When the problem is to determine the total flux or flux density produced by magnetomotive forces impressed at various places, that is, the calculation as in (b), the procedure is not so straightforward even if leakage fluxes are neglected. In certain simple combinations of paths, graphical methods are applicable. These are illustrated in examples that follow. In complicated combinations of paths, a successive-approximations method leads rapidly to a solution. For these complicated combinations the magnetomotive force required to produce an assumed value of flux ϕ_1 is first calculated. If the calculated magnetomotive force does not approach the impressed magnetomotive force within reasonable limits, a second trial value ϕ_2 is chosen, larger or smaller than ϕ_1, depending upon whether the magnetomotive force applied is more or less than the amount required to equal the magnetic potential drop produced by the assumed flux ϕ_1. After a few tries and with the application of reasonable judgment in the choice of ϕ_1, ϕ_2, and so on, a solution is obtained.

When the magnetic structure is assembled through stacking laminations punched from thin sheet material, the volume occupied by the stacked laminations does not truly represent the volume of iron that conducts the flux. A region whose relative permeability is essentially that of air exists between the laminations because of the presence of irregularities or scale on the surfaces of the sheets, because of a thin coat of insulating varnish sometimes deliberately applied to avoid contact between the sheets and so to reduce eddy-current loss as explained in Ch. V, or because of burrs on the edges of the sheets which are caused by the punching operation. This region conducts negligible flux on account of its low relative permeability; so, to allow for its effect in decreasing the net volume of iron, the customary procedure is to express the effective cross-sectional area of iron as equal to the cross-sectional area of the stack times a factor called the *stacking factor*. The stacking factor, defined as the ratio of cross-sectional area of iron to cross-sectional area of stack, ranges between about 0.95 and 0.90, for lamination thicknesses between 0.025 inch and 0.014 inch respectively. For thinner laminations, about 0.001 to 0.005 inch thick, owing to greater difficulties in clamping and in reducing burrs, and because the insulating film is proportionately thicker, the stacking factor ranges from about 0.4 to 0.75, but can be improved by special manufacturing procedures. The flux density in the iron is then equal to the total flux ϕ divided by the product of the stacking factor and the cross-sectional area of the stack.

The details of the methods involved in calculations of this kind are best illustrated by the solution of a few numerical examples.

3. ILLUSTRATIVE EXAMPLE OF CALCULATIONS FOR MAGNETIC STRUCTURE OF UNIFORM MATERIAL AND CROSS–SECTIONAL AREA

The magnetic structure shown in Fig. 2 is similar to that of a core-type alternating-current transformer. It is made of annealed 4 per cent silicon transformer sheet-steel laminations (designated as transformer 72) 0.025 inch thick, stacked into a pile 3 inches thick. The stacking factor is 0.95 owing to nonmagnetic scale on the surfaces of the laminations. The exciting winding has 200 turns. The problem is to compute the current required in the exciting winding to produce a core flux of 500,000 maxwells. Leakage flux is to be neglected.

Solution: The cross-sectional area of the stack is uniform (except at the corners) and, for the dimensions given, is 6 sq in. The equivalent cross-sectional area of iron equals the stack area multiplied by the stacking factor; that is, 6 × 0.95 or 5.7 sq in. Hence, except at the corners, the average flux density $\mathfrak{B}$ is 500/5.7, or 87.5 kilolines/sq in. If this flux density is taken to be uniform throughout the core — that is, if leakage flux is neglected and no attempt is made to investigate the exact effect of the corners of the core on the flux distribution — then, from curve 3 of Fig. 1a, the magnetizing force or magnetic potential difference per unit length of core required to establish a flux density of 87.5 kilolines/sq in. is approximately 25 amp-turns/in.

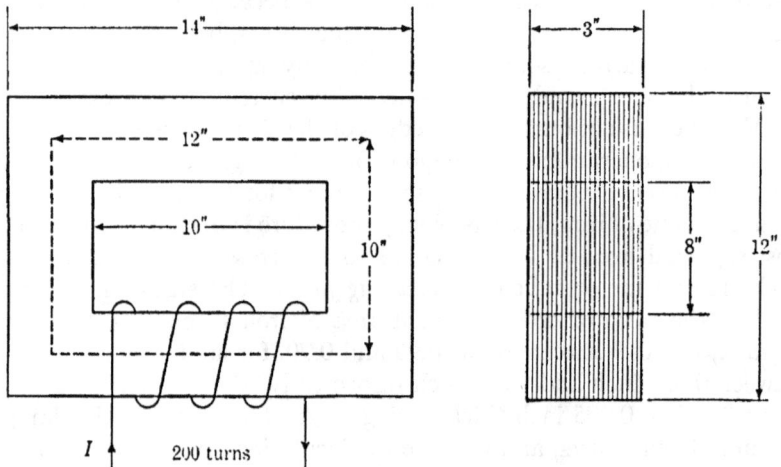

Fig. 2.   Magnetic structure of a core-type transformer.

The precise computation of the flux distribution at the corners is difficult. As an approximation, the mean length of the magnetic circuit is assumed to be 44 in. as shown by the dotted lines in Fig. 2. The magnetic potential difference U between the faces of a hypothetical cut in the circuit, between which is lumped the applied magnetomotive force, is approximately 25 × 44, or 1,100 amp-turns. Since the coil has 200 turns, the exciting current is 1,100/200, or 5.5 amp, approximately.

For this simple illustration, in which the magnetic field intensity is assumed to be uniform throughout the length of the core, calculation of the total flux when the magnetomotive force is specified is an equally simple matter. For example, the flux produced by an exciting current of 11 amp is calculated as follows:

Solution: The magnetomotive force for the specified exciting current is 11 × 200, or 2,200 amp-turns. Since the cross-sectional area of the core is assumed to be uniform, the magnetizing force is uniform and equal to 2,200/44, or 50 amp-turns/in. From Fig. 1a the flux density established by this value of $\mathcal{H}$ is approximately 93 kilolines/sq in. The total flux is therefore 93 × 5.7, or 530 kilolines, or 530,000 maxwells approximately.

4. ILLUSTRATIVE EXAMPLES OF CALCULATIONS FOR MAGNETIC STRUCTURE OF UNIFORM MATERIAL AND NONUNIFORM CROSS-SECTIONAL AREA

In the magnetic structures of electrical apparatus, portions of the magnetic circuit often are of different cross-sectional areas. Sometimes these differences are the result of structural-strength or stiffness considerations or are to provide a section of core that is readily saturated. The analysis of such magnetic circuits may often become exceedingly

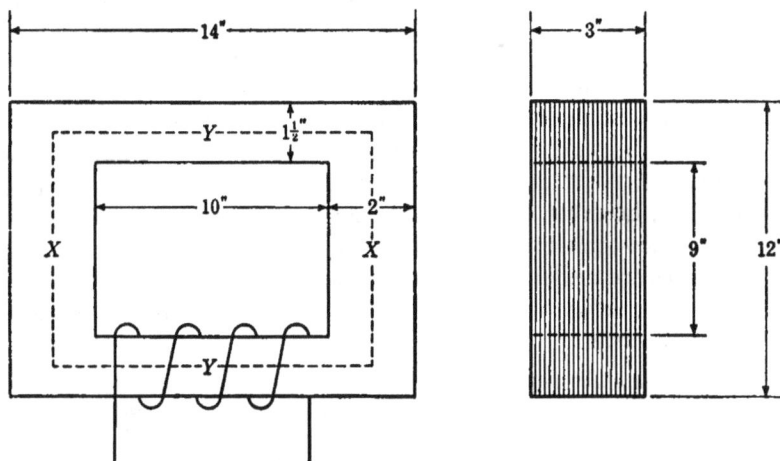

FIG. 3. Illustrative magnetic structure. The legs X and Y have different cross-sectional areas.

complex. The transformer core shown in Fig. 3 is chosen as an illustration, not because it is typical of any particular device but because it is less confusing and equally satisfactory to consider as a first example of a simple structure. This core is made of annealed 4 per cent silicon transformer sheet-steel punchings 0.014 inch thick stacked into a pile 3 inches thick. The stacking factor is 0.91 for the particular thickness of lamination and method of clamping the core. The problem is to calculate the ampere-turns necessary to establish a flux of 410,000 maxwells in the core. The leakage flux is to be neglected.

Solution: The cross-sectional area of the iron portion of the legs X is $2 \times 3 \times 0.91$, or 5.46 sq in., and of the legs Y is $1.5 \times 3 \times 0.91$ or 4.10 sq in. The flux density in the iron is 410/5.46 or 75 kilolines/sq in. for legs X and 410/4.10 or 100 kilolines/sq in. for legs Y. From curve 3 of Fig. 1a, the magnetizing force is 8.6 amp-turns/in. for legs X, and 108 amp-turns/in. for legs Y. The mean length of the flux paths as shown by the dotted lines in Fig. 3 is 21 in. total for two legs X, and 24 in. total for two legs

Y. The sum of the magnetic potential differences for the two legs X is 8.6 × 21 or 180 amp-turns, and that for the two legs Y is 108 × 24 or 2,590 amp-turns. The total ampere-turns required for the whole magnetic circuit are therefore 2,770.

When a specified magnetomotive force acts on a core made of portions of different materials or cross sections, the problem of calculating the fluxes produced is not always simple. If the core consists of more than two different portions in which the relationships between $\mathfrak{B}$ and $\mathfrak{K}$ are nonlinear, a direct solution for the flux becomes impractical, because, on account of the nonlinearity of each portion, the division of the magnetic potential among the many different portions of the circuit cannot be computed directly. When a study of the magnetic characteristics of the structure involving only a few values of magnetomotive force is contemplated, a successive-approximations process usually leads to a solution in a reasonable time, and by its use even relatively complicated problems can be solved with a sufficient degree of accuracy, provided reasonable judgment is exercised and the method of analysis is systematized. If, however, the core consists of not more than two different nonlinear series-connected portions, such as X and Y in Fig. 3, and, if leakage fluxes are neglected, a direct graphical method of solution can be used. Both methods are illustrated, using a magnetomotive force of 2,700 ampere-turns.

Solution by successive approximations: If the results of the previous calculations were not known, the value to assume for the flux density for the first approximation would necessitate an estimate of the probable magnetic potential difference between the ends of each core portion. Since the cross-sectional area of the iron portion of the Y legs, Fig. 3, is considerably smaller than the cross-sectional areas of the X legs, the flux density is much larger in the Y legs than in the X legs. The Y legs therefore constitute the " neck of the bottle." If the flux density is large enough for saturation effects to be pronounced, most of the magnetomotive force of the exciting coil may be required to overcome the reluctance of the two Y legs in series. As a first approximation, all the magnetomotive force is taken for the Y legs. The resulting magnetic potential gradient in them is therefore 2,700/24, or 112 amp-turns/in. The flux density is then about 100 kilolines/sq in. from Fig. 1a. Since the density is in the region of saturation, the assumption for the distribution of the magnetomotive force between the X and Y legs is nearly correct. Actually, not all the magnetomotive force is absorbed by the Y legs, and the flux density in the Y legs is somewhat less than 100 kilolines/sq in.

As a second approximation, a flux density of 99 kilolines/sq in. is assumed for the Y legs. This flux density requires a magnetizing force of about 96 amp-turns/in. The magnetic potential difference required for the Y legs is then 96 × 24 or 2,300 amp-turns. Since the total flux is the same for both legs, the densities in the respective legs are inversely proportional to the areas. Hence the flux density in the X legs is (4.10/5.46)99 or 74.3 kilolines/sq in., neglecting leakage. At this density, the X legs require a magnetizing force of about 8.3 amp-turns/in., or a magnetic potential difference of 8.3 × 21 or 174 amp-turns. The magnetomotive force required by the whole circuit to produce the assumed flux of 406 kilolines is therefore 2,474 amp-turns.

If the second approximation is not sufficiently close, a third approximation is necessary, and so on until a sufficiently close approximation to the specified magneto-motive force is obtained. A second approximation, using 99.5 kilolines/sq in. for the Y legs, gives 2,450 amp-turns for the Y legs, 74.8 kilolines/sq in. and 180 amp-turns for the X legs, and a total of 408 kilolines and 2,630 amp-turns for the circuit. To attempt to obtain too close an agreement with the specified magnetomotive force is a waste of time. Assumptions made about the effects of the corners, disregard of leakage flux, use of the normal magnetization curve which ignores the effects of hysteresis, inability to read the curves very accurately, deviation of the properties of individual samples of iron from the average curve, and deviation from the assumed stacking factor — all these factors tend to introduce errors. Hence an agreement within about 5 per cent is usually satisfactory.

If the explanatory remarks are eliminated from this calculation and the procedure is reduced to its essentials, it seems relatively simple, even though a " cut-and-try " solution is used.

TABLE II

Data from Fig. 1b annealed sheet steel		Data for portion X		Data for portion Y	
$\mathcal{H}$ amp-turns/ in.	$\mathcal{B}$ kilolines/ sq in.	$U_X = 21\mathcal{H}$ amp-turns	$\phi_X = 5.46\mathcal{B}$ kilolines	$U_Y = 24\mathcal{H}$ amp-turns	$\phi_Y = 4.10\mathcal{B}$ kilolines
1	16.0	21	87	24	65
2	36.2	42	197	48	147
4	57.7	84	315	96	237
6	67.8	126	370	144	278
8	73.8	168	403	192	303
10	77.0	210	421	240	316
15	82.1	315	458	360	337
20	85.1	420	465	480	349
40	92.1	840	503	960	378
60	94.8	1260	517	1440	389
80	97.2	1680	531	1920	398
120	101.0	2520	551	2880	414

Direct graphical solution: This graphical method is used in principle in a large number of steady-state circuit problems. It is used extensively in vacuum-tube circuit studies where the vacuum tube behaves as a nonlinear element with respect to the conduction of current in a manner analogous to the way the magnetic structure behaves as a nonlinear element with respect to magnetic flux.

The procedure is first to determine the relationship between the total flux and the total magnetic potential difference for each of the two nonlinear portions of the circuit. In this example, the two legs X and the two legs Y form one portion each. A linear abscissa scale is required; hence the data of Fig. 1b for the core material used are first tabulated as in Table II. For legs X, the flux ϕ_X is $5.46\mathcal{B}$, and the magnetic potential difference U_X is $21\mathcal{H}$, since the cross-sectional area of the iron in the legs X is 5.46 sq in. and the length of legs X totals 21 in. Likewise for legs Y, ϕ_Y is $4.10\mathcal{B}$, and U_Y is $24\mathcal{H}$. The curves of ϕ_X as a function of U_X and ϕ_Y as a function of U_Y are plotted, as shown in Fig. 4, in such a manner that the abscissa axis of the curve for the legs X has increasing values of U from left to right; whereas the abscissa axis of the curve for the legs Y has increasing values of U from right to left and its zero of abscissas is located at the point where U equals 2,700 on the plot for the legs X. In other words, the plot for legs Y is turned end for end and called a *negative magnetization curve* and its origin is put at the point where U equals 2,700 on the plot for the legs Y. The point 2,700 is chosen because it is equal to the applied magnetomotive force. If any other value of total magnetomotive force is applied, the origin of the curve for the legs Y which has been turned end for end is placed at the point corresponding to this value.

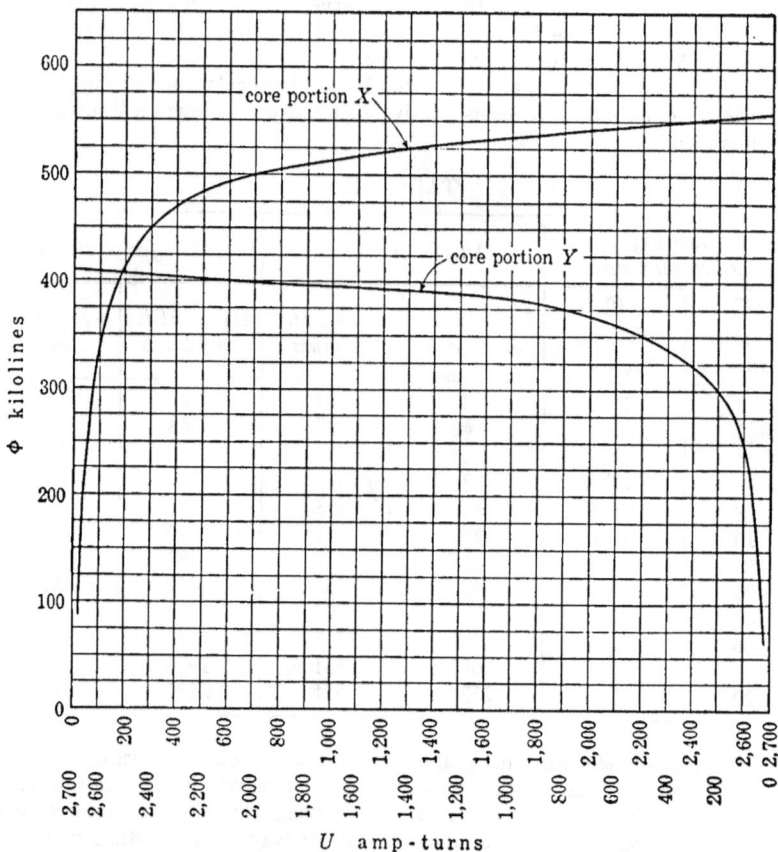

Fig. 4.   Graphical solution for example of Art. 4.

Since the same total flux ϕ is present in both the legs X and Y, the unique solution for the assigned value of F is the value of ϕ corresponding to the co-ordinates of the point of intersection of the two curves. From Fig. 4, the value of ϕ at this point is 410 kilolines, which is substantially in agreement with the result obtained by the successive-approximations method. Also the value of U_X is about 180 amp-turns and U_Y about 2,520 amp-turns.

The explanation shows that a plot of ϕ as a function of F gives a magnetization curve for the portions X and Y combined. Thus, the two magnetization curves are reduced to a single curve for the equivalent core structure.

The graphical method of solution involves preliminary work that is seldom necessary for the successive-approximations method, which can be applied to the solution of more complicated magnetic structures. For problems in which ϕ is desired for only one or two values of F, the successive-approximations method, combined with good judgment, may require the shorter time. If ϕ is required for several different values of F, the graphical method is probably preferable, because when the preliminary work is done the additional time required to obtain ϕ for the specified values of F is small and the total time is probably less than is required by the successive-approximations method. Finally, when a complete magnetization curve is desired, a large number of solutions are probably necessary, and the direct calculation of F for different assumed values of ϕ covering the range desired offers probably the quickest method. Familiarity with the different aspects of each available method provides the information on which to choose the best method for the various problems encountered. These matters are mentioned again in connection with problems involving air gaps.

5. OCCURRENCE OF AIR GAPS IN FERROMAGNETIC CIRCUITS

The magnetic circuits of large numbers of devices contain air gaps in series with the ferromagnetic flux path: for mechanical or constructional reasons, to modify the magnetic behavior of the circuit in order to make the device function in a particular manner, or to produce a magnetic field in air in a desired region. In all rotating machines, for example, air gaps occur between the rotor and stator portions for mechanical reasons and frequently other air gaps occur within the rotor or stator magnetic circuits because of limitation inherent in the construction of an economical design. Also, air gaps are often introduced into iron-core inductors in order to make the inductance of the element essentially independent of the current in the coil throughout a prescribed working range, but at the same time to make the inductance larger than if the inductor had the same coil and only an air core.

Whenever an air gap is inserted in series in a magnetic circuit of ferromagnetic material, an increase in the reluctance of the magnetic path is produced. With certain devices — for example, transformers — the cores, in order to avoid this increase in reluctance, may be assembled from hollow rectangular or ring laminations that are free from air gaps, but, because of the additional complications of winding a coil on a closed magnetic core, such cores are used only when the undesirable effects of the air gap at the joints of the laminations more than outweigh the additional cost of the closed-core winding. Many current transformers used for measurement, which must have low-reluctance magnetic paths, contain laminated ring cores free from air gaps. Also, certain high-quality transformers in communication circuits use cores assembled from compressed powdered Permalloy rings for reasons of performance; the coils are machine wound on a toroidal coil winder.

The cores of most transformers are assembled from separate strips of lamination sheet interleaved or, in other words, stacked like bricks in the manner shown in Fig. 5. The coils are wound separately and slipped

Courtesy General Electric Co.

Assembled core and coils of a luminous-tube transformer; 110-v primary, 15,000-v, 30-ma secondary. The parallel paths and air gaps make this magnetic circuit difficult to calculate.

over the leg of the core before the laminations for one end are assembled. Many audio-frequency transformer cores are made of E-punchings closed by straight punchings, the middle leg of the E being short, so as to leave an air gap between it and the straight punching. Typical lamination shapes are shown in Fig. 12, p. 17. Even with an interleaved construction small air gaps are introduced between the ends of the laminations, and their effect on the total reluctance of the magnetic path must be considered in the design. Recently methods have been developed on a commercial scale for making the core for distribution transformers by threading a continuous strip of material in a roll through the coils after they are wound.* This method gives a core whose reluctance is essentially only that of the iron, and in which the flux paths are kept in the direction

---

* This manufacturing process and an alternative one for accomplishing a similar result are illustrated in Figs. 6 and 8, pp. 290 and 293.

of the grain. Another method consists in winding ribbon spirally on a mandrel, impregnating it with an insulating binder, and then cutting open the core thus formed, so that form-wound coils can be slipped on the core.* Such a core has air gaps, but the flux paths are kept in the direction of the grain. The flux paths can also be kept in the direction of the grain by a special scheme of assembling sheets having 45-degree interlocking corners. With high-voltage transformers, in particular, present practice still adheres to the laminated-interleaved construction, because the difficulties involved in forming a well-insulated high-voltage coil on a closed core assembled from punched rectangular sheets as shown in Figs. 2 and 3 are such as to render the procedure prohibitively expensive if not wholly impractical.

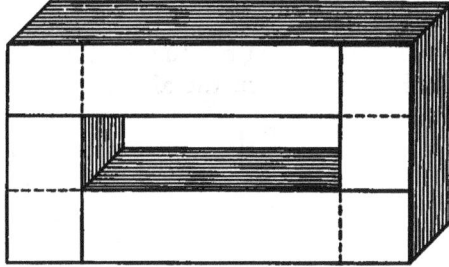

FIG. 5. Construction of a core-type transformer core.

6. FRINGING AND LEAKAGE FLUXES

When an air gap is inserted in a magnetic circuit, the flux spreads out, or fringes, around the gap as shown by the flux lines in Fig. 6, and the flux density in the gap assumes a nonuniform distribution. The flux that terminates near the edges of the gap is called *fringing flux*. Owing to this spreading of the flux, the apparent reluctance of the gap is not that of an air space of the same dimensions as the gap.

Since the permeability of iron is usually at least several hundred times the permeability of air, the reluctance of even a short air gap usually is large compared to the reluctance of the iron portion of the magnetic circuit. Relatively large magnetic potential differences may therefore exist between parts of the core not immediately adjacent to the gap. These potential differences produce flux in the air between those parts of the core, called *leakage flux*, which often is appre-

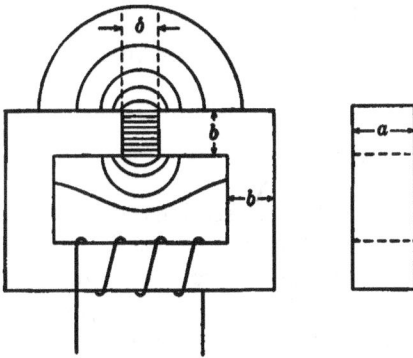

FIG. 6. Magnetic circuit of an iron-core reactor with an air gap showing fringing and leakage fluxes.

* See Figs. 9 and 10, p. 294.

ciable and causes considerable increase in the flux density in certain parts of the core. The design of some electromagnets is such that substantially less than half the flux set up by the magnetizing coil actually crosses the air gap.

The air gaps in rotating machines introduce additional complications. In Fig. 7 is shown the air gap of a typical direct-current dynamo. For reasons that have to do with the operating characteristics of the machine the air-gap clearance usually is larger under the pole tips than in the center of the pole, and the armature windings usually are embedded in slots punched in the armature laminations. The flux that traverses the gap fringes at the pole tips and at the edges of the teeth on the rotor, and because of the large reluctance of the air gap the flux that goes from pole to pole without entering the armature core — that is, the pole leakage flux — is often as much as 25 per cent of the flux in the core of the field pole.

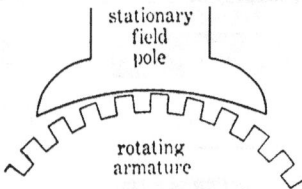

Fig. 7.   Air gap of a typical d-c dynamo.

The calculation of the effects of the fringing and leakage fluxes must be performed accurately in the design of numerous pieces of apparatus. Such calculation involves a three-dimensional magnetic field problem which can be solved approximately by the field mapping methods discussed briefly in Arts. 7 to 11, Ch. i, of the volume on electric circuits in this series. Such methods are usually laborious, and the solution obtained applies only to the particular gap and core used. Approximate methods for evaluating the fringing and leakage flux effects have therefore been developed. Some of these are presented in Arts. 7 to 12.

7. METHODS OF CALCULATION FOR SHORT AIR GAPS

When the air gap is short compared with its cross-sectional dimensions and has parallel faces, magnetic-circuit calculations can usually be performed with a precision approximating the limits of reliability of most magnetic data. As for the other magnetic-circuit computations, the method yields the total flux and average flux density for the air gap. The determination of actual flux distribution in the gap, the distribution of fringing and leakage fluxes, is a field-mapping problem. The procedure here used is first to neglect the effect of leakage flux. The effect of fringing is then taken into account for computation of total flux through replacing the actual parallel-face gap with its fringing by a parallel-face gap assumed to have no fringing but a reluctance equivalent to that of the actual gap. If the cross-sectional dimensions of the core are the same on both faces of the gap, the equivalent gap is assumed to have a length δ equal to the actual gap, but to have a cross-sectional area

$$A = (a + \delta)(b + \delta), \qquad [3]$$

where a and b are the cross-sectional dimensions of the actual core faces, Fig. 6. In other words, the cross-sectional dimensions of the equivalent gap are the original dimensions increased by the gap length δ.

If one of the faces of the gap has a cross-sectional dimension much larger than the corresponding dimensions of the other face, a correction of 2δ should be used. The quantity 2δ follows from a consideration of images, because the configuration of the fringing flux lines when one face is large and the other small is similar to that for a gap of twice the length with cross-sectional dimensions for both parallel faces equal to those of the smaller face.

Experience shows that these rules ordinarily give satisfactory results if the correction applied does not exceed about one-fifth of the cross-sectional dimension to which it is applied.[2]

Within the region of the equivalent gap the flux density is assumed to be uniform. Therefore, between the core faces,

$$\mathcal{H} \text{ (oersteds)} = \mathcal{B} \text{ (gausses)} \qquad [4]$$

$$\mathcal{H} \text{ (amp-turns/cm)} = \frac{1}{0.4\pi} \mathcal{B} = 0.796\mathcal{B} \quad \text{(gausses)} \qquad [5]$$

$$\mathcal{H} \text{ (amp-turns/in.)} = \frac{1}{0.4\pi} \times \frac{2.54}{(2.54)^2} \times 10^3\mathcal{B}$$

$$= 313\mathcal{B} \text{ (kilolines/sq in.)} \qquad [6]$$

$$\mathcal{H} \text{ (praoersteds)} = 10^7\mathcal{B} \text{ (webers/sq m).} \qquad [7]$$

Since the flux density is assumed to be uniform, the total flux is

$$\phi = \mathcal{B}A. \qquad [8]$$

If the flux or flux density is known, the magnetizing force $\mathcal{H}_a$ in the air gap can be calculated from Eqs. 4, 5, 6, or 7, or read from the curve for air shown in Figs. 1a and 1b. The abscissa values for the curve for air on these figures should be multiplied by 200. Likewise, the magnetizing force $\mathcal{H}_s$ for the steel portion of the core can be read from the curves of

---

[2] For short gaps having nonparallel faces, chamfered edges, teeth, or other complicating geometry, semi-empirical rules exist for obtaining equivalent lengths and areas. A toothed gap is used in the example of Art. 12. The manner of computing the permeance of many sorts of air paths is outlined in Herbert C. Roters, *Electromagnetic Devices* (New York: John Wiley & Sons, 1942), Ch. v.

Fig. 1a or 1b. The magnetomotive force for the circuit is

$$F = \mathfrak{K}_a \ell_a + \mathfrak{K}_s \ell_s,$$ [9]

where ℓ_a and ℓ_s are the lengths of air gap and steel paths respectively.

If the total applied magnetomotive force is known, a direct analytical solution for the flux is not possible because of the nonlinear properties of the core, but a successive-approximations solution, as indicated in Art. 4, can readily be used. Since the air gap usually requires most of the exciting ampere-turns, an estimate of the flux to be assumed for the first approximation can be obtained through considering that all the ampere-turns are required to overcome the reluctance of the air gap. Unless this flux gives a value for the flux density greater than the value at about the knee of the magnetization curve for the magnetic material, the actual flux is then somewhat — usually only a few per cent — less than this value. However, a direct graphical method of solution similar to that outlined also in Art. 4 is sometimes useful as a means for determining the flux. For series-connected air and steel paths

$$F = U_a + U_s,$$ [10]

where the subscripts a and s refer to the air and the steel respectively. Then, if the effects of leakage, though not necessarily fringing, are neglected the flux in the steel is equal to the flux that crosses the air gap, or

$$\phi_a = \phi_s.$$ [11]

The solution is the value of flux for which U_a and U_s satisfy Eq. 10.

A graphical solution that satisfies Eqs. 10 and 11 may be had through superposing on a plot of ϕ_s as a function of U_s a plot of the ϕ_a as a function of U_a, with the air-gap plot turned end-for-end about its ordinate axis. The origin of co-ordinates of the plot for the air gap is placed at the point along the abscissa axis of the plot for the steel portion that corresponds to the total applied magnetomotive force F. The ordinate and abscissa co-ordinates of the intersection of the two curves are respectively the total flux ϕ_s and the magnetic potential drop in the steel U_s. The construction for the graphical solution is illustrated by the curves of Fig. 8.

In Fig. 8, the relation between ϕ_s and U_s is obtained from the normal induction curve of the material. The magnetization curve of the air gap is obtained from the calculated reluctance of the gap,

$$\mathfrak{R}_a = \frac{\ell_a}{\mu_0 A_a},$$ [12]

where μ_0 is a constant, ℓ_a is the length of the gap, and A_a is the apparent cross-sectional area; that is, the cross-sectional area of the equivalent gap as given by Eq. 3 to allow for fringing. The relation between ϕ_a and U_a is

$$\phi_a = \frac{U_a}{\mathcal{R}_a} = \frac{\mu_0 A_a}{\ell_a} U_a. \qquad [13]$$

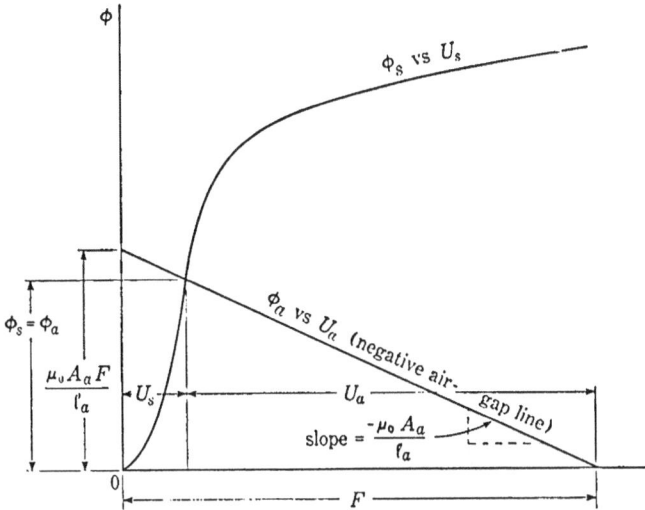

FIG. 8.   Construction for graphical solution for flux in a circuit having a steel path of uniform cross-sectional area in series with an air gap.

The location of the magnetization line for the air gap on Fig. 8 is readily determined from the fact that it intersects the ϕ axis at

$$\phi = \frac{\mu_0 A_a}{\ell_a} F. \qquad [14]$$

The value of ϕ given by Eq. 14 is the air-gap flux that would result if the magnetomotive force F were applied wholly to the air gap. The slope of the line is the negative of the permeance (the reciprocal of the reluctance) of the equivalent air gap. Therefore the line is often called the negative of the air-gap line, abbreviated *negative air-gap line*. The intersection of the negative air-gap line with the curve for the steel is then the desired solution to Eqs. 10 and 11, since this point is the only point that satisfies these equations simultaneously.

Since the curve of flux as a function of magnetic potential difference usually must be derived from the normal magnetization curve of the core material — that is, $\mathcal{B}$ as a function of $\mathcal{H}$ — the direct use of the magnetization curve is frequently more convenient. If the ordinates of Fig. 8 are multiplied by $1/A_s$ and the abscissa is multiplied by $1/\ell_s$, the curve for ϕ_s as a function of U_s becomes $\mathcal{B}_s$ as a function of $\mathcal{H}_s$, and the co-ordinates of the point of intersection with the *modified negative air-gap line* become the flux density in the steel $\mathcal{B}_s$ and the magnetizing force for the steel $\mathcal{H}_s$. The values of $\mathcal{B}_a$ and $\mathcal{H}_a$ are not given directly by the modified negative air-gap line. For the point of intersection, the values are,

$$\mathcal{B}_a = \frac{\mathcal{B}_s A_s}{A_a}, \qquad [15]$$

and

$$\mathcal{H}_a = \frac{F}{\ell_a} - \frac{\mathcal{H}_s \ell_s}{\ell_a}. \qquad [16]$$

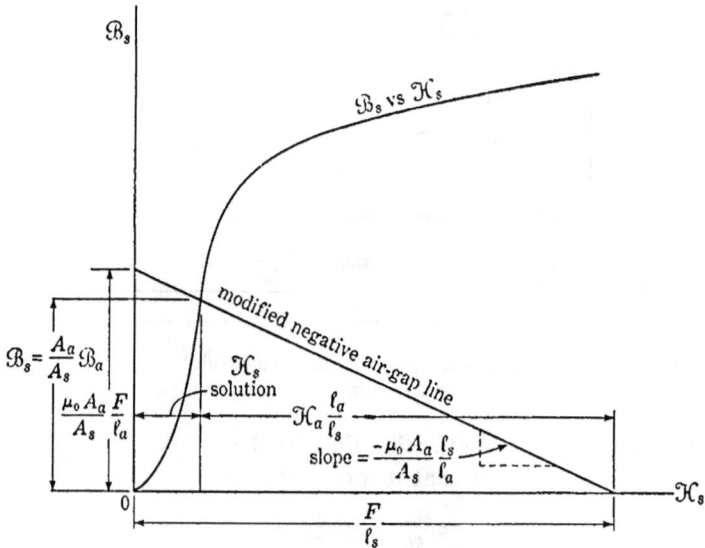

FIG. 9. Modification of Fig. 8 for direct use of $\mathcal{B}_s$ vs. $\mathcal{H}_s$ curve.

The negative air-gap line intersects the $\mathcal{B}_s$ and $\mathcal{H}_s$ axes at

$$\mathcal{H}_s = \frac{F}{\ell_s}, \qquad [17]$$

and

$$\mathcal{B}_s = \frac{\mu_0 A_a}{A_s} \frac{F}{\ell_a}. \qquad [18]$$

Figure 9 shows the plots for the modified construction.

A graphical solution as indicated by Fig. 8 or 9 is useful when a study of the resultant flux as a function of the air-gap length is to be per-formed. The absolute magnitude of the slope of the negative air-gap line is inversely proportional to the length of the gap ℓ_a. Thus, in Fig. 8

the intersections with the ϕ_s curve, of a group of straight lines which fan out from the point F on the abscissa axis, give the solutions for ϕ_s for the air-gap lengths corresponding to the slopes of the respective lines. Corresponding lines can be drawn on Fig. 9 to give the corresponding values of $\mathfrak{B}_s$.

The graphical construction indicated by Figs. 8 and 9 may not be the most useful when the air gap is fixed, and the solution for the flux as a function of magnetomotive force is desired. For these conditions, use of a curve of flux as a function of magnetomotive force for the particular core assembly may be more convenient. Such a curve serves only for the values of core and gap dimensions and core material assumed for the problem. The curve can be obtained through performing a process termed *shearing* the curve of the steel into the line for the air gap. On the assumption

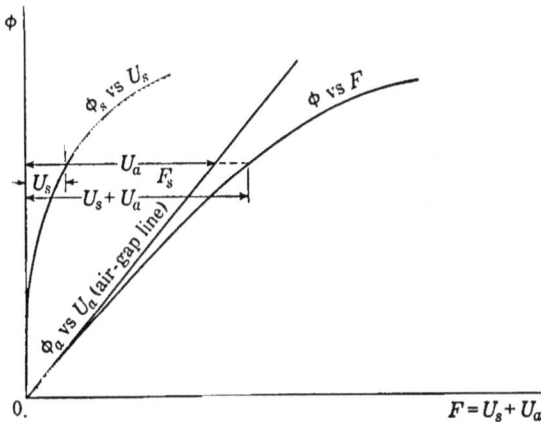

Fig. 10.   Illustrating the shearing of the flux curve for steel into air-gap line.

that leakage can be neglected, the procedure is as illustrated in Fig. 10. The curve for the steel portion of the core is plotted as a function of U_s from Fig. 1b, for the magnetic material used. The values of the abscissas are equal to $\mathcal{H}_s\ell_s$, and the values of ordinate are equal to $\mathfrak{B}_s A_s$. Likewise, ϕ_a is plotted as a function of U_a, where the values of the abscissas are $\mathcal{H}_a\ell_a$ and the values of the ordinates are $\mathfrak{B}_a A_a$. The curve for the gap and steel combined is then obtained through plotting arbitrary values of flux ϕ as a function of the corresponding sum of U_s and U_a, or F.

The methods of solution indicated in Figs. 8, 9, and 10 apply not only to magnetic problems but also generally to any series-connected circuit composed of two passive elements, either of which may be nonlinear or linear, and a specified impressed force; or to a parallel-connected combination of two passive elements and an impressed flux or current. As an example of this generality, the analogy between the electric circuit in which a fixed electromotive force is applied in series with a linear and nonlinear resistance, and the magnetic circuit in which a fixed magnetomotive force is impressed on a series-connected air gap and ferromagnetic material, may be considered. The graphical construction indicated by Fig. 8 or 10 can be applied equally well to a connection such as a battery with an electromotive force E in series with a resistance R and a load having a characteristic specified graphically as current I as a function of voltage V. Here E is analogous to F, I is analogous to ϕ_s, V is analogous to U_s, and the voltage across R is analogous to U_a. On the plot of I versus V, the *negative resistance line* can be plotted beginning at E on the abscissa axis with a slope $-1/R$. The ordinate of the intersection of the negative resistance line and the plot of I versus V then is the value of current in the circuit. For shearing — that is, to obtain a curve of I versus E — the *resistance line* is plotted from the same origin as the I versus V line, and voltage drops of the series elements corresponding to arbitrarily selected values of I are plotted.

Thus far, only a graphical procedure as indicated by Fig. 10 has been mentioned for determining the relationship between flux and total magnetomotive force for the series magnetic circuit, or the volt-ampere characteristic for the series electric circuit. However, the fact that the data may equally well be tabulated to show U_s and U_a for equal values of ϕ, and the values of U_s and U_a may then be added in pairs to give the total F, should not be overlooked. Unless plots of the curves are desired for other reasons, the tabular procedure is probably quicker than the plotting procedure, since the time required to plot the curves is saved.

8. Illustrative Example of Calculations for a Simple Ferromagnetic Circuit with a Short Air Gap

The magnetic structure shown in Fig. 6 is similar to that frequently used for the core assembly of a reactor except that the coil would be placed on the leg containing the air gap. It is made of 4.25 per cent silicon transformer sheet-steel laminations 0.014 inch thick stacked into a pile 2 inches thick (dimension a). The stacking factor is 0.90 for this structure. Dimension b is 2.5 inches. The air-gap length δ is 0.10 inch. The mean length of the steel part of the magnetic circuit is 30 inches. The problem is to find the magnetomotive force necessary to establish a total flux of 250 kilolines.

Solution: On the basis of the approximation indicated by Eq. 3 for a short air gap, the equivalent cross-sectional area of the gap is $(2.0 + 0.1)(2.5 + 0.1)$, or 5.45 sq in. The average flux density in the gap therefore equals 250/5.45, or 45.9 kilolines/sq in. From the curve for air in Fig. 1a, the magnetizing force for the gap with this average flux density is 72×200, or 14,400 amp-turns/in. Since the length of the gap is 0.1 in., the ampere-turns required for the gap are 1,440 approximately.

The net cross-sectional area of the steel is 5 × 0.9, or 4.5 sq in. The flux density in the steel is then 250/4.5, or 55.5 kilolines/sq in. If curve 3, Fig. 1a, is taken to represent the properties of this steel the magnetizing force in the laminations is about 3.7 amp-turns/in. The calculated magnetic potential drop required for the laminations is therefore 3.7 × 30 or 111 amp-turns. The magnetomotive force required for the complete magnetic structure is therefore 1,550 amp-turns. The significance of the third and fourth figures is, of course, very doubtful.

Solution involving a rough correction for leakage: The foregoing solution takes into account the increase in flux density in the iron produced by the fringing flux. Actually there is a considerable amount of leakage flux which further increases the flux density in the iron. This leakage flux is a function of the length of the gap, the shape of the pole faces, the size, shape, and location of the magnetizing coils, and, to a lesser degree, of several other factors. Experiments with reactors of the type considered in this problem having a coil of medium proportions placed over or near the air gap show that the flux in the iron is larger than that obtained by use of the fringing correction. A rough correction which gives more nearly the correct value of total flux involves the use of 2δ instead of δ when finding the equivalent air-gap area. This correction may be called a *fringing and leakage* correction.

The equivalent air-gap area, using the 2δ correction, is $(2.0 + 0.2)(2.5 + 0.2)$, or 5.95 sq in. The average gap density is 250/5.95, or 42 kilolines/sq in. The magnetizing force for the gap is 65 × 200, or 13,000 amp-turns/in., and the magnetic potential drop in the gap is 1,300 amp-turns. The flux density and potential drop in the steel are the same as before, giving a total of about 1,400 amp-turns.

Even though the length of the laminated part of the magnetic structure is 300 times the length of the air gap, the laminated part requires only 111/1,400, or about 8 per cent of the ampere-turns for the complete circuit. This condition means that even though the iron laminations have a nonlinear magnetization curve the complete magnetic circuit has an overall magnetization curve, which, for many purposes, may be considered to be essentially a straight line up to and slightly beyond 250 kilolines. Thus, provided the flux does not appreciably exceed 250 kilolines, the magnetizing coil may, for many applications, be regarded as having constant inductance.

The presence of the magnetic core material is highly effective in increasing the flux, however, as may be seen by comparing the total flux in this structure with the total flux in the core of a coil distributed uniformly over a nonmagnetic core of the same shape as the laminated iron core. To a first approximation

$$\mathfrak{B}_{air\ core} = \frac{0.4\pi \times 1,400 \times 2.54^2}{30 \times 2.54} \qquad [19]$$

$$= 150 \text{ lines/sq in.}$$

If the core area is assumed equal to 5 square inches, then

$$\phi_{air\ core} = 150 \times 5 = 750 \text{ lines,} \qquad [20]$$

and

$$\frac{\phi_{iron\ core}}{\phi_{air\ core}} = \frac{250,000}{750} = 330, \tag{21}$$

which is essentially the same as the ratio of the length of the magnetic circuit to the length of the air gap, namely, 300. This comparison represents a rather fictitious situation, however, because the flux does not take the same path in air if the iron is absent. If the comparison is made on the basis of total flux set up by the iron-core arrangement as contrasted with an air-core coil using the same length of wire of the same size as the iron-core coil, the contrast is less favorable for the iron-core arrangement. For example, if the wire is wound to give a coil of the best possible Q,* the ratio represented by Eq. 21 is of the order of 20. A proper comparison of the effectiveness of the iron-core and air-core coils cannot be made unless the basis for comparison can be set in terms of what the coils are designed to accomplish. The sheet-steel core effectively conducts and concentrates the flux or, in another sense, effectively concentrates the magnetomotive force of the whole coil across the air gap without regard to the location of the winding on the core (except for leakage effects) and produces a region of substantially constant flux density within the air gap. With the simple air-core coil, the flux paths cannot be controlled, and no extensive region of substantially uniform flux density exists.

If the total magnetomotive force is specified, a direct solution for the flux is not possible because of the nonlinear properties of the core material. A successive-approximations solution or a graphical solution, as outlined in Art. 4, can readily be performed. Since the successive-approximations procedure is the same as that illustrated by the solution given in Art. 4, and for a more complicated circuit in Art. 9, only the solution by the graphical procedure is given here. For this illustration the core is again the one shown in Fig. 6 with the same dimensions and material used in the preceding illustration, and with the applied magnetomotive force of 1,400 ampere-turns found for the preceding illustration.

Solution: The curve of flux as a function of magnetic potential difference is first drawn using the data of Fig. 1a. The values used for the magnetic potential difference are thirty times the magnetizing force since the equivalent length of core is 30 in., and the values used for the flux are 4.5 times the flux density because the net core area is 4.5 sq in. This curve is shown in Fig. 11 (similar to Fig. 8). From Eq. 14 the intersection of the negative air-gap line with the ordinate axis of Fig. 11 is at

$$\phi_a = \frac{\mu_0 A_a}{\ell_a} F, \tag{22}$$

* A circular coil of copper wire having a square winding cross section and a mean diameter 3.02 times the depth of the winding.

or, in mixed English units,

$$\phi = \frac{\text{amp-turns} \times \text{area of equivalent gap in sq in.}}{313 \times \text{length of gap in inches}} \quad \text{kilolines.} \qquad [23]$$

$$= \frac{1,400 \times 5.95}{313 \times 0.10} = 266 \text{ kilolines.} \qquad [24]$$

The negative air-gap line intersects the F axis at 1,400 amp-turns. The solution to the problem, as given by the co-ordinates of the point of intersection of the curve and

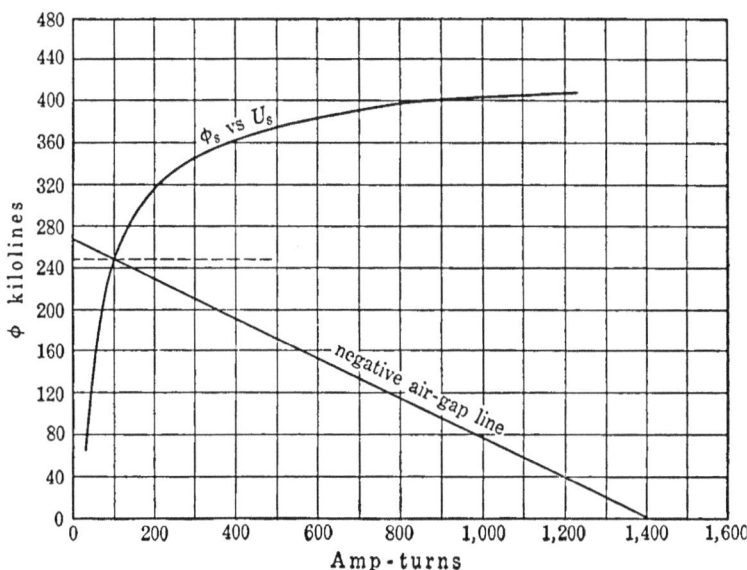

FIG. 11.   Graphical solution for flux for example of Art. 8.

the air-gap line of Fig. 11, is approximately 250 kilolines, agreeing with the data for the previous solution.

The solution indicated by Fig. 12, similar to Fig. 9, involves $\mathcal{B}$ and $\mathcal{H}$, instead of ϕ and F. In Fig. 12, the curve of $\mathcal{B}$ as a function of $\mathcal{H}$ for the steel is plotted from the data of Fig. 1a. The intersection of the negative air-gap line with the $\mathcal{B}_s$ axis is at

$$\mathcal{B} = \frac{\mu_0 A_a F}{A_s \ell_a}. \qquad [25]$$

The value used for A_a should be the cross-sectional area of the equivalent gap; that is, the actual gap face area corrected for fringing and leakage.   Placing numbers in Eq. 25 gives

$$\mathcal{B} = \frac{1 \times 5.95 \times (2.54)^2 \times 1,400 \times 1.257}{4.5 \times (2.54)^2 \times 0.10 \times 2.54} = 9,150 \text{ gausses.} \qquad [26]$$

When converted to the system of units used in the problem, $\mathcal{B}$ is 59,000 lines/sq in.

The intersection of the negative air-gap line with the $\mathcal{H}_s$ axis is at F/ℓ_s, or at

$$\frac{F}{\ell_s} = \frac{1,400}{30} = 46.7 \text{ amp-turns/in.} \qquad [27]$$

The solution given by the co-ordinates of the point of intersection of the two curves of Fig. 12 is 55.5 kilolines/sq in. for $\mathcal{B}_s$. The total flux ϕ is approximately 250 kilolines, a figure which agrees with the solution obtained from Fig. 11.

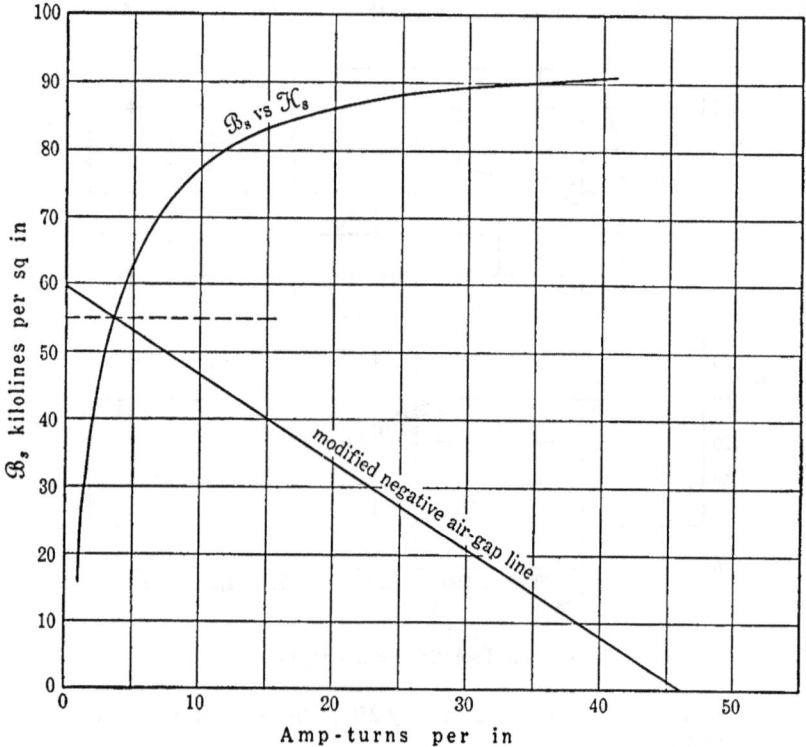

Fig. 12.   Graphical solution for flux density for example of Art. 8.

The factors which should influence the choice of method used in any particular problem are, of course, the same as those mentioned in Art. 7.

9. ILLUSTRATIVE EXAMPLES OF CALCULATIONS FOR SERIES–PARALLEL IRON AND SHORT AIR–GAP PATHS

A type of core construction used frequently for certain transformers in which a relatively small magnetic coupling (that is, large magnetic leakage) between primary and secondary coils is desired is similar to that shown in Fig. 13. The coils are located on the legs X and Z, and a fraction of the flux set up by their magnetomotive forces is by-passed or leaks through the leg Y, thus decreasing the magnetic coupling. Because of the nonlinear properties of the core material, the fraction of the flux by-passed through the leg Y varies with the amount of magnetic saturation.

For this illustration, a flux of 380,000 maxwells is to be set up in the leg X of Fig. 13 by a coil wound on this leg. The ampere-turns required of this coil are to be determined. The magnetic circuit is assembled from 4.25 per cent silicon transformer sheet-steel laminations 0.014 inch

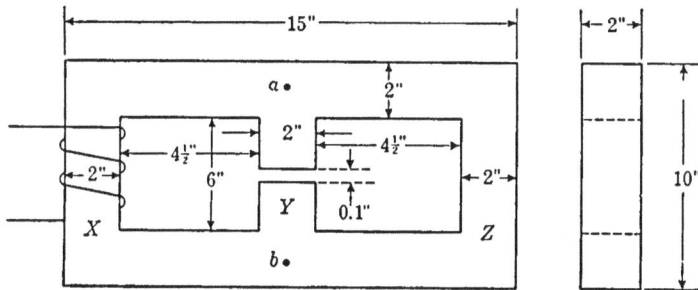

FIG. 13.   Magnetic structure with an air gap and series-parallel paths.

thick. The stacking factor is 0.90 for this structure. As in the previous examples, magnetic leakage is to be neglected and the flux density is to be assumed uniform for the cross-sectional areas of each component part of the magnetic structure. The paths aXb and aZb of the structure are each assumed to have a constant cross-sectional area and a mean length of 21 inches. The mean length of core in the leg Y is assumed equal to 8 inches.

Solution: The cross-sectional area of the steel in each part of the core is equal to $2 \times 2 \times 0.9$, or 3.6 sq in. The cross-sectional area of the air gap, corrected for fringing, is 2.1×2.1, or 4.4 sq in. The flux density in the X leg is 380/3.6, or 105.5 kilolines/sq in. The component of magnetic potential difference required to set up this flux density in the path aXb is calculated from curve 3 of Fig. 1a and the mean length of the path aXb. The magnetizing force for this density is 175 amp-turns/in.; hence the magnetic potential drop in the steel is 175×21, or 3,680 amp-turns.

The total flux of 380 kilolines in the path aXb divides between the Z and Y legs in such a way that the magnetic potential drop from a to b in Fig. 13 is the same through either the aZb or aYb path. Because of the different nonlinear properties of these flux paths, further direct calculation is not possible. The division of the flux cannot be computed by direct calculation; hence a successive-approximations method is used. Of the several variations of this method, the following is simple and effective:

A flux in the air gap or the Y leg is assumed. For this assumed value the magneto-motive force from a to b necessary to produce the assumed flux is calculated. This magnetomotive force also acts on the path aZb, and, since the cross section of the path is assumed to be uniform, the flux in it can be calculated simply. This flux is added to the assumed flux in the Y leg. If the magnitude of the flux assumed in the Y leg is correct, the sum of the flux in the Z and Y legs should equal the desired flux in the X leg. If the assumption is not correct, the procedure must be repeated with a different assumed value.

Unless the flux density in the Z leg nearly saturates the steel, probably the path aZb effectively short-circuits the path aYb, because of the relatively large reluctance of the gap. For the first trial, a small flux in the Y leg, for example 80 kilolines, is assumed. The corresponding flux density in the air gap is 80/4.4, or 18.2 kilolines/sq in. From Eq. 6 or the air-gap curve in Fig. 1a or 1b, the corresponding magnetizing force is approximately 5,700 amp-turns/in. Since the air-gap length is 0.1 in., the magnetic potential drop for the air gap is 570 amp-turns. An inspection of the magnetization curve for sheet steel in Fig. 1a or 1b shows that the magnetic potential drop for the laminations in the Y leg at the corresponding flux density of 80/3.6 or 22 kilolines/sq in. is negligible compared with the 570 amp-turns for the air gap. Hence, if the flux in the Y leg is the assumed value of 80 kilolines, the magnetic potential drop from a to b is about 570 amp-turns. This magnetic potential drop acting on the Z leg produces a magnetizing force of 570/21, or 27 amp-turns/in. approximately, which from the data of Fig. 1a or 1b establishes a flux density of 88 kilolines/sq in. The corresponding flux in the Z leg therefore is 88 × 3.6 or 317 kilolines. The flux in the X leg then is 397 kilolines. Since this flux is substantially more than the specified value of 380 kilolines, a second trial must be made using a smaller value of flux in the Y leg.

TABLE III

Part of Circuit	Area sq in.	Length, in.	kilolines	kilolines sq in.	amp-turns in.	amp-turns
Leg X	3.6	21	380 (specified)	105.5	175	3680
First Approximation Gap	4.4	0.1	80 (assumed)	18.2	5700	570 } (In parallel)
Leg Z	3.6	21	317	88	27	570 }
			ϕ gap + ϕ_z =	397 (too large)		
Second Approximation Gap	4.4	0.1	70 (assumed)	15.9	4800	480 } (In parallel)
Leg Z	3.6	21	310	86.2	22.9	480 }
			ϕ gap + ϕ_z =	380		

Since most of the total flux is in the Z leg, and gives a flux density for which saturation effects are important, the magnetic potential drop from a to b in the second trial must be reduced considerably more than in direct proportion to the error in the result obtained from the first approximation. The magnetic potential drop from a to b is directly proportional to the assumed flux in the Y leg because the gap reluctance predominates; hence, the estimate of the flux in the Y leg must be reduced by considerably more than in direct proportion to the error in the Z-leg flux as calculated from the first approximation.

Therefore, for the next approximation, the flux in the Y leg is assumed to be 70 kilolines. The air-gap density is therefore 70/4.4 or 15.9 kilolines/sq in. The magnetizing force for the air gap given by a procedure similar to that used for the first approximation is 4,800 amp-turns/in. For the 0.10 in. air-gap length, the magnetic potential drop is 480 amp-turns. Since for this low density in the laminations of the

Y leg, the magnetic potential drop is negligible compared with the magnetic potential drop for the air gap, the magnetic potential drop from *a* to *b* is taken as 480 amp-turns. This magnetic potential drop is across the *Z* leg also. Hence the magnetizing force in the *Z* leg is 480/21 or 22.9 amp-turns/in. From Fig. 1a or 1b, this magnetizing force establishes a flux density of about 86.2 kilolines/sq in. The flux in the *Z* leg therefore is 86.2 × 3.6 or 310 kilolines. The sum of the fluxes in the *Y* and *Z* legs gives 380 kilolines for the flux in the *X* leg.

In all but the simplest magnetic circuit problems, a systematic tabulation of the calculations saves a considerable amount of time. A tabulation such as that shown in Table III could be used to advantage in solving Example 4.

The magnetomotive force to be provided by the coil equals the sum of the magnetic potential drop for the paths *aXb* and *aZb*. This sum is 4,160 amp-turns.

10. CALCULATIONS FOR LONG AIR GAPS AND FOR LEAKAGE FLUX PATHS

The procedure outlined for short gaps is reasonably satisfactory provided the smaller dimension of the cross-sectional area of the gap is not less than about five times the length of the gap. However, the gap length often is comparable to or even greater than the cross-sectional dimensions of the gap. In circuits having long gaps, the leakage fluxes may reach values several times as large as the useful gap fluxes. The calculation of fluxes for long gaps and the calculation of leakage fluxes must be approached from the point of view of the field rather than the circuit. However, if a computation of the total flux without detailed computation of its distribution is sufficient, certain useful procedures are available. Actual design procedures involve much judgment and experience, because the variety in size, shape, and proportion among magnetic structures requires the adaptation and refinement of general rules for each special application. The details of these procedures are too involved to be discussed in this text, but certain important general principles can be presented.

The magnetic flux existing between two surfaces of ferromagnetic material sufficiently separated to require analysis as a long-gap or leakage-flux problem depends primarily upon the areas of the surfaces, their average curvature, and the magnetic potential difference between them. It is much less dependent upon the distance of separation and relative orientation of the surfaces, because most of the reluctance of the flux path is near the surfaces. The conditions are somewhat analogous to the problem of ground resistance discussed on p. 65 of the volume on electric circuits in this series.

In design procedure, the actual magnetic structure is represented by a structure consisting of portions having about the same surface area and average surface curvature but having a geometry more easily dealt with mathematically. Spheres, hemispheres, cylinders, or portions of cylinders are found most convenient, because the field distribution around them is comparatively easy to determine. The actual procedure is similar to that used in determining the electric field distribution around charged conducting surfaces.

In the design of a magnetic structure to provide a certain flux density at the coil of an instrument, the flux density desired determines the magnetic potential difference required across the air gap in which the coil is located. Hence, in order to keep leakage fluxes at a minimum, the surface areas not contributing flux to the gap and the average difference of magnetic potential between them should be made the minimum permissible within whatever other restriction practical design may impose. In general, this arrangement is accomplished through keeping the source of magnetomotive force (coil or permanent-magnet material) as close to the gap as possible. With some experience as a guide, estimates of required magnetomotive force within a tolerance of 10 per cent should be attainable. Estimates of the ratio of leakage flux to useful flux should not be in error by more than 20 per cent. Such values are close enough for many design purposes.

11. ILLUSTRATIVE EXAMPLE OF CALCULATION OF LEAKAGE FLUX FOR THE MAGNETIC CIRCUIT OF AN OSCILLOGRAPH

The magnetic structure of Fig. 14 represents a simplification of the magnetic circuit of a three-element oscillograph. The small soft iron blocks *A*, *B*, *C*, *D* each have a square cross-sectional area of 0.50 square inch and a length of 1 inch. Each air gap is 0.10 inch long. The cast steel yoke has a square cross-sectional area of 2.0 square inches and an equivalent length of 18 inches. The magnetizing coil has 2,000 turns and carries 6 amperes.

By the successive-approximations method, leakage being neglected, the flux in the circuit is computed to be 65 kilolines. The problem presented here is to estimate the correctness of this result by computing the amount of leakage flux and thus to obtain an approximate comparison of the coil flux and air-gap flux.

Solution: The parallel portions of the yoke are treated as parallel cylinders. The axial separation is taken as

$$2a = 4.3 + 1.4 = 5.7 \text{ in.} \qquad [28]$$

The radii of the cylinders are computed so as to make the circumference of the cylinder equal to the perimeter of the cross section of the yoke:

$$r = \frac{4 \times 1.4}{6.28} = 0.89 \text{ in.} \qquad [29]$$

The axial length of a cylinder is taken as equal to the axial length of one of the parallel sides of the yoke:

$$\ell = \frac{18 - 4.3 - 1.4 - 1.4}{2} + 1.4 = 7.0 \text{ in., or } 18 \text{ cm.} \tag{30}$$

The reluctance between the parallel cylinders is[3]

$$\mathcal{R} = \frac{\ln \dfrac{5.7}{0.89}}{18\pi} = \frac{1.8}{18\pi} = \frac{1}{10\pi} \quad \text{cgs units.} \tag{31}$$

The leakage flux

$$\phi_\ell = \frac{0.4\pi NI}{\mathcal{R}} = 39.4NI \approx 39NI \text{ lines,} \tag{32}$$

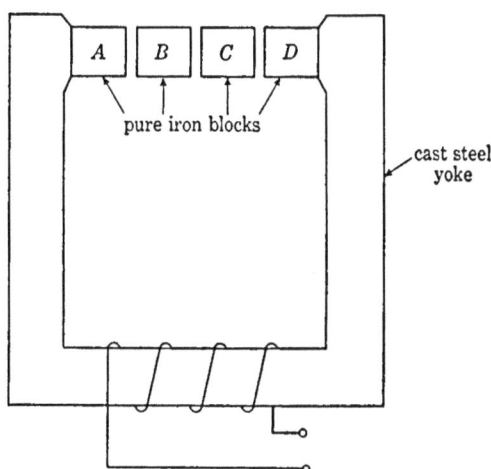

FIG. 14. Equivalent magnetic circuit of three-element oscillograph.

if NI is the average ampere-turns magnetic potential drop between the cylinders. The procedure now is to assume values of air-gap flux, to compute the corresponding magnetic potential difference across the midpoints of the parallel cylinders, and to compute the leakage flux, the coil flux, and the reluctance drop in the yoke in the path through the coil between the midpoints of the parallel cylinders. The procedure is repeated until the total reluctance drop of the circuit equals closely enough the magnetomotive force of the coil.

As a matter of interest, 65 kilolines is taken as the first value of the air-gap flux. The corrected air-gap area is

$$A_a = (0.707 + 0.1)(0.707 + 0.1) = 0.65 \text{ sq in.} \tag{33}$$

The air-gap flux density is 65/0.65 or 100 kilolines/sq in.; the flux density in the blocks is 65/0.50 or 130 kilolines/sq in.; the flux density in the yoke at the air-gap ends is 65/2.0 or 32.5 kilolines/sq in., which is assumed to be constant for half the axial length of the hypothetical cylinders, or 3.5 in. The air gaps require $160 \times 200 \times 0.3$

[3] By analogy to Eq. 104b, p. 51 of the volume on electric circuits in this textbook series.

or 9,600 amp-turns. The blocks require 610 × 4 or 2,440 amp-turns. The part of the yoke considered requires 8 × 7 or 56 amp-turns. These figures total slightly more than 12,000 amp-turns, merely because the precision of the solution which resulted in the 65 kilolines with leakage flux neglected was considered satisfactory when the computed reluctance drop exceeded the available magnetomotive force by about 200 in 12,000 amp-turns. Hence for all practical purposes the 12,000 amp-turns can be taken across the midpoints of the cylinders in this first approximation. The leakage flux is then 39 × 12,000 or 468,000 lines. The flux density in the part of the yoke extending through the coil then is (468 + 65)/2.0 or 278 kilolines/sq in. Obviously this figure is absurd, but it serves to indicate the serious influence of leakage flux.

As a second trial, the air-gap flux is taken as 30 kilolines. For this assumption, the air-gap flux density is 30/0.65 or 46.2 kilolines/sq in.; the flux density in the blocks is 30/0.50 or 60 kilolines/sq in.; the flux density in the air-gap ends of the yoke is 30/2.0 or 15 kilolines/sq in. The air gaps require 14,800 × 0.3 or 4,430 amp-turns; the blocks require 4 × 4 or 16 amp-turns; the air-gap ends of the yoke require 5 × 7 or 35 amp-turns. The magnetic potential drop across the midpoints of the cylinders is about 4,480 amp-turns. The leakage flux is 39 × 4,480 or 175,000 lines, and the flux density in the part of the yoke extending through the coil is (175 + 30)/2.0 or 103 kilolines/sq in. The reluctance drop within this part of the yoke is 98 × 11 or 1,080 amp-turns. The total of 5,560 amp-turns is much too small. Further trials lead to the result of 38 kilolines of air-gap flux, 222 kilolines of leakage flux, and 260 kilolines of core flux, a result vastly different from the figure of 65 kilolines obtained through neglecting leakage.

Insufficient experimental work has been done towards verification of this method of calculating leakage or long air-gap flux to permit an estimate of the precision of the method. Perhaps the axial length of the cylinders should be increased to allow for end effects. However, the basic idea seems to be sound, and use of the method probably at least gives results that are likely to be substantially closer to the truth than results obtained when no leakage correction is made. Further experimental work may provide general rules for adjusting the dimensions and spacing of the hypothetical cylinders to provide known tolerances for the results obtained.

12. Calculation of the Magnetization Curve of a Direct-Current Dynamo

Although the methods and simplifying assumptions outlined in preceding articles yield for many problems results well within the accuracy with which most ferromagnetic data are obtainable, numerous problems exist in which the shapes of the core structures make refinement to these methods and the use of sound engineering judgment essential for a reliable solution. Devices in which such core shapes are encountered are rotating dynamo machines, certain precise measuring instruments, special magnetic relays, and so forth. Because of the importance of the rotating dynamo machine in engineering practice, and the costs involved in the construction of the larger units, considerable attention has been given during the past decade or so to the problem of the precise prediction of the performance of a machine of proposed design, with the result that methods for treating the problem are reasonably well established. To illustrate certain of the more important phases in the treatment of this problem, the calculation of the magnetization curve of a direct-current dynamo is briefly outlined. The application to other problems of the principles involved in this illustration should then readily be made.

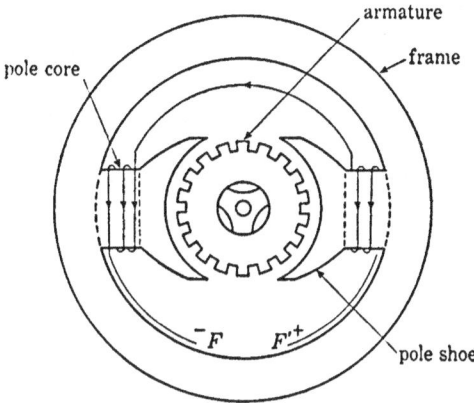

FIG. 15.   Magnetic circuit of a d-c dynamo.

Figure 15 shows the magnetic circuit of a typical two-pole dynamo. The center-slotted or sprocket-shaped member is the rotor or armature, and carries the rotor winding in the slots. It is usually assembled from laminated sheet-steel punchings having 1 or 2 per cent of silicon. The outer member is the field structure, the cylindrical portion being the yoke or frame which frequently is either of cast iron, cast steel, or cold-rolled steel. The protruding portions of the field structure are the poles, the cores usually being of laminated sheet steel.

With a direct voltage applied to the field windings at the terminals FF', the steady value of the field current is determined entirely by the magnitude of the applied voltage and the resistance of the field circuit. For the polarity shown, the field winding produces a magnetomotive force in the direction to establish a magnetic flux from left to right through the field poles, air gaps, and armature. The flux path is then completed through the frame. The magnetic circuit is redrawn schematically in Fig. 16, the flux paths being indicated by the light lines carrying arrowheads.

The degree of accuracy with which the constants of the magnetic circuit of a machine may be calculated depends largely on the amount of time and effort that can be afforded for the work. If a field map of the complete magnetic circuit is drawn, which invariably involves trial and error and therefore is time consuming, a fairly accurate result is possible. The labor involved makes this procedure impracticable for the design of the less important machines, but field plots to obtain the flux distribution for certain critical portions of the magnetic circuits such as the teeth, air gap, and leakage paths are frequently drawn as part of the design of the larger machines or those of which a particular performance is desired. Frequently, approximation of the actual magnetic circuit by one which may be more readily calculated is desirable.

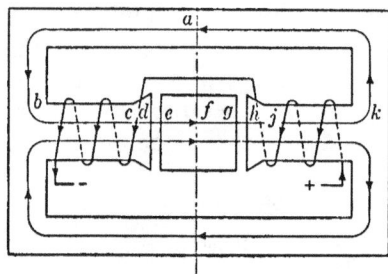

FIG 16. Schematic equivalent of Fig. 15.

In order to determine the portions of the magnetic circuit whose properties predominate in the design and to serve as a guide to justifiable approximations, a graph of the relative magnetic potential of various points around the circuit is sketched in Fig. 17. For this purpose a plane perpendicular to the paper in Fig. 16 cutting the yoke midway between the two field poles and passing through the center of the armature at f is considered as the plane of zero magnetic potential, owing to the symmetry of the structure. In a machine having more than one pair of poles, such a plane would pass midway between adjacent poles. The drop in magnetic potential U around the circuit in the direction of the flux is first considered. The drop from a to b, namely U_{ab}, is

$$U_{ab} = \int_a^b \mathcal{H} \cdot d\ell. \qquad [34]$$

If leakage flux is considered negligible for the present, and the cross-sectional area of the path is constant, magnetic field intensity $\mathcal{H}$ is constant from a to b, so that

$$U_{ab} = \mathcal{H}_{ab}\ell_{ab}, \qquad [35]$$

where ℓ_{ab} denotes the effective length from a to b. The curve of the variation of magnetic potential from a to b is shown by the dot-dash line in Fig. 17. From b to c and c to d, similar conditions hold but the magnetic field intensity in each portion of the core is probably different because of different material or different flux density. Hence, the curves for these portions have different slopes. The air gap offers a similar drop, but because of the high reluctance of the gap the curve from d to e has an extremely large slope. If the plot is completed for the remainder of the path, returning to *a*, the total reluctance drop is found to be equal to *a–A* on the figure.

For the drop *a–A* to exist, an equal rise in magnetic potential must be somewhere in the circuit. This potential rise is established by the magnetomotive force of the field windings, the magnitude of the contribution from each being half of *a–A*, or *c–F* for each winding. The dotted curve

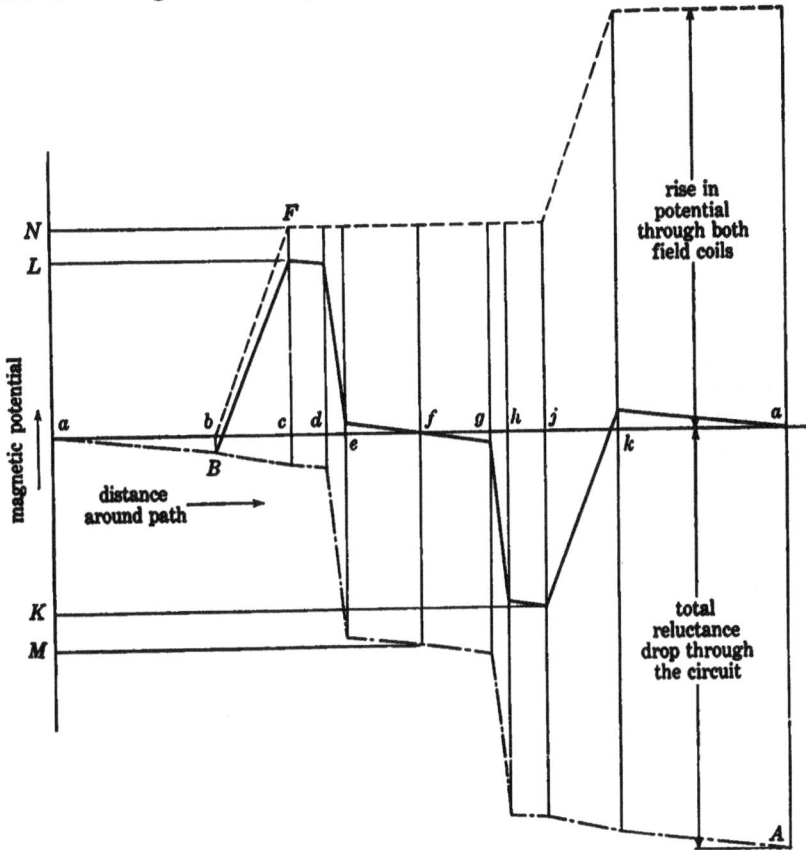

FIG. 17.   Plots of magnetic potentials for magnetic circuit of Fig. 16.

shows the rise in potential given by the windings, and the solid curve shows the sum of the rises and drops around the circuit. The solid curve thus gives the actual magnetic potential at each point with respect to the point *a*.

Several important facts are deduced from a study of the plot of Fig. 17. The first is that points *a* and *f* are at the same magnetic potential; that is, the reluctance drop around the half of the circuit *abcdef* is balanced by the magnetomotive force of the coil on the pole between *b* and *c*. The coil on the pole between j and k supplies the magnetomotive force for the remaining portion *fghjka*. Therefore, since the two paths are alike, all calculations may be made on the one section from a to f.

The resultant curve in Fig. 17 also shows that all points on the yoke, portion *bak*, are at nearly the same potential, the maximum potential difference from b to k being only $2(bB)$. Therefore the flux leakage from b to k through the air (or between any two points on the yoke) is relatively small. For the remainder of the circuit, however, the leakage is not everywhere small. Figure 17 shows that, from b to c and from j to k, the potential difference increases until it reaches a value equal to L-K between the tops of the pole shoes. Since the magnetic potential difference from one pole shoe to the next, that is, U_{dh}, is almost as large as that at the junction of the pole shoe and pole core, and the distance between the tips of adjacent pole shoes is relatively small, the leakage flux from pole shoe to pole shoe is likely to be appreciable.

In practical calculations, the leakage flux from one shoe to the next accounts for most of the leakage flux which is of sufficient magnitude to enter into the calculation. In a new design its magnitude may be estimated from previous experience with similar machines or from a field map. This leakage flux is often as much as 10 or 20 per cent of the useful flux from the pole to the armature. In practice, the ratio of the flux through the pole core to the useful flux across the air gap is used to specify the leakage and is called the *coefficient of dispersion*. This ratio is usually determined for rated full-load flux, and, although it varies slightly with the degree of saturation, it is assumed to remain constant within the normal range of excitation. Once the coefficient of dispersion is known, the flux in all parts of the circuit and the reluctance drop required for each section can be determined.

One problem of interest to the designer is the determination of the curve of generated voltage as a function of field current, that is, the magnetization curve or some point on it. As shown in any text on direct-current machinery,[4] the relation between generated voltage and the air-gap flux is

$$E_a = \frac{Z}{a}\frac{n}{60}p\phi \times 10^{-8} \qquad [36]$$

where

　　$Z \equiv$ total number of active inductors on the armature,

　　$p \equiv$ number of field poles on the machine,

[4] For example: Royce G. Kloeffler, Jesse L. Brenneman, Russell M. Kerchner, *Direct-Current Machinery* (New York: The Macmillan Company, 1934); Alexander S. Langsdorf, *Principles of Direct-Current Machines* (5th ed., New York: McGraw-Hill Book Company, Inc., 1940).

$a \equiv$ number of current paths through the armature winding,

$n \equiv$ speed in revolutions per minute,

$\phi \equiv$ magnetic flux crossing the air gap from one pole shoe to the armature, in lines or maxwells,

$E_a \equiv$ voltage generated between armature terminals.

For a given machine, all factors except E_a, ϕ, and n are fixed. Therefore

$$E_a = K\phi n, \qquad [37]$$

where

$$K \equiv \frac{Zp10^{-8}}{a60}. \qquad [38]$$

From Eq. 37, ϕ can be calculated for each value of E_a. This value of ϕ is the *useful* flux entering the armature from one field pole. It is therefore the flux in the air gap and in the armature teeth.

The pole core carries not only the air-gap flux but also the leakage flux and therefore has a flux through it equal to ϕ times the coefficient of dispersion. In the frame or yoke a–b, the flux is equal to one half that in the field pole core. Thus the flux in each portion of the magnetic circuit may be calculated for any specified value of E_a. With the flux in each portion of the circuit known, the flux density and the reluctance drop for each portion can be calculated. The most difficult part of this calculation is the determination of the proper area and length to use for the air gap. Since, in most problems, this one section of the circuit accounts for more than 80 per cent of the total reluctance drop, the drop in it must be determined as accurately as possible.

A justifiable simplification of the problem is to assume that the pole face is an equipotential surface. This simplification is permissible because the distribution of the flux in the iron portions of the circuit is fairly uniform over each cross section perpendicular to the field, and because the low reluctance of the ferromagnetic path makes any slight nonuniformity in distribution at any cross section have little effect on the reluctance drop. The same assumption may sometimes be made for the armature surface under a pole, but in this treatment the armature surface at the roots of the teeth is considered as an equipotential surface. The flux distribution between these equipotential surfaces, however, is nonuniform and therefore requires special analysis.

Figure 18 shows a section of a pole and armature. In a consideration of the reluctance drop in the air gap, two difficulties are encountered. First, the slots in the armature cause the length of the air gap to be different at different points around the surface. Second, these slots make difficult the definition of area of tooth face so as to take account of fringing. The problem is further complicated by the practice of making the air gap longer at the pole tip than at the center of the pole, the reason for which is found in any textbook on direct-current machinery.[5]

An effective method for taking account of the effect of the slots is that devised by F. W. Carter.[6] This assumes that the tops of the teeth and the pole face are parallel planes and that the slots are infinite in depth. These assumptions fit most problems reasonably accurately if the slots are considered in small groups or singly, since the curvature of the surfaces is slight and the depth of the slot so large in comparison with the air-gap length that very little flux penetrates

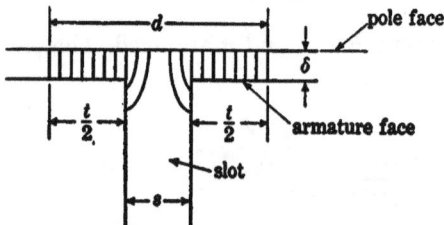

FIG. 18.   Dimensions of a toothed air gap, used in Carter's formulas.

deep into the slot. Carter showed that the flux crossing an air gap of length δ, Fig. 18, over a section d that embraces the slot s is the same as the flux crossing a gap of the same length but width $(d - \sigma s)$ where σ is given by

$$\sigma = \frac{2}{\pi}\left\{\tan^{-1}\frac{s}{2\delta} - \frac{\delta}{s}\ln\left[1 + \left(\frac{s}{2\delta}\right)^2\right]\right\}.\qquad[39]$$

Thus the presence of the slot decreases the flux to a value $\dfrac{d - \sigma s}{d}$ of that calculated for a smooth surface of width d and without a slot. Since the permeability of air is constant, $\mathcal{B}$ in the gap is proportional to $\mathcal{H}$; hence the same result is obtained through multiplying the air-gap length δ by the reciprocal of the foregoing relation to give an equivalent air-gap length δ' and by assuming the equivalent air gap to be uniform over the entire width d. If d is taken as the width of a slot plus one tooth width, the coefficient by which δ is multiplied becomes

$$\frac{d}{d - \sigma s} = \frac{t + s}{t + s - \sigma s}\qquad[40]$$

$$= \frac{t + s}{t + s(1 - \sigma)} = C,\qquad[41]$$

so that

$$\delta' = C\delta.\qquad[42]$$

---

[5] Royce G. Kloeffler, Jesse L. Brenneman, Russell M. Kerchner, *Direct-Current Machinery* (New York: The Macmillan Company, 1934); Alexander S. Langsdorf, *Principles of Direct-Current Machines* (5th ed., New York: McGraw-Hill Book Company, Inc., 1940).

[6] F. W. Carter, " Air-Gap Induction," *Electrical World and Engineer, 38* (1901), 884–888.

The factor C, as defined by this equation, is called Carter's coefficient. Equation 39 shows that σ is dependent only on the ratio s/δ. The curve of Fig. 19 is a plot of this equation, and from it the value of σ may be read for different values of the ratio of slot width to air-gap length.

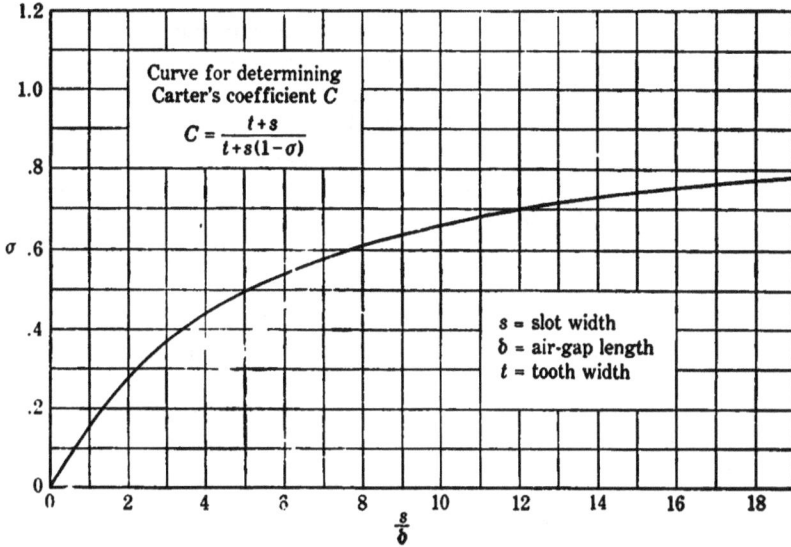

Curve for determining Carter's coefficient C

$$C = \frac{t+s}{t+s(1-\sigma)}$$

s = slot width
δ = air-gap length
t = tooth width

FIG. 19.   Curve for obtaining Carter's coefficient for toothed air gap.

The treatment thus far takes account only of the fringing of the flux at the armature teeth. Because of the fringing at the pole shoe and the variation of air-gap length usually introduced, an appraisal of the net reluctance of the air gap can best be made with the aid of a rough field map of the region. Since the construction is usually symmetrical about the pole axis, only the region between a pole axis and the plane midway between adjacent poles need be considered.

A sufficiently refined map is usually obtained if the armature surface included within this region is divided into eight or ten equal sections. Frequently the width of each section can be made equal to a slot pitch, as is shown by the dashed lines in the sketch of Fig. 20. The slots need not be shown in the plot, however, because their effect is taken into account by the coefficient C. The flux lines to the center of each section, the slots being again disregarded, are next drawn by the flux-plotting methods outlined in Art. 10, Ch. I, of the volume on electric circuits in this textbook series. The length of each flux line is then measured, and is considered as the gap length δ of the corresponding section. The slot width is then divided by this value of δ, and the value of σ for each section is determined from the curve of Fig. 19. The value of the coefficient C

for each section and finally the length of the equivalent air gap are then determined from Eqs. 41 and 42 respectively.

If the original gap sections 1, 2. 3 $\cdots n$ of length δ_1, δ_2, $\delta_3 \cdots \delta_n$, respectively, are now increased to correspond to the equivalent gap sections of lengths δ_1', δ_2', $\delta_3' \cdots \delta_n'$, and a smooth curve is drawn through the extremities of the lines, an armature having a smooth surface with an air gap whose apparent reluctance is equivalent to the original armature is obtained. The broken-line curve drawn through the tips of the teeth in Fig. 20 shows the approximate form of this equivalent armature surface.

The equivalent gap is of nonuniform length. Its apparent reluctance can now readily be obtained from a new flux plot drawn between the pole face and the equivalent armature face, both faces being assumed to be equipotential surfaces. However, to simplify the work by eliminating the need for the flux plot, the gap of nonuniform length is replaced by a gap of uniform length equal to δ_1' at the center of the pole and of circumferential width which gives the same reluctance as the actual gap.

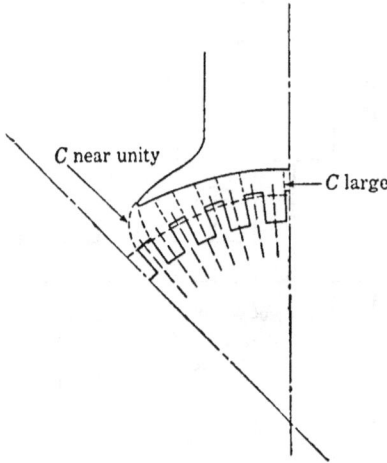

FIG. 20. Flux plotting in air gap to obtain equivalent smooth-faced gaps.

An understanding of the procedure for determining the circumferential width of this equivalent gap of uniform length δ_1' can be obtained through considering the flux path of one section of Fig. 20. Since the magnetic potential difference of the air gap is the same for every path and the cross-sectional areas of the paths are equal, the flux of each path is inversely proportional to the length of the path. If the path length is decreased from δ_n' to δ_1', the path width must be decreased in the ratio δ_1'/δ_n' in order to keep the reluctance fixed. The total width of the equivalent air gap of uniform length δ_1' therefore must be reduced by the factor

$$\frac{1}{n}\left[\frac{\delta_1'}{\delta_1'} + \frac{\delta_1'}{\delta_2'} + \frac{\delta_1'}{\delta_3'} + \cdots + \frac{\delta_1'}{\delta_n'}\right] = \frac{\delta_1'}{n}\Sigma\frac{1}{\delta'} = P, \qquad [43]$$

where n is the number of flux paths of equal width and P is called the *pole constant*. As derived, it is equivalent also to the ratio of the reluctance of a gap of uniform length δ_1' and of pole-pitch width (the arc distance between center lines of adjacent poles) to the reluctance of the actual gap. This ratio is the same as the ratio of flux of the actual gap to flux of the gap of uniform length δ_1' and pole-pitch width, for equal magnetic potential drops. If the actual gap were of uniform length and so short that negligible fringing were present at the edges of the pole shoe, the pole constant would equal the ratio of pole arc to pole pitch.

To systematize the work, a tabular form of solution as indicated by Table IV is suggested.

TABLE IV

Path	δ	s/δ	σ	C	$\delta' = \delta C$	$1/\delta'$
1						
2						
3						

An approximation frequently used is to compute Carter's coefficient and δ_1' at the center of the pole, then draw the pole shoe, and locate the armature as a circle of such radius that the air gap is δ_1'. The flux paths are then plotted and their lengths are measured. The sum of the reciprocals of these lengths is then used as $\Sigma 1/\delta'$ in Eq. 43.

The effective axial length of the air gap is equal to the actual axial length of the poles corrected for fringing at the ends and for the effect of ventilating ducts. The correction for fringing is usually considered as an elongation equal to $2\delta_1$, and the correction for ventilating ducts as a shortening equal to $\sigma_d D$, where D is the width of the duct and σ_d is determined from the curve of Fig. 19 for the ratio of D/δ_1 encountered. Thus

$$\text{Effective axial length} = \text{actual length} + 2\delta_1 - \sigma_d D. \qquad [44]$$

The dimensions to be used for the equivalent air gap of a dynamo having p poles of axial length ℓ and an armature of diameter d are, for gap length:

$$\ell_a = \delta_1', \qquad [45]$$

and for gap area:

$$A_a = \frac{\pi d P}{p} (\ell + 2\delta_1 - \sigma_d D). \qquad [46]$$

After the equivalent gap dimensions have been determined and an estimate of pole tip leakage flux has been made by use of a coefficient of dispersion based upon experience, by the method of Art. 11, or by flux mapping, the remainder of the problem of calculating the magnetization curve is identical with the procedures outlined in preceding articles.

PROBLEMS

1. A cast-iron ring of uniform circular cross section 5 in. inside diameter and 7 in. outside diameter is wound with a coil of 500 turns.

What values of flux are produced by the following values of current in the coil: (a) 1 amp? (b) 2 amp? (c) 4 amp? Does the flux double when the current is doubled?

What direct currents are necessary to produce the following values of flux in the iron: (d) 50 kilolines? (e) 25,000 maxwells? (f) 10^{-4} weber?

What values of flux are produced by the following values of magnetizing force from the coil: (g) 60 amp-turns per in.? (h) 30 oersteds? (i) 25,000 praoersteds?

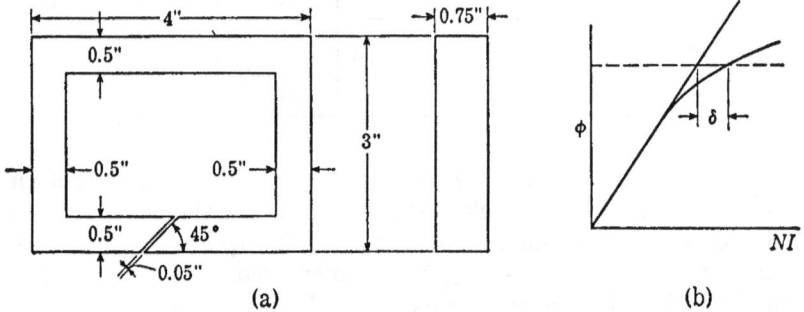

Fig. 21. Magnetic circuit for Prob. 2.

2. In the design of audio-frequency inductances, the continuity of the ferromagnetic circuit is broken with an air gap in order to make the magnetization curve of

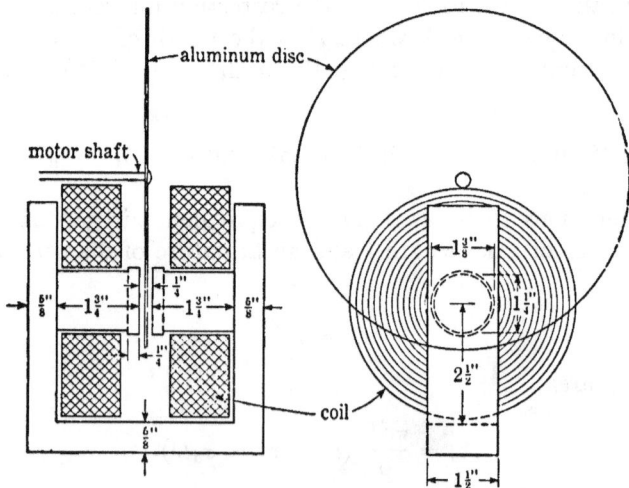

Fig. 22. Eddy-current brake of Prob. 3.

the circuit more nearly linear. The core is made of Radio A-3 (Transformer-72) steel laminations with a stacking factor of 0.9; the dimensions are shown in Fig. 21a.

What is the maximum flux ϕ_{max} at which this core can be worked if the deviation δ of the magnetization curve is to be kept less than 5% of the value NI would have if the magnetization curve were a straight line as shown in Fig. 21b?

3. An eddy-current brake used in the laboratory to measure the mechanical output of a motor is shown in Fig. 22. The torque produced on the motor shaft is a function of the air-gap flux density and of the speed of the aluminum disc which rotates between

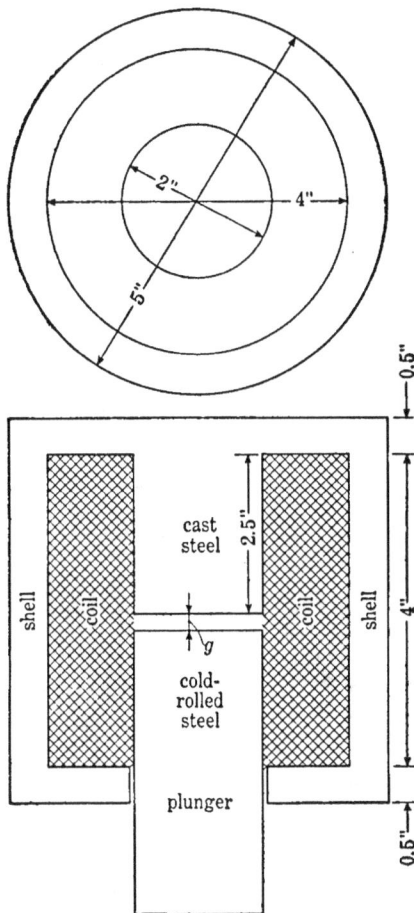

FIG. 23. Ironclad plunger magnet of Prob. 4.

the poles of the magnet. The frame which forms the magnetic circuit is solid cold-rolled steel. The two coils of 700 turns each are connected in series.

Find the approximate flux density in the air gap when the coil current is 2.5 amp.

The speed of the disc is such that the demagnetizing magnetomotive force produced by the currents in the disc is 2000 amp-turns.

4. The ironclad-solenoid magnet shown in Fig. 23 is used for tripping circuit breakers, operating valves, and in other applications in which a relatively large force is applied to a member that moves a relatively small distance. When the coil current is zero, the plunger drops against a stop such that the air gap g is 0.50 in. When the

coil is energized by a direct current of sufficient magnitude, the plunger is raised until it hits another stop set so that g is 0.1 in. The plunger is so supported that it can move freely in the axial direction. The air gap between the shell and the plunger can be assumed to be uniform and 0.01 in. long.

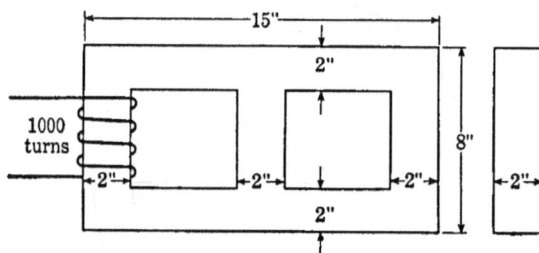

FIG. 24.   Magnetic core for Prob. 5.

If the coil has 1000 turns and carries 2 amp, what are the flux densities between the working faces of the shell and the plunger for gaps g of 0.10, 0.20, and 0.50 in.?

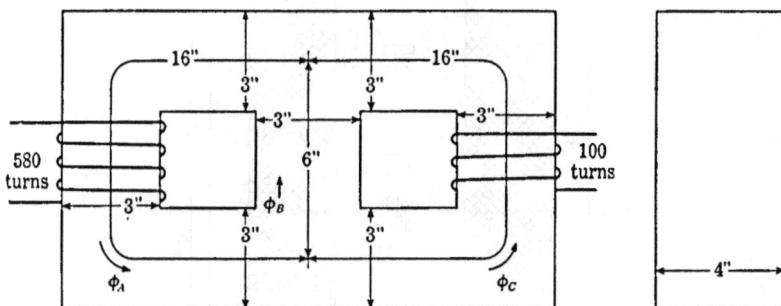

FIG. 25.   Magnetic core for Prob. 6.

5. The magnetic circuit of Fig. 24 is made of electrical-grade sheet steel having a stacking factor of 0.9. How many amperes are required in the coil to produce a flux of 0.0014 weber in the right-hand leg?

6. The magnetic core shown in Fig. 25 is made of transformer steel (curve 3 of Fig. 1). The stacking factor is 0.9. When the fluxes are $\phi_A = 1,000$, $\phi_B = 400$, and $\phi_C = 600$ kilolines, how many amperes must each coil carry and in what direction?

7. If the current entering the top terminal of the left-hand coil in Fig. 25 of Prob. 6 is 0.20 amp, what must be the current in the right-hand coil and in what direction in order that of the flux ϕ_C shall be zero?

8. The magnetic circuit of a small special dynamo is shown in Fig. 26. The rotating member (armature) is made of pure iron; the yoke is made of cast steel and has a square cross section.

How many ampere-turns are required of the winding on the yoke to give an average flux density of 10,000 gausses in the air gap?

9. Develop a formula for the permeance of the leakage path between two long parallel cylinders of ferromagnetic material. The radius of each cylinder is r and the

thickness of structure 0.5"

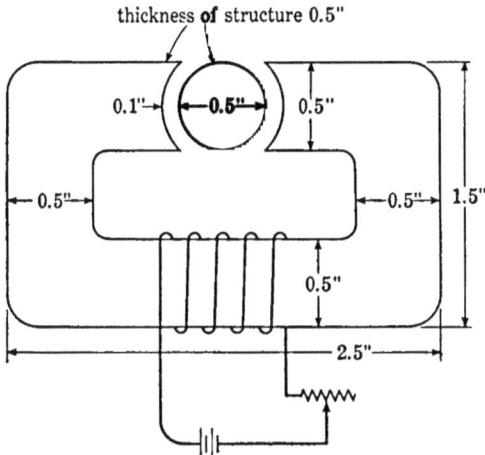

FIG. 26. Magnetic circuit of small special dynamo for Prob. 8.

distance between the axes of the cylinders is $2a$, which is large compared to r. The permeance is to be expressed in kilomaxwells per ampere-turn for each inch of cylinder length.

CHAPTER IV

Permanent Magnets and Energy in the Magnetic Field

Improvements made during recent years in the magnetic properties of magnetically hard materials have made possible a considerable extension in the practical utilization of these materials. In some of the newer applications, permanent magnets are used in place of electromagnets with their attendant energy loss in the exciting winding. In many of the other applications, devices such as radio loudspeakers, electrical meters and

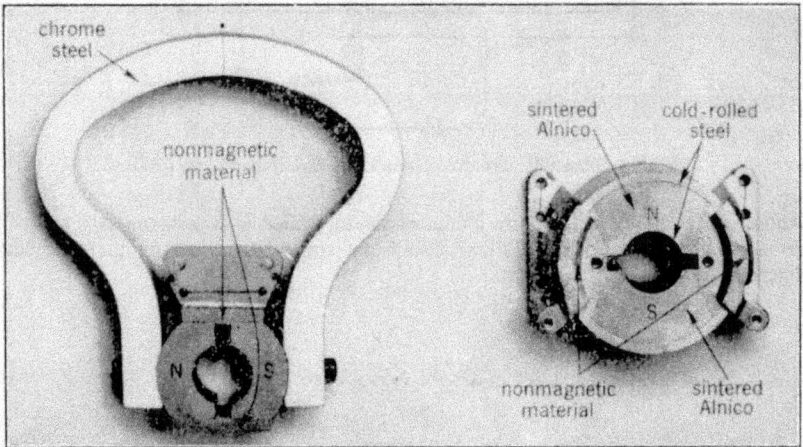

Courtesy General Electric Co.

A radical change in the form and size of the magnetic circuit becomes necessary when sintered Alnico is used in modern electrical measuring instruments in place of chrome steel.

instruments, fractional horsepower motors and generators, and electric clocks have not only been made more compact but have been radically changed in appearance by use of the new alloys possessing greater magnetic energy per unit weight or per unit volume.

All materials which exhibit the properties desired of permanent magnets have one characteristic in common; namely, they are so constituted that the magnetic moments of their domains can be reoriented only with comparatively great difficulty. The initial magnetization of these materials requires very large magnetomotive forces, which fortunately need to be applied only for a relatively short time. Normal magnetization curves and hysteresis loops for typical permanent-magnet materials are

shown in Figs. 15a and 15b of Ch. I. The crystalline state that results in
the condition of magnetic hardness is brought about by metallurgical
treatment which frequently also produces physical hardness or brittleness in the material.[1] The brittleness is very noticeable in the aluminum-
nickel-cobalt alloys such as Alnico, which cannot be worked satisfactorily,
and hence are usually cast initially into the desired shape and ground
where close fitting is necessary.

A method commonly used to magnetize groups of circular or U-shaped
magnets simultaneously consists in placing a large conductor through the
center of several of them and subjecting the conductor to a current surge
having a maximum value of several thousand amperes. During the operation, the open end of each magnet — that is, its air gap — is bridged by
a piece of ferromagnetic material so that nearly all the applied magnetomotive force serves to overcome the reluctance of the magnet itself.
When the bridging piece is removed, the flux decreases by an amount
depending on the dimensions of the magnet and gap, the retentivity,
the coercivity, and the shape of the demagnetization curve of the material. If the air gap is reclosed, the flux may increase somewhat, but it
never again reaches the value it had before the bridging piece was initially
removed. However, subsequent opening and closing of the gap produce
no appreciable permanent change in the flux. Only a cyclic variation
occurs, and the magnet is considered to have reached its permanent
state. When the magnet is a straight bar, it may be magnetized by being
placed between the poles of a suitable electromagnet, or inside a solenoid.

The appropriate length and cross-sectional dimensions for a permanent
magnet which is to maintain a specific magnetic field in a given air gap
can be predetermined to a degree of approximation satisfactory for most
purposes.[2] The dimensions depend on (a) the length and cross-sectional
area of the gap, (b) the flux density desired in the gap, (c) the magnetic
properties of the particular material, and (d) certain other factors whose
significance is discussed in this chapter.

1. MAGNETIC CIRCUITS INVOLVING PERMANENT MAGNETS

The magnetic properties of an alloy are usually given in the form of a
demagnetization curve obtained experimentally by a method which is
equivalent to testing a specimen of the material formed into a closed
magnetic circuit of uniform cross section and without an air gap. A
magnetomotive force large enough to carry the flux density to saturation
is applied to the specimen and then the descending portion of the curve

---

[1] J. Q. Adams, " Alnico — Its Properties and Possibilities," *G.E. Rev., 41* (1938), 518–523.
[2] S. Evershed, " Permanent Magnets in Theory and Practice," *I.E.E.J., 58* (1920), 780–
837, and *63* (1925), 725–821.

is found as described in Art. 5 of Ch. I. The curves of Fig. 1 represent the relation between the flux density and the magnetizing force during the process of magnetization. The intercept on the $\mathcal{B}$ axis is the retentivity $\mathcal{B}_r$ and the intercept on the horizontal axis is the coercivity $\mathcal{H}_c$. The

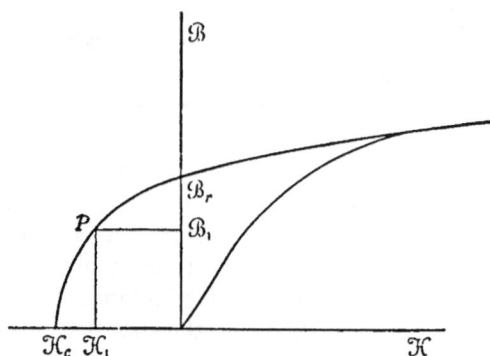

FIG. 1.   Curve of a permanent-magnet material magnetized to saturation and then demagnetized.

portion of the descending curve included between these two intercepts is called the *demagnetization curve;* the operating points associated with a permanent-magnet material usually lie on or near this curve. The demagnetization curves for several typical alloys used as permanent-magnet materials are shown in Fig. 2.

The method of finding the flux or flux density in the air gap of a permanent magnet can be best illustrated by reference to a simple magnetic circuit of uniform cross-sectional area A_m and length l_m with a short air gap of length l_a. In this introductory discussion, leakage and fringing are neglected. With the air gap temporarily bridged by a piece of soft iron, a current of sufficient magnitude to saturate the magnetic circuit is sent through the magnetizing winding as shown in Fig. 3. After the magnetomotive force has been reduced to zero, the flux density remains at the retentivity value $\mathcal{B}_r$ as long as the magnetic circuit is completed by the soft iron bridge. When the bridge is removed, the magnetomotive force applied to the circuit is still zero, and the flux density is further reduced until it reaches a value such that the magnetic potential rise U_m or $\mathcal{H}_m l_m$ clockwise in the magnet from the minus pole to the plus pole is equal to the magnetic potential drop U_a across the air gap. This condition is represented by the operating point $P(-\mathcal{H}_1,\mathcal{B}_1)$ in Fig. 1. A specific computation will indicate just how this point is found for a magnet of specified dimensions.

The magnet of Fig. 3 is to be made of 36 per cent cobalt steel with a cross-sectional area of 0.25 square inch and a length of 6 inches. The length of the air gap is 0.05 inch. What is the air-gap flux?

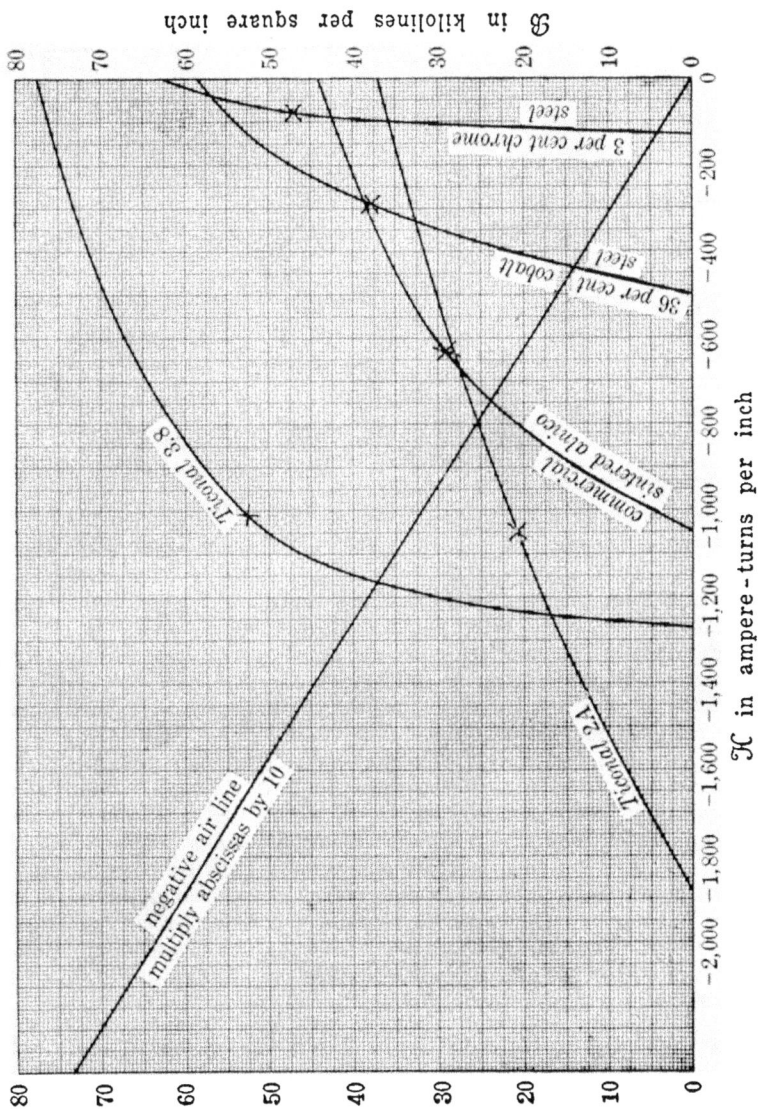

FIG. 2. Demagnetization curves for typical permanent-magnet materials. Crosses indicate points at which (BH) is a maximum.

The method of solution is the one described in Art. 4 of Ch. III and may consist of a direct graphical process in which the curve of flux versus magnetic potential rise $\mathcal{H}_m \ell_m$ in the magnet is plotted as shown in Fig. 4. This curve is the same-as the one for cobalt steel in Fig. 2 except that the abscissa scale of Fig. 2 is multiplied by the length of steel path, 6 in., and the ordinate scale is multiplied by the cross-sectional

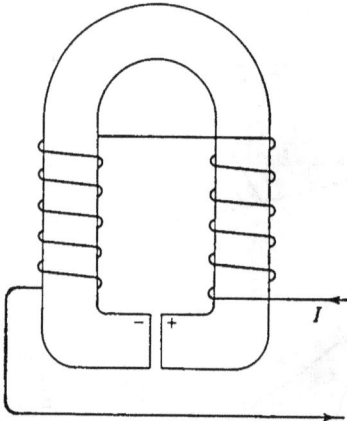

Fig. 3. A permanent magnet with its magnetizing winding.

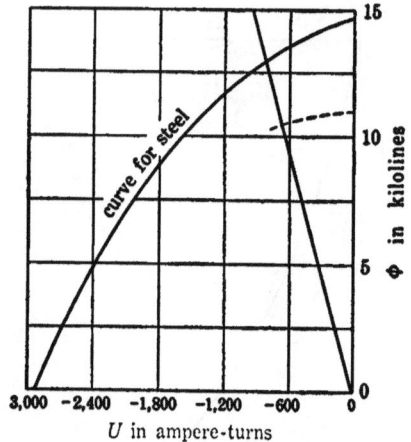

Fig. 4. Graphical solution for finding the flux in the air gap of a permanent magnet.

area of the steel, 0.25 in. The negative air-gap line is drawn with a negative slope equal to the permeance of the air gap, which in mixed English units is (3.192 × 0.25)/0.05 or 15.96 lines/amp-turn. The intersection of the negative air-gap line with the curve gives the flux in the magnetic circuit as 13 kilolines.

The solution may also be carried out by the cut-and-try process, in this way avoiding the necessity for the construction of additional curves. Without the aid of the preceding solution, a first guess of 50 kilolines/sq in. would seem reasonable, since the air gap is short and does not carry the density far down the curve. Corresponding to 50 kilolines/sq in., the magnetizing force in the cobalt steel is 165 amp-turns/in. The magnetic potential rise in the cobalt steel is then 6 × 165 or 990 amp-turns. In the air gap, the magnetizing force is 15,700 amp-turns/in., and the potential drop is 0.05 × 15,700 or 785 amp-turns, which does not check the 990 amp-turns in the steel. A larger flux density will increase the potential drop in the air gap and decrease the rise in the steel. When the final estimate is taken as 52 kilolines/sq in., the rise in the steel is 6 × 140 or 840 amp-turns. The drop in the air gap is 0.05 × 16,300 or 815 amp-turns, which is a good enough check.

Near the beginning of this article it was pointed out that some of the newer alloys are superior to the older material such as tungsten and cobalt steels. It should be interesting to find what the flux would be if the magnet of the previous example were made of Alnico instead of cobalt steel. The negative air-gap line is the same as before, and the curve for the Alnico portion is like the Alnico curve in Fig. 2 except that the abscissas are multiplied by 6 and the ordinates divided by 4. The useful part of this curve is shown dotted in Fig. 4, and it intersects the negative air-gap line at a flux of 10.5 kilolines. This is a poorer magnet, then, when made of Alnico than when made of cobalt steel. The fault is in the improper use of Alnico in the new design, as will be shown in the next article.

2. DESIGN FOR MINIMUM AMOUNT OF MAGNETIC MATERIAL

The relatively high cost of permanent-magnet materials and the necessity of conserving space in many applications make desirable a design which requires the least possible volume of material. A magnet of the form shown in Fig. 3, having a length ℓ_m and cross-sectional area A_m, has a volume $\ell_m A_m$. The magnetic potential rise in the magnet $\mathcal{H}_m \ell_m$ equals the drop in the air gap, or

$$\ell_m = \frac{U_a}{\mathcal{H}_m}, \qquad [1]$$

and, when leakage is neglected, the flux ϕ_a in the air gap is equal to the flux in the magnet, from which

$$A_m = \frac{\phi_a}{\mathcal{B}_m} . \qquad \blacktriangleright[2]$$

The volume of magnetic material is given by the product of Eqs. 1 and 2 as

$$\ell_m A_m = \frac{U_a \phi_a}{\mathcal{H}_m \mathcal{B}_m} . \qquad [3]$$

The air-gap drop U_a can be expressed in terms of the air-gap dimensions and ϕ_a as follows:

$$U_a = \mathcal{H}_a \ell_a = \frac{\mathcal{B}_a \ell_a}{\mu_a} = \frac{\phi_a \ell_a}{\mu_a A_a}, \qquad [4]$$

and the substitution of this value of U_a into Eq. 3 gives

$$\ell_m A_m = \frac{\ell_a \phi_a^2}{\mu_a A_a \mathcal{H}_m \mathcal{B}_m} = \frac{\phi_a^2}{\mathcal{P}_a \mathcal{H}_m \mathcal{B}_m}, \qquad [5]$$

where $\mathcal{P}_a$ is the permeance of the air gap.

▶ For a magnetic circuit with an air gap of specified area A_a and length ℓ_a and having a specified air-gap flux ϕ_a, the volume of magnetic material is least when the *energy product* $\mathcal{H}_m \mathcal{B}_m$ is a maximum.◀

Reference to the demagnetization curves of Fig. 2 shows that the energy product is zero at each of the intercepts and should reach a maximum value at some point on the curve. For each of the materials, the point of $(\mathcal{B}_m \mathcal{H}_m)_{max}$ is indicated with a small cross. With these values of $\mathcal{B}_m$ and $\mathcal{H}_m$ the optimum dimensions of the magnet can now be found from Eqs. 1 and 2. By the use of Eqs. 4, Eq. 1 becomes

$$\ell_m = \frac{\mathcal{B}_a \ell_a}{\mu_a \mathcal{H}_m} = \frac{\phi_a \ell_a}{\mu_a A_a \mathcal{H}_m} = \frac{\phi_a}{\mathcal{P}_a \mathcal{H}_m}. \qquad \blacktriangleright[6]$$

The relationships just developed take into account the fact that the air gaps and the magnet may be of different cross-sectional areas as they are in the designs of Figs. 7 and 8. The equations also allow for a fringing correction at the air gap.

The magnets of Art. 1 will now be redesigned to give the same air-gap flux as before, 13 kilolines. First, for the cobalt-steel magnet the optimum flux density $\mathcal{B}_m$ (corresponding to the point marked x on the curve of Fig. 2) is 37.7 kilolines/sq in. and $\mathcal{H}_m$ is 290 amp-turns/in. From Eqs. 2 and 6,

$$A_m = \frac{13}{37.7} = 0.35 \text{ sq in.} \qquad [7]$$

$$\ell_m = \frac{13{,}000 \times 0.05}{3.192 \times 0.35 \times 290} = 2.01 \text{ in.} \qquad [8]$$

The volume of magnetic material has been reduced from 6×0.25 or 1.5 cu in. to 2.01×0.35 or 0.7 cu in. by the better choice of operating point on the demagnetization curve.

The best value of $\mathcal{B}_m$ for the Alnico magnet is, from Fig. 2, 29 kilolines/sq. in. and the corresponding value for $\mathcal{H}_m$ is 630 amp-turns/in. The area is 13/29 or

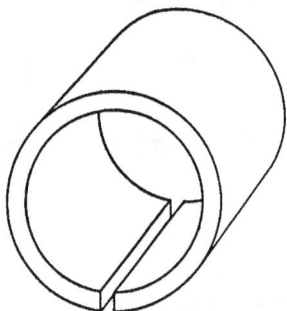

FIG. 5. Small cylindrical ring of cobalt steel.

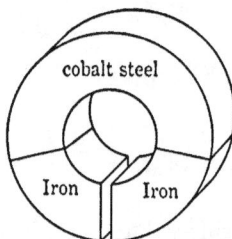

FIG. 6. Cobalt-steel magnet with soft-iron poles.

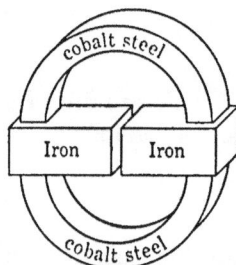

FIG. 7. Magnet with two cobalt-steel sections in parallel.

0.449 sq in. and the length of magnet is $(13{,}000 \times 0.05)/(3.192 \times 0.449 \times 630)$ or 0.72 in. The new volume of Alnico is 0.449×0.72 or 0.32 cu in., and this is only about one-fifth as much as was used in the original poor design which produced only 10.5 kilolines of air-gap flux as compared with 13 for this design.

The question now arises how pieces of magnetic material of these optimum dimensions can be utilized practically in a magnetic circuit. Though a cylindrical ring of cobalt steel 2 in. in mean circumference and 0.35 sq in. in cross-sectional area can be formed as indicated in Fig. 5, such a magnet may not have the desired flux distribution for the particular application. A square or round cross section for the pole pieces can be provided by the introduction of some soft iron or steel into the circuit as shown in Figs. 6 and 7. In these designs, the flux density in the iron is so low that the magnetic potential drop in it can be neglected. In Fig. 6, the length and cross section of the cobalt steel correspond to the optimum values. The diameter of the ring is chosen which is consistent with the practical requirements of the application. When the design of Fig. 7 is used, the cross-sectional area of the cobalt steel is one half the computed value, and the length of each half is 2 in.

If the particular application requires a flux density of 52 kilolines/sq in. rather than a total flux of 13 kilolines in the air gap, the soft-iron poles of Figs. 6 and 7 can be tapered to give an air-gap area of 0.25 sq in. instead of 0.35 sq in. If the total flux is kept to 13 kilolines, the cross-sectional area of cobalt steel should be the same as before, but its length has to be increased somewhat because of the decrease in the permeance of the air gap.

For the new design using Alnico, the arrangement of Fig. 6 can still be used, but the magnet material and the soft iron should be interchanged to reduce leakage as shown in the next article.

3. Fringing and leakage[3]

The methods discussed in Arts. 6 to 11 of Ch. III for taking fringing and leakage into account for electromagnets can be used for permanent magnets. For a short air gap, this cross-sectional area should be taken somewhat larger than the area of the pole faces, as indicated in Art. 7 of Ch. III. When the air-gap length is more than one-fifth of the shorter pole dimension, or when leakage between other parts of the magnetic circuit becomes significant, the principles of Art. 10. Ch. III must be applied. An example shows best how these principles are utilized in a specific design.

The magnetic circuit of Fig. 8 will be designed to give a flux density of 30 kilolines per square inch between the poles. The instrument which is to utilize this magnetic circuit requires a 0.20-inch air gap with a pole area of about 0.75 square inch. The permanent-magnet material will be used in the construction of the poles T, and these will be supported by a soft-iron yoke Y in which the magnetic potential drop will be negligible at the flux density required. Reference to the demagnetization curves of Fig. 2 shows that cobalt steel has its optimum working point at just under a flux density of 40 kilolines per square inch, but, because of its relatively low magnetizing force, the poles would have to be longer than they would if they were made of Alnico, Ticonal 2A, or Ticonal 3.8. In Alnico, the design would use the least volume of material if it provided a flux density of about 28 kilolines per square inch in the Alnico, the necessary air-gap density being produced by adding some tapered soft-iron pole pieces P as indicated in Fig. 9. The fringing between the conical surfaces is so great, however, that the air-gap requirements could be met with greater economy of material if an operating point on the Alnico curve were chosen somewhat higher than the point indicated by

[3] A. Th. van Urk, " The Use of Modern Steels for Permanent Magnets," *Philips Tech. Rev.*, *5* (1940), 29–35.

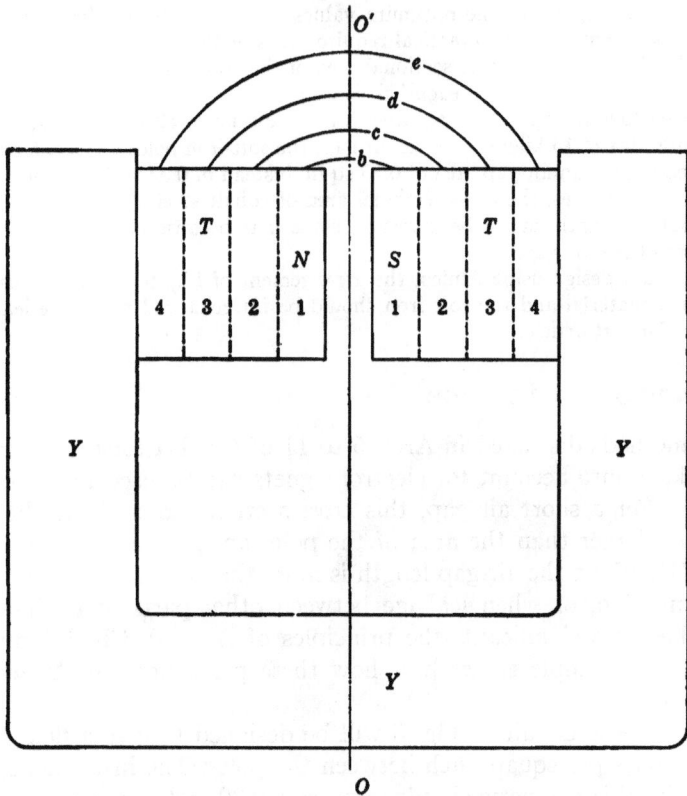

(a) the poles are sectionalized for the computation
of leakage fluxes *b*, *c*, *d* and *e*.

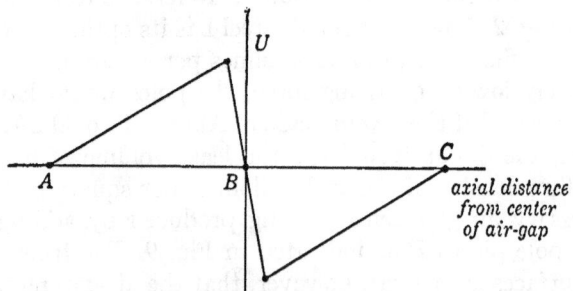

(b) magnetic potential distribution along pole axis

Fɪɢ. 8.   Permanent magnet with Ticonal poles *T*, soft-iron yoke *Y*.

the cross. Another possibility is to look for a material better suited to work at higher flux densities. Some types of cast Alnico[4] (not shown on the curve sheet) could be used. A material known as Ticonal 3.8 (titanium, cobalt, nickel, aluminum, and iron) is evidently better than any of the others represented in Fig. 2, and will be used in this design. Cylin-

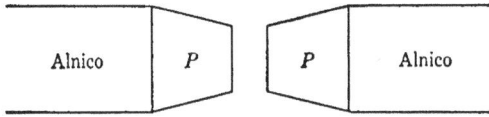

Fɪɢ. 9.   Alnico poles with tapered soft-iron pole pieces.

drical poles T, Fig. 8, one inch in diameter (area 0.785 square inch) are chosen, and their length will be determined in the process of calculating the flux and magnetic potential relations.

Figure 8a includes a sketch showing approximately the directions of some of the flux lines which enter the cylindrical surfaces of the pole pieces. A plane passing through the center line OO' perpendicular to the axis of the poles is a plane of zero magnetic potential. The distribution of potential along the pole axis is shown in Fig. 8b to the same horizontal scale as is used in Fig. 8a.

The potential drop across the air gap is

$$U = 0.20 \times 9,400 = 1880 \text{ amp-turns,} \tag{9}$$

since from Fig. 2 $\mathcal{H}$ in air corresponding to a flux density of 30 kilolines per square inch is 9,400 ampere-turns per inch. Equation 9 gives the magnetic potential difference to be overcome by the permanent magnets and is one of the most important factors in the determination of the pole length.

The flux in the Ticonal magnets increases with the distance from the air gap on account of the considerable amount of useless flux entering the cylindrical surfaces of the poles. An accurate and theoretically sound computation of this extraneous flux could be made, although it would involve a great amount of labor, if the construction of a flux map were feasible for this configuration of magnetic potentials. A good approximation should be possible after examining by means of iron filings the flux paths in the vicinity of the adjacent poles of two bar magnets placed end to end and separated by an air gap. Such an examination reveals that the paths can be fairly closely approximated by portions of concentric circles with centers at the midpoint of the air gap, as indicated at b, c, d, and e

[4] B. M. Smith, " Alnico, Properties and Equipment for Magnetization and Test," *G.E. Rev.*, *45* (1942), 210–213.

of Fig. 8a. Flux paths of this nature will be assumed and the extraneous flux computed.

Consider two equal cylindrical poles of radius r with parallel plane faces of opposite polarity as shown in Fig. 10. The reluctance of a thin spherical shell of thickness t and of average radius a will first be found.

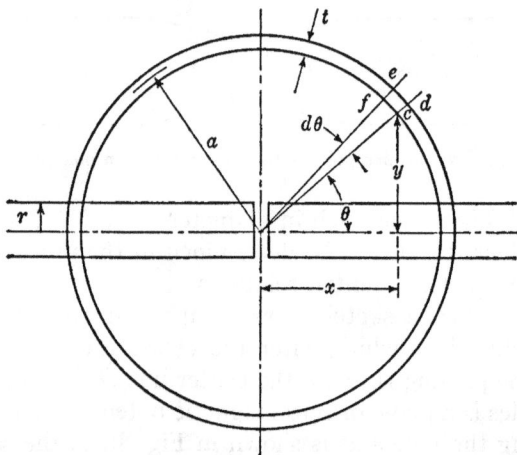

FIG. 10.   Cross section of cylindrical poles with fringing path.

The differential element is a zonal shell made by revolving the differential area *cdef* about the axis of the magnets. The length of the shell in the direction of flux is $ad\theta$, and its cross section is $2\pi yt$ perpendicular to the flux. The reluctance of this element is

$$d\mathcal{R} = \frac{ad\theta}{\mu 2\pi yt},$$ [10]

and, since

$$d\theta = \frac{dy}{\sqrt{a^2 - y^2}},$$ [11]

the total reluctance of the thin shell is

$$\mathcal{R} = \frac{a}{\pi \mu t} \int_r^a \frac{dy}{y\sqrt{a^2 - y^2}}$$

$$= -\frac{1}{\pi \mu t} \left[\cosh^{-1} \frac{a}{y} \right]_r^a$$

$$= \frac{1}{\pi \mu t} \cosh^{-1} \frac{a}{r},$$ [12]

and the permeance is the reciprocal of the reluctance or

$$\mathcal{P} = \frac{\pi \mu l}{\cosh^{-1} \dfrac{a}{r}}. \tag{13}$$

This result will now be used in the design of the magnet shown in **Fig. 8.**

The magnetic potential drops in the poles are found through division of the poles into quarter-inch lengths and then introduction of the leakage flux into each of these short sections. The flux density in the first sections would be about the same as in the air gap if it were not for the great concentration of fringing flux at the pole edges.

Applying the fringing correction, the flux density $\mathcal{B}_1$, in the first half of each first section, is taken as $30 \times (1.2)^2$ or 43.2 kilolines/sq in. The corresponding value of $\mathcal{H}$ in the Ticonal is, from Fig. 2, 1,125 amp-turns/in. The drop in magnetic potential from the air gap to the midpoints of the two inside sections is $0.25 \times 1,125$ or 280 amp-turns. The potential difference along the path b between these midpoints is

$$U_b = 1,880 - 280 = 1,600 \text{ amp-turns.} \tag{14}$$

The permeance of leakage path b is found through substitution into Eq. 13 of the values

$$r = 0.50 \text{ in.} \tag{15}$$

$$a = \sqrt{(0.225)^2 + (0.50)^2} = 0.548 \text{ in.} \tag{16}$$

$$l = 0.25 \times \frac{0.225}{0.548} = 0.103 \text{ in.} \tag{17}$$

The inverse hyperbolic cosine of a/r (1.096) is 0.435.

$$\mathcal{P}_b = \frac{3.14 \times 3.19 \times 0.103}{0.435} = 2.38 \text{ lines/amp-turn.} \tag{18}$$

The flux in path b is

$$\phi_b = 1,600 \times 2.38 = 3,800 \text{ lines.} \tag{19}$$

This leakage flux enters the cylindric surfaces of the poles in the first section. The sum of this leakage flux and the flux entering the air-gap ends of the first section is the total flux

$$\phi_2 = 3,800 + 43,200 \times 0.785 = 37,700 \text{ lines} \tag{20}$$

entering the second sections. The density in the second sections is 37.7/0.785 or 48 kilolines/sq in. The corresponding value of $\mathcal{H}$ is 1,080 amp-turns/in. The potential drop between the midpoints of the first sections and the midpoints of the second sections is $0.50 \times 1,080$ or 540 amp-turns. The potential drop along path c is then

$$U_c = 1,600 - 540 = 1,060 \text{ amp-turns.} \tag{21}$$

For path c the permeance is

$$\mathcal{P}_c = \frac{3.14 \times 3.19 \times (0.25 \times 0.69)}{0.85} = 2.04 \text{ lines/amp-turn.} \tag{22}$$

The leakage flux in path c is

$$\phi_c = 1,060 \times 2.04 = 2,160 \text{ lines,} \tag{23}$$

and this will enter the cylindric surfaces of the second sections, making the total flux ϕ_3 entering the third sections $37,700 + 2,160$ or $39,860$ lines. This gives a flux density in the third sections of 51 kilolines/sq in. The corresponding value of $\mathcal{H}$ is 1,030 amp-turns/in. The drop from the midpoints of the second to the midpoints of the third sections is $0.5 \times 1,030$ or 515 amp-turns.

$$U_d = 1,060 - 515 = 545 \text{ amp-turns.} \tag{24}$$

$$\mathcal{P}_d = \frac{3.14 \times 3.19 \times (0.25 \times 0.825)}{1.17} = 1.77 \text{ lines/amp-turn.} \tag{25}$$

$$\phi_d = 545 \times 1.77 = 960 \text{ lines.} \tag{26}$$

$$\phi_4 = 39,860 + 960 = 40,800 \text{ lines.} \tag{27}$$

$$\mathcal{B}_4 = 40.8/0.785 = 52 \text{ kilolines/sq in.} \tag{28}$$

$$\mathcal{H}_4 = 1,020 \text{ amp-turns/in.} \tag{29}$$

The remaining potential drop from the midpoints of the third sections to the yoke is only 545 amp-turns. This requires $545/1,020$ or 0.535 in. of Ticonal for both poles. The leakage flux over the remaining $(0.535/2) - 0.125$ or 0.143 in. of cylindric surface is so small that it will not be computed.

The total length of Ticonal (both poles) is the total length along which potential drops were computed; this sum is $0.25 + 0.50 + 0.50 + 0.535 = 1.79$ in.

The total extraneous flux is $(43,200 - 30,000)/0.785 + 3,800 + 2,160 + 960 = 23,700$ lines, and this is $23.7 \times 100/38.2$ or 62% of the flux between pole faces.

The importance of these methods of estimating the effect of extraneous flux certainly does not lie in the accuracy with which they can predict correct results but merely in their ability to convey a notion of the approximate magnitudes involved. A designer depends much more upon his experience with previous circuits and the principles of model theory explained in Ch. VII than upon calculations similar to those made in this article.

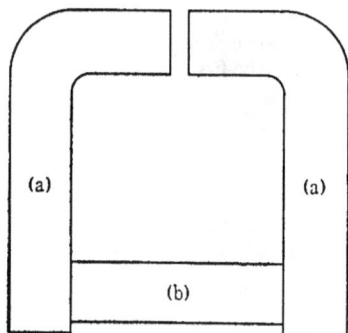

Fig. 11. Permanent magnet having excessive leakage: (a) soft iron, (b) magnet material.

If the Ticonal magnets are used as the vertical members instead of the pole pieces of the circuit of Fig. 8, the leakage flux is much greater. The worst design such a magnet can have from the standpoint of efficient use of magnetic material is of the type shown in Fig. 11, in which the permanent-magnet section is as far away from the air gap as possible. Here the surfaces of the poles and vertical members have the maximum potential difference, and the leakage flux is several times the useful flux.

FIG. 12.   Rear view of a watthour-meter magnet assembly.   The Alnico pole pieces are cast integral with the cold-rolled-steel base plate; this unit is then resistance welded to the cold-rolled-steel yoke which controls the air gap and acts as a magnetic shield against extraneous demagnetizing forces.

Very often where the design involves a large amount of leakage flux, as in the watthour-meter magnet of Fig. 12, the most economic use of material is obtained through increasing the cross section as the distance from the air gap is increased, thus allowing all the permanent-magnet material to be worked at the optimum flux density. The flux required at the ends of the magnet, called the main flux, is first determined and then the length of the magnet is computed as though there were no leakage flux, this design usually being made on the basis of optimum flux density. The main flux occupies a path of constant cross section and the leakage occupies the additional material which may be termed the *leakage shell*, as shown in Fig. 13.

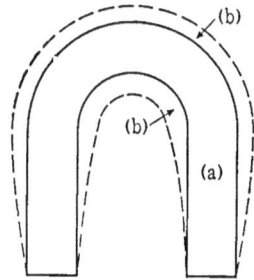

FIG. 13.   Diagram showing increase in cross-sectional area to maintain constant flux density: (a) main path, (b) leakage shell.

4. STABILIZATION OF PERMANENT MAGNETS

The foregoing criteria of design for permanent magnets are reasonably satisfactory when the magnets are to be designed on the basis of minimum volume of alloy and are subjected only to the demagnetizing influence of a fixed gap. In many applications, this ideal situation does not exist. For example, the magnetic material may be subjected to a demagnetizing influence such as that caused by alternating current in the coil of a telephone receiver or in a conductor near the damping magnet in a watthour meter. These currents set up alternating magnetomotive forces.

After the original magnetization process and the removal of the soft-iron bridge across the air gap, an operating point P_1 is determined at the intersection of the negative air-gap line and the curve of flux versus magnetic potential rise in the alloy as shown in Fig. 14. If now a demagnetizing magnetomotive force F is applied to the circuit, the new operating point must be such that, in the direction of the flux, the potential rise in the alloy is balanced by the potential drop across the air gap and the negative magnetomotive force F. The new flux is ϕ_2, corresponding to a rise in potential U_2 in the alloy and an air-gap drop $U_2 - F$. When the demagnetizing influence is removed, the recovery of flux versus potential rise in the alloy does not ascend along the original demagnetization curve but along a curve P_2P_3 corresponding to the incremental permeability of the material at point P_2. After this half cycle of demagnetizing magnetomotive force the operating point is at P_3. A reversal of F, which now becomes a magnetizing magnetomotive force, carries the operating point up the minor loop to P_4 where the flux is ϕ_4, the rise

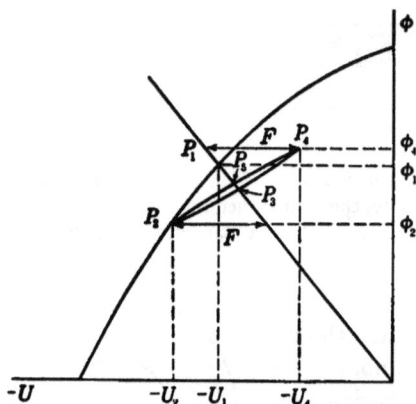

Fig. 14. Diagram showing the demagnetizing effect of an alternating magnetomotive force on a permanent magnet.

in potential is U_4 in the alloy, and the air-gap drop is $U_4 + F$. The removal of the magnetizing magnetomotive force F results in operation at P_5. A complete cycle of alternating magnetomotive force leaves the magnet reduced in strength to the condition shown at P_5. A second cycle like the first one causes the operating point to descend approximately to P_2 and then to retrace the minor loop.

▶ If a permanent magnet is demagnetized by the application of an alternating magnetomotive force of maximum value F, it will not thereafter suffer further significant demagnetization when subjected to alternating magnetomotive forces smaller than F.◀

Operation subsequent to the " alternating-current knockdown," as it is called, is stabilized in the small region in the neighborhood of P_3 and P_5. Magnets like those shown in Fig. 12 are used to produce the braking torque in watthour meters. They are subjected to an alternating-current knockdown of considerable magnitude and are also put through definite temperature cycles and vibration tests.

The foregoing discussion indicates that when such factors as alternating fields are taken into consideration, the criteria of design should be modified. The length and cross-sectional area of the magnet can be increased to correspond to operation at a point halfway between P_3 and P_5 with the required air-gap flux, but this would not necessarily be the optimum operating point to give minimum volume of alloy. The new operating point can be transferred to the demagnetization curve of the material,

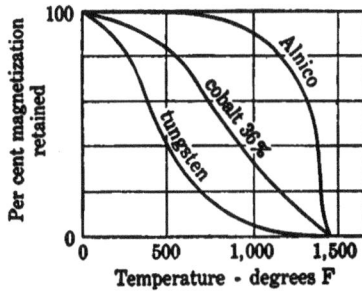

(a) resistance to effects of heating

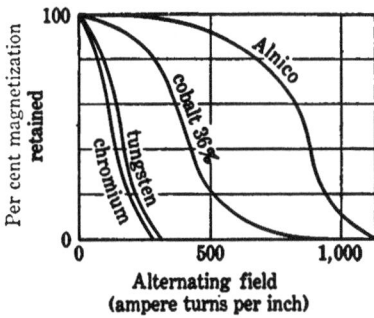

(b) resistance to effects of stray field   (c) resistance to effects of vibrations

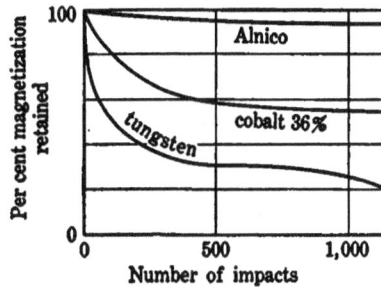

Fig. 15.   Effects of temperature (a), alternating field (b), and vibration (c) on permanent magnets.

and then a locus of similar points as the dimensions of the magnet are varied can be determined by graphical construction. A procedure similar to that outlined for the original curve on the assumption that no demagnetizing influence is present may then be utilized to obtain the optimum point on the new curve. The curves of Fig. 15a, b, and c indicate that the effect upon the strength of a magnet seems to be essentially the same whether the demagnetizing agent is a magnetomotive force, a mechanical vibration, or a thermal agitation.

The use of a length greater than, and a cross-sectional area less than, that indicated by the maximum product of $\mathfrak{B}$ and $\mathcal{H}$ produces a magnet which is less affected by a given magnitude of demagnetizing agent than one whose length and cross-sectional area correspond to the point of $(\mathcal{B}\mathcal{H})_{max}$. That is, the per cent variation of $\mathcal{B}$ produced by a given variation in $\mathcal{H}$ is less when the operating point is chosen on the higher part of the demagnetization curve. For this reason, the older magnets were not designed for optimum ratio of dimensions, but rather for greater permanency of flux. The iron *keeper* or magnetic shunt placed across the poles of a magnet not in use is effectively a device for moving the magnetic state along the line P_2P_4 extended nearly to the $\mathcal{B}$ axis so that demagnetizing influences are less likely to carry it back to and down the demagnetization curve, thus causing a permanent weakening. Keepers are not, in general, required with the newer alloys because of the greater stability of magnetization which these alloys exhibit as indicated by the relative flatness of their demagnetization curve. The curves of Fig. 15a, b, and c illustrate the relative stability characteristics of typical permanent-magnet alloys.

In the large-scale production of a magnetic device, variations in the composition of the alloy, in the effectiveness of welding the parts of the magnetic circuit together, in the magnetization process, and in the assembly of the finished product would seem to make good quality control impossible. A test of 3 per cent of 319,000 meter magnets[5] of the type shown in Fig. 12 reveals that 1 per cent of them had a flux deviation of 5 per cent from the average value. This high attainment in quality control and the permanency of the magnets after manufacture constitute scientific achievements of as great importance as the other elements of the design process.

5. ENERGY IN THE MAGNETIC FIELD

In the discussion of the fundamental properties of electric and magnetic fields given in Ch. I of the volume on electric circuits, the existence of each field was shown to indicate a storage of energy. As demonstrated by the classical work of Maxwell, both fields are mutually dependent and, whenever they are established, energy is delivered to the region in which they exist. Under certain conditions, which depend upon the properties of the medium of the region, the energy put into the region when the fields are established is returned to the original system when the fields are removed. In vacuum, the process is completely reversible. Examples of energy-storage phenomena in the form of energy stored in inductors or capacitors are mentioned frequently throughout this series of textbooks.

In every physical piece of apparatus in which electric and magnetic fields exist, the energy storage-and-release processes are not completely reversible. A certain amount of energy is always dissipated in the form

---

[5] J. H. Goss, " Permanent Magnets," *Metals and Alloys, 15* (1942), 576–582

of heat within the medium permeated by the field, and though the magnitude of this energy is negligibly small in certain mediums it is often appreciable in others. Although the relations for the energy stored in a magnetic field could be developed for the more general problem, it is equally satisfactory for the purpose of this treatment, and less confusing, to consider a simple configuration such as a toroidal magnetic circuit. Consider the core dimensions in Fig. 16 to be such that the thickness T of the core is small compared with the mean diameter D of the toroid, or, in other words, that the lengths of all concentric flux paths are essentially the same. The core is assumed to be made of a material that has essentially infinite resistivity, so that the circulatory currents present in the material as a result of the electromagnetic forces developed within the material by a changing flux are negligible. These circulatory currents, called eddy currents and discussed in more detail subsequently in Ch. V, should be small, because when they occur energy is dissipated by them in the core material in the form of heat, and cannot be returned to the exciting circuit which supplies the energy.

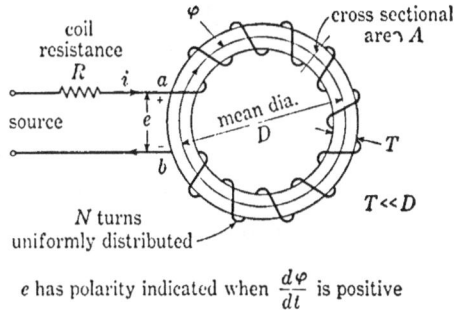

FIG. 16. Combination of electric and magnetic circuits for the study of energy relations of Art. 5.

If the exciting winding is distributed uniformly around the core, magnetic leakage is negligible, so that all the flux φ may be considered to link each of the N turns of the exciting winding. Also, if the resistance of the coil is considered negligible or assumed to be lumped outside the terminals a and b of Fig. 16, the potential difference appearing between a and b is that generated as a result of any rate of change of flux $d\varphi/dt$. Therefore, if at any given instant there is a current i in the coil and, as a result of the magnetomotive force which it produces, a flux φ in the core, the instantaneous voltage e is given by

$$e = N\frac{d\varphi}{dt}, * \qquad\qquad [30]$$

where e and φ have the senses indicated in Fig. 16. The instantaneous power input p_{ab} at terminals a and b is

$$p_{ab} = ei = Ni\frac{d\varphi}{dt}. \qquad\qquad [31]$$

---

* No units are indicated in this article because the equations are written so that any self-consistent unrationalized system of units can be used. Thus the equations are true in aes, in aem, and in mks systems of units.

This power is, for the conditions stipulated, absorbed wholly in the magnetic field, and, depending upon the properties of the medium, it is either all stored and hence recoverable, or part stored and part dissipated and hence only partially recoverable.

For reasons that are more apparent from the subject matter of Ch. V it is useful to express p_{ab} in terms of $\mathcal{H}$, $\mathcal{B}$, and the dimensions of the core rather than in terms of the turns in the particular coil, the current in it, and the core flux φ. Since the thickness of the core is small compared with the diameter of the toroid, the flux density may be assumed to be uniform over the cross section. Therefore,

$$\varphi = A\mathcal{B} \qquad [32]$$

and

$$\frac{d\varphi}{dt} = A\frac{d\mathcal{B}}{dt}, \qquad [33]$$

where A is the constant cross-sectional area of the core. Also,

$$\mathcal{H} = \frac{4\pi Ni}{l}, \qquad [34]$$

or

$$Ni = \frac{\mathcal{H}l}{4\pi}, \qquad [35]$$

in which l equals πD, the mean length of the core.

Substituting Eqs. 33 and 35 into Eq. 31 gives

$$p_{ab} = \frac{\mathcal{H}lA}{4\pi}\frac{d\mathcal{B}}{dt} = \frac{\mathcal{V}}{4\pi}\mathcal{H}\frac{d\mathcal{B}}{dt}, \qquad [36]$$

where $\mathcal{V}$ is lA, the volume of the core.

In any small time dt, the energy input dW into the magnetic field is

$$dW = p_{ab}\,dt = \frac{\mathcal{V}}{4\pi}\mathcal{H}\frac{d\mathcal{B}}{dt}\,dt = \frac{\mathcal{V}}{4\pi}\mathcal{H}d\mathcal{B}. \qquad [37]$$

Equation 37 shows that the energy dW is independent of the length of time required for $\mathcal{B}$ to change by the amount $d\mathcal{B}$. That is, dW depends only on the amount of the change in $\mathcal{B}$ and not upon the rate at which it changes. In general, the energy W put into the field as $\mathcal{B}$ is changed from any value $\mathcal{B}_1$ to any other value $\mathcal{B}_2$ is

$$W = \frac{\mathcal{V}}{4\pi}\int_{\mathcal{B}_1}^{\mathcal{B}_2}\mathcal{H}d\mathcal{B}, \qquad [38]$$

or the energy input per unit volume is

$$w = \frac{1}{4\pi}\int_{\mathcal{B}_1}^{\mathcal{B}_2}\mathcal{H}d\mathcal{B}. \qquad [39]$$

In order to perform the indicated integrations, it is necessary to know $\mathcal{H}$ as a function of $\mathcal{B}$ for the particular change in flux density under consideration. For example, if $\mathcal{B}$ is undergoing a cyclic change, and the region permeated by the field is a ferromagnetic material, the magnetization curve of the core material throughout a cycle is needed. The general subject of cyclic changes in $\mathcal{H}$ and $\mathcal{B}$ is treated in the next chapter. At this point several interesting results can be derived through assuming the permeability of the medium to be constant.

If the permeability is assumed constant, hysteresis is thereby assumed negligible, and

$$\mathcal{H} = \frac{\mathcal{B}}{\mu}. \tag{40}$$

Equation 39 then becomes

$$w = \frac{1}{4\pi\mu} \int_{\mathcal{B}_1}^{\mathcal{B}_2} \mathcal{B}d\mathcal{B} = \frac{\mathcal{B}_2^2 - \mathcal{B}_1^2}{8\pi\mu}. \tag{41}$$

Thus the energy absorbed per unit volume of the magnetic field when a flux density $\mathcal{B}$ is established from zero is

$$w = \frac{\mathcal{B}^2}{8\pi\mu}. \qquad \blacktriangleright[42]$$

This energy is *stored* in the magnetic field and it is returned to the exciting circuit when the flux density is reduced to zero. As is shown in the next chapter, the energy is not returned in full to the exciting circuit when the field permeates any medium that exhibits hysteresis phenomena.

If an air gap is cut in the toroidal magnetic circuit of Fig. 16, the flux densities in the iron and in the air are essentially the same, unless the air gap is long. At the usual working flux densities, the permeability of iron is likely to be of the order of 2,000 times that of air. Hence, since the stored energy per unit volume is inversely proportional to the permeability, the stored energy per unit volume in the air gap is about 2,000 times that in the iron. If the volume of the air gap is even as small as 1 per cent of the volume of the iron, the energy stored in the air gap predominates. This condition makes it relatively simple to obtain approximate expressions for the magnetic forces acting on the iron faces of an air gap, which are the subject of the next article.

6. MAGNETIC PULL

Consider the magnetic circuit shown in Fig. 17. As is well known, the two faces of iron bounding the air gap attract each other when the core is magnetized. From the results of the preceding section, an approximate expression for the magnetic forces acting on the two faces of the gap can be derived. In many practical problems, the approximation represented by the expression is a good one, as can be appreciated when the assumptions on which the expression is based are considered.

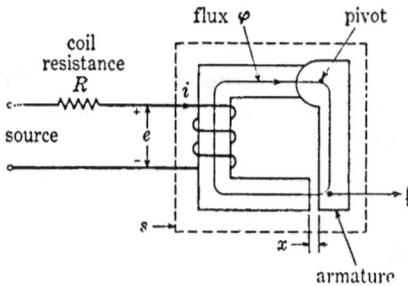

coil resistance R

source e

flux φ pivot

i

f

s

x

armature

FIG 17. Magnetic circuit with variable air gap for the study of force of attraction between poles.

If the air gap is short compared with its cross-sectional dimensions, the fringing flux is small; and if the gap has parallel faces, most of the flux goes straight across the gap and the flux density is substantially uniform. The iron part of the flux path is assumed to have a uniform cross section.

The force f required to displace the armature against the force of magnetic attraction can be calculated by applying the law of conservation of energy to the system s enclosed by the dotted line. If any changes in total energy occur in this system they must be consistent with the equation for conservation of energy, namely,

$$\begin{pmatrix}\text{Electrical}\\\text{energy}\\\text{input}\end{pmatrix}+\begin{pmatrix}\text{Mechanical}\\\text{energy}\\\text{input}\end{pmatrix}=\begin{pmatrix}\text{Increase}\\\text{in stored}\\\text{energy}\end{pmatrix}+\begin{pmatrix}\text{Energy irrevers-}\\\text{ibly converted}\\\text{into other forms}\end{pmatrix} \quad [43]$$

Consider Eq. 43 term by term as the air gap is increased by an amount dx in a time dt by a force f. The electrical energy input dW_e is

$$dW_e = ei\,dt = N\frac{d\varphi}{dt}i\,dt = Ni\,d\varphi, \quad [44]$$

in which e is the potential difference appearing across the terminals of the coil as a result of the flux change $d\varphi/dt$, the change being caused by the increasing reluctance of the flux path, and $d\varphi$ is the net change in the flux during the time dt.

The mechanical energy input dW_m is

$$dW_m = f\,dx. \quad [45]$$

The energy W_s stored in the system s is in magnetic form. The increase dW_s is composed of two components, first the increase dW_{si} in the iron, and second the increase dW_{sa} in the air gap. By Eq. 37

$$dW_{si} = \frac{\mathcal{V}_i}{4\pi}\mathcal{H}_i d\mathcal{B}_i, \quad [46]$$

in which $\mathcal{V}_i,\ \mathcal{H}_i,$ and $\mathcal{B}_i$ are respectively the volume of iron, $\mathcal{H}$ in the iron, and $\mathcal{B}$ in the iron. The increase dW_{sa} results both from a change in the air-gap flux density $\mathcal{B}_a$ and in the air-gap volume $\mathcal{V}_a$. It can be obtained best through differentiating the expression for the total energy stored in the air gap (w of Eq. 42, multiplied by $\mathcal{V}_a$) with respect to x. Thus

$$\frac{dW_{sa}}{dx} = \frac{d}{dx}\left(\frac{\mathcal{V}_a\mathcal{B}_a^2}{8\pi\mu_0}\right) = \frac{\mathcal{B}_a^2}{8\pi\mu_0}\frac{d\mathcal{V}_a}{dx} + \frac{2\mathcal{V}_a\mathcal{B}_a}{8\pi\mu_0}\frac{d\mathcal{B}_a}{dx}, \qquad [47]$$

μ_0 being the permeability of air in the system of units used for the calculations. Since $\mathcal{V}_a = A_a x$, A_a being the area of the air gap, the differential energy storage in the air gap when the gap is lengthened by dx is the expression in Eq. 47 multiplied by dx; that is,

$$dW_{sa} = \frac{dW'_{sa}}{dx}dx = \frac{\mathcal{B}_a^2 A_a}{8\pi\mu_0}dx + \frac{\mathcal{V}_a\mathcal{B}_a}{4\pi\mu_0}d\mathcal{B}_a. \qquad [48]$$

Since $\mathcal{B}_a/\mu_0 = \mathcal{H}_a$, the magnetizing force in the air gap, Eq. 48 becomes

$$dW_{sa} = \frac{\mathcal{H}_a\mathcal{B}_a A_a}{8\pi}dx + \frac{\mathcal{V}_a}{4\pi}\mathcal{H}_a d\mathcal{B}_a. \qquad [49]$$

The fourth term of Eq. 43, namely the energy irreversibly converted to other forms, includes hysteresis and eddy-current conversion of electrical or magnetic energy into heat which are effects assumed in the statement of the problem to be negligible. Hence the fourth term is assumed to be zero, and represents a reasonable approximation in most cases.

Equation 43 can now be written in mathematical terminology by use of the foregoing expressions for the various terms. This gives

$$dW_e + dW_m = dW_{si} + dW_{sa} + 0, \qquad [50]$$

or

$$Ni\,d\varphi + f\,dx = \frac{\mathcal{V}_i}{4\pi}\mathcal{H}_i d\mathcal{B}_i + \frac{\mathcal{H}_a\mathcal{B}_a A_a}{8\pi}dx + \frac{\mathcal{V}_a}{4\pi}\mathcal{H}_a d\mathcal{B}_a. \qquad [51]$$

Equation 51 may appear to be involved but, if $Ni\,d\varphi$ is expressed in terms of the volumes, the magnetizing forces, and the flux densities in the iron and in the air gap, a simple result is obtained. Note that

$$4\pi Ni = U_i + U_a = \mathcal{H}_i \ell_i + \mathcal{H}_a x, \qquad [52]$$

or

$$Ni = \frac{1}{4\pi}(\mathcal{H}_i \ell_i + \mathcal{H}_a x), \qquad [53]$$

in which U_i and U_a are the magnetic potential drops for the iron and air paths respectively, and ℓ_i is the length of the flux path in iron. Also

$$d\varphi = A_i d\mathcal{B}_i = A_a d\mathcal{B}_a. \qquad [54]$$

Multiplying Eq. 53 by Eq. 54 gives an expression for the first term of Eq. 51,

$$Ni\,d\varphi = \frac{1}{4\pi}\,(A_i \ell_i \mathcal{H}_i d\mathcal{B}_i + A_a x \mathcal{H}_a d\mathcal{B}_a)$$

$$= \frac{1}{4\pi}\,(\mathcal{V}_i \mathcal{H}_i d\mathcal{B}_i + \mathcal{V}_a \mathcal{H}_a d\mathcal{B}_a). \qquad [55]$$

Substitution of Eq. 55 into Eq. 51 gives an expression involving ℓ and the field parameters, namely,

$$\frac{1}{4\pi}\,(\mathcal{V}_i \mathcal{H}_i d\mathcal{B}_i + \mathcal{V}_a \mathcal{H}_a d\mathcal{B}_a) + \ell\,dx = \frac{1}{4\pi}\,(\mathcal{V}_i \mathcal{H}_i d\mathcal{B}_i$$

$$+ \mathcal{V}_a \mathcal{H}_a d\mathcal{B}_a) + \frac{\mathcal{H}_a \mathcal{B}_a A_a}{8\pi}\,dx, \qquad [56]$$

which on cancellation of equal terms and division by dx gives

$$\ell = \frac{\mathcal{H}_a \mathcal{B}_a A_a}{8\pi}, \qquad [57]$$

which can be written

$$\ell = \frac{\mathcal{B}_a^2 A_a}{8\pi\mu_0}. \qquad [58]$$

Equation 58 is the desired expression for the force of attraction between the adjacent faces of an air gap in a magnetic circuit.

When expressed in English units, Eq. 58 becomes

$$\ell = 0.0139\mathcal{B}_a^2 A \qquad \blacktriangleright[59]$$

where

$\ell \equiv$ force in pounds,
$\mathcal{B}_a \equiv$ flux density in the gap in kilolines per square inch,
$A \equiv$ cross-sectional area of the gap in square inches.

Before the limitations of Eq. 58 are discussed, it is worth while to consider the physical significance of the various terms of Eq. 56. The terms in parentheses on the left represent the electrical energy input. They are balanced by identical terms on the right giving the component of energy-storage change that would result from changing the flux density without altering the air-gap length. The $\ell\,dx$ term, representing the mechanical input, is therefore equal to the remaining term on the right, that representing the energy in the additional volume of air gap $A_a\,dx$ introduced by lengthening the gap by dx.

If the generality of Eq. 58 is examined it is noticed that usually the chief source of inaccuracy is the assumption of uniform flux density in the air gap. For gaps whose length is small compared with the dimensions of the faces the approximation involved is a good one. For other cases the calculation of an accurate result is a complex problem requiring somewhat more elaborate methods.

The derivation of Eq. 58 assumes that the flux path in iron has uniform cross section. This restriction does not limit the generality of the result since the assumption was made merely to simplify the expressions. If in Eqs. 46 and 53 the single term for iron is replaced by a summation of terms, obtained by dividing the iron into separate parts each of uniform cross section, and writing a term for each, the final result is Eq. 58 because the same additional terms appear on both sides of the equation corresponding to Eq. 56 and therefore cancel each other.

Thus Eq. 58 is applicable to the faces of any air gap in an iron core provided the flux density can be assumed uniform over the air gap and provided the eddy-current loss and hysteresis loss in the iron are negligible. It also gives the correct result for systems containing permanent magnets with or without electrical windings but the proof is not given here. It is reasonable however that this should be so since Eq. 58 for the force involves only conditions in the air gap and nothing concerning the iron or winding when the losses are neglected.

PROBLEMS

1. Two U-shaped tungsten steel permanent magnets, each having a cross section of $\frac{1}{2}$ in. $\times$ 1 in. and a mean length of 7.3 in., are placed with their flat pole faces in contact, and then magnetized to saturation. The magnets are then moved apart to form a $\frac{1}{16}$ in. air gap in each leg. Find the air-gap flux density. Leakage is not to be taken into account, but the fringing correction is to be applied.

The magnets are next separated completely, and it is found that the flux density at the midpoint of the U, that is, the yoke flux density, is reduced to 2.5 kilogausses. If the magnets are then brought back to the $\frac{1}{16}$ in. gap position, what is the yoke flux density?

The data for the demagnetization curve are:

$\mathfrak{B}$ (kilogausses)	10.5	9.8	9.5	8.3	6.0	2.9	0
$\mathfrak{H}$ (oersteds)	0	11	16	25.5	32	35	37

The recovery incremental permeability, defined as $\Delta\mathfrak{B}/\Delta\mathfrak{H}$, is 50 gausses per oersted at the working point for the open magnet.

2. A piece of Alnico V in the shape of a right circular cylinder 1 in. in diameter and 1 in. long is held by screws between two soft-iron pole pieces to form a magnet, as shown in Fig. 18. The Alnico is magnetized to saturation by placing the assembly with the armature in position between the poles of a powerful electromagnet. Following the magnetization process, the armature is carefully forced away from the pole faces, leaving two air gaps of 0.01 in. each. What is the approximate flux density in the air gap? Allow 0.004 in. for the length of both air gaps at the junctions of the Alnico and the soft iron.

3. If the magnet described in Art. 3 must be designed with the Ticonal in the vertical members, what would be the required volume of Ticonal? The air-gap dimensions

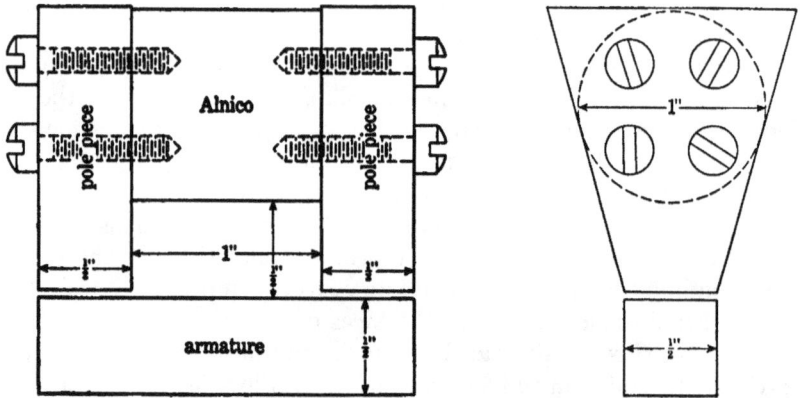

FIG. 18. Alnico magnet having large leakage.

and air-gap flux are to be the same in both designs. The distance between the vertical members is to be 2 in.

4. A strap of chrome steel $\frac{3}{4}$ in. $\times \frac{3}{16}$ in. in cross-sectional dimensions is bent into the shape shown in Fig. 19. A piece of soft iron is placed across the $\frac{1}{16}$-in. air gap while

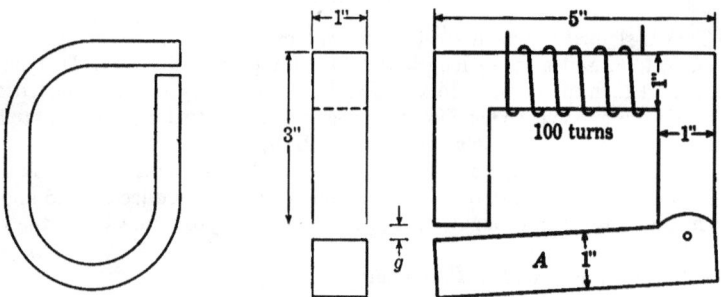

FIG. 19. Permanent
magnet made from
strap steel.

FIG. 20. Relay magnet of Prob. 6.

the strap is magnetized to saturation. The length of the flux path in the steel is 8 in. How much flux, including fringing flux, is in the air gap after the removal of the soft-iron bridge?

(a) Neglect leakage.

(b) Explain whether the methods described in the text are adequate for the computation of leakage flux around this structure.

5. (a) What is the force exerted on the plunger of the magnet of Fig. 23, Ch. III, for each of the air-gap lengths g of Prob. 4, p. 95.

(b) How many foot-pounds of work can this magnet perform per stroke?

(c) If the actual copper of the winding occupies half the gross winding space, and if the solenoid dissipates approximately 0.006 w per degree C temperature rise for each square inch of outside shell area, what is the temperature rise of the coil if it is continuously energized?

6. Figure 20 represents the magnetic circuit of a d-c circuit breaker. The core material is cast steel. A force of 10 lb is required to start the motion of armature *A* when the average gap length at *g* is 0.10 in. An air gap of constant length 0.010 in. is at the hinge end of the armature.

(a) How many amperes are required in the trip coil to start the armature when the gap length at *g* is 0.10 in.?

(b) With the number of amperes found in (a) what is the magnetic pull on the armature when the gap length at *g* is 0.01 in.?

CHAPTER V

Losses in Magnetic Cores Containing Time-Varying Fluxes

The foregoing chapters have indicated the extensive use of ferro-magnetic materials in the construction of both direct-current and alter-nating-current apparatus. In magnetic devices operating with constant flux, no heating occurs in the core materials. A direct-current lifting magnet or a direct-current relay, for example, has almost no energy loss in its magnetic circuit unless it is energized and de-energized very fre-quently. Direct-current-motor and generator armatures, synchronous-motor and generator armatures, induction motors (both rotor and stator), power and audio-frequency transformers, iron-core choke coils, and devices actuated by alternating currents have alternating fluxes in their magnetic circuits, and these fluxes give rise to currents which produce heat in the iron or steel.

The losses that occur in the material arise from two causes: (a) the tendency of the material to retain magnetism or to oppose a change in magnetism, often referred to as magnetic hysteresis; and (b) the I^2R heating which appears in the material as a result of the voltages and consequent circulatory currents induced in it by the time variation of flux. The first of these contributions to the energy dissipation is known as *hysteresis loss* and the second as *eddy-current loss*. The hysteresis loss is the result of the tendency for the $\mathfrak{B}(\mathfrak{H})$ characteristic of the material to involve a loop when the material is subjected to a cyclic magnetizing force. Typical loops are shown in Figs. 14a and 15b of Ch. I, and in Fig. 1 of this chapter. The distinction between hysteresis and hysteresis loss is important. The phenomenon known as hysteresis is the result of the material's property of retaining magnetism or opposing a change in magnetic state. The hysteresis loss is the energy converted into heat because of the hysteresis phenomenon and, as usually interpreted, is associated only with a cyclic variation of magnetomotive force. This interpretation is the result of the extensive engineering use of the material under cyclic magnetizing forces, and the relatively large importance of loss data representative of this manner of use. The *eddy-current loss* is produced by the currents in the magnetic material, and these currents are caused by the electromotive forces set up by the varying fluxes. The sum of the hysteresis and eddy-current losses is called the *total core loss*.

1. HYSTERESIS LOSS

The occurrence of hysteresis loss is a matter intimately associated with the phenomenon whereby energy is absorbed by a region which is permeated by a magnetic field. If the region is other than a vacuum, only

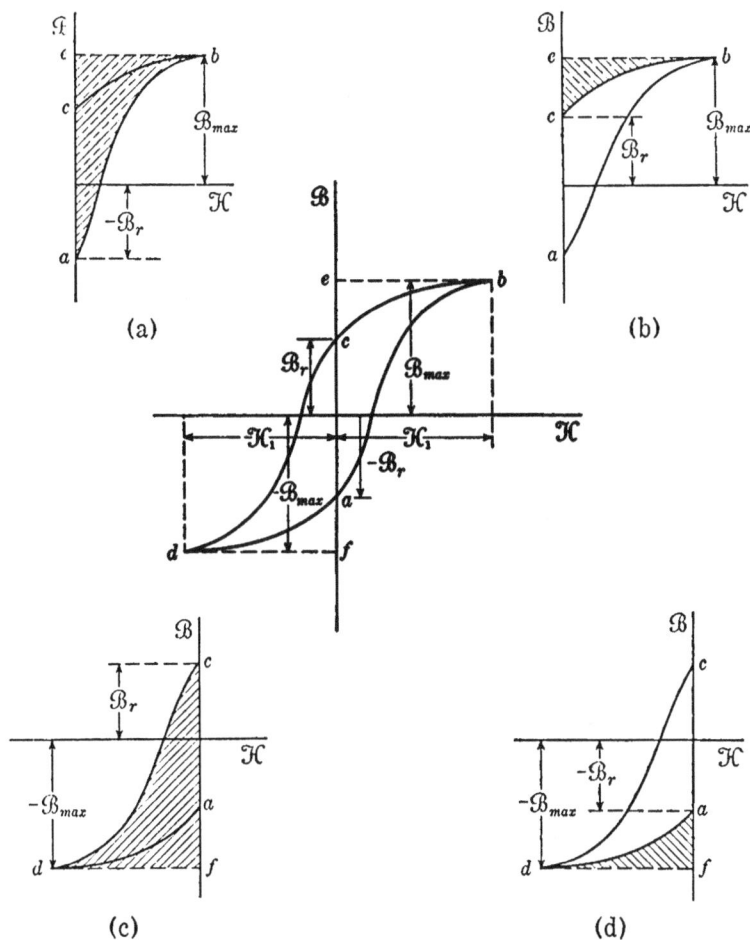

FIG. 1. Hysteresis loop. Shaded areas in (a) and (c) show energy absorbed; in (b) and (d) energy returned by steel.

a portion of the energy taken from the electric circuit is stored and wholly recoverable from the region when the magnetizing force is removed. The rest of the energy is converted into heat as a result of work done on the material in the medium when it responds to the magnetization. In Art. 5 of Ch. IV, it is shown that when the flux density in a region is increased from a value $\mathcal{B}_1$ to a value $\mathcal{B}_2$, energy is absorbed by the region. The magnitude of the energy absorbed per unit volume is given by Eq. 39, Ch. IV, as

$$w = \frac{1}{4\pi} \int_{\mathcal{B}_1}^{\mathcal{B}_2} \mathcal{H}d\mathcal{B}. \qquad [1]$$

The integral of Eq. 1 is proportional to the area bounded by the $\mathcal{B}(\mathcal{H})$ curve for the region, the $\mathcal{B}$ axis, and the lines parallel to the $\mathcal{H}$ axis representing constant $\mathcal{B}_1$ and $\mathcal{B}_2$ respectively. Hence, its magnitude depends on the values of $\mathcal{B}_1$ and $\mathcal{B}_2$ and the shape of the curve between $\mathcal{B}_1$ and $\mathcal{B}_2$. If the flux density is decreased from any specified value to a smaller value, the algebraic sign of w is negative and energy is given up by the material.

When the region considered consists of ferromagnetic material, the magnetization curve between any two values $\mathcal{B}_1$ and $\mathcal{B}_2$ corresponding to decreasing values of $\mathcal{H}$ is different from the curve corresponding to increasing values of $\mathcal{H}$. As indicated in Art. 4, Ch. I, the values of flux density in a ferromagnetic material are larger for a given magnetizing force $\mathcal{H}$ when $\mathcal{H}$ is decreasing than when $\mathcal{H}$ is increasing, even though, for a cyclic variation in $\mathcal{H}$, the extreme values of $\mathcal{B}$ are the same for each cycle when the material reaches its steady-state condition. On account of the difference in the two curves, which, for the cyclic condition, actually form the two sides of a closed loop, it follows that the energy absorbed by the material when the flux density is increased from $\mathcal{B}_1$ to $\mathcal{B}_2$ is larger than the energy returned when the flux density is decreased from $\mathcal{B}_2$ to $\mathcal{B}_1$. The difference in these energies is the magnitude of the *hysteresis loss*. By the graphical evaluation of the integral of Eq. 1 over a complete cycle of magnetization, the energy loss *per cycle* caused by magnetic hysteresis can be determined.

As an illustration of the graphical integration process, the energy stored in the toroid of Fig. 13a, Ch. I, is to be considered. The exciting coil is carrying an alternating current so that the magnetizing force is reversing cyclically between the limits $+\mathcal{H}_1$ and $-\mathcal{H}_1$. The relation between $\mathcal{B}$ and $\mathcal{H}$ is as shown by the hysteresis loop in Fig. 1. During the part of the cycle ab, the magnetic energy absorbed by the core per unit volume is

$$w_1 = \frac{1}{4\pi} \int_{-\mathcal{B}_r}^{\mathcal{B}_{max}} \mathcal{H}d\mathcal{B} = \frac{1}{4\pi} \times (\text{area } abea \text{ shown thus } \text{\it{\char"2592}}\ \text{Fig. 1a}). \quad [2]$$

The unit area is taken as a rectangle of width equal to one unit of $\mathcal{H}$ on the graph and of height equal to one unit of $\mathcal{B}$. If the $\mathcal{H}$ scale is

K' units of $\mathcal{K}$ per inch and the $\mathcal{B}$ scale K'' units of $\mathcal{B}$ per inch, then

$$w_1 = \frac{K'K''}{4\pi} \times \text{(area } abea \text{ in square inches)}.$$

The area can be determined through counting squares, by the use of a planimeter or by other methods described in Art. 5b, Ch. XIII, of the volume on electric circuits.

During the part of the cycle bc, the energy absorbed magnetically per unit of volume is, by Eq. 1,

$$w_2 = \frac{1}{4\pi} \int_{\mathcal{B}_{max}}^{\mathcal{B}_r} \mathcal{K} d\mathcal{B}. \tag{3}$$

Since $\mathcal{B}_r < \mathcal{B}_{max}$ and $\mathcal{K}$ is positive, this integral is negative; that is, energy is being given up by the magnetic field and returned to the exciting circuit. The energy absorbed is then

$$w_2 = -\frac{K'K''}{4\pi} \times \text{(area } bceb \text{ shown thus } \blacksquare \text{ in Fig. 1b).} \tag{4}$$

Similarly, during the part of the cycle cd, the energy absorbed magnetically is equal to

$$w_3 = \frac{K'K''}{4\pi} \times \text{(area } cdfc \text{ shown thus } \blacksquare \text{ in Fig. 1c).} \tag{5}$$

During the part of the cycle da, energy is given up by the magnetic field and returned to the electric circuit. The absorbed energy is therefore negative and given by

$$w_4 = -\frac{K'K''}{4\pi} \times \text{(area } dafd \text{ shown thus } \square \text{ in Fig. 1d).} \tag{6}$$

The net energy w_h absorbed by the magnetic field per unit of volume for one complete cycle is

$$w_h = w_1 + w_2 + w_3 + w_4 =$$
$$\frac{K'K''}{4\pi} \times \text{(area of the hysteresis loop } abcda\text{).} \tag{7}$$

This energy is dissipated as heat in the material each cycle. The dissipation is called the hysteresis loss. Its occurrence has an important effect on the efficiency, the temperature rise, and hence the rating of many electromagnetic devices.

Although the area of a closed hysteresis loop indicates how much energy is dissipated in the core per unit volume per cycle because of hysteresis, it does not indicate at what part of the cycle the dissipation occurs. For example, during the part of the cycle *ab*, an amount of energy w_1, Eq. 2, is absorbed by a unit volume of the core. However, this analysis does not indicate how much of this absorbed energy is dissipated as heat during this part of the cycle, nor how much of it is stored, to be dissipated later during the part of the cycle *bc*.

If a volume $\mathcal{V}$ of magnetic material, throughout which the flux distribution is everywhere uniform and for which the hysteresis loop is known, is subject to a cyclic change at a frequency of *f* cycles per second, the rate at which energy is dissipated because of hysteresis (*hysteresis power loss*) is

$$P_h = \frac{\mathcal{V}f}{4\pi} \times \text{(area of the loop)}. \qquad [8]$$

As in all the equations of this section any consistent system of units can be used. In aem units, P_h is expressed in ergs per second, $\mathcal{V}$ in cubic centimeters, $\mathcal{H}$ in oersteds, and $\mathcal{B}$ in gausses. In mks units, P_h is expressed in watts, $\mathcal{V}$ in cubic meters, $\mathcal{H}$ in praoersteds, and $\mathcal{B}$ in webers per square meter. In either case, the area of the hysteresis loop is expressed as the product of the $\mathcal{H}$ and $\mathcal{B}$ units employed. From Eq. 8 the hysteresis loss in watts is

$$P_h = 10^{-5}\mathcal{V}fK'K'' \times \text{(area of loop in sq in.)}, \qquad \blacktriangleright[9]$$

when the flux density is in kilolines per square inch, the magnetizing force in ampere-turns per inch, and the volume in cubic inches. The constant K' is the number of ampere-turns per inch corresponding to 1 inch on the magnetizing-force scale, and K'' is the number of kilolines per square inch corresponding to 1 inch on the flux-density scale. If the volume is made equal to 1 cubic inch, Eq. 9 gives the hysteresis power loss per cubic inch as

$$p_h = 10^{-5}fK'K'' \times \text{(area of loop in sq in.)}. \qquad \blacktriangleright[10]$$

The hysteresis loss per cycle can be calculated by means of the foregoing relations if the hysteresis loop for the given maximum flux density $\mathcal{B}_{max}$ is known, but the manner in which this loss varies as a function of $\mathcal{B}_{max}$ can be determined only through repeating the calculation for hysteresis loops having various values of $\mathcal{B}_{max}$. Empirically, Steinmetz[1] found from the results of a large number of measurements that the area of the normal hysteresis loop of specimens of various irons and steels commonly used in the construction of electromagnetic apparatus of his day was approximately proportional to the 1.6th power of the maximum flux density throughout the range of flux densities from about 1,000 to 12,000 gausses. As a result of research on ferromagnetic materials, numer-

[1] C. P. Steinmetz, " On the Law of Hysteresis," *A.I.E.E. Trans.*, *9* (1892), 3–51.

ous magnetic steels having widely varying properties have been made available since Steinmetz performed his measurements. The exponent 1.6 fails nowadays to give the area of the loops with a sufficient degree of accuracy to be useful. The empirical expression for the energy loss per unit volume per cycle is more properly given as

$$w_h = \eta \mathcal{B}_{max}^n, \qquad\qquad \blacktriangleright[11]$$

where n and η have values that depend on the material. Equation 11 should be used with caution since the value for n, which may range between 1.5 and 2.5 for present-day materials, may not be constant for a given material. For some materials. an expression of the form of Eq. 11 is not sufficiently accurate to be generally useful. These constants hence should be evaluated for a certain range of $\mathcal{B}_{max}$ and then subsequently used for values of $\mathcal{B}_{max}$ only within this range.

If Eq. 11 is written in its logarithmic form,

$$\log w_h = n \log \mathcal{B}_{max} + \log \eta, \qquad\qquad [12]$$

a straight-line relationship between $\log w_h$ and $\log \mathcal{B}_{max}$ is indicated. From test data, several values of $\log w_h$ can be plotted as ordinates corresponding to the different values of $\log \mathcal{B}_{max}$ as abscissas. These points should lie along a straight line having a slope equal to the exponent n and having an intercept on the vertical axis equal to $\log \eta$. Obviously two points would be sufficient to determine the values of n and η, but, if several points are used, the straightness of the curve joining them indicates how well the resulting Eq. 11 fits the data in the range under consideration. If the points do not lie along a straight line, the constant-exponent type of equation is not appropriate.[2]

Any convenient system of units can be used for w_h and $\mathcal{B}_{max}$ in Eq. 11 if the corresponding value of the coefficient η is used. The total hysteresis loss in a volume $\mathcal{V}$ in which the flux density is everywhere uniform and varying cyclically at a frequency of f cycles per second can then be expressed empirically as

$$P_h = \eta \mathcal{V} f \mathcal{B}_{max}^n. \qquad\qquad [13]$$

Thus far, this article has treated only the symmetrical hysteresis loop in which $\mathcal{B}$ ranges between equal positive and negative values, and in which there are no re-entrant loops such as at *efe* in Fig. 13b, Ch. I. In all instances, however, the hysteresis loss per cycle per unit volume can be calculated from Eq. 10 by use of the actual hysteresis loop. If the curve has re-entrant loops, their areas are added to that of the main loop, thus counting these areas twice. If a planimeter is used to determine

[2] P. G. Agnew, " A Study of The Current Transformer with Particular Reference to Iron Loss," *Bul. Nat. Bur. Stand.*, *7* (1911), 423–474.

the area by tracing continuously around the curve as it actually would be swept out in the $\mathfrak{B}(\mathfrak{H})$ plane, the areas of the re-entrant loops are automatically added to that of the main loop. When the loop is unsymmetrical or contains minor loops, the empirical relations of Eqs. 11 and 13 do not apply.[3]

The quantitative analysis of the loss given thus far assumes that the flux experiences a variation only in magnitude along a fixed direction throughout a magnetic circuit such as, for example, a transformer core. The direction of the magnetization is usually approximately along the length of the core. Its magnitude continuously increases in one direction until it reaches a maximum; it then decreases, passes through zero,

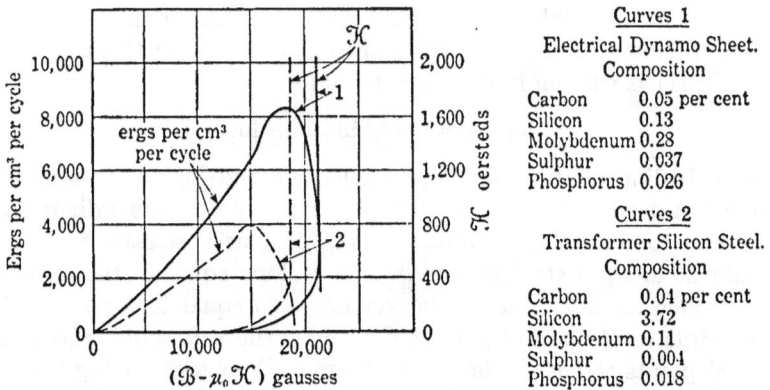

Curves 1

Electrical Dynamo Sheet.
Composition

Carbon	0.05 per cent
Silicon	0.13
Molybdenum	0.23
Sulphur	0.037
Phosphorus	0.026

Curves 2

Transformer Silicon Steel.
Composition

Carbon	0.04 per cent
Silicon	3.72
Molybdenum	0.11
Sulphur	0.004
Phosphorus	0.018

FIG. 2. Rotational hysteresis loss and magnetization curves. (*From reference 5.*)

reverses, and repeats the variation in the opposite direction. The magnetization in these instances throughout any volume is always along a fixed axis in space, and the hysteresis loss occurring in this process is called *alternating hysteresis loss.* However, a loss can occur on account of another mode of variation of the state of magnetization. If the magnitude of the field is kept constant but its direction is changed, by change in the direction of $\mathfrak{H}$ with respect to the body of material or vice versa, a hysteresis loss also occurs. This is known as *rotational hysteresis loss.* The magnitude of the loss occurring in a body of material subjected to rotational variation of flux is materially different from that occurring in the same body subjected to a magnitude alternation of $\mathfrak{B}$ even though $\mathfrak{B}_{max}$ is the same for both conditions. Figure 2 indicates the form of the variation of this loss with $\mathfrak{B}_{max}$ for two typical grades of sheet steel. The rotational hysteresis loss tends toward zero as saturation is approached.

Many devices are used in engineering in which the magnetization not

---

[3] J. D. Ball, " The Unsymmetrical Hysteresis Loop," *A.I.E.E. Trans.,34* (1915), 2693–2715.

only changes cyclically in magnitude and sign but also changes direction in space and thus involves both alternating and rotational hysteresis. The variation of the flux in the armature core of a rotating machine is an example of this condition, and is indicated in Fig. 3. The short arrows show approximately the magnitude and direction of the flux density at a series of points slightly below the roots of the teeth under a pole. Under adjacent field poles, the flux plot is similar but the direction of magnetization is opposite to that for a corresponding point under the pole which is shown. As the armature rotates

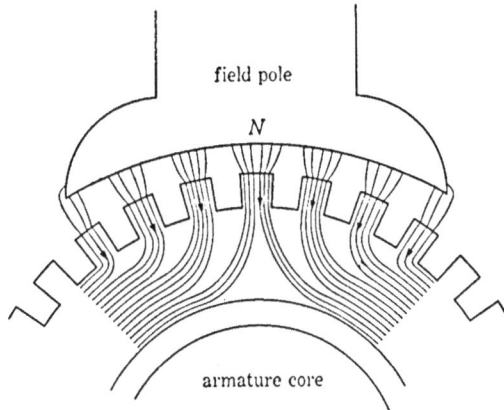

Fig. 3. Approximate no-load flux distribution in the armature iron under a pole of a rotating machine.

past the field poles, the flux density changes in magnitude and direction in every portion of the core material. In the teeth, the *direction* of magnetization is essentially normal to the circumference of the armature, but in any portion of the remainder of the core the magnetization *successively takes on all the directions indicated in the figure* in portions of the core similarly located with respect to the teeth. The change in the direction — that is, the rotation of magnetization in these portions of the core — is a cyclic phenomenon and the second half cycle is repeated under the next pole where the magnetization is in the opposite sense. The hysteresis taking place in the teeth is essentially alternating hysteresis, but that taking place in the remainder of the core under these conditions is a combination of rotational and alternating hysteresis. Since the manner in which the magnetization is changing in this portion of the core is different from that in which it is changing in the teeth, the two phenomena do not follow the same laws.

Rotational hysteresis was carefully investigated by F. G. Bailey[4] in 1895, and is still being given careful study[5] in the hope of explaining the phenomenon. The rotational hysteresis loss at flux densities commonly used in electrical machinery is several times that occurring in an equivalent alternating field. However, at very high saturations, above 18,000

[4] F. G. Bailey, " The Hysteresis of Iron and Steel in a Rotating Magnetic Field," *Phil. Trans., 187* (1896), 715–746.

[5] F. Brailsford, " Rotational Hysteresis Loss in Electrical Sheet Steels," *I.E.E.J., 83* (1938), 566–575.

to 20,000 gausses, the loss decreases and rapidly approaches a small value as indicated by the curves of Fig. 2. When the flux varies in magnitude as well as in direction, the loss is less than that occurring when the flux variation is only rotational with the same value of maximum flux density. In the usual type of electrical machine, a relatively small part of the loss occurs in the iron subjected to the rotating field, and the error produced through using the losses computed for an alternating field is therefore generally not large.

2. EDDY–CURRENT LOSS AND SKIN EFFECT

Whenever the magnetic flux in a medium is changing, an electric field appears within the medium as a result of the time variation of the flux. The line integral of this electric field $\mathcal{E}$ taken around any closed path that bounds the flux is given by the Faraday induction law as

$$\oint_{abcd} \mathcal{E} \cdot d\ell = -d/dt \int \mathcal{B} \cdot n \, ds \qquad [14]$$

where *abcda* is the circuital path bounding the area crossed by the flux φ or $\int \mathcal{B} \cdot n \, ds$. When the medium is conducting, a current* is set up around this path by the induced electromotive force e resulting from the line integral of the electric field. These currents are called *eddy currents*. Their presence results in an energy loss in the material proportional to $i^2 R$, called *eddy-current loss*, the energy being absorbed from the circuit that sets up the field and being dissipated as heat in the medium.

Since the flux density in ferromagnetic materials is usually relatively large, and since the resistivity of the materials is not extremely large, the induced electromotive forces, the eddy currents, and the eddy-current loss may become appreciable if means to minimize them are not taken. This loss is of considerable importance in determining the efficiency, the temperature rise, and hence the rating, of electromagnetic apparatus in which the flux density varies.

To illustrate the conditions typical of those that occur in an iron core, the thin metal slab shown in Fig. 4 is considered to be permeated by an alternating flux φ. From Eq. 14, the electromotive force e induced around a boundary *abcda* of an area through which a flux is changing is given by

$$e = -\frac{d\varphi}{dt}. \qquad [15]$$

This voltage acting around the circuit *abcda* causes a current i to circulate around the boundary and to set up a magnetomotive force in such a

---

* Displacement currents are not considered in this analysis, since their effects are here negligible.

direction as to oppose any change in φ. The effect of such currents is to screen or shield the material from the flux, and to bring about a smaller flux density near the center of the slab than at the surface. For a specified total flux varying periodi-
cally, the maximum flux density at the center is smaller than the value obtained from dividing the total maximum flux by the area. Another way to describe this effect is to say that the total flux tends to be crowded toward the surface of the slab. This phenomenon is known as *skin effect*. A similar skin-effect phenomenon occurs in an electric conductor that has a vary-
ing current,[6] even when it is composed of material having unit relative permeability. In such a conductor, the electric current density is largest at the surface. Since magnetic and electric skin effects are similar in nature, they are subject to the same type of analysis. The gen-
eral solution is somewhat com-
plicated mathematically, and

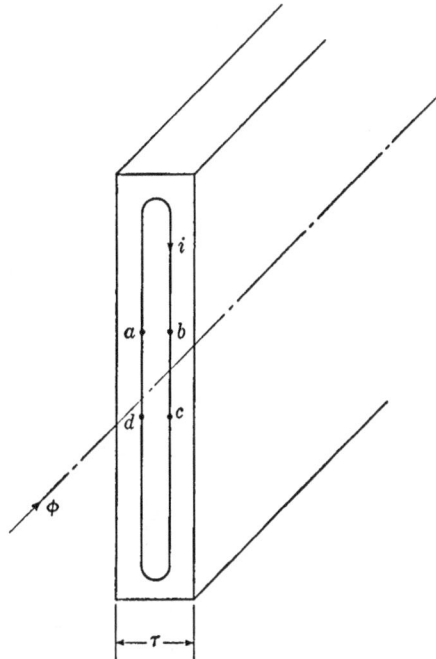

FIG. 4.   Cross section of lamination showing a current path.

treatment of it is reserved for inclusion in the reference volume: *The Theory of Transmission Lines, Wave Guides, and Antennas.* However, an analysis of eddy-current loss that arbitrarily ignores skin effect is useful and relatively simple, and gives results that are sufficiently accurate for many applications, especially in devices having laminated cores. This simpler analysis is given below together with a criterion from the more accurate analysis which indicates whether in any given problem the latter should be performed.

The simpler analysis is developed for a thin plane slab of electrically conducting material having a thickness τ as shown in Fig. 4. In this slab is assumed a uniformly distributed magnetic field whose magnitude is varying with time and whose direction is everywhere parallel to the arrow. The assumption of uniform magnetic-field distribution means that the magnetomotive forces of the eddy currents have negligible effect

[6] L. F. Woodruff, *Electric Power Transmission and Distribution* (2nd ed.; New York: John Wiley & Sons, 1938), Ch. iii.

on the flux distribution, and that the current paths such as *abcda* are symmetrical about the center line through zero as shown. Also, because of the great height as compared with the thickness, the voltage gradient is practically uniform along the vertical current paths except near the top and bottom of the slab. For this reason, any horizontal slice of unit height not too near the top or the bottom has practically the same configuration of voltage gradients and current densities as any other horizontal slice. The portion of the slab considered is shown in Fig. 5; it is a

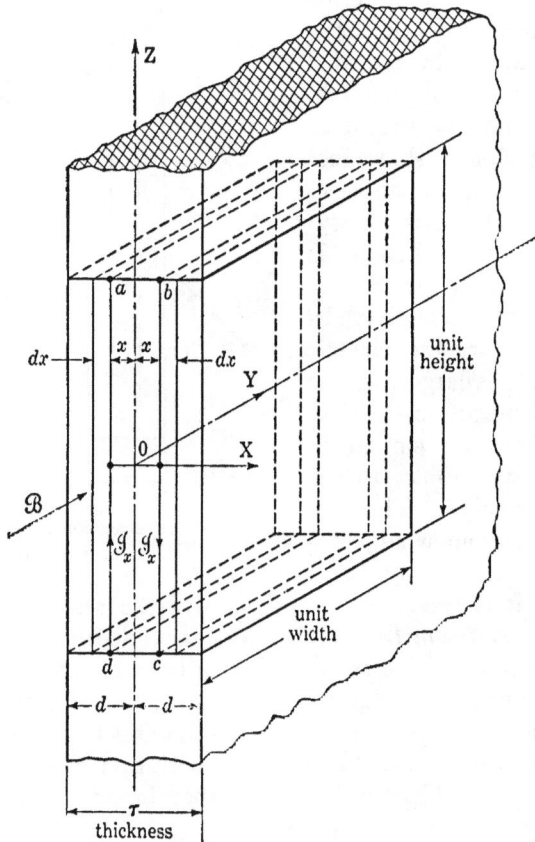

Fig. 5.   Unit element of lamination for calculation of eddy-current loss.

rectangular parallelepiped which has unit height, unit width, and thickness τ and is symmetrical about the OY axis which passes through the center of the slab. The narrow face containing points a, b, c, and d is normal to the direction of the flux. The *decrease* in the magnitude of the flux density with time through the area *abcd* in the direction shown induces a voltage around the path in the direction *abcd*.

The application of the Faraday induction law to the path *abcda* in the *XZ* plane normal to the direction of $\mathscr{B}$ gives

$$\oint \mathscr{E}_z \cdot d\ell = \frac{-d}{dt} \int \mathscr{B} \cdot n \, ds, \qquad [16]$$

where $\mathscr{E}_z$ is the vertical voltage gradient at a horizontal distance x from the *YZ* plane. In accordance with the above argument, the value of the line integral $\oint \mathscr{E}_z \cdot d\ell$ taken around the closed loop *abcda* is $2\mathscr{E}_z$, since the parallelepiped is of unit height. The surface integral $\int \mathscr{B} \cdot n \, ds$, evaluated over the plane *abcd*, is $(\mathscr{B}) \times (2x) \times (1)$; hence Eq. 16 can be written

$$2\mathscr{E}_z = -\frac{d}{dt}(2\mathscr{B}x). \qquad [17]$$

If the conducting material has a volume resistivity ρ, the current density $\mathscr{J}_x$ along *bc* or *da* is

$$\mathscr{J}_x = \frac{\mathscr{E}_z}{\rho} = -\frac{1}{\rho}\frac{d}{dt}(\mathscr{B}x) \qquad [18]$$

$$= (-x/\rho)d\mathscr{B}/dt \qquad [19]$$

since x is not a function of t. Along the two planes parallel to the extended faces and containing lines *bc* and *da* respectively the *instantaneous* power loss per unit volume is

$$\mathscr{J}_x^2\rho = \frac{x^2}{\rho}\left(\frac{d\mathscr{B}}{dt}\right)^2. \qquad [20]$$

This power loss occurs at the distance x from the central *YZ* plane of the slab. The instantaneous power loss in the differential slab dx thick is

$$\mathscr{J}_x^2\rho \, dx = \frac{1}{\rho}\left(\frac{d\mathscr{B}}{dt}\right)^2 x^2 \, dx. \qquad [21]$$

The instantaneous loss in the volume of slab having unit width and unit height and thickness $\tau = 2d$ is

$$2\int_0^d \mathscr{J}_x^2\rho \, dx = \frac{2}{\rho}\left(\frac{d\mathscr{B}}{dt}\right)^2 \int_0^d x^2 \, dx = \frac{2}{3}\frac{d^3}{\rho}\left(\frac{d\mathscr{B}}{dt}\right)^2. \qquad [22]$$

A unit cube of laminated material made up of similar slabs contains $1/2d$ such volumes; hence the instantaneous eddy-current loss per unit cube of laminated material with perfect insulation between the slabs so that no current exists across the lamination is

$$\frac{1}{2d}\left[\frac{2}{3}\frac{d^3}{\rho}\left(\frac{d\mathscr{B}}{dt}\right)^2\right] = \frac{d^2}{3\rho}\left(\frac{d\mathscr{B}}{dt}\right)^2. \qquad [23]$$

Equation 23 gives the instantaneous power loss caused by the time variation of $\mathcal{B}$. In alternating-current-machinery practice, the variation of $\mathcal{B}$ is usually sinusoidal. If $\mathcal{b}$ is its instantaneous value,

$$\mathcal{b} = \mathcal{B}_{max} \cos \omega t,$$ [24]

from which

$$\frac{d\mathcal{b}}{dt} = -\omega \mathcal{B}_{max} \sin \omega t,$$ [25]

$$\left(\frac{d\mathcal{b}}{dt}\right)^2 = \omega^2 \mathcal{B}_{max}^2 \sin^2 \omega t,$$ [26]

and therefore the instantaneous power loss is

$$\frac{d^2 \omega^2 \mathcal{B}_{max}^2}{3\rho} \sin^2 \omega t.$$ [27]

Since the *average* value of a sine-squared function over any integral number of cycles, or over any long-time interval is one-half the maximum, the *average* eddy-current power loss per unit volume when the flux density is varying sinusoidally at a frequency f is

$$p_e = \frac{d^2 2\pi^2 f^2 \mathcal{B}_{max}^2}{3\rho} = \frac{\pi^2 f^2 \tau^2 \mathcal{B}_{max}^2}{6\rho},$$ [28]

where τ is the thickness of the individual slab, or lamination.

In a magnetic circuit containing a volume $\mathcal{V}$ of laminated core material subjected to the same magnetic conditions as the foregoing unit volume, the average eddy-current power loss is

$$P_e = \mathcal{V} p_e = \frac{\pi^2 f^2 \tau^2 \mathcal{B}_{max}^2}{6\rho} \mathcal{V}.$$ [29]

When $\mathcal{V}$ is expressed in cubic meters, f in cycles per second, τ in meters, $\mathcal{B}_{max}$ in webers per square meter, and ρ in ohms per meter cube, P_e is expressed in watts in Eq. 29.

The assumption that $\mathcal{B}$ is uniform throughout the entire lamination requires that the magnetomotive force of the eddy currents have negligible effect in crowding the flux away from the central plane of each lamination toward its surfaces. In the more accurate analysis, this crowding effect is taken into account and shows that the eddy-current loss is equal to

$$P_e = \frac{\mathcal{V} f \mathcal{B}_{max}^2}{12\mu} (\alpha d)^2 \left[1 - \frac{6}{945} (\alpha d)^4 \cdots \right],$$ [30]

where μ is the static permeability of the material and is assumed constant and αd is given by

$$\alpha d = \pi \tau \sqrt{\frac{2\mu f}{\rho}} \; ; \tag{31}$$

hence

$$P_e = \frac{\pi^2 f^2 \tau^2 \mathfrak{B}_{max}^2}{6\rho} \mathcal{V} \left[1 - \frac{6}{945} (\alpha d)^4 \cdots \right]. \tag{32}$$

When the value of αd is near unity the higher-order terms of Eq. 32 are negligible and the loss as given by Eq. 32 is identical with that given by Eq. 29. As an indication of the effect of the higher-order terms, the loss is computed for a laminated material having a relative static permeability of 2,500 — that is, an actual permeability 2,500 times that of air — a resistivity of 25×10^{-6} ohm-centimeter, and a thickness of 0.014 inch (29-gauge). In mks units,

$$\alpha d = \pi \times 2.54 \times 14 \times 10^{-5} \sqrt{\frac{2 \times 2,500 \times 10^{-7} \times f}{25 \times 10^{-8}}}$$

$$\approx 0.05 \sqrt{f} \; . \tag{33}$$

Since the first term involving αd inside the brackets of Eq. 32 is equal to only 0.01 when αd is 1.12, the error involved in Eq. 29 is less than 1 per cent for frequencies up to about 500 cycles per second and for the assumed values of ρ, μ, and τ.

The relation given by Eq. 29 should not be considered as an accurate means for determining the true eddy-current loss of any particular core, but rather as one showing the way in which the eddy-current loss depends on the various factors. Provided αd is not much greater than unity, this loss is proportional to the square of the frequency, the square of the thickness of lamination, and the square of the flux density, and is inversely proportional to the resistivity of the material. The loss for any specific material is preferably written as

$$P_e = k_e f^2 \tau^2 \mathfrak{B}_{max}^2 \mathcal{V} \tag{34}$$

and, although theoretically

$$k_e = \frac{\pi^2}{6\rho} , \tag{35}$$

the effects of finite volume of material, low resistance between laminations, and air gaps within the core make the eddy-current calculation more accurate if k_e is determined from power measurements performed on a sample of the material and used with Eq. 34.

An important deduction can be made from Eq. 20. For any core, the instantaneous voltage e generated magnetically in a winding on the core is proportional to $d\mathcal{B}/dt$. The instantaneous eddy-current loss given by Eq. 20 is proportional to $(d\mathcal{B}/dt)^2$ and thus to e^2. Therefore the average eddy-current loss is proportional to the average of e^2. But the average of the instantaneous voltage squared is by definition the square of the root-mean-square voltage. Thus, regardless of the waveform of the generated alternating voltage, the loss P_e is proportional to the square of the root-mean-square value of the voltage within the limits of the validity of Eq. 20. It must not be implied, however, that in every instance in engineering practice P_e is merely a function of e^2 or $(d\mathcal{B}/dt)^2$. For example, in the armatures of direct-current machinery, the flux variations are far from sinusoidal and the flux density $\mathcal{B}$ undergoes a rotational variation of direction as well as an alternating variation of magnitude, which is not considered in the analysis.

3. Total core loss

The total power loss occurring in iron cores subjected to an alternating magnetizing force is the sum of the hysteresis and the eddy-current losses. From Eqs. 10 and 28, the total power loss p_c per unit volume is expressed by

$$p_c = p_h + p_e = 10^{-5} f K' K'' (\text{area of loop}) + \frac{\pi^2 f^2 \tau^2 \mathcal{B}_{max}^2}{6\rho}, \qquad [36]$$

where the symbols have the significance previously given. If the core material is such that the hysteresis loss follows the empirical relation given by Eq. 13, this loss can be written as

$$p_c = \eta f \mathcal{B}_{max}^n + \frac{\pi^2 f^2 \tau^2 \mathcal{B}_{max}^2}{6\rho}. \qquad [37]$$

If the average flux density is the same throughout the volume $\mathcal{V}$ of core, the total loss P_c in this volume is

$$P_c = \mathcal{V} p_c. \qquad [38]$$

Devices in which ferromagnetic materials carry alternating fluxes practically always have associated electric circuits which interlink the magnetic circuits. Transformers and iron-core reactors, for example, have laminated or powdered ferromagnetic cores around which are wound the turns of one or more coils of wire. The core losses are related to the electromotive force induced in such a coil by the changing flux. Equation 19, Art. 7 of Ch. VI, gives the maximum flux ϕ_{max} in terms of the effective value of the induced electromotive force E in a coil of N turns as

$$\phi_{max} = \frac{E}{4.44 f N} \qquad [39]$$

when the flux and consequently the electromotive force vary sinusoidally. If the flux density is uniform over the cross-sectional area A of the core,

$$\mathcal{B}_{max} = \frac{\phi_{max}}{A} = \frac{E}{4.44fNA}. \qquad [40]$$

For a given transformer or reactor, the number of turns and the core area are fixed by the design. Then

$$\mathcal{B}_{max} = K\frac{E}{f} \qquad [41]$$

which, substituted into Eq. 37, gives

$$p_c = \eta f\left(\frac{KE}{f}\right)^n + \frac{\pi^2 f^2 \tau^2 K^2 E^2}{6\rho f^2}$$

$$= K_1 \frac{E^n}{f^{n-1}} + K_2 E^2. \qquad [42]$$

Equation 42 applies only when the waveform is sinusoidal. Although the hysteresis loss is dependent on the maximum flux density and is independent of the waveform of the flux as long as the hysteresis cycle is symmetrical and without loops, the relation between the maximum value of the flux density and the effective value of the generated electromotive force does depend upon the waveform. Thus, when expressed in terms of effective electromotive force, the hysteresis loss is correctly given by the first term in the right-hand member of Eq. 42 only when the waveform is sinusoidal.

In contrast, the second term in the core-loss expression, Eq. 42, gives the correct eddy-current loss regardless of the waveform provided the frequencies involved in the nonsinusoidal wave are not high enough to produce a considerable skin effect. When the flux wave is made up of components, each of these components induces eddy currents in the core. The eddy-current loss produced by each harmonic component in the flux is proportional to the square of the same harmonic component of the electromotive force generated in the winding. Then if E_1, E_3, E_5, $\cdots$ are the effective values of the fundamental and harmonic components of generated electromotive force, the total eddy-current loss per unit volume is, according to the second term of Eq. 42,

$$p_e = K_2(E_1^2 + E_3^2 + E_5^2 + \cdots). \qquad [43]$$

But the sum of E_1^2, E_3^2, E_5^2, $\cdots$ equals the square of the effective value E of the generated electromotive force. Note also that the eddy-current loss, when expressed in terms of E, is independent of frequency.

Changes in temperature such as are encountered in practice have a

negligible effect on hysteresis loss. The eddy-current loss decreases somewhat with increase in temperature. For a given flux variation, the eddy-
current loss is inversely proportional to the resistivity of the core material
as indicated in Eq. 29. The resistivity increases with temperature.

Since Eqs. 36 and 37 are derived on the basis of several assumptions
that in practice may not be fulfilled, they commonly yield numerical
results that are too small, sometimes by a factor of 2 or more. The
important use of these equations is therefore not so much for the calculation of the loss in particular cases as for showing the functional relation
between the loss and the variables. They serve effectively as guides to
the analysis of experimental data and also indicate the possible ways of
modifying the loss. Since the validity of the assumptions depends on
the conditions of use of the materials, a restatement of these assumptions
should be serviceable. The derivation of the hysteresis-loss term in
Eq. 36 assumes that:

(a) Each lamination is homogeneous magnetically; that is, each element of its volume has the same magnetic characteristics.

(b) The flux density is uniform throughout each lamination; that is,
the effect of the eddy currents on the flux distribution is negligible.

Furthermore the empirical expression for the hysteresis term of Eq. 37
is subject to the additional assumptions that:

(c) The hysteresis loop is of the normal symmetrical shape with no
re-entrant loops. Provided this condition is satisfied, no restriction
is placed on the manner in which $\mathfrak{B}$ varies with time throughout a
cycle of magnetization.

(d) The material, the range of maximum flux density, and the manner
of flux-density variations are such that an empirical exponent n
can be used with reasonable accuracy.

The derivation of the eddy-current loss term of Eq. 36 or 37 assumes
that:

(a) The material is magnetically and electrically homogeneous. In
practice, this condition is not perfectly fulfilled, since such factors
as grain size, the direction of the grain produced by rolling, and
the relatively poorer magnetic properties of the surface layers
have an appreciable effect, especially in thin laminations.

(b) The thickness of the laminations is constant and very small compared with their other dimensions. This condition is usually
realized in practice.

(c) The flux density is uniform throughout the thickness of the lamination; that is, the eddy-current magnetomotive force is negligible
compared with the magnetizing magnetomotive force acting on

the core. Equation 32 indicates the order of error involved, which is usually small for thin laminations of high-resistivity materials subjected to cyclic variations at less than 5,000 cycles per second, but may be very large under other conditions.

(d) The volume of core involved is subjected to a uniform field so that at any given instant the flux density is the same in the different laminations.

(e) The laminations are perfectly insulated from each other. This assumption is seldom fulfilled in commercial apparatus on account of the considerable pressures under which the laminations are clamped together.

(f) The flux density undergoes a sinusoidal time variation and is always directed parallel to the plane of the lamination. The assumption of a sinusoidal time variation is not a restriction, however, since it was shown that the factor $(f\mathcal{B}_{max})^2$ can be replaced by E^2 times a constant, where E is the root-mean-square voltage induced in a coil linked by the alternating core flux which may have any waveform.

Table I gives representative data for core loss and resistivity for typical samples of selected core materials.

TABLE I*

REPRESENTATIVE PROPERTIES OF MAGNETICALLY SOFT MATERIALS

Material	Resistivity, Microhm-cm	Hysteresis Loss, ergs per cm^3 per cycle	Total Core Loss, watts per pound 29-gauge, $f = 60$ cps
Permalloy (78.5% Ni) .	16	200	0.34†
Permalloy (78.5% Ni 3.8% Cr)	65	200	. . . . .
Hipernik (50% Ni) . . .	35	220	0.20†
Mumetal	25	200	. . . . .
Perminvar (45–25) . . .	19		. . . . .
Perminvar (7–70) . . .	80		. . . . .
Permendur	7	12,000	. . . . .
Permendur with 2% V .	26	6,000	. . . . .
Silicon steel (4.25% Si) .	60	1,340†	0.60†
Silicon steel (1.0% Si) .	24	2,640†	1.17†
Armco or Norway iron .	11	5,000†	excessive

* From same source as data in Table I, Ch. I.

† These values correspond to $\mathcal{B}_m = 10,000$ gausses; the others to saturation.

4. Reduction of total core loss; design considerations

With a core of given configuration and stated conditions under which it is to be used, the hysteresis loss can be made small by the use of a core material which has a hysteresis loop of small area. The eddy-current loss can be reduced by the use of thin sheets of a material which has a high resistivity and insulating the sheets from each other. Powdered Permalloy dust,[7] which has a small hysteresis loop, is frequently mixed with an insulating binder and compressed into a solid of the core shape desired. Such a material then often has a resistivity as high as 50 ohm-centimeters. The grades of silicon steel in frequent use have a resistivity ranging between 20 and 60 microhm-centimeters, for a silicon content ranging between 1 and about 4 per cent. As a rule, the various desired properties can be attained to a high degree through proper alloying and heat treatment of the material, the details of which are beyond the scope of this discussion.

Theoretically, the eddy-current losses can be made small through using very thin laminations, provided the laminations are adequately insulated from each other and provided no bolts, clamps, or other solid pieces of metal come within the region in which the varying magnetic field exists. However the use of very thin laminations increases the manufacturing cost and the size of the apparatus by decreasing the stacking factor. If the decrease in the stacking factor is appreciable in transformer cores, the coefficient of coupling, which is desired as near unity as possible, may be reduced to an undesirable extent. The allowable lamination thickness is also determined by the economic balance between the manufacturing cost when thin laminations are used and the cost of the energy lost or other reduction in performance caused by eddy currents.

The hysteresis loss is also related to the grain size, in general the loss being smaller as the grain size is increased. However, if a sheet is to have much strength it must be at least several crystals thick. The photomicrographs shown in Fig. 6 indicate the relative grain size for several samples of silicon steel. The steel shown in Fig. 6a is United States Steel electrical grade with 1 per cent silicon, hot rolled, and not annealed. Figure 6b shows the same steel annealed. The resistivity is 24 microhm-centimeters and the total core loss 1.17 watts per pound. Figure 6c shows United States Steel motor grade with 2.5 per cent silicon content, annealed. The resistivity is 40 microhm-centimeters, and the core loss is 1.01 watts per pound. The transformer grade, shown in Fig. 6d, has 4 per cent silicon and is annealed; its resistivity ranges from 50 to 62 microhm-centimeters and its core loss is 0.7 watt per pound.

[7] W. J. Shackelton and I. G. Barber, " Compressed Powdered Permalloy," *A.I.E.E. Trans.,* *47* (1928), 429–436.

For the materials which are commercially available at the present time, and in view of the present relative costs, a good balance among these various factors is usually found in use of annealed-silicon-steel laminations of 0.012 inch to 0.028 inch thick in the cores of apparatus

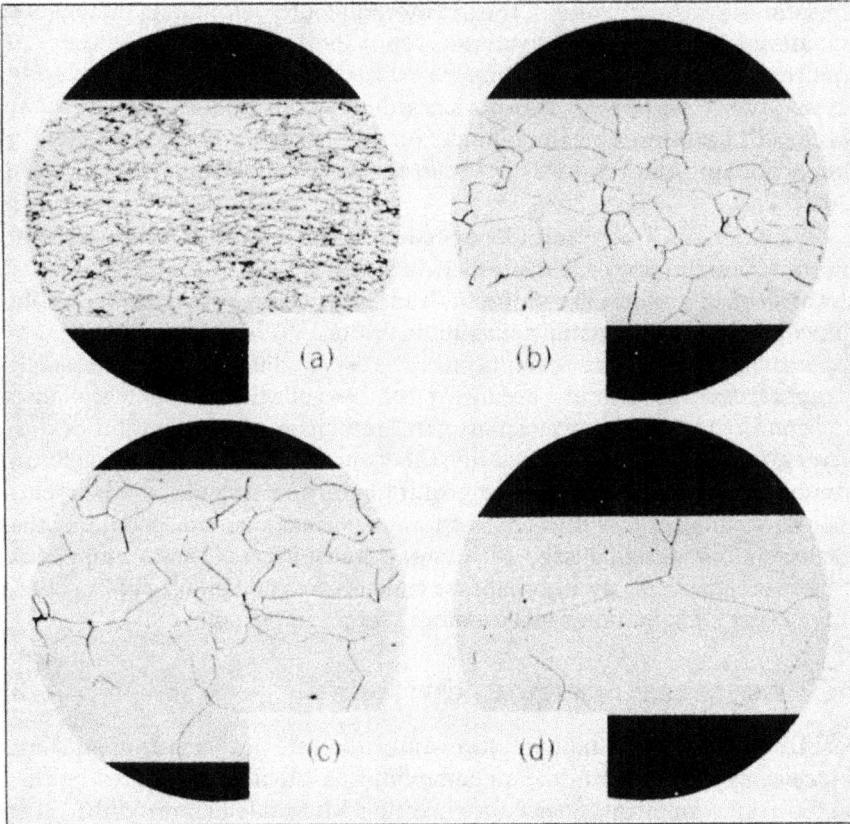

Courtesy Carnegie-Illinois Steel Corp.

FIG. 6. Photomicrographs showing grain size of steels containing different amounts of silicon. (a) Electrical grade hot rolled, not annealed. (b) Electrical grade annealed. (c) Motor grade, annealed. (d) Transformer grades annealed.

operated at commercial power frequencies. A thickness of 0.014 inch is perhaps the most common in the United States. For radio-frequency apparatus, laminations of 0.001 inch to 0.003 inch thick or powdered iron or Permalloy cores are commonly used. Silicon steel has the desirable properties of relatively small bulk cost, comparatively high resistivity, a fairly small hysteresis loop, and large (though not constant) permeability and is readily stamped into laminations of any desired form. The laminations are usually heat treated or at least annealed after being stamped to relieve the strains introduced during the stamping.

Since, at the frequencies encountered in power transmission, eddy-current losses can easily be reduced to a fairly small value by using thin laminations of high-resistivity material, whereas the reduction for hysteresis losses involves the greater difficulty of finding a magnetic material having a small hysteresis loop, the hysteresis loss at these frequencies is usually large compared with the eddy-current loss. In 60-cycle-per-second power transformers the hysteresis loss is usually of the order of two-thirds to three-fourths of the total core loss. A method for apportioning the total loss between the two components is described in Art. 7.

The division of the total loss between the hysteresis and eddy-current components in power transformers is relatively easy to control, since the waveform is essentially sinusoidal and the voltage and flux variations have essentially constant maximum values. With communication apparatus, however, the conditions are very different. The frequency ranges over wide limits because of the peculiarities of the waveforms encountered, which in effect are the intelligence. The amplitudes also vary over wide limits. As a rule, the question of losses in small communication transformers is important more on account of their effect on the technical performance of the transformer than on account of their effect in dissipating energy. With large transformers in power amplifiers, however, particularly in broadcast transmitters and similar applications, the energy dissipation aspect assumes larger importance.

5. Measurement of total core loss

In order that the manufacturers and the users of ferromagnetic sheet steels may have a mutual understanding as to the properties of these steels, the American Society for Testing Materials has prescribed a set of definite procedures to be used in testing for total core losses. The test used for the commercial sheet steels at power frequencies is described here in detail. For a description of methods of obtaining losses at low flux densities or at 1,000 cycles per second, the student should refer to the latest edition of the A.S.T.M. circular A–34.

The Epstein frame used in the commercial test is shown in Fig. 7. It is constructed in such a way that its primary and secondary windings are placed one over the other in four long solenoidal sections along the sides of a square. The winding length of each of the sections is 42 centimeters, and the coils are wound on forms having a square cross section of inside dimensions 4 × 4 centimeters. The secondary coil of 150 turns per section is first wound uniformly over the length of the form and then the primary of the same number of turns is wound on the outside. The four primary windings are connected in series aiding, and the four secondary windings are similarly connected.

The test specimens consist of 10 kilograms (22 pounds) of steel strips 50 centimeters (19$\frac{11}{16}$ inches) in length and 3 centimeters (1$\frac{3}{16}$ inches) in width. These strips are assembled into stacks of 2.5 kilograms each, forming four bundles to be placed inside the solenoidal coils. One half the strips are to be cut with their length parallel to the direction of rolling and the other half at right angles to this direction. The like samples are

Fig. 7. Epstein core-loss apparatus in Electrical Measurements Laboratory at the Massachusetts Institute of Technology.

placed on opposite sides of the square. If the steel is to be used in an application like the " spirakore " and " hipersil " transformers, in which the grain direction plays an important part in the action of the device, the strips should all be cut so that in the test the flux has the same relation to the direction of rolling as it has in the finished product.

The diagram of connections for the core-loss test is shown in Fig. 8. The primary coils of the Epstein frame are designated by P and the secondary coils by S. For the 60-cycle-per-second test, the generator should be capable of providing a sine-wave voltage variable up to approximately 160 volts at 60 cycles per second. The voltage should be adjustable by field control or else a low-impedance autotransformer should be interposed for voltage adjustment. If any possibility of a varying frequency exists, a frequency meter F must be connected across the supply line.

The ammeter A measures the root-mean-square value of the current in the primary winding. This primary current is the transformer exciting current described in Ch. VI and is nonsinusoidal. If the waveform of the core flux and of the electromotive forces induced in the windings is to retain approximately a sinusoidal form, the IZ drops in the generator and in the leads between the generator and the primary winding must be kept low. If these conditions are met, the effective value of the voltage induced in the secondary is a measure of the maximum flux density in the core as indicated by the relationship from Eq. 39

$$E = 4.44Nf\phi_{max} = 4.44NfA\mathfrak{B}_{max}.$$ [44]

Suppose, for example, that a 60-cycle-per-second core-loss measurement is to be made at a maximum flux density of 10,000 gausses in a 4 per cent silicon-steel specimen. The cross-sectional area of steel, 6.5 square centimeters, is found through dividing the total mass, 10,000 grams, by

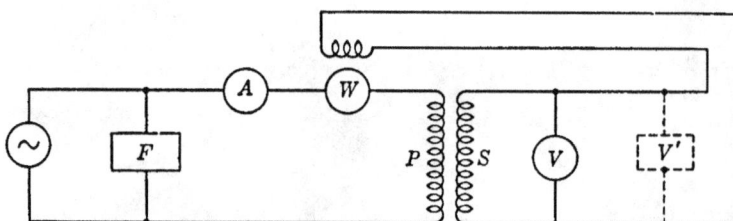

FIG. 8.   Diagram of connections for Epstein core-loss test.

the density, 7.5 grams per cubic centimeter, and by the total length of core, 200 centimeters. The required maximum flux density will be obtained in the sample when the root-mean-square voltmeter V reads

$$E = 4.44 \times 600 \times 60 \times 6.5 \times 10,000 \times 10^{-8} = 103.8 \text{ volts}.$$ [45]

The total core loss is now indicated by the wattmeter W, and this reading in watts, corrected for instrument losses, divided by the mass of the specimen (10 kilograms or 22 pounds) gives the result in watts per kilogram or per pound. For any other flux density the necessary reading of V can be computed from Eq. 44, and the corresponding loss read from the wattmeter.

The cross-sectional area of the specimen is to be computed as indicated above, a density of 7.7 grams per cubic centimeter being used for steels with a silicon content of two per cent or less and a density of 7.5 if the silicon content is greater than two per cent.

It is often impossible to produce a sinusoidal variation of flux density in the specimen, either because a sine-wave generator of adequate capacity is not available or because the flux density desired in the test is so high that a large exciting current is required. When the waveform of the flux density is distorted, the effective value of induced electromotive force does not measure the value of $\mathfrak{B}_{max}$. Another type of voltmeter, sometimes called a flux voltmeter,[8] can be used under these conditions. The

[8] G. Camilli, " A Flux Voltmeter for Magnetic Tests," *A.I.E.E. Trans.,* *45* (1926), 721-728.

design of this instrument is based on the following mathematical discussion.

The instantaneous electromotive force induced in the secondary winding is

$$e = N \frac{d\phi}{dt}.$$ [46]

The integration of this equation over a half-period of the electromotive-force wave,

$$\int_{t_1}^{t_2} e \, dt = N \int_{-\phi_{max}}^{+\phi_{max}} d\phi = 2N\phi_{max} = 2NA\mathcal{B}_{max},$$ [47]

shows that the maximum flux or flux density regardless of the waveform is proportional to the average value of a positive (or negative) half wave of induced electromotive force, or

$$\mathcal{B}_{max} = \frac{e_{av}(t_2 - t_1)}{2NA},$$ [48]

where $t_2 - t_1$ is the constant half-period $1/2f$ of the wave. Since the d'Arsonval type of direct-current voltmeter measures the average value, one of these instruments used in conjunction with a full-wave rectifier[9] constitutes a satisfactory flux voltmeter. For use in the Epstein test, it is connected in parallel with the root-mean-square voltmeter as shown by the dotted lines of Fig. 8 and is calibrated to read the same as the root-mean-square instrument when the waveform is sinusoidal. When the waveform is nonsinusoidal, the instrument will read $1.11e_{av}$ since 1.11 is the *form factor*, or the ratio of root-mean-square voltage to average voltage for a sine wave. The value to be held on the flux voltmeter for any specified value of $\mathcal{B}_{max}$ is, from Eq. 48,

$$E' = 1.11e_{av} = 4.44fNA\mathcal{B}_{max}$$ [49]

as before, but, if the flux wave is nonsinusoidal, the reading V of the root-mean-square voltmeter is larger than the reading V' of the flux voltmeter. The form factor of the flux wave is 1.11 times the ratio V/V'.

The current coils of the instruments used in this test should have exceptionally low resistance[10] and the voltage coils should have exceptionally high resistance in order that they may not absorb too much power. The wattmeter should be of the low-power-factor type.

The data for the curves given in the next article were obtained from the standard Epstein test.

[9] Article 2b, Ch. VI, of the volume on engineering electronics describes this type of rectifier.
[10] Also see article: B. M. Smith and C. Concordia, " Measuring Core Loss at High Densities," *E.E.*, *51* (January, 1932), 36–38.

6. EXPERIMENTAL LOSS CURVES, EXPERIMENTAL COEFFICIENTS

For all calculations in which a precise knowledge of core loss is necessary, the only reliable data are obtained experimentally on samples of the actual material to be used. If ranges of frequency and flux density are to be considered, these data are conveniently expressed by curves of total core loss such as those of Figs. 9 and 10, which apply to a grade of silicon transformer steel containing about four per cent silicon.

In Fig. 9, the total loss in watts per pound is given as a function of frequency for several different values of maximum flux density $\mathfrak{B}_{max}$. The same data are shown as a function of $\mathfrak{B}_{max}$ for several values of frequency in Fig. 10, and, on an enlarged scale in Fig. 20, Ch. VI, for the range of flux densities and frequencies used in power applications. From these curves, the total core loss per pound can be obtained for a wide variety of conditions. The scales are logarithmic.

The forms of the curves on log-log scales are suggestive of relatively simple expressions to represent the loss. The curves of p_c as a function of $\mathfrak{B}_{max}$ in Fig. 10 for various frequencies are almost parallel straight lines, indicating empirically that the total loss varies as some power of $\mathfrak{B}_{max}$ over the range shown. A comparison of the data as indicated by the dotted lines on Fig. 9 with those given by the curves shows that the total loss can be assumed to vary as some power of the frequency f from about 30 to 500 cycles per second with an error not greater than about 10 per cent. Thus, an expression of the form

$$p_c = Cf^n(\mathfrak{B}_{max})^m \qquad [50]$$

is suggested as an empirical expression for the total core loss per unit volume, where C, n, and m depend upon the properties of the particular material. The measurement of the geometrical slopes on Figs. 9 and 10, where the abscissa and ordinate scales are the same, gives the approximate values of n and m respectively as

$$1.36 \quad \text{and} \quad 1.71,$$

for f in cycles per second and $\mathfrak{B}_{max}$ in gausses.

For a maximum flux density of 10,000 gausses and a frequency of 60 cycles per second, the value of the coefficient C for the units used and for 29-gauge 14-mil-thick material is

$$C = \frac{0.60}{60^{1.36}10,000^{1.71}} = \frac{0.60}{261 \times 6.91 \times 10^6} = 3.32 \times 10^{-10},$$

or

$$p_c = 3.32 \times 10^{-10} f^{1.36} \mathfrak{B}_{max}{}^{1.71} \quad \text{watts per pound} \qquad [51]$$

for this particular steel and gauge. This form of relation is often useful for analytical work in which a range of flux densities and frequencies is

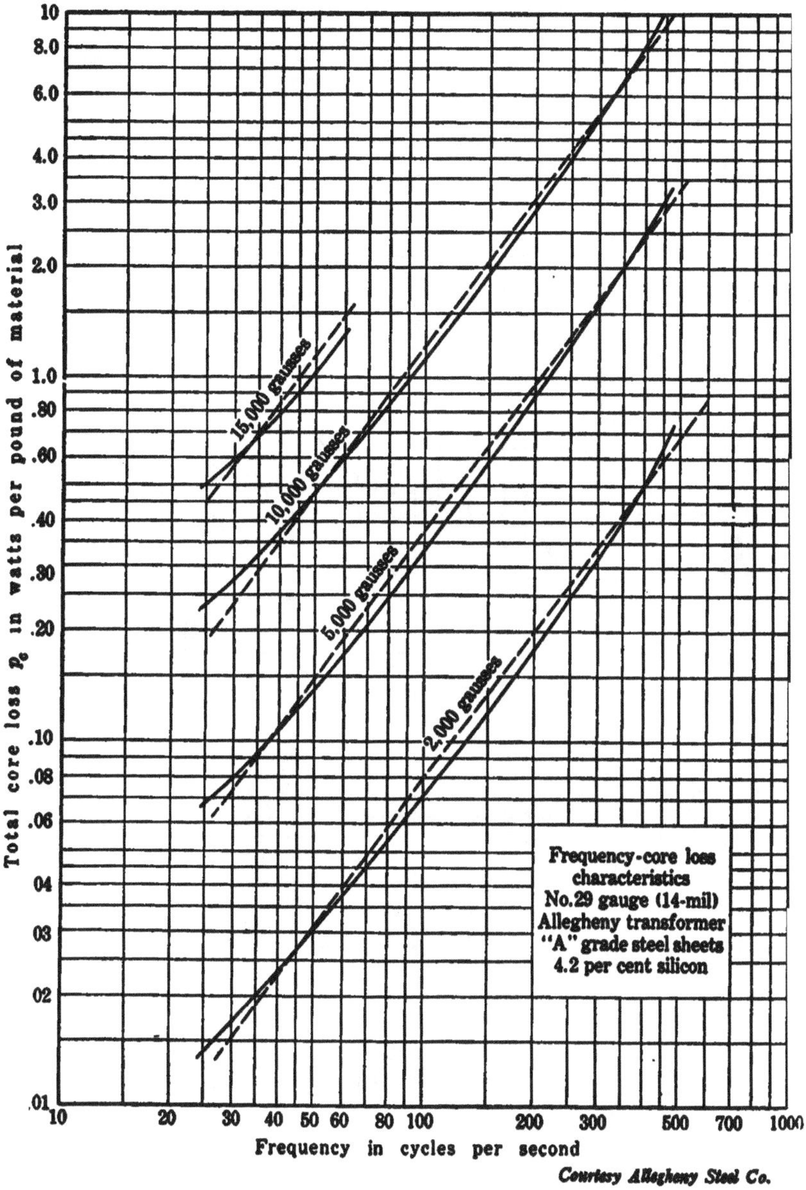

10
8.0
6.0
4.0
3.0
2.0
1.0
.80
.60
.40
.30
.20
.10
.08
.06
04
03
02
.01
Total core loss p_c in watts per pound of material
15,000 gausses
10,000 gausses
5,000 gausses
2,000 gausses
Frequency-core loss characteristics
No. 29 gauge (14-mil)
Allegheny transformer
"A" grade steel sheets
4.2 per cent silicon
10
20
30
40
50
60
80
100
200
300
500
700
1000
Frequency in cycles per second
Courtesy Allegheny Steel Co.

Fig. 9.

flux density in kilolines per square inch

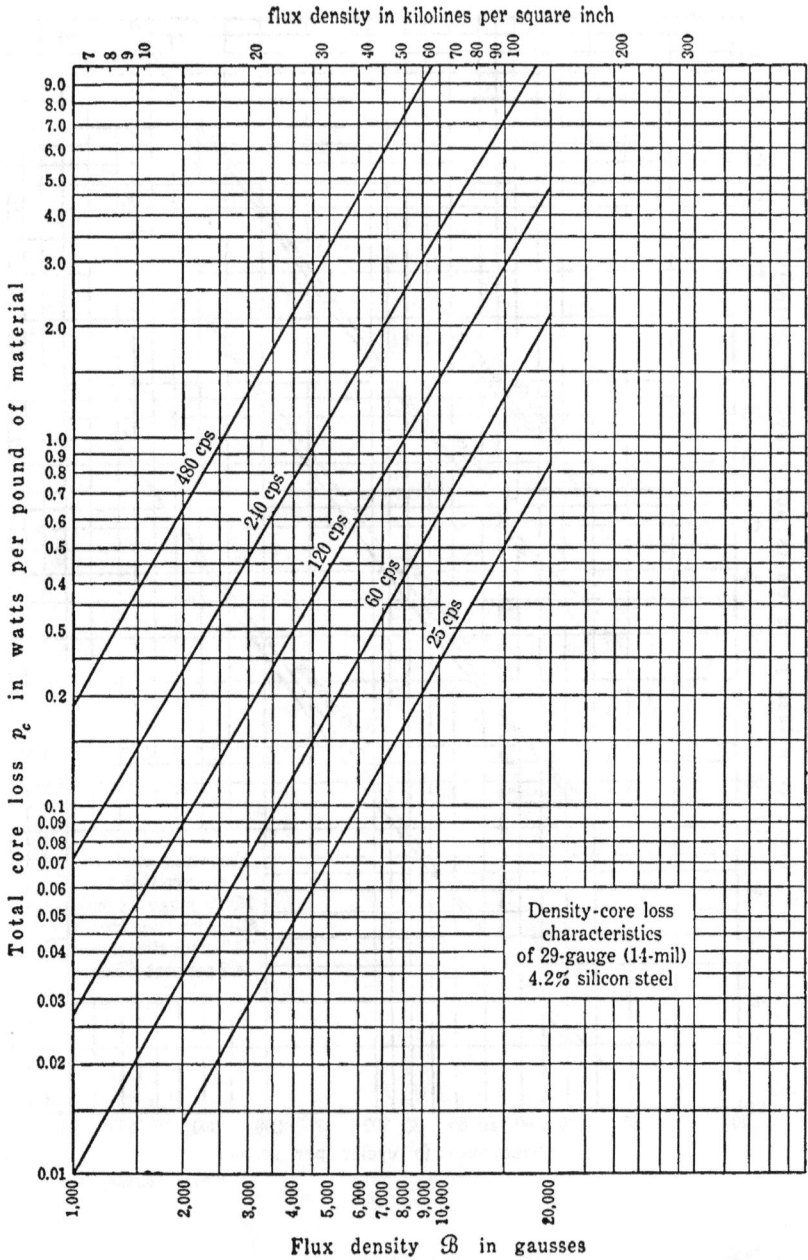

Density-core loss characteristics of 29-gauge (14-mil) 4.2% silicon steel

Flux density $\mathcal{B}$ in gausses

Courtesy Allegheny Steel Co

FIG. 10.

being considered. The significance of the third figure given for C, n, and m is of course, doubtful, and the value given for p_c must be interpreted with appropriate recognition of the approximations inherent in the procedure used.

An expression of the form of Eq. 50 fits the experimental curves of some materials better than others. It is purely empirical, and its applicability can be determined most readily from log-log plots such as Figs. 9 and 10, from which the values of the exponents can also be derived if the curves are found to be approximated sufficiently closely by straight lines.

As an indication of the effect of lamination thickness on total core loss, the curves of Fig. 11 are given for a single flux density of 10,000 gausses and a frequency of 60 cycles per second. Since the United States standard gauge is logarithmic in character, an arithmetic scale in gauge number is equivalent to a reversed log scale of thickness. In the figure as drawn a factor of two in the thickness, a difference of six gauge numbers, occupies the same distance on the abscissa scale as a factor of eight on the ordinate scale. The ratio of the distances on the thickness and watts-per-pound scales that correspond to the same interval of logarithm or factor is $\dfrac{\log 8}{\log 2}$ or three. A geometrical slope of $-\frac{2}{3}$ as shown by the dotted line therefore represents a loss varying directly as the square of the thickness. For the material shown and for 10,000 gausses and 60 cycles per second, the curves indicate that, as the thickness is reduced, the total loss decreases less rapidly than the square relation theoretically applicable to the eddy-current component.

The loss data obtained from tests must be used with caution, since the conditions of the test are not likely to be duplicated exactly in commercial utilization of the materials. For example, in commercial apparatus the punchings are often subjected to considerable pressure in order to improve the stacking factor, to obtain mechanical rigidity, and to reduce the noise from vibrating laminations. The resulting stress, the electrical conducting paths formed between adjacent laminations by punching burrs which puncture the surface of the lamination insulation, and the short-circuited paths formed by the clamps in poorly designed apparatus are conditions not duplicated in the Epstein test, and all tend to increase the losses above the test values. The necessary correction factors to use with the standard test data can be obtained only from the results of measurements performed on apparatus of construction similar to that in which the material is to be used.

7. SEPARATION OF HYSTERESIS AND EDDY–CURRENT LOSSES

Although the total iron loss is usually an item of major concern in the design of alternating-current electromagnetic apparatus, an intelli-

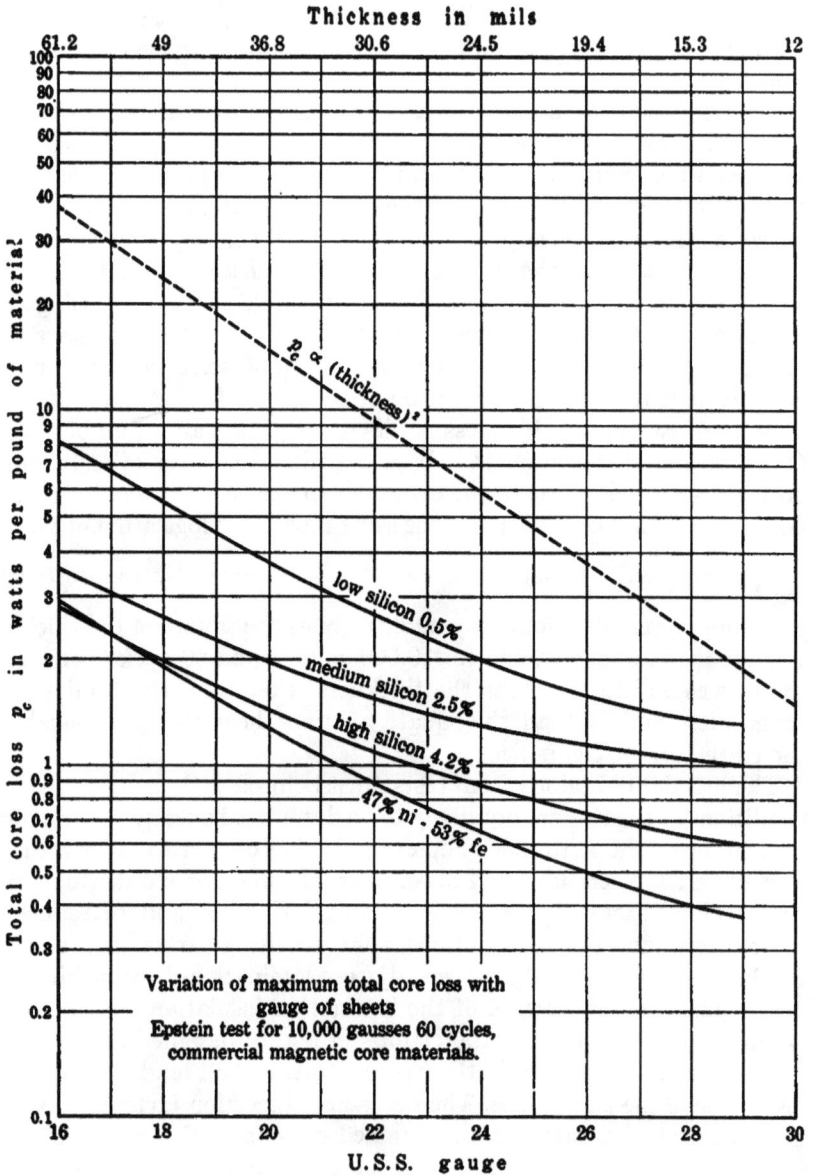

Fig. 11.

gent attempt to reduce the total loss can be made only when the relative magnitudes of the hysteresis and eddy-current components are known. If the eddy-current component predominates and thinner laminations are feasible, a reduction in the total loss can be effected by using them. However, if the hysteresis loss predominates, it is futile to change the lamination thickness. As a preliminary step toward the reduction in the total loss, a simple means of determining the division of the total iron loss between the two components is therefore desirable.

Any direct measurement of the power loss in the iron necessarily gives the total loss, but the division into the two components can be determined through the different ways in which the two are related to the variables. Equations 36 and 37 indicate that, when the flux density is a sinusoidal function of time, each component of loss is a different function of frequency and maximum flux density. In addition, the eddy-current loss is a function of lamination thickness and of the resistivity; but, since these quantities cannot be changed in an actual sample of material, they are not available as variables.

A simple method of separating the hysteresis and eddy-current components of core loss depends on the fact that the hysteresis component varies linearly and the eddy-current component varies as the square of the frequency. Stated differently, the hysteresis component of the *total energy per cycle*, P_c/f, is independent of frequency and the eddy-current component is a linear function of frequency. If the loss is measured at a given constant maximum flux density and the frequency is varied, then a plot of the *loss per cycle* as a function of the frequency, if it follows the theoretical relation

$$P_c/f = \frac{\mathcal{U}}{4\pi} \oint_{loop} \mathcal{H}\, d\mathcal{B} + \left(\frac{\mathcal{U}\pi^2 \tau^2 \mathcal{B}_{max}^2}{6\rho} \right) f = K_1 + K_2 f, \qquad [52]$$

should be a straight line. The ordinate where the line intercepts the P_c/f axis at zero frequency gives the hysteresis loss per cycle K_1, and the slope of the line is the parenthetical coefficient of f in Eq. 52, namely

$$K_2 = \frac{\mathcal{U}\pi^2 \tau^2 \mathcal{B}_{max}^2}{6\rho}. \qquad [53]$$

The procedure is performed experimentally as follows: A generator having essentially a sinusoidal waveform is connected to a low-resistance coil wound around a suitable sample of iron such as a ring or an Epstein test specimen; and the total power input, the voltage, the frequency, and the current are measured by suitable instruments. The power input to the iron is the total wattmeter reading less any corrections such as instrument I^2R losses and the I^2R loss in the coil resistance, the latter correction not being necessary with the Epstein method of measurement.

If Eq. 52 is to be used, $\mathcal{B}_{max}$ must be kept constant; hence the root-mean-square voltage E_φ generated in the coil by the $d\varphi/dt$ in the iron must be kept proportional to the frequency. Thus, if the IR drop in the coil is made small through use of a low-resistance winding, and this is usually done, the applied voltage E is with little error taken equal in magnitude to the voltage E_φ. A series of test points for constant E_φ/f throughout a range of frequencies extending as near to zero frequency as practicable, gives the desired data. A plot of P_c/f extrapolated to zero frequency then gives the hysteresis loss per cycle.

These measurements can be performed on test samples or on transformers or reactors, or in a modified form on rotating machines.

PROBLEMS

1. A 60 $\sim$ core-type transformer is designed for 11,000-v operation on its high-tension side. The laminated steel core has a cross-sectional area of 72.2 sq in. and the stacking factor is 0.9. The high-tension winding has 910 turns.

   (a) At what maximum flux density does the core operate?
   (b) If the length of the core is increased by 10%, how much is the flux changed?
   (c) If the number of turns in the winding is reduced 10%, at what maximum flux density does the core operate?
   (d) The inside coils of this transformer are circular and just fit over the square core. If the square core is changed to one of cruciform cross section which just fits inside the coil, at what flux density does the new core operate? The winding has 910 turns.

   Note: The maximum ratio of the area with a cruciform cross section to the area of a core with a square cross section both of which fit into the same circle is $\sqrt{5} - 1$.

2. The high-voltage winding of a power transformer consists of two coils which may be connected in series or in parallel. When these coils are connected in parallel across 22,000 v at 60 $\sim$, the no-load current is 0.077 amp and the power is 371 w, 28% of which is eddy-current loss. What are the approximate no-load current and power when the coils are connected in series across a 22,000-v 30 $\sim$ source?

3. When the secondary winding of an iron-core transformer is open and a 60 $\sim$ sinusoidal voltage which has an effective value of 500 v is impressed on the primary, the primary power is 1,200 w. When a 30 $\sim$ sinusoidal voltage which has an effective value of 500 v is impressed on the secondary with the primary open, the secondary power is 500 w. The ratio of the number of primary to secondary turns is 0.5.

   If the leakage reactance and resistance drops in the windings at no load are neglected, what are the eddy-current loss and the hysteresis loss when operating with a sinusoidal 60 $\sim$ voltage of 500 v effective value impressed on its primary winding?

4. A transformer core has a mean length of 70 in. and a uniform gross cross section of 30 sq in. The stacking factor is 0.90; the density of the iron is 0.272 lb per cu in. The primary winding has 94 turns. When this transformer is operated at no load, the measured core loss, that is, the power input minus the primary copper loss, for different sinusoidal impressed voltages and frequencies is given in the following table.

Test	V		P
A	678.7 v	60 ∿	904.0 w
B	282.8	25	344.4
C	484.8	60	420.0
D	202.0	25	159.4
E	291.0	60	177.3
F	121.2	25	64.25

The primary resistance drop and the magnetic leakage may be neglected. When the effects of these are negligible, the impressed voltage is equal to the voltage induced in the primary winding by the core flux.

(a) What are the maximum flux densities in kilolines per square inch and the core loss in watts per pound of core for the conditions corresponding to tests A, B, C, D, E, and F? The results are to be tabulated.

(b) What are the eddy-current and the hysteresis losses in watts per pound of core for each test?

(c) What are the Steinmetz coefficient and exponent? The 60 ∿ data are to be used in finding these.

5. A magnetic core is to be built of standard L-shaped punchings, as shown in Fig. 12, with the joints in successive layers staggered as indicated by the solid and dotted lines. The stack is $\frac{3}{4}$ in. high. The stacking factor is 0.88.

The core material is to be a high-silicon steel having a density of 0.274 lb per cu in. and a resistivity of 60 microhm-cm. Manufacturer's tests taken with $\mathfrak{B}_{max} = 10$ kilogausses show that the hysteresis loss is 1,340 ergs per cu cm per cycle and the eddy-current loss is 0.114 w per lb at 60 ∿ for 29-gauge sheets.

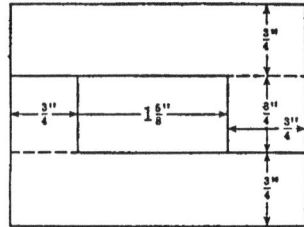

FIG. 12. Laminated transformer core built of L-sections for Prob. 5.

(a) Predict the total core loss this core will have at 200 cycles for $\mathfrak{B}_{max} = 10$ kilogausses if the laminations are 29 gauge (0.014 in. thick).

(b) By what per cent would the core losses be increased if the same net weight of iron were used under the same conditions of flux density and frequency as in part (a) but if the material were 26 gauge (0.0188 in. thick)?

6. An iron-core 1,000-kva transformer is designed to operate with an effective primary voltage of 2,300 v, 60 ∿. At rated primary voltage and frequency the maximum flux density in the core is 11,000 gausses if the voltage is sinusoidal and the core loss is 1.02 w per lb. The mean length of the core is 95 in. Its gross cross-sectional area, which is uniform, is 278 sq in. The density of the iron in the core is 0.272 lb per cu in. and the stacking factor is 0.9. The Steinmetz exponent is 1.6. With a sinusoidal impressed voltage which gives the same maximum flux density when the frequency is 25 ∿ the core loss is 0.383 w per lb.

If a 60 ∿ 2300 v (rms) voltage which has the following waveform

$$v = V_1 \sin \omega t + 0.1 V_1 \sin 3\omega t$$

is impressed on the primary winding of the transformer, what is the core loss?

CHAPTER VI

Alternating-Current Excitation Characteristics of Iron-Core Reactors and Transformers

A coil constructed and used primarily for its inductance property is known as a reactor, choke coil, or retardation coil. When intended for use at audio frequencies and below, such a coil usually has a so-called iron core of laminated silicon steel or other ferromagnetic material. At these frequencies the major effect of adding an iron core to a coil is a great increase in the inductance.

From an engineering point of view, this increase in inductance is very important, for use of an iron rather than a nonmagnetic core in a reactor which is to have a specified inductance often permits reduction of the coil dimensions, the winding resistance, the reactor weight, or all three. A further important advantage of the iron core is that the flux is confined almost wholly to the iron path, and therefore for a given inductance the dispersion of the magnetic field into the region surrounding the reactor is less than with an air core. As a result, the magnetic coupling between an iron-core reactor and adjacent circuits can be kept relatively small.

The use of iron, however, introduces secondary phenomena that may have undesirable effects. The core losses inherent in iron subjected to time-varying magnetization may affect the circuit adversely or cause heating of the core, and limit the usefulness of the device. The nonlinear magnetization characteristics of the core, moreover, cause the inductance to be a variable dependent on the flux, and thus introduce complications in the analysis of circuits containing iron-core coils. Although in numerous applications neither the core loss nor the variable inductance causes trouble, in others these factors are important, and skillful engineering is necessary to minimize their results or to predict the behavior of circuits affected by them. These and other matters important in the application of iron-core reactors and transformers form the subject of this chapter.

1. GENERAL PROPERTIES OF IRON–CORE REACTORS

The total losses in an iron-core reactor comprise the effective-resistance loss I^2R, and the hysteresis and eddy-current losses in the core. The *alternating-current effective resistance R* exceeds the direct-current resistance of the winding because of skin effect and other causes discussed in Art. 3b of this chapter. When the impedance of a reactor is measured, the real component of the impedance — called the *apparent resistance* —

is found to be greater than the effective resistance of the winding. By definition, the apparent resistance equals the total power dissipated in the reactor divided by the square of the coil current, and hence must be greater than the effective resistance of the winding whenever core losses are present. The distinction between the *effective resistance* R and the *apparent resistance* R_a should be carefully noted. The effective resistance accounts for the loss in the winding only, whereas the apparent resistance accounts for the total loss in the reactor.

As the frequency is increased, the advantages of the iron core become less marked. The increase in core loss with frequency may cause the apparent resistance to become excessive. The screening effect of the eddy currents may be so great at high frequencies that the apparent alternating-current permeability of the core may be appreciably less than its value for low frequencies. Hence the apparent inductance may be decreased.

In many applications of inductance coils, particularly in resonant circuits, the ratio of inductive reactance ωL to apparent resistance R_a should be as large as possible. In spite of the increase in apparent resistance due to core loss and the decrease in apparent inductance due to the screening effect of eddy currents, the ratio $\omega L/R_a$ can be made larger with an iron core than with an air core throughout the frequency range up to near supersonic frequencies. To achieve this condition at the higher frequencies, the maximum flux density must be kept small, and thin laminations of high resistivity or powdered cores must be used to keep the core losses and the screening effect of the eddy currents small. At radio frequencies, air-core reactors are commonly used. In these high-frequency applications, a relatively small inductance L produces a large inductive reactance ωL, so that a large inductance usually is not needed.

2. IRON CORES IN TRANSFORMERS

Iron cores are almost universally used in communication transformers for the audio range of frequencies. In such a transformer the ratio between the internal voltage of the power source connected to its primary and the voltage supplied to the load connected to its secondary should be nearly constant throughout the frequency range over which the transformer is operated. As is shown in Ch. XVIII, if this ideal is to be approached over a wide frequency range, the coefficient of coupling between the windings should be extremely close to unity, and the primary inductance should be large for satisfactory performance at low frequencies. The desired results can be obtained with a ferromagnetic core. An unfavorable consequence of the use of an iron core is the distortion due to the nonlinear magnetic properties of the core, which is discussed in Art. 5.

In power-system transformers, iron cores are also universally used; the high self-inductance reduces the no-load or exciting current to a reasonably small value, and the high coefficient of coupling results in small voltage regulation under load. An iron core, however, introduces core losses which have an important effect on the efficiency and temperature rise of the transformer. The nonlinear magnetization characteristics of iron also cause the waveform of the exciting current to be nonsinusoidal even when the flux varies sinusoidally, as is shown in Art. 8. In some circumstances, the harmonics thus introduced in the exciting current may be troublesome.

When only a single winding of an iron-core transformer is energized, the transformer behaves as an iron-core reactor. The discussion of this chapter, therefore, applies not only to problems concerning iron-core reactors but also to the important problem of determining the behavior of an iron-core transformer at no load.

3. Assumptions

The assumptions described below are fundamental to all discussion of iron-core reactors and transformers in this text.

3a. *Capacitance.* — Except when specifically noted to the contrary, the effects of the distributed capacitances of the windings are neglected. They may markedly affect the characteristics at high frequencies and during rapidly changing transient conditions, but their effects usually are negligible at power-system frequencies and at sufficiently low audio frequencies.

3b. *Resistance.* — When a conductor is carrying alternating current, its effective resistance may be appreciably greater than when it is carrying steady direct current; that is, the heat loss per ampere may be appreciably greater with alternating current. The increase in the loss is due to the nonuniform current density caused by the varying magnetic field produced within the conductor by its own current and by the currents in neighboring conductors. These phenomena are often called *skin effect* and *proximity effect.* The extra loss due to them increases with the frequency of the current and with the size of the conductor, and is reduced when large conductors are stranded and the strands are suitably transposed.

When the load current delivered by a transformer is increased, the eddy-current and hysteresis losses in the core and in portions of the structure near the windings usually increase, even when the principal core flux does not change. These extra losses — commonly called the *stray load losses* — are caused by the increase in leakage fluxes which occurs when the load is increased. They depend more nearly on the currents in the windings than on the magnitude of the principal core flux. Since the stray load losses vary approximately as the squares of the currents in the windings, they are approximately taken into account if the windings are assumed to have effective alternating-current resistances greater than the effective resistances due to skin effect and proximity effect alone. In a properly designed transformer the stray load losses are small, and the effective resistances of the windings are often only slightly greater than their direct-current resistances. In the following theory, unless specifically noted to the contrary, the effective resistances are assumed to be constant, independent of the current and of the frequency.

4. RELATIONS BETWEEN INDUCED VOLTAGE, FLUX, AND CURRENT

The relation between the instantaneous value v of the voltage at the terminals of a coil and the instantaneous current i in the coil is given by the equation

$$v = Ri + \frac{d\lambda}{dt}, \qquad \blacktriangleright[1]$$

where R is the resistance of the winding and λ is the instantaneous flux linkage. Any consistent system of units may be used. This equation should be clearly understood, since it is basic to the theory of all electromagnetic apparatus. The following brief review is a reminder of the significance of positive values of current, flux, and voltage in Eq. 1. In many transformer problems, particularly those involving interconnection between two or more windings, these positive directions must be kept clearly in mind.

If the current flowing in a coil appears clockwise to an observer looking along the axis of the coil, then by experiment the direction of the flux produced by the current is away from the observer. That is, the direction of positive current is that of rotation of a right-hand screw which is progressing in the direction of positive flux, as shown in Fig. 1. When the current is alternating, the arrows show merely the directions of *positive values* of the current and flux.

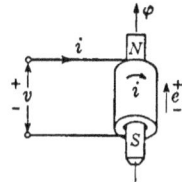

FIG. 1.   Positive directions of current, flux, and voltages.

While the flux linking a winding is changing, an electromotive force equal to the time rate of change of the flux linkage is generated in the winding and tends to send current through the winding in a direction to oppose the change in flux. Therefore, if the flux is positive and increasing, the electromotive force e induced in the coil is in the direction indicated in Fig. 1, since a current produced by this electromotive force would tend to prevent the flux from increasing. If the flux is positive and increasing, the derivative of the flux linkage is positive. Hence the electromotive force induced in the coil in the direction shown in Fig. 1 is

$$e = +\frac{d\lambda}{dt}.\qquad\qquad\blacktriangleright[2]$$

When the flux is alternating, the direction merely shows the significance of positive values of e. Note that e is an electromotive force or rise in potential in the left-hand-screw direction about positive flux.

If the resistance of the winding is R, the component voltage drop due to the resistance is an instantaneous fall in potential Ri in the direction of the current. Hence if i is the instantaneous value of the current in the positive or right-hand-screw direction about positive flux, and v is the instantaneous fall in terminal potential in this same direction, as shown by the polarity markings in Fig. 1, then

$$v = Ri + e = Ri + \frac{d\lambda}{dt},\qquad\qquad\blacktriangleright[3]$$

as in Eq. 1.

If a direct voltage V_{dc} is applied to an iron-core reactor, the steady-state value I_{dc} of the current is not affected by the dimensions or quality of the magnetic core, but is determined entirely by the direct-current resistance of the winding and is

$$I_{dc} = \frac{V_{dc}}{R}.\qquad\qquad[4]$$

The dimensions and quality of the core and the number of turns in the winding, however, determine the value of the core flux.

5. Alternating Applied Voltage

Quite a different situation arises when an alternating voltage is impressed on a series circuit containing an iron-core reactor or transformer.

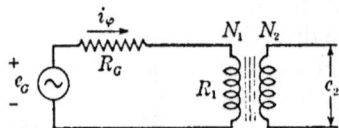

Fig. 2. Transformer connected to a generator whose internal resistance is R_G.

Consider the simple series circuit of Fig. 2, which shows a generator of internal resistance R_G and instantaneous internal electromotive force e_G connected to the primary terminals of an iron-core transformer. The equivalent of this arrangement is common in communication circuits, so that the behavior of this circuit gives useful information about the general nature of some of the effects produced in communication systems by the use of ferromagnetic materials in transformer and reactor cores. Assume that the transformer secondary circuit is open. The voltage equation for the circuit of Fig. 2 is

$$e_G = (R_G + R_1)i_\varphi + \frac{d\lambda_1}{dt}, \qquad [5]$$

where

R_1 is the effective resistance of the primary winding,
i_φ is the instantaneous current,
λ_1 is the instantaneous primary flux linkage.

In an iron-core transformer at no load, the flux linkage produced by the core flux usually is greater than 99 per cent of the total flux linkage. If all the flux φ is assumed to link all N_1 primary turns, as it would if the flux were entirely confined to the core, the primary flux linkage λ_1 equals $N_1\varphi$, and the counter electromotive force e_1 induced in the primary is

$$e_1 = N_1 \frac{d\varphi}{dt}. \qquad [6]$$

Then Eq. 5 can be written

$$e_G = (R_G + R_1)i_\varphi + N_1 \frac{d\varphi}{dt}. \qquad \blacktriangleright[7]$$

The core flux also links all N_2 turns of the secondary. Since nearly all the flux is confined to the core, the voltage e_2 induced in the open-circuited secondary is, very nearly,

$$e_2 = N_2 \frac{d\varphi}{dt}. \qquad [8]$$

In communication circuits the waveform of the generator voltage e_G is often directly related to the intelligence that is being transmitted, and the output voltage e_2 therefore should have as nearly as possible the same waveform as the generator voltage e_G. In general, the signal voltage e_G is a complicated function of time; much can be learned, however, concerning the behavior of the circuit by study of its performance when the generator voltage e_G is a sine wave. Accordingly, in the following discussion the generator voltage e_G is assumed to be a known sine wave.

The flux φ is produced by the magnetomotive force of the current i_φ, which is therefore called the *exciting current*. A flux-current characteristic is shown in Fig. 3. The abscissas are the instantaneous values of the exciting current, and the ordinates are the corresponding instantaneous values of the core flux. The loop shown in Fig. 3 is symmetrical about the origin. In communication circuits, direct current is often present in the winding, in addition to the alternating exciting current produced by the alternating signal voltage; and in these circumstances the loop is displaced from the origin.*

The flux density in the iron is proportional to the core flux, and the magnetizing force is approximately proportional to the exciting current. Hence the flux-current loop has approximately the same shape as the hysteresis loop for the core material at the corresponding maximum flux density. This conclusion neglects the magnetic effect of the eddy currents, the uncertainties introduced by the corners of the core, and the magneto-motive force required by air gaps which may purposely be inserted in the core or which may unavoidably be present owing to the joints between laminations. The magnetic effect of the eddy currents generally is relatively small at frequencies of about 60 cycles per second but grows larger with increasing frequency. To simplify the presentation of fundamental principles, eddy currents are neglected here. The flux-current characteristic with alternating current then equals the static characteristic measured with a ballistic galvanometer.[1]

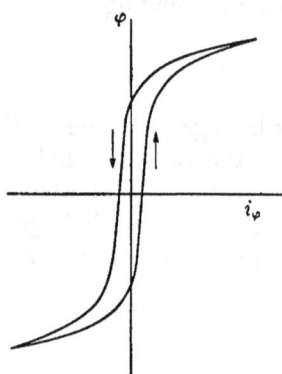

Fig. 3. Flux-current loop.

According to Eq. 7, the exciting current must adjust itself so that the sum of the instantaneous resistance drop $(R_G + R_1)i_\varphi$ plus the instantaneous counter electromotive force e_1 due to the flux set up by the exciting current equals the instantaneous value of the generator voltage e_G. When both the resistance drop and the counter electromotive force are important, the solution of Eq. 7 is indeed difficult, for several reasons. In the first place, the hysteresis loop relating the exciting current and the flux is nonlinear and, in fact, is not even single valued. Since the hysteresis loop cannot be expressed in analytical form, any attempt at a theoretical solution must be based on graphical methods. A further difficulty arises from the fact that the shape of the hysteresis loop depends on the amplitude of the loop. Hence it is not possible even to determine on which of a family of hysteresis loops the core is operating until the peak value of either the current or the flux is known. If a family of experimentally determined flux-current loops is available, Eq. 7 can be solved for the instantaneous current by step-by-step methods or by mechanical means

---

* The effects of superposed alternating- and direct-current excitations are discussed in Arts. 14 and 15.

[1] See T. Spooner, *Properties and Testing of Magnetic Materials* (New York: McGraw-Hill Book Company, Inc., 1927), 220.

such as the differential analyzer.[2] However, such methods of solution are laborious, and, whenever possible, direct experiment — either on the actual circuit or on a suitable model — is preferable.

Figure 4 shows oscillograms of the exciting current i_φ and of the voltage e_2 induced in the open-circuited secondary of an iron-core transformer excited through a series resistance from a source of essentially sinusoidal

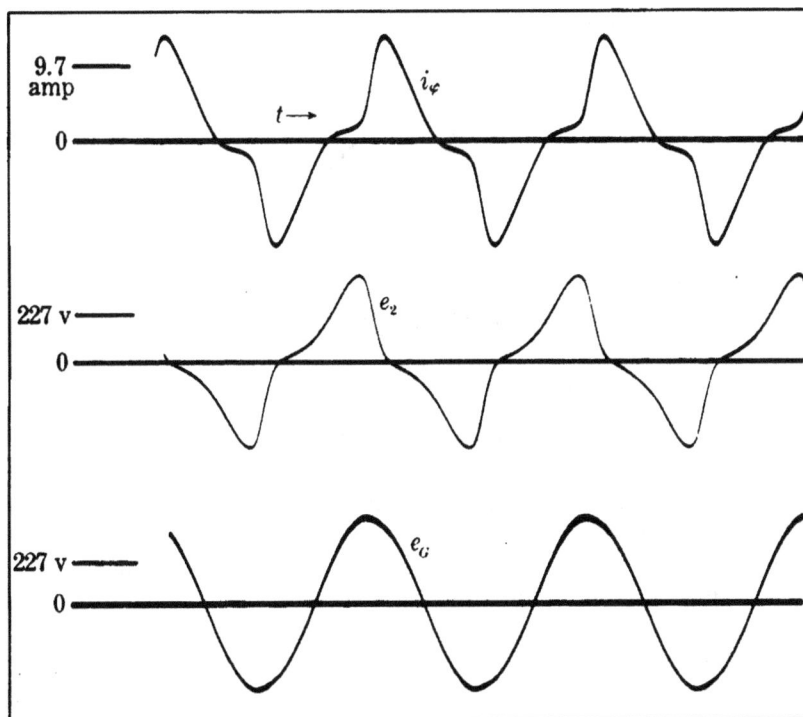

Fig. 4.   Oscillograms obtained from the circuit of Fig. 2.

voltage e_G, as in the circuit of Fig. 2. Because of the magnetic nonlinearity of the core, the exciting current is nonsinusoidal even when the generator voltage e_G is a sine wave. Hence the resistance drop $(R_G + R_1)i_\varphi$ is non-sinusoidal and has the same waveform as the exciting current. This resistance drop, calculated from the oscillogram of i_φ and the measured value of $(R_G + R_1)$, is shown in Fig. 5, which also shows the secondary induced voltage e_2 obtained from the oscillogram and plotted to the same scale as the resistance drop. Since this transformer has the same number of primary and secondary turns, and since magnetic leakage is

[2] The differential analyzer is described in the volume on electric circuits (1940), Ch. XIII, Art. 10.

negligible at no load, the primary induced voltage e_1 very nearly equals the secondary induced voltage e_2. The waveform of the induced voltage is markedly nonsinusoidal. However, the sum of the nonsinusoidal resistance drop $(R_G + R_1)i_\varphi$ and the nonsinusoidal induced voltage e_1 closely approximates a sinusoid, as shown by the broken line in Fig. 5. This result must occur, since, according to Eq. 7, the sum of these quantities equals the generator voltage e_G, which is essentially sinusoidal. That is, the waveforms of the induced voltage and of the exciting current must adjust themselves to produce harmonics in the induced voltage

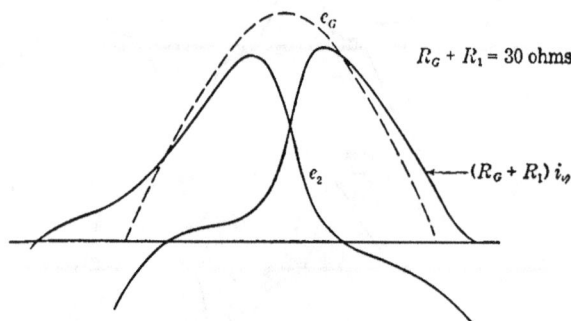

FIG. 5.   Curves obtained from the oscillograms of Fig. 4.

which are equal and opposite to the harmonics caused by the nonsinusoidal resistance drop $(R_G + R_1)i_\varphi$ due to the exciting current. Thus when the circuit resistance is large, the waveform of the secondary voltage e_2 may differ considerably from the waveform of the generator voltage e_G. In communication circuits this distortion may be serious, particularly for strong signals at low frequencies, for which condition the exciting current may be large. This distortion also occurs if, when a power-system transformer is tested at no load, the applied voltage is regulated by means of an adjustable series resistance.

Although it is difficult to calculate the performance of the circuit of Fig. 2 when both the resistance drop and the counter electromotive force in the transformer are important, the two limiting conditions for which either one of these component voltages is negligible can be calculated by relatively simple approximate methods. These solutions are given below.

6. SINUSOIDAL EXCITING CURRENT

When the counter electromotive force induced in the transformer is negligible compared with the resistance drop, an exciting current whose waveform is nearly sinusoidal results from a sinusoidal generator voltage e_G in the circuit of Fig. 2. Thus, if the effective or rms value of the sinusoidal generator voltage is E_G, the effective value of the current is approximately

$$I_\varphi = \frac{E_G}{R_G + R_1}. \tag{9}$$

If the exciting current is assumed to be sinusoidal in waveform, and of known amplitude $\sqrt{2}I_\varphi$, the approximate waveforms of the flux and of the voltage induced in the transformer can be determined from the appropriate flux-current loop by a simple graphical method, as follows.[3]

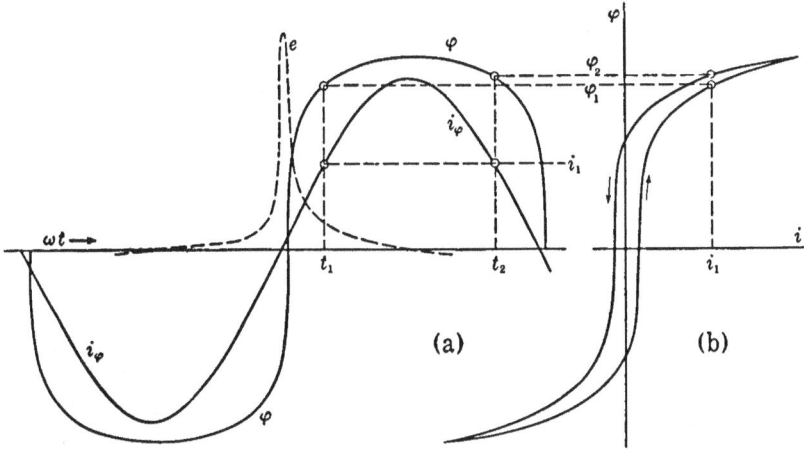

Fig. 6. Graphical construction for the determination of the waveforms of flux and induced voltage when the exciting current is a known sinusoidal function of time.

Figure 6a shows a sinusoidal curve of exciting current i_φ, and Fig. 6b shows the corresponding symmetrical loop relating the flux and current, determined by means of a ballistic galvanometer. If the magnetic effect of the eddy currents is neglected, this loop is also the flux-current characteristic with alternating current. At time t_1, Fig. 6a, the exciting current is increasing and has an instantaneous value i_1. From Fig. 6b, the corresponding instantaneous value of the flux is φ_1. At time t_2, Fig. 6a, the exciting current also has a value i_1 but is decreasing, and the corresponding instantaneous value of the flux in Fig. 6b is φ_2. These and other instantaneous values of the flux, determined in like manner and plotted with respect to time, give the waveform of the flux φ shown in Fig. 6a. When the hysteresis loop is symmetrical, as in Fig. 6b, the waveforms of the positive and negative half-cycles are identical. The flux

[3] For discussion of an analytical solution, see E. Peterson, "Harmonic Production in Ferromagnetic Materials at Low Frequencies and Low Flux Densities," *B.S.T.J.*, **7** (1928), 762–796.

variation, therefore, contains only odd harmonics.* If the hysteresis loop were unsymmetrical, as it would be if direct current were present in the winding, the flux variation would contain even harmonics also. Note that the flux lags the exciting current, because of hysteresis. The flux wave, Fig. 6a, rises rapidly, shows a relatively flat top which is slightly lopsided, and then falls rapidly.

Since the induced voltage is proportional to the slope $d\varphi/dt$ of the flux wave, the induced voltage wave can be obtained through graphically determining this slope, and is shown by the broken line in Fig. 6a. This voltage rises to a sharp peak corresponding to the steep portion of the flux wave. Generally, a flux wave which has a flatter top than a sine wave gives rise to an induced-voltage wave which is more sharply peaked than a sinusoid. Also, the harmonics in the induced voltage are relatively greater than those in the flux. For example, if the flux variation is

$$\varphi = \phi_{1max} \sin \omega t + \phi_{3max} \sin 3\omega t, \qquad [10]$$

the induced voltage e is

$$e = N \frac{d\varphi}{dt} = N(\omega\phi_{1max} \cos \omega t + 3\omega\phi_{3max} \cos 3\omega t). \qquad [11]$$

Thus the third-harmonic component of the induced voltage is relatively three times as great as the third-harmonic component of the flux.

In communication circuits the harmonics in the induced voltage constitute distortion. Figure 6a, therefore, represents conditions that are avoided, rather than conditions that actually occur. The distortion is reduced if the transformer is designed to operate at maximum flux densities for which the hysteresis loop is not too badly distorted by magnetic nonlinearity, and if the inductance of the transformer is made sufficiently large compared with the internal resistance of the tube with which it is to be used. In power circuits the induced voltages are usually nearly sinusoidal in waveform, as in Art. 7 below. If, however, for any reason the induced voltage is nonsinusoidal, the waveform of the voltage may have an important effect on the core loss.

7. Voltage induced by a sinusoidal flux

As shown in Art. 5, calculating the performance of the circuit of Fig. 2 is difficult when both the resistance drop and the induced counter electromotive force in Eq. 7 must be taken into account. In Art. 6, the approximate solution is given for the limiting condition when the induced counter electromotive force is negligible compared with the resistance drop. A simple approximate solution can also be obtained, as follows,

* See Art. 9c.

for the limiting condition when the resistance drop is negligible compared with the induced voltage. This solution is of great importance in determining the design and performance of iron-core reactors and transformers.*

The impedances of the transmission lines and feeders in a power system are usually so small that the impedance drops due to the exciting currents of the connected transformers are but a fraction of a per cent of the generator voltage. Hence the problem of determining the no-load characteristics of a power-system transformer usually reduces to that illustrated in Fig. 7, where the voltage v applied to the terminals of the excited winding is practically independent of the exciting current i_φ. The voltage equation is

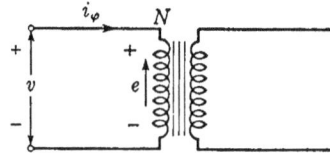

FIG. 7.   Transformer with one winding excited from a source of sinusoidal voltage.

$$v = Ri_\varphi + e, \tag{12}$$

where R is the effective resistance of the excited winding and e is the induced counter electromotive force.

Resistance of the usual transformer winding is so small that the resistance drop due to the exciting current is negligible compared with the terminal voltage. In a power-system transformer at no load, normal voltage, and normal frequency, the effective value of the resistance drop in the excited winding is usually less than 0.1 per cent of the effective value of the applied voltage. For such a condition it is sufficiently accurate to neglect the resistance drop and equate the terminal voltage v to the counter electromotive force e induced by the changing flux φ; thus,

$$v = e = N\frac{d\varphi}{dt}, \tag{13}$$

where N is the number of turns in the excited winding. The flux must adjust itself so that the instantaneous counter electromotive force generated by the changing flux very nearly equals the applied voltage, and the exciting current accordingly must adjust itself to produce this flux. If the waveform of the applied voltage is sinusoidal, the waveform of the flux also is very nearly sinusoidal. This condition is usually approxi-

---

* When a sinusoidal voltage is suddenly applied to an iron-core transformer, the instantaneous peak value of the transient flux density may be about twice its normal steady-state peak value; and because of the then saturated condition of the core, the corresponding peak value of the transient exciting current may be many times its normal steady-state value. For discussions of these exciting-current transients, see the volume on electric circuits (1940), 699–705; L. F. Blume, editor, *Transformer Engineering* (New York: John Wiley & Sons, 1938), 23–36. The following discussion is confined to steady-state conditions.

mated in power-system transformers. If

$$\varphi = \phi_{max} \sin \omega t,$$ [14]

by differentiation,

$$e = \omega N \phi_{max} \cos \omega t.$$ [15]

The induced voltage leads the flux by 90 degrees, as in Fig. 8a, in which *e* is the electromotive force or rise in potential in the left-hand-screw direction about positive flux, as explained in Art. 4. If the voltage and

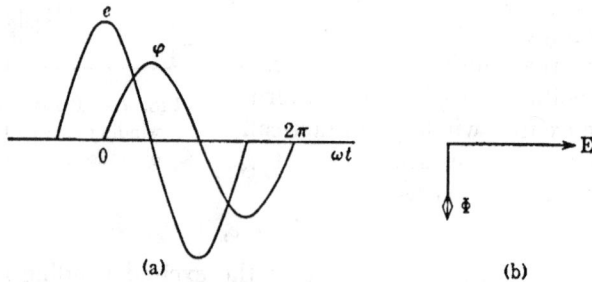

(a)   (b)

Fig. 8.   Phase relations of flux and induced voltage.

flux are represented by the vectors E and Φ, the vector diagram is in Fig. 8b.

If the frequency is *f*, the relation between the maximum values of the voltage and flux is

$$E_{max} = \omega N \phi_{max} = 2\pi f N \phi_{max}.$$ [16]

The effective value *E* of the generated voltage is

$$E = \frac{2\pi}{\sqrt{2}} f N \phi_{max}$$ [17]

or

$$E = 4.44 f N \phi_{max}.$$ ▶[18]

Any consistent system of units can be used in this equation. Thus if ϕ_{max} is expressed in webers, *E* is obtained in volts; or if ϕ_{max} is expressed in maxwells, *E* is obtained in abvolts. If ϕ_{max} is expressed in maxwells, the right-hand side of Eq. 18 must be multiplied by 10^{-8} to obtain *E* in volts.

▶ Equation 18 is one of the most important relations in the theory of all alternating-current electromagnetic apparatus, since it relates the design of the winding (*N* turns) to the magnetic loading of the core (ϕ_{max}) when the electric operating conditions (*E* and *f*) are specified. It is the general relation between the *effective or rms value* of the voltage generated by a sinusoidally varying flux and the *maximum value* of the flux.◀

In the many instances in which the resistance drop is negligible, the voltage generated by the flux very nearly equals the terminal voltage. Therefore, when a sinusoidal voltage is impressed at the terminals of a winding, the maximum value of the core flux is determined by the effective value and the frequency of the applied voltage, and the number of turns in the winding. That is, from Eq. 18,

$$\phi_{max} = \frac{E}{4.44fN}.$$ ▶[19]

When E is in volts, ϕ_{max} is in webers. If E is in volts and ϕ_{max} is desired in maxwells, multiply the right-hand side of Eq. 19 by 10^8.

▶ The value of the flux is *independent* of the dimensions and quality of the magnetic core as long as the resistance drop is negligibly small compared with the terminal voltage. The dimensions and quality of the core, however, determine the value of the exciting current required to establish the core flux.◀

8. WAVEFORM OF THE EXCITING CURRENT FOR SINE–WAVE FLUX

Because of the peculiar shape of the hysteresis loop of most magnetic core materials, the waveform of the exciting current is not sinusoidal when the flux varies sinusoidally. Figure 9 shows oscillograms of the sinusoidal voltage e induced in the open-circuited secondary, and of the nonsinusoidal exciting current i_φ in the primary, when the transformer of Fig. 10 is excited by a sinusoidal voltage. Since this transformer has the same number of primary and secondary turns, and since magnetic leakage is negligible at no load, the voltage e measured in the open-circuited secondary very nearly equals the voltage induced in the primary. This induced voltage very nearly equals the impressed voltage.

Although it is often unimportant, the waveform of the exciting current must be reckoned with in numerous problems. For example, distortion introduced by the exciting current is important in communication transformers. Harmonics in the exciting current of power-system transformers often have an important effect on inductive interference between adjacent power lines and communication circuits. Sometimes the characteristics of a polyphase bank of transformers are markedly affected by harmonics in the exciting current. Such harmonics may occasionally be

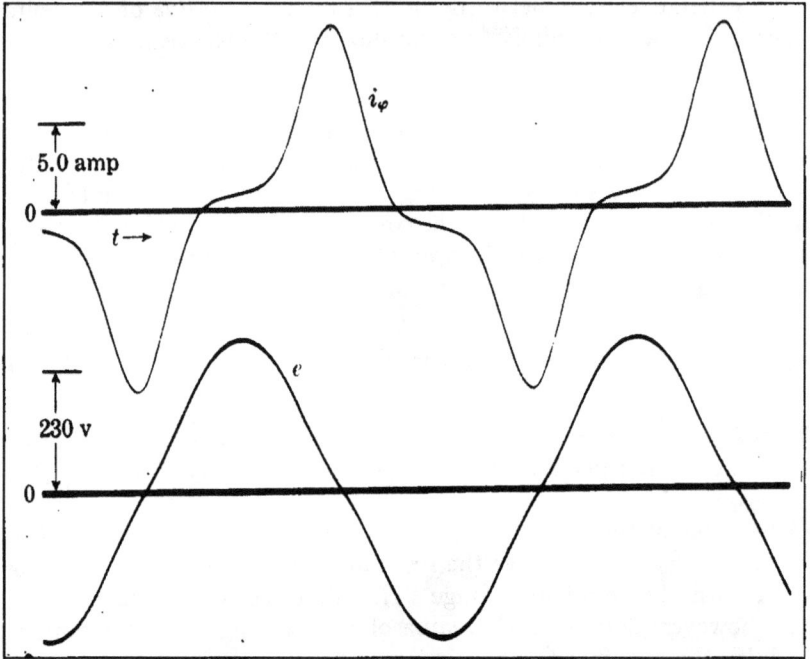

FIG. 9. Exciting-current oscillogram for the transformer of Fig. 10 with rms induced voltage of 200 volts at a frequency of 60 cycles per second.

greatly exaggerated by resonance with the capacitances of transmission lines or cables.

Figure 11a shows the sinusoidal curve of induced voltage e, replotted from the oscillogram of Fig. 9, and the corresponding sinusoidal flux

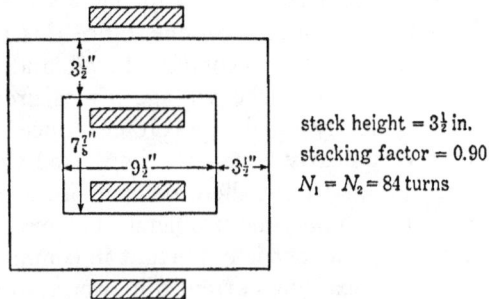

stack height = $3\frac{1}{2}$ in.
stacking factor = 0.90
$N_1 = N_2 = 84$ turns

FIG. 10. Core of a small experimental transformer.

wave φ, lagging the induced voltage by 90 degrees, as in Fig. 8. From Eq. 19 and the data given with Figs. 9 and 10, the amplitude of this flux wave is

$$\phi_{max} = \frac{200}{4.44 \times 60 \times 84}$$

$$= 0.00895 \text{ weber or } 895,000 \text{ maxwells.} \qquad [20]$$

From the core dimensions and the stacking factor of 0.90, the maximum flux density in the iron is

$$\mathfrak{B}_{max} = \frac{\phi_{max}}{\text{net area}} \qquad [21]$$

$$= \frac{895,000}{3.50 \times 3.50 \times 0.90} = 81,200 \text{ lines/sq in.} \qquad [22]$$

$$= \frac{81,200}{6.45} = 12,600 \text{ gausses, or } 1.26 \text{ webers/sq m.} \qquad [23]$$

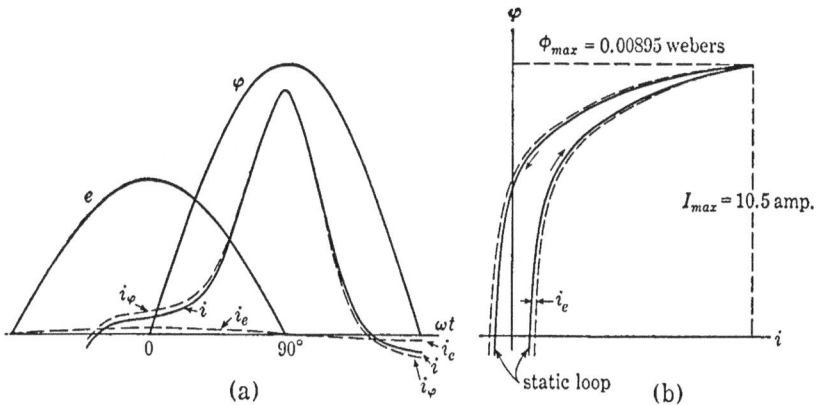

FIG. 11.   Graphical construction for determination of the exciting current of the transformer of Fig. 10.

This is approximately the flux density commonly used in 60-cycle power transformers. The solid-line curve labeled *static loop* in Fig. 11b is half of a symmetrical flux-current loop for this same maximum flux, determined by means of a ballistic galvanometer. From this loop and the known time variation of the flux, the corresponding time variation of the current i, shown in Fig. 11a, can be determined in a manner similar to that illustrated in Fig. 6. By symmetry, the waveform of the negative half-cycles is identical with that of the positive half-cycle shown. If the magnetic effect of the eddy currents is negligible, the current i is the exciting current.

The effect of the eddy currents can be approximately accounted for in the following manner. As in the analysis of Art. 2, Ch. V, the eddy currents are proportional to the time rate of change of the flux density and are therefore sinusoidal in waveform and instantaneously proportional to the electromotive force induced in the exciting winding. The direction of the eddy currents is such as to oppose any externally applied magnetomotive force tending to change the flux; that is, while the time derivative of the flux is positive, the eddy currents are circulating in the laminations in the left-hand-screw direction about positive flux. Hence a sinusoidal component of magnetomotive force must be supplied by the exciting winding in a positive or right-hand-screw direction, in order to offset the effect of the eddy currents and to permit the flux variation to remain unchanged at the value which induces the necessary counter electromotive force in the winding. This component of the exciting current is in phase with the counter electromotive force e, as shown by the broken-line sinusoidal wave i_e in Fig. 11a. The current i_e supplies the power absorbed by the eddy-current loss in the core. Hence the effective value I_e of the eddy-current component of the exciting current is

$$I_e = \frac{P_e}{E},$$ [24]

where E is the effective value of the counter electromotive force and P_e is the eddy-current loss. The eddy-current loss can be determined from the measured values of the core loss at various frequencies and constant flux density, as described in Art. 7, Ch. V. From data obtained in this manner the eddy-current loss in the transformer to which Figs. 9, 10, and 11 apply is 36 watts when the induced voltage is 200 volts and the frequency is 60 cycles per second, as in Fig. 11. Hence, in this example

$$I_e = \frac{36}{200} = 0.18 \text{ amp.}$$ [25]

The amplitude of the sinusoidal eddy-current component i_e of the exciting current in Fig. 11a is therefore $\sqrt{2}I_e$, or 0.25 ampere. The resulting exciting current i_φ is

$$i_\varphi = i + i_e$$ [26]

and is shown by the broken-line curve in Fig. 11a. In Fig. 12 is shown a comparison of the calculated exciting current and the oscillogram of Fig. 9. The agreement is within the experimental accuracy.

The effect of the eddy currents on the flux-current loop is interesting. Since the eddy currents in the core oppose the change in flux, the component i_e in the exciting current which offsets their effect is in a direction to aid the change in flux. Hence, if this offsetting component, proportional to $d\varphi/dt$, is added to the abscissas i of the static loop in Fig. 11b, a broader dynamic loop is obtained, as shown by the broken line in Fig. 11b. The area of the static loop is proportional to the hysteresis loss per cycle, but the area of the dynamic loop is proportional to the total core loss per cycle.

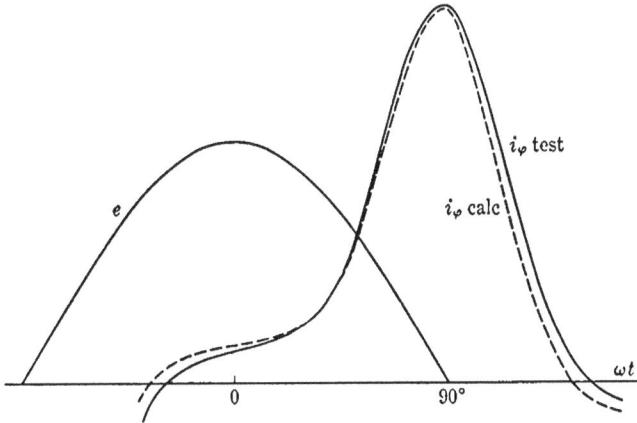

FIG. 12.   Comparison of calculated and measured exciting-current waves.

9. GENERAL PROPERTIES OF FOURIER SERIES

Since the steady-state exciting current is a periodic function and can therefore be resolved into a Fourier series, a few matters of general interest in the Fourier analysis of nonsinusoidal currents and voltages are discussed here. Detailed treatment of the Fourier methods of analysis is given elsewhere in this series.[4]

Any single-valued, periodic function can be expressed in a Fourier series comprising a constant term and an infinite number of sine and cosine terms whose frequencies are integer multiples of the fundamental frequency. Hence, if $f(t)$ is a periodic function of time,

$$f(t) = \frac{a_0}{2} + a_1 \cos \omega t + a_2 \cos 2\omega t + \cdots$$
$$+ b_1 \sin \omega t + b_2 \sin 2\omega t + \cdots, \qquad \blacktriangleright [27]$$

where ω is the angular frequency of the function. The a's and b's are constant coefficients and are given by the definite integrals

$$a_n = \frac{\omega}{\pi} \int_{-\pi/\omega}^{\pi/\omega} f(t) \cos n\omega t \, dt \qquad [28]$$

$$= \textit{twice} \text{ the average value of } f(t) \cos n\omega t \qquad [29]$$

---

[4] See the reference volume, *The Mathematics of Circuit Analysis*, Ch. VII.

$$b_n = \frac{\omega}{\pi} \int_{-\pi/\omega}^{\pi/\omega} f(t) \sin n\omega t \, dt \tag{30}$$

$$= \textit{twice} \text{ the average value of } f(t) \sin n\omega t. \tag{31}$$

In these equations, n indicates the order of the harmonic component, and the averages are taken over a complete period. The constant a_0 is the value of a_n for n equals zero; that is,

$$a_0 = \frac{\omega}{\pi} \int_{-\pi/\omega}^{\pi/\omega} f(t) \, dt, \tag{32}$$

and therefore the constant term $a_0/2$ in Eq. 27 is

$$\frac{a_0}{2} = \text{the average value of } f(t). \tag{33}$$

Alternative forms of the Fourier series can be obtained through combining sine and cosine terms of the same frequency; thus

$$f(t) = \frac{a_0}{2} + c_1 \cos (\omega t + \alpha_1) + c_2 \cos (2\omega t + \alpha_2) + \cdots \quad \blacktriangleright[34]$$

where

$$c_n = \sqrt{a_n^2 + b_n^2} \tag{35}$$

$$\alpha_n = -\tan^{-1}\left(\frac{b_n}{a_n}\right). \tag{36}$$

The Fourier series for the time functions with which this text is concerned usually converge fairly rapidly, so that relatively few terms are required to represent the function with engineering accuracy.

If the function $f(t)$ is given in analytical form, the coefficients of the Fourier series can be determined from Eqs. 28 to 33. Many methods have been developed for evaluating the coefficients when the function $f(t)$ is given in graphical form[5] — for example, as an oscillogram such as Fig. 9. Although a quantitative analysis requires determination of the numerical values of the coefficients, a relatively simple investigation of the waveform of $f(t)$ reveals a number of useful facts regarding the general character of the Fourier series which represents the time function. Some of these general matters are discussed below.

9a. *The Constant Component.* — The constant term $a_0/2$ in Eqs. 27 and 34 is the average value of $f(t)$, and, if $f(t)$ represents a current or voltage, $a_0/2$ is its direct component. A direct component of current may be produced by a direct electromotive force, or by rectifying action in any

[5] A discussion of graphical methods is given in the reference volume, *The Mathematics of Circuit Analysis*, Ch. VII.

circuit whose volt-ampere characteristic for one direction of current is not the negative of its characteristic for the opposite direction.

▶ In electric-circuit problems, examining the circuit for possible causes of a direct-current component usually suffices, therefore, to determine the presence of any constant component in $f(t)$. If no direct-current component is present, the areas of the positive and negative portions of each cycle are equal.◀

9b. *Even and Odd Functions.** — The cosine functions are even functions of their arguments; that is,

$$\cos (-x) = \cos x. \tag{39}$$

The sine functions, however, are odd; that is,

$$\sin (-x) = -\sin x. \tag{40}$$

The cosine part of the series, Eq. 27, hence is an even function, and the sine part odd.

▶ Therefore, if the function $f(t)$ is an odd function, its Fourier series, Eq. 27, contains only the sine components; but if $f(t)$ is an even function, the sine components are not present. If $f(t)$ is neither even nor odd, all components may be present in Eq. 27.◀

For example, consider the triangular wave shown in Fig. 13. If the origin is taken at 0, this triangular wave is an odd function, and its Fourier series then contains only sine terms. On the other hand, if 0′ is the origin, the wave represents an even function, and only cosine terms are present in

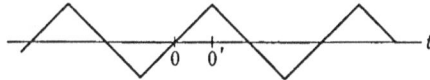

FIG. 13. Saw-toothed wave.

its Fourier-series expansion. However, the origin in Fig. 9 cannot be chosen so as to make the exciting current either an odd or an even

---

* An *even* function is defined as one for which the values of the function for negative values of the argument are the same as for equal positive values of the argument. An *odd* function is one for which the values of the function for negative values of the argument are the *negatives* of the values for equal positive values of the argument. Thus, if the function $f(t)$ has the property that

$$f(-t) = f(t), \tag{37}$$

it is said to be an *even* function. But if

$$f(-t) = -f(t), \tag{38}$$

the function is *odd*.

The reader should note the distinction between the meanings of the terms " even " and " odd " as used to describe functions having the forms of symmetry, Eqs. 37 and 38, with respect to the origin, and the same terms as applied to the order of the harmonic components. For example, the third-harmonic cosine term $a_3 \cos 3\omega t$ in Eq. 27 is an *even function* but an *odd-harmonic* component of $f(t)$.

function, and therefore the Fourier series, Eq. 27, for the exciting current must have both sine and cosine terms.

9c. *Even and Odd Harmonics.* — Frequently the periodic function $f(t)$ satisfies the condition

$$f\left(t \pm \frac{T}{2}\right) = -f(t), \qquad [41]$$

where T is the *period* of the function and is the time interval between successive equal values of the function. Equation 41 states that the sequence of values of the function throughout any half-period is the negative of the values encountered throughout the preceding or succeeding half-period; that is, the waveforms of positive and negative half-cycles are identical except for sign.

▶ The Fourier series for a function of this kind contains only odd-harmonic components — a fact which can be proved through showing that the odd harmonics satisfy Eq. 41, but that the even harmonics do not, and therefore cannot exist if the function itself satisfies Eq. 41.◀

Consider the nth-harmonic cosine term in Eq. 27. The value of this harmonic for $t \pm (T/2)$ is

$$a_n \cos\left[n\omega\left(t \pm \frac{T}{2}\right)\right] = a_n \cos\left(n\omega t \pm n\omega \frac{T}{2}\right). \qquad [42]$$

But the relation between the period T and the frequency f is

$$T = \frac{1}{f}, \qquad [43]$$

whence

$$n\omega \frac{T}{2} = \frac{n\omega}{2f} = \frac{n2\pi f}{2f} = n\pi. \qquad [44]$$

Therefore

$$a_n \cos\left[n\omega\left(t \pm \frac{T}{2}\right)\right] = a_n \cos\left(n\omega t \pm n\pi\right). \qquad [45]$$

If n is odd, the phase angle $n\pi$ equals an odd number of half-periods, and therefore

$$a_n \cos\left(n\omega t \pm n\pi\right) = -a_n \cos n\omega t, \qquad [46]$$

whence

$$a_n \cos\left[n\omega\left(t \pm \frac{T}{2}\right)\right] = -a_n \cos n\omega t. \qquad [47]$$

Similarly, for the nth-harmonic sine component, when n is odd,

$$b_n \sin\left[n\omega\left(t \pm \frac{T}{2}\right)\right] = -b_n \sin n\omega t. \qquad [48]$$

Thus when n is odd, the sine and cosine harmonic components satisfy Eq. 41, and therefore the odd-harmonic components can be present in the Fourier series for a function which satisfies this equation.

If n is even, however, the phase angle $n\pi$ in Eq. 45 equals an integer number of whole periods, and therefore

$$a_n \cos (n\omega t \pm n\pi) = +a_n \cos n\omega t, \qquad [49]$$

or

$$a_n \cos \left[n\omega \left(t \pm \frac{T}{2} \right) \right] = +a_n \cos n\omega t. \qquad [50]$$

Similarly, when n is even,

$$b_n \sin \left[n\omega \left(t \pm \frac{T}{2} \right) \right] = +b_n \sin n\omega t. \qquad [51]$$

Hence the even-harmonic components cannot be present in the Fourier series for a function that satisfies Eq. 41, since they do not satisfy this equation.

The oscillograms of Figs. 4 and 9 show examples of waves whose positive and negative half-cycles are identical except for sign. Consequently these functions have the property expressed in Eq. 41. The Fourier series for these currents and voltages contain only odd-harmonic components. Because the waves are neither even nor odd functions, both sine and cosine terms are present.

9d. *Power.* — The expression for the instantaneous power absorbed by a circuit element through which an instantaneous fall in potential $v(t)$ occurs in the direction of the current $i(t)$ is

$$p(t) = v(t)i(t). \qquad \blacktriangleright[52]$$

Over any time interval Δt, the average power P_{av} is

$$P_{av} = \frac{1}{\Delta t} \int_0^{\Delta t} v(t)i(t)\, dt. \qquad \blacktriangleright[53]$$

These equations are entirely general expressions and can be evaluated explicitly when the time functions $v(t)$ and $i(t)$ are known. For steady-state conditions, the instantaneous power $p(t)$ is a periodic function, and the average value in Eq. 53 may be taken over any integer number of periods of the instantaneous power.

When both the current and the voltage are sinusoidal functions of time and have the same frequency, the instantaneous power is the sum of a constant component and a sinusoidal component whose frequency is twice that of the current and voltage.[6] The double-frequency term rep-

---

[6] See the volume on electric circuits (1940), Ch. IV, Arts. 9 and 10, pp. 275–277.

resents an oscillation of power, but the average power due to this component is zero. The constant component is the average power, and its value is

$$P_{av} = VI \cos \theta, \qquad \blacktriangleright[54]$$

where V and I are the effective values of the sinusoidally varying voltage and current and θ is the phase angle between them.

Frequently the power must be determined when the current and voltage are nonsinusoidal functions of time. The following discussion assumes that the current and voltage are both periodic functions of the same period. They can then be expressed by Fourier series. For the following analysis the form of the Fourier series given in Eq. 34 is convenient; thus

$$v(t) = V_{dc} + \sqrt{2}V_1 \cos (\omega t + \psi_1) + \sqrt{2}V_2 \cos (2\omega t + \psi_2) + \cdots [55]$$

$$i(t) = I_{dc} + \sqrt{2}I_1 \cos (\omega t + \alpha_1) + \sqrt{2}I_2 \cos (2\omega t + \alpha_2) + \cdots, [56]$$

where

$$V_{dc}, I_{dc} \text{ are the direct components,}$$

$$V_1, V_2, \cdots I_1, I_2 \cdots \text{ are the } \textit{effective values} \text{ of the}$$
$$\text{harmonic components.}$$

According to Eq. 35, the effective value of each harmonic component equals the square root of the sum of the squares of its sine and cosine components. These equations can be expressed in compact form as follows:

$$v(t) = V_{dc} + \sum_k \sqrt{2}V_k \cos (k\omega t + \psi_k) \qquad [57]$$

$$i(t) = I_{dc} + \sum_n \sqrt{2}I_n \cos (n\omega t + \alpha_n). \qquad [58]$$

The instantaneous power is

$$p(t) = v(t)i(t)$$

$$= V_{dc}I_{dc} + V_{dc}\sum_n \sqrt{2}I_n \cos (n\omega t + \alpha_n)$$

$$+ I_{dc}\sum_k \sqrt{2}V_k \cos (k\omega t + \psi_k)$$

$$+ [\sum_k \sqrt{2}V_k \cos (k\omega t + \psi_k)][\sum_n \sqrt{2}I_n \cos (n\omega t + \alpha_n)]. \qquad [59]$$

When the last term on the right-hand side of Eq. 59 is expanded, a typical term is

$$[\sqrt{2}V_k \cos (k\omega t + \psi_k)][\sqrt{2}I_n \cos (n\omega t + \alpha_n)]$$

$$= 2V_kI_n \cos (k\omega t + \psi_k) \cos (n\omega t + \alpha_n). \qquad [60]$$

By means of the trigonometric identity

$$\cos x \cos y = \tfrac{1}{2}[\cos (x + y) + \cos (x - y)], \qquad [61]$$

Eq. 60 can be converted to

$$[\sqrt{2}V_k \cos (k\omega t + \psi_k)][\sqrt{2}I_n \cos (n\omega t + \alpha_n)]$$
$$= V_k I_n \cos [(k + n)\omega t + \psi_k + \alpha_n]$$
$$+ V_k I_n \cos [(k - n)\omega t + \psi_k - \alpha_n]. \qquad [62]$$

Thus a typical term in the expansion of the last term on the right-hand side of Eq. 59 can be expressed as the sum of two cosine components whose frequencies are, respectively, the sum and difference of the frequencies of the harmonic components in the voltage and current. The last term on the right-hand side of Eq. 59 is the summation of all such terms.

▶ Therefore the instantaneous power $p(t)$ contains harmonic components whose frequencies are the sums and differences of the frequencies of all the harmonic components in the voltage and current. ◀

The average power P_{av} is the average value of the instantaneous power taken over a complete cycle. In Eq. 59, the first term is constant, and therefore its average value is $V_{dc}I_{dc}$. The second and third terms represent harmonic components whose frequencies are integer multiples of the fundamental frequency. These terms represent harmonic oscillations in the power caused by the interaction of direct and alternating components, but the average value of each one of these components for an integer number of cycles is zero. Equation 62 shows that when k is not equal to n, the power resulting from the product of a harmonic voltage of one frequency and a harmonic current of a different frequency — as in the last term of Eq. 59 — is the sum of two alternating components, and their average values therefore are zero. When k equals n, however, the last term in Eq. 62 becomes

$$V_n I_n \cos (\psi_n - \alpha_n) = V_n I_n \cos \theta_n, \qquad ▶[63]$$

where $\theta_n = \psi_n - \alpha_n$, and is the phase angle between the nth-harmonic components of voltage and current. The average value of the last term in Eq. 59 is therefore zero for harmonic components of voltage and current of different frequencies, but is

$$\sum_n V_n I_n \cos \theta_n$$

for all the harmonic components of voltage and current of the same frequency. Consequently the average power is

$$P_{av} = V_{dc}I_{dc} + \sum_n V_n I_n \cos \theta_n. \qquad ▶[64]$$

▶ The average power thus equals the sum of the average powers due to components of voltage and current of the *same frequency*. The average power due to the direct components is $V_{dc}I_{dc}$, and that due to each harmonic is $V_n I_n \cos \theta_n$, just as for sine-wave voltages and currents, Eq. 54. A voltage of one frequency and a current of a different frequency produce alternating components of *instantaneous* power, but the *average* power resulting from them is zero.◀

9e. *Effective Values.*[7] — By definition, the effective value of a periodically varying current $i(t)$ is the square root of the average value of $[i(t)]^2$. Because of the way in which it is computed, the effective value is often called the root-mean-square (rms) value. Expressed as an equation, the effective value I is

$$I = \sqrt{\text{average}[i(t)]^2} = \sqrt{\frac{1}{T} \int_0^T [i(t)]^2 \, dt}, \qquad \blacktriangleright [65]$$

where T is the period of $i(t)$. Similarly, the effective value V of a periodically varying voltage $v(t)$ is

$$V = \sqrt{\text{average}[v(t)]^2} = \sqrt{\frac{1}{T} \int_0^T [v(t)]^2 \, dt}. \qquad \blacktriangleright [66]$$

▶ Note that the average of the square of the current is entirely different from the square of the average current.◀

Equations 65 and 66 are general expressions for the effective values. When the current or voltage is a sinusoidal function of time, these equations give the well-known result that the effective value is $1/\sqrt{2}$ of the amplitude.

It is frequently necessary to determine the effective value of a nonsinusoidal current or voltage which is expressed as a Fourier series of harmonic components. For example, let

$$i(t) = I_{dc} + \sum_n \sqrt{2} I_n \cos (n\omega t + \alpha_n) \qquad [67]$$

where I_{dc} is the direct-current component and I_n is the *effective value* of the nth-harmonic component. The square of $i(t)$ is

$$[i(t)]^2 = I_{dc}^2 + 2I_{dc}\sum_n \sqrt{2} I_n \cos (n\omega t + \alpha_n)$$
$$+ [\sum_n \sqrt{2} I_n \cos (n\omega t + \alpha_n)]^2. \qquad [68]$$

The average value of the first term in Eq. 68 is I_{dc}^2. The second term represents a number of alternating components whose average values

[7] Effective values and their measurement are also discussed in this series in the volume on electric circuits (1940), Ch. IV, Arts. 11 and 12, pp. 278–282.

are zero. The last term represents the cross products of all the harmonic terms. For example, a typical term in the expansion of the last term of Eq. 68 is the product of the kth and nth harmonics of the current, and, as in Eqs. 60 to 62, this term can be expressed as

$$[\sqrt{2}I_k \cos{(k\omega t + \alpha_k)}][\sqrt{2}I_n \cos{(n\omega t + \alpha_n)}]$$
$$= I_kI_n \cos{[(k + n)\omega t + \alpha_k + \alpha_n]}$$
$$+ I_kI_n \cos{[(k - n)\omega t + \alpha_k - \alpha_n]}. \qquad [69]$$

The last term in Eq. 68 is the summation of all such terms. Equation 69 shows that the cross products of harmonics of different frequencies (when k is not equal to n) produce alternating components whose average values are zero. When k equals n, however, the last term in Eq. 69 is I_n^2, and hence the average value of the last term in Eq. 69 is the sum of the squares of the effective values of all the harmonic components. Consequently the effective value I is

$$I = \sqrt{I_{dc}^2 + \sum_{n} I_n^2}. \qquad \blacktriangleright[70]$$

▶ The effective value of a periodic current or voltage which is expressed as a Fourier series therefore equals the square root of the sum of the squares of the effective values of all its harmonic components.◀

10. Fourier analysis of the exciting current for sine–wave flux

Because of the peculiar shape of the hysteresis loop, the exciting current for sine-wave flux is a lopsided peaked wave which can be determined from the appropriate flux-current loop for the core by means of the graphical construction shown in Fig. 11. This computed curve is indicated by the broken-line curve i_φ in Fig. 11, and Fig. 12 shows that the computed exciting current is in reasonably close agreement with the actual waveform determined from the oscillogram of Fig. 9. Because of its peculiar waveform, the exciting current can conveniently be expressed as a Fourier series of harmonic components.

The time variation of the exciting current consists of a fundamental component and a series of odd-harmonic components. No even harmonics are present when the hysteresis loop is symmetrical, since positive and negative half-cycles are identical except for sign.* Note that the exciting current is neither an even nor an odd function, and consequently when the exciting current is expressed as a Fourier series — as in Eq. 27 — both sine and cosine terms are present.† Thus if the origin is chosen, as

---

* See Art. 9c.
† See Art. 9b.

in Fig. 11a, so that the flux variation is a sine function

$$\varphi = \phi_{max} \sin \omega t, \qquad [71]$$

then the exciting current is

$$i_\varphi = \sqrt{2}[I_1' \sin \omega t + I_3' \sin 3\omega t + I_5' \sin 5\omega t + \cdots$$
$$+ I_1'' \cos \omega t + I_3'' \cos 3\omega t + I_5'' \cos 5\omega t + \cdots]. \qquad [72]$$

The $\sqrt{2}$ is introduced in Eq. 72 in order that the I's should be the effective or rms values of the component currents. When the time variation of the exciting current is given in graphical form, the numerical values of the harmonic components can be found by graphical methods.[8] The effective ampere values of the principal harmonic components of the calculated exciting current i_φ, Fig. 11, are given in the following table.

Sine Components	Cosine Components
$I_1' = +4.50$	$I_1'' = +0.87$
$I_3' = -2.10$	$I_3'' = -0.15$
$I_5' = +0.50$	$I_5'' = +0.05$
$I_7' = -0.10$	$I_7'' = \quad 0$

The fundamental current consists of a sine component

$$i_1' = \sqrt{2}I_1' \sin \omega t \qquad [73]$$

in phase with the flux, and a cosine component

$$i_1'' = \sqrt{2}I_1'' \cos \omega t \qquad [74]$$

in phase with the counter electromotive force $\sqrt{2}E \cos \omega t$. The average power P_c absorbed by the core is

$$P_c = EI_1'', \qquad [75]$$

since I_1'' corresponds in phase and frequency to the induced voltage E. Hereafter this component current I_1'' accounting for the total core loss P_c is designated by I_c and is called the *core-loss component* of the exciting current or, briefly, the *core-loss current*. That is,

$$I_c = I_1''. \qquad [76]$$

Hence, from Eq. 75,

$$I_c = \frac{P_c}{E}. \qquad \blacktriangleright[77]$$

[8] A discussion of graphical methods for the determination of the harmonic components is given in the reference volume, *The Mathematics of Circuit Analysis*, Ch. VII. A method especially suitable for the analysis of the exciting current is described in an article by J. Albert Wood, Jr., "A Graphical Method of Wave Analysis," *Cornell University Engineering Experiment Station*, Bulletin No. 22 (July, 1936).

The fundamental sine component I_1' contributes nothing to the average power absorbed by the core, since it is in quadrature with the induced voltage. None of the other components in the current contributes to the average power absorbed by the core, since their frequencies differ from the frequency of the induced voltage.*

Plotting a curve of i_φ^2 with respect to time, and determining the average value of i_φ^2 from this curve, will give I_φ^2, the square of the effective value I_φ of the exciting current i_φ. The effective value also can be expressed in terms of the harmonic components as

$$I_\varphi = \sqrt{(I_1')^2 + I_c^2 + (I_3')^2 + (I_3'')^2 + \cdots}. \qquad [78]\dagger$$

If harmonics above the seventh are neglected, the calculated effective value of the current i_φ in Fig. 11a is

$$I_\varphi = 5.07 \text{ amp.} \qquad [79]$$

The measured effective value of the exciting current is 5.35 amperes.

The waveform of the exciting current is sharply peaked, and hence the effective value of the current does not equal the maximum value divided by $\sqrt{2}$, but is considerably less than this value. For the current in Fig. 11a,

$$\frac{I_{\varphi(rms)}}{I_{\varphi(max)}} = \frac{5.07}{10.5} = 0.484. \qquad [80]$$

The resultant fundamental component $I_{\varphi 1}$ of the exciting current is

$$I_{\varphi 1} = \sqrt{I_c^2 + (I_1')^2} \qquad [81]$$

$$= 4.59 \text{ amp.} \qquad [82]$$

This is 90.5 per cent of the exciting current. The principal harmonic is the third. The resultant third-harmonic current is

$$I_{\varphi 3} = \sqrt{(I_3')^2 + (I_3'')^2} \qquad [83]$$

$$= 2.10 \text{ amp.} \qquad [84]$$

This is 41.5 per cent of the exciting current. The third harmonic is largely responsible for the sharp peak in the waveform of the current. Figure 14 shows the fundamental and third-harmonic components. Their sum, which approximates the actual exciting current of Fig. 11a, is shown by the broken line in Fig. 14. The most obvious discrepancy is the two points of inflection at about -15 degrees and $+15$ degrees in the resultant curve of Fig. 14, and is due to the omission of the higher harmonics, the principal one being the fifth-harmonic sine component i_5'. In general, if the maximum flux density is increased, the exciting

* See Art. 9d.
† See Art. 9e.

current becomes more sharply peaked, as is illustrated in Fig. 15, which shows the superposed results of four oscillograms of the exciting current of the transformer of Fig. 10. With increasing flux density, the per cent third harmonic increases, and the per cent fundamental decreases.[9] When the waveform of the exciting current produces serious distortion,

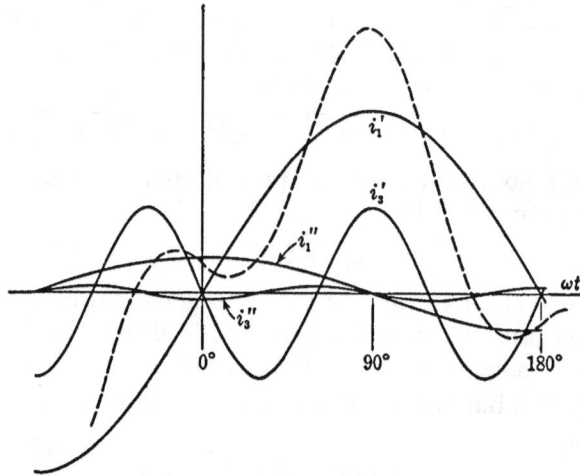

Fig. 14.   Fundamental and third-harmonic components of the exciting current.

as it may in communication circuits, this distortion is reduced if operation at high flux densities is avoided.

10a. *Effect of an Air Gap.* — The exciting current may be made more nearly sinusoidal by insertion of a small air gap in the core. For example, suppose an air gap of 0.100 inch — only about 0.2 per cent of the mean length of the magnetic circuit — is cut in the core of Fig. 10. If the impressed voltage is still 200 volts, as in the preceding numerical example, the maximum core flux must still be 0.00895 weber (Eq. 20) as long as the resistance drop in the winding is still negligible. Hence, except for fringing, the maximum flux density in the air gap is

$$\mathcal{B}_{max} = \frac{\phi_{max}}{\text{area}}$$
$$= \frac{0.00895}{3.5 \times 3.5 \times 6.45 \times 10^{-4}} = 1.13 \text{ webers/sq m.} \qquad [85]$$

[9] Data concerning the variation of the exciting-current harmonics with flux density are given in the following references:

L. F. Blume, editor, *Transformer Engineering* (New York: John Wiley & Sons, 1938), 21.

O. G. C. Dahl, *Electric Circuits, Theory and Applications.* Vol. I: *Short-Circuit Currents and Steady-State Theory* (New York: McGraw-Hill Book Company, Inc., 1928), 226.

J. W. Butler and E. B. Pope, " The Effect of Overexciting Transformers on System Voltage Wave Shapes and Power Factor," *A.I.E.E. Trans.*, *60* (1941), 49–53.

The maximum value of the magnetomotive force required by the gap is

$$F_{gap} = \frac{\mathfrak{B}_{max} \times \text{gap length}}{\mu_0}$$

$$= \frac{1.13 \times 0.100 \times 2.54 \times 10^{-2}}{10^{-7}} = 28{,}800 \text{ pragilberts.} \quad [86]$$

The maximum value of the component added to the exciting current by insertion of the gap is

$$\frac{F_{gap}}{4\pi N} = \frac{28{,}800}{4\pi \times 84} = 27.2 \text{ amp.} \quad [87]$$

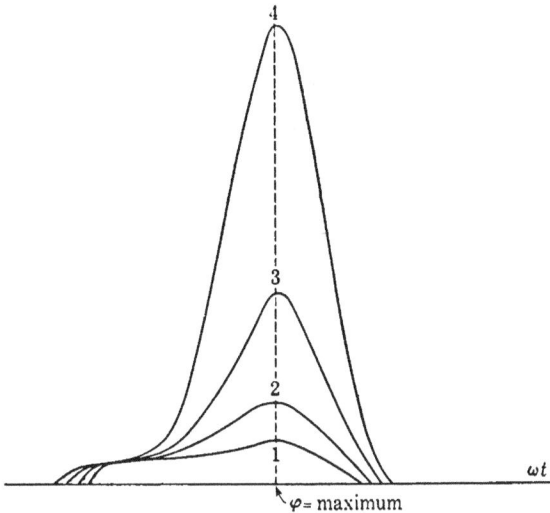

Fig. 15. Effects of maximum flux density on the waveform of the exciting current. The four curves are for the following approximate maximum flux densities (in gausses): (1) 7,500; (2) 10,000; (3) 12,500; (4) 15,000.

This component current is sinusoidal, since the flux varies sinusoidally. Hence its effective value I_{gap} is $27.2/\sqrt{2}$ or 19.3 amperes. Since this added component current is in phase with the flux, it adds directly to the fundamental sine component I_1' of the exciting current. The other components of the exciting current are practically unchanged, since conditions in the iron portion of the magnetic circuit are essentially unaltered. The air gap therefore increases the effective value of the exciting current to

$$\sqrt{(I_1' + I_{gap})^2 + I_c^2 + (I_3')^2 + (I_3'')^2 + \cdots} = 24.0 \text{ amp.} \quad [88]$$

Although the harmonics are unchanged in ampere value, they are a much smaller percentage of the increased exciting current; the third harmonic, for example (Eq. 84), now is only 8.7 per cent of the exciting current. The waveform of the exciting current therefore is more nearly sinusoidal.

Air gaps often are inserted in the cores of iron-core reactors in order to reduce the inconstancy of the inductance caused by the iron. Since they may reduce the inductance itself, a suitable compromise must be struck between constancy of the inductance and bulk of a reactor designed to have a specified inductance. When both direct and alternating currents flow in a reactor winding, an air gap is often inserted in the core in order to prevent saturation due to the flux set up by the direct current. These conditions are discussed in Arts. 14 and 15. Air gaps may also be used to obtain a large value of the ratio of inductance to apparent resistance. This matter is discussed in Art. 5, Ch. VII.

11. VECTOR REPRESENTATION OF THE EXCITING CURRENT

The exciting current of a transformer is usually small, that of a power-system transformer, for example, usually being 4 to 8 per cent of the full-load primary current. Frequently the peculiar waveform of this small current can be neglected and the exciting current can be treated as if it were a sine-wave current. Vector methods then can be used. This simplification usually leads to sufficiently accurate results in calculations involving the distribution of power and reactive volt-amperes in power-system analysis. In many other problems, however, to neglect the peculiar waveform of the exciting current is to ignore basic phenomena of primary importance. Some of these problems are mentioned in the introduction to Art. 8.

In the following analysis the flux is assumed as varying sinusoidally, a condition usually approximated very closely in power-system transformers. The flux, the induced voltage, and the fundamental sine and cosine components of the exciting current then are all sine waves and can therefore be correctly represented by vectors, as in the vector diagram of Fig. 16. This vector diagram, however, entirely ignores the effect of the harmonics in the exciting current. The harmonics contribute nothing to the power, since their frequencies differ from the frequency of the induced voltage,[*] but they do increase the effective value of the exciting current.

▶ Thus the effect of the harmonics on the rms value of the exciting current is similar to that of a component current in quadrature with the voltage; that is, they increase the effective value in the same way that the reactive component of a sine-wave current increases the effective value of the current. ◀

* See Art. 9d.

Let I_m be the effective value of the reactive current including the harmonics. Then

$$I_m = \sqrt{(I_1')^2 + (I_3')^2 + (I_3'')^2 + \cdots} \qquad \blacktriangleright[89]$$

That is, the effective value of the total reactive current equals the square root of the sum of the squares of all the components in the exciting current except the core-loss component I_c. In the example of Art. 10, the fundamental component I_1' of the reactive current is 4.50 amperes, but the total reactive current is 5.00 amperes. The total reactive current I_m is usually called the *magnetizing current*. If its peculiar waveform is

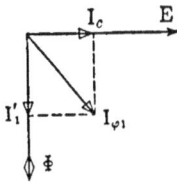

FIG. 16. Vector diagram of fundamental components of the exciting current.

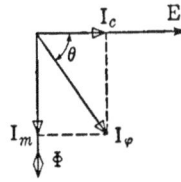

FIG. 17. Vector diagram of equivalent sine-wave components of the exciting current.

neglected, the magnetizing current can be represented by a vector I_m in quadrature with the induced voltage and therefore in phase with the flux, as in Fig. 17. Since the effective value I_φ of the exciting current equals the square root of the sum of the squares of all its harmonic components, all of which except the core-loss current I_c are included in the magnetizing current, then

$$I_\varphi = \sqrt{I_c^2 + I_m^2} \qquad \blacktriangleright[90]$$

$$= 5.07 \text{ amp.} \qquad [91]$$

Since the harmonic components combine in the same way in which quadrature sinusoidal components are added, the effective value of the exciting current I_φ is given correctly by the magnitude of the vector sum of the core-loss component I_c and the magnetizing component I_m in Fig. 17. The waveforms of the magnetizing and exciting currents, however, are far from sinusoidal.

$\blacktriangleright$ The vectors I_m and I_φ in Fig. 17 merely represent equivalent sine-wave currents having the same effective values as the actual nonsinusoidal currents and producing the same average power as these currents. That is, the length of the vector I_φ, Fig. 17, is the effective value of the exciting current, and the phase angle θ of this equivalent sine-wave current is given by

$$\cos \theta = \frac{P_c}{EI_\varphi} \qquad\qquad [92]$$

where P_c is the core loss and E and I_φ are effective values, as read on ordinary alternating-current instruments.

To determine the vector diagram of Fig. 17 by tests using ordinary alternating-current instruments is hence a simple matter. Because the exciting current often is small, vector representation by means of equivalent sine waves is often satisfactory.

Since the magnetizing current consists of all the components in the exciting current i_φ except the core-loss component i_c, the instantaneous value i_m of the magnetizing current is

$$i_m = i_\varphi - i_c. \qquad\qquad [93]$$

The waveforms of the exciting current and its two components — the core-loss current and the magnetizing current — are shown in Fig. 18.

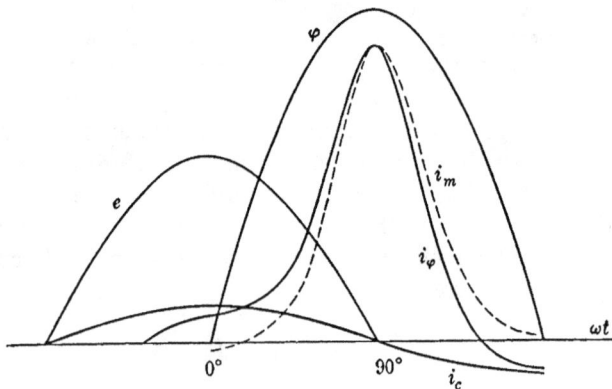

FIG. 18.   Waveforms of the magnetizing and core-loss components of the exciting current.

In analysis of a situation where the exciting current is combined with a second current which varies sinusoidally, representation of the exciting current by an equivalent sine wave is not strictly proper. Rigorously, the fundamental component of the exciting current should be combined vectorially with the second current; the square root of the sum of the squares of the resulting fundamental current and all the harmonics should then be taken in order to combine the harmonic components with the resulting fundamental current. This rigorously correct procedure is rather laborious and requires a knowledge, not often available, of the harmonic components of the exciting current. Hence the equivalent sinusoidal exciting current is often treated as if it were a true sine wave which may be added vectorially to other sine-wave currents. However, it should be appreciated that the vector treatment of the exciting current in combination with other currents is an approximation which may not always be justifiable.

12. ALTERNATING–CURRENT EXCITATION CHARACTERISTICS

The alternating-current magnetic characteristics of core materials frequently are plotted as in Figs. 19 and 21. These curves, together with the core-loss curves of Fig. 20, are useful in comparing the magnetic qualities of core materials and in predicting an approximate value of the rms exciting current of a power-system transformer.* All three of these figures are for typical 4.25 per cent silicon steel laminations of good quality, tested with a sine-wave flux. The magnetization characteristics of Figs. 19 and 21 are the average results of measurements on unbroken ring samples and represent the range of performance to be expected within the limits of commercial power frequencies and sheet gauges. The core-loss curves of Fig. 20 are for 29-gauge laminations tested in an Epstein frame, as described in Art. 5, Ch. V. Figure 19 shows the relation between the *maximum* value of the sine-wave flux density and the *effective* value of the nonsinusoidal magnetizing force in rms ampere-turns per inch. Its use in computing the rms magnetizing component of the exciting current for a specified core flux is similar to the use of the direct-current magnetization curve in analogous direct-current problems. Figure 21, showing the relation between the maximum flux density and the rms reactive volt-amperes per pound, presents essentially the same information as Fig. 19 but in a form which sometimes may be more useful. Note that the reactive volt-amperes required to excite cores of the same material to the same maximum flux density depend solely on the frequency and the weights of the cores. Since this fact also holds for the core loss, the no-load power factor of a transformer is dependent solely on the frequency and the maximum flux density, and the total excitation in rms volt-amperes EI_φ is determined by the frequency, the maximum flux density, and the weight of the core and is independent of the number of turns in the exciting winding.

The curves of Figs. 19, 20, and 21 must be used with considerable caution, and empirical correction factors must be applied to take into account the effects of stacking pressures, intersheet eddy currents, grain alignment, corners of the core, and joints in the magnetic circuit. A butt joint is equivalent to a small air gap whose length is slightly greater

---

* Core-loss curves covering a wide range of frequency and flux density are given in Figs. 9 and 10, Ch. V.

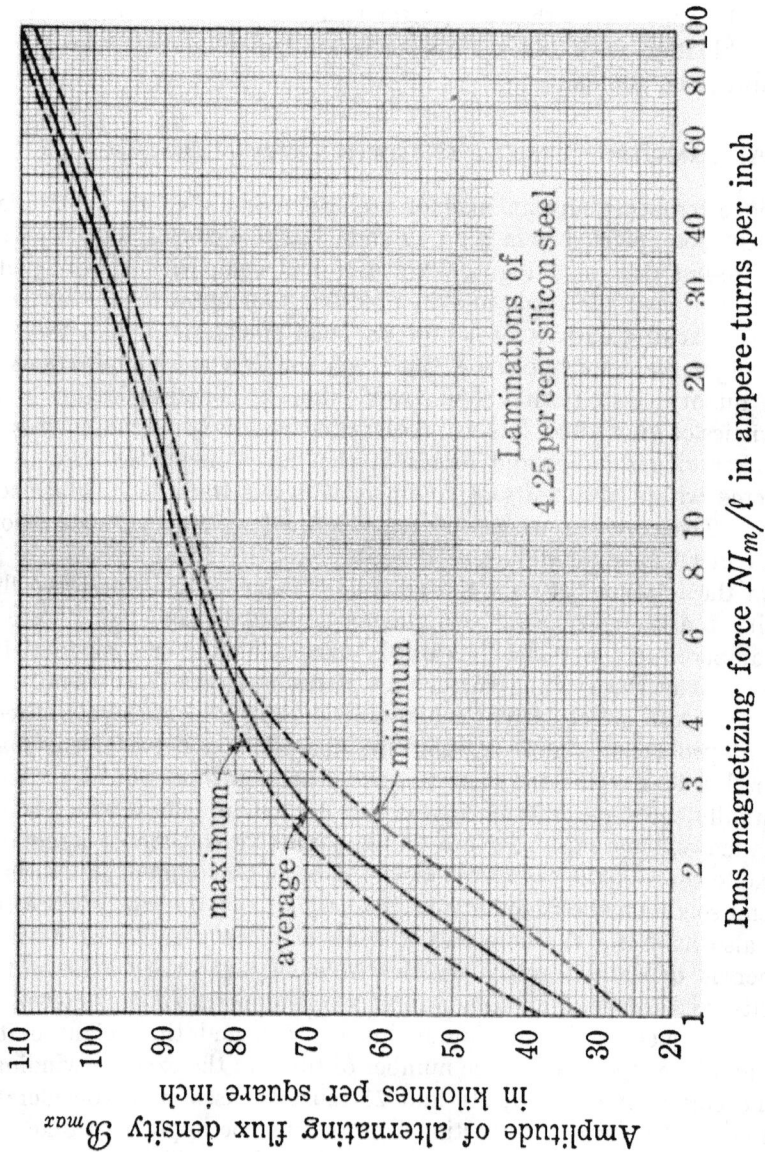

Rms magnetizing force NI_m/ℓ in ampere-turns per inch

Amplitude of alternating flux density $\mathcal{B}_{max}$ in kilolines per square inch

Laminations of
4.25 per cent silicon steel

maximum

average

minimum

Courtesy Allegheny Ludlum Steel Corp.

Fig. 19.   Rms magnetization curve for sine-wave flux.

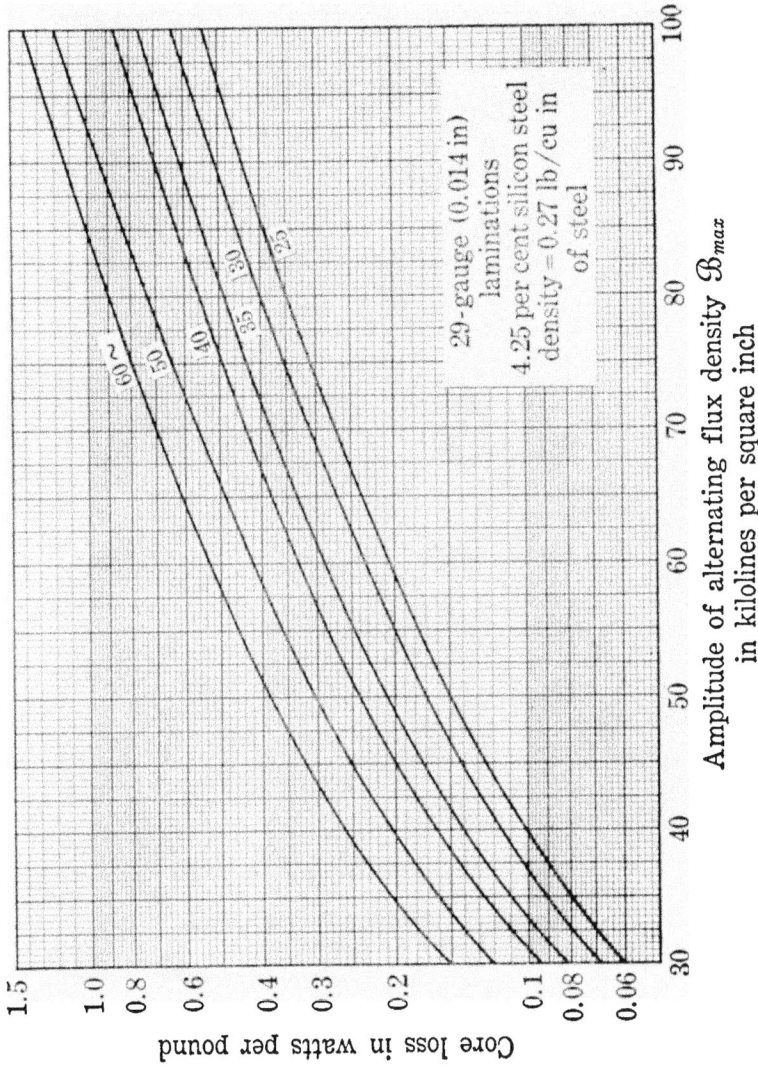

FIG. 20.   Core-loss characteristics for sine-wave flux.

29-gauge (0.014 in)
laminations
4.25 per cent silicon steel
density = 0.27 lb/cu in
of steel

Amplitude of alternating flux density $\mathcal{B}_{max}$
in kilolines per square inch

Core loss in watts per pound

Courtesy Allegheny Ludlum Steel Corp.

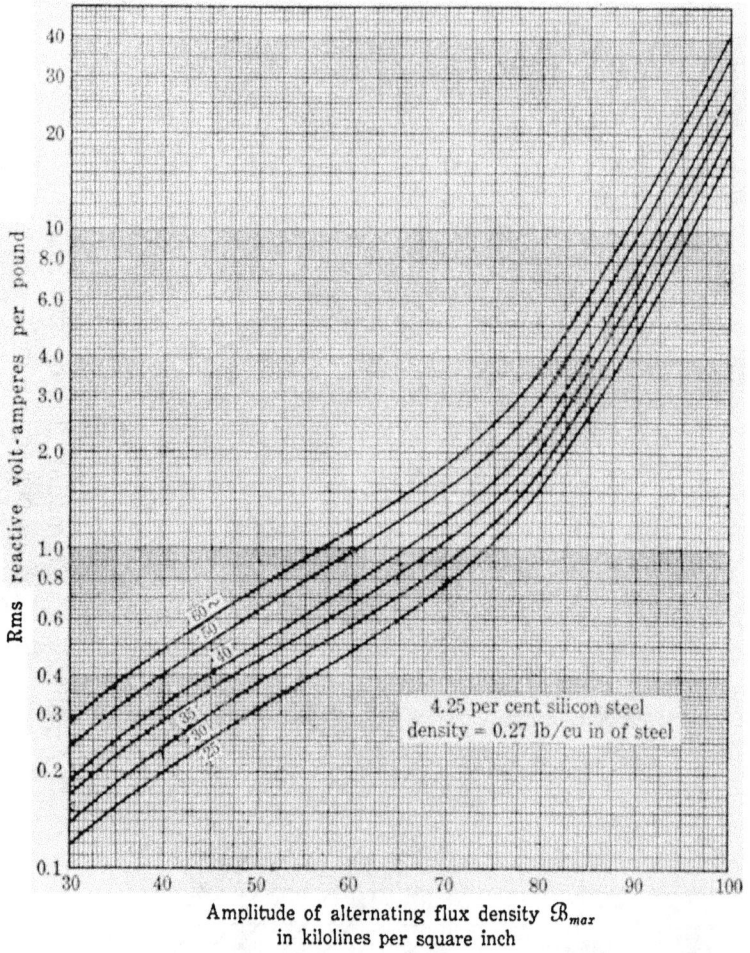

Amplitude of alternating flux density $\mathfrak{B}_{max}$
in kilolines per square inch

Courtesy Allegheny Ludlum Steel Corp.

FIG. 21. Reactive volt-ampere characteristics for sine-wave flux.

than the thickness of the paper insulation in the joint. Lap joints such as that at the corner of a core, shown in Fig. 22a, exhibit the effects of saturation. For simplicity, however, consider the straight lap joint of Fig. 22b, which is also shown enlarged, edgewise of the laminations, in Fig. 22c. Most of the flux crosses the joint from layer to layer, as indicated in Fig. 22c. Hence the flux density is more than normal in portions of the laminations near the joint, and the relation between the flux density and the magnetomotive force required by the joint is therefore nonlinear.[10]

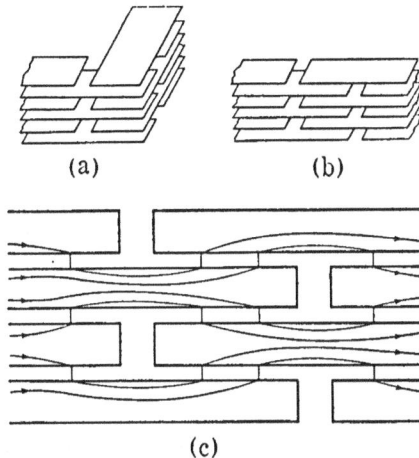

(a)   (b)

(c)

FIG. 22. Flux paths in a lap joint. (a) Lap joint at the corner of a core; (b) straight lap joint; (c) enlargement of (b) showing approximate flux paths.

12a. *Illustrative Example of Iron-Core Reactor Design.* — Consider an iron-core reactor with a rectangular core of 29-gauge 4.25 per cent silicon steel laminations whose characteristics are given in Figs. 19, 20, and 21. The core dimensions are shown in Fig. 23. The core is constructed of L-shaped laminations with lap joints. The winding has 145 turns. The problem is to determine the core loss and rms exciting current when a voltage of 220 volts at a frequency of 60 cycles per second is applied to the reactor.

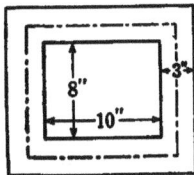

stack height = 3 in.
stacking factor = 0.90

FIG. 23. Core of reactor for illustrative example of Art. 12a.

Solution: Although the magnetomotive force required by the joints may be important, the effects of the joints are neglected in the following solution.

The mean length of the magnetic path (center line in Fig. 23) is 48 in. Therefore the net volume of steel in the core is

$$48 \times 3 \times 3 \times 0.90 = 389 \text{ cu in.} \qquad [94]$$

[10] A saturation curve for lap joints is shown in L. F. Blume, editor, *Transformer Engineering* (New York: John Wiley & Sons, 1938), 10. According to this saturation curve, each lap joint requires a magnetomotive force equal to roughly ten times the rms ampere-turns per inch in the iron over the range of flux densities commonly used in the design of power-system transformers (50–90 kilolines per square inch). The magnetomotive force required by the joints is affected by many variables, however, such as manufacturing tolerances, stacking pressures, and other factors which depend on the skill with which the core is assembled.

The density of 4.25 per cent silicon steel is 0.27 lb/cu in., so that the weight of the core is

$$0.27 \times 389 = 105 \text{ lb.} \qquad [95]$$

From Eq. 19, the maximum value ϕ_{max} of the core flux is

$$\phi_{max} = \frac{E}{4.44fN} = \frac{220}{4.44 \times 60 \times 145} \qquad [96]$$

$$= 0.00570 \text{ weber or } 570 \text{ kilolines.} \qquad [97]$$

Hence the maximum flux density $\mathcal{B}_{max}$ is

$$\mathcal{B}_{max} = \frac{\phi_{max}}{\text{net area}}$$

$$= \frac{570}{3 \times 3 \times 0.90} = 70.3 \text{ kilolines/sq in.} \qquad [98]$$

From Fig. 20, at 60 $\sim$ and 70.3 kilolines/sq in., the core loss is 0.70 w/lb, and therefore the total core loss is

$$105 \times 0.70 = 74 \text{ w.} \qquad [99]$$

From Fig. 21, at 60 $\sim$ and 70.3 kilolines/sq in., the reactive power is 1.85 vars/lb, and therefore the total reactive power is

$$105 \times 1.85 = 194 \text{ vars.} \qquad [100]$$

The core-loss and magnetizing components of the exciting current now can be computed; thus

$$I_c = \frac{74}{220} = 0.34 \text{ amp.} \qquad [101]$$

$$I_m = \frac{194}{220} = 0.88 \text{ amp.} \qquad [102]$$

The rms value I_φ of the exciting current is then given by

$$I_\varphi = \sqrt{(0.34)^2 + (0.88)^2} = 0.94 \text{ amp.} \qquad [103]$$

The rms magnetizing current I_m also can be computed from Fig. 19. From this curve, at a maximum flux density of 70.3 kilolines/sq in., the rms magnetizing force is 2.7 rms amp-turns/in., and, since the mean length of the magnetic circuit is 48 in., the magnetomotive force is 48 $\times$ 2.7 rms amp-turns. The winding has 145 turns, and therefore the rms magnetizing current I_m is

$$I_m = \frac{48 \times 2.7}{145} = 0.89 \text{ amp.} \qquad [104]$$

Consequently the rms exciting current I_φ is

$$I_\varphi = \sqrt{(0.34)^2 + (0.89)^2} = 0.96 \text{ amp.} \qquad [105]$$

The two methods for computing exciting current yield substantially the same results. Note that the effective values of the magnetizing and exciting currents do not differ greatly. Often this difference is ignored and the rms exciting current is assumed to equal the rms magnetizing current.

13. EQUIVALENT CIRCUITS

In circuit analysis, an iron-core reactor is often represented by an approximate equivalent circuit consisting of lumped-parameter elements so connected that their impedance characteristics approximately represent the effects of the reactor as a circuit element. When the effects of magnetic nonlinearity on the waveforms of current and voltage can be neglected, either a series or a parallel combination of a resistance and an inductance can be found, at any rms applied voltage V and angular frequency ω, which will absorb the same average power P and draw the same rms current I as the reactor. These equivalent circuits are

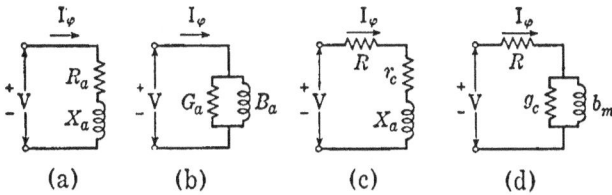

(a)   (b)   (c)   (d)

FIG. 24.   Equivalent circuits for an iron-core reactor.

shown in Figs. 24a and 24b. If they are equivalent to the reactor, their constants must be related to the voltage, current, and power taken by the reactor, as follows:

In Fig. 24a,   In Fig. 24b,

$$R_a = \frac{P}{I^2} \qquad \blacktriangleright[106] \qquad\qquad G_a = \frac{P}{V^2} \qquad \blacktriangleright[108]$$

$$X_a = \sqrt{\left(\frac{V}{I}\right)^2 - R_a^2} \qquad \blacktriangleright[107] \qquad\qquad B_a = -\sqrt{\left(\frac{I}{V}\right)^2 - G_a^2} \qquad \blacktriangleright[109]$$

where

R_a is the apparent resistance,
X_a is the apparent reactance,
G_a is the apparent conductance,
B_a is the apparent susceptance of the reactor.

Because of the magnetic nonlinearity of the core, the parameters may not be constant but may depend on the voltage and frequency at which they are measured.

In the circuits of Figs. 24a and 24b, R_a and G_a produce a power loss equal to the total power loss P in the reactor, consisting of the core loss P_c plus the copper loss I^2R, where R is the effective resistance of the winding. Thus, as stated in Art. 1, the apparent resistance R_a is greater than the effective resistance R of the winding. The copper and core losses vary in different ways with changes in the voltage and frequency and with changes in the design of the windings and the core. Hence, to indicate the component losses by separate resistances is usually desirable, as in Figs. 24c and 24d, in which R is the effective resistance of the winding, and r_c or g_c accounts for the core loss. In the theory of transformers it is also desirable to resolve the apparent reactance X_a into a component due to leakage flux and a component due to core flux. Although in a loaded transformer magnetic leakage is important, it is usually very small in an iron-core reactor or in a transformer at no load. Hence the apparent reactance X_a practically equals the reactance due to the core flux.

In spite of the peculiar magnetic properties of iron, an iron-core reactor or transformer can often be represented with sufficient accuracy by means of an equivalent circuit whose parameters are assumed to be constant and independent of voltage and frequency. If the frequency is constant, assuming that the parameters r_c, X_a, g_c, and b_m in Figs. 24c and 24d are independent of voltage is equivalent to assuming that the current is proportional to the flux, and that the core loss varies as the square of the flux. Both equivalent circuits yield the same results, and whichever is more convenient may be used. In the actual transformer or reactor, the current increases more rapidly than the flux, the eddy-current loss varies as the square of the flux, and the hysteresis loss may often be assumed also to vary as the square of the flux.

If the frequency varies, and if it is assumed that the resistors r_c and g_c are constant and that the circuit elements X_a and b_m have constant inductances, the equivalent circuits of Figs. 24c and 24d give different results. In these circumstances the equivalent circuit of Fig. 24c is unsatisfactory, since the apparent core loss computed from this circuit is approximately proportional to the square of the flux but is independent of frequency. However, the parallel circuit of Fig. 24d, with constant g_c and constant inductance in the magnetizing branch, may be a sufficiently accurate representation of the reactor. With these assumptions, in Fig. 24d the magnetizing current is proportional to the flux, and the total core loss is proportional to the square of the induced voltage. This is the manner in which the eddy-current loss varies in the reactor,* but the hysteresis loss varies as the first power of the frequency. When the amplitude of the alternating flux density is very small, as in many reactors and transformers used in communication circuits, the nonlinearity of the magnetic characteristics often is unimportant, and the assumption of constant inductance frequently is justifiable.

---

* See Art. 2, Ch. V.

▶ Therefore the equivalent circuit of Fig. 24d, with constant g_c and constant magnetizing inductance, can often be used to represent an iron-core reactor over a wide frequency range, if the hysteresis loss is negligible compared with the eddy-current loss and if the amplitude of the alternating flux density is small.[11]◀

At the flux densities commonly used in power apparatus, however, the hysteresis loss is usually greater than the eddy-current loss. Under these conditions it is probably more accurate to assume that the resistances of r_c and g_c, Figs. 24c and 24d, are proportional to frequency and that the circuit elements X_a and b_m have constant inductances. That is, assume that:

In Fig. 24c,   In Fig. 24d,

$$r_c = \omega R' \qquad [110]$$

$$g_c = \frac{1}{\omega R''} \qquad [112]$$

$$X_a = \omega L' \qquad [111]$$

$$b_m = -\frac{1}{\omega L''} \qquad [113]$$

where R', L', R'', and L'' are constants. With these assumptions it can be shown that both equivalent circuits, Figs. 24c and 24d, yield the same results; namely, that the current is proportional to the flux and the core loss is proportional to the frequency and the square of the flux. This result is the correct variation of the hysteresis loss with frequency and may also be approximately the correct variation of the total core loss with flux. However, when the inductances are assumed to be constant, the equivalent circuits all neglect the effects of saturation on the magnetizing current.

▶ Equivalent circuits give sinusoidal currents when the waveform of the applied voltage is sinusoidal, and hence the equivalent circuits are applicable only when it is sufficiently accurate to represent the current by an equivalent sine wave. Fortunately this can often be done.◀

14. CHARACTERISTICS OF IRON SUBJECTED TO SUPERPOSED DIRECT AND ALTERNATING MAGNETIZING FORCES

Thus far, discussion has concerned the behavior of ferromagnetic materials under conditions where the materials are subjected to steady or direct magnetizing forces, and also when they are subjected to alternating magnetizing forces. In practice, however, the materials are often

---

[11] For further discussion of this equivalent circuit, see L. B. Arguimbau, " Losses in Audio-Frequency Coils," *Gen. Rad. Exp.*, *11*, No. 6 (November, 1936), 1–4; P. K. McElroy and R. F. Field, " How Good Is an Iron-Cored Coil?" *Gen. Rad. Exp.*, *16*, No. 10 (March, 1942), 1–12.

subjected to superposed alternating and steady magnetizing forces. For example, in many useful control circuits, superposed alternating- and direct-current excitation of iron cores is often used to secure a desired result. Also, in many circuits associated with electronic apparatus there is of necessity a unidirectional component of current in addition to an alternating component. In all such applications it is important to know the manner 'n which the iron responds to the combined excitations. Frequently the problem is to determine the effect that direct current in the windings of an iron-core inductor has on its apparent inductance.

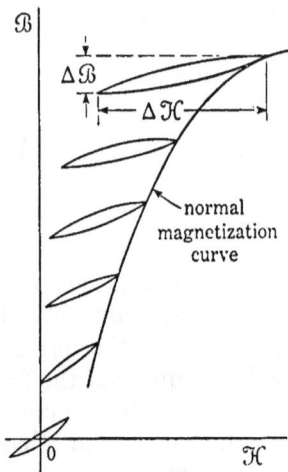

In this article certain characteristics of iron subjected to superposed direct and alternating excitations are briefly presented. A new term, called *incremental permeability*, is introduced and the instances where it can be used effectively are explained.

When both unidirectional and alternating magnetizing forces are applied simultaneously by an exciting winding, the resulting hysteresis loop is unsymmetrical and different for different values of the unidirectional magnetizing force. A family of typical loops is shown in Fig. 25, in which the steady component of the magnetizing force $\mathcal{H}$ is different for each loop but the amplitude of the alternating component of flux density $\mathcal{B}$ is the same.[12]

Fig. 25. Displaced hysteresis loops due to superposed direct and alternating magnetizing forces.

An important characteristic of the iron is its apparent average or *incremental permeability* for the alternating component of magnetizing force. This incremental permeability is approximately proportional to the average slope $\Delta\mathcal{B}/\Delta\mathcal{H}$ of the small loops. Its magnitude is calculated from the relation

$$\text{Incremental permeability} = \frac{d\mathcal{B}}{d\mathcal{H}} \approx \frac{\Delta\mathcal{B}}{\Delta\mathcal{H}} \qquad [114]$$

or, when sinusoidal variations are assumed, the incremental permeability μ_{ac} may be defined as

$$\mu_{ac} = \frac{\mathcal{B}_{rms}}{\mathcal{H}_{rms}}. \qquad \blacktriangleright[115]$$

The flux density $\mathcal{B}_{rms}$ is the rms value of the alternating component of flux density which induces the alternating voltage in the coil, and the

[12] J. D. Ball, " The Unsymmetrical Hysteresis Loop," *A.I.E.E. Trans.*, *34*, Part 2 (1915), 2693–2715.

magnetizing force $\mathcal{H}_{rms}$ is the component of the equivalent sine-wave magnetizing force that is in phase with the flux.

▶ Since the alternating components of $\mathcal{B}$ and $\mathcal{H}$ do not have the same waveform, the permeability μ_{ac} defined by Eq. 115 is an average or equivalent permeability. The term has meaning only in the sense that μ_{ac} is the ratio between the rms values of the alternating components of $\mathcal{B}$ and $\mathcal{H}$.◀

▶ Two qualitative generalizations can be drawn from the shapes and sizes of the loops of Fig. 25. The first is that the incremental permeability μ_{ac} is markedly less than the slope of the normal magnetization curve at the operating point corresponding to the unidirectional component of $\mathcal{H}$. That is, the apparent permeability of the material to cyclic changes in magnetizing force cannot be determined by measurement of the slope of a normal magnetization curve. The second generalization is that the larger the direct magnetizing force is made, the smaller the incremental permeability becomes. This generalization is illustrated in Fig. 25 by the increased cyclic $\mathcal{H}$ range required for the same cyclic $\mathcal{B}$ range when the unidirectional component of $\mathcal{H}$ is increased, and is shown more strikingly by the curves of incremental permeability given in Fig. 26.◀

Any exact mathematical statement of these generalizations is impracticable because of the nonlinear nature of the relations involved. Approximate methods of analysis using data expressed in graphical form, or experiments on models or full-scale apparatus, offer the only practicable means for determining the performance of iron-core reactors having superposed alternating and direct excitations. The problem is essentially the same as that discussed earlier in this chapter, except that here the direct magnetizing force $\mathcal{H}_{dc}$ is an additional variable. The previous discussion concerning nonlinearity, waveforms, equivalent sine waves, and so forth applies here almost directly for any given value of $\mathcal{H}_{dc}$, with the additional consideration that the hysteresis loop is now displaced from the origin and is unsymmetrical.

A useful way of describing the properties of iron subjected to superposed direct and alternating magnetizations is by means of curves of the incremental or effective permeability μ_{ac} as a function of the maximum value of the alternating component of flux density $\mathcal{B}_{max}$ with $\mathcal{H}_{dc}$ as a parameter. A typical family of such curves is shown as Fig. 26. The data for Fig. 26 were taken on a core of L-shaped punchings, stacked interleaved. The data must be regarded only as typical in form, and quantitatively applicable only to the particular material when used under magnetic conditions similar to those applying when the data were obtained. Even when the same material is used in a different form of

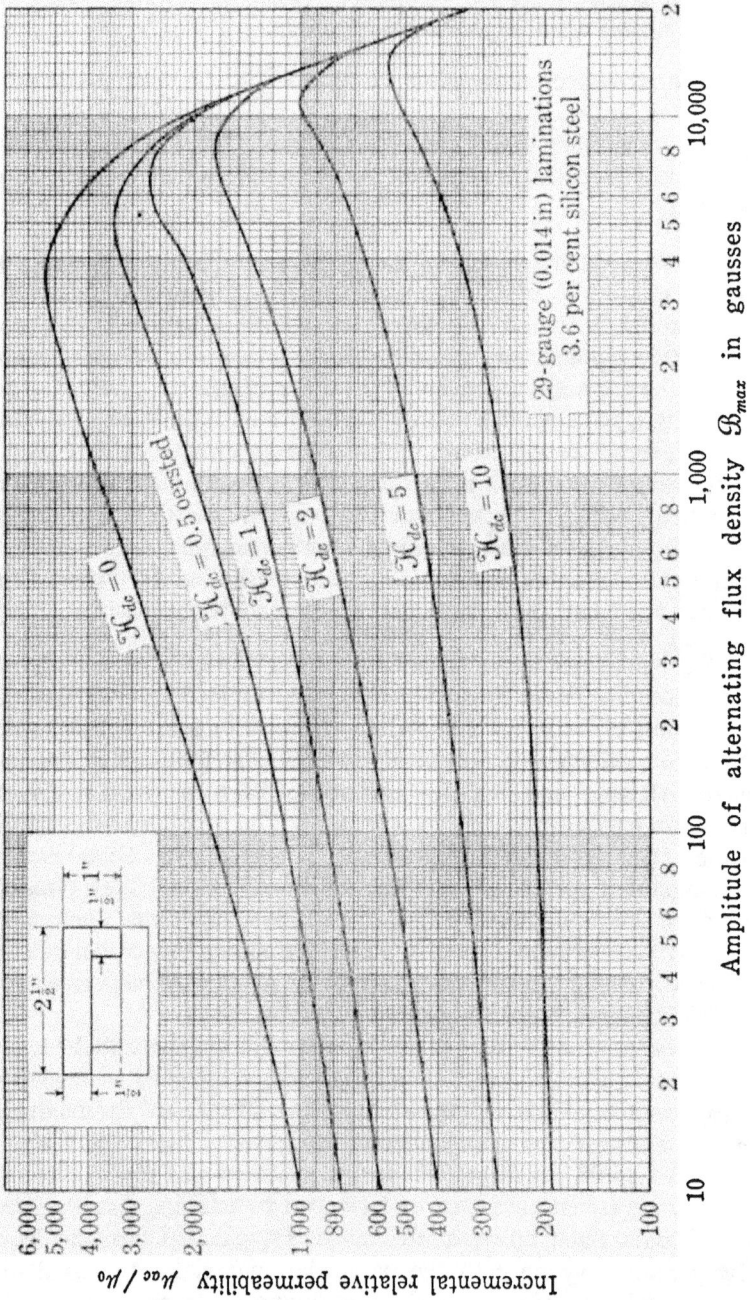

Amplitude of alternating flux density $\mathcal{B}_{max}$ in gausses

Courtesy Allegheny Ludlum Steel Corp.

Fig. 26.   Incremental permeability characteristics.

punching, or when differently stacked, as with butt instead of interleaved joints, the data are somewhat different. Regardless of these limitations, however, the data, if properly interpreted, are useful in any approximate design calculation. The use of incremental permeability as defined by Eq. 115 for the calculation of the performance of reactors is treated in Art. 15.

Where a large incremental permeability is desired (as for example in audio-frequency transformers used in vacuum-tube amplifiers) the direct component of magnetizing force, though not easily avoided, is undesirable, because it tends to decrease the inductances of the transformer windings and hence their coefficient of coupling. In many applications of rectifier filter chokes, a large incremental permeability is desirable with a large direct magnetizing force. When the direct magnetizing force is increased, the decrease in incremental permeability decreases the apparent inductance of the element and gives rise to the term " swinging choke." As shown in the volume on electronics, swinging chokes are used effectively and economically in certain rectifier filter circuits. Sometimes the change in incremental permeability is made to serve a useful function in a large number of control devices. The attractive feature of these devices comes largely from the ease with which the effective or incremental permeability of the core material can be varied by varying the direct magnetizing force.

In applications in which waveforms are important, a different definition of alternating-current permeability may be desirable. If the discussion is limited to situations in which a sinusoidal voltage is applied to the reactor and the resistance of the coil is negligible, the variation of flux density is essentially sinusoidal. The permeability may then be defined as the ratio of the sinusoidal component of $\mathfrak{B}$ to the *fundamental component* of $\mathfrak{K}$ that is in phase with $\mathfrak{B}$. One or more of a kind of " harmonic " permeability, defined as the ratio of the fundamental component of $\mathfrak{B}$ to any specified harmonic component of $\mathfrak{K}$, may also be found useful. Note that since the displaced hysteresis loop is unsymmetrical, even as well as odd harmonics are present in the magnetizing force. Such harmonic permeabilities are important when an exact reproduction of signals is desired, but the measurement and use of them are not discussed here.

When an inductor is used in a tuned circuit, a large ratio of reactance to apparent resistance is usually desired, and, if the losses in the iron form a substantial part of the total loss, their effect is important, since the ratio of reactance to resistance decreases as the apparent resistance increases. In other applications the contribution of the core loss to the apparent alternating-current resistance of the reactor may have small relative importance, as in rectifier filter circuits where the direct-current resistance of the winding is important because it increases the regulation of the direct-current source. Significant data regarding the loss in iron subjected to superposed alternating and direct excitations can be obtained best by experiment, with magnetic conditions in the sample similar to those for which data are desired.

15. IRON–CORE REACTORS WITH SUPERPOSED DIRECT AND ALTERNATING EXCITATIONS

Superposed alternating and direct excitation of the iron occurs in two principal forms of iron-core reactor. In one form the direct excitation is used to control the effective reactance, and the direct and alternating magnetomotive forces act on different magnetic circuits, parts of which are common. A typical reactor of this form with series-connected alternating-current coils is shown in Fig. 27. When the direct excitation is zero, the alternating magnetic flux traverses only the outer path if the alternating-current coils are similar and connected either in series or in parallel with their magnetomotive forces aiding with respect to this path. When the direct magnetomotive force is applied, it sets up a unidirectional flux that traverses the entire core as shown, and, to a first approximation, at any given instant of the alternating magnetization cycle adds to the alternating flux in one outer leg and subtracts from the alternating flux in the other outer leg. Actually, because of saturation effects, the amount subtracted from one leg does not equal the amount added to the other leg. In other words, linear superposition is not applicable here.

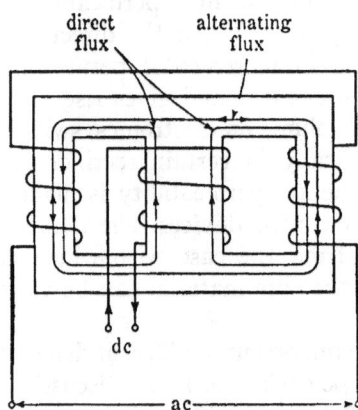

FIG. 27. Three-leg reactor with d-c and a-c windings.

Experiment shows that the apparent reactance of the alternating-current winding in such a reactor can be varied between wide limits merely by change in the magnitude of the direct current. Consequently, reactors using this principle find many applications in problems of control. The analysis of such reactors is performed in a manner similar to that given in part (a) of this article.

The second form of reactor in which the excitations are superposed applies the two magnetomotive forces by the same coil, and the paths for both the alternating and the direct fluxes are therefore the same. When the flux path in iron has substantially uniform cross section throughout its length, the approximate performance of reactors of this form can be determined rather easily from magnetic data such as those of Fig. 26. In part (a) of this article, the theory of such an analysis is developed as needed in the consideration of a particular numerical example.

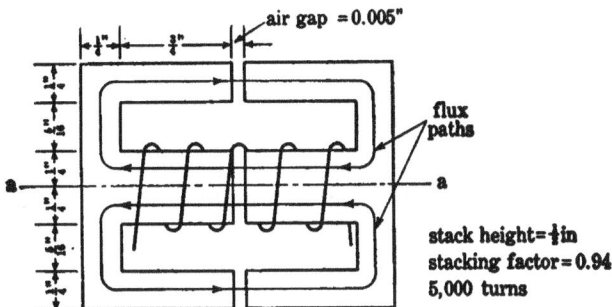

FIG. 28.   Core of inductance coil.

15a. *Illustrative Example of Calculation of Apparent Inductance.* — Suppose it is desired to find the apparent inductance of an iron-core choke or inductance coil similar to that used frequently as a coupling element in a vacuum-tube circuit. A typical core consists of laminations stacked $\frac{1}{2}$-inch high and assembled as shown in Fig. 28. A material fre-

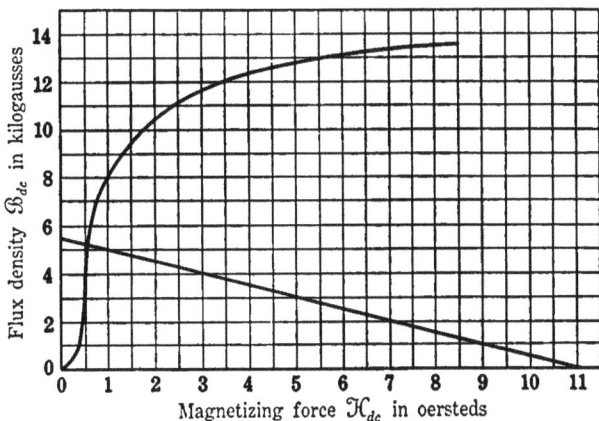

FIG. 29.   Normal d-c magnetization curve for the steel of Fig. 26, and graphical construction for the determination of $\mathcal{H}_{dc}$ in Art. 15a.

quently used is a silicon steel with about 4 per cent silicon, for which the normal magnetization curve is given by Fig. 29 and the incremental permeability curves are given by Fig. 26. The stacking factor for the assembly as indicated is assumed to equal 0.94. Assume that the coil has 5,000 turns and is wound on the center leg of the core.

For illustration, consider that the coil carries 0.020 ampere direct current, and that the apparent inductance is desired when a voltage of 50 volts rms at 120 cycles per second is applied to the coil. There are no windings on the outer legs.

Solution: The general procedure for the solution is first to determine the direct magnetizing force $\mathcal{H}_{dc}$ in the steel. The alternating maximum flux density corresponding to the alternating component of voltage is then computed. From these data the apparent incremental permeability of the steel is determined from Fig. 26. The apparent inductance is then computed on the assumption that the magnetic circuit is linear within the range of the superposed alternating magnetomotive force.

To determine the value of $\mathcal{H}_{dc}$, the total direct magnetizing force must be divided between the air gaps and the steel for the condition that the flux across the air gap equals the flux in the iron. For this calculation the core may be divided into two parts in parallel magnetically, as shown by the plane *aa* perpendicular to the paper in Fig. 28. The magnetic dimensions for one half can be computed and dealt with alone, except that the total flux linking the coil is then twice that in either half of the core.

The graphical solution for the direct magnetizing force $\mathcal{H}_{dc}$ is given by the curves of Fig 29. The method used for the solution is identical with that illustrated by the solution of the second example in Art. 8, Ch. III. The numerical values used in Fig. 29 for this solution are obtained as follows.

The direct magnetomotive force F_{dc} is

$$F_{dc} = 5,000 \times 0.020$$

$$= 100 \text{ amp-turns or 126 gilberts.} \tag{116}$$

To compute the length ℓ_s of the flux path in the steel core, assume that the flux rounds the corners in the steel on a mean radius of $\frac{1}{8}$ inch. Then

$$\ell_s = 4\left(\frac{3}{4}\right) + 2\left(\frac{5}{16}\right) + \frac{\pi}{4}$$

$$= 4.4 \text{ in., or 11.2 cm.} \tag{117}$$

The cross-sectional area A_s of the steel is

$$A_s = (0.94 \times 0.25 \times 0.50)$$

$$= 0.117 \text{ sq in., or 0.76 sq cm.} \tag{118}$$

The total length ℓ_a of the two air gaps in series with the flux path is

$$\ell_a = 0.010 \text{ in., or 0.0254 cm.} \tag{119}$$

The equivalent cross-sectional area A_a of the gap is determined by the effect of fringing. To allow for fringing, each gap may be assumed to have effective cross-sectional dimensions greater than the steel dimensions by the gap length.* While in this illustration this refinement does not affect the area by an amount great enough to be numerically significant, the refinement is included for completeness in principle. Thus the equivalent area of the outer-leg gap is

$$A_a = (0.25 + 0.005)(0.5 + 0.005)$$

$$= 0.129 \text{ sq in., or 0.83 sq cm.} \tag{120}$$

---

* See Eq. 3, Ch. III.

The equivalent area of the center-leg gap is

$$(0.50 + 0.005)(0.50 + 0.005) = 0.255 \text{ sq in.,}$$

and therefore the equivalent area on each side of the central plane *aa*, Fig. 28, is

$$\frac{0.255}{2} = 0.1275 \text{ sq in., or } 0.82 \text{ sq cm.}$$

Although this equivalent area is slightly less than the equivalent area of the outer-leg gap, subsequent calculations are based on the assumption that the combined effect of both gaps is the same as the effect of a single gap whose area A_a is 0.83 sq cm, Eq. 120, and whose length ℓ_a is 0.0254 cm, Eq. 119.

The location of the negative air-gap line is determined by Eqs. 17 and 18, Ch. III. Hence the intersection of this line with the $\mathcal{H}$ axis is

$$\frac{F_{dc}}{\ell_s} = \frac{126}{11.2} = 11.2 \text{ oersteds,} \qquad [121]$$

and its intersection with the $\mathcal{B}$ axis is

$$\frac{\mu_0 A_a F_{dc}}{A_s \ell_a} = \frac{1.00 \times 0.83 \times 126}{0.76 \times 0.0254} = 5,420 \text{ gausses,} \qquad [122]$$

where μ_0 is the permeability of free space.

The co-ordinates of the point of intersection of the magnetization curve of the steel and the negative air-gap line give the values of direct magnetizing force $\mathcal{H}_{dc}$ and the flux density in the steel, in the absence of any alternating magnetizing force. From Fig. 29, the value of $\mathcal{H}_{dc}$ used on entering Fig. 26 is 0.55 oersted approximately.

Before the incremental permeability μ_{ac} can be obtained, the maximum value $\mathcal{B}_{max}$ of the alternating component of flux density in the steel must be calculated. To determine $\mathcal{B}_{max}$, the maximum value ϕ_{max} of the alternating component of flux required to generate 50 v rms at 120 $\sim$ must be computed. From Eq. 19, this is

$$\phi_{max} = \frac{E}{4.44fN} = \frac{50}{4.44 \times 120 \times 5,000}$$

$$= 1.88 \times 10^{-5} \text{ weber or 1,880 maxwells.} \qquad [123]$$

For a stacking factor of 0.94, the net area of steel in the center leg, which carries the total flux, is

$$0.94 \times 0.500 \times 0.500 = 0.235 \text{ sq in.} = 1.52 \text{ sq cm.} \qquad [124]$$

Hence the alternating component of flux density in the steel is

$$\mathcal{B}_{max} = \frac{1,880}{1.52} = 1,240 \text{ gausses.} \qquad [125]$$

From Fig. 26, the incremental relative permeability μ_{ac}/μ_0 corresponding to $\mathcal{H}_{dc}$ of 0.55 oersted and $\mathcal{B}_{max}$ of 1,240 gausses is approximately 2,000. Therefore, in mks units,

$$\mu_{ac} = 2,000 \times 10^{-7}. \qquad [126]$$

To compute the apparent inductance, the incremental flux linkages per ampere in the coil are determined, the reluctance of the magnetic circuit being calculated first. Since the plane *aa*, Fig 28, divides the magnetic circuit into two identical parallel paths, the reluctance is one-half the value for either of these paths. If all quantities are expressed in mks units, the desired inductance is given in henries.

The reluctance $\mathcal{R}_s$ of the steel part of the path *in each half* of the reactor is

$$\mathcal{R}_s = \frac{\ell_s}{\mu_{ac} A_s},$$ [127]

and on substitution of the values of ℓ_s and A_s (Eqs. 117 and 118) converted to mks units,

$$\mathcal{R}_s = \frac{0.112}{2{,}000 \times 10^{-7} \times 0.76 \times 10^{-4}}$$

$$= 0.74 \times 10^7 \text{ mks units, or pragilberts/weber.}$$ [128]

As in the preceding computations, the air gaps in each half of the magnetic circuit are considered to be equivalent to a single gap whose length ℓ_a is given by Eq. 119 and whose area A_a is given by Eq. 120. Therefore the reluctance $\mathcal{R}_a$ of the air gaps *in each half* of the magnetic circuit is

$$\mathcal{R}_a = \frac{\ell_a}{\mu_0 A_a} = \frac{0.0254 \times 10^{-2}}{10^{-7} \times 0.83 \times 10^{-4}}$$

$$= 3.06 \times 10^7 \text{ pragilberts/weber.}$$ [129]

The effective reluctance of each half of the magnetic circuit is the sum of Eqs. 128 and 129, and the combined reluctance $\mathcal{R}$ of the two halves in parallel is

$$\mathcal{R} = \frac{0.74 + 3.06}{2} \times 10^7$$

$$= 1.90 \times 10^7 \text{ pragilberts/weber.}$$ [130]

The apparent inductance L_a in henries is the flux linkages produced per ampere in the coil. One ampere in the coil produces a magnetomotive force of $4\pi(5{,}000)$ pragilberts, resulting in a flux ϕ of

$$\phi = \frac{4\pi \times 5{,}000}{1.90 \times 10^7} = 0.00330 \text{ weber.}$$ [131]

This flux links 5,000 turns; whence the inductance is

$$L_a = \frac{N\phi}{I}$$ ▶[132]

$$= 5{,}000 \times 0.00330 = 16.5 \text{ h,}$$ [133]

which is the desired result.

Note that the determination of $\mathcal{H}_{dc}$ from the normal magnetization curve, and hence the values of μ_{ac} and $\mathcal{R}_s$, as given are not highly accurate. However, the steel term constitutes only about 20 per cent of the total reluctance; so the over-all result is much more precise than is the value for the steel reluctance. In fact, the most uncertain data in this calculation are probably the air-gap lengths, which cannot readily be known to as good accuracy as the other data. Good accuracy is not ordinarily necessary in such calculations, a result that is reliable to within 5 or 10 per cent usually being adequate.

The example given illustrates the calculation of the performance of one particular reactor. In practice, the more usual problem is that of designing the cheapest or smallest reactor to have specified characteristics.

This involves the selection of the best size and shape of core that can be made from available punchings, and the determination of the optimum air gap. In general, the design of such reactors includes calculations of the kind illustrated above, together with extensions or generalizations to determine the effects of varying dimensions on the reactor performance. Model theory as briefly presented in Ch. VII is often very useful in such design.

PROBLEMS

1. A sinusoidal exciting current impressed on one winding of an iron-core transformer is adjusted so that the maximum flux density in the core is 10,000 gausses. Data for the top half of the symmetrical hysteresis loop for this maximum flux density are given in the following table:

$\mathcal{B}$ — gausses	$\mathcal{H}$ — oersteds
0	+0.60
2,000	+0.65
4,000	+0.73
6,000	+0.92
7,000	+1.07
8,000	+1.30
9,000	+1.70
10,000	+2.43
9,500	+1.00
9,000	+0.53
8,000	+0.03
7,000	−0.23
6,000	−0.37
4,000	−0.50
2,000	−0.57
0	−0.60

Assume that the eddy currents and the effects of joints in the magnetic circuit are negligible.

(a) Plot the waveform of the flux. The following scales are suggested:

2,000 gausses per in.
1 oersted per in.
40 degrees per in.

(b) Plot the waveform of the induced voltage.

2. A sinusoidal voltage which results in a maximum flux density of 10,000 gausses is applied to one winding of the transformer of Prob. 1. Plot the resultant $\mathcal{B}(\mathcal{H})$ loop when the rms value of the eddy-current-loss component of the exciting current is 10% of the total rms value of the exciting current.

3. Find the ratio of the exciting admittances, I_φ/E, for the transformer of Prob. 1 for the following two conditions: (1) maximum flux density of 10,000 gausses, applied voltage of the form: $e = E_{1max} \sin \omega t − 0.3 E_{1max} \sin 3\omega t$, (2) maximum flux density of 10,000 gausses, applied voltage of the form: $e = E_{max} \sin \omega t$.

4. If the induced voltage of a reactor is of the form $E_1 \sin \omega t + k E_1 \sin 3\omega t$, how large can k be if there is to be no re-entrance on the hysteresis loop.

(a) If k is positive?

(b) If k is negative?

5. A 240-v 60 $\sim$ transformer is connected to a 240-v 60 $\sim$ sine-wave generator through a line with a resistance of 10 ohms. A variable condenser is connected across the terminals of the transformer as shown in Fig. 30. When the transformer is operated at no load, the rms current indicated by ammeter A is found to be least when the condenser is adjusted to have a capacitance of 21 μf. The rms current in A then is 0.60 amp, and an oscillographic record shows that its harmonic content is given by

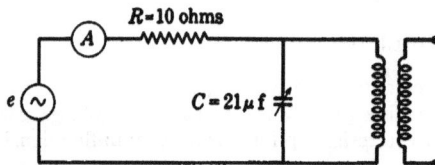

Fig. 30.   Transformer connection, Prob. 5.

$$i = I_1\sqrt{2} \ (\cos \omega t - 2.57 \sin 3\omega t + 0.19 \cos 3\omega t + 0.69 \sin 5\omega t) \qquad [134]$$

if the generator voltage is expressed as

$$e = E\sqrt{2} \cos \omega t. \qquad [135]$$

Neglecting the winding resistance of the transformer, find:

(a) The waveform of the induced voltage in the transformer,

(b) The rms current drawn from the generator in per cent of the rms exciting current in the transformer.

6. In a region where 50 $\sim$ and 60 $\sim$ power systems serve adjacent areas, a lightning storm has damaged a 1,000-kva 11,000 : 2,200-v 50 $\sim$ power transformer. This transformer is to be replaced temporarily by a spare 1,000-kva 12,000 : 2,400-v 60 $\sim$ transformer borrowed from the 60 $\sim$ system. The question has arisen whether the third harmonics in the exciting current will cause inductive interference in a telephone circuit adjacent to the 11,000-v 50 $\sim$ transmission line which would supply the primary of the replacement transformer. As a first step in the solution of the problem, determine the rms value of the third-harmonic component of the exciting current in the 11,000-v 50 $\sim$ line when the 60 $\sim$ unit is used on the 50 $\sim$ system.

The following data concerning the construction of the 60 $\sim$ spare transformer are available:

> Number of turns in the high-voltage winding = 400.
> Number of turns in the low-voltage winding = 80.
> Material of core: 4.25% silicon steel (characteristics of which are given in Figs. 19, 20, and 21).
> Gross cross-sectional area of the core = 172 sq in.
> Stacking factor = 0.90.

The acceptance test of the 60 $\sim$ spare transformer has given the following no-load data:

> Voltage = 2,400 v. applied to low-voltage winding.
> Exciting current = 11.3 amp. in low-voltage winding.
> Core loss = 7.15 kw.
> Frequency = 60 $\sim$.

The values of the third-harmonic component of the exciting current, in per cent of the rms value of the total exciting current, for different values of maximum flux density are given in the following table:

$\mathfrak{B}_{max}$ (kilolines per sq in.)	65	70	75	80	85
Per cent third harmonic	29.5	33.1	36.9	41.0	45.6.

7. A 6,600-v 60 ~ single-phase transformer has a laminated core of 4.25% silicon steel (for which the magnetic characteristics are given in Figs. 19, 20, and 21) assembled with a stacking factor of 0.90. The gross cross-sectional area of the core is 35 sq in., the mean length is 88 in., and there are four lap joints. Each lap joint is to be considered to take ten times as many reactive ampere-turns as are required per inch of core.

If the transformer is to be operated at rated voltage and frequency and with a maximum flux density of 70 kilolines per sq in.:

    (a) How many turns must the 6,600-v winding contain?
    (b) What effective reactive current will be required for the core?
    (c) What will be the core-loss current?
    (d) What will be the total no-load current?
    (e) What rms value of no-load current would the transformer take if a 0.2-in. air gap were inserted in the core? The waveform of the current taken at no load without the air gap is

$$i = I_{1max} [\sin \omega t + 0.35 \sin (3\omega t + \alpha_3) + 0.05 \sin (5\omega t + \alpha_5)]. \qquad [136]$$

8. It is desired to build a 115-v reactor which will have a reactance of 100 ohms at 60 ~ with a loss not to exceed 22 w. A number of stampings are available with dimensions as shown in Fig. 31. The material is 4.25% silicon steel. See Figs. 19, 20, and 21 for magnetic characteristics.

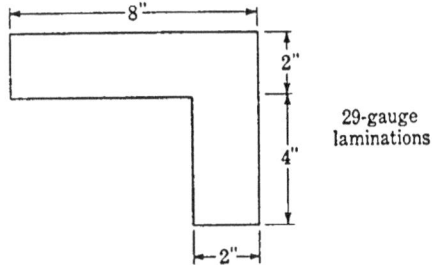

Determine the core dimensions and the number of turns which should be used to meet the specifications. Make suitable assumptions concerning stacking factor and the magnetomotive force required by the joints. Neglect copper loss.

Fig. 31. Core stamping, Prob. 8.

9. A choke coil consists of 1,000 turns of No. 16 wire wound on a 4.25% silicon steel core with dimensions as shown in Fig. 32. See Figs. 19, 20, and 21 for magnetic characteristics. Assume that the mean turn length is 15 in. and that the effective resistance of the wire is 1.01 times its direct-current resistance. Calculate an equivalent circuit of the form shown in Fig. 24d:

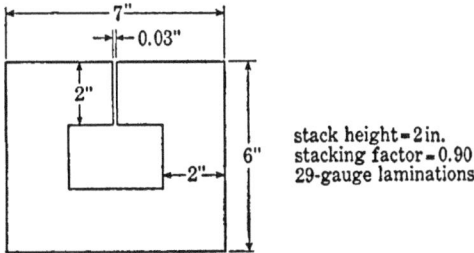

    (a) For an applied voltage of 900 v at 60 ~.

    (b) For an applied voltage of 450 v at 60 ~.

Fig. 32. Core stamping, Prob. 9.

10. Values of core loss and of exciting current at different no-load terminal voltages and at different frequencies are tabulated below for a 10-kva 2,400 : 240-v 60 ∼ single-phase transformer. The effective resistance of the 240-v winding at 60 ∼ is approximately 0.05 ohm.

AT RATED VOLTAGE

Frequency, *cycles per second*	*Exciting Current,* *amperes*	*Core Loss,* *watts*
40	4.32	71.0
50	1.63	54.5
60	1.27	50.0
70	1.16	47.3

AT RATED FREQUENCY

Terminal Voltage, *volts*	*Exciting Current,* *amperes*	*Core Loss,* *watts*
180	1.02	32.0
220	1.16	44.5
260	1.45	61.0
300	1.90	78.5

(a) Determine the parameter values for equivalent circuits of the type shown in Figs. 24c and 24d from the data taken at rated voltage and frequency.

(b) From the equivalent circuits found in part (a), keeping the parameters constant, plot curves of core loss and magnetizing current versus voltage and frequency. Compare these curves with similar curves plotted from the original data.

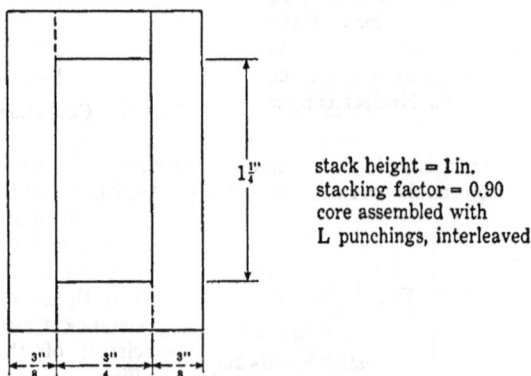

stack height = 1 in.
stacking factor = 0.90
core assembled with
L punchings, interleaved

FIG. 33.   Core assembly for reactor, Prob. 11.

11. The cross section of a laminated reactor core made of 29-gauge 3.6% silicon steel is shown in Fig. 33. See Figs. 26 and 29 for magnetic data. A reactor is to be designed with this core, to be used in a small power-rectifier filter. The maximum

direct current which the reactor will have to carry is 60 ma. The alternating voltage applied to the reactor will have an rms value of 140 v, and a frequency of 120 ∿.

(a) What is the smallest number of turns that can be used to give an inductance of at least two henries when I_{dc} equals 60 ma?

(b) What will be the inductance of the reactor when I_{dc} equals 15 ma if it is wound with the number of turns computed in part (a)?

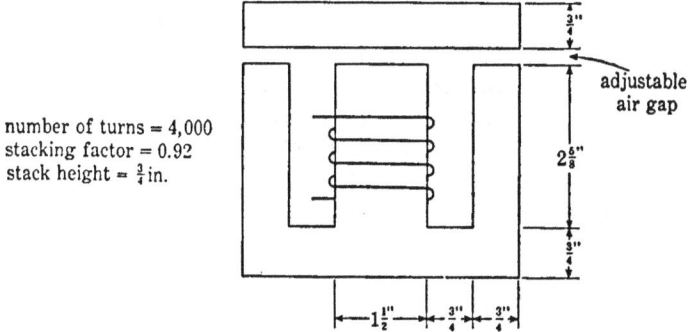

number of turns = 4,000
stacking factor = 0.92
stack height = $\frac{3}{4}$ in.

adjustable air gap

FIG. 34.   Reactor with adjustable air gap, Prob. 12.

12. A special reactor for laboratory use is constructed with a variable air gap as shown in Fig. 34. See Figs. 26 and 29 for magnetic data.

(a) To what length should the air gap be adjusted to meet the voltage, current, and inductance specifications of Prob. 11a?

(b) What will be the inductance of the reactor, with the air gap adjusted as in part (a), at the voltage and current specified in Prob. 11b?

CHAPTER VII

Model Theory and Design of Iron-Core Reactors

Because magnetic core materials are nonlinear, the most practicable way of determining the characteristics of apparatus embodying them is usually experimental. Analysis, while often valuable, is largely empirical and must therefore be verified by actual experimental data. By the use of model theory, however, the experimental data obtained on one unit can be made to apply rigorously to all geometrically similar units, regardless of size, provided certain similarity conditions are observed. This is a very important conception pertaining to nonlinear systems in general, though only the application to iron-core apparatus is considered here.

Broadly stated, the basic condition for electrical similarity between two reactors (or other apparatus) that have the same shape but different size is that the values of the variables characterizing the state of the non-linear medium shall be the same at corresponding points of the two reactors. In other words, if one made three-dimensional plots of the fields in and about two similar reactors, and then reduced or enlarged one plot to the same size as the other, the two would be congruent. This matter will be made more apparent by consideration of a few specific examples.

1. GENERAL RELATIONS

First consider a reactor whose characteristics are known as a function of the flux density $\mathfrak{B}$ at some part of the core. In the following discussion, this reactor is called the *prototype*. For simplicity, the applied voltage and flux density are assumed to be sinusoidal in waveform and of constant frequency. Throughout the following analysis, the frequency is assumed to be so low that the effects of distributed capacitances of the winding are negligible. This assumption is made to simplify the discussion and focus attention on the main points of the argument, the general principles of model theory as applied to nonlinear devices. The reader should realize, however, that the effects of capacitances may be extremely important at high frequencies and that the analysis presented in this chapter must be expanded to include capacitive effects when model theory is applied to high-frequency problems.

Assume that the following values indicated with the subscript 0 have been observed on the prototype reactor. Any consistent unrationalized system of units may be used.

A_{s0} ≡ cross-sectional area of magnetic material at any convenient section of the core

ℓ_{s0} ≡ mean length of flux path,

N_0 ≡ number of turns in the winding,

R_0 ≡ resistance of the winding,

V_0 ≡ rms value of the applied voltage,

I_0 ≡ rms value of the current,

P_0 ≡ average power absorbed by the reactor.

From these measured values, the following quantities can be computed:

Z_{a0} ≡ apparent impedance of the reactor

$$= \frac{V_0}{I_0} \qquad [1]$$

$\cos \theta_{a0}$ ≡ apparent power factor

$$= \frac{P_0}{V_0 I_0} \qquad [2]$$

R_{a0} ≡ apparent resistance

$$= Z_{a0} \cos \theta_{a0} = \frac{P_0}{I_0^2} \qquad \blacktriangleright[3]$$

X_{a0} ≡ apparent reactance

$$= Z_{a0} \sin \theta_{a0} = \sqrt{Z_{a0}^2 - R_{a0}^2} \qquad \blacktriangleright[4]$$

L_{a0} ≡ apparent inductance

$$= \frac{X_{a0}}{\omega}. \qquad \blacktriangleright[5]$$

Many reactor applications require a large ratio of reactance to resistance. This ratio is commonly denoted by the symbol Q and may be thought of as a quality factor.[1] In a high-Q coil, relatively little loss is associated with the desired inductance. The Q of the prototype reactor is

$$Q_{a0} \equiv \frac{X_{a0}}{R_{a0}} = \frac{\omega L_{a0}}{R_{a0}} = \omega T_{a0}, \qquad \blacktriangleright[6]$$

where

$$T_{a0} \equiv \frac{L_{a0}}{R_{a0}} \equiv \text{the time constant.} \qquad [7]$$

In order to study the effects of changes in design the effects of the core must be separated from those of the winding. Article 13, Ch. VI, shows

[1] See the volume on electric circuits (1940), p. 321.

that the reactor can be represented by an equivalent circuit comprising the winding resistance in combination with either a series or a parallel arrangement of a resistance and an inductance, as shown in Fig. 1. In these circuits,

$$E_0 \equiv \text{rms value of the voltage induced by the flux.}$$

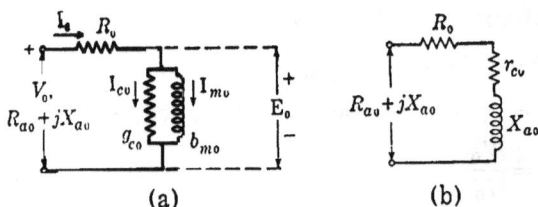

FIG. 1. Equivalent circuits for an iron-core reactor.

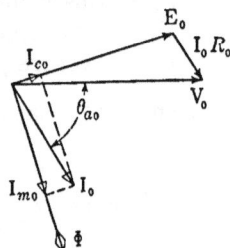

FIG. 2. Vector diagram for an iron-core reactor.

The vector diagram is shown in Fig. 2, from which, vectorially

$$\mathbf{E_0} = \mathbf{V_0} - \mathbf{I_0}R_0. \qquad [8]$$

The core loss can be determined from the measured power input; thus

$$P_{c0} \equiv \text{core loss}$$

$$= P_0 - I_0^2 R_0. \qquad [9]$$

The current can now be resolved into its core-loss and magnetizing components, and the parameters of the equivalent circuit of Fig. 1a can be determined as follows:

$$I_{c0} \equiv \text{core-loss component of the current}$$

$$= \frac{P_{c0}}{E_0} \qquad [10]$$

$$g_{c0} \equiv \text{core-loss conductance}$$

$$= \frac{I_{c0}}{E_0} = \frac{P_{c0}}{E_0^2} \qquad [11]$$

$$I_{m0} \equiv \text{magnetizing component of the current}$$

$$= \sqrt{I_0^2 - I_{c0}^2} \qquad [12]$$

$$b_{m0} \equiv \text{magnetizing susceptance}$$

$$= -\frac{I_{m0}}{E_0}. \qquad [13]$$

The negative sign in Eq. 13 indicates that b_{m0} is an inductive susceptance. In Fig. 1a, the vector admittance of the parallel circuit which represents the effects of the core as viewed from the winding — hereafter called the vector admittance of the core — is

$$Y_{\varphi 0} \equiv \text{vector admittance of the core}$$
$$= g_{c0} + j b_{m0}. \tag{14}$$

The relations between the parameters of the equivalent circuits, Figs. 1a and 1b, are:

$$Z_{\varphi 0} \equiv \text{vector impedance of the core}$$
$$= \frac{1}{Y_{\varphi 0}} \tag{15}$$

$$r_{c0} \equiv \text{equivalent series resistance of core loss}$$
$$= \text{real part of } Z_{\varphi 0}$$
$$= \frac{g_{c0}}{g_{c0}^2 + b_{m0}^2}. \tag{16}$$

Note that r_{c0} does *not* equal $1/g_{c0}$.

$$X_{a0} \equiv \text{equivalent series reactance}$$
$$= \text{imaginary part of } Z_{\varphi 0}$$
$$= \frac{-b_{m0}}{g_{c0}^2 + b_{m0}^2}. \tag{17}$$

The magnetic conditions in the core are as follows:

$$\lambda_0 \equiv \text{rms value of the alternating flux linkage}$$
$$= \frac{E_0}{\omega} \tag{18}$$

$$\phi_0 \equiv \text{rms value of the alternating flux}$$
$$= \frac{\lambda_0}{N_0} \tag{19}$$

$$\mathcal{B}_0 \equiv \text{rms value of the flux density at the cross section } A_{s0}$$
$$= \frac{\phi_0}{A_{s0}} \qquad \blacktriangleright[20]$$

$$F_0 \equiv \text{rms value of the magnetomotive force}$$
$$= 4\pi N_0 I_{m0} \tag{21}$$

$$\mathcal{K}_0 \equiv \text{rms magnetomotive force per unit length}$$
$$= \frac{F_0}{l_{s0}}. \qquad \blacktriangleright[22]$$

Equations 19 and 20 assume that all the flux is confined to the core and that the flux density is uniform over the cross section A_{s0}. If the flux were entirely confined to the core and if the flux density at all cross sections were constant, the flux density at every point would equal $\mathcal{B}_0$, and $\mathcal{H}_0$ would be the magnetic field intensity or magnetizing force. However, even if the flux density varies from point to point, magnetic conditions in the core can be completely specified in terms of an average $\mathcal{B}_0$ at any convenient cross section and an average $\mathcal{H}_0$, and therefore the following analysis includes the effects of magnetic leakage and is applicable to cores of any arbitrary shape.

▶ From the test results on the prototype, Eqs. 1–22, the performance of any other reactor geometrically similar to the prototype can be predicted for operating conditions which give the same magnetic conditions at corresponding points.◀

Several applications of these principles are illustrated in the following articles.

2. Effects of changes in the winding

Next consider a reactor with a core identical to that of the prototype and a coil having the same weight and geometrical arrangement, but wound with a times as many turns, necessitating smaller wire, of course. The new reactor is to be operated at the same frequency as the prototype. Though winding capacitances are neglected in the following treatment, their effects may be important. An increase in turns results in increases in both the inductance and the capacitances, and therefore the frequency at which capacitive effects become important is lowered. Even though capacitive effects may be negligible in the prototype, they may be important at the same frequency in the new reactor if the winding turns are greater than in the prototype. If the values pertaining to the new reactor are designated by subscript 1, for similarity

$$\mathcal{B}_1 = \mathcal{B}_0 \qquad \blacktriangleright[23]$$

$$\mathcal{H}_1 = \mathcal{H}_0 \qquad \blacktriangleright[24]$$

and since the cores are identical

$$\phi_1 = \phi_0 \qquad [25]$$

$$F_1 = F_0 \qquad [26]$$

$$P_{c1} = P_{c0}. \qquad [27]$$

But

$$N_1 = aN_0, \qquad [28]$$

and therefore, for the same flux,

$$\lambda_1 = a\lambda_0 \qquad [29]$$

$$E_1 = aE_0. \qquad \blacktriangleright[30]$$

For the same core loss,

$$I_{c1} = \frac{P_{c1}}{E_1} = \frac{P_{c0}}{aE_0} = \frac{I_{c0}}{a} \qquad [31]$$

$$g_{c1} = \frac{I_{c1}}{E_1} = \frac{I_{c0}}{a^2 E_0} = \frac{g_{c0}}{a^2}. \qquad [32]$$

For the same magnetomotive force, but *a* times as many turns,

$$I_{m1} = \frac{I_{m0}}{a} \qquad [33]$$

$$b_{m1} = -\frac{I_{m1}}{E_1} = -\frac{I_{m0}}{a^2 E_0} = \frac{b_{m0}}{a^2}. \qquad [34]$$

Therefore **the vector admittance of the core is**

$$\mathbf{Y}_{\varphi 1} = g_{c1} + jb_{m1} = \frac{\mathbf{Y}_{\varphi 0}}{a^2}. \qquad [35]$$

Hence the impedance of the core is

$$Z_{\varphi 1} = a^2 Z_{\varphi 0} \qquad [36]$$

and its series resistance and reactance components are

$$r_{c1} = a^2 r_{c0} \qquad [37]$$

$$X_{a1} = a^2 X_{a0}. \qquad \blacktriangleright[38]$$

For the same weight of copper and the same geometrical arrangements and space factors of the windings, the new winding has *a* times the length of wire but the cross-sectional area of the wire must be $1/a$ times the cross-sectional area of the wire in the prototype. Therefore the winding resistance of the new reactor is

$$R_1 = a^2 R_0. \qquad [39]$$

From Fig. 1b, the apparent resistance is the sum of the winding resistance and the series core-loss resistance, whence

$$R_{a1} = R_1 + r_{c1} = a^2(R_0 + r_{c0}) = a^2 R_{a0}. \qquad \blacktriangleright[40]$$

From Eqs. 38 and 40, the quality factor Q_{a1} of the new reactor is

$$Q_{a1} \equiv \frac{X_{a1}}{R_{a1}} = \frac{a^2 X_{a0}}{a^2 R_{a0}} = Q_{a0}. \qquad \blacktriangleright[41]$$

▶ Thus the reactor with a times the turns has a^2 times the apparent reactance and a^2 times the apparent resistance. The ratio Q_a of reactance to resistance, however, is the same as in the prototype.◀

Since both the core-loss and magnetizing components of the current are $1/a$ times the prototype values, the current in the new reactor is

$$I_1 = \frac{I_0}{a}. \qquad\qquad ▶[42]$$

The copper loss in the winding of the new reactor is

$$I_1^2 R_1 = \left(\frac{I_0}{a}\right)^2 a^2 R_0 = I_0^2 R_0. \qquad ▶[43]$$

▶ Thus, for a fixed geometrical arrangement of winding and a constant weight of copper, the I^2R loss in the winding is independent of the number of turns as long as magnetic conditions in the core are unaltered.◀

The resistance drop in the winding of the new reactor is

$$I_1 R_1 = \frac{I_0}{a} a^2 R_0 = a I_0 R_0 \qquad [44]$$

By Eq. 30, the induced voltage E_1 equals aE_0, and therefore the applied voltage, which equals the vector sum of the induced voltage and the winding resistance drop, is

$$V_1 = aV_0. \qquad\qquad ▶[45]$$

▶ Thus for similarity conditions, Eqs. 23 and 24, the new reactor must be operated at an applied voltage a times the value for the prototype. The current in the new reactor is $1/a$ times the current in the prototype, but the power and volt-amperes in the two reactors are the same.◀

3. EFFECTS OF CHANGES IN LINEAR DIMENSIONS

As a second modification, consider a reactor which has the same number of turns as the prototype, but has all linear dimensions except lamination thickness (which is unchanged) k times those of the prototype. The new reactor is to be operated at the same frequency as the prototype. If the new values are designated by subscript 2, for similarity

$$\mathcal{B}_2 = \mathcal{B}_0 \qquad\qquad ▶[46]$$

$$\mathcal{H}_2 = \mathcal{H}_0. \qquad\qquad ▶[47]$$

Since the cross-sectional area of the core of the new reactor is k^2 times that of the prototype,

$$\phi_2 = k^2 \phi_0,$$

and, since the mean length of the flux path in the new reactor is k times that of the prototype,

$$F_2 = kF_0. \tag{48}$$

Equation 48 assumes that the magnetomotive force required by joints in the core is proportional to the linear dimensions — an assumption that may be erroneous. Therefore Eq. 48 may be inaccurate if the joints are important. The core volume of the new reactor is k^3 times that of the prototype, and therefore for the same frequency, flux density, and lamination thickness

$$P_{c2} = k^3 P_{c0}. \tag{49}$$

Equation 49 neglects the effects of changes in stacking pressures, intersheet eddy currents, and other more or less indeterminate variables.

The winding turns in the two reactors are the same, or

$$N_2 = N_0, \tag{50}$$

and therefore for a flux $k^2\phi_0$, the flux linkage is

$$\lambda_2 = k^2\lambda_0 \tag{51}$$

and the induced voltage is

$$E_2 = k^2 E_0. \qquad \blacktriangleright[52]$$

From Eqs. 49 and 52,

$$I_{c2} = \frac{P_{c2}}{E_2} = \frac{k^3 P_{c0}}{k^2 E_0} = kI_{c0} \tag{53}$$

$$g_{c2} = \frac{I_{c2}}{E_2} = \frac{kI_{c0}}{k^2 E_0} = \frac{g_{c0}}{k}. \tag{54}$$

For a magnetomotive force kF_0 and the same number of turns, the magnetizing current in the new reactor is

$$I_{m2} = kI_{m0}, \qquad \blacktriangleright[55]$$

and the magnetizing susceptance is

$$b_{m2} = -\frac{I_{m2}}{E_2} = -\frac{kI_{m0}}{k^2 E_0} = \frac{b_{m0}}{k}. \tag{56}$$

Therefore the vector admittance of the core is

$$Y_{\varphi 2} = g_{c2} + jb_{m2} = \frac{Y_{\varphi 0}}{k}, \tag{57}$$

the impedance of the core is

$$Z_{\varphi 2} = kZ_{\varphi 0}, \tag{58}$$

and its resistance and reactance components are

$$r_{c2} = kr_{c0} \qquad [59]$$

$$X_{a2} = kX_{a0}. \qquad \blacktriangleright[60]$$

For similar geometrical arrangements and the same space factors, the new winding has k times the length of wire but the cross-sectional area of the wire in the new winding is k^2 times the cross-sectional area of the wire in the prototype. Therefore the winding resistance of the new reactor is

$$R_2 = \frac{kR_0}{k^2} = \frac{R_0}{k}. \qquad [61]$$

The apparent resistance is

$$R_{a2} = R_2 + r_{c2} = \frac{R_0}{k} + kr_{c0}. \qquad \blacktriangleright[62]$$

The quality factor Q_{a2} of the new reactor is

$$Q_{a2} = \frac{X_{a2}}{R_{a2}} = \frac{kX_{a0}}{\dfrac{R_0}{k} + kr_{c0}} = \frac{X_{a0}}{\dfrac{R_0}{k^2} + r_{c0}}. \qquad \blacktriangleright[63]$$

In the k-times reactor there are certain rather fundamental differences from the prototype. The quality factor is higher for k greater than unity.

▶ Thus the Q_a of a fixed shape of reactor can be increased if the reactor is made larger.◀

Note, however, that Q_a is increased only because of reduction of winding resistance; that is

$$\frac{X_{a2}}{R_2} = k^2 \frac{X_{a0}}{R_0}. \qquad [64]$$

The ratio of reactance to series core-loss resistance is the same as in the prototype; that is

$$\frac{X_{a2}}{r_{c2}} = \frac{kX_{a0}}{kr_{c0}} = \frac{X_{a0}}{r_{c0}}. \qquad \blacktriangleright[65]$$

▶ Consequently there is a definite limit to the increase in Q_a that can be made merely through increasing size.◀

From Eqs. 52 and 55,

$$E_2 I_{m2} = k^3 E_0 I_{m0}. \qquad \blacktriangleright[66]$$

▶ Thus for operation at the same frequency and flux density, the reactive volt-amperes required to excite the k-times reactor equal k^3 times the reactive volt-amperes required by the prototype. In a certain sense, therefore, the " rating " of the k-times reactor is k^3 times that of the prototype.◀

Summarizing broadly the changes produced in similar reactors shows that changing the winding turns (constant weight of winding) merely changes the resistance and inductance without altering their ratio. Changing size, however, alters the time constant. If a change in time constant without a change of inductance is wanted, the time constant is secured by a change in size, following which a suitable alteration in turns gives the desired inductance.

These two changes, in turns and in size, are the only ones that can be made while the reactor is kept geometrically similar to the prototype. However, the effect of not too great changes in stack height of core can be computed with good accuracy by modification of the foregoing principles.

In the foregoing analysis, no mention is made of superposed direct-current excitation. However, if the direct-current field intensity is maintained at the prototype value, just as the alternating field intensity is maintained, the changes in turns or size do not alter conditions in the core material, and the same similarity relations apply.

4. ENERGY CONSIDERATIONS

The discussion in the preceding articles is based on the circuit theory of the iron-core reactor; that is, the reactor is considered as a magnetic circuit interlinked with an electric circuit. The theory of the reactor can also be developed from the basis of field theory, the reactor being regarded as a magnetic field interlinked with an electric field. The principal phenomena associated with the magnetic field, most of which is confined to the core, are the storage of energy and the conversion of energy to heat caused by the core losses. Similar phenomena are also associated with the electric field due to the current and voltage in the winding. Although the energy stored in the electric field (in the distributed capacitances of the windings) may be very important at high frequencies, it is neglected in the following treatment, which is concerned primarily with phenomena associated with the magnetic core. The magnetic and electric fields are related to each other through the magnetomotive-force equation, the induced-voltage equation, and the dimensions and properties of the materials of the core and windings.

Many of the important characteristics of an iron-core reactor can be derived in terms of the density of energy stored in the magnetic field, the density of core loss, the density of copper loss, and the volumes of the core and windings. This concept is particularly useful as a basis for the design not only of iron-core reactors but also of most electromagnetic apparatus, since certain broad generalizations can be made through considering merely the masses of materials involved and the flux densities and current densities at which they are operated.

In order to simplify the argument so that the important points stand out clearly, the following discussion neglects the nonlinear magnetic characteristics of iron. That is, the permeability of the core is assumed to be constant, whence the inductance is constant and the waveform of the exciting current for sine-wave flux is sinusoidal. It is also assumed that the flux is entirely confined to the core and that the core is proportioned so that the flux density is the same throughout the iron. In spite of these assumptions, however, it should be evident that the principal results of the argument apply to an iron-core reactor of any shape, and that, when magnetic leakage and nonlinearity are important, their effects can be determined from tests on a model. In addition to the symbols already defined, let

$\mathcal{J}$ $\equiv$ rms value of the current density in the winding,

ρ $\equiv$ volume resistivity of the copper in the winding,

μ_s $\equiv$ permeability of the steel core,

A_{cu} $\equiv$ net area of copper in a cross section of the winding,

ℓ_{mt} $\equiv$ mean length of turn in the winding,

$\mathcal{V}_{cu}$ $\equiv$ volume of copper in the winding,

$\mathcal{V}_s$ $\equiv$ volume of the steel core,

W_{av} $\equiv$ average value of the magnetic energy stored in the core,

w_{av} $\equiv W_{av}/\mathcal{V}_s \equiv$ average stored energy per unit core volume,

p_c $\equiv P_c/\mathcal{V}_s \equiv$ core loss per unit core volume.

Since the primary purpose of a reactor is to absorb reactive power, a knowledge of the physical nature of reactive power is needed for a clear understanding of the phenomena associated with the behavior of iron-core reactors. It is shown elsewhere in this series[2] that "the reactive power can be interpreted physically as the amplitude of the power oscillation resulting from the component of current in quadrature with the voltage." Since this power oscillation is due to interchange of energy between the source and the magnetic field, it follows that the reactive power is related to the magnetic stored energy. The relation can be derived from the well-known fact[3] that the energy stored in a constant inductance L is $\frac{1}{2}Li^2$ when the instantaneous current is i. With a steady-state alternating current, the stored energy varies cyclically between zero and a maximum value. The average value of the stored energy is

$$W_{av} = \tfrac{1}{2}L(\text{average value of } i^2). \qquad [67]$$

---

[2] See the volume on electric circuits (1940), p. 298.

[3] See the volume on electric circuits (1940), p. 22.

But the average value of i^2 is the square of the rms value, and consequently

$$W_{av} = \tfrac{1}{2}LI^2, \qquad\qquad [68]$$

where I is the rms value of the current. Multiplication of both sides of Eq. 68 by 2ω gives

$$2\omega W_{av} = \omega LI^2 = I^2 X, \qquad\qquad \blacktriangleright[69]$$

where X is the inductive reactance. But $I^2 X$ is the reactive power.

▶ The reactive power absorbed by a constant inductance therefore equals 2ω times the average value of the magnetic stored energy. This is a useful concept of the physical nature of the reactive power in an inductive circuit.◀

Article 5, Ch. IV, shows that the energy stored in a magnetic field per unit volume is $\ell^2/8\pi\mu$, where ℓ is the instantaneous value of the flux density and μ (assumed to be constant) is the permeability of the medium in which the field exists. If the flux density is alternating, the average value of the stored energy per unit volume is

$$w_{av} = \frac{\text{average value of } \ell^2}{8\pi\mu}. \qquad\qquad [70]$$

But the average value of ℓ^2 is the square of the rms value. Therefore

$$w_{av} = \frac{\mathcal{B}^2}{8\pi\mu}, \qquad\qquad [71]$$

where $\mathcal{B}$ is the rms value of the flux density.

Thus if magnetic leakage is neglected, if the flux density is assumed to be the same in all portions of the core, and if the permeab ity of the core is assumed to be constant, the average value of the energy stored in the magnetic field of a reactor is

$$W_{av} = w_{av}\mathcal{V}_s = \frac{\mathcal{B}^2}{8\pi\mu_s}\mathcal{V}_s. \qquad\qquad \blacktriangleright[72]$$

From Eqs. 69 and 72, the reactive power absorbed by an iron-core reactor can be expressed as

$$I^2 X_a = 2\omega W_{av} = 2\omega w_{av}\mathcal{V}_s, \qquad\qquad [73]$$

where X_a is the apparent reactance. But from Eq 71,

$$2\omega w_{av} = \frac{4\pi f\mathcal{B}^2}{8\pi\mu_s} = \frac{f\mathcal{B}^2}{2\mu_s}, \qquad\qquad [74]$$

whence

$$I^2 X_a = \frac{f\mathcal{B}^2}{2\mu_s}\mathcal{V}_s. \qquad\qquad \blacktriangleright[75]$$

Equation 75 shows several important facts relating to the design of iron-core apparatus. Usually the frequency f s specified by the operating requirements, the optimum value of the rms flux density $\mathfrak{B}$ is determined by the magnetic properties of the core (such as magnetic saturation and harmonic generation), and the permeability μ_s then is fixed by the flux density.

▶ For specified f, $\mathfrak{B}$, and μ_s, the reactive power absorbed by a reactor is proportional to the volume of the core and is independent of the design of the winding. These facts are very useful in the design of a reactor to absorb a specified amount of reactive power at a given frequency, since the volume of the core can be determined as soon as the flux density is decided upon. Thus, from Eq. 75,

$$\mathcal{V}_s = \frac{2\mu_s(\text{reactive volt-amperes})}{f\mathfrak{B}^2}. \qquad ▶[76]$$

Although the reactive power is independent of the arrangement of the winding when magnetic leakage is negligible, the design of the winding does influence the quality factor Q_a, as is shown subsequently. ◀

From the series circuit of Fig. 1b, the quality factor Q_a is

$$Q_a = \frac{X_a}{R_a} = \frac{X_a}{R + r_c}. \qquad [77]$$

On multiplication of the numerator and denominator of Eq. 77 by I^2, Q_a can be expressed as

$$Q_a = \frac{I^2 X_a}{I^2 R + I^2 r_c}. \qquad [78]$$

In Eq. 78, $I^2 X_a$ is the reactive power, $I^2 R$ is the copper loss in the winding, and $I^2 r_c$ is the core loss. With the aid of Eq. 73, the quality factor therefore can be interpreted physically as

$$Q_a = \frac{2\omega(\text{average stored energy})}{\text{copper loss} + \text{core loss}}. \qquad ▶[79]$$

In the following analysis it is convenient to deal with the *loss ratio;* that is, the reciprocal of the quality factor, or

$$\frac{1}{Q_a} = \frac{R_a}{X_a} = \frac{\text{copper loss} + \text{core loss}}{2\omega(\text{average stored energy})}. \qquad ▶[80]$$

If quality factors for the winding and for the core are defined by

$$Q_w = \frac{X_a}{R} = \frac{I^2 X_a}{I^2 R} \qquad [81]$$

$$= \frac{2\omega W_{av}}{\text{copper loss}} \qquad ▶[82]$$

and

$$Q_c = \frac{X_a}{r_c} = \frac{I^2 X_a}{I^2 r_c} \tag{83}$$

$$= \frac{2\omega W_{av}}{\text{core loss}}, \qquad \blacktriangleright[84]$$

then the loss ratio of the reactor, Eq. 80, can be expressed as

$$\frac{1}{Q_a} = \frac{1}{Q_w} + \frac{1}{Q_c}. \qquad \blacktriangleright[85]$$

Thus the variables determining the quality factors for the winding and for the core can be investigated separately and the results can be combined as in Eq. 85.

The quality factor for the core can be expressed as

$$Q_c = \frac{2\omega w_{av} \mho_s}{p_c \mho_s} = \frac{2\omega w_{av}}{p_c}, \qquad \blacktriangleright[86]$$

where p_c is the core loss per unit volume and w_{av} is the average stored energy per unit volume. If the thickness of the laminations is specified, the core loss per unit volume is determined by the frequency, flux density, and magnetic properties of the core material — the same variables that determine the stored energy per unit volume. The core loss per unit volume is independent of the size of the core if intersheet eddy currents are negligible.

▶ Thus for fixed lamination thickness, the quality factor Q_c of the core is determined by the frequency and flux density and is independent of the volume of the core.◀

The quality factor for the winding, however, depends on a number of variables. The flux density, permeability, and mean length ℓ_s of the core determine the magnetomotive force that must be produced by the winding. Thus

$$4\pi N I_m = \mathcal{K}\ell_s = \frac{\mathcal{B}\ell_s}{\mu_s} \tag{87}$$

or

$$N I_m = \frac{\mathcal{B}\ell_s}{4\pi\mu_s}. \tag{88}$$

In addition to the reactive magnetizing current I_m, the winding must carry a core-loss current I_c, and, since these component currents are in quadrature, the resultant current I is given by

$$I^2 = I_m^2 + I_c^2 = I_m^2 \left[1 + \left(\frac{I_c}{I_m} \right)^2 \right] \qquad [89]$$

$$= I_m^2 \left[1 + \left(\frac{EI_c}{EI_m} \right)^2 \right], \qquad [90]$$

where E is the voltage induced by the flux. But EI_c is the core loss, and EI_m is the reactive power. Consequently, in Eq. 90,

$$\frac{EI_c}{EI_m} = \frac{1}{Q_c}, \qquad [91]$$

and, from Eqs. 90 and 91, the relation between the total current and its magnetizing component is

$$I = I_m \sqrt{1 + \frac{1}{Q_c^2}}. \qquad [92]$$

If Eq. 92 is multiplied by the number of turns N and Eq. 88 is substituted in the result,

$$NI = \text{ampere-turns} = \frac{\mathcal{B}\ell_s}{4\pi\mu_s} \sqrt{1 + \frac{1}{Q_c^2}}. \qquad [93]$$

This equation gives the relation between the magnetomotive force that must be developed by the winding and the magnetic conditions in the core.

The ampere-turn product NI equals the total current that links the magnetic circuit. The number of turns need not be considered in the following analysis, however, since the magnetomotive force can be regarded as the result of a current sheet NI, rather than as the result of the current I in N turns. This current sheet can be expressed conveniently in terms of the current density $\mathcal{J}$ and the net area A_{cu} of copper in a cross section of the winding; thus

$$NI = \text{ampere-turns} = \mathcal{J} A_{cu}. \qquad [94]$$

The copper loss can be expressed in terms of the current density; thus

$$\text{Copper loss} = \mathcal{J}^2 \rho \mathcal{V}_{cu}. \qquad [95]$$

Equation 95 is the general expression for the copper loss in a volume $\mathcal{V}_{cu}$ of copper the resistivity of which is ρ and the current density in which is $\mathcal{J}$. The current density can be expressed in terms of the current sheet $\mathcal{J} A_{cu}$; thus

$$\mathcal{J} = \frac{\mathcal{J} A_{cu}}{A_{cu}} = \frac{\text{ampere-turns}}{A_{cu}}. \qquad [96]$$

The volume of copper can be expressed as

$$\mathcal{V}_{cu} = \ell_{mt} A_{cu},$$ [97]

where ℓ_{mt} is the mean length of turn. Substitution of Eqs. 96 and 97 in Eq. 95 gives

$$\text{Copper loss} = \left(\frac{\text{ampere-turns}}{A_{cu}}\right)^2 \rho \ell_{mt} A_{cu}$$ [98]

$$= (\text{ampere-turns})^2 \frac{\rho \ell_{mt}}{A_{cu}}.$$ [99]

Note that $\rho \ell_{mt}/A_{cu}$ is the resistance of a solid strap of copper whose length equals ℓ_{mt} and whose cross-sectional area equals A_{cu}.

If Eq. 93 is substituted in Eq. 99, the copper loss can be expressed in terms of the magnetic conditions in the core; thus

$$\text{Copper loss} = \left(1 + \frac{1}{Q_c^2}\right)\left(\frac{\mathcal{B}\ell_s}{4\pi\mu_s}\right)^2 \frac{\rho \ell_{mt}}{A_{cu}}.$$ ▶[100]

The quality factor for the winding can then be expressed as

$$Q_w = \frac{\text{reactive power}}{\text{copper loss}},$$ [101]

and, from Eqs. 75 and 100,

$$Q_w = \frac{\dfrac{f\mathcal{B}^2}{2\mu_s} A_s \ell_s}{\left(1 + \dfrac{1}{Q_c^2}\right) \dfrac{\mathcal{B}^2 \ell_s^2}{16\pi^2 \mu_s^2} \dfrac{\rho \ell_{mt}}{A_{cu}}}$$ [102]

$$= \frac{8\pi^2 f}{1 + \dfrac{1}{Q_c^2}} \left(\frac{A_{cu}}{\rho \ell_{mt}}\right)\left(\frac{\mu_s A_s}{\ell_s}\right)$$ ▶[103]

Note that in Eq. 103

$$\frac{\mu_s A_s}{\ell_s}$$

is the permeance of the core, and

$$\frac{A_{cu}}{\rho \ell_{mt}}$$

is the conductance of a solid strap whose length equals ℓ_{mt} and whose cross-sectional area equals A_{cu}.

▶ Equation 103 shows that the quality factor Q_w for the winding depends on the over-all dimensions of core and coils. It is independent of the number of turns in the winding.◀

Although the flux density does not appear explicitly in Eq. 103, μ_s and Q_c are determined by the flux density and therefore Q_w is influenced by the magnetic conditions in the core. From Eq. 103, the best geometrical arrangement for a high-Q coil is one in which the area-to-length ratios for the flux path and for the current path are both as large as possible. For a fixed shape of reactor, an increase in all linear dimensions by a factor k results in an increase in the A/ℓ ratios for both coils and core of k^2/k or k, and therefore increases Q_w by a factor k^2.

▶ Thus, as pointed out in Art. 3, a higher quality factor Q_w for the winding can be secured from any specified shape through an increase in size. However, the quality factor Q_c for the core depends solely on the magnetic conditions in the core and is independent of the volume and disposition of magnetic material. For fixed $\mathfrak{B}$ and f, Q_c is unchanged by increase in size. Therefore the quality factor Q_c for the core may become the limiting feature in determining the combined quality factor Q_a of the reactor. ◀

5. Effects of an Air Gap

Insertion of an air gap in the magnetic circuit of an iron-core reactor alters the characteristics of the reactor in several ways. Except for reactors which are subjected to superposed direct and alternating magnetizing forces, the insertion results in a decrease in the inductance. If the reactor is operated with superposed direct and alternating magnetizing forces, however, the insertion of a small air gap may actually result in an increase in the incremental inductance, due to the increase in incremental permeability caused by the reduction of the unidirectional component of flux. The air gap also reduces the effects of the nonlinear magnetic characteristics of the core. Thus the insertion of the gap causes the inductance to be more nearly constant with changes in flux density, and also results in a reduction of harmonic distortion.* A further effect of inserting the air gap is a change in the quality factor of the reactor. As shown subsequently, the insertion results in an increase in the quality factor for the magnetic circuit, but a decrease in the quality factor for the winding.

5a. *Reduction of Nonlinear Effects by Means of an Air Gap.* — A problem arising frequently in the design of iron-core reactors for operation over a wide range in flux densities is the determination of the length of air gap that must be inserted in the core in order to reduce the variations in inductance to within specified tolerances. The following analysis assumes that both the flux and the current are sinusoidal in waveform

* See Art. 10a, Ch. VI.

and that the core-loss component of the current is so small compared with the magnetizing component that the total current substantially equals its magnetizing component. The inductance L can then be expressed as

$$L = \frac{N\phi}{I} = \frac{4\pi N^2 \phi}{4\pi N I} \qquad [104]$$

$$= 4\pi N^2 \frac{\phi}{F}, \qquad [105]$$

where ϕ is the rms value of the flux and F is the rms value of the magnetomotive force. But F/ϕ is the reluctance $\mathcal{R}$ of the magnetic circuit, whence

$$L = \frac{4\pi N^2}{\mathcal{R}}. \qquad [106]$$

The reluctance $\mathcal{R}$ is the series combination of the reluctances of the steel path and of the air gap, or

$$\mathcal{R} = \mathcal{R}_s + \mathcal{R}_a = \frac{l_s}{\mu_s A_s} + \frac{l_a}{\mu_a A_a}, \qquad [107]$$

where subscript s refers to the steel and a to the air gap.

When the flux density is changed, the inductance also changes because of changes in the permeability of the steel. For a specified range in flux density, the maximum and minimum values of the permeability are known, and therefore the maximum variation in inductance for this range in flux density is

$$\Delta L = 4\pi N^2 \left(\frac{1}{\mathcal{R}_{min}} - \frac{1}{\mathcal{R}_{max}} \right) \qquad [108]$$

$$= 4\pi N^2 \left(\frac{\mathcal{R}_{max} - \mathcal{R}_{min}}{\mathcal{R}_{max}\mathcal{R}_{min}} \right), \qquad [109]$$

where $\mathcal{R}_{max}$ and $\mathcal{R}_{min}$ are the maximum and minimum values of the reluctance. If most of the reluctance is in the air gap,

$$\mathcal{R}_{max}\mathcal{R}_{min} \approx \mathcal{R}_a^2, \qquad [110]$$

and therefore

$$\Delta L \approx 4\pi N^2 \left(\frac{\mathcal{R}_{max} - \mathcal{R}_{min}}{\mathcal{R}_a^2} \right). \qquad [111]$$

An approximate expression for the inductance is

$$L \approx \frac{4\pi N^2}{\mathcal{R}_a}. \qquad [112]$$

Thus the maximum *fractional* change in inductance for the specified range in flux density is found through division of Eq. 111 by Eq. 112;

thus
$$\frac{\Delta L}{L} \approx \frac{\mathcal{R}_{max} - \mathcal{R}_{min}}{\mathcal{R}_a} \qquad [113]$$

$$\approx \frac{\dfrac{l_s}{A_s}\left(\dfrac{1}{\mu_{s\,min}} - \dfrac{1}{\mu_{s\,max}}\right)}{\dfrac{l_a}{\mu_a A_a}} \qquad [114]$$

$$\approx \frac{l_s}{l_a}\frac{A_a}{A_s}\left(\frac{\mu_a}{\mu_{s\,min}} - \frac{\mu_a}{\mu_{s\,max}}\right) \qquad [115]$$

$$\approx \frac{l_s}{l_a}\frac{A_a}{A_s}\frac{\mu_a}{\mu_{s\,min}}\left(1 - \frac{\mu_{s\,min}}{\mu_{s\,max}}\right), \qquad [116]$$

where $\mu_{s\,max}$ and $\mu_{s\,min}$ are the maximum and minimum values of the permeability of the steel for the specified range in flux density.

Equation 116 suggests a method of approach to the problem of designing a reactor whose inductance must not vary by more than a specified amount when the operating conditions vary. Thus, if δ is the tolerance on the fractional change in inductance,

$$\delta = \frac{\Delta L}{L}, \qquad [117]$$

and from Eq. 116 the ratio of air-gap length to length of the steel path can be expressed in terms of δ; thus

$$\frac{l_a}{l_s} = \frac{1}{\delta}\frac{A_a}{A_s}\frac{\mu_a}{\mu_{s\,min}}\left(1 - \frac{\mu_{s\,min}}{\mu_{s\,max}}\right). \qquad \blacktriangleright[118]$$

Usually the first step in design is to decide upon the operating range in flux density. As soon as this is done, $\mu_{s\,min}$ and $\mu_{s\,max}$ can be determined. The area ratio A_a/A_s in Eq. 118 can be estimated through assuming a stacking factor and a correction for fringing. The minimum value of the ratio l_a/l_s required to satisfy the specified tolerance δ can then be computed from Eq. 118.

5b. *Effects of an Air Gap on Quality Factor.* — Consider the effects of changing the length of an air gap in the magnetic circuit of a specified reactor. The loss ratio of the reactor can be expressed as

$$\text{Loss ratio} = \frac{1}{Q_a} = \frac{\text{core loss} + \text{copper loss}}{\text{reactive volt-amperes}} \qquad [119]$$

$$= \frac{P_c + I^2 R}{E I_m} \qquad [120]$$

$$= \frac{P_c + I_c^2 R + I_m^2 R}{E I_m}, \qquad [121]$$

where

$P_c \equiv$ core loss,
$E \equiv$ rms value of the induced voltage,
$I \equiv$ rms current and equals $\sqrt{I_c^2 + I_m^2}$,
$I_c \equiv$ rms core-loss component of the current,
$I_m \equiv$ rms magnetizing component of the current,
$R \equiv$ resistance of the winding.

Let the frequency and rms value of the flux be maintained constant (by adjustment of the applied voltage as the air gap is changed), and assume that the waveform of the flux is sinusoidal; also neglect the effects of changes in magnetic leakage caused by changes in air-gap length. With these assumptions, the rms value of the flux density in the core is constant. The core loss P_c and the induced voltage E are therefore constants. Since

$$I_c = \frac{P_c}{E}, \qquad\qquad [122]$$

the core-loss current I_c also is constant. Hence, in Eq. 121 only the reactive magnetizing current I_m is affected by changes in the gap length. The magnetizing current must adjust itself to produce the same flux in spite of changes in reluctance due to changes in the air gap.

Equation 121 shows that the loss ratio is affected in two ways. The first effect of increasing the gap length is an improvement in the quality factor Q_c for the magnetic circuit. Thus the loss ratio of the magnetic circuit (including the air gap) is

$$\frac{1}{Q_c} = \frac{P_c}{EI_m}, \qquad\qquad [123]$$

and therefore the increased magnetizing current results in a lower loss ratio; that is, an improved Q_c. This result should be expected, since core loss occurs when pulsating energy is stored in iron, but no loss occurs when it is stored in air. The way to increase Q_c is therefore to store more energy in air; in other words, to insert an air gap in the magnetic circuit or to increase the length of an existing one. On the other hand, the second effect of the increased magnetizing current is a reduction of the quality factor Q_w for the winding. The loss ratio for the winding is

$$\frac{1}{Q_w} = \frac{I^2 R}{EI_m} = \frac{I_c^2 R + I_m^2 R}{EI_m}, \qquad\qquad [124]$$

and, since the magnetizing current I_m usually is considerably greater than the core-loss current I_c, the copper loss is approximately proportional to the square of the magnetizing current. Thus when the magnetizing current is increased by a lengthening of the gap, the copper loss increases more than does the reactive power. Therefore the loss ratio for the winding is increased and the quality factor Q_w is reduced by an increase in gap length.

The loss ratio for the reactor is the sum of the loss ratios for the magnetic circuit and winding, one of which is reduced while the other is increased by the increase in the gap length. To what extent, then, should the magnetizing current be increased by the insertion of an air gap when a minimum over-all loss ratio is desired? From Eq. 121,

$$\text{Over-all loss ratio} = \frac{P_c + I_c^2 R}{EI_m} + \frac{RI_m}{E}, \qquad [125]$$

and for constant frequency and flux density, I_m is the only variable when the gap length is changed. To determine the value of I_m that results in a minimum loss ratio, Eq. 125 is differentiated with respect to I_m and the derivative set equal to zero; thus

$$0 = -\frac{P_c + I_c^2 R}{EI_m^2} + \frac{R}{E} \qquad [126]$$

or

$$I_m^2 R = P_c + I_c^2 R. \qquad [127]$$

Therefore if the frequency and rms value of the flux density are maintained constant, the minimum loss ratio is obtained when the gap length is adjusted so that the copper loss due to the magnetizing current equals the sum of the core loss and the copper loss due to the core-loss current.

6. MODEL THEORY APPLIED TO ENTIRE NETWORKS

In many applications of nonlinear circuit elements in networks, the effects of the nonlinear element on the entire network must be taken into account. In many control circuits using nonlinear elements, for instance, the nonlinearity is the source of the desired action of the circuit. One useful application of model theory is in the design of nonlinear networks which are to have, for example, certain voltage and current values and are to behave in a specified manner with respect to time. The voltage and current values and the time constants in the desired network may be large and the apparatus therefore expensive. A prototype network, similar in arrangement to the desired network but having smaller voltage and current values and smaller time constants, can be set up and adjusted experimentally until it has the proper characteristics. The full-size network can then be designed through suitably scaling up the dimensions and parameters. Suppose that such a prototype network has been set up and adjusted to behave in the desired manner. A new network is to be designed to behave in a similar manner, but with the following changes in scale:

(1) The time intervals between corresponding events in the derived network are to be k_t times as great as in the prototype.

(2) All voltages in the derived network are to be k_v times as great as in the prototype.

(3) All currents in the derived network are to be k_i times as great.

The scale factors k_t, k_v, and k_i may be chosen arbitrarily. What changes must be made in the circuit parameters and in the design of the reactor to make the derived network behave similarly to the prototype? This question can be answered by comparison of the differential equations for corresponding branches of the two networks,[4] as shown in the following analysis.

6a. *Relations among the Variables.* — First consider the effects of the desired scale factors on the relations among the variables in the two networks. Let

t = time,

e_s = instantaneous source electromotive force,

v = instantaneous voltage drop through a circuit branch,

i = instantaneous current,

q = instantaneous charge.

Let values pertaining to the prototype be designated by subscript 0 and those pertaining to the derived network by subscript 3. The variables and parameters in both networks are expressed in the same system of units. According to the desired time scale factor, corresponding instants of time are when

$$t_3 = k_t t_0. \tag{128}$$

The times are measured from reference points when corresponding events take place in the two networks — for example, when corresponding switches are closed. According to the desired voltage and current scale factors, at corresponding instants of time

$$e_{s3}(t_3) = k_v e_{s0}(t_0) \tag{129}$$

$$v_3(t_3) = k_v v_0(t_0) \tag{130}$$

$$i_3(t_3) = k_i i_0(t_0). \tag{131}$$

The relation between corresponding instantaneous values of the charges and of the rates of change of corresponding currents in the two

[4] An alternative approach is through dimensional analysis. For example, see Frederick E. Fowle, editor, *Smithsonian Physical Tables* (8th ed. [revised]; Washington: The Smithsonian Institution, 1933), p. xxxiii.

networks now can be determined from Eqs. 128 and 131. By differentiation of Eq. 128,

$$\frac{dt_3}{dt_0} = k_t \qquad\qquad [132]$$

or

$$dt_3 = k_t dt_0, \qquad\qquad [133]$$

and, by differentiation of Eq. 131,

$$\frac{di_3}{di_0} = k_i \qquad\qquad [134]$$

or

$$di_3 = k_i di_0. \qquad\qquad [135]$$

The functional notation is omitted from Eqs. 134 and 135 and from subsequent equations, with the understanding that the symbols for the variables mean their values at the corresponding instants of time specified by Eq. 128. Equations 133 and 135 state that an increment of time dt_0 in the prototype network corresponds in the derived network with a time increment dt_3 which is k_t times dt_0, and that the resulting increment di_3 of current in the derived network is k_i times the corresponding increment di_0 in the prototype. The time derivative of current in the prototype can be expressed as

$$\frac{di_3}{dt_3} = \frac{di_0}{dt_0}\frac{di_3}{di_0}\frac{dt_0}{dt_3}, \qquad\qquad [136]$$

and, on substitution of Eqs. 132 and 134 in Eq. 136,

$$\frac{di_3}{dt_3} = \frac{k_i}{k_t}\frac{di_0}{dt_0}. \qquad\qquad [137]$$

▶ At corresponding instants of time, the rate of change of current in the derived network is k_i/k_t times as great as that in the prototype, because the current increment is k_i times as great, but the time increment is k_t times as great.◀

The charge in the derived network is

$$q_3 = \int i_3 dt_3, \qquad\qquad [138]$$

in which $\int i_3 dt_3$ is understood to mean $\int_0^{t_3} i_3 dt_3 + q_3(0)$, where $q_3(0)$ is the charge when t_3 equals 0. If i_3 is expressed in terms of the corresponding value of i_0 (Eq. 131) and dt_3 is expressed in terms of dt_0 (Eq. 133), Eq. 138 becomes

$$q_3 = k_i k_t \int i_0 dt_0 = k_i k_t q_0. \qquad\qquad [139]$$

▶ The charge in the derived network is $k_i k_t$ times the corresponding instantaneous charge in the prototype, because the current is k_i times as great and the current exists for k_t times as long. ◀

6b. *Relations between Corresponding Linear Parameters.* — The relations between corresponding parameters in the linear portions of the two networks now can be determined by comparison of the voltage equations for two corresponding branches. Thus a typical branch equation for the derived network is

$$v_3 = L_3 \frac{di_3}{dt_3} + R_3 i_3 + \frac{q_3}{C_3} , \qquad [140]$$

and for the prototype at the corresponding instant of time

$$v_0 = L_0 \frac{di_0}{dt_0} + R_0 i_0 + \frac{q_0}{C_0} , \qquad [141]$$

where L, R, and C are linear parameters. If Eq. 141 is multiplied by k_v and the variables are expressed in terms of the corresponding variables in the derived network — as in Eqs. 130, 131, 137, and 139 — then Eq. 141 becomes

$$k_v v_0 = v_3 = \frac{k_v k_t}{k_i} L_0 \frac{di_3}{dt_3} + \frac{k_v}{k_i} R_0 i_3 + \frac{k_v}{k_i k_t} \frac{q_3}{C_0} . \qquad [142]$$

Comparison of Eqs. 140 and 142 shows that the coefficients of corresponding terms in the two differential equations must be equal. For similar conditions with the desired time, voltage, and current scale factors, the relations between corresponding parameters in the linear portions of the two networks therefore must be

$$L_3 = \frac{k_v k_t}{k_i} L_0 \qquad \qquad ▶[143]$$

$$R_3 = \frac{k_v}{k_i} R_0 \qquad \qquad ▶[144]$$

$$C_3 = \frac{k_i k_t}{k_v} C_0. \qquad \qquad ▶[145]$$

Note that

$$\frac{L_3}{R_3} = k_t \frac{L_0}{R_0} \qquad \qquad [146]$$

and

$$R_3 C_3 = k_t R_0 C_0. \qquad \qquad [147]$$

That is, the time constants of the derived circuit are k_t times the corresponding time constants of the prototype.

6c. *Relations between the Derived and Prototype Reactors.* — Two changes can be made — turns and size — if strict geometrical similarity is maintained between the derived reactor and the prototype. Instead of restricting the analysis to geometrically similar reactors as in Arts. 2 and 3, the following analysis assumes that magnetic leakage can be neglected.* The method of analysis presented here merely illustrates a line of reasoning that can be applied under a variety of assumptions. The assumption of no leakage permits the cross-sectional area of the core to be changed independently of the core length, subject to the following restrictions when the cross-sectional area is not the same throughout the length: First, the cross-sectional areas, if changed, must all be changed in the same ratio; second, the lengths of parts of the core, if changed, must all be changed in the same ratio. The length and area ratios may be different, however. This statement includes any air gaps that may be in the magnetic circuit.

The one condition that cannot be altered if the similarity relations between the derived reactor and the prototype are to hold is that the $\mathfrak{B}$ and $\mathfrak{H}$ conditions must be the same at corresponding points in the two cores. Thus if

$\mathfrak{l}$ ≡ instantaneous flux density,

$\mathfrak{h}$ ≡ instantaneous magnetic field intensity,

then

$$\mathfrak{l}_3(t_3) = \mathfrak{l}_0(t_0) \tag{148}$$

$$\mathfrak{h}_3(t_3) = \mathfrak{h}_0(t_0). \tag{149}$$

The voltages induced in the reactors at corresponding instants of time are

$$e_3 = N_3 A_3 \frac{d\mathfrak{l}_3}{dt_3} \tag{150}$$

$$e_0 = N_0 A_0 \frac{d\mathfrak{l}_0}{dt_0}, \tag{151}$$

where N is the number of turns and A is the cross-sectional area at the point where the flux density is $\mathfrak{l}$. According to the desired voltage scale, the voltage induced in the derived reactor must be k_v times the voltage induced in the prototype at corresponding instants, and therefore the quotient of Eqs. 150 and 151 is

$$\frac{e_3}{e_0} = k_v = \frac{N_3 A_3 d\mathfrak{l}_3/dt_3}{N_0 A_0 d\mathfrak{l}_0/dt_0}. \tag{152}$$

* The assumption of no leakage could equally well have been used as a point of departure for reactor model theory similar to that given in Arts. 2 and 3, with somewhat different results.

But since ℓ_3 equals ℓ_0, the quotient of the derivatives in Eq. 152 is

$$\frac{d\ell_3/dt_3}{d\ell_0/dt_0} = \frac{dt_0}{dt_3} = \frac{1}{k_t} . \tag{153}$$

Consequently, Eq. 152 becomes

$$k_v = \frac{N_3 A_3}{k_t N_0 A_0} \tag{154}$$

or

$$N_3 A_3 = k_v k_t N_0 A_0. \qquad \blacktriangleright[155]$$

▶ To maintain the same $\mathscr{B}$ and $\mathscr{H}$ conditions in the core, the turns-area product $N_3 A_3$ in the derived reactor must be $k_v k_t$ times the turns-area product $N_0 A_0$ in the prototype, because the voltage in the derived network is k_v times as great, and the time rate of change of the flux density is $1/k_t$ times as great.◀

The reactor design must also be changed so that for the same $\mathscr{B}$ and $\mathscr{H}$ conditions the current is changed by the desired current scale factor k_i. The relations between the currents and the magnetic field intensities in the two reactors are

$$4\pi N_3 i_3 = \mathscr{H}_3 \ell_3 \tag{156}$$

$$4\pi N_0 i_0 = \mathscr{H}_0 \ell_0, \tag{157}$$

where ℓ is the core length. From the ratio of Eqs. 156 and 157,

$$\frac{N_3 i_3}{N_0 i_0} = \frac{\mathscr{H}_3 \ell_3}{\mathscr{H}_0 \ell_0} . \tag{158}$$

But for corresponding instants of time $\mathscr{H}_3$ equals $\mathscr{H}_0$ (Eq. 149) and from the desired current scale (Eq. 131),

$$\frac{i_3}{i_0} = k_i. \tag{159}$$

Therefore Eq. 158 reduces to

$$k_i \frac{N_3}{N_0} = \frac{\ell_3}{\ell_0} \tag{160}$$

or

$$\frac{\ell_3}{N_3} = k_i \frac{\ell_0}{N_0} . \qquad \blacktriangleright[161]$$

▶ To maintain the same $\mathscr{B}$ and $\mathscr{H}$ conditions in the core, the length-to-turns ratio ℓ_3/N_3 in the derived reactor must be k_i times the length-to-turns ratio ℓ_0/N_0 in the prototype when the current is k_i times as great.◀

Three changes can be made in the design of the reactor — changes in turns N, in core length l, and in core area A — but only the two conditions expressed by Eqs. 155 and 161 must be satisfied for these changes to result in the same $\mathfrak{B}$ and $\mathfrak{K}$ conditions at corresponding times and with specified time, voltage, and current scale factors. Thus one additional condition can be chosen arbitrarily — for example, the turns may be kept constant. The voltage change could then be brought about through altering the core area as in Eq. 155, and the current change through altering the core length as in Eq. 161.

Noting the relations between the stored energies in the two systems is worth while. It can be shown that the ratio of the stored energies W_3 and W_0 is

$$\frac{W_3}{W_0} = \frac{\frac{1}{2}C_3 v_3^2}{\frac{1}{2}C_0 v_0^2} = \frac{\frac{1}{2}L_3 i_3^2}{\frac{1}{2}L_0 i_0^2} \qquad [162]$$

$$= \frac{\text{volume of core 3}}{\text{volume of core 0}}. \qquad [163]$$

One restriction that may not be obvious is implicit in Eqs. 148 and 149. The $\mathfrak{B}(\mathfrak{K})$ relation must be the same in both prototype and derived reactors. Hence the actual $\mathfrak{B}(\mathfrak{K})$ characteristic, including the effect of eddy currents, must be the same for both. If the time scale is changed, the effects of the eddy currents are altered, and therefore the $\mathfrak{B}(\mathfrak{K})$ characteristics of the two reactors are not identical unless the lamination thickness is also suitably changed at the same time. Practically this restriction may usually be unimportant, but it should be noted in order to avoid difficulty from this source in unusual cases.

Sometimes difficulty may be encountered because the time constants of small inductive apparatus are inherently smaller than those of large apparatus unless an appropriate value of k_l is used.

▶ The concept of similar nonlinear circuits is a very important one that has many practical applications to problems that would be very difficult, economically or analytically, to solve without it. Whenever nonlinear apparatus of any kind, electrical or otherwise, is to be developed or designed, model methods should receive serious consideration.◀

PROBLEMS

1. A 500-kva 11,000-v 60 $\sim$ transformer takes 3.35 amp and 2,960 w at no load when rated voltage is applied at rated frequency. Another transformer has a core with all its linear dimensions $\sqrt{2}$ times as large as the corresponding dimensions of the core of the first transformer and with magnetic properties identical with those of the first transformer. The primary copper loss and the leakage-reactance voltage drop at no load may be neglected in both transformers.

If the same number of turns are used for both the primary windings, what no-load current and no-load power will the second transformer take when a voltage of 22,000 v at 60 ∼ is impressed on its primary winding?

2. **The following data apply to a shell-type reactor with a winding of 1,000 turns:**

Frequency	Terminal Voltage	Current	Power
60 ∼	180 v	0.0624 amp	1.13 w
60	160	0.0239	0.91
60	140	0.0145	0.71
60	120	0.0106	0.54

Assuming that the copper loss and the effective resistance drop are negligible find the number of turns and the scaling factor k to give a geometrically similar reactor with a power loss of 10 w and a current of 0.30 amp at a terminal voltage of 120 v, 60 ∼.

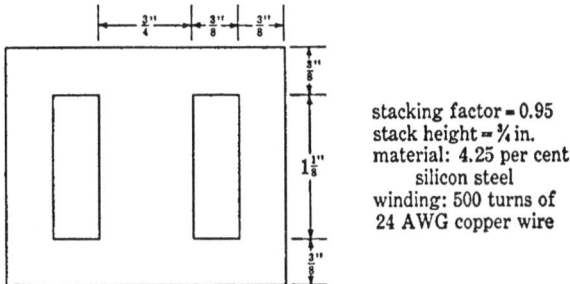

stacking factor = 0.95
stack height = ¾ in.
material: 4.25 per cent
  silicon steel
winding: 500 turns of
24 AWG copper wire

FIG. 3.   Cross-section of reactor core, Prob. 3.

3. A cross section of the core of a reactor is shown in Fig. 3. Core and winding data are given with the figure. Magnetic data for the core material are given in Figs. 19, 20, and 21, pp. 190–192.

(a) Calculate the quality factor of the reactor for an impressed sinusoidal voltage of 50 v at 60 ∼.

(b) Calculate the quality factor of a geometrically similar reactor operating at the same flux density and with the same effective inductance, but at a voltage of 150 v at 60 ∼.

4. In the reactor of Prob. 3 an air gap is to be cut of such length that the reactor will have not greater than a 5% change in inductance, referred to its maximum value, when a sinusoidal applied voltage is varied from 25 to 60 v at 60 ∼.

(a) Find the minimum permissible length of the air gap.

(b) Give complete specifications for a reactor with an inductance of 1 h at 50 volts and a 5% change in inductance over the range 20.8 to 50 v at 60 ∼.

5. Find the length of air gap to be cut in the core of the reactor of Prob. 3 to give a maximum quality factor if the flux density is to be held at the value used in that problem. Compare the value of quality factor obtained with the air gap to that obtained in Prob. 3.

6. A nonlinear circuit employing an iron-core inductor which is used in some control systems is shown in Fig. 4. Such a circuit can be designed to have a critical voltage, below which only a small current flows, but above which the current increases very rapidly.

Experimentally obtained data of current versus voltage for such a circuit operating on 60 ∼ are given below:

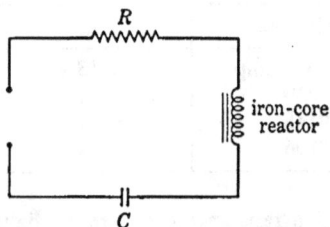

Current	Voltage
0.02 amp	10 v
0.02	50
0.02	80
0.12	95
0.60	100
0.80	102
1.00	107
1.20	116
1.40	128

Fig. 4. Nonlinear circuit to give critical voltage control, Prob. 6.

In the circuit from which these data were obtained the resistance R was 67 ohms and the capacitance C was 25 μf. Data on the iron-core inductor are as follows:

Mean length of magnetic circuit = 7.4 in.
Center-leg area of core = 1.3 sq in.
Winding: 480 turns.

A circuit of this type is to be built to actuate a voltage regulator on a 50 ∼ system whenever the voltage reaches 117 v. A small relay, possessing very low inductance, is to be connected in series with the other elements shown in Fig. 4. This relay, which trips at a current of 1 amp, will set the regulator in operation.

Determine the values of R and C which, when used with a suitable inductor, will give most sensitive operation in the application described. For the condition that the inductor for the new control circuit is to be built of punchings identical with those in the experimental reactor, determine its core area and number of turns. Assume that the magnetic relay has no effect on the electrical characteristics of the circuit.

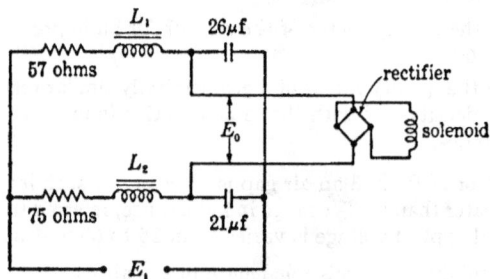

Fig. 5. Voltage-selective nonlinear bridge circuit, Prob. 7.

7. Two nonlinear circuit branches of the type discussed in Prob. 6 when combined in a bridge arrangement give a circuit with exceptional voltage selectivity. Since the output from such a circuit can be used to close a magnetic contactor, the circuit is suitable for use as a voltage-selective relay. A complete relay circuit is shown in Fig. 5.

Voltage data obtained from this circuit at a frequency of 60 $\sim$ are given below:

E_1 — volts	E_0 — volts
50	0
90	1
100	2
109	10
110	88
111	12
120	3
130	2

Two of these selective circuits are to be connected in parallel on a 1,000 $\sim$ system to control two different relays. The input voltages E_1 at which the circuits are sensitive are to be 80 and 120 v. The maximum voltage output E_0 of each bridge circuit should be 60 v. If the iron-core reactors are all geometrically similar to the model, specify the turns ratio a and the linear scaling factor k for each reactor. Specify also the resistances and capacitances. Assume that the magnetic relays have no effect on the electrical characteristics of the circuits.

CHAPTER VIII

Thermal Properties and Heat Flow

In general, all losses in electric power apparatus are converted into heat, in the ratio of 3,413 British thermal units for each kilowatt-hour of energy so converted. When a steady operating temperature is reached, the heat carried away by the combined effects of conduction, convection, and radiation must equal the heat produced by conversion from electrical and mechanical forms of energy. The usual and in fact almost universal criterion of power output rating of electrical machinery is the temperature rise. The maximum safe continued load is that at which the steady temperature is at the highest safe operating point. Since heating, rather than electrical or mechanical considerations directly, determines the permissible output of a machine, a careful study of thermal properties and of heat flow is a very important element in design.

1. EFFECTS OF HEAT ON MATERIALS

In electrical machinery, the insulating materials suffer most from overheating, but the effect of heat on the current-conducting copper and on the magnetic material is not negligible.

The resistance of copper through a large temperature range varies in proportion to its excess in temperature over minus 234.5 degrees centigrade. As a consequence, the ohmic copper loss at 40 degrees centigrade above ordinary room temperature is about 15 per cent higher than that at room temperature.

Core losses in the magnetic circuit are slightly reduced as the temperature is raised the normal amount — approximately 40 degrees centigrade — associated with full-load operation.

Insulating materials in common use in electrical machinery and equipment have been classified by the American Institute of Electrical Engineers into the four groups listed in Table I.

For purposes of standardization, the following values of maximum "hottest-spot" temperatures have been established by the American Institute of Electrical Engineers:

Class O	90 C
Class A	105 C
Class B	130 C
Class C	No limit selected

Standard limiting ambient temperature of the air is chosen as 40 degrees centigrade, which is near the maximum encountered in most parts of the United States.

TABLE I

CLASSIFICATION OF INSULATING MATERIALS[1]

Class *Description of Material*

O   Class O insulation consists of cotton, silk, paper, and similar organic materials when neither impregnated nor immersed in a liquid dielectric.

A   Class A insulation consists of: (1) cotton, silk, paper, and similar organic materials when either impregnated or immersed in a liquid dielectric; (2) molded and laminated materials with cellulose filler, phenolic resins, and other resins of similar properties; (3) films and sheets of cellulose acetate and other cellulose derivatives of similar properties; and (4) varnishes (enamel) as applied to conductors.

B   Class B insulation consists of mica, asbestos, fiber glass, and similar inorganic materials in built-up form with organic binding substances. A small proportion of Class A materials may be used for structural purposes only.

C   Class C insulation consists entirely of mica, porcelain, glass, quartz, and similar inorganic materials.

[1] From " American Institute of Electrical Engineers Standards," No. 1: *General Principles upon Which Temperature Limits are Based in the Rating of Electrical Machinery and Apparatus* (New York: American Institute of Electrical Engineers, 1940), 6.

The deterioration of organic insulation, Classes A and B, under high temperatures is gradual and is a relatively smooth function of the temperature. It shows up principally in a drying out and charring of the material, effecting a brittleness and loss of mechanical strength rather than loss of dielectric strength directly. After very severe charring or carbonizing, the dielectric strength may be seriously impaired, but failure is more generally associated with mechanical failure of the insulation caused by vibration or short-circuit mechanical stresses.

V. M. Montsinger[2] by repeated tests has shown that the rate of mechanical deterioration doubles for each 8-degree centigrade increase in temperature. Deterioration is more rapid with oil impregnation than without it. Class A insulation in oil was found to lose half its tensile strength after 18 weeks at 110 degrees centigrade, after 40 weeks at 100 degrees, and after 85 weeks at 90 degrees. Although the presence of oil or other impregnating material may hasten the deterioration of the fibrous insulation, it has the compensating advantages of greatly increasing the heat conductivity and of increasing the dielectric strength.

Hot oil in the presence of air or oxygen will slowly oxidize and form a

[2] V. M. Montsinger, "Loading Transformers by Temperature," *A.I.E.E. Trans.*, *49* (April, 1930), 776–790.

sludge. In oil-cooled apparatus such as transformers, this sludge by sticking to surfaces interferes with convection and the free flow of heat away from its sources. Oxidation and sludge formation are combated by the use of inert gas over the oil and by continual filtering and inspection, with replacement when necessary.

2. CONDUCTION OF HEAT

Heat conduction through the volume of a substance is very nearly proportional to the temperature gradient. The steady conduction of heat through any volume is a field phenomenon analogous mathematically to many others, such as the passage of dielectric flux through a non-conducting dielectric or of magnetic flux through a field. If heat is generated within the volume of the material which conducts it, the dielectric field with space charge is analogous. The simplest example of heat flow is the one in which the direction of flow is everywhere parallel, and here the rate of heat flow per degree conducted through a plate or sheet of material is equal simply to the product of thermal conductivity times section area, divided by thickness. In equation form, the temperature difference resulting from a given rate of heat flow is

$$\theta_2 - \theta_1 = \frac{Pd}{Ak}, \tag{1}$$

where

θ_2 and θ_1 are the steady temperatures on the two sides,
P is the rate of heat flow,
d is the thickness in the direction of heat flow,
A is the section area in a plane perpendicular to the direction of heat flow,
k is the thermal conductivity.

For simple radial flow between radii r_2 and r_1 of a cylindrical shell,

$$\theta_2 - \theta_1 = \frac{P_u}{2\pi k} \ln \frac{r_2}{r_1}, \tag{2}$$

where P_u is the rate of heat flow per unit length of the cylinder.

It is convenient and general practice in dealing with thermal problems associated with electrical machinery to use units of watts, inches, and degrees centigrade. The element of time is included in the watt unit, which is a *rate* of flow of energy, rather than a unit of energy or heat simply.

Table II presents approximate values of thermal conductivities, resistivities, and specific thermal capacities of materials commonly used in electric power equipment. The thermal conductivity of a material is

TABLE II

APPROXIMATE THERMAL PROPERTIES OF MATERIALS

Material	*Resistivity* Deg. C inch /watt	*Conductivity* Watts /deg. C inch	*Specific thermal capacity* Watt-sec /cu inch deg. C
Copper	0.11	9.0	58
Aluminum	0.30	3.3	40
Wrought iron	0.50	2.0	64
Sheet carbon steel	0.90	1.1	64
Cast iron	1.0	1.0	64
Silicon steel (longitudinal)	2.3	0.44	64
Silicon steel sheets (transverse)	15–50	0.07–0.02	64
Mica (transverse)	140	0.007	35
Varnished cambric	200	0.005	25
Rubber	250	0.004	..
Pressboard, oiled	250	0.004	25
Brick	400	0.0025	..
Pressboard, dry	400–500	0.002–0.0025	..
Untreated cloth or felt	600	0.0016	..

the rate of heat flow through a unit cube per unit temperature difference between two opposite faces. From Eq. 1,

$$k = \frac{P}{A \dfrac{(\theta_2 - \theta_1)}{d}}, \qquad [3]$$

where $(\theta_2 - \theta_1)/d$ is the temperature gradient. From Eq. 3, the dimensions of thermal conductivity based on units of watts, inches, and degrees centigrade are

$$[k] = \frac{[\text{watts}]}{[\text{in.}]^2 \dfrac{[\text{deg. C}]}{[\text{in.}]}} = \frac{[\text{watts}]}{[\text{deg. C}][\text{in.}]} \qquad [4]$$

Thermal resistivity is the reciprocal of thermal conductivity and can be expressed in degrees centigrade inch per watt. The specific thermal capacity of a material is the heat energy stored in unit volume for unit temperature rise and can be expressed in watt-seconds (or joules) per cubic inch per degree centigrade.

The thermal conductivity is greatly increased in fibrous material with impregnation, as has been stated. The relatively new fibrous glass insulation, without impregnation, could successfully withstand an astonishingly high temperature, but the advantage of it would be largely nullified by its low thermal conductivity. With impregnation, its safe maximum operating temperature is set by the impregnating material rather than by the fiber.

Transverse conduction of heat through stacks of thin steel laminations encounters its principal thermal resistance in the varnish and air spaces between the laminations. Any variations in thickness of varnish and looseness of packing will have corresponding effects on the transverse resistivity. In general, the data given in Table II are to be taken as only approximate.

For the general two-dimensional flow of heat, such as that in a long slotted field or armature, field mapping by curvilinear squares may be used.* In many practical problems there is a generation of heat throughout the conducting material. This complicates the problem, and a trial and error method of solution is often used. Sometimes the volume is arbitrarily divided into layers and the drops through the layers are added. The heat losses in the layers are added successively to the total heat flow.

The entire calculation of heat flow in geometrically complicated machines is somewhat too involved a problem of detailed design to be illustrated completely here. Fuller treatments will be found in some of the references listed in the bibliography.

3. RADIATION OF HEAT

Radiated heat is energy in the form of infra-red electromagnetic waves. Radiation is not of much importance in the cooling of high-power electrical machinery, but, for small devices having poor ventilation characteristics, its effect is appreciable. The cooling of the anodes in some thermionic tubes is largely dependent on radiation.

Clean air and of course vacuum sustain radiation with negligible loss. Glass, particularly that containing ferrous oxide FeO, has considerable absorption; water has even more.

The fundamental law of heat radiation from a surface of temperature θ degrees centigrade is

$$P = Ke(273 + \theta)^4, \qquad [5]$$

known as Stefan's law of emission. Here

P is the rate of emission,
K is a constant depending primarily on the size of the surface,
$273 + \theta$ is the absolute temperature on the centigrade scale,
e is the emissivity coefficient of the surface material.

---

* In a region where there are no sources, Laplace's equation describes the field. For a brief discussion of Laplace's equation and the general principles of field mapping, see the volume on electric circuits, Arts. 6a and 7, Ch. I.

The range of e is from unity for the ideal black body to about 0.025 for a polished mirror surface.

TABLE III

APPROXIMATE THERMAL EMISSIVITY COEFFICIENTS

Material	Emissivity e
Ideal black body	1.00
Lampblack	0.98
Rough insulating material	0.90
Oxidized iron	0.75
Oxidized copper	0.65
Oxidized aluminum	0.15
Polished aluminum	0.05
Polished copper	0.04
Polished silver	0.025

In view of the fact that radiant heat from a rough, irregular, or convoluted body may reach other parts of the same body, a precise analytical determination of radiation effects would be a stupendously complicated problem dependent on the exact geometry of the body in question and all its surroundings, as well as their thermal coefficients. For many purposes, however, a simple approximate method will suffice.

The Stefan law, for radiated heat expressed in watts per square inch, is

$$p = 37e \left(\frac{\theta + 273}{1,000} \right)^4 \text{ watts/sq in.} \tag{6}$$

The body which is radiating heat at this rate will be receiving heat by radiation from its surroundings. If the temperature of the body is θ degrees centigrade, and it is completely surrounded by a much larger enclosure of temperature θ_1, the *net* radiation of heat will be

$$p = 37e \left[\left(\frac{\theta + 273}{1,000} \right)^4 - \left(\frac{\theta_1 + 273}{1,000} \right)^4 \right] \text{ watts/sq in.} \tag{7}$$

In the cooling of ordinary electrical machinery, the difference between the temperature of the warm machine and that of its enclosure (that is, the inner walls and ceiling) is a small fraction, perhaps one-tenth, of the absolute temperature. If $\theta_a + \Delta$ is the absolute temperature of the device, and θ_a that of the surrounding walls, the net radiation is

$$p = 37e10^{-12}[(\theta_a + \Delta)^4 - \theta_a^4] \tag{8}$$
$$= 37e10^{-12}(4\theta_a^3\Delta + 6\theta_a^2\Delta^2 + 4\theta_a\Delta^3 + \Delta^4)$$

If Δ is small compared with θ_a,

$$p \approx 37e10^{-12}(4\theta_a^3\Delta) \text{ watts/sq in.} \tag{9}$$

The net radiated watts per square inch for small temperature differentials is therefore approximately

$$p \approx 0.148e \left(\frac{273 + \theta}{1,000} \right)^3 \Delta \text{ watts/sq in.}, \qquad [10]$$

where Δ is the temperature rise in degrees centigrade above ambient temperature. The variation of p with Δ is approximately linear for small temperature rises. If a moderate ambient temperature of 25 degrees centigrade is assumed, the net radiation per square inch is found to be

$$p \approx 0.0039e\Delta \text{ watts/sq in.} \qquad [11]$$

A rise of 40 degrees centigrade would thus dispose of about one-sixth watt per square inch by radiation with an emissivity near unity.

For very large rises in temperature above that of the surroundings, the heat absorbed by the hot body through reception of radiation from its surroundings may be negligible, and the total heat radiated away taken as the net radiated heat.

Between parallel surfaces of emissivity coefficients e_1 and e_2, the effective emissivity coefficient e can be shown to be

$$e = \frac{1}{\dfrac{1}{e_1} + \dfrac{1}{e_2} - 1} \qquad [12]$$

For a small radiating body of surface area A_1 and emissivity e_1 surrounded by a larger surface of area A_2 and emissivity e_2, the effective emissivity e is given by

$$e = \frac{1}{\dfrac{1}{e_1} + \dfrac{A_1}{A_2} \left(\dfrac{1}{e_2} - 1 \right)}. \qquad [13]$$

Note that when A_1/A_2 equals 0, e equals e_1; and when A_1 equals A_2, this expression becomes that for parallel surfaces.

According to A. D. Moore,[3] the increases in radiation obtained with square slots are those given in Table IV. Convoluted surfaces in general afford greater radiation than flat surfaces of the same projected area.

4. CONVECTION OF HEAT

Though air is an extremely poor conductor of heat, any hot body immersed in a cooler atmosphere will set up natural convection currents which materially assist in cooling. The amount of heat dissipated by

[3] A. D. Moore, *Fundamentals of Electrical Design* (New York: McGraw-Hill Book Company, Inc., 1927), 116.

TABLE IV

APPROXIMATE EFFECT OF SQUARE SLOTS ON RADIATION OF HEAT

Emission Coefficient of Plain Surface	Effective Coefficient* with Square Slots
0.25	0.48
0.50	0.75
0.75	0.89
1.00	1.00

* This is the effective emission coefficient to be used with the projected area.

natural convection in air is of the same order of magnitude as that dissipated by radiation, for temperature differences such as exist ordinarily in electrical machinery. However, by the use of forced currents of air or other gases, or by the use of fluids such as oil or water having much higher thermal capacity (by volume) than air, the dissipation of heat may be increased many times over.

The quantity of heat carried away by natural convection in air is a nonlinear function of the size, shape, surface material, condition, and orientation of the hot body; its temperature and that of the surrounding air; and the character of the surroundings especially insofar as they may affect the free flow of air currents. It is obvious that general convection constants cannot exist, since the relations are nonlinear. Table V will, however, serve to convey a general idea of the quantities involved.

The total rate of dissipation varies as the 1.25 power of temperature rise in each case. For the first position,

$$p = 0.00157\Delta^{1.25} \text{ watts/sq in.,} \quad\quad [14]$$

in which Δ is the temperature rise in degrees centigrade. For the second (vertical) position,

$$p = 0.00123\Delta^{1.25} \text{ watts/sq in.} \quad\quad [15]$$

and, for the third,

$$p = 0.00081\Delta^{1.25} \text{ watts/sq in.} \quad\quad [16]$$

In general, smaller bodies are capable of dissipating more heat per square inch because of the smaller effect on local air temperature by other parts of the heated body. The air to which the center of a three-foot-square hot plate is exposed is hot because of the large adjacent hot surface; a minute surface of the same temperature would have much cooler air available to carry away the heat, other things being equal.

Heat carried away under forced ventilation varies almost linearly with velocity, for velocities up to about 5,000 feet per minute, and more slowly thereafter. In long tubes or cooling ducts the air is cooler at

TABLE V

HEAT DISSIPATION IN WATTS PER SQUARE INCH BY NATURAL
CONVECTION FROM A THREE–FOOT–SQUARE STEEL
OR ALUMINUM PLATE[4]

Deg. C Rise	Horizontal Facing Upward		Vertical		Horizontal Facing Downward	
	Per Deg.	Total	Per Deg.	Total	Per Deg.	Total
0	..	0	..	0	..	0
10	0.0027	0.027	0.0021	0.021	0.0014	0.014
20	0.0033	0.066	0.0025	0.050	0.0017	0.034
30	0.0036	0.108	0.0029	0.087	0.0019	0.057
40	0.0039	0.156	0.0031	0.124	0.0021	0.084
50	0.0041	0.205	0.0033	0.165	0.0022	0.110
60	0.0043	0.258	0.0034	0.204	0.0023	0.138
70	0.0045	0.315	0.0035	0.245	0.0024	0.168
80	0.0047	0.376	0.0036	0.288	0.0024	0.192
90	0.0048	0.432	0.0037	0.333	0.0025	0.225
100	0.0049	0.490	0.0038	0.380	0.0025	0.250

[4] From E. Griffiths and A. H. Davis, " The Transmission of Heat by Radiation and Convection," *Special Report 9* (London: H. M. Stationery Office, 1931). See also A. D. Moore, *Fundamentals of Electrical Design* (New York: McGraw-Hill Book Company, Inc., 1927), 119.

entrance and so the heat dissipation per square inch of duct surface falls off as the duct length is increased, other things being equal. Owing to the nonlinear variation of dissipation with shape, size, and length of duct, as well as velocity, usable cooling data require several families of curves for proper presentation. These may be found in various books on electrical machine design. A rough expression giving at least the order of magnitude of dissipation in ducts is 0.01 watt per square inch per 1,000 feet per minute per degree centigrade above average air temperature.

5. Cooling of Electrical Apparatus

From the foregoing discussion and data, it may be seen that the total heat dissipation by natural means, both radiation and convection, is of the order of one-half watt per square inch for a rise of 40 degrees centigrade in devices such as relays or small enclosed motors. This figure is worth remembering as a rough basis for conservative design of apparatus depending on natural cooling. With organic insulation, dissipation at a rate exceeding one watt per square inch would be very likely to cause trouble if long continued. On large machines, the heat dissipation per square inch would be even less if no special provisions for cooling were made.

In electromagnetic machinery, current densities and flux densities are of the same order of magnitude in large and small machines, and therefore losses vary about as the volume, or the cube of the linear dimensions. The surface, varying as the square of linear dimensions, increases more slowly with size than do the losses. Consequently, the larger the machine, the more difficult in general is the design of an adequate cooling system.

The cooling problem grows in importance likewise with size and rating of electron tubes.[5] The problem is of a different nature here, however, because at least the cathode has to work at a temperature of several hundred degrees to provide the necessary electron emission. Heating rather than cooling has to be applied to it. There is no organic insulation in the ordinary tube, and so higher temperatures are permissible, although for some applications the thermal noise developed in the tube makes desirable an anode temperature as low as practicable. The change in size and shape with heating affects the performance of tubes in certain critical circuits. Water-cooled anodes are common in high-power tubes. In general, however, the cooling of electronic devices, while as vital to successful operation, is proportionately less costly of solution than the cooling of power machinery.

5a. *Transformers.** — Transformers of 100 kva or less operating at low voltage may be cooled satisfactorily by natural air circulation around the windings and around the outer case.

In larger sizes, and very often also in small sizes down to 10 kva, the transformer case is filled with oil, which both promotes cooling and increases the dielectric strength. Smaller clearances may be used in an oil-filled transformer, and a more compact design can be attained. The oil is so effective in removing heat from the windings and core that the principal thermal problem in the oil-immersed self-cooled transformer is the disposal of heat from the case. A larger case may be used, but from 30-kva sizes upward it is usual to increase the cooling surface by corrugation, or by the use of external radiators attached to the tank and carrying the circulating oil.

In very large transformers, water is sometimes circulated in copper coils immersed in the oil. This method of cooling requires large quantities of water, but the first cost is usually lower than for transformers equipped with the large radiators that would otherwise be required.

Sometimes the oil is pumped through external tubing immersed in water. This method has the advantage that scale formation inside the tubes is avoided. Any scale from impurities in the water will form on the outside, where it can be cleaned off with relative ease.

---

[5] The effects of heating on the rating of electron tubes are discussed in the volume on electronics, Art. 2, Ch. IV.

* For further discussion of cooling of transformers, see Art. 3, Ch. XI.

The safe capacity of self-cooled oil-immersed transformers may be increased very considerably by the addition of blowers to force air over the radiator surfaces. For example, the receiving-end transformers on the transmission line running from Boulder Dam to Los Angeles* are rated 48,750 kva each when self-cooled, and 65,000 kva with blowers. The increase in rating is $33\frac{1}{3}$ per cent.

5b. *Rotating Machines.* — The armatures of motors and generators are made of sheet steel laminations or punchings and are divided into sections, usually two to three inches long, separated by radial ventilating ducts. The width of the ducts is usually $\frac{3}{8}$ to $\frac{1}{2}$ inch, and the spacing is provided by small I-beams spot-welded to a lamination on one side. Rotation of the machine provides a considerable draft of air, and usually fan blades are attached to at least one end of the rotor to increase the circulation.

Tremendous quantities of cooling air must be blown through very large turbo-alternators. For example, a 50,000-kw unit operating at 98 per cent efficiency would have losses of 1,000 kw, which would require 60,000 cubic feet of air per minute, on the assumption that the air temperature rises 30 degrees centigrade in passing through the machine. To pass this huge quantity of air through the ducts and gaps, hurricane velocities must be used, and the friction losses are very large.

To reduce these friction losses and to promote better cooling, hydrogen has now practically displaced air as the cooling medium in new large high-speed generators and synchronous reactors. In such machines complete enclosure is necessary, so that a few additional precautions against leakage make it feasible to use hydrogen instead of air. The cooling medium, whether air or hydrogen, is ordinarily circulated through cooling and washing units and returned to the machine.

Hydrogen has about the same specific heat *per unit volume* as air, although it is about 14 times lighter. Its viscosity is less, and the energy required to accelerate it to a given velocity is very much less. The over-all windage losses are only about ten per cent as much for hydrogen as for air. Hydrogen is a better conductor of heat.

The enclosing jackets around hydrogen-cooled machines are normally designed with sufficient strength to prevent any external damage due to an explosion of the contained hydrogen when mixed with air in the proportions to give the most explosive mixture. In one respect the use of hydrogen instead of air promotes safety, because the hydrogen would not support combustion of the organic insulating material.

It seems likely also that deterioration with age would be somewhat retarded because of the elimination of slow oxidation. Conclusive evidence

---

* These transformers are shown in Fig. 4, p. 274.

of this is not available, because of the short time hydrogen cooling has been in use.

6. TRANSIENT HEATING

A knowledge of the transient heating characteristics of electrical machines and of conductors is necessary for determination of their short-time overload capabilities, and also of the length of time that short-circuit currents of various magnitudes may be allowed to flow in them without causing serious damage to the insulation. A related problem is the determination of heating under cyclical loads.

Large short-circuit currents produce such rapid heating that heat dissipation is of very minor importance. If dissipation is neglected altogether, and the assumption is made that all the ohmic heating remains in the conductor, a direct calculation can be made of the time t required to bring a conductor from an initial temperature θ_0 up to a temperature θ, due to a steady current I amperes starting at time $t_0 = 0$. The heating rate in watts is I^2R, where R is the resistance of the conductor in ohms. Since the resistance of copper varies as $234.5 + \theta$ over the range considered, the heating rate for a unit length of conductor is

$$\frac{I^2\rho_0}{A} \times \frac{234.5 + \theta}{234.5 + \theta_0} \text{ watts,} \qquad [17]$$

where ρ_0 is the resistivity at θ_0 degrees centigrade and A is the section area. If G designates the thermal capacity per unit volume of the conductor material, AG is the thermal capacity of unit length of the conductor. The rate of temperature rise is proportional to the rate at which heat is developed, and inversely proportional to the thermal capacity, whence

$$\frac{d\theta}{dt} = \frac{I^2\rho_0}{A^2G} \times \frac{234.5 + \theta}{234.5 + \theta_0}. \qquad [18]$$

Integration of this equation gives

$$t = \frac{A^2G}{I^2\rho_0} (234.5 + \theta_0) \ln \frac{234.5 + \theta}{234.5 + \theta_0} \text{ sec.} \qquad [19]$$

For copper, at $\theta_0 = 20$ C,

$$t = 2.13 \times 10^{10} \frac{A^2}{I^2} \ln \frac{234.5 + \theta}{234.5 + \theta_0} \text{ sec,} \qquad [20]$$

where A is in square inches and I in amperes. If it is preferred to express A in circular inches (circular mils $\div 10^6$) and to use common logarithms,

$$t = 3.03 \times 10^{10} \frac{A^2}{I^2} \log_{10} \left(\frac{234.5 + \theta}{234.5 + \theta_0} \right) \text{ sec.} \qquad [21]$$

Example. How long would it take for a piece of No. 0000 copper wire to heat from 20 C to 200 C if it carries 10,000 amperes?

Solution: The area of No. 0000 is 0.212 circular inch.

$$t = 3.03 \times 10^{10} \frac{0.212^2}{10,000^2} \log_{10} \frac{434.5}{254.5} \tag{22}$$

$$= 13.6 \log_{10} 1.70 \tag{23}$$

$$= 3.1 \text{ sec.} \tag{24}$$

The justification for omitting from consideration any loss of the heat developed during this short period may be checked through comparing the watts of heat dissipated at 200 degrees with the I^2R power converted into heat. For the specific problem, this method of treatment is quite satisfactory, because of the small ratio of heat dissipated to heat developed. On the other hand, the same wire carrying a current of only 1,000 amperes (one-tenth as much as before) would by no means heat up to 200 degrees in 310 seconds as indicated by the formula, because the dissipation of heat would be a controlling factor.

In electrical machinery, the transient heating characteristics are usually expressed through statement of the overload capacity for a period of one-half hour, one hour, or two hours, following long continuous operation at full load. A typical machine, for example, having a rise of 40 degrees centigrade under continuous full load, might carry 125 per cent of full load for two hours with a 55-degree total rise above ambient temperature.

Many electrical machines are called upon to carry a cyclical load. Heating may not always be the limiting factor if the loads are very short and severe, as voltage drop, torque, or — in direct-current machines — commutation may control. For short-time cycles of load variation, the heating may be calculated as that due to a hypothetical constant load current equal to the root-mean-square of the actual load current over the load cycle, plus such constant losses as core loss, field copper loss, and windage. For this method to be appropriate, the load cycle must be only a small fraction of the thermal time constant of the machine. For such a condition, the cyclical temperature variation will be only a small fraction of the total rise above the ambient temperature and may be neglected.

Examples of short-time duty cycles are those on generators and transformers for electric welding; on motors driving machine tools such as planers and shapers, where reversing is required; and in elevator service, mine hoists, and rolling mills.

For load cycles which cover a time comparable to or greater than the thermal time constant of a machine, thermal capacity as well as thermal conductance and dissipation must be considered, since there will be considerable cyclic temperature variation.

Reference to a single " thermal time constant " of a machine, although convenient, actually implies a considerable oversimplification. The various parts of a machine do not come up to their final respective temperatures under constant load along the same or even similar curves.

Long load cycles occur in power system generators, transformers, and cables (24-hour cycle), and in motors driving large boring mills and lathes requiring perhaps hours to complete a cut. Transformers supplying power for furnaces or electrochemical baths or arcs may have long load cycles. There is as yet no standardized method of rating machines on such cycles,* and considerable interest is evinced in recent technical papers on this question. Suggested further reading may be found in the references listed in the bibliography.

* The proposed American standards for rating power and distribution transformers are discussed in Art. 8, Ch. XIV.

Part II

TRANSFORMERS

CHAPTER IX

General Design and Cost Considerations

Thus far, consideration has been given to various types of field and circuit phenomena individually — electric, dielectric, magnetic, and thermal. The more important types of electrical machinery and other apparatus are treated hereafter. The theories from which their electrical characteristics can be computed are developed through the combination of the fundamental concepts of the electric, dielectric, magnetic, and thermal circuits contained in them. Transformers are discussed in the remainder of this volume, and electronic devices in the next, leading the student to further study in the fields of power, communications, or industrial measurements and control.

The reader who is following the continuity of treatment as arranged in this series may well pause and meditate, at this transition point, on the general principles which have guided the development of electrical machinery and other equipment throughout the thousands of man-years of effort spent in the design of each type, and which continue to guide the further development.

1. ELECTRICAL MACHINERY

The best design is the best compromise among the several contending elements which must or should enter into it. In electrical machinery these are:

(1) The electric circuit (or circuits),
(2) The magnetic circuit,
(3) The dielectric circuit (insulation),
(4) The thermal circuit,
(5) The mechanical circuit (means for supporting the various mechanical stresses),
(6) Cost balance,
(7) Manufacturing facility,
(8) Ease of maintenance and repair,
(9) Acoustical characteristics (reasonably quiet operation).

Cost is of the essence in good engineering and is a controlling factor in electrical machine design. Practically all existing machines can easily be redesigned to have higher efficiencies and better operating characteristics, at the expense of higher cost of manufacture; but, when the existing design has been properly made in the light of existing technical knowledge, the incremental cost of improved efficiency or other character-istics will be greater than the value of that improvement. Specifically, the efficiency of a machine of a certain design is best if the total of initial cost and present value of losses throughout the expected useful life is a minimum, other things being equal. Price competition may force the design to a point of lower cost and lower efficiency, but such economy is unsound.

The choice of materials is naturally influenced by costs, and the con-tinuing development of improved materials keeps design in a state of flux, as does increasing technical knowledge. Copper is superior to all other metals except silver in conductivity, and so is used almost exclu-sively in machine windings, since the cost of silver is some 50 times greater. Aluminum is usually competitive with copper on a cost basis (per mho-foot) but because of its greater size for a given conductance it is not ordinarily suitable for windings, since the space available must be parceled out jealously to magnetic and insulating material and to cooling means — ducts and so on — as well as to the conductors them-selves. Squirrel-cage rotors of induction motors are often made of cast aluminum, for reasons that may be appreciated later.

At the high flux densities used in electric power machinery, iron and soft steel are the best magnetic materials, on the basis both of first cost and of performance. At low flux densities, nickel-iron alloys (the "permalloy" group) and very pure iron are superior, and so find appli-cation in communication transformers and choke coils. To reduce the effects of eddy currents induced by time-varying flux, the magnetic circuit is commonly built of varnished silicon-steel laminations, whose thickness is that determined by the optimum compromise between ex-pense of manufacture and gain in performance. Thickness of 0.014 inch is commonly used in power-system transformers. Pole faces and poles carrying principally direct flux but with some superposed ripple usually have much thicker laminations. High-speed steam-turbo-alternator rotors are made of solid steel forgings, because of the high centrifugal stresses and in spite of the ripple losses. Here, incidentally, the entire design is affected by the high speed, and the machine is made relatively long and thin to reduce centrifugal stresses and windage losses. The characteristics of the prime mover, demanding high speed for good efficiency, affect the electrical design most importantly.

The mechanical strength of the various parts of a machine usually should be many times greater than normal load forces. Sufficient rugged-ness must be provided to withstand short-circuit forces, inertia forces due to sudden stalling of load or to quick reversing, and the hard knocks which must always be expected in practice. Rotating machines which carry no external load, such as synchronous reactors and converters, are normally supplied with much lighter shafts than are motors of the same kilovolt-ampere rating.

In the quantity manufacture of electrical machinery, effort is made to utilize standardized individual piece designs in the maximum practical number of machines. The same motor lamination design, for example, may be used in dozens of motors of different speeds, voltages, powers, phases, and even types. An order for a special motor of moderate size would be filled not by the prohibitively expensive method of making a complete new design, but by the selection first of an appropriate frame size, and then of laminations, preformed winding coils, and other standard parts for assembly into a machine to meet the specifications.

Very often the designer must be guided by two or more objectives, and a relative evaluation of them must be made. An induction motor, for example, has a performance limited by its maximum torque and by its stator heating. Decision must be made as to how many times full-load torque should be provided as the maximum torque. A synchronous reactor may be called upon to take either leading or lagging reactive power. Decision must be made as to the proper ratio of the two capacities.

Rotating machines usually carry windings in slots, and the magnetic flux of the machine finds the most restricted portion of its iron path in the teeth about these slots. Utmost care is required in the evaluation of flux densities at the narrowest portions of the teeth, because of the rapidity with which core loss and magnetizing force go up with flux density above the knee of the saturation curve. If the flux density is low in the teeth, the design is poor, because there is then a great waste in material and weight in other parts of the magnetic circuit where the density is still lower. There is but a relatively narrow range for good design.

It is not the intent here to set forth even briefly the details of electrical machine design, but merely to point out some of the background principles which guide them. Knowledge and appreciation of these principles serve to provide perspective in the study of the theory of operation and of the performance characteristics.

2. POWER SYSTEMS

Proper design of a power system, utilizing individual units of machines and equipment, is guided by the same fundamental principles as is the design of the units themselves. Certain definite requirements are imposed by the characteristics of load devices. The electric power supplied by a system should be characterized by constant or nearly constant voltage, dependability, balanced voltage (if polyphase), sinusoidal waveform, and constant frequency. In addition to these requirements imposed by the loads, inductive interference with communication circuits must be combated, and the entire system must be designed to fulfill its requisite functions at minimum over-all annual cost. Hydroelectric generating stations need to be supplemented by steam plants if periods of low water are to be expected. Steam plants need to be located within reach of adequate cooling-water facilities and cheap fuel transportation. Selection of generator size and number of stand-by units; choice of frequency, voltage, phases, type of distribution system; and determination of types and degree of lightning protection and protective relaying are among the problems which a power-system designer must handle.

3. ELECTRONIC DEVICES

Small low-power electronic and other electrical devices and elements are used in great numbers in communication, control, and other circuits. In such equipment, the dissipation of heat may be no problem at all, and power consumed may be of negligible importance except for its effect on general operation. For such devices, including receiver-type vacuum tubes, small capacitors, relays, choke coils, loudspeakers, and so on, the design criteria include a balance between incremental cost and quality of performance. In high-quality apparatus of small size, the consideration of cost may well be kept in the background during the development of the design, until the production problem is considered. This fact does not hold for large heavy apparatus. High-power transmitter-type tubes require care in the provision for anode cooling, the circulation of water through the anode being a common expedient. The use of the best materials for the cathode and for other parts, sturdy mounting to minimize vibration and microphonics, and precise spacing and location of the various electrodes, while all of importance, do not add materially to the cost of the tube when large quantities are involved.

The power limitation on a tube may be set by its voltage capacity, by the breakdown strength between its plate or anode and its other electrodes, or, in gas tubes, by its inverse-voltage breakdown strength.[1] Current limitation may be by anode heating or by cathode emission modified by space-charge effects. At ultrahigh frequencies, electron transit time is an important consideration and in conventional tubes may require a drastic modification of the size and arrangement of electrodes. For a tube of any specific power rating, the choice by the designers of the plate voltage and current ratings represents a compromise arrived at by consideration of relative incremental costs in the tube itself, and also the costs of power sources and of other circuit elements and devices to be used in conjunction with the tubes. For example, a 5-watt

---

[1] The rating of electronic tubes is discussed in the volume on electronics, Art. 2, Ch. IV.

tube designed for 500 volts and 10 milliamperes is much more economical than one rated at 50 volts and 100 milliamperes, because the large cathode emission, the large capacitances required in typical external circuits, and the low-efficiency rectifier associated with the latter would add much more to the costs than the value of the small saving due to smaller spacing and insulation breakdown strength.

There are so many types even of conventional electron tubes, including combinations of two or more in one envelope, that any brief statement of design principles can be only of the most general kind. Starting from some tentative design, any change which increases the effectiveness of the tube in any way is justified unless its advantage is nullified by some other decrease in effectiveness, or unless the cost of the changed design outweighs the gain in tube characteristics.

A great number of tube designs are on the market. One important consideration in the design of an entire series of tubes is the proper balance between, on the one hand, standardization on a relatively few designs in order to obtain maximum benefits of mass production and minimum inventory requirements and, on the other hand, production of a relatively large number of designs in order to make available tubes designed individually and specifically for a narrower and more specialized field of application. Under receiving-tube types, for example, there are necessarily rectifiers, indicator tubes, converters, voltage amplifiers, detectors, and power amplifiers. Approximately 30 tubes in all these categories are sufficient for most radio receiver applications, although several hundred are manufactured.

4. ELECTRONIC AND OTHER CIRCUITS

The analyses of many circuits and devices utilizing electronic and other low-power control elements are characterized by considerable complexity, more especially when full equivalent circuits are considered. The designer of these devices usually finds it sufficient to select from the commercially available units those which are most suitable, and to arrange his circuit to accomplish the desired results in the simplest and most direct way. To a limited degree, experimental methods of design may profitably be used, since the complete theoretical advance analysis of a circuit may be very tedious or almost impossible, and therefore costly. The cost of well-directed cut-and-try experimental procedures, however, may be relatively small because of the low costs of the elements themselves and the experimental assembly of them.

In signal and control circuits, safety may be the all-important criterion. In railway signaling, for example, failure of a mechanism, of a circuit, or of a power supply must result in a danger signal to a locomotive engineer rather than a green light. Circuit failure may be reduced by the use of double contacts in relays, and even parallel wiring. In the design of circuits for control and other special purposes, there may be a conflict between first cost and maintenance cost. By the use of one relay, for example, to perform several functions, the number of relays in a circuit may be materially reduced; but the adjustments of the multipurpose relay will probably be more critical, and the training of maintenance men more difficult.

It is not uncommon to find that a particular type of device may be repeatedly redesigned with the utmost care for some specific purpose, without being able to fulfill adequately its intended function. This situation is often the forerunner of the *invention* of an entirely new method of accomplishing the desired result. If the invention is an important one, it will usually pass through a series of design developments and refinements. Examples are the art of " loading " telephone transmission lines, so long advocated by Heaviside and later advanced by M. I. Pupin and G. A. Campbell; the invention of the thermionic triode amplifier, displacing the electromechanical " repeater "; and the velocity-modulation tube for ultrahigh-frequency generation to avoid trouble from and in fact to make use of the electron transit time, which is a serious limitation in tubes of conventional type.

Compromises must be made continually in engineering practice, and the best engineer is he who can best recognize and evaluate the pertinent forces bearing on design, application, and operation, and with imagination, precision, practicality, and resourcefulness proceed to work out the best compromise.

CHAPTER X

Transformers: General Principles

In its simplest form, a transformer consists of two conducting coils having mutual inductance. The *primary* is the winding which receives electric power, and the *secondary* is the one which may deliver it. The coils usually are wound on a laminated core of magnetic material, or of compressed powdered alloy, and the transformer is then known as an *iron-core transformer*. Sometimes, as in many radio-frequency transformers, there is no magnetic core; the transformer then may be described as an *air-core transformer*. In other types, as in the induction coil, the core may consist of a bundle of fine iron wire, but the return path for the magnetic flux is in air. Other examples of this open-core construction are the small transformers used in subscribers' telephone sets and some types of radio-frequency transformers.

1. ELEMENTARY TRANSFORMER THEORY

When displacement currents due to the capacitances of the windings are neglected, the fundamental principles from which the theory of transformers is developed are expressed in the well-known voltage equations:

$$v_1 = R_1 i_1 + \frac{d\lambda_1}{dt} = R_1 i_1 + e_1 \tag{1}$$

$$v_2 = R_2 i_2 + \frac{d\lambda_2}{dt} = R_2 i_2 + e_2, \tag{2}$$

where the subscripts 1 and 2 refer to the primary and secondary windings, and

v_1 and v_2 are the instantaneous terminal voltages,
i_1 and i_2 are the instantaneous currents,
R_1 and R_2 are the effective resistances,
λ_1 and λ_2 are the instantaneous flux linkages,
e_1 and e_2 are the instantaneous voltages induced by the time-varying flux linkages.

Any consistent system of units may be used. In these equations, the positive directions of all voltages are chosen as falls in potential in a right-hand-screw direction about an assumed positive direction of flux, as indicated by the + and − signs in Fig. 1. It is convenient also to assume the positive directions of both primary and secondary currents in this same right-hand-screw direction about positive flux, as indicated by the arrows i_1 and i_2 in Fig. 1. For these positive directions the algebraic signs in the voltage equations are as in Eqs. 1 and 2.*

To make use of Eqs. 1 and 2, it is necessary to find the relations between the flux linkages and the currents in the windings. If the perme-

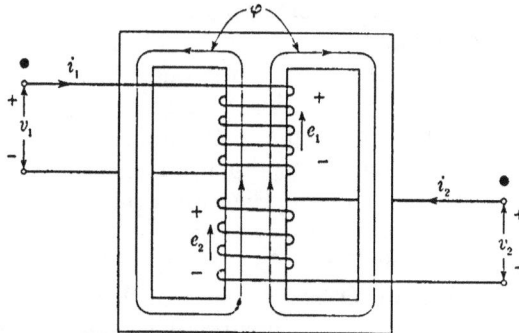

FIG. 1. Schematic diagram of a transformer, showing positive directions of the currents and voltages in Eqs. 1 and 2.

ability of the core is constant, the flux linkages are proportional to the currents producing them and, consequently, by the Principle of Superposition, the flux linkages can be expressed as the sum of the components produced by each current acting alone. That is,

$$\lambda_1 = L_1 i_1 + M i_2 \tag{3}$$

$$\lambda_2 = L_2 i_2 + M i_1, \tag{4}$$

where L_1 and L_2 are the self-inductances of the windings and M is the mutual inductance. In these equations, $L_1 i_1$ is the component flux linkage with winding 1 produced by its own current, and $M i_2$ is the component flux linkage with winding 1 produced by the current in the other winding. Similarly, $L_2 i_2$ and $M i_1$ are the self- and mutual components of the flux linkage with winding 2. The inductances L_1, L_2, and M are the constants of proportionality relating the component flux linkages and the currents producing them. Then, according to the classical theory of linear coupled circuits,[1] the voltage equations can be written as

$$v_1 = R_1 i_1 + L_1 \frac{di_1}{dt} + M \frac{di_2}{dt} \tag{5}$$

$$v_2 = R_2 i_2 + L_2 \frac{di_2}{dt} + M \frac{di_1}{dt}. \tag{6}$$

* With regard to these algebraic signs, Art. 4, Ch. VI, should be reviewed.

[1] Coupled circuits are discussed in this series in the volume on electric circuits (1940), Art. 10, Ch. VI, p. 383

The permeability of the core of an iron-core transformer is not constant and therefore its inductances are not constant; their values depend on the instantaneous magnetic conditions in the core. Thus it is evident that Eqs. 3, 4, 5, and 6 with constant inductances do not apply rigorously to an iron-core transformer, although they are often used as the basis of the analysis of such transformers and the results so obtained, if properly interpreted, usually agree satisfactorily with experimental results. This method of analysis is discussed in Ch. XVII, but for the present it is advisable to adopt a different method of attack which is particularly well suited to the analysis of iron-core transformers and does not involve superposition of component fluxes — a process that requires considerable justification when nonlinear magnetic cores are dealt with.

In order to obtain a physical concept of the behavior of an iron-core transformer and a first approximation to a theory of its behavior, assume that all the flux is confined to the high-permeability magnetic core and therefore links all the turns of both windings. The effects of magnetic leakage are discussed subsequently. According to this assumption, the flux linkages with the primary and secondary windings are

$$\lambda_1 = N_1\varphi \qquad\qquad [7]$$

$$\lambda_2 = N_2\varphi, \qquad\qquad [8]$$

where N_1 and N_2 are the numbers of turns in the primary and secondary windings and φ is the instantaneous value of the core flux produced by the combined magnetomotive forces of the primary and secondary currents. That is, the flux in Eqs. 7 and 8 is the resultant flux produced by the combined action of both the primary and secondary magneto-motive forces acting simultaneously, instead of the sum of the components due to each current acting separately, as in the linear circuit theory of Eqs. 3, 4, 5, and 6. Then Eqs. 1 and 2 can be written as

$$v_1 = R_1 i_1 + N_1 \frac{d\varphi}{dt} = R_1 i_1 + e_1 \qquad\qquad [9]$$

$$v_2 = R_2 i_2 + N_2 \frac{d\varphi}{dt} = R_2 i_2 + e_2, \qquad\qquad [10]$$

where e_1 and e_2 are the voltages induced by the time-varying core flux.

First consider conditions when the secondary circuit is open and an alternating voltage of constant amplitude and frequency is impressed on the primary terminals. According to Eq. 9, the flux must adjust itself so that the sum of the primary resistance drop and the counter electromotive force induced by the time-varying flux exactly balances the impressed voltage, and the primary current which produces the flux must adjust itself to satisfy these conditions. Since the no-load resistance drop in the primary winding usually is very small, the primary induced voltage very nearly equals the impressed voltage. The core flux induces a voltage in the secondary winding also, and, since most of the flux is confined to the core, the ratio of the primary voltage to the secondary voltage very nearly equals the ratio of the number of turns in the primary to the number of turns in the secondary. Thus this simple static device is capable of changing the voltage at which the power of an alternating-current source is available.

If the secondary is now connected to a load, a secondary current results. The core is now acted upon by the magnetomotive force of the secondary current, but, in spite of the secondary magnetomotive force, the core flux must remain practically unchanged, since by Eq. 9, this flux must still induce a counter electromotive force in the primary differing from the primary impressed voltage only by the primary resistance drop, which usually is small even under load. Hence when current flows in the secondary, the primary current changes so as to counteract the magnetomotive force of the secondary current.

It is often convenient to consider the primary current as the sum of an *exciting component* i'_φ and a *load component* i'_L. That is,

$$i_1 = i'_\varphi + i'_L. \tag{11}$$

The exciting current i'_φ is the component of the primary current that is sufficient by itself to produce the flux required to induce the counter electromotive force in the primary, and equals the no-load current for a no-load condition for which the core flux is the same as under the load condition. The load component i'_L of the primary current produces a magnetomotive force that opposes and exactly balances the magnetomotive force of the secondary current. If the positive directions of both primary and secondary currents are taken in the same directions about the core — as are i_1 and i_2 in Fig. 1 — then the relation between the secondary current i_2 and the load component i'_L of the primary current is

$$N_1 i'_L = -N_2 i_2. \tag{12}$$

Therefore, when the secondary is connected to a load, the current taken by the load causes a compensating change in the primary current.

2. THE IDEAL TRANSFORMER

The modern iron-core transformer has so nearly approached perfection that in many problems it may be considered a perfect transforming device. In the simplest form of the theory of the transformer it is assumed that:

(1) The resistances of the windings are negligibly small;
(2) The core loss is negligibly small;
(3) The entire magnetic flux links all the turns of both windings;
(4) The permeability of the core is so high that a negligibly small magnetomotive force produces the required flux;
(5) The capacitances of the windings are negligibly small.

That is, the transformer is assumed to have characteristics approximating those of an ideal transformer, with no losses, no magnetic leakage, and no exciting current.[2]

According to Assumptions 1 and 3, for an ideal transformer Eqs. 9 and 10 reduce to

$$v_1 = e_1 = N_1 \frac{d\varphi}{dt} \qquad [16]$$

$$v_2 = e_2 = N_2 \frac{d\varphi}{dt}, \qquad [17]$$

where φ is the resultant flux produced by the simultaneous action of the primary and secondary currents. Hence

$$\frac{v_1}{v_2} = \frac{N_1}{N_2} \qquad \blacktriangleright [18]$$

Thus, for an ideal transformer, the instantaneous terminal voltages are proportional to the numbers of turns in the windings, and their waveforms are identical. Also, when the windings of Fig. 1 are traced from their dot-marked terminals to their unmarked terminals, the core is encircled in the same direction by both windings, and therefore for an ideal transformer the primary and secondary terminal voltages are in

---

[2] The ideal transformer is discussed on the basis of the coupled-circuit equations in the volume on electric circuits (1940), Art. 11, Ch. VI, pp. 384–389, where it is shown that, according to Assumption 3, the ratios between the inductances of an ideal transformer are

$$\frac{L_1}{M} = \frac{M}{L_2} = \frac{N_1}{N_2} \qquad [13]$$

$$\frac{L_1}{L_2} = \left(\frac{N_1}{N_2}\right)^2. \qquad [14]$$

The coefficient of coupling k, defined as

$$k \equiv \frac{M}{\sqrt{L_1 L_2}} \qquad [15]$$

is a useful quantity in coupled-circuit theory. From Eqs. 13 and 14, the coefficient of coupling of an ideal transformer is unity. According to Assumption 4, the inductances of an ideal transformer are all infinitely large; but the above ratios, Eqs. 13 and 14, are maintained among them.

phase when their positive directions are taken in the directions shown by the + and − signs in Fig. 1. That is, at any instant the dot-marked primary terminal is actually of the same relative polarity as the dot-marked secondary terminal.

▶ Note that the performance of a transformer depends on time-varying flux, and therefore in the steady state a transformer operates on alternating voltage only.◀

According to Assumptions 2 and 4, the net magnetomotive force required to produce the resultant flux is zero. The net magnetomotive force is the resultant of the primary and secondary ampere-turns, and hence, if the positive directions of both primary and secondary currents are taken in the same directions about the core, as in Fig. 1,

$$N_1 i_1 + N_2 i_2 = 0. \qquad [19]$$

That is, for an ideal transformer, the exciting current is zero, and therefore from Eq. 11 the primary current equals its load component, and Eq. 12 reduces to Eq. 19. From Eq. 19, for an ideal transformer,

$$\frac{i_1}{i_2} = -\frac{N_2}{N_1}. \qquad ▶[20]$$

The minus sign in Eq. 20 indicates that the currents produce opposing magnetomotive forces. When used as an approximation for an actual transformer, Eqs. 19 and 20 do not apply to steady direct currents which may exist in the windings because of external causes.

Note that the arrows i_1 and i_2 in Fig. 1 indicate only the positive directions of the currents, both assumed in the right-hand-screw direction about positive flux, for convenience in setting up the equations. The arrows do not mean that while the current i_1 is in its positive direction the secondary current is also in its positive direction. Since the primary and secondary currents produce opposing magnetomotive forces, the right-hand primary and secondary currents in an ideal transformer are in phase opposition, as shown by the minus sign in Eq. 20. That is, while the primary current is entering the dot-marked primary terminal, the secondary current is leaving the dot-marked secondary terminal and entering the corresponding terminal of the load.

When Eq. 18 is multiplied by Eq. 20,

$$\frac{v_1 i_1}{v_2 i_2} = -1. \qquad [21]$$

That is, for an ideal transformer, the instantaneous powers on the primary and secondary sides are numerically equal. The minus sign shows that, while the secondary winding is delivering power to the load, the primary winding is absorbing power from the source.

Dividing Eq. 18 by Eq. 20 gives

$$\frac{v_1/i_1}{v_2/i_2} = -\left(\frac{N_1}{N_2}\right)^2 \qquad [22]$$

or

$$\frac{v_1}{i_1} = -\left(\frac{N_1}{N_2}\right)^2 \frac{v_2}{i_2} . \qquad [23]$$

If a resistance load R_L is connected to the secondary, as shown in Fig. 2, the instantaneous current through the load is in the same direction as the instantaneous fall in potential across the load. If the secondary terminal voltage is v_2, Fig. 2, the current through the load in the direction of the fall in potential across the load is i_L. Note that, in Fig. 2,

$$i_L = -i_2. \qquad [24]$$

Hence

$$v_2 = R_L i_L = -R_L i_2 \qquad [25]$$

or

$$\frac{v_2}{-i_2} = R_L \qquad [26]$$

From Eqs. 23 and 26,

$$\frac{v_1}{i_1} = \left(\frac{N_1}{N_2}\right)^2 R_L. \qquad [27]$$

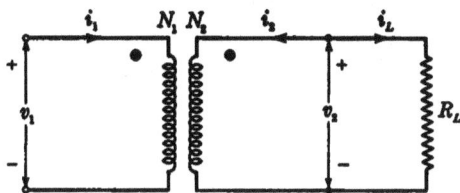

FIG. 2.   Simplified diagram of a transformer. The dot-marked terminals are of the same relative polarity and correspond to the dot-marked terminals in Fig. 1.

Thus, on the primary side, the combination of the load and the transformer is equivalent to a resistance

$$R_L' = \left(\frac{N_1}{N_2}\right)^2 R_L \qquad \blacktriangleright[28]$$

connected in the primary circuit. This result can be extended to an ideal transformer with an impedance Z_L connected to its secondary terminals. Hence, if a transformer is interposed between a load and a generator, the apparent impedance which the load presents to the generator can be transformed to a value different from the actual impedance of the load.

▶ Therefore it may be said that a transformer is a device which transforms alternating voltage, or alternating current, or impedance. It may also serve to insulate one circuit from another or to isolate direct current, at the same time maintaining alternating-current continuity between the circuits.◀

3. Uses of transformers

In the distribution of electric power, safety demands that the voltage at which the power is supplied to the consumer must not exceed a few hundred volts. For household use, 120 volts is standard in most localities in America. Assume there is a demand for 500 kw of single-phase power at 120 volts, unity power factor, at a point 10 miles from the generating station. If the amount of copper in the transmission wires is adjusted so that the transmission loss (I^2R) is 75 kw, or the efficiency is 87 per cent, about 75,000,000 pounds of copper are required for transmission at 120 volts, but only 7,500 pounds for transmission at 12,000 volts. If copper cable costs 20 cents per pound, the cost of the transmission conductors alone is $15,000,000 in the first case and but $1,500 in the second. A 500-kva transformer to reduce the line voltage from 12,000 volts to 120 volts costs about $1,800. Its full-load efficiency is about 98.5 per cent.

The alternating-current system of transmission and distribution has come into almost universal use largely because the transformer makes possible the operation of different parts of the system at their most suitable voltages. Other important factors in favor of the alternating-current system are the desirable qualities of the synchronous generator and of the induction motor. It is no exaggeration to say, however, that, without the simplicity, reliability, and high efficiency of the transformer, the enormous growth of electric transmission and distribution systems during the past fifty years would have been impossible.

In the field of electrical communications also the transformer is indispensable. Among other uses, it makes possible a maximum transfer of power from one section of a circuit to another. For example, consider a linear (Class-A) vacuum-tube amplifier supplying power to a loudspeaker. For alternating current, the output of a linear amplifier can be considered as a generator of internal electromotive force μe_g in series with the dynamic plate resistance r_p of the tube, where μ is the amplification factor of the tube and e_g is the alternating signal voltage impressed between its grid and cathode.[3] Consider an amplifier tube having an amplification factor of 5 and a plate resistance of 2,000 ohms, supplying power to a dynamic-type loudspeaker which, as a first approximation, may be considered as a pure resistance load R_L of 10 ohms. If the 10-ohm loudspeaker were directly connected in series with the plate circuit of the tube and if the effective value μE_g of the alternating signal voltage were 100 volts, the power delivered to the loudspeaker would be

$$I^2R_L = \left(\frac{\mu E_g}{r_p + R_L}\right)^2 R_L = \left(\frac{100}{2,010}\right)^2 \times 10 = 0.025 \text{ watt.} \quad [29]$$

---

[3] The theory of linear amplifiers is discussed in this series in the volume on electronics.

As is well known, the power delivered to a resistance load is greatest when the resistance of the load equals the internal resistance of the generator (both load and generator being without reactance, and the internal electromotive force and resistance of the generator being constant). Hence, if distortion were unimportant, the maximum power which could be obtained from the amplifier tube whose internal electromotive force is 100 volts and whose internal resistance is 2,000 ohms would occur for a load resistance of 2,000 ohms and would be

$$\left(\frac{100}{4,000}\right)^2 \times 2,000 = 1.25 \text{ watts.} \tag{30}$$

In an actual amplifier, considerations of distortion often require that the load resistance be two or three times the plate resistance of the tube. In this example, if the loudspeaker were connected to the plate circuit of the tube through an ideal transformer whose turns ratio N_1/N_2 equals $\sqrt{500}$, by Eq. 28 the load would appear as a resistance of 5,000 ohms, or $2.5r_p$, connected in the plate circuit of the tube. The power delivered to the loudspeaker for the same signal strength would be

$$\left(\frac{100}{7,000}\right)^2 \times 5,000 = 1.02 \text{ watts} \tag{31}$$

or about 41 times greater than without the transformer. The loss in an actual transformer is small compared with this great gain in power.

Transformers are probably the most widely used of all electromagnetic apparatus. They range greatly in size and must be designed to meet the demands of a variety of operating conditions. For example, those used in communication circuits usually operate with widely varying voltage and frequency and should produce as little difference as possible between the waveforms of the voltages or currents on the primary and secondary sides. On the other hand, power-system transformers usually operate at nominally constant voltage and frequency, and high efficiency usually is of more importance than in communication transformers, since the amount of power involved usually is much greater.

Transformers used in communication circuits are often classified according to the frequency range over which they are used, as *radio-frequency, intermediate-frequency,* and *audio-frequency transformers*. Within each of these frequency classifications, they are also classified by purpose as input, interstage coupling, and output transformers. Generally an *input transformer* is used to connect a microphone or other signal generator to the grid-cathode terminals of a vacuum tube in such a manner as to supply to the grid as high a voltage as possible consistent with reasonable frequency-response characteristics.* The power delivered is

* Frequency-response characteristics are discussed in Ch. XVIII.

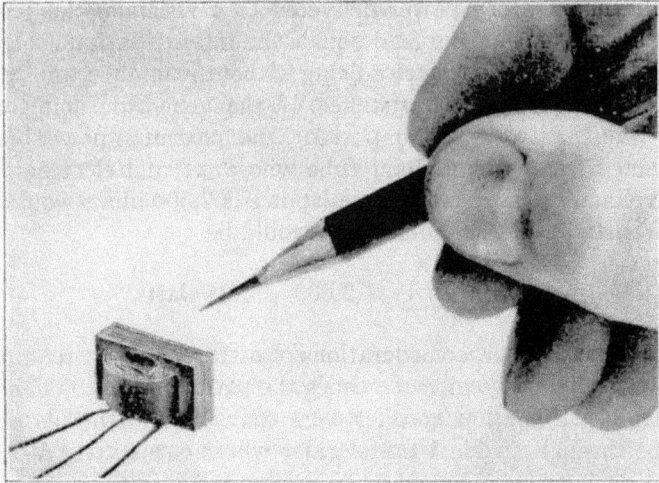

Courtesy United Transformer Corp.

FIG. 3

Courtesy Westinghouse Electric and Manufacturing Co.

FIG. 4

FIGS. 3 and 4. Contrasts in transformer sizes and applications. The tiny unit shown in Fig. 3 is the core-and-coil assembly of an autotransformer used in the circuit of a hearing-aid device. It has an incremental inductance of 20 henries. Figure 4 shows a 3-phase 275,000-v bank of autotransformers installed at the receiving end of the transmission line from Boulder Dam to Los Angeles. Each unit has blower fans mounted at the rear for cooling the oil as it circulates through the external radiators. The rating of each unit with blowers in operation is 65,000 kva.

often but a few microwatts. *Interstage coupling transformers* are sometimes used in vacuum-tube amplifiers to interconnect the plate circuit of one tube with the grid circuit of the succeeding tube and are generally required to perform the same function as input transformers. *Output transformers* are usually required to operate over a specified frequency range and to supply maximum power, without distortion, to a load such as a loudspeaker, transmission line, or radio antenna. The power delivered may be a fraction of a watt to several hundred kilowatts. *Modulating transformers* serve to superpose an audio-frequency signal on a high-frequency carrier wave. In certain radio transmitting stations, the

FIG. 5. A variety of communication transformers of typical forms.

modulating transformer weighs several tons. *Filament transformers* are in almost universal use to supply power to the filaments, or heaters, of vacuum tubes in communication and control systems. Transformers are used in " phantom " telephone circuits to permit carrying three simultaneous audio-frequency signals over two pairs of wires, and in " simplex " circuits to permit simultaneous operation of a telegraph signal and of a telephone signal on one pair of wires; in this service, transformers are called *repeating coils*. Transformers separate the alternating, or " voice," current from the direct current in a local-battery telephone transmitter circuit and transmit only the " voice " current on the lines; in this service, transformers are called *induction coils*.*

Power-system transformers are commonly classified as power transformers or distribution transformers according to whether they are ordinarily used in the high-power generating stations and substations, or in the distribution networks. The familiar transformers, mounted

* See Ch. XX for further discussion of transformer applications in telephone circuits.

overhead on poles or underground in vaults, which form the last voltage-changing link between the central-station system and the ultimate consumer of electrical energy are known as *distribution transformers*. They compose a major element in the total value of transformers manufactured.

FIG. 6. Rectifier-tube filament transformer for radio broadcast service. The tube is mounted in the insulated socket supported by the transformer. The over-all height of the unit is 11 in.

Distribution transformers are built in standard voltage, frequency, and kilovolt-ampere ratings up to and including 500 kva.* They are manufactured in quantities, and in the smaller ratings usually are available

* Standard ratings of distribution transformers are listed in Appendix A.

from manufacturers' stocks. Large transformers whose ratings are above 500 kva are known as *power transformers*. Single-phase units have been constructed with continuous ratings of 65,000 kva and two-hour ratings of 80,000 kva with auxiliary blowers, for operation in Y-connected three-phase banks at line-to-line voltages of 275,000 volts. One of these three-

Fig. 7. Two 37½-kva distribution trans-formers mounted on an overhead plat-form. The bank is used to transform from a 3-phase 4-wire 4,160-v distri-bution circuit to a 5-wire 120/240-v 2-phase system.

Fig. 8. Distribution transformers in a public-utility company's stockroom. Some of the transformers are used, and others are new.

phase banks is shown in Fig. 4. Each transformer weighs more than 180 tons.[4] For use in high-voltage laboratories, transformers have been con-structed with voltage ratings of over a million volts.

Many other transformers are used in power systems for special pur-poses.[*] For example, *constant-current* distribution transformers are used to supply nominally constant current to series-connected street-lighting systems. *Voltage-regulating* transformers of many types are used for voltage control of individual circuits or parts of a system. *Instrument transformers*, whose secondary voltages or currents are very nearly directly proportional to their primary voltages or currents, are used to actuate measuring instruments and control relays. Instrument trans-

[4] W. G. James and F. J. Vogel, " Power Transformers for 287.5 Kv Service," *A.I.E.E. Trans.*, 55 (May, 1936), 438–444.

[*] Some of these special applications are discussed in Ch. XIX.

formers are called *potential transformers* or *current transformers* in accordance with their use for measurement of voltage or current.

In addition to their uses in power and communication systems, transformers are used in many domestic and special-purpose applications,

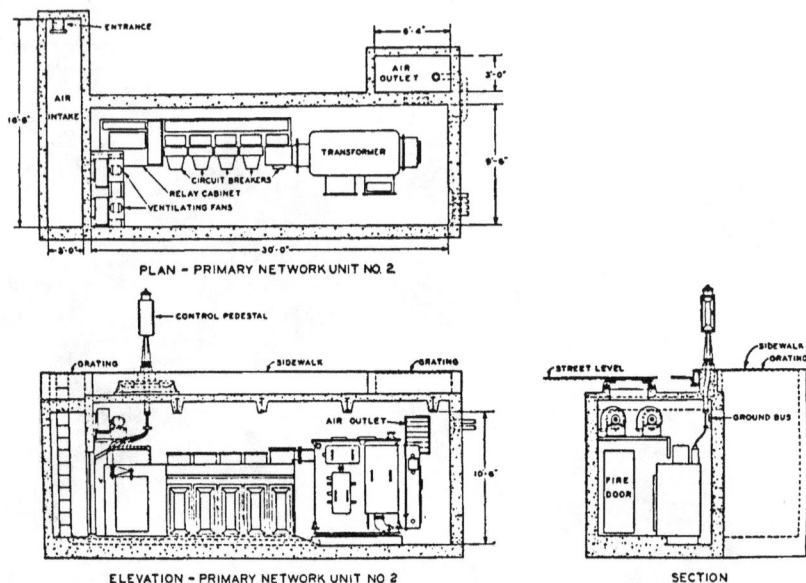

PLAN - PRIMARY NETWORK UNIT NO. 2

ELEVATION - PRIMARY NETWORK UNIT NO 2   SECTION

FIG. 9. An underground primary network installation. The 1,500-kva 60 ~ oil-immersed transformer receives power from a 13,800-v 3-phase circuit and delivers power to a 3-phase 4-wire Y-connected 2,300/4,000-v distribution network. The transformer is equipped with automatic tap-changing equipment for control of the output voltage. The vault contains automatic circuit breakers in the transformer output circuit and in each of four outgoing 4,000-v circuits, an operating battery with its associated controls, and ventilating fans. When an automatic operation results from trouble in the vault or on an outgoing circuit, signals are transmitted both to the control pedestal at the curb and by telephone wires to a central point where an operator is in attendance.

such as for energizing door bells, thermostat circuits, luminous tubes, and electric fences; for spark-plug and oil-burner ignition coils; and for many other purposes.

4. EVOLUTION OF TRANSFORMERS

Several factors have influenced the development and improvement of transformers. The physical properties of the materials of which they are constructed — particularly insulation and magnetic core materials — have been improved greatly. Experience has led to more effective use of the available materials and to better methods of assembling them. The use of oil as an insulating and cooling medium has made possible the construction of large, high-voltage power transformers. More exact methods have been developed for computing the details of design. The foundations upon which much of this progress rests are the principles of electric and magnetic fields and of electromagnetic induction.

The same principles apply to the tiny transformer in the radio receiver and to the giant in the power system. It should not be concluded, however, that a knowledge of these fundamental principles alone is sufficient for the analysis of transformers, since it is rarely possible or desirable to apply the fundamentals exactly. Available mathematical methods are either inadequate or too complex to enable one to deal with the actual configurations of the electric and magnetic fields present in a transformer, and therefore it has been necessary to devise approximate methods of analysis to overcome these difficulties. The choice of the approximate method appropriate to the analysis of any specific problem

Courtesy General Electric Co.

Fig. 10. The change in style and size of 69,000-v 400-amp. current transformers resulting from improvements in design, materials, and production methods. The transformer at the left was built in 1929; that at the right in 1940.

should always depend not only upon a thorough understanding of the physical theory but also, wherever possible, upon experience.

5. TRANSFORMER PROBLEMS

Although many problems involving the applications of transformers can be solved satisfactorily if the transformer is considered as a perfect device, it is apparent that many other problems arise from the departure of the characteristics in one or more ways from those of an ideal transformer. The satisfactory solution of such problems is important in design and important also because these characteristics of the transformer often markedly affect the behavior of the electrical network of which a transformer is a part. A number of these problems are discussed briefly below, and the remainder of this book is devoted largely to the solution of some of them.

5a. *Losses.* — The efficiency of a transformer is determined by the copper losses in its windings and the hysteresis and eddy-current losses in its core. The cost of these losses is a matter of great importance in most transformers whose outputs are above a few watts. Since increasing the efficiency usually results in a more expensive transformer, the design for the best over-all economy primarily depends on the proper balance between the annual cost of the losses and the annual capital costs of the transformer. Although the cost of the losses is unimportant in many small communication transformers, the losses may be important in another way, in that they affect the amplification, distortion, and frequency-response characteristics of the circuit. Hence it is important for the designer to be able to predict the losses and for the engineer who uses transformers to understand their effects and the manner in which they vary with operating conditions. In the analysis of problems involving power losses, usually it is sufficiently accurate to neglect magnetic leakage and exciting current.

5b. *Cooling.* — The losses develop heat within the windings and core; the effectiveness with which it is dissipated determines the temperature rise at a given load and hence the life of the insulation.* The cost of the transformer and its life determine the annual depreciation charges. Since reducing the temperature rise for a given load increases the cost of the transformer, the maximum safe load or kilovolt-ampere rating is determined primarily by that balance between the life of the transformer and its initial cost which results in the lowest annual capital cost. Thus the losses give rise to many economically important thermal problems. Although the thermal problems are relatively simple in small transformers, they become exceedingly complex for large power transformers, since, with increase in size, the losses increase more rapidly than the surface area from which they must be dissipated as heat energy. It is interesting to note that in the design of a large transformer, increasing the efficiency above the minimum acceptable value actually may result in a lower cost, because of the reduction in the cost of the means of cooling.

5c. *Magnetic Leakage.* — If there were no magnetic leakage, the ratio between the primary and secondary terminal voltages would differ from the turns ratio only by the relatively small resistance drops in the windings, as in Eqs. 9 and 10. However, magnetic leakage contributes an additional reactive component to the voltage drop through the trans-

---

* See Art. 1, Ch. VIII.

former and increases the departure of the voltage ratio from the ideal. Since this voltage drop is inductive, it not only depends on load but also increases with the frequency, and the voltage ratio of a communication transformer at high frequencies therefore differs from its voltage ratio at lower frequencies. The determination of these frequency-response characteristics is an extremely important problem in communication circuits.

Although the frequency of a power system is substantially constant, the load varies, and therefore the secondary voltage of a power-system transformer varies even if the primary voltage is maintained constant. This undesirable voltage regulation is largely determined by magnetic leakage. On the other hand, magnetic leakage has a beneficial effect in reducing the excessive currents resulting from accidental short circuits on a power system. These short-circuit currents produce large electro-magnetic forces on the windings and hence affect the mechanical design of power-system transformers.

Thus magnetic leakage introduces a number of important problems in communication and power circuits. In the analysis of these problems, it is usually permissible to neglect the exciting current and core loss, and it is often also permissible to neglect the resistances of the windings.

5d. *Exciting Current.* — The exciting current gives rise to a number of problems which are discussed in Ch. VI and which need only be mentioned briefly here. In communication circuits, the exciting current causes a voltage drop in the internal impedance of the source to which the primary of the transformer is connected, and thus causes the primary terminal voltage to differ from the internal voltage of the source. This effect is particularly important at low frequencies when the exciting current may be relatively large. Hence the low-frequency end of the frequency-response characteristic is largely determined by the exciting current, and in addition to this effect — which would be present even if the magnetic properties of the core were linear — the nonlinearity of the core introduces harmonics in the voltage waveform. Therefore the exciting current is important in problems involving frequency response and harmonic distortion in communication circuits.

In power systems, the exciting current usually decreases the power factor and therefore increases the current required to supply a given kilowatt load, increases the copper losses in the transmission lines and generators, increases the necessary kilovolt-ampere capacity of the generators, and increases the voltage regulation. In addition to these undesirable effects, harmonics in the exciting currents of power-system transformers may cause serious inductive interference in adjacent communication circuits.

Hence the exciting currents of both power and communication transformers are kept as small as is consistent with reasonable cost. In the analysis of problems involving the exciting current, it is frequently permissible to neglect magnetic leakage, and in power-system problems the primary-circuit impedances also usually are neglected.

5e. *The Electric Field.* — Several classes of problems have to do with the distribution of the electric field in the vicinity of the windings. This field causes the windings to have self- and mutual capacitances, and capacitances to ground and to adjacent circuits. In communication circuits, the effects of these capacitances may be large at sufficiently high frequencies, and may cause the transformer to behave in a completely different manner from that which would be predicted if the capacitances were ignored.

In transformers for operation at more than a few hundred volts, the design of the insulation is an important problem. The voltage gradient in the insulation under normal operating conditions affects the life of the insulation and hence is an important factor in determining the normal voltage rating. Under abnormal conditions, the voltage stresses due to surges set up by lightning and other causes may be very much greater than during normal operation. It is particularly important that power-system transformers be designed to withstand these abnormal stresses.

5f. *Manufacturing Problems.* — In addition to these problems pertaining to the electrical, thermal, and mechanical characteristics of transformers, a large number of manufacturing problems are of at least equal importance to the designer. The end in view should be the attainment of the best compromise between quality of product and its cost. Some of the major factors contributing to reduced cost are quantity production by machine processes, standardization of parts, development of standard designs which are suitable for a variety of general-purpose applications, and elimination of waste materials.

5g. *Other Considerations.* — Some other factors which must be considered by the designer and the application engineer are the cost of installation, maintenance, and repairs. Transportation may have an important effect on the design, especially of large units. Sometimes size or weight, magnetic noise, appearance, and fire hazard must be given primary consideration.

The discussion that follows is devoted principally to the analysis of the electrical behavior of the transformer, since this is of fundamental importance and determines the effect of the transformer on the network of which it is a part. To the designer, however, the thermal, dielectric, mechanical, and manufacturing problems are of major importance.

PROBLEMS

1. In each of the diagrams of Fig. 11, the instantaneous polarity of the primary voltage of a transformer (assumed to be ideal) is shown for a particular time in the cycle. Indicate: (1) the polarity of the instantaneous voltage induced in the secondary and (2) the directions of the instantaneous currents existing in the primary and secondary when each secondary winding is connected to a resistance load.

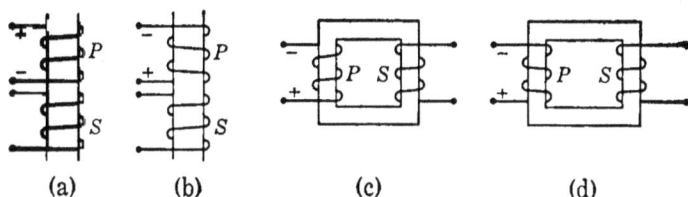

(a) (b) (c) (d)

Fig. 11.   Transformer windings with polarity markings, Prob. 1.

2. Refer to the diagrams of Fig. 11; let the positive direction of the primary terminal voltage be as marked. Then, using the conventions established in this chapter, indicate the nominal positive directions for: (1) flux in the core, (2) secondary terminal voltage, and (3) primary and secondary currents i_1 and i_2.

3. With the use of the conventions of this chapter for positive directions of voltage and of current flow in a transformer, draw to the scales of 5 amp per in. and 50 v per in. a vector diagram for an ideal transformer supplied with a sinusoidally varying primary voltage of 240 v. The ratio of primary turns to secondary turns is 2 : 1, and the load is an impedance of $8.0 + j5.3$ ohms. Include in the diagram vectors representing rms values of primary and secondary terminal voltages, primary and secondary induced electromotive forces, primary and secondary currents, and a vector showing the phase of the core flux. Indicate in a sketch of the transformer the positive directions.

4. In the ideal transformer of Fig. 12, at a time when the instantaneous primary voltage equals $+100$ v and is increasing, what are the instantaneous magnitudes

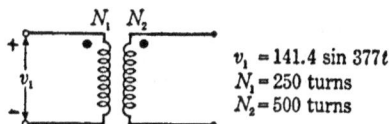

$$v_1 = 141.4 \sin 377t$$
$$N_1 = 250 \text{ turns}$$
$$N_2 = 500 \text{ turns}$$

Fig. 12.   Transformer circuit, Prob. 4.

and directions of the currents and the instantaneous magnitude and polarity of the secondary voltage:

(a) For a resistance load of 100 ohms?

(b) For an inductive load of impedance $50 + j86.6$ ohms?

(c) For a capacitive load of impedance $0 - j100$ ohms?

5. In an electric substation a single-phase 66,000 : 2,400-v transformer is connected to a short 2,400-v line supplying a single-phase load. The receiving end of the line is connected to the load through a 2,400 : 480-v transformer. The line has a resistance of 0.40 ohm and a reactance of 0.60 ohm (both conductors). The load may be con-

sidered as equivalent to an impedance of $0.30 + j0.34$ ohm. Assume that both transformers may be considered ideal.

(a) Determine the total impedance of line and load as viewed from the high-voltage side of the 66,000 : 2,400-v transformer.

(b) In this system, a power factor of unity is desired on the 66,000-v side of the single-phase circuit. A shunt capacitor is to be connected at the second transformer either on its 2,400-v side or on its 480-v side in order to secure this result.

Find: (1) the capacitance necessary if the capacitor is to be connected on the 2,400-v side and (2) the capacitance necessary if the capacitor is to be connected on the 480-v side.

(c) The cost of 480-v capacitors in dollars equals 1.16 times the kva rating. The costs of 2,400-v capacitors at various kva ratings are given by the following data:

Kva	Cost in dollars
270	690
360	780
540	960
720	1,140
1,080	1,500
1,260	1,680

On the assumption that capacitors of any desired kva rating can be purchased, find the costs of a suitable 2,400-v capacitor and of a suitable 480-v capacitor.

6. Engineers of a public utility company are making a study to determine whether distribution transformers used to carry a certain type of load should be of 7.5-kva or of 10-kva rating. The typical load being studied is equivalent to 9 kva during peak periods and is present about 900 hours per year. The load during the remainder of the year is negligible. Since the peak load is of short duration, the 7.5-kva transformer is capable of carrying it.

The transformer losses include the core loss, which is constant throughout the year, and the copper loss, which in a transformer is proportional to the square of the kva load. Data on the losses are:

Core loss
   7.5-kva transformer: 52 w
   10-kva transformer: 57 w

Copper loss
   7.5-kva transformer: 136 w at rated load
   10-kva transformer: 188 w at rated load

The cost of losses is 0.9 cent per kwhr.

The annual cost of taxes, insurance, and miscellaneous overhead charges is 7% of the original delivered cost of the transformer. Depreciation may be charged in equal annual amounts over the life of the transformer. The investment in the transformer becomes less each year, so that interest should be charged on the remaining investment — not on the original cost. The average annual charge for interest is then given by

$$\text{Interest} = \frac{r}{2}\left[2C - \frac{(n-1)(C-S)}{n}\right] \text{ dollars,} \qquad [32]$$

where

      C is the original delivered cost of transformer, dollars,

      n is the number of years of expected life of the transformer,

      S is the salvage value of the transformer after n years,

      r is the annual rate of interest on the investment, and may be assumed as 4%.

The expected life of the 10-kva transformer is 22 years, while that of the 7.5-kva transformer is about 5 years less, because of its higher operating temperature. The delivered costs are

7.5-kva transformer $ 95.00

10-kva transformer $112.00

and the salvage value is 50 cents per kva. The cost for labor and miscellaneous materials used to install a transformer is $20.00 for either rating, and should be treated as an increase in first cost.

Compute the approximate total annual costs of owning and operating each transformer for this load.

CHAPTER XI

Physical Features of Transformers

The service which a transformer is to meet dictates the physical features of its construction. With communication transformers, the major service requirement usually is fidelity of signal reproduction over a wide range of frequency and voltage. In such transformers, the materials and proportions of the core and the arrangement of the windings are governed by the necessity for obtaining electric and magnetic characteristics that result in satisfactory frequency-response characteristics and tolerable distortion. Heating is of minor importance in many communication transformers. Quite different matters, however, influence the design of power-system transformers. Their desired electrical characteristics are high efficiency, low voltage regulation, and high dielectric strength. With them, heating is a major factor in the determination of their physical characteristics.

1. Cores

The two basic types of transformer assembly are the *core type*, in which a single core is linked by two groups of windings, as in Fig. 1, and

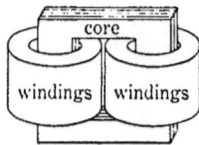

FIG. 1. Core-type transformer.

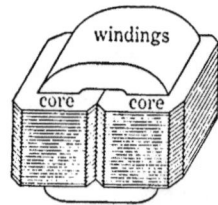

FIG. 2. Shell-type transformer.

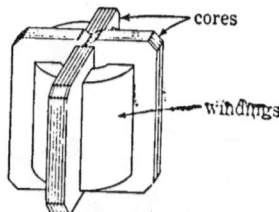

FIG. 3. Distributed-shell-type transformer.

the *shell type*, in which a single group of windings is linked by at least two flux components that are present in parallel magnetic circuits, as in Fig. 2. A modification of the shell type, called the *distributed shell type*, shown in Fig. 3, is common in certain ratings of distribution transformers. The core type is commonly used in large high-voltage distribution transformers and in power transformers. The cross section of a core is often square or rectangular in small transformers, but in large units the circular opening in the coil is used more effectively through building up the laminations in layers of varying width, making a stepped circular core, as in Figs. 4 and 5.

In addition to the necessity of obtaining desirable electrical characteristics, a number of practical considerations, such as cost of construction and repairs, space requirements, cooling, insulation, and mechanical strength, influence the choice of a type of core construction. Since the core usually serves as the frame, it must be mechanically strong. Under normal load, the mechanical stresses in a transformer are but little more than those due to the weight of its parts. Under short-circuit conditions, however, the electromagnetic stresses may become enormous, since they are proportional to the squares of the currents in the windings. If rated voltage is maintained at the primary terminals of a normal power-system transformer whose secondary is short-circuited, the steady-state currents usually are 10 to 25 times full-load current and the electromagnetic stresses 100 to 625 times their full-load values. The transient currents may be even greater. Transformers have been wrecked by these forces. In very large transformers, the short-circuit forces tending to separate the primary and secondary windings may be as great as 1,500,000 pounds.[1]

The cores of power-system transformers are made of suitable annealed steel laminations. In the traditional types of core shown in Figs. 1, 2, and 3, silicon steel having about 4 per cent silicon is generally used, for this material provides a good compromise in cost, workability, low hysteresis and eddy-current losses, and high permeability at relatively high flux densities. The cores are usually operated at maximum alternating flux densities between 65 and 90 kilolines per square inch. In small transformers, the oxide scale on the laminations may be sufficient insulation to prevent intersheet eddy currents, but in larger cores the laminations are varnished or coated with a thin layer of steel-adherent glass. Although the eddy-current loss can be further reduced by the use of thinner laminations, the cost of laminations and of core construction then increases and the stacking factor is reduced. In normal 60-cycle transformers with 29-gauge (0.014 inch) sheets, the eddy-current loss is only about one-fifth the total core loss. Little therefore would be gained through using thinner laminations to reduce this relatively small component of the total core loss. When the laminations are clamped together, the clamping bolts or rivets must not form eddy-current paths.

[1] K. K. Paluev, "Power Transformers with Concentric Windings," *A.I.E.E. Trans.*, 55 (June, 1936), 649–659.

FIG. 4. The two main legs and bottom yoke of a 55,000-kva core-type power transformer. The stepped-circular core construction and ventilating ducts are shown in the cross-sectional insert. This core is part of a single-phase transformer used in one of the 3-phase banks at the sending end of the transmission line from Boulder Dam to Los Angeles. They raise the generator voltage of 16,320-v to 287,500-v.

Courtesy General Electric Co.

FIG. 5. The core and coils of a 180-kw coupling transformer used in the fourth stage of the audio-frequency amplifier of broadcasting station WLW. The input winding consists of two separate parts, interlaced symmetrically with respect to the single output winding, and supplied from the outputs of two separate vacuum tubes. Each input winding is rated 10,200-v peak, and the output winding is rated 6,000-v peak. The assembly shown weighs 27,000 lb., of which the sheet steel in the magnetic circuit weighs 25,700 lb.

Courtesy Westinghouse Electric and Manufacturing Co.

The traditional core constructions represented in Figs. 1, 2, and 3 have three serious objections. First, there are inherently two or more air gaps in these forms of core, no matter what shapes of punching are used. Second, considerable hand labor is involved in assembling or stacking the cores, since the laminations must be interleaved at each joint. Third, there are always certain places in the core where the flux goes across the direction of rolling of the metal, thus increasing the core loss and exciting current. In spite of these difficulties, these three designs are still widely used for power transformers and distribution transformers of the larger sizes.

In recent years, improvements in the technique of manufacture have resulted in cores in which the magnetic flux is almost parallel to the direction of rolling of the steel at all points in the core, and in which air gaps are greatly reduced and assembly labor is saved. In Fig. 6, a shell-type construction is achieved through forming each of the two cores from a single continuous steel ribbon, spot-welded at the ends to hold the ribbon tight. After annealing, the cores are wound through the opening in preformed coils by the ingenious manufacturing process shown in the figure.[2] In Fig. 7 steel strips are bent in a folding machine and interleaved to form two D-shaped cores which are annealed after folding. The method of winding the coils around the core is shown in Fig. 7. In Fig. 8 ribbons of different widths are wound on a form, making a continuous cruciform core which is then annealed. The coils are wound around the core, by means of the clever motor-driven collapsible coil-forming arrangement shown in Fig. 8.

Another important development in the design and construction of transformer cores is the use of Hipersil, a silicon steel having at common excitations about one-third higher permeability than is possessed by the usual silicon steels.[3] Hipersil must be employed in such a manner that the flux is parallel to the direction of rolling of the strip. This requirement is fulfilled in small distribution transformers through winding the core in a continuous strip on a rectangular mandrel. The core is annealed on the mandrel, and then cut as indicated in Fig. 9 to permit the coils to be placed in the core opening to form a shell-type transformer. In the larger ratings, single straight strips of Hipersil are assembled with the joints cut at 45-degree angles and overlapped in alternate layers, as shown in Fig. 10. With this construction, the flux paths at the corners of the core are more nearly parallel to the direction of rolling of the steel than in the conventional type of lap joint.

[2] E. D. Treanor, " The Wound-Core Distribution Transformer," *A.I.E.E. Trans.*, *57* (November, 1938), 622–625.

[3] J. K. Hodnette and C. C. Horstman, " Hipersil, a New Magnetic Steel and Its Use in Transformers," *The Westinghouse Engineer*, *1* (August, 1941), 52–56.

(a)

(b)

(c)

(d)

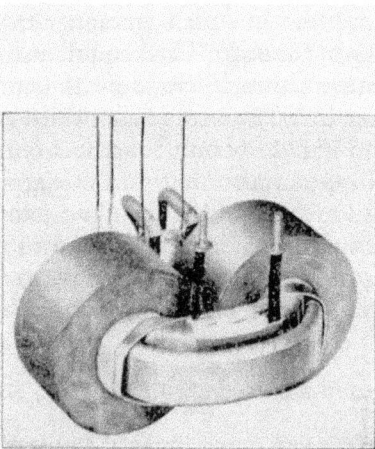

(e)

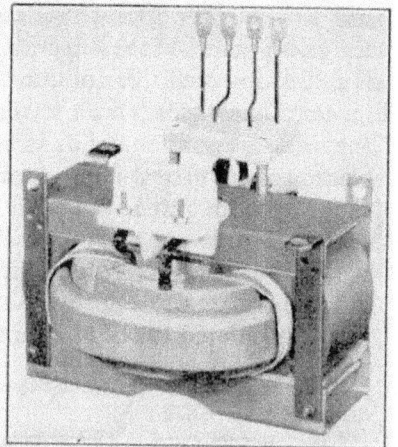

(f)

Courtesy General Electric Co.

FIG. 6

In communication circuits, input transformers usually operate with a relatively feeble alternating flux density, frequently only a fraction of a gauss. A core material having high permeability at low field strengths is desired for these transformers. A relatively large direct current may be present in the primary winding, which may produce a constant component of flux of several thousand gausses density. A core material having high incremental permeability at relatively high polarizations is then desired. Since hysteresis introduces distortion, the hysteresis loss should be low. Since eddy currents produce a screening effect and therefore reduce the effective alternating-current permeability, the eddy currents should be minimized by use of thin laminations of a core material having high resistivity. Sometimes the cores of communication transformers are molded from compressed powder, as described in Art. 6, Ch. I. Toroidal cores manufactured in this manner are shown in Fig. 20, Ch. I. Often constancy of permeability over the operating range is desirable. Various grades of silicon steels and permalloy[4] are commonly used. Since the magnetic properties of some of these special alloys are adversely affected by mechanical stresses, care must be taken in manufacturing and in the design of the structure to avoid stresses in the laminations.

When flat-punched cores of power-system transformers are assembled, the laminations are usually laid down with the joints between them overlapping in successive layers, as shown in Fig. 22a, Ch. VI. This system gives a rigid core and reduces the reluctance introduced by the breaks in the laminations at each corner. Sometimes the laminations in successive layers are not interleaved, giving a butt joint which has a small air gap and greater reluctance than the lap joint. Since the butt joint is easier to assemble than the lap joint, it is frequently used in the cores of audio-frequency transformers, where mechanical strength is not

[4] C. II. Crawford and E. J. Thomas, "Silicon Steel in Communication Equipment," *A.I.E.E. Trans., 54* (December, 1935), 1348–1353.

A. G. Ganz and A. G. Laird, "Improvements in Communication Transformers," *A.I.E.E. Trans., 54* (December, 1935), 1367–1373.

G. W. Elmen, "Magnetic Alloys of Iron, Nickel and Cobalt," *A.I.E.E. Trans., 54* (December, 1935), 1292–1299.

FIG. 6. Several steps in the construction of a small wound-core Spirakore distribution transformer. The core is first wound from a continuous strip of cold-rolled steel, and placed between two power-driven rollers, as in (a), near the already wound coils. The outside turn of the core is then threaded through the coils and temporarily cleated to the next turn, as in (b). In (c), the steel has all been unwound from its roller, the inside turn of the core is clamped to the coil insulation, the inner roller is lifted out, and the outside turn is freed to permit tightening. The power-driven rollers (top right in each view) are placed against the core, and pull it together by friction. This operation forms the tight assembly shown in (d), which is held together by spot-welding the outer layer to the adjacent layer. After the second core has been assembled by the same procedure, the core and coils appear as in (e), while the final core-and-coil assembly, with terminals and frame, is shown in (f).

important. In an audio-frequency transformer with a relatively large direct current in its primary winding, the introduction of a small air gap in the core reduces the flux due to the direct current and may thus increase the alternating-current incremental permeability so greatly that the effective reluctance of the core to the alternating component of flux is actually decreased by the addition of the air gap.

In a punched-lamination shell-type transformer, the core is assembled

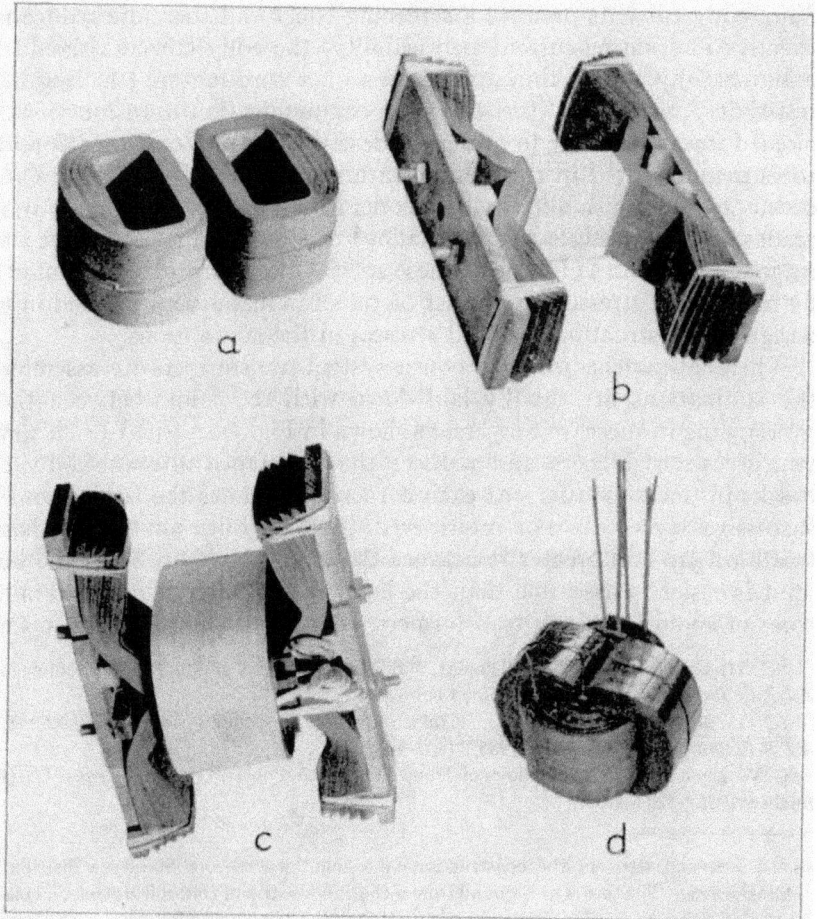

FIG. 7. Four steps in assembling the cores and coils of a Bent-Iron distribution transformer. The two core sections are first formed in a folding machine and annealed, appearing as in (a). The laminations are then opened and tied to a frame (b). Next the coils are wound around the core, through rotating the frame, as in (c). The complete assembly, after dipping in an insulating compound and baking, is shown in (d).

around the form-wound coils, as shown in Fig. 11. When a damaged shell-type coil is to be replaced, the whole core must be taken apart. In a core-type transformer, the core is assembled as a U, as in Fig. 4, the form-wound coils are slipped over the vertical legs, and the top yoke is then added. Only the top yoke need be removed when a damaged coil is to be replaced.

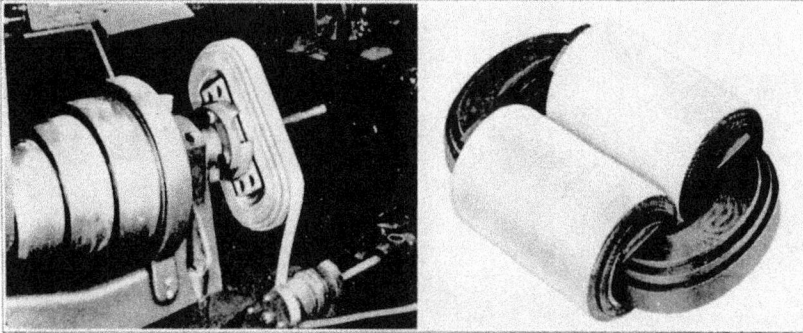

      (a)   (c)

Courtesy Line Material Co.

      (b)   (d)

FIG. 8. Various steps in the production of a Round-Wound distribution transformer. The core is constructed from continuous strips of steel wound on a form, as in (a), and then annealed. To wind the coils (b), the core is placed in a machine which rotates the coil form through a pair of split gears. When the coil is completely wound, the gears are removed. The final core-and-coil assembly appears in (c). The cutaway view (d) of the complete transformer shows how the various parts are mounted. The oil level is high enough to immerse the core and coils completely, but not to cover the porcelain bushings.

Courtesy Westinghouse Electric and Manufacturing Co.

FIG. 9.   Core and coils of a wound-core distribution transformer using Hipersil core material. The cores are wound on a form from a continuous strip, then sawed into two parts to permit the insertion of the coils, as indicated by the unassembled core at the left.

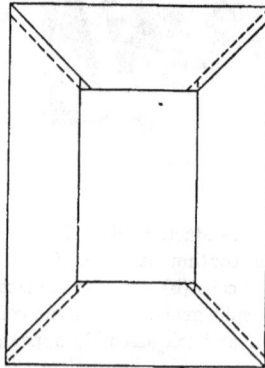

FIG. 10.   Core of Hipersil laminations showing modified lap joints with corners cut diagonally, to promote flux flow in the direction of the grain.

Courtesy Carnegie-Illinois Steel Corp., and Wagner Electric Corp.

Fig. 11.   Stacking the laminations in a shell-type transformer.

2. WINDINGS

The windings usually consist of form-wound coils wrapped with insulating tape, vacuum-treated, impregnated with insulating varnish, and baked. In small low-voltage transformers, round wire is used, but in large transformers the conductors are usually rectangular. If the cross-section of a solid conductor is large or the frequency is high, the effective resistance of the conductor to alternating current may be appreciably larger than its direct-current resistance. To reduce the extra loss which is due to nonuniform current distribution within the conductor, large

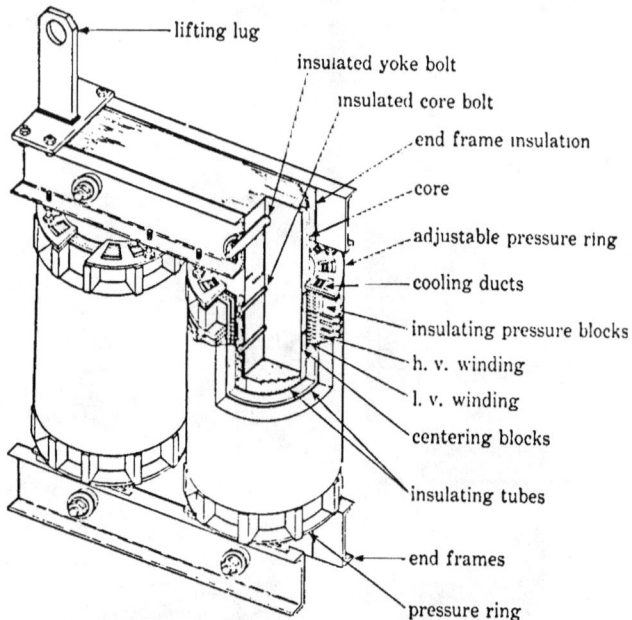

lifting lug
insulated yoke bolt
insulated core bolt
end frame insulation
core
adjustable pressure ring
cooling ducts
insulating pressure blocks
h. v. winding
l. v. winding
centering blocks
insulating tubes
end frames
pressure ring

FIG. 12. Cutaway line drawing of a core-type air-cooled distribution transformer with stepped-circular core and concentric cylindrical low- and high-voltage windings.

conductors are usually subdivided into strands lightly insulated from each other and suitably transposed throughout the winding, as in Fig. 13. If strands of the same size are transposed so that each strand links the same total flux, the total current divides equally among the strands and the copper loss is reduced to a minimum.

In the core-type transformer, the primary and secondary windings are each divided into two equal parts, one of which is placed on each vertical leg of the core. The purpose of this subdivision of the windings is to reduce magnetic leakage between the primary and secondary. The windings are usually concentric with the low-voltage winding next to the core, and are separated from the core and from each other by insulating barriers. In transformers having stepped circular cores, circular coils which are mechanically strong and easy to insulate can be used, as in Fig. 12. Each of the two low-voltage coils may be wound as a continuous helix, as in Fig. 13; but, if the voltage per coil exceeds a few thousand volts, the winding is usually subdivided. Circular disc coils are then commonly used. Figure 14 shows a large circular disc coil on the winding machine. A complete winding consists of a number of these ring coils stacked on each other, as shown in Fig. 15. The discs are usually separated by wooden spacers in order to facilitate cooling.

In the distributed-shell-type transformer, the windings are usually concentric, with the high-voltage winding sandwiched between two halves of the low-voltage winding. This type of winding is seldom used for voltages above 5,000 volts. The concentric sandwiched arrangement may also be used in low-voltage core-type transformers. The sandwiched arrangement of the windings gives less magnetic leakage than the simple concentric arrangement, but has higher manufacturing cost.

In small low-voltage shell-type transformers, either the straight concentric or the concentric sandwiched interleaving of the windings may be used. These arrangements are common in audio-frequency transformers. In large shell-type transformers, thin pancake coils, shown in Fig. 16, are generally used. They are assembled in a stack, interleaving high- and low-voltage coils. The coils are separated by spacers to facilitate cooling and insulation. In order to use the core opening efficiently in the wound-core transformer, Fig. 6, the high-voltage winding is made wider than the low-voltage winding, and is placed in the middle of the opening. The low-voltage winding is in two parts, narrower than the high-voltage winding, and fills most of the remainder of the opening. The complete windings then have a cruciform cross section.

3. COOLING AND INSULATION*

In very small transformers, the surface area is relatively large compared with the volume. Cooling by radiation and natural convection is usually sufficient to keep the operating temperature below the maximum which the insulation can stand without seriously reducing its life. As the size of an object increases, however, the volume increases as the cube, but the surface area increases only as the square of the linear dimensions. Thus for a given loss per unit volume of working parts, the heat which must be dissipated per unit surface area increases proportionally to the linear dimensions. Hence, as the size increases, either the surface area

---

* The effects of temperature on insulating materials are discussed in Art. 1, Ch. VIII, and a brief discussion of cooling of electrical apparatus is given in Art. 5, Ch. VIII.

FIG. 13. A helical winding under construction. This is a partly wound low-voltage coil for a 3-phase 3,000-kva 60 ~ 4,600 : 460-v oil-immersed power transformer.

Courtesy General Electric Co.

Courtesy Allis-Chalmers Manufacturing Co.

FIG. 14.   Winding a disc-type coil for the 230,000-v winding of a 40,000-kva single-phase transformer.

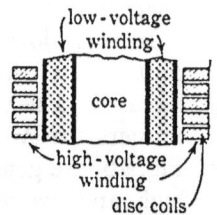

FIG. 15. Section of concentric winding composed of circular-disc high-voltage coils.

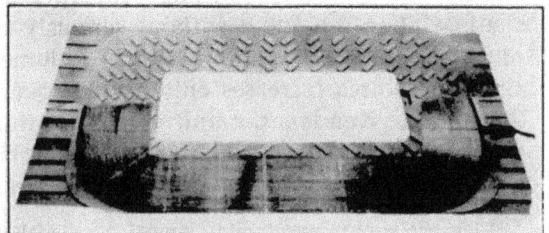

FIG. 16. A "pancake" coil used in shell-type high-voltage transformers. The insulating barrier is partly cut away to show the coil.

Courtesy Westinghouse Electric and Manufacturing Co.

must be increased or artificial means must be provided for accelerating the heat dissipation. Often these two means of facilitating the cooling are combined. Furthermore, as the linear dimensions increase, the distances from the internal parts to the surface increase, and the windings and core must be provided with suitable ventilating ducts.

3a. *Air Cooling.* — In some installations, the fire risk from a transformer must be minimized, and conventional oil cooling is therefore undesirable. Examples are transformers used inside buildings to transform the voltage of 600-, 480-, and 240-volt circuits to 120 volts for lighting purposes. Air-cooled transformers, through which air circulates by convection, are well adapted to this purpose. The dimensions of a transformer having convected air cooling are somewhat larger than those of an oil-cooled transformer of the same rating.

3b. *Air-Blast Cooling.* — When the voltage is below about 40,000 volts and when it is important to keep down the weight and space requirements, the transformer may be cooled by a blower. A common application of this type of cooling is in the transformers used in alternating-current locomotives and multiple-unit railway cars.

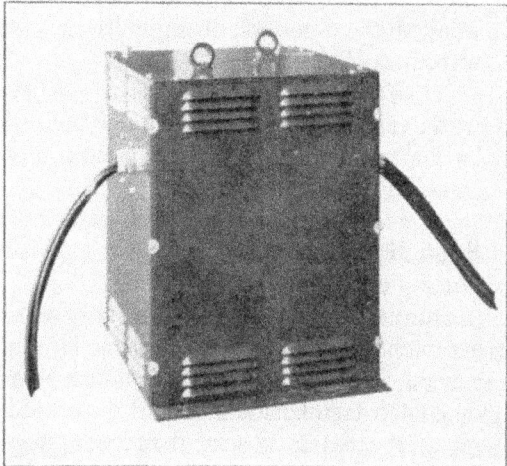

Courtesy General Electric Co.

FIG. 17. An air-cooled transformer for use inside buildings, to provide lighting service from a higher-voltage wiring system. This transformer is rated 50-kva single-phase 240/480-v primary to 120/240-v secondary.

3c. *Transformer Oil.* — A generally more satisfactory method of cooling is to immerse the transformer working parts in oil, which serves the twofold purpose of facilitating the removal of heat from the core and windings, and at the same time has valuable insulating properties. The oil should have high dielectric strength, low viscosity, low freezing point, and high flash point and should be free from corrosive acids, alkalies, and sulphur. The oil should not oxidize or sludge. Unfortunately, the presence of very small amounts of moisture or suspended particles seriously affects the dielectric strength of the oil so that in large transformers special means are provided to prevent moisture from entering it.

3d. *Oil-Immersed, Self-Cooled.* — Most distribution transformers and many power transformers are of this type. In small sizes a smooth tank provides sufficient surface for cooling the oil. In intermediate sizes the outside walls of the tank are provided with fins, or corrugations, or with vertical tubes through which the oil circulates downward as it cools. In large sizes radiators are massed all over the available surface.

3e. *Oil-Immersed, Forced-Air-Cooled.* — Thermostatically controlled blowers or a number of small fans may be provided to blow cool air through the radiators. Transformers of this type have two ratings; the smaller the self-cooled rating with natural convection, the larger its forced-air-cooled rating.

3f. *Oil-Immersed, Water-Cooled.* — Where a suitable supply of cooling water is available and space must be conserved, the hot oil may be cooled by water circulating through a coil of copper tubing placed around the inside of the tank.

3g. *Forced-Oil Cooling.* — The hot transformer oil may be pumped through an external cooler, which is usually water-cooled. This method of cooling is common in Europe.

Standard transformer oil is fairly volatile, and if it vaporizes there is great chance of a serious explosion. Even if it does not explode, the oil can burn, producing an intense flame and heat. Therefore conventional oil-cooled transformers should be operated outdoors, or if indoors they must be located in special fireproof vaults that conform to fire underwriters' specifications. This objection to transformer oil is overcome by the use of special noninflammable liquid compounds.

3h. *Pyranol, Inerteen, or Chlorextol.* — Air cooling for transformers to be used inside buildings has been mentioned. The limitations of air cooling are serious, however, when the voltage is high so that dust gathering on the windings might be harmful, or when small space is an important requirement. Chemical compounds, bearing the trade names Pyranol, Inerteen, or Chlorextol, have been developed to replace ordinary transformer oil. These compounds are nonvolatile, noncombustible, and nonexplosive, and are sufficiently thin to circulate freely around the windings. They have high dielectric strength and thus, like standard transformer oil, serve both to insulate and to cool the windings.

4. TANKS

Transformers using liquid cooling necessarily have their cores and windings enclosed in leakproof tanks, which are of welded steel construction and may be round, oval, elliptical, or rectangular. Allowance must be made for the expansion and contraction of the oil caused by temperature changes. With distribution transformers, common practice is to use an air-tight tank with sufficient air space between the cover and

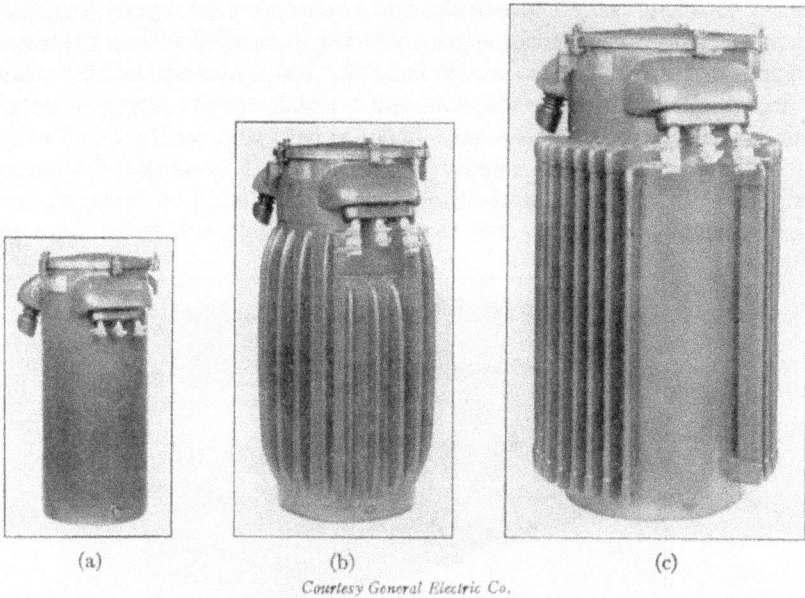

FIG. 18. Three 2,400 : 120/240-v 60 ~ single-phase distribution transformers, showing the effects of rating on the provisions made for heat dissipation. Transformer (a) is rated 15 kva, and has a smooth tank; transformer (b) is rated 75 kva, with a corrugated tank; while (c), of 100-kva capacity, is cooled by external tubes welded to the main tank. The illustrations are in approximately correct scale proportion. The over-all top-to-bottom dimension of transformer (c) is 71 in.

FIG. 19. A 3-phase 1,500-kva 13,200 : 4,330-v transformer equipped with auxiliary fans for cooling the oil. When the fans are in operation, the transformer rating is increased to 2,000 kva.

the oil to allow the oil to expand and compress the trapped air. With some larger transformers the space above the oil is filled with dry nitrogen the pressure of which is maintained slightly above atmospheric by means of an automatic pressure-operated valve which admits nitrogen from a cylinder of dry compressed gas. There is also a pressure relief valve. Large transformers more usually are allowed to " breathe." A common method is to mount a horizontal drum called an " oil conservator " above the tank and connected to the tank through a U tube, as shown

Courtesy General Electric Co.

FIG. 20.   A single-phase 625-kva 25 ∼ 11,000 : 460-v transformer designed for service on a multiple-unit electric railway car. Pyranol, cooled by forced-air ventilation, is used for insulation and heat dissipation. The high-voltage bushing appears at the left, the low-voltage terminals at the right.

in Fig. 27. The oil completely fills the tank and partly fills the expansion drum. On the top of the drum there is a breather opening into the atmosphere. This breather may be equipped with a chemical filter for removing moisture and oxygen from the air admitted to the conservator. The conservator is equipped with a sump for occasional removal of sludge and moisture. In the event of a short circuit within the transformer, gas bubbles may be rapidly formed; hence the cover is provided with a projecting pipe closed by a thin relief diaphragm which breaks under excessive pressure and thus prevents the tank from bursting. The danger of an explosion resulting from hot gas bubbles is eliminated if oxygen is excluded from the top of the tank.

5. BUSHINGS

With moderate voltages the leads are brought out of the top of the transformer case through porcelain bushings or through wiping-sleeve type joints, which are wiped to the lead sheaths of underground cables. For high voltages there are two common types of bushings, the oil-filled and the condenser type, which are similar in external appearance but differ in their principle of operation.

The oil-filled bushing consists of a conducting rod surrounded by a number of thin concentric cylinders of insulating material separated by treated hardwood spacers. The rod and insulating cylinders are supported inside two hollow porcelain cones, as shown in Fig. 21. The open spaces in the bushing are filled with insulating oil. The outsides of the insulators

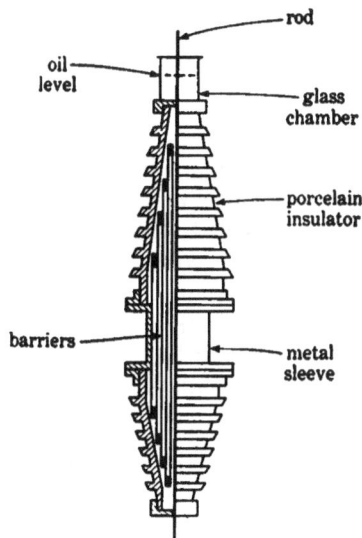

Fig. 21. Principal parts of an oil-filled bushing.

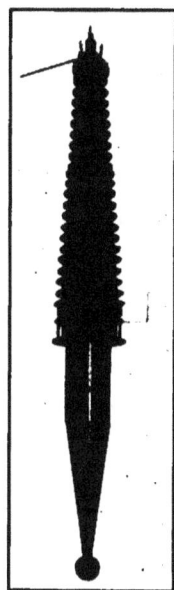

Courtesy *Westinghouse Electric and Manufacturing Co.*

Fig. 22. A 288,000-v condenser-type bushing.

are made with petticoats to increase the creepage distance between the terminals and the grounded tank. The lower cone projects below the oil level and requires shorter creepage distance than the upper cone, which is exposed to the weather. The bushing is proportioned so that the voltage gradient is nearly uniform along its surface.

The condenser bushing, shown in Fig. 22, is made of alternate layers of insulation and metal foil wrapped around a central conducting rod. The upper part of this internal assembly is mounted inside a hollow porcelain cone, as in the oil-filled bushing. The lower end projects into the transformer oil. The purpose of the foil is to bring about a fairly uniform voltage gradient within the insulation. If the foil were omitted, the voltage gradient between the rod and the sleeve would vary approximately inversely as the radial distance from the center of the rod, and hence the gradient in the insulation next to the rod would be very much greater than near the sleeve, unless the rod diameter were very large. The metal foil breaks up the insulation into a number of condensers in series. The voltages across condensers in series are inversely proportional to their capacitances. If the layers of insulation are all of the same thickness, the capacitances of the condensers are proportional to their surface areas. If the capacitances are equal, the voltages across the layers of insulation are equal. Hence if the axial lengths of the layers vary inversely as their diameters, the maximum insulation stress at the inner fibers of each layer is approximately the same, and the material is used safely and efficiently.

6. PROTECTIVE DEVICES

Transformers used in power-system or industrial applications need to be installed in such a manner that they are protected against excessively high voltages and dangerous overloads; they must be arranged for easy maintenance; and the installation must not constitute a serious fire or accident hazard. Two sources may cause excessive voltages across the terminals. The first cause, lightning, produces traveling waves of very steep wave fronts, which travel in both directions from the point of incidence on wire lines and are reflected from ends and junctions, causing a series of sharp voltage peaks. When one of these high-voltage waves reaches the terminals of a transformer, the insulation may break down between turns, or between a winding and the tank. The obvious way to minimize the effect of these waves is to connect a device between the power line and the ground which provides a heavy-current-carrying path around the transformer, and thus harmlessly dissipates the energy in the wave. The devices used for this purpose are called *lightning arresters*. They are manufactured in a great variety of forms and materials, but they all have the property of nonlinear resistance. The active element is a chemical substance, assembled in powder, pellet, or pressed block form, whose electrical resistivity becomes lower as the voltage gradient across the element rises. In most lightning arresters, one or more air gaps are placed in series with the nonlinear resistance element. The gaps break down when the high-voltage wave front arrives, and they cause negligible resistance in series with the main elements. However, when the wave has nearly passed, the arc in the air gaps goes out at a time when the arc current is zero, and thereafter no conduction current flows through the arrester until another steep wave comes along.

The second cause of dangerously high voltage is a surge which results from a fault in the power system or from a switching error. The cure for this type of high voltage is also to use lightning arresters. Care must be taken to be sure that the arrester selected for any application will success-

FIG. 23. A 37,000-v lightning arrester, cut away to show the internal assembly. The lower portion of the unit consists of porous ceramic blocks, whose current-voltage characteristic is nonlinear. Although negligible current flows at normal voltage, the blocks conduct 1,500 amp at 2.1 times normal voltage. The upper portion of the arrester consists of small series gaps. (Other types of arresters are illustrated on pp. 672–673 of the volume on electric circuits.)

Courtesy Westinghouse Electric and Manufacturing Co.

Courtesy Westinghouse Electric and Manufacturing Co.

FIG. 24. Three 50-kv lightning arresters protecting a 3-phase bank of single-phase transformers in an outdoor substation.

fully withstand the continuously maintained power-line voltage across it during a line-to-ground fault on another phase. This voltage is, for many three-phase systems, much greater than the normal voltage across the arrester.

Not only must transformers be provided with suitable surge-voltage protection, but also they must be automatically disconnected from the

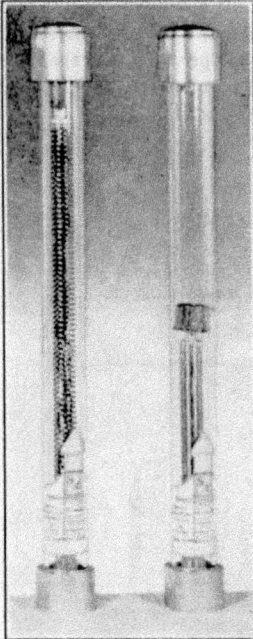

Courtesy Schweitzer and Conrad Co.

FIG. 25. A 34,500-v 30-amp fuse before and after blowing. This fuse is filled with an arc-extinguishing liquid, and has an interrupting capacity of 6,400 amp.

FIG. 26. Three fuses, containing solid arc-extinguishing material, installed in a 66,000-v circuit. These fuses are rated to interrupt 7,000 amp at 66,000 v.

power source if a dangerously high winding current is present, either from an abnormal overload, or as a result of internal failure of the transformer insulation. This overcurrent protection is provided either by fuses or by automatic circuit breakers which open the circuit before the heating effect of the overcurrent becomes serious. The usual overcurrent protection, either by fuses or by circuit breakers, is adjusted with an "inverse-time characteristic," which means that the circuit is opened after a shorter period of time for high currents than for lower currents.

Fuses are available for protection of transformers in ratings up to 138,000 volts. In high-voltage fuses, the main problem is to extinguish the arc rapidly. This is done in a variety of ways, including the blowout action resulting from the expansion of gas in a tube containing the hot arc, the rapid separation under spring tension of the terminals on either side of a fusible link, and the quenching of the arc in a liquid such as carbon tetrachloride.

To make a transformer installation reasonably safe from an accident standpoint, all bare terminals should be well out of reach, or protected by grounded metal or insulating screens. The metal tanks should be well grounded, and high-voltage warning signs should be conspicuously displayed if the transformers are in a location accessible to the public or to employees not concerned with maintenance of them. Transformers having bare high-voltage terminals should always be placed in a fenced and locked enclosure or in a locked room, and, if they contain inflammable cooling oil, they should be mounted above a bed of crushed stone, or else surrounded by a low concrete wall that will prevent leaking oil from spreading during a fire.

A very slight explosion risk is associated with transformers cooled by ordinary oil. A sustained arc inside the tank will vaporize oil, and, when enough oil vapor has gathered, an explosion may follow unless a pressure relief device is provided. It is not advisable, therefore, to allow people to stand or work near large power transformers as a regular practice.

A recent development in distribution transformers is the so-called CSP, or "completely self-protecting," transformer. This contains its own lightning protection and fuses with indicators which show whether the fuses have blown or are in operation.

All distribution, power, or instrument transformers must be installed with suitable disconnecting devices which can be opened to permit safe access to the electrical parts for inspection, maintenance, or repair. Distribution transformer fuses are usually built into a porcelain housing, called a *cutout*. The fuses are pulled to disconnect the transformer. However, for large power transformers a separate switch is provided which de-energizes the fuses or circuit breaker as well as the transformer. An open-air disconnect switch must always be placed in series with an oil circuit breaker, so that a workman can see with his own eyes that the circuit is open before he starts work on the transformer. If there is a possibility of two-way feed, disconnect switches must be provided for both windings.

Many indoor potential transformers are provided with fuse mountings and fuses in series with the high-voltage windings, to protect the line against short circuits resulting from a failure within the transformer. When not furnished with its own fuses, a potential transformer should

Fig. 27. An open-type substation used to energize a 13,800-v and a 4,800-v system from a 66,000-v transmission line. The main transformer bank in the foreground consists of three single-phase 500-kva units. The 66-kv lightning arresters are mounted at the rear of the structure, slightly above center, and the 13.8-kv arresters at the right side, just below center. Liquid-filled fuses protect the 66-kv circuits. The outgoing 13,800-v line

Fig. 28. A 600-kva 3-phase 13,200 : 4,330-v CSP (" completely self-protecting ") transformer, which has all necessary auxiliary devices built in at the factory. The various compartments mounted on the front of the transformer are: upper left, mechanism for tap-changing under load; upper right, 4,330-v circuit breaker; lower left, tap-changer control and oil-temperature indicator; lower right, overload, reverse power, and reclosing relays. The necessary lightning arresters are mounted on top of the main housing. When all doors are closed and the case grounded, there are no live parts exposed below the top cover.

---

at the right is controlled and protected by an oil circuit breaker. A 650-kva 3-phase transformer (partly visible behind the right-hand main transformer) steps down from 13,800 v to 4,800 v and supplies a distribution system. The two small distribution transformers at the right supply power for use in the station itself.

Each transformer bank has its own disconnecting device for manual opening during repairs or maintenance work. The 66-kv air-break disconnect switches on top of the structure are used for sectionalizing the high-voltage line in either direction, and may be opened while carrying normal load.

The entire steel structure, all metal tanks, and the safety fence are connected by heavy conductors to a grounding network beneath the substation. The ground wire for the 66-kv line is electrically connected to the grounded structure at the upper right.

always be installed with separate fuses in series with the high-voltage winding. The fuses used to protect potential transformers should always be equipped with or placed in series with current-limiting resistors, which

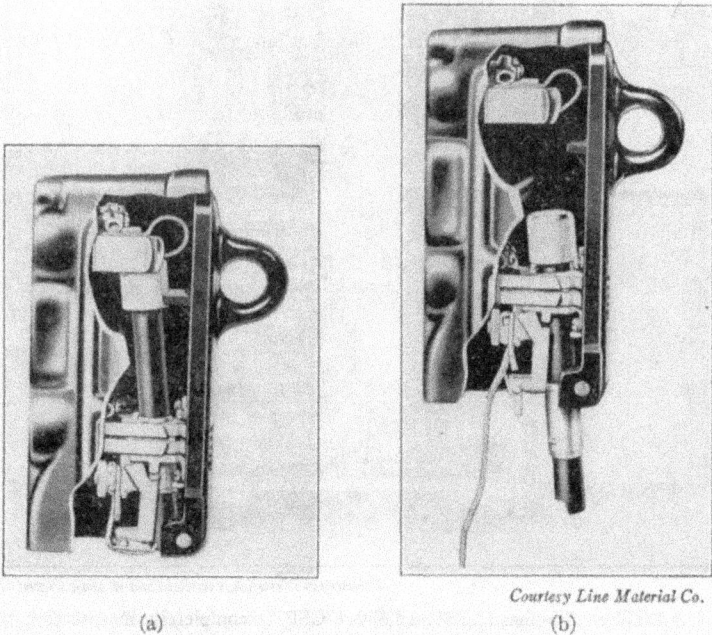

(a)   (b)

Fig. 29. A primary cutout, used both as a fuse and as a disconnecting device for distribution transformers. The fuse is contained within the fiber cylinder supported by the hinged cover, as in (a). To open the circuit, the hinged cover is pulled to the right. If the fuse blows, the broken end drops out of the cylinder and the top spring drives the cylinder down so that it projects below the porcelain housing, indicating the blown condition, as in (b).

will prevent the formation of destructive power arcs within the fuses when the fuses blow.

7. SHIELDING

Even though lightning arresters are provided to protect a transformer, the local voltage between turns may be excessive at certain places in the windings. In other words, the distribution of voltage gradient may not be uniform. One method of improving this situation is to provide a carefully designed metallic shield which is so located with respect to the windings that the voltage gradients are kept within safe values for all windings. This system is known as electrostatic shielding.

In communication transformers, electrostatic shielding is commonly

Courtesy General Electric Co.

FIG. 30. The assembled core and coils of a 55,000-kva shielded-winding power transformer. This is the transformer whose core is shown in Fig. 4, p. 288. The high-voltage line lead at right has not been enclosed in its final casing.

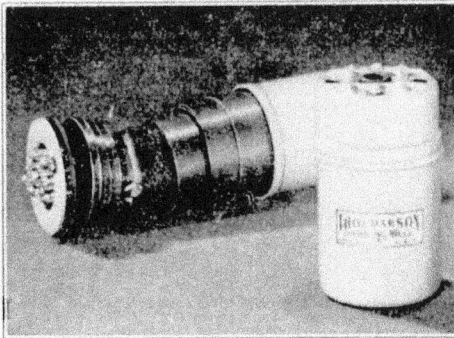

Courtesy Thordarson Electric Manufacturing Co.

FIG. 31. Small audio-frequency transformer mounted in a case with three high-permeability shields.

employed to prevent electrostatically induced voltages from arising in one winding as a result of a voltage present on the other, and to prevent similar static interference between the transformer windings and other nearby circuit elements. This electrostatic shielding also performs the important function of definitely fixing the winding-to-ground and interwinding capacitances. Magnetic shielding of communication transformers is often provided in the form of a steel case, to minimize electromagnetic interference between the transformer and other circuit elements.

CHAPTER XII

Magnetic Leakage in Transformers

In the simplest theory given in Art. 2, Ch. X, the transformer is assumed electrically perfect. A more comprehensive theory of its electrical characteristics, however, must take into account at least approximately the following imperfections which occur in iron-core transformers:

(1) The windings have resistance;
(2) There is magnetic leakage;
(3) An exciting current is required to produce the flux;
(4) There are hysteresis and eddy-current losses in the core.

In some problems involving high frequencies, it is also necessary to take into account the capacitances of the windings, but in the following analysis these capacitances are neglected. Then, as in Eqs. 1 and 2, Ch. X, the voltage equations are

$$v_1 = R_1 i_1 + \frac{d\lambda_1}{dt} \qquad \blacktriangleright [1]$$

$$v_2 = R_2 i_2 + \frac{d\lambda_2}{dt} \qquad \blacktriangleright [2]$$

where

v_1 and v_2 are the instantaneous primary and secondary terminal voltages,
i_1 and i_2 are the instantaneous currents,
R_1 and R_2 are the effective resistances of the windings,
λ_1 and λ_2 are the instantaneous flux linkages resulting from the effects of both currents.

Before these equations can be solved, it is necessary to determine the relations between the flux linkages and the currents, taking into account magnetic leakage and the magnetic properties of the core. The remainder of this chapter is devoted to the discussion of these matters and of the simplifying approximations necessary to obtain a useful solution.

1. MAGNETIC LEAKAGE AT NO LOAD

Magnetic leakage has an important effect on the load characteristics of all alternating-current machines. Although the magnetic leakage in an iron-core transformer usually is very small at no load, a discussion of the magnetic field produced by the current in one winding is an aid to a clear understanding of the important and more complicated condition when there are currents in both windings.

Consider the two air-core coils shown in Fig. 1, constituting an elementary transformer. If the instantaneous current in coil 1 is i_1 and coil 2 is open-circuited, the general character of the magnetic field produced by i_1 is shown by the roughly drawn field map of Fig. 1a. Since

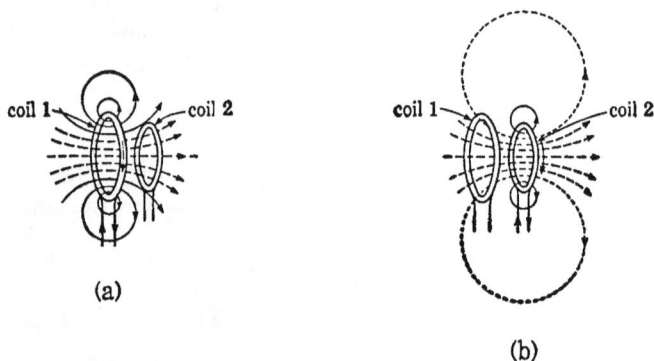

(a)

(b)

FIG. 1.   Magnetic fields of compact, air-core coils; (a) current in coil 1 only; (b) current in coil 2 only.

the turns of coil 1 are concentrated in a compact coil, practically all the flux produced by coil 1 links all N_1 turns of coil 1. Let φ_{11} be the total instantaneous flux linking coil 1. A portion of this flux also links coil 2, as shown by the broken lines in Fig. 1a. Let φ_{21} be this mutual flux. The difference between the total flux linking coil 1 and the mutual flux which also links coil 2 is the leakage flux φ_{l1} of coil 1 with respect to coil 2. That is,

$$\varphi_{l1} \equiv \varphi_{11} - \varphi_{21}. \qquad [3]$$

This leakage flux is shown by the unbroken flux lines in Fig. 1a.

The magnetic field produced when coil 2 is similarly excited by an instantaneous current i_2 and coil 1 is open-circuited is shown in the roughly drawn field map of Fig. 1b. If the total flux linking coil 2 is φ_{22} and the mutual flux produced by i_2 is φ_{12} — shown by the broken lines in Fig. 1b — the leakage flux φ_{l2} of coil 2 with respect to coil 1 is

$$\varphi_{l2} \equiv \varphi_{22} - \varphi_{12}. \qquad [4]$$

This leakage flux is shown by the unbroken flux lines in Fig. 1b.

In the compact coils of Fig. 1, there is a fairly clear-cut demarcation between the leakage and mutual fluxes. The windings of an actual transformer are not such simple compact coils, however, so that it is necessary to take into account the effects of the magnetic field within the regions occupied by the windings. For example, consider the roughly drawn field maps of Fig. 2 which show, in cross section, the core and coils of a core-type transformer with concentric windings. In Fig. 2 there is current in only the inner winding 1, as indicated by the dots and crosses representing the heads and tails of arrows pointing in the direction of the current i_1. The principal components of the flux produced by i_1 are shown in Fig. 2a. Most of the flux is confined to the core and therefore links all

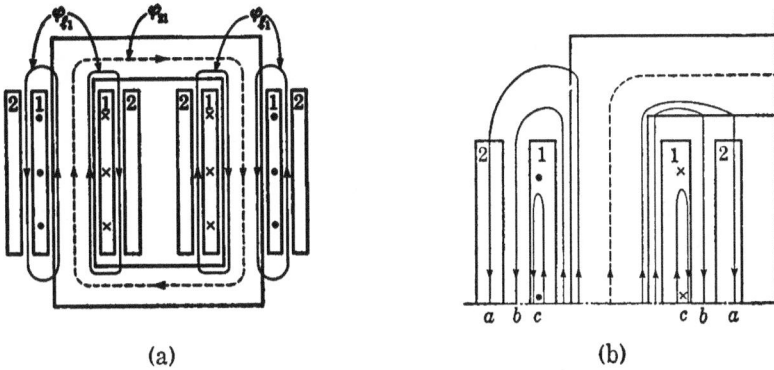

(a)   (b)

FIG. 2. Magnetic field due to current in the inner winding of a core-type transformer. The principal fluxes are shown in (a), and (b) shows the general character of the field in the upper, left-hand quadrant of (a).

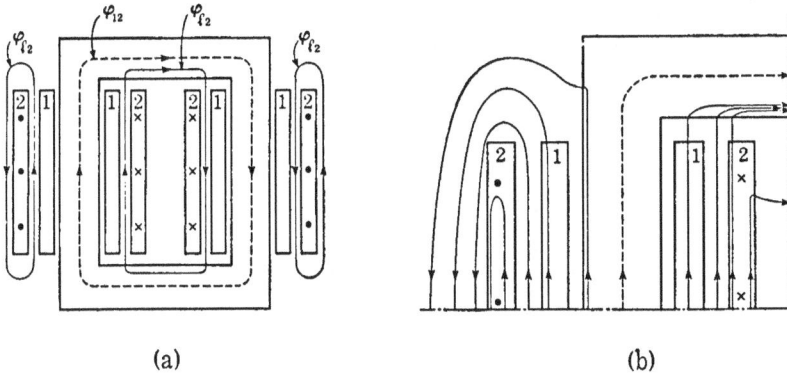

(a)   (b)

FIG. 3. Magnetic field due to current in the outer winding of Fig. 2.

the turns of both windings, as shown by the broken lines in Fig. 2. Additional flux, not entirely confined to the core, is shown by the light unbroken flux lines. Figure 2a shows the approximate paths of this air-borne flux, and Fig. 2b shows a roughly drawn map of the magnetic field in the upper, left-hand quadrant of Fig. 2a. An examination of Fig. 2b shows that some of the air-borne flux, such as lines *a* and *b*, links all the turns of winding 1, while another portion of it, such as lines c, links only a fraction of the turns of winding 1. Lines such as b and c link the turns of winding 1 only, but lines such as a also link some of the turns of winding 2. Figure 2b shows that, because of the partial flux linkages produced by fluxes, such as a and c, whose paths lie among the turns of the windings, the distinction between leakage and mutual fluxes in an actual transformer is less evident than in the compact coils of Fig. 1.

Because of the ease with which the leakage and mutual fluxes can be visualized in compact coils, it is convenient to simplify the picture of the magnetic field in a transformer by introducing the concept of equivalent fluxes in the following manner. Let λ_{11} be the flux linkage with winding 1 produced by the current in winding 1. A small contribution to the flux linkage λ_{11} is produced by flux, such as that shown in lines c, Fig. 2b, that links only a fraction of the turns of the winding. Let φ_{11} be an equivalent flux considered as linking all N_1 turns of winding 1 and producing a flux linkage $N_1\varphi_{11}$ equal to the actual flux linkage λ_{11}. Then

$$\varphi_{11} \equiv \frac{\lambda_{11}}{N_1}. \qquad [5]$$

The flux φ_{11} is the average flux linkage with winding 1 per turn. A portion of this flux also links winding 2. Let λ_{21} be the flux linkage with winding 2 produced by the current in winding 1. A small contribution to the flux linkage λ_{21} is produced by flux, such as that shown in lines a, Fig. 2b, that links only a fraction of the turns of winding 2. Let φ_{21} be an equivalent flux considered as linking all N_2 turns of winding 2 and producing a flux linkage $N_2\varphi_{21}$ equal to the actual flux linkage λ_{21}. Then

$$\varphi_{21} \equiv \frac{\lambda_{21}}{N_2}. \qquad [6]$$

The flux φ_{21} is the average mutual flux linkage with winding 2 per turn.

The difference between the average flux φ_{11} linking winding 1 and the average mutual flux φ_{21} which also links winding 2 is the leakage flux $\varphi_{\ell1}$ of winding 1 with respect to winding 2. That is,

$$\varphi_{\ell1} \equiv \varphi_{11} - \varphi_{21} \qquad [7]$$

$$\equiv \frac{\lambda_{11}}{N_1} - \frac{\lambda_{21}}{N_2}. \qquad [8]$$

Thus the actual flux distribution of Fig. 2b is equivalent to the simplified picture of Fig. 2a, in which the magnetic field produced by i_1 is represented by an equivalent mutual flux φ_{21}, shown by the broken lines in Fig. 2a, linking all the turns of both windings, and a leakage flux $\varphi_{\ell1}$ linking all the turns of winding 1 but none of the turns of winding 2, as shown by the light unbroken lines in Fig. 2a. The magnetic field of an actual transformer then may be visualized in terms of equivalent leakage and mutual components similar to those of the simple, compact coils of Fig. 1. When these equivalent fluxes are expressed as average values of the flux linkages per turn, as in Eqs. 6, 7, and 8, they correctly take into account the partial flux linkages produced by the magnetic field within the winding spaces.

Similarly, if there is a current i_2 in winding 2 while winding 1 is open-circuited, the flux distribution is approximately as shown in Fig. 3b, p. 315. The average flux φ_{22} linking winding 2 is

$$\varphi_{22} \equiv \frac{\lambda_{22}}{N_2}, \tag{9}$$

where λ_{22} is the flux linkage with winding 2 produced by its own current. The average mutual flux φ_{12} linking winding 1 is

$$\varphi_{12} \equiv \frac{\lambda_{12}}{N_1}, \tag{10}$$

where λ_{12} is the flux linkage with winding 1 produced by the current in winding 2. The leakage flux of winding 2 with respect to winding 1 is

$$\varphi_{l2} \equiv \varphi_{22} - \varphi_{12} \tag{11}$$

$$\equiv \frac{\lambda_{22}}{N_2} - \frac{\lambda_{12}}{N_1}. \tag{12}$$

The equivalent fluxes φ_{12} and φ_{l2} are shown in Fig. 3a.

Most of the mutual flux in an iron-core transformer is confined to the core, and because of the magnetic nonlinearity of the iron is not proportional to the magnetomotive force producing it. The leakage flux, however, is in air for a considerable portion of the length of its path. Hence the reluctance of the iron portion of the leakage-flux paths is small compared with the reluctance of the paths in air.

▶ Therefore, in spite of the magnetic nonlinearity of the iron, the leakage flux is very nearly directly proportional to the current producing it. This important property of the leakage flux greatly simplifies the analytical treatment of iron-core transformers.◀

2. Voltage Equations and Flux Distribution under Load

When there are currents in both windings, the resultant flux linkages λ_1 and λ_2 in Eqs. 1 and 2 are produced by the combined effects of both currents and are in part due to the partial linkages of the magnetic field within the winding spaces. As an aid to visualizing the phenomena, let

φ_1 and φ_2 be equivalent fluxes linking all the turns of the windings with which they are associated and producing flux linkages $N_1\varphi_1$ and $N_2\varphi_2$ equal to the actual resultant flux linkages λ_1 and λ_2. That is, let

$$\varphi_1 \equiv \frac{\lambda_1}{N_1} \qquad [13]$$

$$\varphi_2 \equiv \frac{\lambda_2}{N_2}. \qquad [14]$$

Then Eqs. 1 and 2 can be expressed as

$$v_1 = R_1 i_1 + N_1 \frac{d\varphi_1}{dt} \qquad \blacktriangleright[15]$$

$$v_2 = R_2 i_2 + N_2 \frac{d\varphi_2}{dt}. \qquad \blacktriangleright[16]$$

The equations are rewritten in this form since the equivalent fluxes can be visualized more readily than can the actual magnetic field. Though most of the flux links both windings, there are also leakage fluxes linking one winding but not the other. The fluxes φ_1 and φ_2 in Eqs. 15 and 16 are the total fluxes linking the windings and include both leakage and mutual components. The problem is to determine the relations among these resultant fluxes and the currents.

2a. *Component Fluxes.* — When there are currents in both windings, the resultant magnetic field depends on the instantaneous values of both currents. In general, therefore, determining the distribution of the magnetic field is then more difficult than when there is current in but one winding. For this reason, it is convenient to resolve the resultant fluxes into the components produced by each current alone.

For the present, consider a transformer having a core of magnetic material whose permeability is constant. With this assumption, the Principle of Superposition can be applied, and the resultant flux linking each winding can be expressed as the sum of the components due to each current acting by itself.* These components are:

(1) The leakage flux due to the current in the winding;
(2) The component mutual flux due to the current in the winding;
(3) The component mutual flux due to the current in the other winding.

---

* It is unnecessary to postulate a constant-permeability core if the component fluxes are regarded as the components due to each current acting *in the presence of the other current.* However, it is probably easier to visualize the component fluxes if constant permeability is assumed, and the results of investigating this simple example readily can be extended to apply to an iron-core transformer, as shown subsequently.

Thus the resultant fluxes φ_1 and φ_2 in Eqs. 15 and 16 can be expressed as

$$\varphi_1 = \varphi_{l1} + \varphi_{21} + \varphi_{12} \qquad [17]$$

$$\varphi_2 = \varphi_{l2} + \varphi_{12} + \varphi_{21}. \qquad [18]$$

The fluxes with double subscripts are the components produced by a single current acting by itself. That is,

φ_{l1} and φ_{l2} are the component leakage fluxes produced by each current,

φ_{21} is the component mutual flux produced by i_1,

φ_{12} is the component mutual flux produced by i_2.

The paths of these component fluxes and their positive directions are shown in Figs. 2 and 3, p. 315, and the resultant magnetic field at any instant can be regarded as the result of superposing these component fields with proper consideration for their instantaneous magnitudes and directions, as shown in part (b) of this article.

The sum $\varphi_{l1} + \varphi_{21}$ of the first and second components on the right-hand side of Eq. 17 is the total flux φ_{11} produced by the current in the primary winding alone. Hence Eq. 17 can be written as

$$\varphi_1 = \varphi_{11} + \varphi_{12}. \qquad [19]$$

Thus, by superposition, the resultant flux φ_1 linking the primary can be expressed as the sum of the flux φ_{11} produced by the primary current alone plus the mutual flux φ_{12} linking the primary but produced by the secondary current alone. Similarly, in Eq. 18, $\varphi_{l2} + \varphi_{12}$ is the total flux φ_{22} linking the secondary due to its own current alone, and φ_{21} is the mutual flux linking the secondary but due to the primary current alone. Hence, by superposition, Eq. 18 can be written as

$$\varphi_2 = \varphi_{22} + \varphi_{21}. \qquad [20]$$

This method of combining the component fluxes leads to the familiar coupled-circuit equations in terms of the self- and mutual inductances of the windings, discussed in Chs. XVII and XVIII. The purpose of introducing Eqs. 19 and 20 at this time is merely to point out that the classical coupled-circuit equations can be derived from the same basic theory as the method of analysis discussed in the remainder of this chapter.

The component fluxes in Eqs. 17 and 18 can also be combined in another way which is particularly convenient for the analysis of iron-core transformers. Note that the resultant mutual flux produced by the combined action of both primary and secondary currents is the sum of the components due to each current acting alone. If φ is this resultant mutual flux,

$$\varphi = \varphi_{21} + \varphi_{12}. \tag{21}$$

Hence Eqs. 17 and 18 can be written as

$$\varphi_1 = \varphi_{l1} + \varphi \qquad \blacktriangleright [22]$$

$$\varphi_2 = \varphi_{l2} + \varphi. \qquad \blacktriangleright [23]$$

That is, the resultant flux linking each winding can be expressed as the sum of the leakage flux due to the current in the winding alone plus the resultant mutual flux due to the combined magnetomotive forces of both primary and secondary currents acting simultaneously. Combining the components in Eqs. 17 and 18 in this way leads to the method of analysis discussed in Arts. 3 and 4.

2b. *Resultant Flux Distribution.* — Before entering into a detailed discussion of the analytical treatment that can be developed from Eqs. 22 and 23, the student should have a clear understanding of the significances of the component fluxes in these equations. Particularly, it is important that he have no misconceptions of the manner in which these component fluxes are related to the actual flux distribution in a loaded transformer, since a considerable amount of loose thinking on this subject has been done.[1] Therefore the following discussion will describe, by means of roughly drawn field maps, the manner in which the magnetic field varies with time in a loaded transformer.

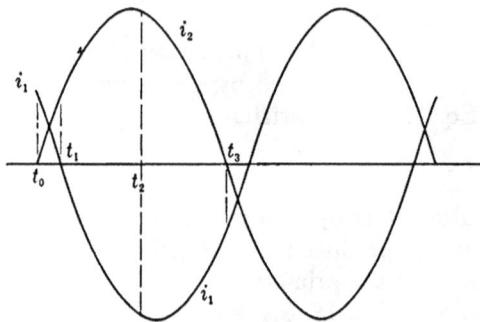

FIG. 4.   Instantaneous primary and secondary currents.

Consider a loaded transformer having the same number of turns in its primary and secondary windings. As shown in Art. 1, Ch. X, the secondary current produces a magnetomotive force that tends to oppose the magnetomotive force of the primary current, and therefore if the positive directions of both currents are chosen in the same direction about positive flux, the primary and secondary currents are approximately, though not exactly, equal in magnitude and opposite in phase, as shown in Fig. 4.

[1] It is interesting to read critically the diverse ideas of a number of writers on the subject of flux distribution in transformers, expressed in the discussions of and in the papers by K. B. McEachron, " Magnetic Flux Distribution in Transformers," *A.I.E.E. Trans.*, *41* (1922), 247-261; O. G. C. Dahl, " Separate Leakage Reactance of Transformer Windings," *A.I.E.E. Trans.*, *44* (1925), 785-791; A. Boyajian, " Resolution of Transformer Reactance into Primary and Secondary Reactances," *A.I.E.E. Trans.*, *44* (1925), 805-810.

The magnitudes and phase relation of the primary and secondary currents depend on the nature of the load as well as on the characteristics of the transformer.

At time t_0, Fig. 4, the instantaneous secondary current i_2 is zero, and the resultant flux distribution at this instant is the same as it would be if the secondary circuit were open. If the primary is the inner winding 1 of Figs. 2 and 3, p. 315, the flux distribution at time t_0 is as shown in Fig. 2. A small fraction of a cycle later, at time t_1, the primary current i_1 is zero, and the magnetic field is then as shown in Fig. 3. During the first part of the time interval from t_0 to t_1, the instantaneous primary current is greater than the instantaneous secondary current, and the primary current has the major influence on the flux distribution; that is, the air-borne leakage flux links the primary and is directed downward in the air spaces between the windings on the left-hand leg of the core, as in Fig. 2. During the latter part of the time interval from t_0 to t_1, the instantaneous secondary current is greater than the instantaneous primary current, and the secondary current has the major influence on the flux distribution; that is, the air-borne leakage flux links the secondary and is directed upward in the air spaces between the windings on the left-hand leg of the core, as in Fig. 3. Thus during the time interval from t_0 to t_1, the flux distribution goes through a transition from the form of Fig. 2 to that of Fig. 3; the flux in the air spaces between the windings (directed downward at t_0) decreases, reverses its direction, and then increases, simultaneously shifting its location in the iron from paths that link the primary (at t_0) to paths that link the secondary (at t_1). During the whole of this time interval, the instantaneous primary and secondary currents are relatively small and the leakage flux therefore is small.

During the time interval from t_1 to t_2, Fig. 4, the currents in the windings produce opposing magnetomotive forces, and the magnetomotive force of the positive secondary current i_2 predominates. Therefore the resultant mutual flux φ is in the direction of the magnetomotive force of i_2, as shown by the broken lines in Fig. 5a which also shows by unbroken lines the component leakage fluxes $\varphi_{\ell 1}$ and $\varphi_{\ell 2}$ produced by the individual currents. Thus Fig. 5a shows the component fluxes of Eqs. 22 and 23 during the time interval from t_1 to t_2. These fluxes are merely *components*, however, and do not exist as lines of force in a map of the resultant magnetic field. The actual flux distribution is the result of the combining of these components. For example, the resultant flux φ and the primary leakage flux $\varphi_{\ell 1}$ are components of the flux in the vertical legs of the core, and these components are in opposite directions, as shown in Fig. 5a. The resultant flux in the vertical legs is the difference between these oppositely directed components. That is, the primary magnetomotive force, opposing that of the secondary, diverts some of the core flux

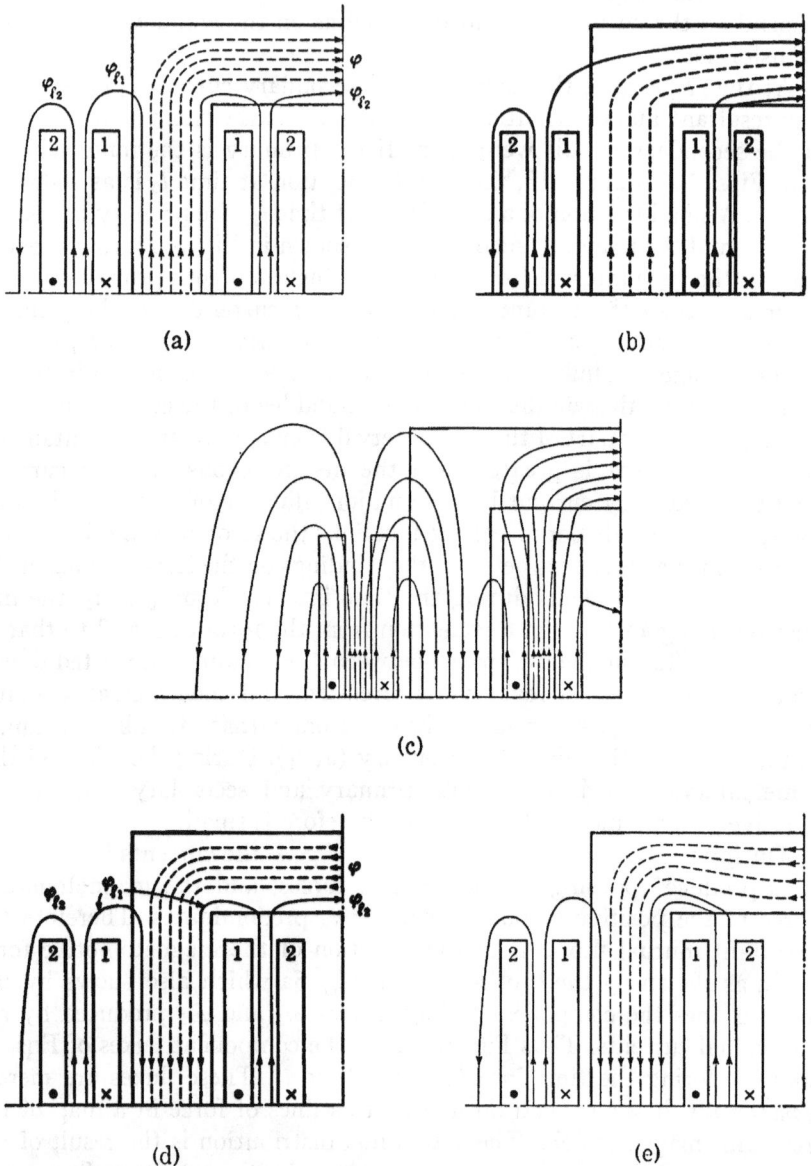

FIG. 5. Magnetic field in the upper, left-hand quadrant of the transformer of Fig. 2 due to opposing primary and secondary magnetomotive forces. (a) Component fluxes and (b) resultant field when i_2 is greater than i_1. (c) Resultant field when i_2 equals i_1. (d) Component fluxes and (e) resultant field when i_1 is greater than i_2.

and forces it into the air spaces between the windings, as shown in the roughly drawn field map of Fig. 5b. Note that the lines of flux instantaneously linking both windings, shown by the broken lines in Fig. 5b, do not represent the resultant mutual component φ, shown by the broken lines in Fig. 5a, and that the air-borne flux, shown by the unbroken lines in Fig. 5b, links the secondary but not the primary.

As time increases from t_1 to t_2, Fig. 4, the resultant mutual flux decreases until at time t_2 the primary and secondary magnetomotive forces are equal and opposite, and the net magnetomotive force around the core is zero. The resultant mutual flux therefore is zero at time t_2, if the effects of hysteresis and eddy currents are neglected. The flux distribution for this condition is approximately shown in the rough field map of Fig. 5c. During the interval from t_1 to t_2 the currents increase and the air-borne flux increases from a very small value at t_1 to a much larger value at t_2.

During the time interval from t_2 to t_3, Fig. 4, the primary and secondary currents are in opposite directions, but now the magnetomotive force of the negative primary current i_1 predominates. The resultant mutual flux φ is therefore in the direction of the magnetomotive force of i_1, as shown by the broken lines in Fig. 5d, which also shows in unbroken lines the component leakage fluxes $\varphi_{\ell 1}$ and $\varphi_{\ell 2}$ produced by each current. Thus Fig. 5d shows the component fluxes of Eqs. 22 and 23 during the time interval from t_2 to t_3, but it should be remembered that the fluxes shown in Fig. 5d are merely *components* and cannot be identified with any of the lines of force in a map of the resultant magnetic field. The resultant flux distribution is the combination of the components and is shown in the approximate field map of Fig. 5e. Note that the flux lines instantaneously linking both windings, shown by the broken lines in Fig. 5e, do not represent the resultant mutual component φ of Fig. 5d, and that most of the air-borne flux, shown by the unbroken lines in Fig. 5e, links the primary but does not link the secondary.

As time increases from t_2 to t_3, Fig. 4, the instantaneous currents decrease, until finally at t_3 there is current in the primary winding only, as at t_0, except that the primary current is now in the opposite direction. The flux distribution at t_3 is then similar but opposite in direction to that shown in Fig. 2.

The important points of this discussion can be summarized briefly as follows:

▶ The flux distribution in a transformer depends not only on the geometrical arrangement of its core and windings, but also on the *instantaneous* magnitudes and directions of the currents. Thus during the half-cycle from t_0 to t_3, the flux distribution is continuously changing, assuming successively the forms shown in Figs. 2, 3, and 5.◀

▶ The mutual and leakage fluxes in Eqs. 22 and 23 are merely *components*, and cannot be identified with any of the lines of force in a map of the resultant magnetic field, except when current exists in only one winding, as in Figs. 2 and 3.◀

▶ When there are instantaneous currents in both windings, most of the air-borne flux links whichever winding has the greater instantaneous magnetomotive force, and the lines of force that link one winding without linking the other do not represent the component leakage flux of this winding with respect to the other.◀

3. LEAKAGE INDUCTANCES

When the resultant fluxes in Eqs. 15 and 16 are expressed as the sum of the components in Eqs. 22 and 23, the transformer voltage equations become

$$v_1 = R_1 i_1 + N_1 \frac{d\varphi_{\ell 1}}{dt} + N_1 \frac{d\varphi}{dt} \qquad [24]$$

$$v_2 = R_2 i_2 + N_2 \frac{d\varphi_{\ell 2}}{dt} + N_2 \frac{d\varphi}{dt}. \qquad [25]$$

Each terminal voltage can hence be expressed as the sum of a resistance drop, a voltage induced by leakage flux, and a voltage induced by the resultant mutual flux φ. The component leakage fluxes $\varphi_{\ell 1}$ and $\varphi_{\ell 2}$ induce voltages in only the winding with which each is associated, but the resultant mutual flux φ links both windings and induces in them voltages whose ratio equals the turns ratio, as in an ideal transformer.

In spite of the magnetic nonlinearity of the iron core, the component leakage fluxes are very nearly directly proportional to the currents producing them, since the paths of the component fluxes are in air for a considerable portion of their lengths.

▶ Hence the components of the self-inductances of the windings due to the leakage fluxes are very nearly constant, so that it is convenient to introduce inductance parameters to account for the voltages induced by the leakage fluxes.◀

The component of the self-inductance of winding 1 due to the leakage flux $\varphi_{\ell 1}$ of winding 1 with respect to winding 2 is defined as the *leakage inductance* of winding 1 with respect to winding 2. Thus the leakage inductance $L_{\ell 1}$ of winding 1 with respect to winding 2 is the leakage-flux linkage per unit current, or

$$L_{\ell 1} \equiv \frac{N_1 \varphi_{\ell 1}}{i_1}. \qquad ▶[26]$$

Similarly, the leakage inductance L_{l2} of winding 2 with respect to winding 1 is

$$L_{l2} \equiv \frac{N_2 \varphi_{l2}}{i_2}.$$ ▶[27]

Note that leakage inductance is a property of one winding with respect to another. If the transformer has more than two independent windings, the leakage inductance of the primary with respect to the secondary generally differs from the leakage inductance of the primary with respect to the third or tertiary winding.* With a two-winding transformer, however, there is no ambiguity if the leakage inductance of the primary with respect to the secondary is called, briefly, the primary leakage inductance. This abbreviated terminology is used throughout the remainder of the discussion of two-winding transformers.

If the voltages induced by the primary and secondary leakage fluxes are expressed in terms of the leakage inductances, Eqs. 24 and 25 become

$$v_1 = R_1 i_1 + L_{l1} \frac{di_1}{dt} + N_1 \frac{d\varphi}{dt} = R_1 i_1 + L_{l1} \frac{di_1}{dt} + e_1$$ ▶[28]

$$v_2 = R_2 i_2 + L_{l2} \frac{di_2}{dt} + N_2 \frac{d\varphi}{dt} = R_2 i_2 + L_{l2} \frac{di_2}{dt} + e_2$$ ▶[29]

where e_1 and e_2 are the voltages induced by the resultant mutual flux φ. Note that the resistances and leakage inductances in Eqs. 28 and 29 are essentially constant parameters and that the only effect of the nonlinear magnetic properties of the core on the theory expressed in Eqs. 28 and 29 is in the relation between the resultant mutual flux φ and the magnetomotive force required to produce it. This aspect of the theory is discussed in Art. 4.

4. EXCITING AND LOAD COMPONENTS OF THE PRIMARY CURRENT

The resultant mutual flux φ is produced by the combined action of both primary and secondary currents. In the analysis of the two-winding iron-core transformer, it is almost always assumed that the mutual flux is entirely confined to the core, and usually that the leakage flux is so small compared with the resultant mutual flux that the flux in the core equals the resultant mutual flux φ. This assumption is not strictly true, since leakage increases the flux in some portions of the core and reduces it in others, as may be seen from the field maps of Fig. 5. Although it is possible approximately to take into account the effects of the leakage fluxes on the magnetic conditions in the core, this refinement is seldom

---

* Multicircuit transformers are discussed in Ch. XXVII.

necessary. If these secondary effects of magnetic leakage are neglected, the primary and secondary magnetomotive forces are equally effective in producing mutual flux, regardless of the different arrangement of these windings on the core.

According to Eq. 28, the resultant mutual flux φ must adjust itself so that the sum of the primary resistance drop, the primary leakage-inductance drop, and the counter electromotive force e_1 induced by the resultant mutual flux equals the primary impressed voltage; and the primary and secondary currents, whose resultant magnetomotive force produces the resultant mutual flux, must adjust themselves to satisfy these conditions. As shown in Art. 1, Ch. X, it is convenient to resolve the primary current into a *load component* i_L', whose magnetomotive force exactly balances the magnetomotive force of the secondary current, and an *exciting component* i_φ', whose magnetomotive force by itself is sufficient to produce the resultant mutual flux required to satisfy the primary voltage equation. The primary current i_1 is the sum of these components; thus

$$i_1 = i_L' + i_\varphi'. \tag{30}$$

If the positive directions of the primary and secondary currents are both taken in the same direction with respect to the flux, the relation between the secondary current and the load component of the primary current is, by definition,

$$N_1 i_L' = - N_2 i_2 \tag{31}$$

or

$$i_L' = - \frac{N_2}{N_1} i_2. \tag{32}$$

That is, the load component of the primary current is instantaneously proportional to the secondary current, and in the opposite direction with respect to the flux. The secondary current and the load component of the primary current are thus related to each other as are the secondary and primary currents in an ideal transformer.

The relation between the resultant core flux and the exciting current is a flux-current loop similar to that shown in Fig. 3, Ch. VI, p. 162. When it is assumed, as stated above, that the core flux equals the resultant mutual flux φ, this flux-current loop under load is the same as the no-load flux-current loop having the same maximum flux density.

5. An equivalent circuit

Examination of Eq. 28 shows that it applies to a circuit in which the primary terminal voltage v_1 is impressed on the primary resistance and leakage inductance in series with the counter electromotive force e_1

induced in the primary by the resultant mutual flux, as shown by part (a) of Fig. 6. Since the exciting current depends on the resultant mutual flux φ, the exciting current i'_φ can be accounted for if an iron-core reactor is connected in parallel with the voltage e_1 induced by this resultant mutual flux, as in part (b) of Fig. 6. The core loss and excitation characteristics of this reactor are those of the actual transformer, and its winding resistance is zero. The current in the main circuit, Fig. 6, to the right of this reactor is the load component i'_L of the primary current. The primary and secondary voltages e_1 and e_2 induced by the resultant mutual flux are directly proportional to the winding turns, as are the terminal voltages

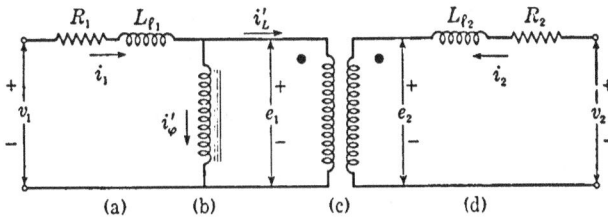

FIG. 6. An equivalent circuit representing an iron-core transformer. The excitation characteristics are represented by the iron-core reactor (b) whose winding has no resistance. The turns ratio of the ideal transformer (c) equals N_1/N_2.

of an ideal transformer. Also the load component i'_L of the primary current is related to the secondary current i_2, Eqs. 31 and 32, as are the primary and secondary currents in an ideal transformer. Hence the turns ratio of the actual transformer can be shown by an ideal transformer, as in part (c) of Fig. 6. The relation between the secondary induced voltage e_2 and the secondary terminal voltage v_2, Eq. 29, is also that shown by the series impedance, part (d) of Fig. 6. Hence on the basis of the assumptions discussed in Art. 4 regarding the factors influencing the excitation characteristics of the core, the circuit of Fig. 6 has electrical characteristics identical with those of the actual transformer.

An advantage of the equivalent circuit is that it shows, probably more clearly than do the equations, the imperfections of an actual transformer. The winding resistances and leakage inductances introduce series impedance drops in the primary and secondary circuits, causing the ratio of the primary to the secondary terminal voltage to differ from the turns ratio. Furthermore, the exciting current causes the ratio of the primary to the secondary current to differ from the negative inverse turns ratio, as shown by the shunt reactor in Fig. 6.

A further advantage of the concept of leakage inductance and of the resulting equivalent circuit is that the effects of the magnetic non-linearity of the iron core are segregated in the shunt reactor representing the excitation characteristics. *Except for this one consideration, the iron-core transformer has essentially the properties of a linear circuit, since the winding resistances and leakage inductances are very nearly constant.* This important consideration is discussed in Art. 6.

6. EFFECTS OF THE IRON CORE AND SIMPLIFYING APPROXIMATIONS

If the operation of an iron-core transformer is to be determined analytically, some simpler relation than the actual flux-current loop must be assumed for the relation between the exciting current and the resultant mutual flux. Since the exciting current usually is small, it may often be treated by approximate methods. There are several alternatives.

(1) The peculiar waveform of the exciting current may be neglected and the exciting current under steady-state alternating-current conditions may be assumed an equivalent sine wave, as in Art. 11, Ch. VI. The core-loss and magnetizing components of the equivalent sinusoidal exciting current may be adjusted with changes in frequency and core flux in accordance with the actual characteristics of the transformer. Thus the nonlinearity of the core may be accounted for to the extent that it affects the rms values of the components of the exciting current, but the harmonics in the exciting current due to the nonlinearity of the core are neglected. For a given operating condition, the iron-core reactor in Fig. 6 representing the excitation characteristics of the core is then equivalent to either a series or a parallel combination of a resistance and an inductance, as in the equivalent circuits of Art. 13, Ch. VI. According to these assumptions, for a given operating condition the transformer is considered to behave as a linear circuit element, but the circuit parameters representing its excitation characteristics change their values when the resultant mutual flux or the frequency varies.

(2) Further simplification results if suitable assumptions are made regarding the circuit parameters representing the excitation characteristics. Thus it is often permissible to assume that the excitation characteristics may be represented by a parallel combination of a *constant* resistance and a *constant* inductance. The effects of these assumptions are discussed in detail in Art. 13, Ch. VI.

(3) In many problems involving the characteristics of a transformer as a circuit element, the core loss may be neglected and the resultant core flux may be assumed to be instantaneously proportional to the magnetizing current. That is, for a given operating condition, it may be assumed that the self- and mutual inductances do not vary cyclically, whence the classical theory of magnetically coupled circuits may be applied to the iron-core transformer. In spite of the magnetic nonlinearity of the core, the characteristics of a transformer depend ultimately on an

essentially linear magnetic leakage and a nonlinear exciting current which, however, is often so small that its peculiarities may be neglected. Hence, at the beginning of the analysis, the transformer may often be considered as an essentially linear circuit element. This alternative is often used in the analysis of communication networks, as in Chs. XVII and XVIII. The effects of magnetic nonlinearity may be taken into account partly by changes in the values of the inductances with changes in the operating conditions.

(4) The third alternative may be simplified further by the assumption that the inductances are constant for changes in the operating conditions.

(5) Since the exciting current often is small compared with the load component of the primary current, the exciting current often may be neglected entirely. This is the simplest alternative and is often used in analysis of the behavior of transformers connected in power networks. It is also applicable to audio-frequency transformers at high audio frequencies. With this assumption the shunt reactor may be omitted from the equivalent circuit of Fig. 6, and the transformer becomes equivalent to the leakage impedances in series with an ideal transformer. This leads to the important concept of equivalent impedance, discussed in Art. 6, Ch. XIII.

PROBLEMS

1. Two very long rectangular coils are arranged concentrically in the same plane with their long sides parallel to each other, as shown in the sectional view of Fig. 7. The width of the outer coil is twice that of the inner coil. Each of the conductors has a diameter $2r$. Assuming that the coils have a length l so great that end effects may be neglected, and neglecting flux linkages within the conductors, determine the mutual inductance, the leakage inductance of the inner coil with respect to the outer coil, and the leakage inductance of the outer coil with respect to the inner coil.

2. For the elementary air-core transformer of Prob. 1, determine the instantaneous leakage fluxes at an instant when the currents in the two coils are equal in magnitude but opposite in direction. Compare these values with the values of leakage flux computed in Prob. 1. How do the instantaneous leakage fluxes at any given time in the cycle compare with the leakage fluxes of each winding with respect to the other?

3. A section through the windings of a special core-type transformer (with coils on one leg only) is shown in the diagrams of Fig. 8. The main windings, numbered 1 and 2, are each rated at 110 v 60 ~ 11.4 amp, and consist of 84 turns wound in three layers of 28 turns each. The windings I, II, III, IV, and V are exploring coils, each of which has a single layer of 21 turns of fine wire.

In a test to determine the relative magnitudes of leakage fluxes at various points about the windings, a voltage of 11 v at 60 ~ is applied first to coil 1 with coil 2 open-

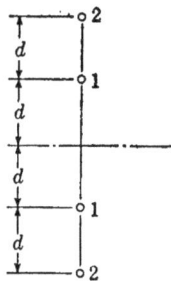

Fig. 7.　Cross-section of plane coils, Probs. 1 and 2.

circuited, and then to coil 2 with coil 1 open-circuited. The no-load current at this voltage equals 125 ma and is observed to be nearly sinusoidal in waveform.

In each part of the test, the exploring coils I–V are successively connected in oppo-

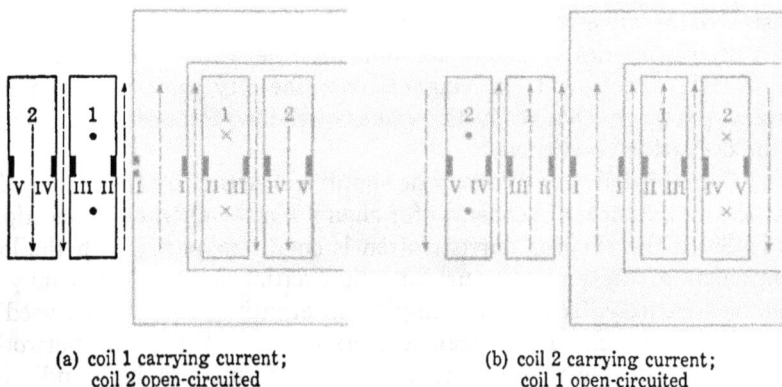

(a) coil 1 carrying current;
coil 2 open-circuited

(b) coil 2 carrying current;
coil 1 open-circuited

FIG. 8.   Cross-section of special transformer with exploring coils, Probs. 3 and 4.

sition each to the next and the difference in potential is measured. The data obtained are tabulated below:

COIL 1 CARRYING CURRENT; COIL 2 OPEN–CIRCUITED

Connection of Exploring Coils	Voltage
II–I	0.0059 v
III–II	0.0001
IV–III	0.0039
V–IV	0.0097

COIL 2 CARRYING CURRENT; COIL 1 OPEN–CIRCUITED

II–I	0.0047
III–II	0.0063
IV–III	0.0050
V–IV	0.0004

On the assumptions that the leakage flux paths may be approximated as shown by the broken-line arrows in Fig. 8, and that the no-load current is entirely magnetizing current, compute the equivalent leakage flux of coil 1 with respect to coil 2, the equivalent leakage flux of coil 2 with respect to coil 1, and the leakage inductances of each winding with respect to the other. List the approximations and assumptions used in arriving at a solution.

4. Sketch the component fluxes and the resultant flux for the transformer of Prob. 3 for a time, intermediate between the times t_0 and t_1 of Fig. 4, when the currents in the two coils are equal.

5. Figure 9 is a cross-sectional view of the core and windings of a 2,400 : 240-v 10-kva 60 $\sim$ shell-type distribution transformer. Four exploring coils (marked I, II, III, IV on the diagram) are wound as shown, each with the same number of turns.

With the 240-v windings open-circuited and a 60 $\sim$ voltage of 240 v applied to the high-voltage winding P, the following data were recorded:

$$\text{Exciting current} = 0.0104 \text{ amp.}$$
$$V_I = 16.2 \text{ v.}$$
$$V_I - V_{II} = 4.03 \times 10^{-3}.$$
$$V_{II} - V_{III} = 0.08 \times 10^{-3}.$$
$$V_{III} - V_{IV} = 8.06 \times 10^{-3}.$$

Find an approximate value for the leakage inductance of the high-voltage winding of the transformer with respect to the low-voltage winding. List and discuss the assumptions and approximations necessary in arriving at a solution.

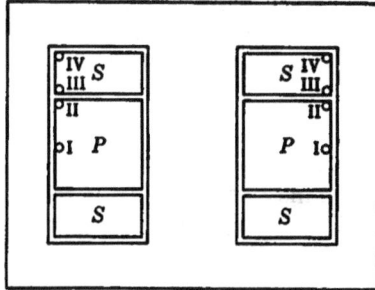

FIG. 9. **Cross-section of transformer with exploring coils, Prob. 5.**

6. Both the primary and secondary windings of a one-to-one-ratio iron-core transformer consist of a pair of wires wound side by side, as shown in Fig. 10. Rated voltage is applied to the primary terminals P_1 and P_2 with the secondary open-circuited, and the primary current is measured.

(a) If P_1 is connected to P_1' and the voltage between P_2 and P_2' is measured with a high-resistance voltmeter, what constant of the transformer can be computed? Is it necessary to know the waveform as well as the rms value of the current?

(b) If P_1' is connected to S_1' and the voltage between P_2' and S_2' is measured with a high-resistance voltmeter, what constant of the transformer can be computed? Is it necessary to know the waveform as well as the rms value of the current?

7. The primary and secondary windings of a one-to-one-ratio iron-core transformer are connected in parallel, with ammeters inserted to measure both the primary and

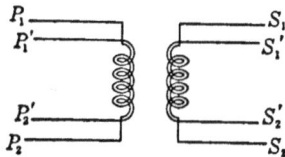

FIG. 10. Windings of transformer with exploring coils, Prob. 6.

the secondary currents. If the resistances of the primary and secondary windings are neglected, what information concerning the transformer constants can be determined? Would the accuracy of the computation be increased if the waveforms of the currents were known? Consider that (a) terminals of like polarity are connected in parallel, or (b) terminals of unlike polarity are connected in parallel.

CHAPTER XIII

Leakage Inductances

The major factors in the theory of the transformer are discussed in Ch. XII, where it is shown that magnetic leakage is important, that the leakage inductances are nearly constant, and that the nonlinearity of the iron core affects only the waveform of the exciting current. Since the exciting current usually is small, this nonlinearity frequently can be neglected, so that the exciting current for sinusoidal flux often may be assumed to be an equivalent sine wave. Vector methods of calculation can then be used.

This chapter develops these general principles into a quantitative theory which can be represented by vector diagrams and equivalent circuits, and which often may be simplified. Tests by which the constants required in the general theory can be measured are shown, and an illustration of the design calculation of these constants is given.

1. VECTOR DIAGRAMS AND AN EQUIVALENT CIRCUIT

Consider an iron-core transformer supplying power to a load, as in Fig. 1. This circuit is typical of many applications of transformers in communication systems. Assume that the internal electromotive force

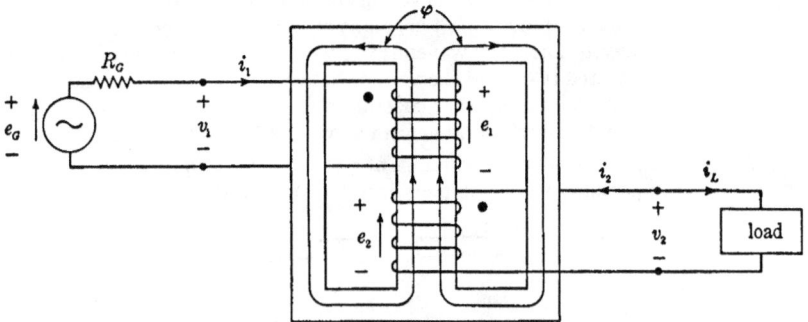

FIG. 1.   Circuit showing a load connected to a generator through a transformer.

e_G of the generator varies sinusoidally, that the generator has constant internal resistance R_G, and that the load has linear characteristics. Except for the effects of the harmonics in the exciting current, the circuit as a whole then has linear characteristics. As shown in Art. 5, Ch. VI, the harmonics in the exciting current produce harmonic voltage drops in the internal impedance of the generator and in the primary resistance and leakage inductance of the transformer. The harmonics hence cause the waveforms of the induced voltages in the transformer to differ from the sinusoidal waveform of the generator electromotive force. If the harmonics in the exciting current and the impedances in the primary circuit are large, distortion may be a serious matter in communication circuits, so that this condition is usually avoided. In power circuits, the exciting current and the primary-circuit impedance are generally so small that the harmonic voltage drops are negligible, and the waveforms of the voltages usually are very nearly sinusoidal. Since an accurate solution that takes into account the harmonics in the exciting current is impractical because of its complexity, the exciting current usually is assumed to be an equivalent sine wave in all problems except those directly relating to the effects of the harmonics. The circuit as a whole can then be analyzed by simple vector methods.

When the currents and voltages are assumed to vary sinusoidally, the voltage equations for the primary and secondary, Eqs. 28 and 29, Ch. XII, can be written in vector form as

$$V_1 = (R_1 + jX_{l1})I_1 + E_1 \qquad \blacktriangleright[1]$$

$$V_2 = (R_2 + jX_{l2})I_2 + E_2, \qquad \blacktriangleright[2]$$

where

V_1 and V_2 are the vectors representing the terminal voltages,

E_1 and E_2 are the vectors representing the voltages induced by the resultant mutual flux,

I_1 and I_2 are the vectors representing the currents,

R_1 and R_2 are the effective resistances of the windings,

X_{l1} and X_{l2} are the leakage reactances; that is,

$$X_{l1} \equiv \omega L_{l1} \qquad \blacktriangleright[3]$$

$$X_{l2} \equiv \omega L_{l2}, \qquad \blacktriangleright[4]$$

L_{l1} and L_{l2} being the leakage inductances.

The positive directions of the currents and voltages in Eqs. 1 and 2 are shown in Fig. 1. The positive directions of the currents i_1 and i_2 are in the right-hand-screw direction about an assumed positive direction of flux, and the voltages are all falls in potential in the same direction. It may also be said that the induced voltages e_1 and e_2 are electromotive forces, or rises in potential, in the left-hand-screw direction about positive flux. The advantage of choosing the positive directions for the currents and voltages in both windings in the same direction relative to the flux is that the algebraic signs in both voltage equations are then the same, as in Eqs. 1 and 2. These voltage equations apply to the transformer ıegardless of the direction of power flow — that is, regardless of whether winding 1 or winding 2 acts as the primary.

A problem frequently encountered in the analysis of power systems is that of determining the voltage that must be impressed at the primary terminals to maintain a specified secondary terminal voltage when the secondary load and power factor are known. The analysis of this problem is a convenient illustration of the vector relations expressed in Eqs. 1 and 2. Therefore, in the following discussion, assume that the secondary terminal voltage and current and the power factor of the load are known, and that the primary terminal voltage, primary current, and power factor on the primary side are to be determined, corresponding to these specified conditions on the secondary side. Also assume that the resistances, leakage reactances, and turns ratio are known, and that data are available giving the core loss and exciting current as functions of the induced voltage. The determination of the resistances and leakage reactances is discussed subsequently, and the core-loss and exciting-current characteristics can be determined by a variable-voltage no-load test, as in Art. 8a of this chapter, or from design calculations, as in Art. 12, Ch. VI. The vector diagrams illustrating the relations expressed in Eqs. 1 and 2 can then be constructed as follows:

Since the secondary is delivering power to the load, the current in the load is in the direction of the fall in potential through the load for the major portion of each half-cycle. If the positive direction of the secondary voltage v_2 is as shown in Fig. 1, the positive direction of the current delivered to the load in the direction of the fall in potential through the load is i_L, Fig. 1. Thus if V_2 and I_L are the vectors representing v_2 and i_L, their phase relation is the power-factor angle of the load. If V_2 is chosen as the reference vector, I_L is as shown vectorially in Fig. 2a, which is drawn for an inductive load of power factor $\cos \theta_L$. In Fig. 1, note that i_L is equal and opposite to i_2. Thus vectorially I_2 is opposite to I_L, as in Fig. 2a. Note that I_L represents the current in the secondary winding in the left-hand-screw direction about positive flux, and that E_2 represents the secondary electromotive force (rise in potential) in the same direction. In writing the fundamental transformer equations it is convenient to assume that both primary and secondary currents are positive in the right-hand-screw direction about positive flux, as I_1 and I_2 in Eqs. 1 and 2. In the analysis of problems in which the direction of power flow is definitely known, however, it may be more convenient to deal with the left-handed secondary current I_L, or $-I_2$, produced by the left-handed secondary electromotive force E_2, since I_L is the current delivered to the load in the direction of the electromotive force E_2 producing it.

According to Eq. 2,

$$E_2 = V_2 - I_2(R_2 + jX_{\ell 2}) \qquad [5]$$

or, in terms of the current I_L delivered to the load,

$$E_2 = V_2 + I_L(R_2 + jX_{\ell 2}). \qquad [6]$$

That is, the internal electromotive force E_2 is the vector sum of the terminal voltage plus the voltage drop due to the internal impedance and the current produced by the internal electromotive force, as in a

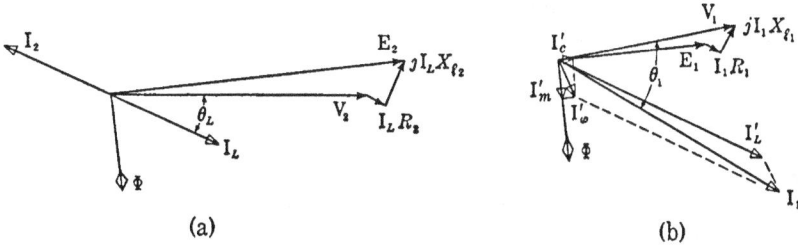

(a)   (b)

FIG. 2.   Vector diagrams for a transformer whose turns ratio N_1/N_2 equals 1/2. The secondary voltages and currents are shown in (a), and the primary voltages and currents in (b).

generator. Equation 6 is shown vectorially in Fig. 2a by the vectors V_2, $I_L R_2$ in phase with I_L, $jI_L X_{\ell 2}$ leading I_L by 90 degrees, and their vector sum E_2.

The voltages E_1 and E_2 are induced by the resultant mutual flux, and their effective values are related to the maximum value of this flux, as in Eq. 18, Ch. VI. The magnitudes of E_1 and E_2 are proportional to the number of turns in the windings, and when the positive directions of these induced voltages both represent electromotive forces in the left-hand-screw direction about positive flux, as in this analysis, they are in time phase and lead the flux by 90 degrees, as in Fig. 8, Ch. VI, p. 168. Accordingly the resultant mutual flux lags the induced voltages by 90 degrees, as shown by the vector Φ in Fig. 2.

As shown in Art. 4, Ch. XII, the primary current can be considered as the sum of a load component that counteracts the magnetomotive force of the secondary current, plus an exciting component that produces the required resultant mutual flux. The magnitude of the load component of the primary current and that of the secondary current are inversely proportional to the number of turns in the primary and secondary windings, and the load component of the primary current is in the opposite direction to the secondary current. As shown in Art. 11, Ch. VI, the exciting current consists of a core-loss component in phase with the induced voltage and a magnetizing component in phase with the flux. If the magnetizing current is treated as an equivalent sine wave, it can be represented vectorially.

The primary current and its components are shown vectorially in Fig. 2b. The load component of the right-handed primary current is I'_L, opposite to the right-handed secondary current I_2, or in phase with the left-handed secondary current I_L. Since in Fig. 2 the turns ratio N_1/N_2 equals $\frac{1}{2}$, the magnitude of I'_L is twice the secondary current. The core-loss component of the exciting current is I'_c in phase with the induced voltage E_1, and the magnetizing current is I'_m in phase with the flux Φ. The exciting current is I'_φ, and the primary current I_1 is the vector sum of the exciting current I'_φ and the load component I'_L.

According to Eq. 1, the primary terminal voltage V_1 is the vector sum of the counter electromotive force E_1 induced by the resultant mutual flux and the primary leakage-impedance voltage drop $I_1(R_1 + jX_{l1})$, as shown in Fig. 2b. The primary power factor is $\cos \theta_1$. Thus from Eqs. 1 and 2 and the excitation characteristics, the complete performance of the transformer can be determined. When the exciting current is treated as an equivalent sine wave, these relations may be represented by the vector diagrams of Fig. 2.

In Fig. 2, voltages V_1 and V_2 differ slightly in phase because of the effects of the leakage-impedance voltage drops, as do the currents I_1 and I_L because of the effects of the exciting current. In these vector diagrams, however, the resistance and leakage-reactance voltage drops and the exciting current are exaggerated. In a typical power-system transformer at rated load, the resistance drops usually are but a fraction of a per cent of rated voltage, the leakage-reactance drops usually are 2 to 4 per cent of rated voltage, and the exciting current usually is 4 to 8 per cent of rated current. Therefore the primary and secondary terminal voltages V_1 and V_2 usually are very nearly in phase, as usually are the right-handed primary current I_1 and the left-handed secondary current I_L. During the major portion of each half-cycle, then, the dot-marked primary and secondary terminals, Fig. 1, are of the same relative polarity, and, for most of the time when current is entering the dot-marked primary terminal, current is simultaneously leaving the dot-marked secondary terminal and entering the corresponding terminal of the load.

As shown in Art. 5, Ch. XII, the electrical characteristics of a transformer may be represented by the equivalent circuit of Fig. 6, Ch. XII, p. 327. When the voltages and currents are assumed to be sinusoidal in waveform, this equivalent circuit becomes that shown in Fig. 3, in which the excitation characteristics of the transformer are represented by the parallel combination of a resistance and a reactance, as in the first alternative of Art. 6, Ch. XII. A series combination also is sometimes used. For small changes in the induced voltage and frequency, the simplifying assumptions discussed in Art. 13, Ch. VI, often are made regarding the constants of the exciting branch, as in the second alternative of Art. 6, Ch. XII.

It is apparent that the fundamental relations, Eqs. 1, 2, 5, and 6, and the vector diagrams of Fig. 2 apply to the equivalent circuit of Fig. 3.

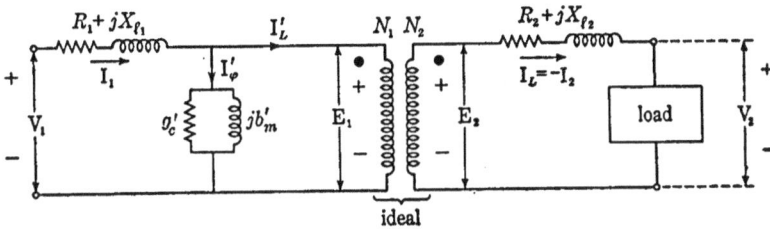

Fig. 3.   An equivalent circuit, assuming an equivalent sine-wave exciting current.

An advantage of the vector diagrams and of the equivalent circuit is that they show, probably more clearly than do the equations themselves, the relations among the currents and voltages expressed in the fundamental equations.

2. TRANSFORMER WITH THE SAME NUMBER OF PRIMARY AND SECONDARY TURNS

When the primary and secondary windings have the same number of turns, the primary and secondary voltages E_1 and E_2 induced by the resultant mutual flux are equal, and the load component I_L' of the right-handed primary current equals the left-handed secondary current I_L. It is therefore convenient to combine the vector diagrams of Fig. 2, as in Fig. 4a, in which E is the left-handed electromotive force induced in either winding by the resultant mutual flux and I_L is the left-handed load current, equal to the load component of the right-handed primary current. Figure 4a shows that the left-handed electromotive force E induced in the secondary is the vector sum of the secondary terminal voltage and the leakage-impedance voltage drop due to the left-handed secondary current I_L. The exciting current I_φ is the vector sum of the core-loss component I_c in phase with the induced voltage E and the magnetizing component I_m in phase with the resultant mutual flux Φ. The primary current I_1 is the vector sum of the load component, which equals the secondary current I_L, and the exciting component I_φ. The primary terminal voltage is the vector sum of the induced counter electromotive force E and the leakage-impedance voltage drop due to the primary current.

When the primary and secondary windings have the same number of turns, the ideal transformer may be omitted from the equivalent circuit of Fig. 3, as in Fig. 4b. On the basis of the assumptions of Ch. XII, this circuit has the same electrical characteristics as the actual transformer.

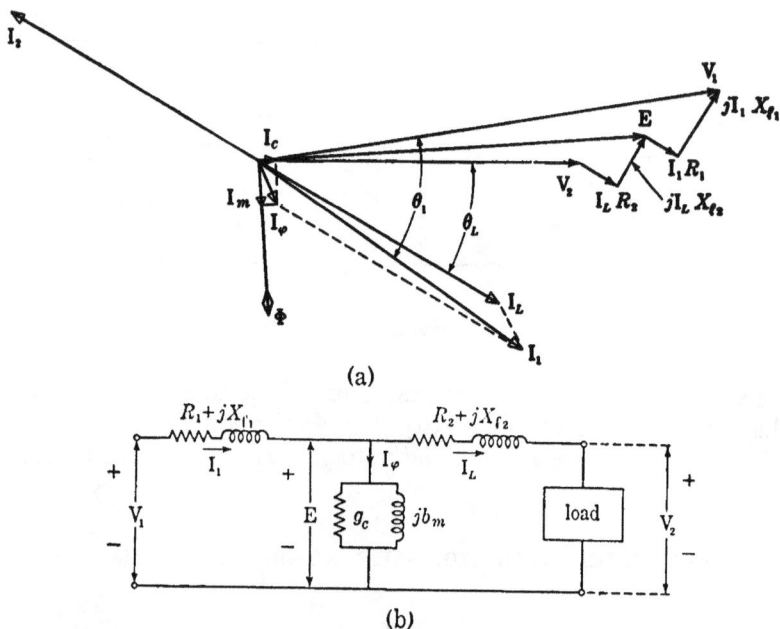

(a)

(b)

FIG. 4. Vector diagram and an equivalent circuit for a transformer with the same number of primary and secondary turns.

In fact, this equivalent circuit can be used to replace the actual transformer in the circuit of which the transformer is a part, provided a single connection between a primary terminal and the secondary terminal of the same relative polarity does not alter conditions in the circuit.

3. RATIO OF TRANSFORMATION

The ratio of the terminal voltages of an ideal transformer exactly equals the turns ratio. In an actual transformer, however, the ratio of the terminal voltages may be a few per cent greater or less than the turns ratio, because of the leakage-impedance voltage drops in the windings. Since these voltage drops depend on the magnitude and power factor of the load, the ratio of the terminal voltages is not a characteristic constant of the transformer alone but depends also on the load. For this reason, it is convenient to define *the ratio of transformation as the turns ratio*, rather than as the ratio of the terminal voltages, since the turns ratio is a definite constant fixed by the transformer alone. It is already apparent that the turns ratio is an important quantity in transformer theory. The turns ratio also equals the ratio of the voltages induced by the resultant mutual flux. That is, if a is the ratio of transformation, or turns ratio, then

$$a \equiv \frac{N_1}{N_2} = \frac{E_1}{E_2}.$$ ▶[7]

The ratio of transformation only approximately equals the ratio of the terminal voltages. The ratio of the voltages given on the nameplate of a power-system transformer is the turns ratio a and may differ by a few per cent from the ratio of the terminal voltages under load.

Since at no load the primary leakage-impedance voltage drop due to the exciting current is very small, the ratio of the terminal voltages at no load very nearly equals the turns ratio, and a very close approximation to the ratio of transformation can be determined by measurement of the primary and secondary terminal voltages at no load. In tests of very small transformers, it is essential that the current required by the voltmeter used to measure the secondary voltage be so small that the leakage-impedance voltage drops due to it are negligible. Other approximate methods for measuring the turns ratio are discussed in Art. 3b, Ch. XVII.

4. Equivalent one–to–one–ratio transformer

For a transformer having the same number of primary and secondary turns, the primary and secondary induced voltages E_1 and E_2 are equal, and the load component I_L' of the right-handed primary current equals the left-handed secondary current I_L. These facts make it convenient to combine the vector diagrams for the primary and secondary, and also permit the omission of the hypothetical ideal transformer from the equivalent circuit, as in Fig. 4. By the use of the following artifices, the same simplifications can be made in the theory of the transformer whose turns ratio is not unity.

Consider the problem discussed in Art. 1, in which the secondary load conditions are given. The vector diagram for the secondary can then be constructed as in Art. 1, Fig. 2a, and is shown by the vectors V_2, I_L, $I_L R_2, j I_L X_{l2}$, E_2, and Φ in Fig. 5a. Since the magnetomotive force $N_1 I_L'$ of the load component I_L' of the right-handed primary current balances the magnetomotive force $N_2 I_L$ of the left-handed secondary current I_L, the relation between these currents is

$$I_L = \frac{N_1}{N_2} I_L' = a I_L'.$$ ▶[8]

Thus the vector I_L represents both the secondary current and the load component of the primary current multiplied by the turns ratio, or aI'_L. If the exciting component of the primary current is also multiplied by the turns ratio a and added vectorially to aI'_L, the result is the primary

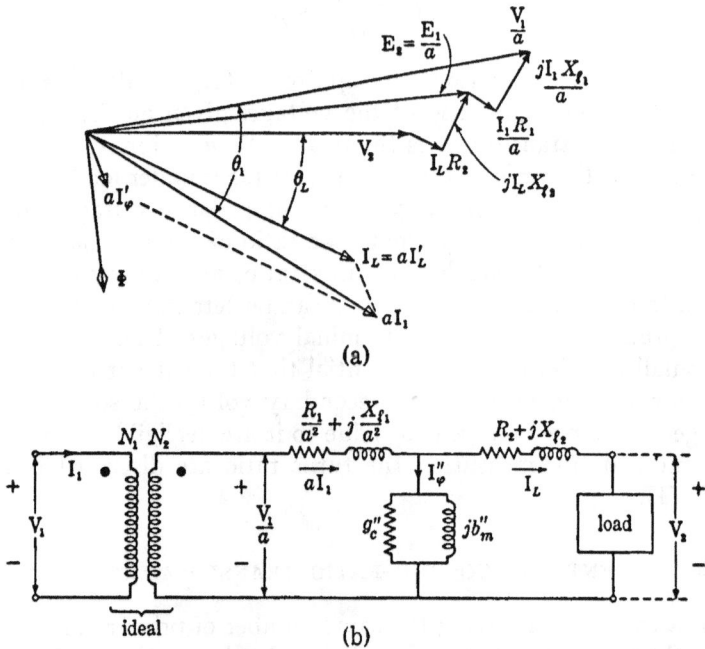

(a)

(b)

Fig. 5. Vector diagram and an equivalent circuit referred to the secondary. The turns ratio N_1/N_2 is denoted by a.

current multiplied by the turns ratio, as indicated by the vector aI_1 in Fig. 5a. The primary current multiplied by the ratio a of primary to secondary turns is called *the primary current referred to the secondary*.

The relation between the primary and secondary induced voltages is

$$E_2 = \frac{N_2}{N_1} E_1 = \frac{E_1}{a}.$$ ▶[9]

Thus the vector E_2 represents both the left-handed secondary electromotive force and the primary counter electromotive force divided by the turns ratio, or E_1/a. If the primary leakage-impedance voltage drop also is divided by the turns ratio a and added vectorially to E_1/a, the result is the primary terminal voltage divided by the turns ratio, as indicated by the vector V_1/a in Fig. 5a. This vector is thus proportional to the primary terminal voltage. The primary terminal voltage divided by the ratio a of primary to secondary turns is called *the primary terminal voltage referred to the secondary*.

The primary leakage-impedance voltage drop divided by the turns ratio a is *the primary leakage-impedance voltage drop referred to the secondary*, and equals

$$\frac{I_1(R_1 + jX_{l1})}{a} = aI_1 \left(\frac{R_1 + jX_{l1}}{a^2}\right), \qquad [10]$$

where aI_1 is the primary current referred to the secondary. The impedance $\dfrac{R_1 + jX_{l1}}{a^2}$ is called *the primary leakage impedance referred to the secondary*, where R_1/a^2 is *the primary resistance referred to the secondary* and X_{l1}/a^2 is *the primary leakage reactance referred to the secondary*. Thus the vector diagram for the primary voltages in Fig. 5a represents Eq. 1 divided by the turns ratio N_1/N_2.

Referring primary currents, voltages, and impedances to the secondary, as above, may be regarded simply as a mathematical manipulation, equivalent to a change in scale that permits primary and secondary quantities conveniently to be represented on the same vector diagram, as in Fig. 5a. The primary currents, voltages, and impedances referred to the secondary, however, also have definite physical significances, understanding which makes it easy to remember the relations between the actual currents, voltages, and impedances of the primary and the corresponding quantities referred to the secondary.

The primary current referred to the secondary is the current in an equivalent primary having the same number of turns N_2 as the secondary and producing the same magnetic effect as the actual primary current I_1 in the actual primary of N_1 turns. For example, if the primary has twice as many turns as the secondary, the magnetically equivalent N_2-turn primary winding then has half the turns of the actual primary. Therefore twice the current is required in the equivalent N_2-turn winding to produce the same magnetic effect as the current in the actual primary. The primary current and its load and exciting components referred to the secondary hence equal twice the values of the actual primary current and its components. It is commonly assumed that a given magneto-motive force in either the primary or the secondary produces the same magnetic effect regardless of the different arrangements of the windings on the core. On this assumption, the load component of the right-handed primary current referred to the secondary equals the left-handed secondary current I_L, and the exciting current referred to the secondary is the exciting current in a winding of N_2 turns and is therefore the exciting current measured on the secondary side at an induced voltage E_2.

Similarly, the primary induced voltage referred to the secondary is the voltage induced in an equivalent primary having the same number of turns as the secondary, and therefore equals the secondary induced voltage E_2.

The primary resistance and leakage reactance referred to the secondary also have definite physical significances; thus, if the magnetically equivalent N_2-turn primary occupies exactly the same space as the actual N_1-turn primary, the leakage impedance of the N_2-turn equivalent winding equals the primary leakage impedance referred to the secondary. For example, assume that the actual primary has twice as many turns as the secondary, or a equals 2. The equivalent winding having the same number of turns as the secondary then has half the turns of the actual primary. For the same geometrical arrangement, the equivalent winding therefore has half the length of copper, but each conductor has twice the cross-sectional area. The space factors of the actual primary and its equivalent winding are here assumed to be the same. Since the resistance is proportional to the length and inversely proportional to the cross-sectional area of conductor, the resistance of the equivalent winding is one-quarter that of the actual primary, or is R_1/a^2, where R_1 is the resistance of the actual primary. But, by definition, R_1/a^2 is the primary resistance referred to the secondary. Note that, since the primary current referred to the secondary equals $a\mathrm{I}_1$, the copper loss $(a\mathrm{I}_1)^2\dfrac{R_1}{a^2}$ in the equivalent winding equals the copper loss $I_1^2 R_1$ in the actual primary. Also, if the equivalent winding has exactly the same geometrical arrangement as the actual primary, the magnetic leakage field is the same for the same magnetomotive force in the actual N_1-turn primary and in its N_2-turn equivalent. In the equivalent winding with half the turns of the actual primary, however, this leakage flux links half as many turns, and twice the current is required to produce it. Since the leakage inductance is the leakage-flux linkage per ampere, the leakage inductance of this equivalent winding is one-quarter that of the actual winding, or its leakage reactance is X_{l1}/a^2, where X_{l1} is the leakage reactance of the actual primary. But by definition, X_{l1}/a^2 is the leakage reactance of the primary referred to the secondary. Note that the expression for the reactive volt-amperes in the leakage field of the equivalent winding is $(a\mathrm{I}_1)^2\dfrac{X_{l1}}{a^2}$, which equals the reactive volt-amperes $I_1^2 X_{l1}$ in the leakage field of the actual primary.

▶ Thus the primary currents, voltages, and impedances referred to the secondary are those in the primary of an equivalent N_2-turn, one-to-one-ratio transformer whose primary has exactly the same space factor and geometrical arrangement as the primary of the actual transformer.◀

When the primary currents, voltages, and impedances are referred to the secondary, the equivalent circuit is as shown in Fig. 5b, in which I_φ'' is the exciting current referred to the secondary side and equals aI_φ', where I_φ' is the primary exciting current and a is the ratio of primary to secondary turns. The core-loss conductance g_c'' and magnetizing susceptance b_m'' are the values referred to the secondary side. Since admittance is the reciprocal of impedance, and since an impedance in the primary circuit is referred to the secondary by division by a^2, the primary exciting admittance is referred to the secondary by multiplication by a^2. That is, if $g_c' + jb_m'$ is the exciting admittance on the primary side, the exciting admittance referred to the secondary is

$$g_c'' + jb_m'' = a^2 g_c' + ja^2 b_m'. \qquad \blacktriangleright[11]$$

Often the hypothetical ideal transformer is omitted from the equivalent circuit. The currents and voltages in the part of the equivalent circuit representing the primary are then these quantities referred to the secondary.

Frequently it is convenient to refer the secondary current, voltages, and impedances to the primary. This may be done by using the reciprocals of the reduction factors used to refer primary quantities to the secondary. Thus the voltage equation for the secondary, Eq. 6, can be multiplied by the turns ratio $N_1/N_2 = a$, giving

$$aE_2 = aV_2 + a(R_2 + jX_{l2})I_L. \qquad [12]$$

The terms appearing in Eq. 12 are *the secondary voltages referred to the primary*. The secondary leakage-impedance voltage drop referred to the primary can be expressed as

$$a(R_2 + jX_{l2})I_L = (a^2 R_2 + ja^2 X_{l2})\frac{I_L}{a}, \qquad \blacktriangleright[13]$$

where $a^2 R_2$ and $a^2 X_{l2}$ are *the secondary resistance and leakage reactance referred to the primary* and I_L/a is *the secondary current referred to the primary*. The secondary current, voltages, and impedances referred to the primary are those in a magnetically equivalent secondary having the same number of turns as the primary and the same space factor and geometrical arrangement as the secondary.

From Eq. 7, the secondary induced voltage referred to the primary, or aE_2, equals the primary induced voltage E_1; also, from Eq. 8, the left-handed secondary current referred to the primary, or I_L/a, equals the load component I_L' of the right-handed primary current. Therefore the vector diagram for the primary may be superposed on the secondary vectors referred to the primary, as in Fig. 6a. Except for a change in scale and in the significance of the vectors, Fig. 6a is identical with Fig. 5a.

When the secondary current, voltages, and impedances are referred to the primary, the equivalent circuit is as shown in Fig. 6b. Often the ideal transformer is omitted; then the current, voltages, and impedances in the part of the equivalent circuit representing the secondary are these quantities referred to the primary. The secondary load also can be

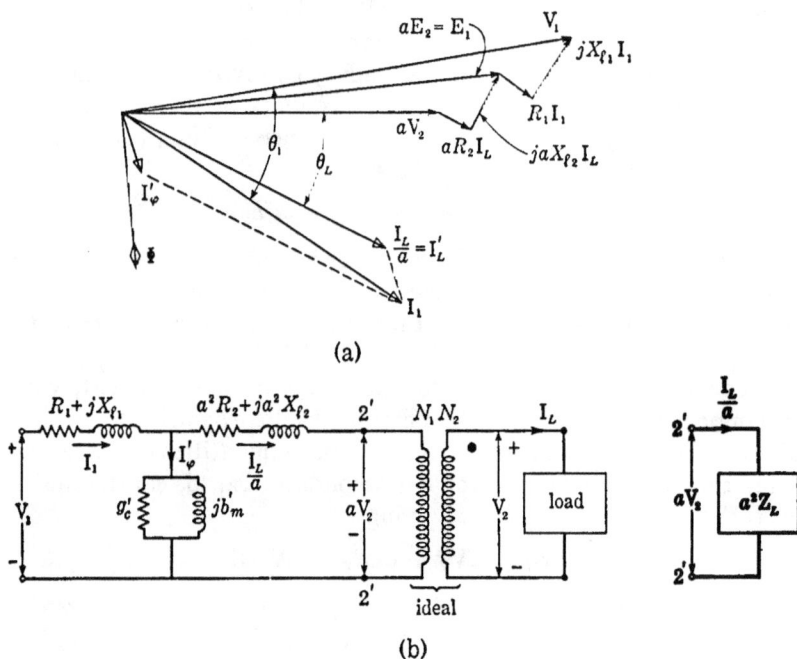

(a)

(b)

FIG. 6. Vector diagram and an equivalent circuit referred to the primary.

referred to the primary. Thus, as shown in Art. 2, Ch. X, the combination of the ideal transformer and its secondary load of impedance Z_L is equivalent to a load of impedance $a^2 Z_L$ connected directly to the terminals $2'$ representing the secondary referred to the primary. The impedance $a^2 Z_L$ is called *the load impedance referred to the primary*.

It is appropriate to conclude this discussion by recapitulating the assumptions on which the vector diagrams and equivalent circuits of Figs. 5 and 6 are based. As shown in Arts. 3 and 4, Ch. XII, these assumptions are:

(1) The effects of the capacitances of the windings are negligible;

(2) The effective resistances and leakage inductances are constant;

(3) The magnetic conditions in the core are determined by the frequency and the resultant mutual flux, and therefore the core loss and exciting current depend on the frequency, magnitude, and waveform of the voltage induced in either winding by the resultant mutual flux;

(4) A given magnetomotive force in either winding produces the same magnetic effect on the core, regardless of the different arrangements of the primary and secondary windings;

(5) It is also assumed in Figs. 5 and 6 that the exciting current may be treated as an equivalent sine wave.

On the basis of these assumptions, Figs. 5 and 6 are "exact" representations of the theory. This theory is about as close an approach to an exact analysis as can be developed without excessive complications, and is more than sufficiently accurate for the solution of most practical problems, except those in which importance attaches to the capacitances of the windings or to the waveform of the exciting current.

5. ILLUSTRATIVE EXAMPLE OF THE "EXACT" VECTOR RELATIONS

The low-voltage secondary winding of a 100-kva 60 ∼ 12,000 : 2,400-volt distribution transformer is supplying its rated kilovolt-amperes at rated secondary voltage to an inductive load of 0.80 power factor. The primary current, primary voltage, and primary power factor are to be determined.

The resistances and leakage reactances of the windings are

$$R_1 = 7.0 \text{ ohms} \qquad R_2 = 0.30 \text{ ohm}$$

$$X_{\ell 1} = 19.0 \text{ ohms} \qquad X_{\ell 2} = 0.75 \text{ ohm}$$

The tabulated results of an open-circuit test, taken on the low-voltage side, are

Applied voltage in volts	Exciting current in amperes	Core loss in watts
2,400	1.50	940
2,500	1.67	1,020
2,600	1.87	1,110

Solution: Since the excitation data are given in terms of the low-voltage secondary side, it is convenient to refer primary quantities to the secondary.

$$\text{Ratio of transformation } a \equiv \frac{N_1}{N_2} = \frac{12,000}{2,400} = 5.00.$$

The primary constants referred to the secondary are

$$\frac{R_1}{a^2} = \frac{7.0}{25.0} = 0.28 \text{ ohm} \qquad [14]$$

$$\frac{X_{t1}}{a^2} = \frac{19.0}{25.0} = 0.76 \text{ ohm.} \qquad [15]$$

Note that the primary resistance and leakage reactance referred to the secondary are of the same order of magnitude as the secondary resistance and leakage reactance, as is usually true.

The vector diagram is given in Fig. 5a. At rated output the load current is

$$I_L = \frac{100,000}{2,400} = 41.7 \text{ amp.} \qquad [16]$$

If the secondary terminal voltage is chosen as the reference vector, the vector expression for the load current is

$$I_L = 41.7\underline{/\cos^{-1}0.80}, \text{lagging} = 41.7\underline{/-36.9°} \text{ amp.} \qquad [17]$$

$$= 41.7 (0.80 - j0.60) = 33.4 - j25.0 \text{ amp.} \qquad [18]$$

The secondary resistance drop is

$$I_L R_2 = (33.4 - j25.0)0.30 = 10.0 - j7.5 \text{ v.} \qquad [19]$$

The secondary leakage reactance drop is

$$jI_L X_{t2} = j(33.4 - j25.0)0.75 = 18.8 + j25.0 \qquad [20]$$

The secondary terminal voltage is

$$V_2 \qquad\qquad = 2,400.0 + j\ 0 \qquad [21]$$

$$E_2 = \text{vector sum} \qquad = 2,428.8 + j17.5 \text{ v.} \qquad [22]$$

$$= 2,429\underline{/0.4°}. \qquad [23]*$$

The data give the resistances and leakage reactances with two significant figures, and therefore the nearest volt is significant in the values of the leakage-impedance voltage drops, Eqs. 19 and 20. In what follows it is assumed that the secondary terminal voltage V_2 also is known to the nearest volt. Tenths of a volt are carried throughout the computations, and the final results are then expressed to the nearest volt referred to the low-voltage side. Although the secondary terminal voltage in an actual circuit seldom would be known to the nearest volt in 2,400, nor would the primary voltage — to be computed — be required with this accuracy, nevertheless four significant figures are kept in these computations, since the results are to be used subsequently in comparison with the results of the illustrative example of Art. 7. Furthermore, it is often the *change* in voltage incident to loading the transformer that is more important than the actual values of the terminal voltages themselves. This

---

* The magnitude of E_2 may be calculated rapidly and accurately from the first two terms of the binomial expansion; thus if A and B are the real and imaginary parts of the complex expression for E_2, when $B \ll A$ the magnitude of E_2 is

$$E_2 = \sqrt{A^2 + B^2} \approx A + \frac{B^2}{2A}. \qquad \blacktriangleright[24]$$

When B is as great as $0.3A$, the error is less than 0.1 per cent.

change in voltage, expressed as a decimal fraction or a per cent of rated voltage, is called the voltage regulation and is discussed in Art. 1, Ch. XIV. The voltage regulation usually is but one or two per cent, so that if this change in voltage is to be computed with two significant figures, as warranted by the accuracy of the leakage-impedance data, four significant figures must be assumed in the values of the terminal voltages.

In order to determine the primary current, it is next necessary to determine the exciting current, which then is added to the load component of the primary current. The exciting current referred to the secondary equals the no-load current measured on the low-voltage side for a no-load condition at which the no-load induced voltage equals 2,429 volts — the value of the secondary induced voltage under load as given by Eq. 23. The open-circuit test data show that, to a first approximation, this exciting current is about 1.5 amperes. Since the leakage impedance of the low-voltage winding is about 0.8 ohm, the leakage-impedance voltage drop for this no-load condition is but 1.5 × 0.8 or 1.2 volts, and therefore is negligible. Thus at no load the induced voltage very nearly equals the applied voltage, and therefore the exciting current for an induced voltage of 2,429 volts under load equals the no-load current for very nearly the same applied voltage — say 2,430 volts. By interpolation from the open-circuit data at 2,430 volts

$$I''_\varphi = 1.55 \text{ amp.} \tag{25}$$

$$P_e = 963 \text{ w.} \tag{26}$$

The no-load power factor is

$$\frac{963}{2,430 \times 1.55} \approx 0.256. \tag{27}$$

That is, the exciting current lags the induced voltage by $\cos^{-1} 0.256$, or 75.2 degrees. Since E_2 is at an angle of $+0.4$ degrees (see Eq. 23), the vector expression for the exciting current is

$$I''_\varphi = aI'_\varphi = 1.55\underline{/-74.8°} \quad = \quad 0.41 - j\,1.50 \text{ amp.} \tag{28}$$

But the load current is, by Eq. 18,

$$I_L = aI'_L \qquad\qquad = \quad \underline{33.4 - j25.0.}$$

The vector sum is

$$aI_1 \qquad\qquad = \quad 33.8 - j26.5 \text{ amp.} \tag{29}$$

$$= \quad 43.0\underline{/-38.1°.} \tag{30}$$

The primary terminal voltage now can be determined by adding the primary resistance and leakage-reactance voltage drops to the induced voltage. Referred to the secondary, these voltage drops are

$$aI_1\frac{R_1}{a^2} = (33.8 - j26.5)0.28 \quad = \quad 9.5 - j\,7.4 \text{ v.} \tag{31}$$

$$jaI_1\frac{X_{t1}}{a^2} = j(33.8 - j26.5)0.76 = \quad 20.2 + j25.7 \tag{32}$$

and by Eq. 22

$$\frac{E_1}{a} = 2{,}428.8 + j17.5 \qquad [33]$$

$$\frac{V_1}{a} = \text{vector sum} \qquad = 2{,}458.5 + j35.8 \text{ v.} \qquad [34]$$

$$= 2{,}459 \underline{/0.8^\circ}. \qquad [35]$$

Hence the results are

$$I_1 = \frac{aI_1}{a} = \frac{43.0}{5.00} = 8.60 \text{ amp.} \qquad [36]$$

$$V_1 = a\frac{V_1}{a} = 5.00 \times 2{,}459 = 12{,}290 \text{ v.} \qquad [37]$$

The primary current lags the primary terminal voltage by

$$\theta_1 = 38.1 + 0.8 = 38.9^\circ.$$

Hence the primary power factor is

$$\cos\theta_1 = 0.778, \text{ lagging current.} \qquad [38]$$

6. SIMPLIFIED EQUIVALENT CIRCUITS; EQUIVALENT IMPEDANCE

Since the exciting current and the leakage-impedance voltage drops usually are small, it is often permissible to make further simplifying approximations. The terminal voltages usually do not differ greatly from the voltages induced by the resultant mutual flux, and therefore very little error usually results if the exciting current is taken as the value corresponding to either terminal voltage — whichever is convenient — rather than as the " exact " value corresponding to the induced voltage. Furthermore, the leakage-impedance voltage drop due to the exciting current usually is very small, and therefore negligible error usually is caused when the leakage-impedance voltage drop is computed on the assumption that the primary and secondary currents are equal when referred to the same side. Consequently the equivalent circuits, Figs. 5b and 6b, often may be simplified by the moving of the shunt branch representing the exciting current from its central position in these " exact " equivalent circuits and connecting it across either the primary or the secondary terminals, whichever is convenient. For example, if the " exact " equivalent circuit of Fig. 5b is altered by the moving of the exciting admittance to the left of the primary leakage impedance, the circuit of Fig. 7a is obtained. The transformer is then represented by an ideal transformer in combination with a shunt admittance and a single series impedance. Evidently the shunt admittance and the series impedance may be moved to the left of the ideal transformer if they are first referred to the primary.

Although the equivalent circuit of Fig. 7a is derived here as an approximate simplification of the " exact " equivalent circuit of Fig. 5b, as a matter of fact a circuit of the form of Fig. 7a but with slightly different parameters is an " exact " equivalent of the transformer. This fact is discussed briefly in Art. 8c of this chapter and again in Art. 5, Ch. XVII.

The series impedance of Fig. 7a is called the *equivalent impedance* of the transformer. The *equivalent resistance* is the sum of the primary and secondary effective resistances both referred to the same side, and the *equivalent reactance* is the sum of the primary and secondary leakage

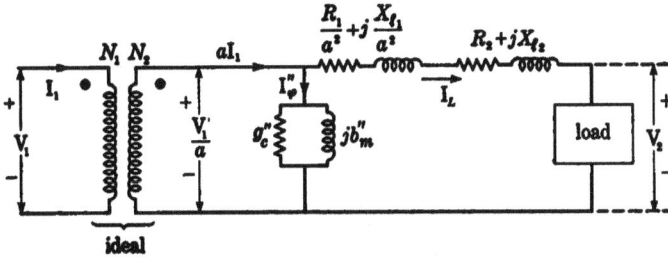

(a)

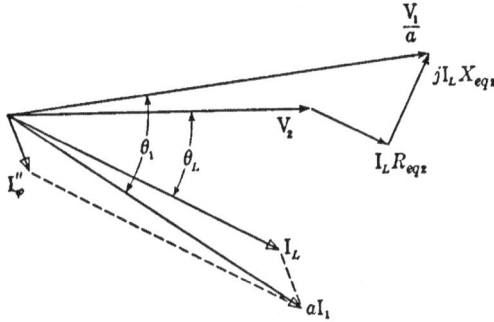

(b)

Fig. 7.   A simplified equivalent circuit and vector diagram.

reactances both referred to the same side. The equivalent reactance is also called the *total leakage reactance* or sometimes the *leakage reactance of the transformer* and should be distinguished from the individual leakage reactances $X_{\ell 1}$ and $X_{\ell 2}$ of its windings with respect to each other. Thus, if a is the turns ratio N_1/N_2, the equivalent impedance Z_{eq2} referred to the secondary is

$$Z_{eq2} \equiv R_{eq2} + jX_{eq2} \qquad \blacktriangleright [39]$$

$$\equiv \frac{R_1}{a^2} + R_2 + j\left(\frac{X_{\ell 1}}{a^2} + X_{\ell 2}\right). \qquad \blacktriangleright [40]$$

Similarly, the equivalent impedance Z_{eq1} referred to the primary is

$$Z_{eq1} \equiv R_{eq1} + jX_{eq1} \qquad\qquad \blacktriangleright[41]$$

$$\equiv R_1 + a^2 R_2 + j(X_{\ell 1} + a^2 X_{\ell 2}). \qquad\qquad \blacktriangleright[42]$$

When the exciting admittance is placed at the left of the primary leakage impedance, as in Fig. 7a, the current in the equivalent impedance is the secondary current, which also equals the load component of the primary current when both currents are referred to the same side and their positive directions are taken as the directions of currents that would produce opposing magnetomotive forces. The primary and secondary leakage-impedance voltage drops then can be combined, as shown in the vector diagram of Fig. 7b. The equivalent-impedance voltage drops in Fig. 7b differ slightly from the sum of the primary and secondary leakage-impedance voltage drops in the " exact " vector diagram of Fig. 5a, but only because the exciting current does not flow in the primary leakage impedance in Fig. 7a; the component of the primary leakage-impedance voltage drop due to the exciting current hence is omitted from Fig. 7b. The exciting current in Fig. 7b is taken as the no-load current corresponding to the primary terminal voltage, and therefore differs slightly, both in phase and magnitude, from the exciting current in the " exact " vector diagram of Fig. 5a.

7. ILLUSTRATIVE EXAMPLE OF THE SIMPLIFIED VECTOR RELATIONS

The results of the illustrative example of Art. 5 are computed below by use of the simplified vector diagram of Fig. 7b.

Solution: From Eqs. 14 and 15 and the data on p. 345, the equivalent resistance and equivalent reactance referred to the secondary are

$$R_{eq2} = \frac{R_1}{a^2} + R_2 = 0.28 + 0.30 = 0.58 \text{ ohm} \qquad [43]$$

$$X_{eq2} = \frac{X_{\ell 1}}{a^2} + X_{\ell 2} = 0.76 + 0.75 = 1.51 \text{ ohms}. \qquad [44]$$

The secondary terminal voltage being used as the reference vector, the vector secondary current is, by Eqs. 17 and 18,

$$I_L = 33.4 - j25.0 \text{ amp}. \qquad [45]$$

The primary terminal voltage referred to the secondary is determined by computation of the vector equivalent-resistance and equivalent-reactance voltage drops referred to the secondary and addition of them to the secondary terminal voltage; thus

$$I_L R_{eq2} = (33.4 - j25.0)0.58 = 19.4 - j14.5 \text{ v}. \qquad [46]$$

$$jI_L X_{eq2} = j(33.4 - j25.0)1.51 = 37.8 + j50.5 \qquad [47]$$

$$V_2 \qquad\qquad = 2{,}400.0 + j\ 0 \qquad\qquad\qquad [48]$$

$$\frac{V_1}{a} = \text{vector sum} \qquad = 2{,}457.2 + j36.0 \text{ v.} \qquad\qquad [49]$$

$$= 2{,}458\underline{/0.8^\circ}. \qquad\qquad\qquad [50]$$

Compare with the value $V_1/a = 2{,}459\underline{/0.8^\circ}$ given by Eq. 35 in the "exact" solution. The error in the approximate solution is only 0.04 per cent. The primary terminal voltage is

$$V_1 = 5.00 \times 2{,}458 = 12{,}290 \text{ v.} \qquad\qquad [51]$$

Note that if only the primary voltage were required, the exciting current need not be known. In many problems, such as those involving voltage regulation, the voltages are important but the actual primary current is relatively unimportant. In the analysis of these problems, the effects of the exciting current are negligible, and the transformer then may be represented very simply by its equivalent impedance in series with an ideal transformer, as in the fifth alternative of Art. 6, Ch. XII.

However, if accurate values of the primary current and power factor are required, the exciting current must be determined. According to the equivalent circuit of Fig. 7a, the exciting current is taken as the no-load current at the primary terminal voltage. Thus the exciting current referred to the secondary is the no-load current at 2,458 volts, Eq. 50. By interpolation from the open-circuit data, p. 345,

$$I_\varphi'' = 1.60 \text{ amp.} \qquad\qquad\qquad [52]$$

$$P_a = 987 \text{ w.} \qquad\qquad\qquad [53]$$

The no-load power factor is

$$\cos\theta_{nl} = \frac{987}{2{,}458 \times 1.60} = 0.251. \qquad\qquad [54]$$

That is, the exciting current lags the applied voltage by an angle θ_{nl}, or 75.5 degrees. Since the angle of V_1 is $+0.8$ degree, the vector expression for I_φ'' is

$$I_\varphi'' = 1.60\underline{/-74.7^\circ} = 0.42 - j\ 1.54 \text{ amp.} \qquad [55]$$

But, from Eq. 45,

$$aI_L' = I_L \qquad\qquad = 33.4 \quad - j25.0$$

$$aI_1 = \text{sum} \qquad = 33.8 \quad - j26.5 \text{ amp.} \qquad\qquad [56]$$

$$= 43.0\underline{/-38.1^\circ}. \qquad\qquad\qquad [57]$$

Compare Eqs. 55–57 with the results of the "exact" analysis, Eqs. 28–30, p. 347. The results are identical to three significant figures.

The primary current lags the primary terminal voltage by

$$\theta_1 = 38.1 + 0.8 = 38.9^\circ,$$

and therefore the primary power factor is

$$\cos\theta_1 = 0.778, \text{ lagging current.} \qquad\qquad [58]$$

This result is identical with the value given in Eq. 38, p. 348.

8. Determination of the Parameters by Tests

The excitation characteristics and the parameters of the equivalent circuits can be determined by means of open-circuit and short-circuit tests.

8a. *Open-Circuit Tests.* — When one winding is open-circuited and an alternating voltage is applied to the other, the voltage induced in the excited winding very nearly equals the applied voltage and the current in the excited winding is the corresponding exciting current referred to that winding. Also the copper loss due to the exciting current usually is negligibly small, and therefore the power input very nearly equals the core loss for the corresponding value of the induced voltage.

With communication transformers, the measurements usually are made by means of a suitable bridge.* Power-system transformers usually are tested by the application of a sine-wave voltage to either winding — whichever is convenient — and measurement of the applied voltage, exciting current, and power input with ordinary alternating-current instruments. If the primary is the excited winding and the exciting current is assumed to be an equivalent sine wave, as shown in Art. 13, Ch. VI, the conductances and susceptances representing the excitation characteristics referred to the primary are

$$g_{oc1} = \frac{P}{V^2} \approx g_c' \qquad \qquad \blacktriangleright [59]$$

$$b_{oc1} = -\sqrt{\left(\frac{I_\varphi}{V}\right)^2 - g_{oc1}^2} \approx b_m', \qquad \qquad \blacktriangleright [60]$$

where

     P, V, and I_φ are the readings of the instruments, corrected if necessary for the effects of the instrument currents,

     g_{oc1} and b_{oc1} are the apparent conductance and apparent susceptance on open circuit,

     g_c' and b_m' are the conductance and susceptance of the exciting admittance referred to the primary.

The error resulting from computing the core-loss conductance and magnetizing susceptance as in Eqs. 59 and 60 usually is entirely negligible, since the power input P very nearly equals the core loss and the applied voltage V very nearly equals the induced voltage.

As mentioned in Art. 3, a close approximation to the ratio of transformation may be determined by measurement of the voltage applied to the excited winding and the voltage of the open-circuited winding. Thus

---

     * For further discussion of impedance measurements with communication transformers, see Art. 3, Ch. XVII.

the ratio of transformation is, very nearly,

$$a \equiv \frac{N_1}{N_2} = \frac{E_1}{E_2} \approx \frac{V_1}{E_{oc2}}$$ ▶[61]

where V_1 is the voltage applied to the primary and E_{oc2} is the voltage of the open-circuited secondary. Equation 61 neglects the primary leakage-impedance voltage drop due to the exciting current. Because of this voltage drop, the primary applied voltage V_1 is very slightly greater than the primary induced voltage, and therefore the ratio V_1/E_{oc2} is very slightly greater than the turns ratio N_1/N_2. Furthermore, the vector voltages V_1 and E_{oc2} are not exactly in phase. That is, the vector ratio V_1/E_{oc2} is a complex number with a very small phase angle.

8b. *Short-Circuit Tests.* — If the secondary winding of a transformer is short-circuited, the equivalent circuit of Fig. 6b reduces to that shown

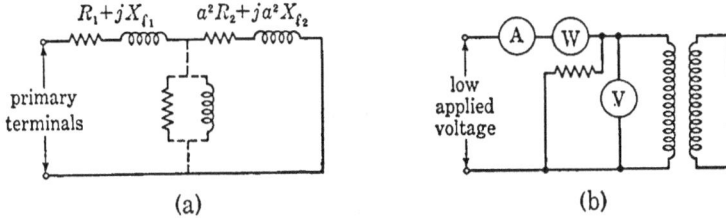

(a)   (b)

Fig. 8.   Short-circuited transformer; (a) equivalent circuit referred to the primary, and (b) connections for the short-circuit impedance test.

in Fig. 8a. If the exciting current is neglected, Fig. 8a shows that the impedance at the primary terminals equals the equivalent impedance of the transformer referred to the primary.

The short-circuit impedance of a communication transformer usually is measured by means of a suitable bridge. With power-system transformers, the test usually is made by the application of a low voltage of such a value that it produces approximately rated current in the excited winding, and measurement of the voltage, current, and power, as shown in Fig. 8b. This applied voltage often is called the *impedance voltage* and, when the current has its rated value, the *full-load impedance voltage*. For power-system transformers, the full-load impedance voltage usually is 3 to 10 per cent of the rated voltage of the excited winding. Thus the core flux during the short-circuit test is only a few per cent of its value for rated-voltage operation. In spite of this fact, however, the leakage reactances, being practically independent of saturation, have essentially the same values for short-circuit as for normal operating conditions.

If the primary current is its rated value and the exciting current is negligible, the secondary current also is its rated value. To produce this rated secondary current, the voltage induced in the secondary must equal the secondary leakage-impedance voltage drop. In a typical power-system transformer this voltage is about 3 per cent of the rated voltage; that is, the core flux is about 0.03 of its value at no load with rated voltage. If magnetic saturation is neglected, the exciting current on short circuit therefore is about 0.03 of its normal value. Since the exciting current at rated voltage usually is about 5 per cent of the rated current, the exciting component of the primary current when the secondary is short-circuited is only about 0.03 × 5 or 0.15 per cent of the primary current and is quite insignificant. Thus the current in the exciting admittance of the equivalent circuit, Fig. 8a, is negligible; and therefore the short-circuit impedance very nearly equals the equivalent impedance. Also, since the core loss varies approximately as the square of the flux, the core loss at short circuit is approximately $(0.03)^2$ or 0.0009 of the normal-voltage core loss. Since the normal-voltage core loss usually is less than the combined primary and secondary copper losses at rated current, the core loss at short circuit usually is less than 0.0009 of the copper loss and thus is entirely negligible compared with the copper loss. Hence, on short circuit, the power input P very nearly equals the equivalent-resistance loss $I^2 R_{eq}$.

If the readings of the voltmeter, ammeter, and wattmeter in Fig. 8b are V, I, and P, the magnitude of the short-circuit impedance is

$$Z_{sc} = \frac{V}{I} \approx Z_{eq} \qquad\qquad \blacktriangleright[62]$$

$$R_{sc} = \frac{P}{I^2} \approx R_{eq} \qquad\qquad \blacktriangleright[63]$$

$$X_{sc} = \sqrt{Z_{sc}^2 - R_{sc}^2} \approx X_{eq}, \qquad\qquad \blacktriangleright[64]$$

where Z_{sc}, R_{sc}, X_{sc} are the short-circuit impedance, resistance, and reactance, and Z_{eq}, R_{eq}, X_{eq} are the equivalent impedance, resistance, and reactance.

Obviously, either winding may be used as the primary. The equivalent impedance is its value referred to the excited winding. If desired, this value may be referred to the other winding. Thus if Z_{eqX} is the equivalent impedance referred to or measured on the low-voltage side, the equivalent impedance Z_{eqH} referred to the high-voltage side is

$$Z_{eqH} = a^2 Z_{eqX}, \qquad\qquad \blacktriangleright[65]$$

where a is the turns ratio N_H/N_X.

Since the voltage drop in a transformer usually is determined quite accurately from its equivalent impedance, it is seldom necessary to know the actual leakage impedances of each winding, except in the analysis of problems concerning the effects of harmonics. Tests for actually measuring the separate leakage impedances of each winding are rather complicated.[1] When an approximate value is required, it is often assumed that the separate leakage impedances of the windings are equal when referred to the same side. Thus if a is the turns ratio N_1/N_2,

$$R_1 + jX_{l1} \approx a^2 R_2 + ja^2 X_{l2} \approx \frac{R_{eq1}}{2} + j\frac{X_{eq1}}{2} \qquad [66]$$

or

$$R_2 + jX_{l2} \approx \frac{R_1}{a^2} + j\frac{X_{l1}}{a^2} \approx \frac{R_{eq2}}{2} + j\frac{X_{eq2}}{2}. \qquad [67]$$

Alternatively, it is sometimes assumed that the effective resistances and leakage reactances are in the same ratio as the direct-current resistances R_{dc1} and R_{dc2} of the windings; that is,

$$\frac{R_1}{R_2} \approx \frac{X_{l1}}{X_{l2}} \approx \frac{R_{dc1}}{R_{dc2}}; \qquad [68]$$

whence

$$R_1 + jX_{l1} \approx \frac{R_{dc1}}{R_{dc1} + a^2 R_{dc2}} (R_{eq1} + jX_{eq1}) \qquad [69]$$

$$R_2 + jX_{l2} \approx \frac{R_{dc2}}{R_{dc1} + a^2 R_{dc2}} (R_{eq1} + jX_{eq1}). \qquad [70]$$

8c. *An " Exact " Equivalent Circuit Determined by Tests.* — If it is assumed that the exciting admittance referred to the primary side equals the open-circuit admittance measured on the primary (as in Eqs. 59 and 60), that the ratio of transformation equals the open-circuit voltage ratio (as in Eq. 61), and that the equivalent impedance referred to the secondary equals the short-circuit impedance measured on the secondary side (as in Eqs. 62, 63, and 64), the equivalent circuit of Fig. 7a becomes that shown in Fig. 9a when the exciting admittance of Fig. 7a is referred to the primary side. Although this equivalent circuit is derived in Art. 6 as an approximate simplification of the " exact " equivalent circuit of Fig. 5b, the circuit of Fig. 9a can be derived in a manner that shows it is an " exact " equivalent of the transformer.

If it is assumed that the exciting current may be treated as an equivalent sine wave and that the exciting admittance is constant for small changes in voltage, the transformer may be considered a linear circuit

---

[1] For a discussion of some of these tests, see O. G. C. Dahl, *Electric Circuits, Theory and Applications*, Vol. I: *Short-Circuit Currents and Steady-State Theory* (New York: McGraw-Hill Book Company, Inc., 1928), Ch. ii.

element, to which Thévenin's theorem applies.[2] According to Thévenin's theorem, the transformer as viewed from its secondary terminals is equivalent to an electromotive force equal to the open-circuit secondary voltage E_{oc2} in series with the impedance Z_{sc2} measured at the secondary terminals with the primary terminals short-circuited, as in the portion to the right of the ideal transformer in Fig. 9a. If the ratio of transformation of the ideal transformer is V_1/E_{oc2}, its primary terminal voltage equals the primary terminal voltage V_1 of the actual transformer.

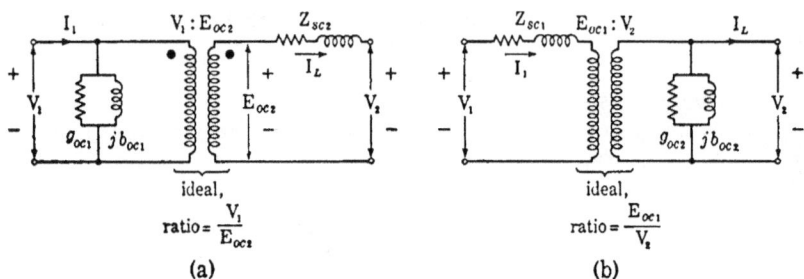

FIG. 9. Two " exact " equivalent circuits.

It is shown in Art. 5, Ch. XVII, that if the open-circuit admittance $g_{oc1} + jb_{oc1}$ is connected across the primary terminals of the ideal transformer, the circuit of Fig. 9a is " exactly " equivalent to the transformer on both its primary and secondary sides.

It is apparent that if the open-circuit measurements are made on the secondary side and the short-circuit measurements on the primary side, the circuit of Fig. 9b also is the " exact " equivalent of the transformer.*

Thus the rather surprising conclusion is reached that the combined effects of a number of approximations give an " exact " result.

The ratio V_1/E_{oc2} is a complex number with a very small phase angle and therefore for complete rigor the ideal transformer, Fig. 9a, must be regarded as one that not only transforms the magnitudes of voltage and current in inverse fashion but also introduces a slight phase shift between its primary and secondary voltages, and an equal and *opposite* phase shift between its primary and secondary currents. However, with power-system transformers, the phase angle on open circuit ordinarily is but a few hundredths of a degree, and therefore the equivalent circuits of Fig. 9 are almost " exact " if the ratio of transformation of the ideal transformer is taken as a real number equal to the magnitude of the open-circuit voltage ratio.

---

[2] Thévenin's theorem is discussed in this series in the volume on electric circuits (1940) Ch. VIII, Art. 15, p. 469.

* For further discussion of equivalent circuits, see Art. 5, Ch. XVII.

9. DETERMINATION OF THE PARAMETERS FROM DIMENSIONS

The determination of the parameters from dimensions and the properties of the materials is, of course, a ı important part of the work of a designer. Although a detailed discussion of design procedure is beyond the scope of this book on general principles, the brief discussion given below serves to illustrate the methods by which the fundamental theory is applied to the designer's problem of determining the transformer circuit constants. The excitation characteristics are described in detail in Ch. VI, and their determination need not be discussed further here. The remainder of this article is devoted to the discussion of the determination of equivalent impedance.*

Consider a one-to-one-ratio transformer supplying power to a load, as in Fig. 10a. A primary and a secondary terminal of the same polarity are shown connected together. If the exciting current is small, the primary current I_1 very nearly equals the oppositely directed secondary current I_L. If these currents were exactly equal, conditions in the primary and

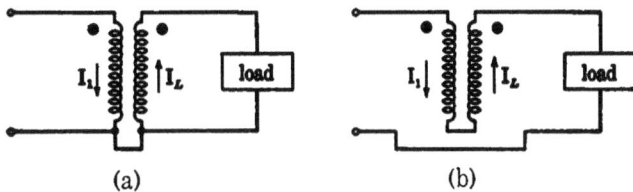

(a)   (b)

FIG. 10. Showing that the equivalent impedance of a 1:1-ratio transformer equals its series impedance as a reactor with primary and secondary magnetomotive forces in opposition.

secondary circuits would be unaltered if the circuit were reconnected as in Fig. 10b. Thus a one-to-one-ratio transformer is equivalent to a simple series impedance, if the exciting current is neglected. From the discussion of Art. 6, it is apparent that this series impedance is the equivalent impedance of the transformer. Therefore the calculation of the equivalent impedance reduces to determining the resistance and reactance of an equivalent one-to-one-ratio transformer connected as a reactor with its primary and secondary magnetomotive forces in opposition.

The direct-current resistance of each winding can be calculated from the number of turns, the mean length of turn, and the cross-sectional area of the conductor. With large transformers or at high frequencies the effective alternating-current resistances may be considerably greater than the direct-current resistances, because of skin effect and stray

* For further sources of information on the determination of equivalent impedance, the reader is referred to the bibliography.

losses. The increase in resistance due to skin effect can be reduced if the conductors are stranded and the strands are suitably transposed. The stray losses due to eddy currents produced by the leakage field are usually calculated by empirical methods. The equivalent resistance is the sum of the primary and secondary effective resistances, both referred to the same side.

The equivalent reactance is determined from the leakage field due to equal and opposing primary and secondary currents. Under these conditions, the general character of the magnetic field in a core-type transformer with concentric, cylindrical windings is shown in Fig. 5c, p. 322. This type of transformer serves as a simple example of the method of approach to the problem of computing equivalent reactance, and illustrates the application of simplifying assumptions to a problem which is very difficult to solve rigorously.

When the primary and secondary windings have the same stack height h, it is commonly assumed that the leakage flux is parallel to the axes of the cylindrical coils throughout the regions *abcd*, Fig. 11a.[3] Compare with Fig. 5c, p. 322. According to this assumption, the flux density is uniform in a direction parallel to the axes of the coils from one end of the coils to the other. It is also assumed that in the regions above and below the coils the flux disperses rapidly, its density is low, and most of the flux finds a return path in the iron core. The assumption is therefore commonly made that the reluctance of the leakage flux paths above and below the coils is zero. This is equivalent to assuming that the coils are surrounded by iron of infinite permeability, as in Fig. 11b. On these assumptions, the equivalent reactance is the reactance of two N-turn coils wound on the simple magnetic circuit of Fig. 11b and connected in series, with their magnetomotive forces in opposition. If the number of turns N in each coil equals the primary turns N_1, the result is the equivalent reactance referred to the primary.

Consider the flux path 1 which threads the primary turns in Fig. 11b. This path closes through the iron to the left of the coils and links the fraction x/t_1 of the primary turns. If the current is I, the magnetomotive force F_x around this closed path is

$$F_x = 4\pi NI \frac{x}{t_1}. \qquad [71]$$

Any consistent unrationalized system of units may be used. Since the permeability of the iron is assumed as infinite and the flux density in

---

[3] When the windings do not have the same stack height, it is possible to resolve them into equivalent component windings in such a way that the method of attack presented here may still be used. See L. F. Blume, editor, *Transformer Engineering* (New York: John Wiley & Sons, 1938), 80.

the air is assumed as uniform, the magnetizing force $\mathcal{H}_x$ along the line *ef*, Fig. 11b, is constant and is

$$\mathcal{H}_x = \frac{F_x}{h} = \frac{4\pi NIx}{ht_1}.$$ [72]

As x increases from zero at the inside of the primary coil to t_1 at its outside, the magnetizing force increases linearly to the value $4\pi NI/h$ at $x = t_1$, as indicated in Fig. 11c.

Within the insulation space between the windings, the magnetomotive

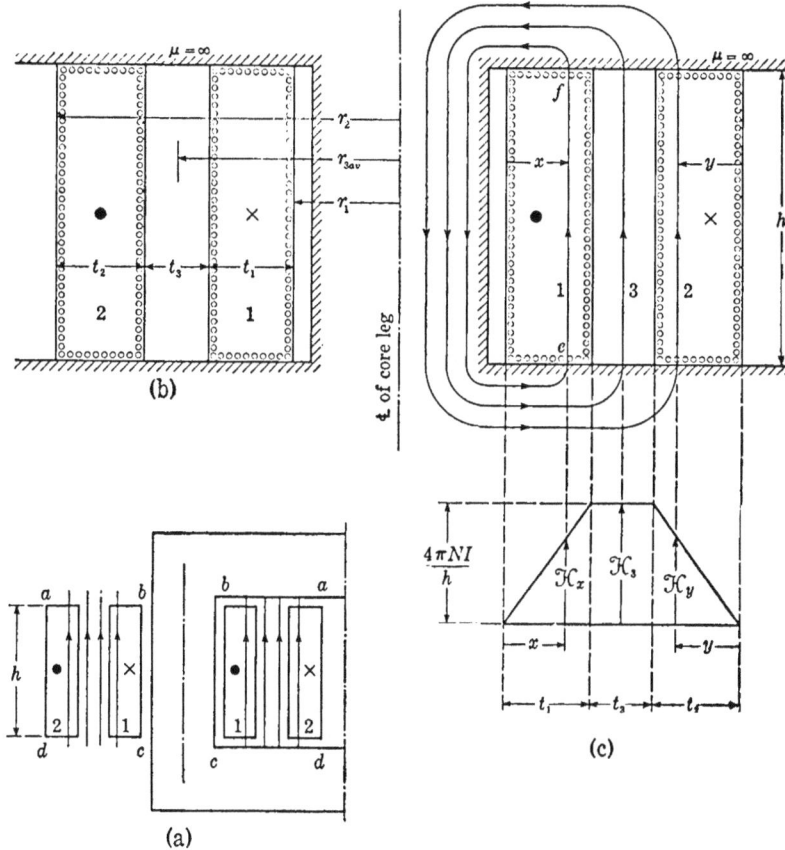

FIG. 11.   One leg of a core-type transformer; (a) arrangement of windings and core, (b) magnetic equivalent, (c) distribution of magnetizing force.

force around a closed path such as 3, Fig. 11b, is the whole magneto-motive force of a single winding. Within this region the magnetomotive force is independent of the radial distance and the constant magnetizing force $\mathcal{H}_3$ in the insulation space is

$$\mathcal{H}_3 = \frac{4\pi NI}{h},$$ [73]

as shown in Fig. 11c. It is evident that the same result is obtained whether the flux path is assumed to close surrounding the primary or the secondary.

Within the space occupied by the secondary turns the magnetizing force falls off uniformly, as shown in Fig. 11c. Consider the flux path 2, Fig. 11b, threading the secondary turns. If the flux path closes to the left and surrounds the primary, as shown, the path surrounds all N primary turns and $\left(\dfrac{t_2 - y}{t_2}\right) N$ secondary turns whose magnetomotive force opposes that of the primary. Note that y is measured from the *outside* of the secondary towards the inside, in order to simplify the mathematics which follows. The magnetomotive force F_y on the path 2 is

$$F_y = 4\pi \left[N - \frac{t_2 - y}{t_2} N \right] I$$

$$= 4\pi NI \frac{y}{t_2},$$ [74]

and the magnetizing force $\mathcal{H}_y$ is

$$\mathcal{H}_y = \frac{4\pi NIy}{ht_2}.$$ [75]

This is the same result that would be obtained if the flux path 2 were assumed to close to the right, surrounding the fraction y/t_2 of the secondary turns.

The equivalent inductance L_{eq} can be computed from the energy stored in the magnetic field. Since the permeability of the iron is assumed to be infinite, this energy is all stored in the cylindrical volume occupied by the copper and insulation between the windings. When the current in the winding is I, this stored energy equals $\frac{1}{2} L_{eq} I^2$.

It is shown in Art. 5, Ch. IV, that the energy put into the magnetic field per unit volume when the flux density is raised from zero to $\mathcal{B}$ is

$$\frac{1}{4\pi} \int_0^{\mathcal{B}} \mathcal{H} d\mathcal{B}.$$

In air, this magnetic energy per unit volume is

$$\frac{1}{4\pi} \int_0^{\mathcal{B}} \frac{\mathcal{B}}{\mu_0} d\mathcal{B} = \frac{\mathcal{B}^2}{8\pi\mu_0} = \frac{\mu_0 \mathcal{H}^2}{8\pi},$$ [76]

where μ_0 is the permeability of free space.

The energy stored in the resultant leakage field of Fig. 11 is the integral of Eq. 76 throughout the volume.

The differential energy dW_x stored in a cylindrical ring of height h, thickness dx, and radius $(r_1 + x)$, in which the magnetizing force is $\mathcal{H}_x$, is

$$dW_x = \frac{\mu_0}{8\pi} \mathcal{H}_x^2 \times \text{(differential volume of the ring)} \qquad [77]$$

$$= \frac{\mu_0}{8\pi} \mathcal{H}_x^2 2\pi (r_1 + x) h \, dx. \qquad [78]$$

From Eq. 72

$$dW_x = \frac{\mu_0}{8\pi} \left(\frac{4\pi N I x}{h t_1} \right)^2 2\pi (r_1 + x) h \, dx$$

$$= \frac{4\pi^2 \mu_0 N^2 I^2}{h t_1^2} (r_1 + x) x^2 \, dx. \qquad [79]$$

Therefore the total energy W_1 stored in the volume occupied by the primary winding is

$$W_1 = \frac{4\pi^2 \mu_0 N^2 I^2}{h t_1^2} \int_0^{t_1} (r_1 + x) x^2 \, dx$$

$$= \frac{4\pi^2 \mu_0 N^2 I^2}{h} \left(\frac{r_1}{3} + \frac{t_1}{4} \right) t_1. \qquad [80]$$

Similarly the energy W_2 stored in the region occupied by the secondary winding is

$$W_2 = \frac{4\pi^2 \mu_0 N^2 I^2}{h t_2^2} \int_0^{t_2} (r_2 - y) y^2 \, dy$$

$$= \frac{4\pi^2 \mu_0 N^2 I^2}{h} \left(\frac{r_2}{3} - \frac{t_2}{4} \right) t_2. \qquad [81]$$

The energy W_3 stored within the space between the windings, where the magnetizing force $\mathcal{H}_3$ is independent of the radial distance, is

$$W_3 = \frac{\mu_0 \mathcal{H}_3^2}{8\pi} \times \text{(volume of space between the windings)}$$

$$= \frac{\mu_0}{8\pi} \left(\frac{4\pi N I}{h} \right)^2 2\pi r_{3av} t_3 h$$

$$= \frac{4\pi^2 \mu_0 N^2 I^2}{h} r_{3av} t_3. \qquad [82]$$

where r_{3av} is the *mean* radius of the insulating cylinder between the windings.

The total stored energy equals $\frac{1}{2}L_{eq}I^2$, whence

$$L_{eq} = \frac{2}{I^2}(W_1 + W_2 + W_3)$$

$$= \frac{8\pi^2\mu_0 N^2}{h}\left[\left(\frac{r_1}{3} + \frac{t_1}{4}\right)t_1 + \left(\frac{r_2}{3} - \frac{t_2}{4}\right)t_2 + r_{3av}t_3\right], \qquad \blacktriangleright[83]$$

where N is the number of turns in the winding to which the equivalent inductance is referred. The dimensions are shown in Fig. 11b. Note that r_1 is the *inside* radius of the inner winding, but r_2 is the *outside* radius of the outer winding. In the mks system, inductance is in henries, dimensions are in meters, and $\mu_0 = 10^{-7}$.

This formula may be simplified if it is noted that $\frac{r_1}{3} + \frac{t_1}{4}$ is approximately $\frac{r_{1av}}{3}$, where r_{1av} is the mean radius of the primary. Similarly $\frac{r_2}{3} - \frac{t_2}{4}$ is approximately $\frac{r_{2av}}{3}$, where r_{2av} is the mean radius of the secondary. Thus

$$L_{eq} \approx \frac{8\pi^2\mu_0 N^2}{h}\left(\frac{r_{1av}t_1}{3} + \frac{r_{2av}t_2}{3} + r_{3av}t_3\right). \qquad \blacktriangleright[84]$$

The formula can also be expressed in terms of the mean length of turn, or volume of copper and insulation, or in various other forms.

Since in an actual transformer there is some reluctance in the region above and below the winding space, the magnetic leakage is not so large as in Fig. 11b and the flux begins to disperse before it has left the winding space, as in Fig. 5c, p. 322. Hence Eqs. 83 and 84 give values that are too large, and empirical correction factors less than unity often are used.

PROBLEMS

1. Each vector diagram of Fig. 12 applies to a transformer. All voltage and induced electromotive force vectors are indicated by an open arrowhead $\longrightarrow$, all current vectors by $\longrightarrow\!\triangleright$, and all flux vectors by $\longrightarrow\!\diamond$. For each of these vector diagrams:
   (a) Draw an equivalent circuit.
   (b) Name the vectors appearing on the diagram and deduce the conventions for positive directions.
   (c) Indicate in the equivalent circuit diagram the positive directions of all the currents, voltages, and electromotive forces represented on the vector diagram, and indicate the direction of power flow.

2. A 5-kva 2,300 : 230-v 60 $\sim$ transformer is being used to supply power to an inductive load. Measured values of terminal voltage, current, and power input on the high-voltage side are, respectively: 2,360 v, 1.96 amp, and 4.16 kw. The magnetizing and core-loss currents corresponding to rated full-load conditions are 0.10 and

0.02 amp. (Neglect the change in exciting current caused by a change in induced voltage.) The primary leakage impedance is $11.5 + j21$ ohms, and the secondary leakage impedance is $0.112 + j0.206$ ohm.

On the assumption that there is no line drop between the secondary terminals of the transformer and the load, determine the power delivered to the load, the load voltage, and the power factor.

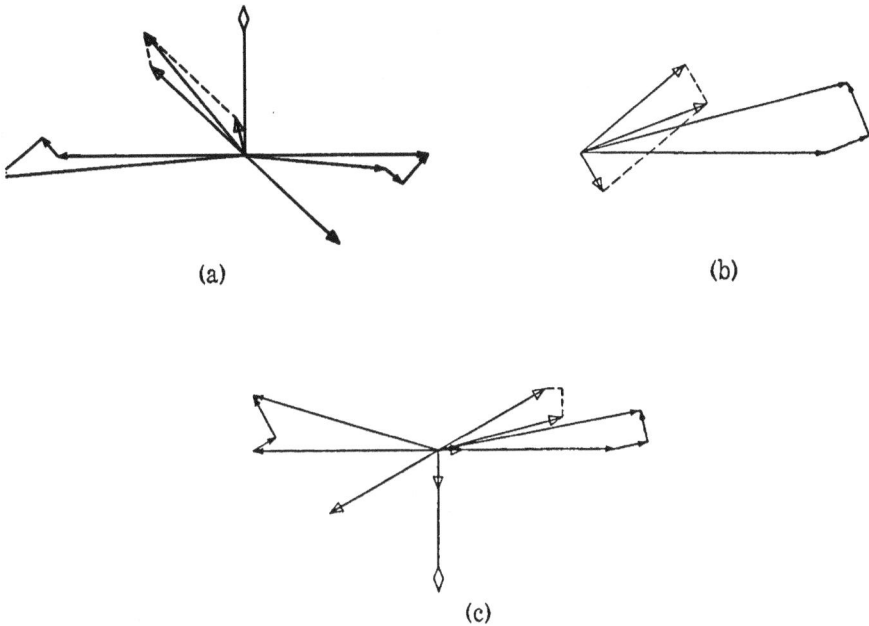

(a)   (b)

(c)

Fig. 12. Vector diagrams of transformers, Prob. 1.

3. The transformer for which the equivalent circuit is shown in Fig. 13 is used to connect a generator with an internal impedance of $10,000 + j0$ ohms to a load of impedance $4 + j0$ ohms.

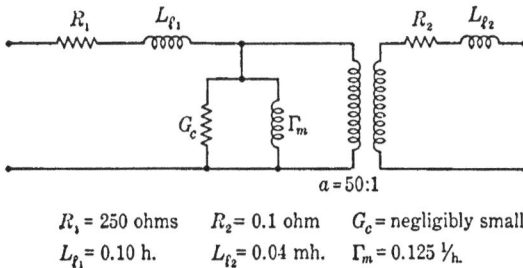

| $R_1 = 250$ ohms | $R_2 = 0.1$ ohm | G_c = negligibly small |
| $L_{\ell_1} = 0.10$ h. | $L_{\ell_2} = 0.04$ mh. | $\Gamma_m = 0.125$ ⅟h. |

Fig. 13. Equivalent circuit of matching transformer, Prob. 3.

If the secondary of the transformer is to be rewound with the same volume of copper but with the correct number of turns so that maximum power will be delivered to the load at a frequency of 1,000 ∼, what should be the new ratio of primary to secondary turns?

4. The following data are given for a 7.5-kva 2,080 : 208-v 60 ∼ transformer when operated at rated frequency and normal induced voltage:

> Primary resistance = 7.5 ohms.
> Primary leakage reactance = 14.0 ohms.
> Secondary resistance = 0.07 ohm.
> Secondary leakage reactance = 0.15 ohm.
> Core-loss current = 1.0 amp.
> Magnetizing current = 2.0 amp.

Determine the voltage that must be impressed on the high-voltage side of this transformer to maintain 208 v across the secondary terminals when the load is 40 amp at unity power factor. Neglect the change in exciting current accompanying a change in load.

5. Compute the primary terminal voltage, current, and power factor for the transformer referred to in the illustrative examples of Arts. 5 and 7 if the exciting admittance is assumed to be connected to the secondary terminals.

6. The following data are given for a 3,333-kva 60 ∼ 55,000 : 6,600-v single-phase transformer:

SHORT–CIRCUIT TEST

(High-voltage winding excited)

Frequency	Voltage	Current	Power
60 ∼	3,800 v	60.6 amp	18,850 w

(a) What is the greatest per cent change that can occur in the secondary terminal voltage between no-load and full-load kva output when the transformer is operated at rated frequency and with the primary voltage held constant at the value which gives rated secondary voltage at full load? Express the change in voltage as a percentage of rated secondary voltage.

(b) What is the greatest per cent change that can occur in the mutual flux under the above conditions?

(c) What is the power factor of the load at which the above conditions occur?

Assume that the primary and secondary resistances and the primary and secondary leakage reactances are equal when referred to the same side.

7. The following data are given for a 1,000-kva 66,000 : 6,600-v 60 ∼ transformer:

	Frequency	Voltage	Current	Power
With low-voltage terminals short-circuited	60 ∼	3,240 v	15.2 amp	7,490 w
With high-voltage terminals open-circuited	60	6,600	9.1	9,300

Assume that the primary and secondary resistances are equal when referred to the same side and that the primary and secondary leakage reactances are similarly equal.

Also assume that the magnetizing current is sinusoidal in waveform and varies as the first power of the voltage corresponding to the mutual flux and that the core loss varies as the square of this voltage.

(a) If an inductive load taking a current of 152 amp at a power factor of 0.80 is connected to the low-voltage terminals of this transformer, what primary voltage is necessary to maintain 6,680 v across the load?

(b) What is the apparent impedance of this transformer at its primary terminals when delivering the above load?

8. The primary and secondary windings of a one-to-one-ratio iron-core transformer are connected in series opposition. When a 60 $\sim$ voltage which causes a current of 10 amp is applied to the outer terminals, the voltage across the primary is 26.7 v and the voltage across the secondary is 23.4 v. The primary and secondary resistances are 0.52 ohm and 0.49 ohm. What are the primary and secondary leakage reactances?

9. Both the primary and secondary windings of a one-to-one-ratio iron-core transformer consist of a pair of wires wound side by side. (See Prob. 6 and Fig. 10, Ch. XII.) The secondary terminals S_1 and S_2 are short-circuited and 5% of rated voltage is applied to the primary terminals P_1 and P_2. If the primary current is measured, what information regarding the constants of the transformer can be obtained using this value of current, and

(a) The voltage $P_2 P_2'$ with P_1 and P_1' connected?

(b) The voltage $S_1' S_2'$?

(c) The voltage $P_2' S_2'$ with P_1' and S_1' connected?

10. A 1,000-kva 66,000 : 11,000-v 60 $\sim$ transformer has a short-circuit impedance of $1.0 + j4.9$ ohms, referred to the low-voltage winding. It is being operated with an impressed terminal voltage of 66,600 v to supply a load through a line with an impedance of $0.8 + j2.0$ ohms. The secondary terminal voltage is 10,750 v, and the secondary load current is 90 amp. Find the power being delivered to the load.

11. Show whether the following equalities are exactly or only approximately true:

(a) $Z_{eq1} = a^2 Z_{eq2}$

(b) $Z_{sc1} = a^2 Z_{sc2}$

(c) $Y_{oc2} = a^2 Y_{oc1}$

(d) $Y_{\varphi}'' = a^2 Y_{\varphi}'$.

12. **The following data apply to a transformer rated at 100 kva 11,000 : 2,300 v and 60** $\sim$:

$$R_1 = 6.21 \text{ ohms.} \qquad R_2 = 0.273 \text{ ohm.}$$

$$X_{p1} = 22.7 \text{ ohms.} \qquad X_{p2} = 0.954 \text{ ohm.}$$

$$g_{c2} = 0.000172 \text{ mho.} \qquad b_{m2} = 0.000681 \text{ mho.}$$

Compare Z_{eq} and Z_{sc}, Z_{φ} and Z_{oc}, and $\dfrac{V_1}{E_{oc2}}$ and $\dfrac{N_1}{N_2}$.

13. Derive an expression for the equivalent inductance of a transformer wound as illustrated in Fig. 14. Let the total number of secondary turns equal the total number of primary turns, and let half the secondary turns be on the inner coil and half on the outer. Make simplifying assumptions corresponding to those of Art. 9.

(a) Given the following values, calculate the leakage inductance of a transformer of the type shown in Fig. 14:

$$l_1 = l_3 = \tfrac{1}{2} \text{ in.}$$

$$l_2 = 1 \text{ in.}$$

$$l_a = l_b = \tfrac{1}{8} \text{ in.}$$

$$r_1 = 2\tfrac{1}{8} \text{ in.}$$

$$h = 6 \text{ in.}$$

$$N_p = N_s = 1{,}000 \text{ turns.}$$

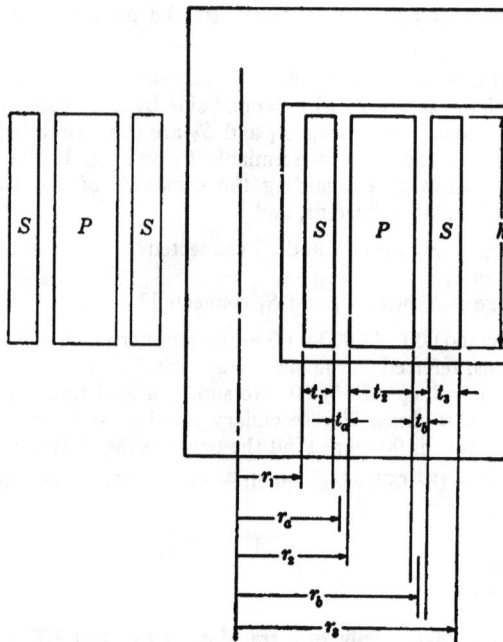

FIG. 14.　Cross-section of transformer, Prob. 13.

(b) Compare this leakage inductance with that of a transformer of the type shown in Fig. 11, for which

$$l_1 = l_2 = 1 \text{ in.}$$

$$l_3 = \tfrac{1}{8} \text{ in.}$$

$$r_1 = 2\tfrac{1}{8} \text{ in.}$$

$$h = 6 \text{ in.}$$

$$N_p = N_s = 1{,}000 \text{ turns.}$$

(The symbols are those used in Fig. 11.)

CHAPTER XIV

Voltage Regulation, Efficiency, and Rating

The general theory of the transformer developed in the two preceding chapters is suitable for the analysis of most problems except those concerned with the capacitances of the windings or the harmonics in the exciting current. Next to be discussed are applications of this theory to the solution of problems of voltage regulation, efficiency, and rating of transformers used in power systems. In the analysis of these and similar problems, it is often convenient to express the currents and voltages as a percentage or decimal fraction of their normal, rated values; power, impedance, admittance, and other quantities can be expressed in similar fashion. Certain advantages of the per unit or per cent system of units are discussed in Arts. 6 and 7.

1. VOLTAGE REGULATION

Since an electric lamp, heating appliance, or motor operates most effectively at rated voltage, it is important that the rated voltage be supplied to these loads and be maintained as nearly constant as possible. Excessively high voltage tends to shorten the life of electrical equipment, whereas subnormal voltage results in low illumination from lamps, improper operation of heating appliances, and subnormal torques from motors. Abnormal voltage — either low or high — therefore results in complaints to the public utility company; on the other hand, the satisfaction derived by the user from rated-voltage operation is a definite stimulus to the sale of electrical energy for additional uses.

Subnormal voltage, furthermore, means reduced sales of energy and a measurable dollar loss to the power company. The energy consumed by motors and heating appliances does not vary materially as a function of applied voltage over the usual range of voltage found in practice, even when the voltage regulation is unsatisfactory. But the power taken by incandescent lamps varies approximately as the lamp voltage raised to the power 1.56, with a corresponding variation in energy consumption. As the voltage drop due to circuit regulation is likely to be most serious during the evening hours, when the load consists mainly of lighting, the exponent 1.56 is a satisfactory guide to the evaluation of revenue loss from subnormal voltage. Suppose that the load delivered by a certain distribution feeder is 500 kw during a peak lasting four hours per day and that the corresponding average voltage delivered is four per cent lower than normal. If voltage-regulating equipment were installed that maintained normal voltage during the peak-load period, the output would be approximately

$$\left(\frac{100}{96}\right)^{1.56} \times 500 = 531 \text{ kw.} \qquad [1]$$

Thus four per cent subnormal voltage results in a loss of sale of 31 kilowatts for four hours daily. At a representative residential rate of $3\frac{1}{2}$ cents per kilowatthour, this loss of revenue is about \$1,585 yearly. It is evident that a utility company has real financial reasons for maintaining proper voltage at the users' premises and that a considerable investment in voltage-regulating equipment — discussed in Ch. XIX — is justified.

In the absence of automatic voltage-regulating equipment, the consumer's voltage varies with the load on the distribution circuit from which he is supplied, because of changes in the impedance drop with fluctuations in the load on the circuit. One component of this impedance drop is that in the transformers.

Consider the 100-kva, 12,000 : 2,400-volt transformer of Arts. 5 and 7, Ch. XIII, delivering its rating at 2,400 volts to an inductive load at 0.80 power factor. In the solution of Art. 7, Ch. XIII, it is shown that the primary impressed voltage required to maintain rated secondary voltage at this load is 12,290 volts, or the primary voltage referred to the secondary is 2,458 volts. This solution neglects the primary leakage-impedance voltage drop due to the exciting current. If this impressed voltage were maintained constant at 12,290 volts and the load were removed, the secondary terminal voltage would rise to 2,458 volts, the primary leakage-impedance voltage drop due to the exciting current being again neglected.

The change in magnitude of the secondary voltage when load is removed and the primary voltage is held constant, expressed as a fraction or a per cent of the magnitude of the secondary voltage under load, is by definition the *voltage regulation* of the transformer for the specified load.[1] That is, with V_1 held constant, by definition,

$$\text{Regulation} \equiv \frac{E_{oc2} - V_2}{V_2}, \qquad \blacktriangleright[2]$$

where E_{oc2} is the open-circuit secondary voltage and V_2 is the secondary terminal voltage under load. Unless otherwise specified, the load is taken as rated kilovolt-amperes at rated secondary voltage and at some specified power factor. A typical value of the power factor of commercial loads is about 0.8, lagging current. Thus at 0.80 power factor, lagging

[1] " American Engineering and Industrial Standards," No. C57: *Proposed American Standards; Transformers, Regulators, and Reactors* (New York: American Standards Association, 1940), paragraph 1.064, p. 11.

current, the voltage regulation of the transformer of Art. 7, Ch. XIII, is

$$\text{Regulation} = \frac{2{,}458 - 2{,}400}{2{,}400} \times 100 = +2.4\%. \tag{3}$$

The regulation computed in Eq. 3 is based on the equivalent circuit of Fig. 7a, Ch. XIII, p. 349, which neglects the primary leakage-impedance voltage drop due to the exciting current, and assumes that the series impedance of the transformer equals its equivalent impedance. However, the error introduced by these approximations is entirely negligible, as can be shown from the results of Arts. 5 and 7, Ch. XIII. In the simple solution of Art. 7, the exciting current is neglected, the primary current referred to the secondary is therefore assumed to equal the secondary load current, and the primary terminal voltage referred to the secondary is found to be $2{,}458\underline{/0.8°}$ volts. Actually the primary current referred to the secondary is the vector sum of the secondary load current and the exciting current I''_φ referred to the secondary. Thus the true value of the primary terminal voltage referred to the secondary is the vector sum of the value computed in Art. 7, Ch. XIII, and the additional primary leakage-impedance voltage drop due to the exciting current. That is,

$$\text{True} \frac{V_1}{a} = 2{,}458\underline{/0.8°} + \left(\frac{R_1}{a^2} + j\frac{X_{l1}}{a^2}\right)I''_\varphi. \tag{4}$$

From the solution of Art. 5, Ch. XIII, this true value is $2{,}459\underline{/0.8°}$.

If the load is removed and the primary terminal voltage is maintained at $2{,}459\underline{/0.8°}$ referred to the secondary, the induced voltage is the primary terminal voltage minus (vectorially) the primary leakage-impedance voltage drop due to the exciting current at no load. This induced voltage referred to the secondary equals the no-load terminal voltage of the secondary, or

$$E_{oc2} = \text{true} \frac{V_1}{a} - \left(\frac{R_1}{a^2} + j\frac{X_{l1}}{a^2}\right) I''_{nl} \tag{5}$$

$$= 2{,}458\underline{/0.8°} + \left(\frac{R_1}{a^2} + j\frac{X_{l1}}{a^2}\right)(I''_\varphi - I''_{nl}), \tag{6}$$

where I''_φ is the exciting current under load and I''_{nl} is the exciting current at no load, both these currents being referred to the secondary. Thus the true magnitude of the secondary terminal voltage at no load differs from the approximate value, 2,458 volts, computed in Art. 7, Ch. XIII, only by the effect on the magnitude produced by the small difference between the exciting currents under load and at no load. The approximate value of the voltage regulation given in Eq. 3 hence does not neglect even so small a quantity as the primary leakage-impedance voltage drop due to the exciting current, but neglects merely the second-order effect of the *change* in this small quantity.

▶ Therefore when the regulation is computed the exciting current is always neglected and the transformer is represented in simple fashion by its equivalent impedance.◀

As a matter of fact, when the equivalent impedance is determined by measurement of the short-circuit impedance on the secondary side, the voltage regulation thus computed is the "exact" value, since by Thévenin's theorem the open-circuit secondary voltage equals the secondary voltage under load plus the voltage drop due to the secondary current and the short-circuit impedance measured on the secondary side, as in the equivalent circuit of Fig. 9a, Ch. XIII, p. 356.

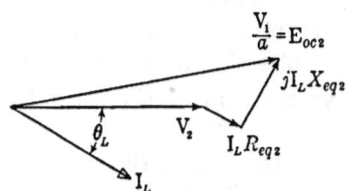

FIG. 1. Vector diagram referred to the secondary.

The vector diagram referred to the secondary is shown in Fig. 1. When the load and power factor are specified, V_2, I_L, and θ_L are known. If the equivalent resistance and equivalent reactance are known, and V_2 is the reference vector, the primary terminal voltage referred to the secondary is given vectorially by

$$\frac{V_1}{a} = V_2 + I_L(\cos \theta_L - j \sin \theta_L)(R_{eq2} + jX_{eq2})$$

$$= V_2 + I_L R_{eq2} \cos \theta_L + I_L X_{eq2} \sin \theta_L$$
$$+ j(I_L X_{eq2} \cos \theta_L - I_L R_{eq2} \sin \theta_L) \qquad [7]$$

$$= A + jB, \qquad [8]$$

where A is the real component and B is the imaginary component of V_1/a in Eq. 7; that is,

$$A = V_2 + I_L R_{eq2} \cos \theta_L + I_L X_{eq2} \sin \theta_L \qquad [9]$$

$$B = I_L X_{eq2} \cos \theta_L - I_L R_{eq2} \sin \theta_L. \qquad [10]$$

Note that θ_L is *positive for lagging current*. This assigned positive direction is convenient because the power factors of commercial loads usually are inductive, and therefore θ_L usually is a positive quantity.

Since B is small compared with A, the magnitude of V_1/a is, almost exactly,

$$\frac{V_1}{a} = \sqrt{A^2 + B^2} = A + \frac{B^2}{2A}.$$ [11]*

Since $B^2/2A$ is the effect of a small component perpendicular to the principal part of V_1/a, the error is entirely negligible if A in the denominator of this term is replaced by its approximate value V_2. Thus

$$\frac{V_1}{a} = A + \frac{B^2}{2V_2}.$$ [12]

The no-load exciting current being neglected, V_1/a equals the secondary terminal voltage E_{oc2} following removal of load. Hence, from Eq. 2,

$$\text{Regulation} = \frac{A + \dfrac{B^2}{2V_2} - V_2}{V_2} = \frac{A - V_2}{V_2} + \frac{1}{2}\left(\frac{B}{V_2}\right)^2.$$ [13]

Substituting Eqs. 9 and 10 in Eq. 13 gives

$$\text{Regulation} = \frac{I_L R_{eq2} \cos\theta_L + I_L X_{eq2} \sin\theta_L}{V_2}$$

$$+ \frac{1}{2}\left(\frac{I_L X_{eq2} \cos\theta_L - I_L R_{eq2} \sin\theta_L}{V_2}\right)^2.$$ ▶[14]

The per cent regulation is 100 times the value given by Eq. 14. The same result is obtained from Eq. 14 whether the high-voltage or the low-voltage winding is assumed to be the secondary. According to the American Engineering and Industrial Standards,[2] the equivalent resistance should be taken as its value at 75 degrees centigrade. The correction for temperature changes is discussed in Art. 3b.

With inductive loads, θ_L is positive and the voltage regulation is positive; that is, the secondary voltage increases when load is

Fig. 2. Vector diagram with leading current.

removed and the primary voltage is held constant. However, with capacitive loads (θ_L is negative and thus $\sin\theta_L$ is negative), V_1/a may be smaller than V_2, as shown in Fig. 2; the regulation may be negative and the secondary voltage may decrease when the load is removed.

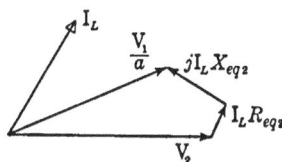

* This expression comes from the first two terms of the binomial expansion. See the footnote on p. 346.

[2] "American Engineering and Industrial Standards," No. C57: *Proposed American Standards; Transformers, Regulators, and Reactors* (New York: American Standards Association, 1940), paragraph 3.035, p. 25.

2. EFFICIENCY

The power efficiency of any machine is, by definition, the ratio of the useful power output to the power input. The efficiency may be determined by simultaneous measurement of the output and input; but, with large machinery especially, such a test is inconvenient and expensive to set up and perform. When the efficiency is high, furthermore, greater accuracy can be obtained if the efficiency is expressed in terms of the power losses; thus

$$\text{Efficiency} \equiv \frac{\text{output}}{\text{input}} = \frac{\text{input} - \text{losses}}{\text{input}} \qquad [15]$$

$$= 1 - \frac{\text{losses}}{\text{input}} = 1 - \frac{\text{losses}}{\text{output} + \text{losses}} \cdot \qquad \blacktriangleright[16]$$

If the losses can be determined, the efficiency can be computed by Eq. 16. When, as in most electrical machinery, the efficiency is high, it can be determined quite accurately in this manner. For example, suppose the total loss in a 100-kva transformer delivering exactly full load at unity power factor is 2.04 kw ± 2 per cent. The input is 102.04 ± 0.041 kw, and the loss as a fraction of the input is

$$\frac{\text{Loss}}{\text{Input}} = \frac{2.04}{102.0} \pm 2\% = 0.0200 \pm 0.0004.$$

Therefore, from Eq. 16,

$$\text{Efficiency} = 1 - 0.0200 \pm 0.0004 = 0.9800 \pm 0.0004.$$

▶ The per cent error in the efficiency resulting from a 2 per cent error in the loss is only 0.04 per cent. This, of course, is a far more accurate value than could be obtained by simultaneous measurement of the input and output. Note also the convenience of Eq. 16. This equation permits the use of a 10-inch slide rule to compute efficiencies above 90 per cent to four or more significant figures, as usually is justified by the accuracy with which the losses are known.◀

The losses in a transformer are the core loss due to the principal core flux, the copper losses in its windings, and the stray losses due to eddy currents induced by the leakage fluxes in the tank, clamps, and various parts of the structure.

2a. *Core Loss.* — In a loaded transformer, almost all the resultant mutual flux is confined to the core. However, since the return path for most of the leakage flux is in the core, the actual flux density in the core is not exactly that determined by the resultant mutual flux alone, as is explained in Art. 4, Ch. XII. Thus the leakage flux may increase the flux density in certain portions of the core while decreasing it in others.[3] It is probably quite accurate, however, to assume that on the average the core flux density is determined by the resultant mutual flux. On this basis the core loss under load equals the no-load core loss for the same induced voltage and frequency.

2b. *Load Loss.* — The losses due to the currents in the windings are the I^2R losses as with direct current plus extra losses caused by the eddy currents induced by the leakage fluxes. When these eddy currents are within the conductors themselves, they produce the phenomenon known as skin effect. When in various structural parts, they produce the stray losses. Since the leakage fluxes inducing these eddy currents are proportional to the currents in the windings, and since any eddy-current loss is proportional to the square of the flux producing it, these extra losses are proportional to the squares of the currents in the windings. Therefore the total loss due to the currents in the windings can be accounted for by use of effective alternating-current resistances greater than the direct-current resistances of the windings, in accordance with the assumptions of Art. 3, Ch. VI. Thus if R_1 and R_2 are the alternating-current effective resistances of the windings, the combined copper loss and stray loss, known collectively as the *load loss*, is $I_1^2R_1 + I_2^2R_2$. Since the copper losses in the windings are the principal components of the load loss, and since the load loss varies with the squares of the currents, as does a true copper loss, the load loss is often called the copper loss, and may be designated by the symbol P_{cu}, although part of the load loss is not associated directly with the copper. Note that the same loss is obtained when the currents and resistances are referred to one winding; thus if the turns ratio N_1/N_2 equals a,

$$I_1^2R_1 + I_2^2R_2 = I_1^2R_1 + \left(\frac{I_2}{a}\right)^2 a^2R_2 = (aI_1)^2\frac{R_1}{a^2} + I_2^2R_2. \quad [17]$$

3. Conventional efficiency

If the losses are determined, the efficiency for any load can be computed by Eq. 16. Since a relatively large error in the value of the losses produces a much smaller error in the computed efficiency, approximations can be made which greatly simplify both the calculations and the test procedure for measuring the losses, and yet cause very little error in the computed efficiency. For standardization, certain test procedures and methods of calculation have been agreed upon.[4] The effi-

---

[3] For further discussion of this point, see L. F. Blume, editor, *Transformer Engineering* (New York: John Wiley & Sons, 1938), 15-16, 67-68.

[4] "American Engineering and Industrial Standards," No. C57: *Proposed American Standards; Transformers, Regulators, and Reactors* (New York: American Standards Association, 1940), paragraphs 2.015 and 2.016, p. 16.

ciency computed in accordance with these standard methods is called the *conventional efficiency* and is sufficiently close to the true value for all practical purposes. A discussion of these standard methods follows.

3a. *Core Loss.* — As shown in Art. 2, the core loss under load very nearly equals the nc-load core loss for the same induced voltage and frequency. Since the exciting current at normal voltage and frequency usually is about 0.05 of the full-load current, the no-load copper loss due to the exciting current is about $(0.05)^2$ or 0.0025 of the full-load primary copper loss. The full-load copper losses in the primary and secondary are approximately equal; hence the no-load copper loss is but $\frac{1}{2} \times 0.0025$ of the total copper loss at full load. In a distribution transformer, the total copper loss at full load usually is approximately twice the core loss at rated voltage and frequency. At no load, therefore, the primary copper loss is approximately only $\frac{1}{2} \times 0.0025 \times 2$, or 0.0025 of the core loss, which is entirely negligible. Also at no load the primary leakage-impedance voltage drop due to the exciting current is approximately 0.05 of its value at rated current. Since this drop at rated current usually is about 0.03 of rated voltage, the primary leakage-impedance voltage drop due to the exciting current at no load is but 0.05×0.03 or 0.0015 of rated voltage and hence is negligible also. Thus the core loss under load is very nearly the no-load power input at a voltage and frequency equal to the induced voltage under load.

▶ Even at full load the induced voltage does not differ greatly from the terminal voltage. In order to simplify the tests and calculations, the core loss under load therefore is taken as the value corresponding to the *secondary terminal voltage* under load, rather than the probably more accurate value corresponding to the induced voltage.◀

For example, consider the 100-kva transformer of Art. 5, Ch. XIII. When it is delivering full load at 0.80 power factor, lagging current, the secondary terminal voltage is 2,400 volts, and the induced voltage, Eq. 23, Ch. XIII, p. 346, is 2,429 volts. Thus if the core loss under load is assumed to be 940 watts corresponding to the secondary *terminal* voltage of 2,400 volts, rather than the probably more accurate value of 963 watts corresponding to the *induced* voltage of 2,429 volts (see the data on p. 345), the error is 23 watts, or 2.4 per cent of the core loss. Since the power input under load is a little more than the 80-kw output, this error in the core loss is slightly less than 23/80,000 or 0.0003 of the input, or the efficiency is about 0.0003 too high. This error is so small that it is scarcely worth the extra time required to obtain the supposedly more accurate value.

For the conventional efficiency the flux is assumed to vary sinusoidally, and therefore the power source for the no-load test should have a sinusoidal electromotive force and relatively small internal impedance. The applied voltage should not be adjusted by means of a variable series impedance, since the harmonic voltage drops due to the harmonics in the exciting current would then distort the voltage waveform.

Since only the relatively small eddy-current component of the core loss varies appreciably with temperature changes, the core loss is commonly assumed to be independent of the temperature.

3b. *Load Loss.* — According to Eq. 17, the load loss due to the currents in the windings is the sum of the effective I^2R losses. If the exciting current is neglected, the primary and secondary currents are equal when referred to the same side, and the load loss is I^2R_{eq}, where I and R_{eq} are referred to the same side.

The error produced by neglecting the exciting current may be estimated for the typical transformer of Art. 5, Ch. XIII. Thus, when the exciting current is included in the primary current, the primary and secondary currents determined in that article are 8.60 amperes and 41.7 amperes, and the load loss is

$$(8.60)^2 \times 7.0 + (41.7)^2 \times 0.30 = 1,040 \text{ watts,}$$

the primary and secondary effective resistances being 7.0 and 0.30 ohms. The equivalent resistance referred to the secondary is 0.58 ohm. Hence, if the exciting current is neglected, the load loss is

$$(41.7)^2 \times 0.58 = 1,010 \text{ watts.}$$

The error in the primary copper loss caused if the exciting current is neglected is about 30 watts. The input is slightly more than 80 kw. Hence the error is about 30/80,000 or 0.0004 of the input. Therefore neglecting the exciting current causes the computed efficiency to be about 0.0004 too high. This error is neglected in computing the conventional efficiency and the load loss is assumed to equal I^2R_{eq}.

▶ As shown in Art. 8b, Ch. XIII, the power input on short circuit is I^2R_{eq}, the exciting current and core loss on short circuit being very small indeed. Thus to compute the conventional efficiency, the load loss is taken as the power input on short circuit for the same current.◀

The load loss should be corrected to its value at 75 degrees centigrade, which is assumed to be the normal operating temperature under continuous full load. Thus if R_{dc1} and R_{dc2} are the direct-current resistances of the windings at the temperature, θ degrees centigrade, of the windings during the short-circuit test, the component of the equivalent resistance at temperature θ due to the direct-current resistances of the windings is $R_{dc1} + a^2R_{dc2}$ referred to the primary, or $(R_{dc1}/a^2) + R_{dc2}$ referred to the secondary. If $R_{dc(\theta)}$ is this equivalent resistance referred to whichever side is convenient, and $R_{eq(\theta)}$ is the effective alternating-current value of the equivalent resistance referred to the same side, the component of the equivalent resistance due to the eddy-current losses in the conductors and in various structural parts is $R_{eq(\theta)} - R_{dc(\theta)}$. The direct-current resistances increase with increasing temperature, and the equivalent direct-current resistance at 75 degrees centigrade is

$$R_{dc(75)} = R_{dc(\theta)} \frac{234.5 + 75}{234.5 + \theta} . \qquad [18]$$

On the other hand, the eddy-current losses decrease with increasing temperature, since they are inversely proportional to the resistivities of the materials in which they exist. Some of these eddy currents are in the copper conductors themselves, whereas others are in the iron and steel structural parts. If the operating temperature of all the structural parts were 75 degrees centigrade and the temperature coefficient of resistance of all the parts were the same as that of copper,* the eddy-current component of the equivalent resistance at 75 degrees centigrade would be

$$[R_{eq(\theta)} - R_{dc(\theta)}] \frac{234.5 + \theta}{234.5 + 75} . \qquad [19]$$

Since the stray losses are a relatively small part of the load loss this temperature correction is considered sufficiently accurate. Thus the equivalent resistance at 75 degrees centigrade is

$$R_{eq(75)} = R_{dc(\theta)} \frac{309.5}{234.5 + \theta} + [R_{eq(\theta)} - R_{dc(\theta)}] \frac{234.5 + \theta}{309.5} . \qquad [20]$$

4. ILLUSTRATIVE EXAMPLE OF REGULATION AND EFFICIENCY COMPUTATIONS

Compute the regulation and efficiency at full load, 0.80 power factor, lagging current, of the 15-kva, 2,400 : 240-volt, 60 $\sim$ distribution transformer to which the following data apply. (Subscript H means high-voltage, subscript X means low-voltage winding.)

Short-circuit test	Open-circuit test
$V_H = 74.5$ v	$V_X = 240$ v
$I_H = 6.25$ amp	$I_X = 1.70$ amp
$P_H = 237$ watts	$P_X = 84$ watts
Frequency = 60 $\sim$	Frequency = 60 $\sim$
$\theta = 25$ C	

Direct-current resistances measured at 25 C

$$R_{dcH} = 2.80 \text{ ohms} \qquad R_{dcX} = 0.0276 \text{ ohm}$$

* The temperature coefficient of resistance of iron and steel varies considerably with composition and heat treatment but is about 0.003 at 20 degrees centigrade, whereas that of copper is 0.00393.

The data given above have been corrected for instrument losses where this correction was necessary.

Solution: The rated current of the high-voltage winding is

$$I_H = \frac{15,000}{2,400} = 6.25 \text{ amp.} \tag{21}$$

The short-circuit test was taken at rated current, as is usually done.
From the short-circuit data,

$$R_{eqH} = \frac{P_H}{I_H^2} = \frac{237}{(6.25)^2} = 6.07 \text{ ohms at 25 C.} \tag{22}$$

For computation of the equivalent resistance at 75 C, the direct-current component of the equivalent resistance must be determined. The ratio of transformation is 10, and from the direct-current resistances

$$R_{dcH} + a^2 R_{dcX} = 2.80 + 2.76 = 5.56 \text{ ohms at 25 C.}$$

The temperature correction factor $309.5/(234.5 + \theta)$ in Eq. 20 is $309.5/259.5 = 1.19$. Therefore, from Eq. 20, at 75 C

$$R_{eqH(75)} = 5.56 \times 1.19 + (6.07 - 5.56)\frac{1}{1.19} = 7.05 \text{ ohms.} \tag{23}$$

The conventional efficiency now can be determined from Eq. 16, as follows:

$$
\begin{aligned}
I^2 R_{eqH(75)} &= (6.25)^2\, 7.05 = &276 \text{ watts} \\
\text{Core loss} &= &\underline{84} \\
\text{Total loss} = \text{sum} &= &360 \\
\text{Output} = 0.80 \times 15,000 &= &12,000 \\
\text{Input} = \text{sum} &= &12,360
\end{aligned}
$$

$$\frac{\text{Losses}}{\text{Input}} = \frac{360}{12,360} = 0.0291.$$

$$\text{Efficiency} = 1 - 0.0291 = 0.9709 \tag{24}$$

Computing the voltage regulation requires that the equivalent reactance be found. From the short-circuit data,

$$Z_{eqH} = \frac{V_H}{I_H} = \frac{74.5}{6.25} = 11.92 \text{ ohms} \tag{25}$$

$$X_{eqH} = \sqrt{Z_{eqH}^2 - R_{eqH}^2}. \tag{26}$$

Note that the value of R_{eqH} at the temperature of the test should be used in determining X_{eqH}. Hence

$$X_{eqH} = \sqrt{11.92^2 - 6.07^2} = 10.27 \text{ ohms.} \tag{27}$$

Since the constants are referred to the high-voltage side, this winding may be considered the secondary when the regulation is computed. Thus in Eq. 14, p. 371,

$$
\begin{aligned}
\cos \theta_L &= 0.800 & \sin \theta_L &= +0.600 \\
I_L R_{eq2} &= 6.25 \times 7.05 & I_L X_{eq2} &= 6.25 \times 10.27 \\
&= 44.1 & &= 64.2.
\end{aligned}
$$

Simplifying the notation, in Eq. 14,

$$IR \cos \theta = 35.3 \qquad\qquad IX \cos \theta = 51.4$$
$$IX \sin \theta = 38.5 \qquad\qquad IR \sin \theta = 26.4$$

$$\text{Sum} = \overline{73.8} \qquad\qquad \text{Difference} = \overline{25.0}$$

$$\frac{73.8}{2,400} = 0.0308 \qquad\qquad \frac{1}{2}\left(\frac{25.0}{2,400}\right)^2 = \text{negligible.}$$

Hence,

$$\text{Regulation} = 0.0308, \text{ or } 3.08 \%. \qquad\qquad [28]$$

5. ENERGY EFFICIENCY; ILLUSTRATIVE EXAMPLE

Although the efficiency of power transfer at full load is an important figure of merit, of even greater importance is the efficiency of energy transfer; that is, the ratio of kilowatthours output to kilowatthours input. This energy efficiency depends not only on the characteristics of the transformer but also on the load cycle.

Since, if the voltage is constant, the core loss is very nearly constant, the energy in kilowatthours consumed daily by the core loss equals the

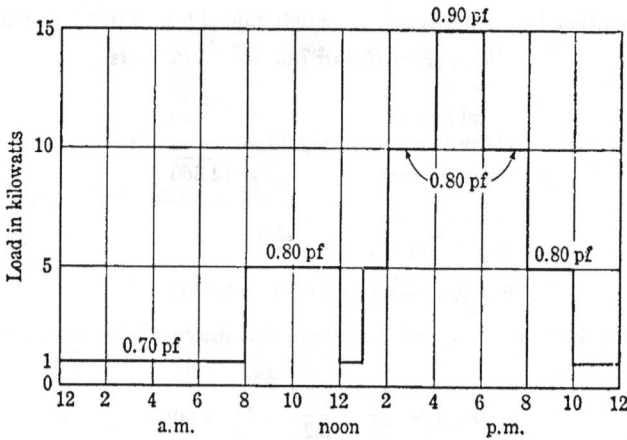

FIG. 3. Simplified daily load curve.

core loss in kilowatts multiplied by the number of hours the transformer is energized daily. The load loss, however, depends on the square of the load current, and therefore is much more difficult to compute, since the load usually fluctuates. The computation of the energy consumed in load loss is discussed in detail in Art. 3, Ch. XVI. If, however, the daily load curve may be considered a step function comprising a number of constant-load periods, as in Fig. 3, the computations are relatively simple.

For example, let it be required to determine the energy efficiency of the 15-kva transformer of Art. 4 operating on the load cycle shown in Fig. 3.

Solution: Examination of Fig. 3 shows that the daily load cycle can be summarized as follows:

Hours	Kilowatts output	Power factor
2	15	0.90
4	10	0.80
7	5	0.80
11	1	0.70

The losses are the constant core loss, which must be supplied continuously, and the load loss, which varies with the square of the current. Thus the energy dissipated in core loss is

$$\frac{24 \times 84}{1,000} = 2.02 \text{ kwhr daily.} \qquad [29]$$

The daily output and load loss can be computed and, for convenience, arranged as in the table below. For computing the load loss the equivalent resistance is taken as the average of its hot and cold values; that is, from Eqs. 22 and 23,

$$R_{eqH} = \frac{6.07 + 7.05}{2} = 6.56 \text{ ohms.} \qquad [30]$$

Hours	Output				I_H	$I_H^2 R_{eqH}$ Kw	Load loss Kwhr
	Kw	Kwhr	P.f.	Kva			
2	15	30	0.90	16.67	6.94	0.316	0.632
4	10	40	0.80	12.5	5.20	0.177	0.709
7	5	35	0.80	6.25	2.60	0.0443	0.310
11	1	11	0.70	1.43	0.595	0.0023	0.025

Total output = 116 kwhr   Total load loss = 1.676 kwhr

The energy efficiency now can be computed from the energy output and losses, as follows:

$$
\begin{aligned}
\text{Total load loss} &= \quad 1.68 \text{ kwhr daily} \\
\text{Total core loss} &= \quad 2.02 \\
\hline
\text{Total loss} &= \quad 3.70 \\
\text{Output} &= 116.00 \\
\hline
\text{Input} &= 119.70
\end{aligned}
$$

Hence

$$\text{Energy efficiency} = 1 - \frac{\text{losses}}{\text{input}} = 1 - \frac{3.70}{119.7} = 0.9691. \qquad [31]$$

6. PER CENT AND PER UNIT QUANTITIES

In the discussion of power apparatus, it is common practice to express the values of the currents and voltages as a per cent or a fraction (per unit) of their nominal full-load values.* The value of the current or voltage expressed as a per cent or a fraction of this base value is defined as the *per cent or per unit value* of the current or voltage.

Frequently this method of expressing the values of currents and voltages conveys more information about the apparatus than does the expression of the currents and voltages in amperes and volts. Thus in Art. 4 it is stated that the exciting current of a 15-kva, 2,400:240-volt distribution transformer is 1.70 amperes, measured on the low-voltage side at rated voltage. To determine the importance of this exciting current, it is necessary to express the current as a fraction or a percentage of the rated current. Thus the rated current of the low-voltage winding is

$$\frac{15,000}{240} = 62.5 \text{ amp.} \tag{32}$$

Hence the exciting current as a fraction of rated current is

$$\frac{1.70}{62.5} = 0.0272. \tag{33}$$

To state that the exciting current is 0.0272 per unit, or 2.72 per cent, immediately conveys information regarding its relative importance in affecting the characteristics of the transformer. There are similar advantages in expressing voltages in per unit or per cent. For example, if the equivalent resistance drop referred to the high-voltage winding of this transformer is 44.1 volts at rated current, this voltage drop is

$$\frac{44.1}{2,400} = 0.0184 \text{ per unit, or } 1.84\%. \tag{34}$$

More information is given regarding the relative importance of the resistance drop by the statement of its per unit or per cent value than by the statement of its value in volts.

Similarly impedances may be expressed conveniently in per unit or per cent values. Thus the normal or base value of impedance is the number of ohms given by the ratio of normal voltage in volts to normal current in amperes, and the value of an impedance expressed as a fraction or per-

---

* The *normal or base values* of the currents and voltages need not be the rated full-load values. In the analysis of power systems involving the interconnection of a number of generators, transformers, and other machines, with various ratings, any convenient base values may be chosen. However, when dealing with a single piece of apparatus, such as a transformer, it is convenient to choose the rated full-load values of current and voltage as the base values.

centage of the normal impedance is defined as the *per unit or per cent value* of the impedance. It can be stated also that the per unit voltage drop at normal current equals the per unit value of the impedance. For example, if the equivalent resistance referred to the high-voltage winding of the 15-kva transformer of Art. 4 is 7.05 ohms, the resistance drop at rated current is

$$6.25 \times 7.05 = 44.1 \text{ v.} \tag{35}$$

Hence the per unit or per cent value of the resistance is given by Eq. 34. At normal current, per unit impedance and per unit impedance-voltage drop are numerically the same.

Admittance may also be expressed in per unit or per cent values. Thus normal admittance is the ratio of normal current in amperes to normal voltage in volts, and the *per unit value* of an admittance is defined as the ratio of its value to the normal value. It also may be said that the per unit current resulting when normal voltage is applied to a circuit element equals the *per unit value* of its admittance. For example, if the magnitude of the exciting admittance of this 15-kva transformer is 0.00708 mho measured on its low-voltage side, the exciting current at normal voltage is

$$240 \times 0.00708 = 1.70 \text{ amp.} \tag{36}$$

Thus the per unit value of the exciting admittance is given by Eq. 33. At normal voltage, per unit admittance and the per unit current taken by the admittance are numerically the same.

Obviously power and volt-amperes may also be expressed in per unit or per cent. For example, if the copper loss in this 15-kva transformer is 276 watts at rated current, it is

$$\frac{276}{15,000} = 0.0184, \text{ or } 1.84\% \tag{37}$$

of the base volt-amperes. Note that at rated current the per unit copper loss, the per unit equivalent resistance drop, and the per unit equivalent resistance are numerically equal.

The relations between the per unit quantities and the volt-ampere-ohm units can be summarized as follows:

$$\text{Per unit current} = \frac{\text{current in amperes}}{\text{base amperes}} \qquad \blacktriangleright[38]$$

$$\text{Per unit voltage} = \frac{\text{voltage in volts}}{\text{base volts}} \qquad \blacktriangleright[39]$$

$$\text{Per unit impedance} = \frac{\text{impedance in ohms}}{\text{base ohms}} \qquad \blacktriangleright [40]$$

$$= \frac{(\text{impedance in ohms})(\text{base amperes})}{\text{base volts}} \qquad [41]$$

$$\text{Per unit admittance} = \frac{\text{admittance in mhos}}{\text{base mhos}} \qquad \blacktriangleright [42]$$

$$= \frac{(\text{admittance in mhos})(\text{base volts})}{\text{base amperes}} \cdot \quad [43]$$

$$\text{Per unit power} = \frac{\text{power in kilowatts}}{\text{base kilovolt-amperes}} \cdot \qquad \blacktriangleright [44]$$

An important advantage of the per unit or per cent system of units is that both large and small transformers of widely different voltage ratings are rather similar when their constants are expressed in per unit or per cent on their own ratings as bases. Thus, although the exciting current of a 2-kva, 2,400-volt distribution transformer is less than 0.1 ampere, whereas the exciting current of a 20,000-kva, 13,800-volt power transformer is about 70 amperes, these exciting currents, when expressed in per unit or per cent, are of the same order of magnitude. Thus it is easy to remember a few per unit or per cent values of important quantities, and these values are approximately correct for both large and small apparatus of normal design. Thus the exciting current of any 60 ∼ power or distribution transformer usually is 4 to 8 per cent, the core loss usually is 0.5 to 2 per cent, the equivalent resistance (that is, the full-load copper loss) usually is 0.5 to 2 per cent, and the equivalent reactance (that is, the full-load reactance drop) usually is 3 to 10 per cent. In general, the per cent exciting current, per cent core loss, and per cent resistance decrease with increasing size, and the per cent reactance increases with increasing voltage.

When the currents, voltages, impedances, and admittances of a transformer are all expressed in per cent or per unit, the transformer can be analyzed as a one-to-one-ratio transformer without the use of any reduction factors to refer primary quantities to the secondary, or vice versa; that is, the per unit value of a current, voltage, impedance, or admittance measured on one side also equals its per unit value referred to the other side. For example, if the secondary induced voltage is E_2 volts, this voltage referred to the primary equals aE_2 volts, a being the turns ratio N_1/N_2. However, since the rated voltage of the primary equals a times the rated secondary voltage, aE_2 volts is the same percentage of rated primary voltage as E_2 volts is of rated secondary voltage. Therefore the per unit value of the secondary voltage also equals its per unit value referred to the primary. The turns ratio which refers voltages on one side to the other side is taken into account in the per cent or per unit system by the different values of the primary and secondary base voltages. Similarly, if the current on the secondary side equals I_2 amperes, this current referred to the primary equals I_2/a amperes. But the ampere rating of the primary equals $1/a$ times the ampere rating of the secondary. Thus I_2/a amperes is the same percentage of the rated primary current as I_2 amperes is of the rated secondary current, or the per unit value of a secondary current also equals its per unit value referred to the primary. Similarly the per unit value of an impedance or admittance referred to the secondary side also equals its per unit value referred to the primary.

There are certain other computational advantages in the use of per unit or per cent values. For example, at rated current the per unit values of resistance R, of resistance drop IR, and of copper loss I^2R are the same numerically. Also the per unit values of reactance X, of reactance drop IX, and of reactive volt-amperes I^2X are the same numerically. Similarly at rated voltage the per unit values of admittance Y, of the current VY, and of the volt-amperes V^2Y taken by an admittance are numerically equal. Since computations are often made at rated current and voltage, these simplifications permit some saving of time. For example, consider Eq. 14 expressing the regulation in terms of the equivalent resistance and reactance drops, the voltage, and the power factor. Note that at rated voltage $I_L R_{eq2}/V_2$ and $I_L X_{eq2}/V_2$ are simply the per unit values of the equivalent resistance and reactance drops. Hence when these drops are expressed in per unit, Eq. 14 reduces to

$$\text{Regulation} = IR_{eq}\cos\theta_L + IX_{eq}\sin\theta_L + \tfrac{1}{2}(IX_{eq}\cos\theta_L - IR_{eq}\sin\theta_L)^2. \ [45]$$

where IR_{eq} and IX_{eq} are the per unit equivalent-resistance and equivalent-reactance voltage drops. At rated current, these voltage drops equal the per unit values of the equivalent resistance and equivalent reactance.

In a three-phase circuit, certain simplifications can be made in the computations by the use of per unit quantities.* In such calculations the factor $\sqrt{3}$ occurs frequently in conversions from line-to-line to line-to-neutral quantities when the computations are carried out in units of volts and amperes. This factor can be omitted from three-phase computations when the per unit system of units is used. For example, if the per unit value of the voltage to neutral of a balanced, Y-connected, three-phase system is 1.1 times normal (or per unit), the line-to-line voltage is also 1.1 times its normal value (or per unit). The normal or base value of the line-to-line voltage, however, equals $\sqrt{3}$ times the

* The use of per unit quantities in three-phase problems is discussed in Art. 2b, Ch. XXII.

normal value of the line-to-neutral voltage. That is, the factor of $\sqrt{3}$ is absorbed by the different volt values of the base voltage for line-to-line and line-to-neutral voltages.

7. ILLUSTRATIVE EXAMPLE OF THE PER UNIT SYSTEM

A 100-kva, 12,000 : 2,400-volt, 60 $\sim$ distribution transformer is connected to a single-phase bus of voltage V_B (nominally 12,000 volts)

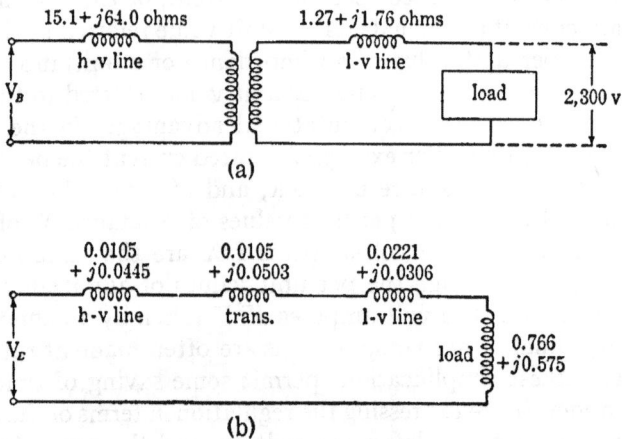

FIG. 4.   Circuit for the illustrative example of Art. 7. (a) Circuit diagram.
(b) Equivalent circuit on a per unit basis.

through a high-voltage line of impedance $15.1 + j64.0$ ohms, and supplies power to an inductive load through a low-voltage line of impedance $1.27 + j1.76$ ohms, as shown in Fig. 4.

Determine the bus voltage V_B necessary to maintain 2,300 volts at the load when the load takes the rated current of the transformer at 0.80 power factor.

The data obtained from a short-circuit test in which the measurements were made on the low-voltage side of the transformer are:

$$V = 123 \text{ volts}$$

$$I = 41.7 \text{ amperes}$$

$$P = 1.05 \text{ kw.}$$

Solution: This problem is conveniently solved by use of per unit quantities. Accordingly, the first step is to express all circuit constants in per unit on the rating of the transformer as a base. If subscripts H and X indicate the high- and low-voltage circuits,

Base $V_H = 12,000$ v.   Base $V_X = 2,400$ v.

$$\text{Base } I_H = \frac{100,000}{12,000} \qquad\qquad \text{Base } I_X = \frac{100,000}{2,400}$$

$$= 8.34 \text{ amp.} \qquad\qquad\qquad = 41.7 \text{ amp.}$$

$$\text{Base } Z_H = \frac{12,000}{8.34} \qquad\qquad \text{Base } Z_X = \frac{2,400}{41.7}$$

$$= 1,440 \text{ ohms.} \qquad\qquad\qquad = 57.5 \text{ ohms.}$$

From the data and the values of the base impedances, the per unit value of the high-voltage line impedance is

$$\frac{15.1 + j64.0}{1,440} = 0.0105 + j0.0445 \text{ per unit impedance,} \tag{46}$$

and of the low-voltage line impedance is

$$\frac{1.27 + j1.76}{57.5} = 0.0221 + j0.0306 \text{ per unit impedance.} \tag{47}$$

Converting the short-circuit data of the transformer to per unit:

$$V = \frac{123}{2,400} = 0.0513 \text{ per unit voltage}$$

$$I = 1.00 \text{ per unit current}$$

$$P = \frac{1.05}{100} = 0.0105 \text{ per unit power.}$$

Since the voltage is the impedance drop at normal current, its per unit value equals the per unit equivalent impedance, and also, since the power is the copper loss at normal current, its per unit value equals the per unit equivalent resistance. Hence the per unit equivalent reactance of the transformer is

$$\sqrt{0.0513^2 - 0.0105^2} = 0.0503 \text{ per unit impedance.} \tag{48}$$

The per unit equivalent impedance of the transformer is

$$Z_{eq} = 0.0105 + j0.0503 \text{ per unit impedance.} \tag{49}$$

The per unit load voltage is

$$\frac{2,300}{2,400} = 0.958. \tag{50}$$

Since the load current is the rated value, the load is equivalent to an impedance whose per unit magnitude equals the per unit load voltage. Since the power factor is 0.800 inductive, the per unit vector load impedance is

$$0.958 (0.800 + j0.600) = 0.766 + j0.575 \text{ per unit impedance.} \tag{51}$$

Since the high-voltage line impedance is only about 0.05 per unit (Eq. 46), and a reasonable value of the exciting current of the transformer (not given) is about 0.05 per unit, the magnitude of the impedance drop in the high-voltage line due to the exciting current is about 0.05 × 0.05 or 0.0025 per unit voltage. Therefore negligible error results if the exciting current is neglected and the transformer is represented by its equivalent impedance in series with the feeders, as in Fig. 4b.

The impedance of the whole circuit viewed from the high-voltage bus is the sum of the four components, Eqs. 46, 47, 49, and 51, and is

$$0.809 + j0.700 = 1.07 \underline{/0} \text{ per unit impedance.}$$

Since the current is the normal value, this per unit impedance also equals the per unit sending-end bus voltage. Hence the bus voltage V_B is

$$V_B = 1.07 \text{ per unit voltage}$$
$$= 1.07 \times 12,000 \text{ or } 12,850 \text{ v.} \qquad [52]$$

8. RATING OF POWER AND DISTRIBUTION TRANSFORMERS

The operating conditions which a transformer is designed to meet compose its *rating*. For a power or distribution transformer, these conditions are the normal frequency, primary and secondary effective voltages, and kilovolt-ampere capacity. In common engineering parlance, the one value most frequently referred to as the rating is the kilovolt-ampere load capacity. The load rating of a transformer is limited by the temperature developed in the insulating materials from heat generated by the transformer losses. The nameplate rating of most types of transformers gives the load the transformer can carry continuously without damage or a shortening of life at altitudes not exceeding 3,300 feet with freely circulating surrounding air having a maximum temperature of 40 degrees centigrade and 24-hour average temperature of 30 degrees centigrade. Excessively high insulation temperatures, resulting from serious overloads, cause deterioration of the insulating materials and shorten the life of the transformer.* On the other hand, if a transformer is overloaded for short periods, the specific heat of its metal parts supplies considerable heat-absorbing capacity, so that temporary overloads do not produce so high a winding temperature as would ultimately be reached if the overload were steadily continued.† The American Standards Association has tentatively approved the curves shown in Figs. 5 and 6 for the short-time overload multipliers which are applicable to the continuous load ratings of transformers manufactured in 1939 or later. If a transformer is operated in surrounding air having a temperature less than 30 degrees centigrade, the following increases in load are permissible, according to the Proposed A.S.A. Standards:[5]

(1) Oil-immersed self-cooled transformers may be loaded continuously one per cent of [above] rated kva for each degree Centigrade that the daily average temperature of the cooling medium (air) is below 30 C.

* See Art. 1, Ch. VIII.
† See Art. 6, Ch. VIII.
[5] " American Engineering and Industrial Standards," No. C57: *Proposed American Standards: Transformers, Regulators, and Reactors* (New York: American Standards Association, 1940), paragraph 11.004(c), p. 87.

(2) An oil-immersed self-cooled, forced-air-cooled transformer may be loaded continuously above its forced-air-cooled rated kva one per cent of its self-cooled rated kva for each degree Centigrade that the daily average temperature of the cooling medium (air) is below 30 C. For example, if the self-cooled rating is 10,000 kva and the forced-air-cooled rating is 13,333 kva, the permissible continuous loading with a daily average air temperature of 0 C would be 13,333 + (0.30 × 10,000) = 16,333 kva with forced-air cooling. This increase amounts to 0.75 per cent of its forced-air-cooled rating for each degree Centigrade that the ambient is below 30 C.

(3) Oil-immersed water-cooled transformers may be loaded continuously one per cent above rated kva for each degree Centigrade that the daily average temperature of the cooling medium (water) is below 25 C.

(4) Continuous loads greater than 130 per cent of rated kva for self-cooled transformers, even though the temperature of the air be lower than 0 C, and continuous loads greater than 125 per cent of rated kva for water-cooled transformers, are not recommended.

The effect of the temperature of the air surrounding the transformer, technically known as *ambient temperature*, is an important element in the operation of transformers. In the northern part of the United States, the peak loads on power transformers generally occur during the winter months. It is then possible to take advantage of the added load capacity obtained in transformers operating in the open air, and thus to keep the investment in transformers lower than would be possible in a warmer climate. During the Christmas season, many power transformers carry extraordinary evening peak loads for a few hours each day for about a week. If it were necessary to install larger transformers at all points to provide capacity for these severe Christmas peak loads, the expense involved would be inordinately high. Of course, some special transformer capacity must be installed, as for decorative lighting of main business thoroughfares, or for residential zones where unusually elaborate Christmas lighting displays are installed.

In many operating utility companies, the trend is toward the practice of overloading distribution transformers during short peak-load periods, within the limits of the curves of Figs. 5 and 6. As an example, one company reports the following improvement in economical use of transformer capacity connected to certain distribution circuits, over a three-year period:

Year	Distribution transformer capacity installed, kva	Aggregate peak kva demand of circuits involved	Utilization factor (col 3/col 2)
1932	74,782	41,175	0.55
1935	70,415	47,224	0.67

In making improvements in transformer utilization, the company's

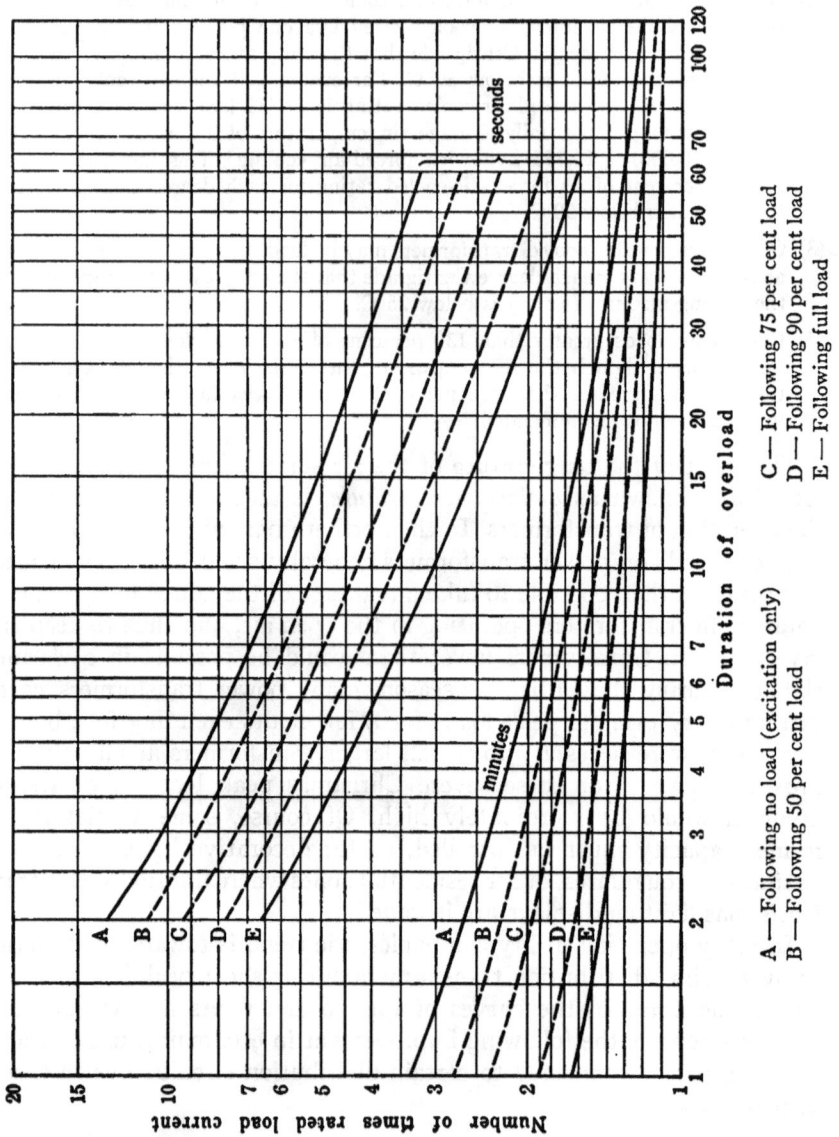

FIG. 5.   Recurrent short-time overloads for oil-immersed transformers larger than 100 kva, self-cooled with 30 C cooling air, or water-cooled with 25 C cooling water. For forced-air-cooled oil-immersed transformers, use 75 per cent of the increase over rated load current shown by these curves. (A recurrent short-time overload is one imposed occasionally, at intervals not less than approximately once every twenty-four hours.)

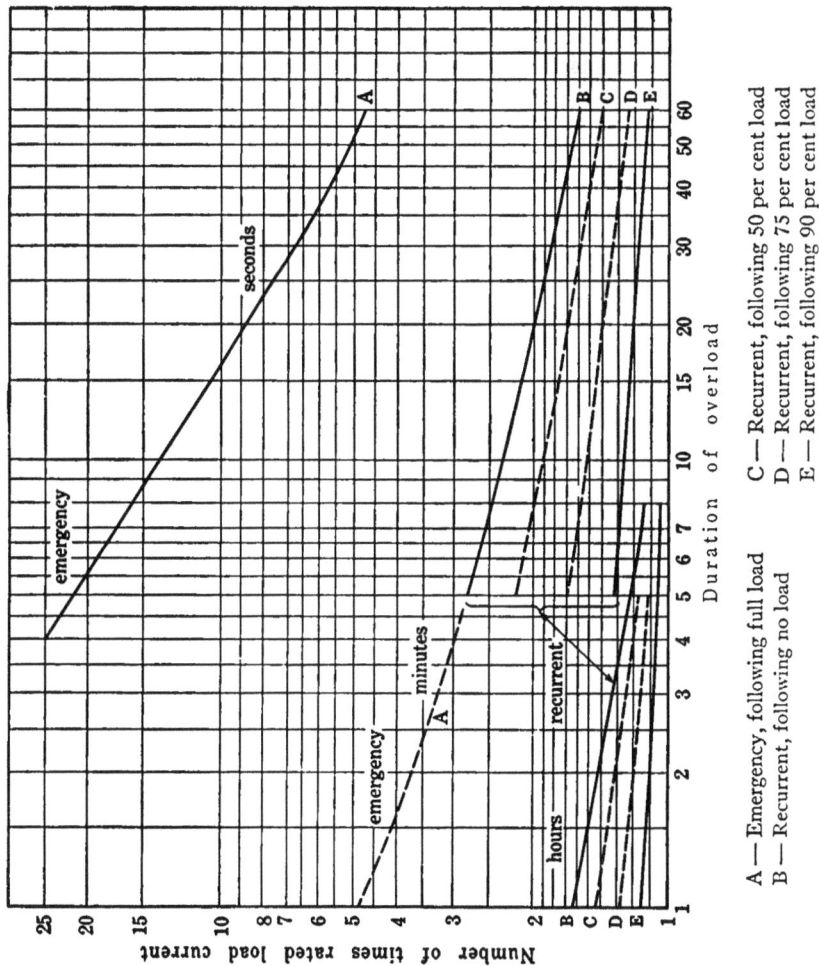

FIG. 6. Short-time overloads for oil-immersed transformers 100 kva and smaller, below 15,000 volts; and for network transformers 500 kva and smaller, self-cooled with 30 C cooling air, or water-cooled with 25 C cooling water. For forced-air-cooled oil-immersed transformers, use 75 per cent of the increase over rated load current shown by these curves. (A recurrent short-time overload is one imposed occasionally, at intervals not less than approximately once every twenty-four hours.)

engineers expect some of the transformers to be overloaded sufficiently so that occasional failures will occur. It is believed to be good economy to have some failures from overload rather than to avoid such failures by operating many transformers with spare capacity.

Two considerations exert a negative influence on the desirability of overloading transformers: voltage regulation and future load growth. Since the impedance voltage drop in a transformer increases with increasing load, the allowable peak load may be limited by voltage drop as well as by heating. During peak loads, the voltage drops in secondary conductors and building wiring are a maximum; hence the simultaneous maximum transformer voltage drop may be a limitation on the quality of service. Concerning load growth, any reasonably certain increase in load that is anticipated within a period of about two years should be considered in the selection of a transformer. For example, in an area where electric ranges and other new appliances are being adopted rapidly, good judgment indicates that a fair margin of transformer capacity should be installed to meet the load increase expected in the near future.

PROBLEMS

1. Show that the full-load regulation formula, Eq. 14, can be put in the form:

$$\text{Regulation} = \frac{V_{sc}}{V_2} \cos (\theta_{sc} - \theta_L) + \frac{1}{2} \left[\frac{V_{sc}}{V_2} \sin (\theta_{sc} - \theta_L) \right]^2, \qquad [53]$$

where V_{sc} is the secondary terminal voltage corresponding to rated current with the primary short-circuited, V_2 is the rated secondary terminal voltage, θ_L is the power factor angle of the load (considered positive for lagging current), and $\theta_{sc} = \cos^{-1} \dfrac{P_{sc}}{V_{sc} I_{sc}}$, where P_{sc} and I_{sc} are respectively the power input and current at the short-circuit condition specified.

2. State whether it is true that the per cent voltage regulation of a transformer is exactly the same, regardless of which winding is the secondary. Prove your conclusion.

3. A 5,000-kva 60 ∼ 69,200 : 19,020-v transformer when operated at rated frequency and rated secondary voltage (19,020 v) with the primary winding open-circuited takes 14,000 w and 3.65 amp. When operated at rated frequency and rated primary current with the secondary winding short-circuited, the transformer takes 32,000 w at 5,220 v.

(a) Compute the per cent voltage regulation of this transformer for an inductive load of 4,600 kw at rated secondary voltage and 0.91 power factor.

(b) Compute the efficiency of the transformer when supplying the load specified in (a). Neglect changes in load loss with temperature.

4. The following data were obtained for a 10-kva 60 ∼ 2,400 : 240-v distribution transformer, tested at 25 C.

	Frequency	*Voltage*	*Current*	*Power*
With high-voltage windings open-circuited	60 ∼	240 v	0.533 amp	63.3 w
With low-voltage terminals short-circuited	60	57.7	4.17	142

$$R_{dc1} = 3.68 \text{ ohms.}$$
$$R_{dc2} = 0.0428 \text{ ohm.}$$

(a) Plot the per cent voltage regulation of this transformer as a function of power factor at rated voltage and kva output for values of power factor ranging from 0.5 lagging to 0.5 leading.

(b) Plot the conventional efficiency of this transformer at rated kva output as a function of power factor for the conditions specified in part (a).

5. A 1,000-kva 60 ∼ 66,000 : 11,000-v transformer has an equivalent impedance of $1.0 + j4.9$ ohms, referred to the low-voltage side, and a no-load loss of 5,500 w at rated terminal voltage.

(a) When this transformer is operated at rated frequency and rated secondary voltage, at what kva output is the efficiency a maximum?

(b) What is the value of the maximum efficiency?

(c) If the transformer delivers its rated kva at rated voltage to a load of variable power factor, for what load power factor is the voltage regulation zero?

6. The transformer for which data are given in Prob. 5 operates at rated secondary voltage on a load cycle represented by the following typical values:

Hours	*Per cent kva load*	*Power factor*
1	125	0.80
2	100	0.84
1	50	0.90
5	12	0.98
15	0	—

(a) What is the energy efficiency of the transformer?

(b) What would be the energy efficiency if the equivalent resistance were 150% of the value given in Prob. 5?

7. The average load demand for a 24-hour period of a 500-kva 60 ∼ 11,000 : 2,300-v single-phase transformer is shown in Fig. 7. When the transformer is tested at open circuit with an applied secondary voltage of 2,300 v at rated frequency, the power input is 2,960 w. When it is tested at short circuit with an applied primary voltage of 345 v at rated frequency, the power input is 3,370 w and the current is 45.5 amp.

Determine the energy efficiency of the transformer, assuming the secondary voltage is maintained at its rated value.

8. A $\frac{1}{2}$-kva 60 ∼ 220 : 110-v transformer has a primary resistance of 0.970 ohm, a primary leakage inductance of 12.8 mh, and an exciting current at normal voltage of 0.100 per unit at 0.100 power factor. The primary is the higher voltage winding. A test with the primary short-circuited gives 0.02 per unit power at 11 v with 100% current.

(a) While the transformer is operating under an unknown inductive load, the primary voltage, current, and power are respectively 200 v, 3 amp, and 500 w. Find the secondary terminal voltage, current, and load power factor.

(b) Find the regulation for full load at unity power factor and rated secondary voltage.

(c) Find the efficiency for the conditions of (b).

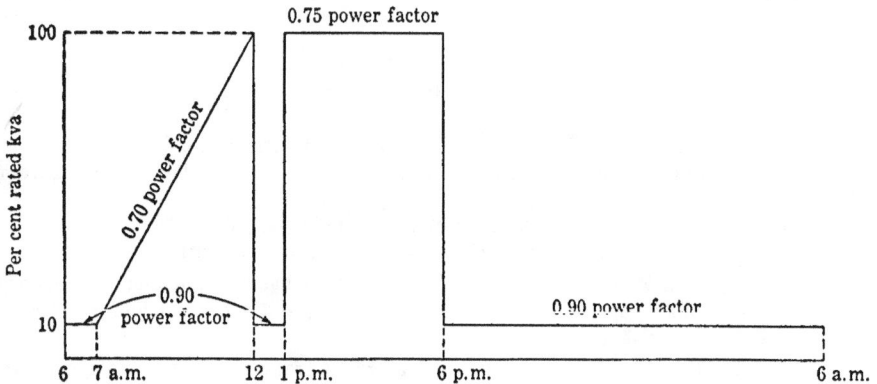

FIG. 7.   Load-demand curve, Prob. 7.

9. An inductive load is supplied with power from a substation through a short distribution line terminated at each end with transformers as shown in Fig. 8. The load consumes 80 kva at 0.90 power factor and 230 v, and the line has an impedance

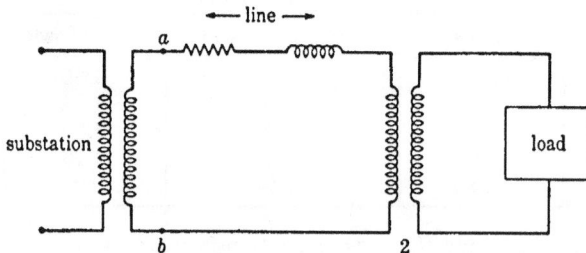

FIG. 8.   Distribution circuit, Prob. 9.

of $0.20 + j0.36$ ohm. Transformer number 1 is rated 150 kva, 60 $\sim$, and 13,800 : 2,300 v. Transformer number 2 is rated 100 kva, 60 $\sim$, and 2,300 : 230 v. Conventional short-circuit tests on the transformers yield the following data:

SHORT–CIRCUIT DATA

	Transformer No. 1	*Transformer No. 2*
Frequency	60 $\sim$	60 $\sim$
Primary terminal voltage	0.060	0.054
Current	1.0	1.0
Power input	0.010	0.012

All values except frequency are per unit. The base for any per unit value is the rating of the transformer to which the value applies.

(a) Determine the per cent voltage regulation at the load.

(b) An additional distribution line is connected at points *a* and *b* which draws a constant load of 60 kva at unity power factor. Determine the per cent change in voltage between *a* and *b* for a change in the original load of zero to 80 kva. The substation voltage is maintained at a constant value which results in 230 v at the original load when this load is 80 kva.

10. The core loss at rated voltage and frequency of transformer number 1 of Prob. 9 is 0.011 per unit; that of transformer number 2 is 0.009 per unit. The bases for the per unit values are the kva ratings of the two transformers. For the load schedule given below, determine the energy efficiency of the distribution system of Prob. 9a.

Hours	*Kva load*	*Power factor*
9 : 00 A.M.–12 : 00 M.	90	0.9
12 : 00 M. – 2 : 00 P.M.	50	0.9
2 : 00 P.M.– 4 : 00 P.M.	90	0.9
4 : 00 P.M.– 5 : 00 P.M.	20	0.95
5 : 00 P.M.– 9 : 00 A.M.	—	—

Assume that the substation voltage is regulated to maintain rated load voltage and that the core losses in the transformers are proportional to the squares of their secondary voltages. Assume that the transformers are continuously energized, and that the load is negligible from 5 : 00 P.M. to 9 : 00 A.M.

CHAPTER XV

Autotransformers

An ordinary two-circuit transformer whose windings 1 and 2 are connected in series, as shown in Fig. 1, is known as an *autotransformer*. Winding 1 is frequently called the *series winding* and winding 2 the *common winding*. An autotransformer can be used to transform power at a voltage V_H to power at a lower voltage V_X, or in the reverse direction. Although the analysis of this chapter is made on the assumption that the transformer is delivering power on the low-voltage side, the conclusions obviously are independent of the direction of power flow.

1. VOLTAGE AND CURRENT RELATIONS

Reference to Fig. 1 shows that the voltage V_X on the low-voltage side is the terminal voltage V_2 of the common winding and that the voltage V_H on the high-voltage side is the vector sum of the terminal voltages of the series and common windings; that is

$$V_X = V_2 \qquad \qquad \blacktriangleright[1]$$

$$V_H = V_1 + V_2. \qquad \qquad \blacktriangleright[2]$$

The voltages V_1 and V_2 differ from the voltages induced by the resultant core flux by only the relatively small leakage-impedance voltage drops in the windings. Since the induced voltages E_1 and E_2 are in time phase, the terminal voltages V_1 and V_2 also are very nearly in time phase and the vector sum of the winding voltages in Eq. 2 very nearly equals the *numerical* sum. The induced voltage E_X appearing in the low-voltage circuit is the voltage E_2 induced in the common winding, and the induced voltage E_H appearing in the high-voltage circuit is the series combination of the voltages induced in the two windings; thus

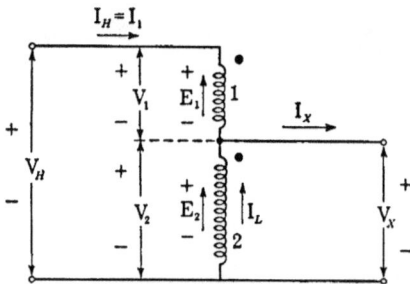

FIG. 1. Connections of an autotransformer.

$$E_X = E_2 \qquad \qquad [3]$$

$$E_H = E_1 + E_2. \qquad \qquad [4]$$

The induced voltages E_1 and E_2 are proportional to the numbers of turns N_1 and N_2 in the series and common windings respectively and are in time phase in the positive directions indicated in Fig. 1, whence

$$E_1 = \frac{N_1}{N_2} E_2, \qquad [5]$$

and from Eqs. 3, 4, and 5

$$E_H = E_2\left(\frac{N_1}{N_2} + 1\right) = \frac{N_1 + N_2}{N_2} E_X. \qquad [6]$$

The ratio of transformation between the high- and low-voltage circuits is E_H/E_X and differs from the ratio of terminal voltages by only the effects of the small leakage-impedance voltage drops; that is,

$$\frac{V_H}{V_X} \approx \frac{E_H}{E_X} = \frac{N_1 + N_2}{N_2}. \qquad [7]$$

Figure 1 also shows that the current I_H in the high-voltage circuit is the current I_1 in the series winding and that the current I_X in the low-voltage circuit is the vector sum of the current I_1 in the series winding and the current I_L in the common winding; that is,

$$I_H = I_1 \qquad \blacktriangleright[8]$$

$$I_X = I_1 + I_L. \qquad \blacktriangleright[9]$$

If the exciting current is neglected, the winding currents produce equal and opposing magnetomotive forces; whence the relation between the oppositely directed winding currents I_1 and I_L is

$$N_1 I_1 \approx N_2 I_L. \qquad [10]$$

The winding currents are very nearly in time phase, and therefore their vector sum in Eq. 9 very nearly equals the *numerical* sum. From Eqs. 8, 9, and 10 the relation between the currents in the external circuits is very nearly

$$I_X \approx I_1\left(1 + \frac{N_1}{N_2}\right) = \frac{N_1 + N_2}{N_2} I_H, \qquad [11]$$

or

$$\frac{I_H}{I_X} \approx \frac{N_2}{N_1 + N_2}. \qquad [12]$$

▶ Structurally, the only difference between an autotransformer and an ordinary two-circuit transformer is that the series winding of the autotransformer must be insulated for the voltage to ground of the high-voltage circuit — a greater voltage than that induced in the winding itself. In all other respects, an autotransformer is simply an ordinary two-winding transformer connected so that its windings are in series.

The internal behavior of an autotransformer is no different from that of an ordinary two-circuit transformer, but because of the method of connection the relations among the voltages and currents in the external circuits and those in the windings are as given in Eqs. 1, 2, 8, and 9.◀

The outstanding features of autotransformers can be derived readily from a comparison of the rating, losses, exciting current, and impedance characteristics of a two-winding transformer when it is connected as an autotransformer with the corresponding characteristic when it is connected as an ordinary two-circuit transformer.

2. RATING

Consider a transformer whose rating as an ordinary two-circuit transformer is 100 kva, 11,500 : 2,300 volts. If its windings are connected as

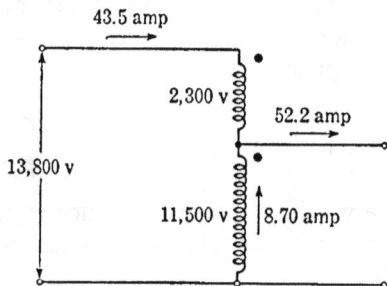

in Fig. 2, this transformer can be used as an autotransformer, provided there is sufficient insulation between the 2,300-volt winding and ground. If the terminal voltage of the 11,500-volt winding is its rated value, the terminal voltage of the 2,300-volt winding is very nearly its rated value, since the leakage-impedance voltage drops usually are very small. Since the winding voltages are very nearly in time phase, the voltage of the high-voltage circuit is very nearly the numerical sum of the winding

FIG. 2. Transformer with 2,300-v and 11,500-v windings connected as a 13,800 : 11,500-v autotransformer.

voltages, or 13,800 volts. Therefore, when connected as an autotransformer, this transformer can be used to interconnect two circuits whose rated voltages are 13,800 volts and 11,500 volts.

The current rating of the 2,300-volt winding is

$$\frac{100,000}{2,300} = 43.5 \text{ amp.} \tag{13}$$

If this rated current flows in the series winding and if the exciting current is negligible, the current in the common winding is its rated value of

$$\frac{100,000}{11,500} = 8.70 \text{ amp.} \tag{14}$$

Since the oppositely directed winding currents are very nearly in phase, the current I_X in the low-voltage external circuit is very nearly the numerical sum of the currents in the series and common windings, or 52.2 amperes.

If the transformer is delivering power to the low-voltage circuit, the kilovolt-ampere output is

$$\frac{11,500 \times 52.2}{1,000} = 600 \text{ kva.} \qquad [15]$$

If the leakage-impedance voltage drops and the exciting current are neglected, the kilovolt-ampere output also equals the kilovolt-ampere input from the high-voltage circuit, or

$$\frac{13,800 \times 43.5}{1,000} = 600 \text{ kva.} \qquad [16]$$

▶ This transformer, whose rating as a two-circuit transformer is only 100 kva, is therefore capable of transforming 600 kva when connected as an autotransformer. ◀

In an ordinary two-circuit transformer, *all* the power delivered at the low-voltage terminals is transferred by electromagnetic induction from the high-voltage side to the low-voltage side. The autotransformer possesses the important advantage that only a *portion* of the power is transferred by electromagnetic induction. In this example, the only change the transformer is required to bring about is a voltage reduction of 2,300 volts and a current increase of 8.70 amperes. The reduction in voltage is the fraction

$$\frac{2,300}{13,800} = \frac{1}{6} \qquad [17]$$

of the input voltage, or in general the fraction

$$\frac{E_1}{E_1 + E_2} = \frac{E_H - E_X}{E_H} \qquad [18]$$

of the voltage on the high-voltage side, where E_1, E_2 are the rated voltages of the windings and E_H, E_X are the rated voltages of the external circuits. When the leakage-impedance voltage drops are negligible, these rated voltages equal the induced voltages.

In general, if the currents and voltages have their full-load values, the volt-ampere rating of the series winding is

$$E_1 I_1 = (E_H - E_X) I_H, \qquad [19]$$

which is also the volt-ampere rating of the common winding. This is the rating of the transformer when it is used as an ordinary two-circuit transformer. The rating as an autotransformer, however, is $E_H I_H$.

Hence these two ratings are in the ratio

$$\frac{\text{Rating as an autotransformer}}{\text{Rating as a 2-circuit transformer}} = \frac{E_H}{E_H - E_X} \qquad \blacktriangleright[20]$$

In this example the rating as an autotransformer is

$$\frac{13,800}{13,800 - 11,500} = 6 \qquad [21]$$

times the rating as a two-circuit transformer. That is, this device, whose rating as a two-circuit transformer is only 100 kva, has a rating of 600 kva when connected as an autotransformer and is therefore equivalent to a 600-kva two-circuit transformer. Obviously the autotransformer is smaller and costs less than the 600-kva two-circuit transformer. The relative sizes of parts, the weight, and the cost of an autotransformer compared with those of a two-circuit transformer for the same application depend on the ratio $(E_H - E_X)/E_H$.

3. Losses and efficiency

Furthermore the efficiency is higher when the transformer is connected as an autotransformer. For example, if the unity-power-factor full-load efficiency of the 100-kva transformer is 0.9825 when it is connected as a two-circuit transformer, its losses are

$$\frac{0.0175 \times 100}{0.9825} = 1.78 \text{ kw.} \qquad [22]$$

When it is connected as an autotransformer, its losses at full load are still 1.78 kw, but these losses are now only

$$\frac{1.78}{601.78} = 0.00296 \qquad [23]$$

of the input. Consequently its unity-power-factor full-load efficiency as an autotransformer is 0.99704. This is a rather close approach to perfection! In general, the ratio between the per cent or per unit losses of a given transformer connected as an autotransformer to its losses as an ordinary two-circuit transformer is the inverse of the ratio of the ratings for these two connections. Thus, by Eq. 20,

$$\frac{\text{Full-load losses in \% of autotransformer rating}}{\text{Full-load losses in \% of 2-circuit transformer rating}} = \frac{E_H - E_X}{E_H} \qquad \blacktriangleright[24]$$

Figure 3 shows the variation of $(E_H - E_X)/E_H$ with the ratio E_H/E_X. Thus when the transformation ratio E_H/E_X between the high-voltage and low-voltage circuits is less than 2 : 1, the fractional change in voltage

$(E_H - E_X)/E_H$ which the autotransformer must bring about is less than $\frac{1}{2}$. Therefore the savings in size and cost, and the increase in efficiency, when an autotransformer is used instead of a two-circuit transformer may be quite important when E_H/E_X is less than about 2, but these advantages of the autotransformer are not so significant for larger values of the ratio of transformation E_H/E_X.

4. EXCITING CURRENT

The exciting current is of less importance when the transformer operates as an autotransformer than when it operates as a two-circuit transformer. If the winding voltages have their rated values at no load, the core flux has its rated value and the total no-load ampere-turns are the

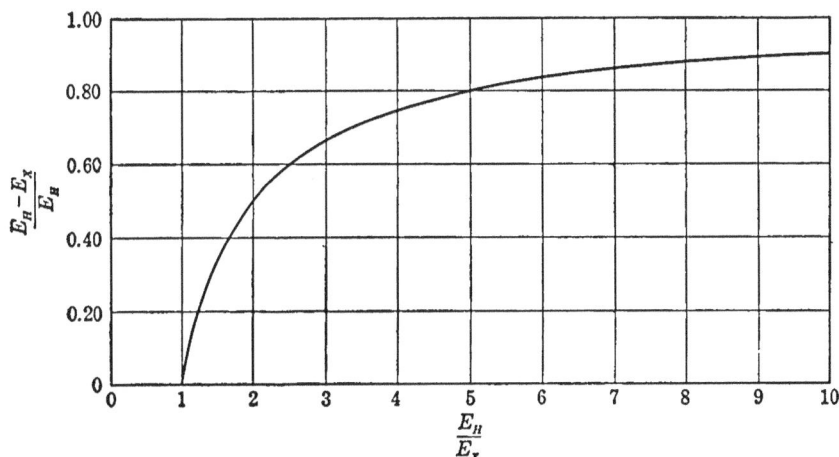

FIG. 3.   Fractional change in voltage $\dfrac{E_H - E_X}{E_H}$ as a function
of the circuit voltage ratio E_H/E_X.

same whether the transformer is connected as an autotransformer or as an ordinary two-circuit transformer. The exciting current varies inversely as the number of turns in which the exciting current flows. Since the rated voltages are proportional to the numbers of turns, the exciting volt-amperes at normal voltage are the same whether the transformer is connected as an autotransformer or as an ordinary two-circuit transformer. For example, if the exciting volt-amperes of the 100-kva transformer of Fig. 2 when operated as a two-circuit transformer are 3 per cent, or 3 kva, its exciting volt-amperes when connected as an autotransformer are still 3 kva. However, this is but 0.5 per cent of its rating of 600 kva as an autotransformer. In general, it follows from Eq. 20 that

$$\frac{\%\,I_\varphi \text{ connected as an autotransformer}}{\%\,I_\varphi \text{ as a 2-circuit transformer}} = \frac{E_H - E_X}{E_H} \qquad \blacktriangleright[25]$$

This relation applies to a given transformer connected as either a two-circuit transformer or an autotransformer. It is only approximately the ratio of the exciting current of an autotransformer to that of a different two-circuit transformer of the same rating, since the percentage of exciting current in normal designs varies somewhat with the size.

Neglecting the exciting current in an ordinary two-circuit transformer usually causes very little error, except in the analysis of problems that are directly concerned with the excitation phenomena, particularly those involving the behavior of the harmonics. Since the exciting current in an autotransformer as a rule is very small indeed, neglecting it usually causes even less error.

5. EQUIVALENT CIRCUITS AND IMPEDANCE PHENOMENA

If the nonlinearity of the excitation characteristics is neglected, an autotransformer can be represented by either of the equivalent circuits of Fig. 4. According to Thévenin's theorem,[1] the autotransformer as

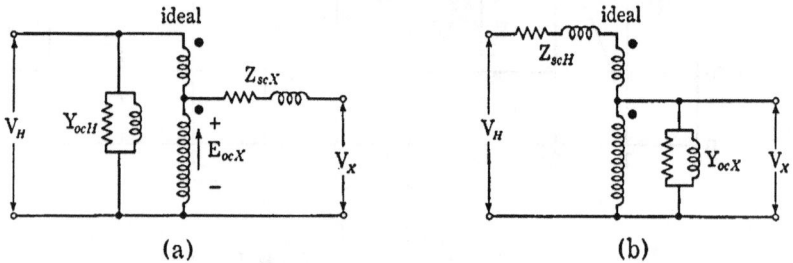

(a)   (b)

FIG 4. "Exact" equivalent circuits for an autotransformer.

viewed from its low-voltage terminals is equivalent to an electromotive force, equal to the open-circuit voltage E_{ocX} measured at the low-voltage terminals, in series with the impedance Z_{scX} measured at the low-voltage terminals with the high-voltage terminals short-circuited, as in the portion to the right of the ideal autotransformer in Fig. 4a. If the ratio of transformation of the ideal autotransformer is V_H/E_{ocX}, the voltage at its high-voltage terminals equals the high-tension voltage V_H of the actual autotransformer. This open-circuit voltage ratio very nearly equals $(N_1 + N_2)/N_2$, where N_1 and N_2 are the numbers of turns in the series and common windings respectively. It can be shown that if the

[1] Thévenin's theorem is discussed in this series in the volume on electric circuits (1940), 469.

open-circuit admittance Y_{ocH} measured on the high-voltage side of the actual autotransformer is connected across the high-voltage terminals of the ideal autotransformer, the circuit of Fig. 4a is an " exact " equivalent of the autotransformer on both its high- and low-voltage sides. Evidently, if the open-circuit measurements are made on the low-voltage side and the short-circuit measurements on the high-voltage side, the circuit of Fig. 4b also is an " exact " equivalent of the autotransformer. When the exciting current is neglected, the " exact " equivalent circuits of Fig. 4

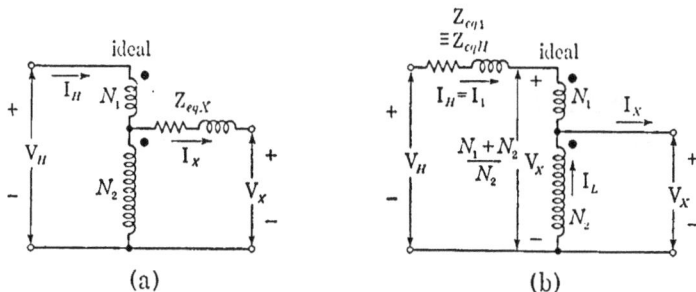

Fig. 5.   Approximate equivalent circuits for an autotransformer.

reduce to the approximate equivalent circuits of Fig. 5. The equivalent circuits are useful for determination of the external behavior of an autotransformer as a circuit element.

Internally, the autotransformer is exactly the same as an ordinary two-circuit transformer, and therefore the equivalent circuits can be derived from two-circuit transformer theory. This alternative derivation is given in the following discussion for the purpose of showing the relations between the autotransformer and the ordinary two-circuit transformer.

If the exciting current is neglected, the relation between the terminal voltages V_1 and V_2 of the windings can be written in terms of the winding currents, the equivalent impedance between the windings, and the ratio of winding turns. Thus, as in two-circuit transformer theory,

$$V_1 = I_1 Z_{eq1} + \frac{N_1}{N_2} V_2, \qquad [26]$$

where $(N_1/N_2)V_2$ is the terminal voltage of the common winding referred to the series winding. Except for the effects of the exciting current, the equivalent impedance Z_{eq1} equals the short-circuit impedance measured at the terminals of the series winding with the common winding short-circuited, as in two-circuit transformer theory. If V_2 is added to both sides of Eq. 26,

$$V_1 + V_2 = I_1 Z_{eq1} + \frac{N_1 + N_2}{N_2} V_2. \qquad [27]$$

In this equation, $V_1 + V_2$ equals the voltage V_H across the high-voltage terminals when the transformer is connected as an autotransformer (see Fig. 1), I_1 is the current I_H in the high-voltage circuit, and V_2 is the voltage V_X across the low-voltage terminals. Thus the equation relating the voltages across the external terminals of an autotransformer can be written as

$$V_H = I_H Z_{eq1} + \frac{N_1 + N_2}{N_2} V_X. \qquad [28]$$

This equation for the autotransformer also applies to the equivalent circuit of Fig. 5b, consisting of an impedance Z_{eq1} in series with an ideal autotransformer. If the exciting current is negligible, the relation between the high-voltage and low-voltage circuit currents of the autotransformer is, by Eq. 12, p. 395,

$$I_H = \frac{N_2}{N_1 + N_2} I_X, \qquad [29]$$

as in the ideal autotransformer of Fig. 5b. Therefore the autotransformer is correctly represented by the equivalent circuit of Fig. 5b.

Equation 29 shows that the equivalent impedance Z_{eqH} of the autotransformer referred to its high-voltage side is the same as the equivalent impedance Z_{eq1} referred to the series winding when the transformer is operated as an ordinary two-circuit transformer; that is,

$$Z_{eqH} \equiv Z_{eq1}. \qquad [30]$$

This impedance can be determined by measurement of the impedance at the terminals of the series winding with the common winding short-circuited, as in testing an ordinary two-circuit transformer. It can also be determined through short-circuiting the low-voltage terminals of the autotransformer and applying a reduced voltage to the high-voltage terminals. Because of the short circuit, the voltage is applied directly to the series winding, and therefore the same impedance is measured in both of these tests.

Sometimes referring the equivalent impedance to the low-voltage circuit may result in a more convenient equivalent circuit. When Eq. 28 is multiplied by $N_2/(N_1 + N_2)$, the result can be expressed as

$$\frac{N_2}{N_1 + N_2} V_H = \frac{N_1 + N_2}{N_2} I_H \left(\frac{N_2}{N_1 + N_2}\right)^2 Z_{eqH} + V_X \qquad [31]$$

$$= I_X Z_{eqX} + V_X, \qquad [32]$$

where

$$Z_{eqX} = \left(\frac{N_2}{N_1 + N_2}\right)^2 Z_{eqH}. \qquad [33]$$

In Eq. 32, Z_{eqX} is the equivalent impedance of the autotransformer referred to its low-voltage side, and $\dfrac{N_2}{N_1 + N_2}\, V_H$ is the terminal voltage on the high-voltage side referred to the low-voltage side. Equation 32 can be represented by an ideal transformer in series with the equivalent impedance Z_{eqX}, as in Fig. 5a. The equivalent impedance Z_{eqX} can be measured by a short-circuit test, with V_H short-circuited and a reduced voltage applied to the low-voltage terminals.

Again consider the 100-kva 11,500 : 2,300-volt two-circuit transformer of Art. 2. Assume that when it is operated as a two-circuit transformer its equivalent-impedance voltage drop at rated full-load current is 6 per cent of its rated voltage. That is, its equivalent-impedance voltage drop $I_1 Z_{eq1}$ referred to the 2,300-volt winding is 138 volts when I_1 is its rated value. When the transformer is operated as a 600-kva autotransformer (as in Fig. 2), the rated voltage on the high-voltage side is 13,800 volts. The equivalent-impedance voltage drop referred to the high-voltage side, still being 138 volts, now is but 1 per cent of the rated high-tension voltage. In general, the full-load impedance drop expressed as a percentage of rated voltage when a transformer is connected as an autotransformer is related to its per cent impedance drop when it is connected as a two-circuit transformer by the ratio

$$\dfrac{\% \text{ full-load impedance drop as an autotransformer}}{\% \text{ full-load impedance drop as a 2-circuit transformer}}$$

$$= \dfrac{E_H - E_X}{E_H} \qquad \blacktriangleright[34]$$

Since the voltage regulation is very nearly proportional to the per cent impedance, a relation like Eq. 34 is approximately true for the voltage regulation at full load and at a specified power factor. That is,

$$\dfrac{\text{Regulation as an autotransformer}}{\text{Regulation as a 2-circuit transformer}} \approx \dfrac{E_H - E_X}{E_H} . \qquad \blacktriangleright[35]$$

Since the short-circuit current when the primary voltage is maintained at its rated value is inversely proportional to the short-circuit impedance, the short-circuit current in an autotransformer is larger than when the same device is connected as a two-circuit transformer. The ratio of the short-circuit currents is

$$\dfrac{\text{S.c. current as an autotransformer}}{\text{S.c. current as a 2-circuit transformer}} = \dfrac{E_H}{E_H - E_X} . \qquad \blacktriangleright[36]$$

When the 100-kva 11,500 : 2,300-volt transformer is used as an ordinary two-circuit transformer. its per unit equivalent impedance is 0.06 and the short-circuit current at rated voltage is 1/0.06, or 16.7 times full-load current. By Eq. 36, when it is connected as a 13,800 : 11,500-volt auto-transformer, its short-circuit current at rated voltage is 6 × 16.7, or 100 times normal current. This is a serious matter, since the destructive forces on short circuit are proportional to the square of the short-circuit current. Largely for this reason, the use of autotransformers is limited to applications where there is sufficient impedance in the primary circuit to limit the short-circuit current to a much smaller value than would be obtained if the primary terminal voltage remained constant during short-circuit conditions.

Note that Eqs. 34, 35, and 36 apply to the *same transformer* connected either as an autotransformer or as a two-circuit transformer. They do not apply to an autotransformer and a different two-circuit transformer of the same rating.

6. Conclusions

It is now apparent that an economically designed autotransformer has some or all of the following advantages over an ordinary two-circuit transformer of the same rating:

(1) Smaller size,
(2) Lower cost,
(3) Greater efficiency,
(4) Smaller exciting current,
(5) Better voltage regulation.

On the other hand, the autotransformer has the following disadvantages:

(1) Larger short-circuit currents,
(2) Conductive connection between low-voltage and high-voltage circuits.

These various characteristics are, of course, subject to considerable variation in design. For example, it is possible to design an autotransformer and an ordinary two-circuit transformer each of which has the same rating and the same efficiency. In such a design the autotransformer is smaller and costs less than the two-circuit transformer. On the other hand, it is possible to design an autotransformer and a two-circuit transformer of the same rating, each to cost the same. In such a design, the autotransformer has the higher efficiency. An economical design is one in which a proper balance is obtained between cost and other fixed charges on one hand, and losses and other operating expenses on the other. Such an economical design is a compromise in which the auto-transformer costs somewhat less than the two-circuit transformer and has a somewhat higher efficiency.

▶ All the desirable features of the autotransformer depend on the ratio of circuit voltages, and little is gained by use of an autotransformer when the circuit voltage ratio E_H/E_X is large.◀

PROBLEMS

1. An autotransformer is to be used to supply a load of 180 amp at 150 v from a 220-v line. If the transformer may be considered ideal:
   (a) What must be the turns ratio between the series and common windings?
   (b) What current will be drawn from the 220-v source?
   (c) What per cent of the power delivered to the load flows conductively?
   (d) What would be the voltage and kva ratings of the transformer if it were used as an ordinary two-winding transformer?

2. Conventional short-circuit and open-circuit tests on a 100-kva 60 ~ 4,400 : 2,200-v two-winding transformer yield the following data:

SHORT-CIRCUIT TEST; HIGH-VOLTAGE WINDING EXCITED

Frequency	Voltage	Current	Power
60 ~	228 v	22.8 amp	1,300 w

OPEN-CIRCUIT TEST; LOW-VOLTAGE WINDING EXCITED

Frequency	Voltage	Current	Power
60 ~	2,200 v	1.08 amp	525 w

If this transformer is connected as a 6,600 : 4,400-v autotransformer:
   (a) What will be its rating?
   (b) What per cent of rated voltage must be applied to the 6,600-v terminals to maintain rated secondary voltage across a load of 250 kva at 0.85 power factor, lagging?
   (c) What will be its efficiency when operating under the conditions of (b)?

3. A 25-kva 60 ~ 660 : 220-v autotransformer, when excited at 27.6 v on the high-voltage side with the low-voltage terminals short-circuited, draws rated primary current with a power input of 320 w. Its core loss at rated voltage is 360 w.
   (a) If the transformer is used to supply a 220-v 25-kva inductive load with a power factor of 0.90, determine its per cent regulation and efficiency.
   (b) If the transformer were reconnected so that its voltage rating was 660 : 440 v, what would be its kva rating and its regulation and efficiency when supplying rated load at the power factor specified in (a)?

4. A diagram of a 10-kva 440 : 220 v 60 ~ autotransformer is shown in Fig. 6. When the terminals *bc* are short-circuited, 19.2 v at 60 ~ applied to the terminals *ab* causes full-load current to flow in the windings. Under this condition, the power taken is 176 w. When rated voltage at 60 ~ is applied to either pair of external terminals with the other pair of terminals open, the power taken is 142 w.
   (a) With 440 v at 60 ~ impressed on the terminals *ac* with terminals *bc* short-circuited, what per cent of full-load current will flow in each winding?

(b) If rated load at 0.75 power factor and rated voltage is supplied from the terminals *bc*, determine the efficiency and per cent regulation.

(c) If the autotransformer is used as a regular 220-v 60 ~ transformer with a ratio of transformation of unity, and a voltage of 220 v at 60 ~ is impressed on the primary with the secondary short-circuited, what per cent of full-load current will flow in each winding?

(d) If the transformer is connected as a regular transformer as in part (c), and a 5-kva load at 0.75 power factor and rated voltage is supplied from its secondary terminals, determine its efficiency and per cent regulation.

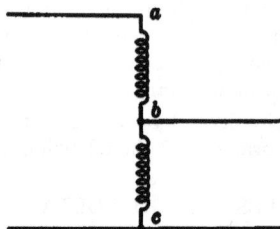

FIG. 6. Connections of an autotransformer, Prob. 4.

5. The autotransformer of Prob. 2 is supplied through a line of impedance $6 + j10$ ohms from one phase of a large three-phase bank of transformers. The secondary terminal voltage of the three-phase bank is 6,720 v, and may be considered constant.

If a fault occurs across the secondary of the autotransformer, causing a dead short circuit, find:

(a) The current in the fault, and

(b) The currents in each winding of the autotransformer.

CHAPTER XVI

Economy in Power-System Transformer Applications

The ownership and operation of any electric power equipment involve expenses of two types: fixed costs, which are inherent in ownership, and are present whether the equipment is used or not; and operating costs, which are related to the manner of use. Both types depend on the design and production methods employed by the manufacturer of the equipment. Although transformers are inherently very efficient, and therefore the costs of operating them are relatively small, they serve well to illustrate the general cost principles which are common to all types of electrical apparatus. Small differences of detail, such as the cost of friction and windage losses found in rotating machines and of arc losses found in electronic devices, should be obvious. In general, three different points of view are important in connection with transformer costs: those of the transformer manufacturer, of the public-utility or traction company, and of the industrial user of transformers.

The transformer manufacturer is faced with several very important economic problems. First, he must strive to keep manufacturing costs as low as is possible, consistent with the high quality expected of his product. The need for reduced manufacturing costs has often led to actual physical improvements in transformers; for example, remarkable reductions in size and weight of high-voltage current transformers have resulted from it, as shown in Fig. 10, p. 279. The manufacturer must also be aware of general changes in the characteristics of transformer loads, because the trend of loads has an important economic relation to the loss characteristics desired in a transformer. Finally, the manufacturer must build transformers so well that their service record will be satisfactory to his customers, thus tending to increase the market for his product.

The largest users of power and instrument transformers are the electric public-utility systems, whose requirements largely dictate the progress of development in the performance characteristics of transformers and transformer accessories. The rapid growth in central-station power sales from 1910 to 1940 has made it desirable to transmit power over long distances from sources of the cheapest energy to centers of load. Long-distance power transmission is economical only when the operating voltage is very high; so the utility systems have asked the manufacturers to furnish transformers of increasingly higher voltage ratings, and have obtained them. The extension of electric service in country areas has

407

required that distribution transformers for rural electric lines have the lowest possible purchase price, and has also resulted in marked improvements in methods of lightning protection for transformers.

Railroads and industrial users of electricity are vitally interested in the economic application of transformers. Almost all industrial enterprises consume large quantities of electrical energy, and usually employ one or more transformer banks to reduce the voltage supplied by the public-utility company to the magnitude used in the plant. When the industrial user owns these step-down transformers, the energy sold to his plant is metered on their high-voltage sides. The cost of transformer losses is important to an industrial user, since these losses are included in the metered energy sold to his plant and therefore cause a charge that must be paid as a definite operating expense. If the industrial plant has its own generating equipment, the capital and generating costs necessary to supply the transformer losses are an element of overhead costs in production expense that should be kept at a minimum.

Some interesting investment studies can be made concerning the method of installing transformers. Should a standard transformer be purchased to be located out of doors, or should it be placed in a fireproof vault inside a building? Should a special transformer containing a noncombustible and nonexplosive insulating liquid be purchased, to be mounted inside the building? Should the outdoor transformer be placed on the ground in a fenced enclosure, or should it be mounted above the ground, supported on one or more poles? Should tap-changers, lightning arresters, oil switches, fuses, disconnecting switches, metering equipments, instrument transformers, cooling fans, and other auxiliary devices be purchased as a single integrated unit, housed together with the transformer (as shown in Fig. 28, p. 309), or should they be installed separately by the user? Should three single-phase transformers or a single three-phase transformer be used for a three-phase bank? Not only are these questions of installation important to the engineer; he is concerned also with the relative operating costs of various suitable transformers. The lowest operating cost that can be achieved in any transformer installation depends largely upon the load cycle imposed on the transformer bank. Hence a knowledge of the probable load characteristics is necessary for the preparation of an estimate of operating costs. The estimates of installed cost and operating cost of the various plans for installation, together with considerations of safety convenience, and appearance, dictate the decision to be made whenever a new transformer substation is to be erected. As a guide to the solution of practical transformer problems, the essential elements of economy in transformer use are treated in the following articles.

1. ELEMENTS OF TRANSFORMER COSTS

A consideration of the detailed costs of manufacturing a transformer is a specialized problem beyond the scope of a textbook on basic principles. It is possible, however, to outline and discuss the general cost elements involved in the use of transformers. A transformer which is owned and operated by a public-utility company may be expected to incur charges of the types indicated in the following outline:

A. Fixed costs (proportional to investment)
   1. Direct (deriving from the installed cost of the transformer and its associated protective and control devices)*
      a. Interest
      b. Taxes
      c. Insurance
      d. Depreciation
      e. Administration
   2. Indirect (fixed costs in other portions of plant, caused by the transformer)
      a. Increase in direct fixed costs in other portions of the plant, resulting from added size of general plant caused by transformer losses and exciting current
      b. Portion of fixed charges on feeder voltage regulators or condensers needed to offset transformer impedance drop

B. **Operating** costs (variable, depending on conditions of use)
   1. Direct (cost of internal transformer losses)
      a. Cost of core loss (practically constant)
      b. Cost of load loss (proportional to square of current)
      c. Cost of energy consumed by auxiliary cooling equipment
      d. Cost of cooling water
   2. Indirect (increase in cost of losses in external circuits caused by the transformer)
      a. Additional copper loss in the circuit and generator supplying the transformer, resulting from transformer exciting current and losses
      b. Portion of losses in feeder voltage regulator needed to maintain constant transformer output voltage

* Note that the installed cost includes charges allocated to the transformer installation from the general overhead expenses of the company owning the transformer, including administration expense, accounting expense, and the direct overhead expenses of installation as described in Art. 2.

3. Maintenance expense
    a. Periodic oil testing
    b. Oil filtering or replacement
    c. Repairs
    d. Periodic load testing
    e. Replacement of inert gas and deoxidizing chemicals

2. DIRECT FIXED COSTS

As a general rule, the direct fixed costs associated with the ownership and operation of a transformer exceed the operating costs. The annual direct fixed costs are usually considered to be proportional to the original cost of the transformer installation. The installed cost includes, in addition to material, labor, and real estate, a number of overhead costs associated with the installation. These are for construction supervision, engineering, tools, interest during construction, employee compensation and public liability insurance, and other less important items.

The annual rates to be charged for interest, taxes, and insurance vary so much with local conditions and company policy that it seems best not to present here any definite percentages to be used in cost studies, but rather to suggest that each case be investigated as an individual problem. For example, in localities where transformers are classed by assessors as tax-exempt personal property, the company owning a transformer pays no taxes on it, although real-estate taxes are paid on the surrounding substation structure; while, if the transformer is mounted on a pole, a special franchise tax is paid on the supporting pole. Many companies do not insure transformers unless they are located in buildings, preferring to consider an occasional failure as a part of the general basis for depreciation rates. Interest charges cannot be avoided, for, even though there are funds available for purchasing and installing a transformer without borrowing, the money spent for the transformer and installation could have been invested in some other way to earn interest.

Depreciation accounting practice varies widely in different companies and in various types of industry. Business risks are greater in manufacturing industries than in public-utility companies; so manufacturers usually write off the investment in equipment, including transformers, more rapidly than do public-utility companies. If a manufacturing process which requires the use of a transformer becomes obsolete, the salvage value of the transformer is determined by the market price for used transformers. On the other hand, a utility company can readily move transformers from place to place, keeping most of them in active service. Therefore a utility company can expect the depreciation rate for transformers to be determined by average service life rather than by obsolescence.

Transformers have, on the average, a fairly long useful life, although individual transformers occasionally fail after a short period of service, because of excessive overload, lightning, or other causes. In a system using a large number of transformers, properly maintained and periodically checked for load conditions, it is not unreasonable to expect an average physical life of about twenty-five years.[1] There exists, therefore, an active trade in used transformers, and industrial managements frequently purchase used equipment in order to reduce the immediate fixed charges. To obtain a reasonable assurance of a satisfactory service life, with a corresponding reduction in fixed charges, the purchaser of a used transformer should obtain a suitable guarantee from the transformer dealer, or should have the condition of the windings and insulating oil carefully inspected.

A portion of the general administrative costs of a company is charged annually as a fixed charge to all the capital of the company, including transformers. Administration costs include a portion of management salaries, fixed capital accounting expenses, engineering records expenses, legal costs, tax and insurance department salaries and expenses, and other similar costs.

3. Direct operating costs

The operating costs of a transformer are direct functions of the manner in which the transformer is used. Except for a part of the maintenance expenses, the operating costs cease when the transformer is removed from service. If the transformer is cooled by external equipment, such as air blowers or circulating cooling water, the cost of operating that equipment is a direct operating cost. Most transformers are continuously energized at approximately rated voltage, so that the core loss is always present. If a transformer is continuously energized, its energy loss from core loss in kilowatthours per year is

Annual energy in core loss = (core loss in kw) × 8,760 kwhr. ▶[1]

The load losses, on the other hand, are proportional to the square of the load current (when the exciting current is negligible), and are more difficult to compute than the core loss. The load losses include the heat loss in the transformer coils and the stray load losses in the iron parts which result from load conditions, as discussed in Art. 2b, Ch. XIV. To conform with common practice, the term *copper loss* is used hereafter in this chapter to mean *load loss*, including both the I^2R losses in the windings, and the stray losses. The known copper loss at full-load current can

---

[1] R. Winfrey, "Statistical Analyses of Industrial Property Retirements," *Iowa State College Engineering Experiment Station Bulletin 125* (December, 1935).

be used in determining the annual energy loss resulting from the variable copper losses, by a method to be described in later paragraphs.

Certain important terms, applied here to transformers only, are of general application to all electric power apparatus as used in service and are essential to discussion of the method. *Demand* is the load on an electrical system or device averaged over a specified interval of time. Common units for demand measurement are kilowatts, kilovolt-amperes, and reactive kilovolt-amperes or kilovars. The concept of demand is very important in power work. Since the magnitude of loads often varies over a wide range, even in a period of a few seconds, it would be of little value to use momentary* values of maximum load for determination of the necessary current-carrying capacity of power apparatus, although momentary peaks are important in their effect on voltage regulation. However, the average value of the load, taken over a suitable time interval, such as 15 minutes, 30 minutes, or an hour, is a measure of the current-carrying requirements of power equipment. Theoretically, the rms value of the load over a specified time interval is a more accurate measure of heat generated by losses than is the average load, but in practice the average value is used to determine the demand because the average value for a time interval of several minutes can be measured by simpler metering arrangements than those required to measure the rms value for the same interval.

If the demand is determined for each interval of a series of consecutive basic demand intervals (such as 24 consecutive one-hour intervals, comprising a 24-hour day), the *rms demand* for the corresponding total elapsed period is the rms value of the basic-interval demands for that period. The term rms demand is introduced here for its usefulness in computing the energy consumed by copper losses in a transformer. Suppose that the following one-hour demands represent the load on a transformer during a 24-hour period:

Demand	Number of hours
10 kva	8
50 kva	6
70 kva	5
100 kva	5

The rms demand for this 24-hour period is

$$\text{Rms demand} = \sqrt{\tfrac{1}{24}[(10^2 \times 8) + (50^2 \times 6) + (70^2 \times 5) + (100^2 \times 5)]} \quad [2]$$

$$= 61.4 \text{ kva.} \quad [3]$$

* The term *momentary* as introduced here means the value which a standard indicating instrument would show as opposed to the *instantaneous* value which would be shown by an oscillographic instrument.

The *maximum demand* of an electrical system or device is the greatest of all the demands which have occurred during a specified period of time. Maximum demand is used for determining the responsibility of any individual load as it contributes to the entire system load, and also for determining the rating of equipment needed to supply a given load.

Load factor of an electrical system or device is the ratio of the average load to the maximum load during a specified period of time, when the

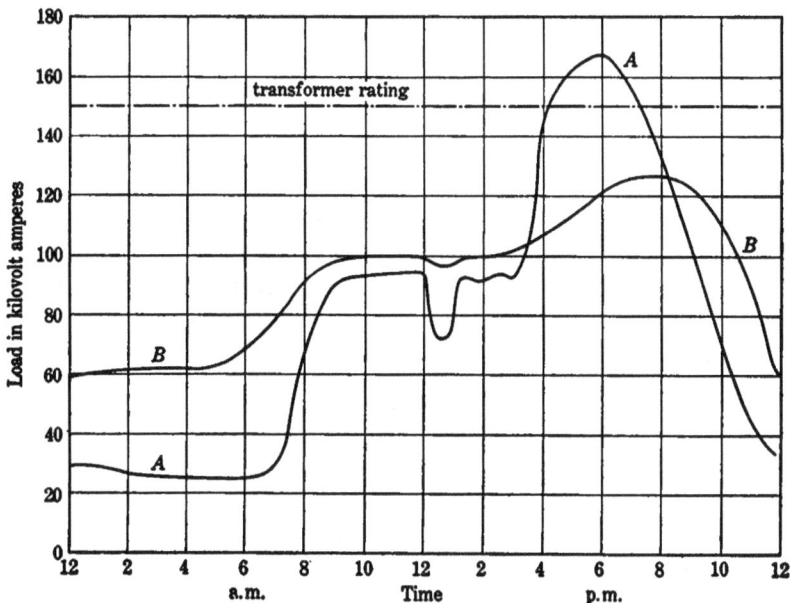

Fig. 1.   Two load curves which illustrate the difference between load factor and rms demand ratio.

average and the maximum loads are measured in the same units. The term load factor is widely used as a measure of the steadiness of a load. If a load is constant for the time interval in question, it has a load factor of 100 per cent. Usually load factor is measured over time intervals of a day, month, or year. Load factor gives no idea of the amount to which a device having a certain rating is loaded; thus, for example, a transformer can operate at 100 per cent load factor, but be loaded to only 70 per cent of its rating.

The following example shows the physical significance of the various terms defined above. In Fig. 1, curves *A* and *B* represent the characteristics of two different loads, measured in kilovolt-amperes, over a 24-hour period. The area under curve *A* is 1,910 kvahr; the area under curve *B* is 2,230 kvahr. If the demand interval is taken as one hour, the area under a curve for one hour divided by 1 gives the one-hour demand for that hour. For example, the area under curve B from 6 to 7 P.M. is 124 kvahr, so the demand for that hour is 124 kva. As an example of a 15-minute demand interval, the area under curve A from 7 to 7:15 A.M. is 8.75 kvahr. The 15-minute demand is then 8.75/0.25, or 35 kva. The maximum demand for a one-hour interval is, for curve A, 167 kva; for curve B, 125 kva. The load factors are:

$$\text{Load factor, curve } A = \frac{1,910/24}{167} \times 100 = 47.7\%; \qquad [4]$$

$$\text{Load factor, curve } B = \frac{2,230/24}{125} \times 100 = 74.3\%. \qquad [5]$$

The load B is steadier than load A and therefore load B has the higher load factor. The load factor, however, indicates not the actual load carried by the transformer, but merely the steadiness of this load. The rms demand, computed in the manner indicated in Eq. 2 and on one-hour basic intervals, is for curve A, 92.0 kva; for curve B, 95.6 kva. The ratio of the daily rms demand to the transformer rating is

$$\frac{\text{Daily rms demand, curve } A}{\text{Transformer rating}} = \frac{92.0}{150} = 0.614; \qquad [6]$$

$$\frac{\text{Daily rms demand, curve } B}{\text{Transformer rating}} = \frac{95.6}{150} = 0.638. \qquad [7]$$

This comparison illustrates the fact that there is no general relation between load factor and the ratio of rms demand to transformer rating. This point should be clearly understood, as there are places in the technical literature where the two terms are incorrectly used in an identical sense. The term load factor has little application in the discussions to follow, but has been introduced in order to bring out this distinction.

In determining an expression for the annual energy loss resulting from transformer copper losses, let

$W_a \equiv$ annual energy loss in kilowatthours due to transformer copper losses,

$P_{cu} \equiv$ known copper loss in kilowatts at rated full-load current,

$p \equiv$ momentary copper loss in kilowatts,

$t \equiv$ time in hours,

$i \equiv$ momentary load current in rms amperes,*

$I_{fl} \equiv$ rated full-load current in rms amperes,

$I_a \equiv$ rms value of the load current in amperes taken over a period of one year, as defined by Eq. 10,

$\lambda \equiv I_a/I_{fl}$.

---

* Note that this is *not* the instantaneous current. See the footnote on p. 412.

The currents i, I_{fl}, and I_a are all referred to the same side of the transformer, or are all expressed in per unit values.

The momentary copper loss p, which is proportional to the square of the momentary load current, may be expressed in terms of the known copper loss at full load and the known full-load current; thus

$$p = P_{cu}\left(\frac{i}{I_{fl}}\right)^2 \qquad [8]$$

The total annual copper loss is

$$W_a = \int_0^{8,760} p \, dt = \frac{P_{cu}}{I_{fl}^2} \int_0^{8,760} i^2 \, dt, \qquad [9]$$

where 8,760 is the number of hours in a year.

By definition of a root-mean-square value, the annual rms load current I_a is

$$I_a \equiv \sqrt{\frac{1}{8,760} \int_0^{8,760} i^2 \, dt}, \qquad [10]$$

whence

$$\int_0^{8,760} i^2 \, dt = 8,760 \, I_a^2. \qquad [11]$$

Substituting Eq. 11 in Eq. 9 gives

$$W_a = 8,760 \left(\frac{I_a}{I_{fl}}\right)^2 P_{cu} = 8,760 \lambda^2 P_{cu} \qquad [12]$$

Equation 12 determines the annual energy loss resulting from the transformer copper losses. If the annual load curve of momentary current were used, the rigorous computation of the annual rms load current I_a would involve some very tedious work, which can be simplified in the following way. As already defined, λ is the ratio of annual rms current to rated full-load current; however, on the assumption that the transformer output voltage is constant, λ is likewise the ratio of annual rms kva demand of the load to the full-load kva rating of the transformer. The annual rms kva demand needed for finding λ can be determined readily from a set of load curves taken for a few representative days throughout the year, by the method of Eq. 2.

The cost of electric service depends not only on the operating costs, but also on the fixed charges. Accordingly, electric rates in general include an energy charge to cover the operating costs, and a demand charge, based on the maximum demand, to cover the fixed charges. Since the losses themselves cause a demand, the cost of the transformer losses includes a demand cost in addition to the energy cost. The maximum demand due to copper loss is

$$\text{Copper-loss maximum demand} = P_{cu} \left(\frac{\text{maximum kva demand}}{\text{transformer kva rating}}\right)^2. \quad [13]$$

The sum of the energy in core loss and that in copper loss determines the total cost of energy resulting from transformer losses. Likewise, the sum of the maximum demand due to core loss and that due to copper loss determines the total cost of demand resulting from transformer losses. If the transformer uses metered power, these two costs are established by the rate paid for electric service at the meter; if the transformer is operated by a public-utility company, the cost of energy is often taken as the actual generating cost of electricity at the power plant, a figure likely to lie between 0.10 cent and 0.75 cent per kilowatthour. Every utility company knows, from experience, the average annual fixed cost per kilovolt-ampere of demand that applies to each operating district of the company; this unit cost is used as the annual demand charge for losses. An annual demand charge of $15.00 per kilovolt-ampere of demand is representative for American conditions.

4. ILLUSTRATIVE EXAMPLE: COMPARISON OF USED AND NEW TRANSFORMERS

An industrial firm purchases power which is supplied from a three-phase 2,400-volt system. Construction of an addition to the factory requires the installation of a 150-kva bank of transformers for reducing the voltage to 240 volts for three-phase distribution throughout the new building. The industrial firm will own and operate the transformers. The following two types of transformer installation are under consideration:

Plan A: Three used transformers (thirteen years old); single-phase, 50 kva each.

Plan B: One new transformer, three-phase, 150 kva.

In either plan, the load to be carried by the transformer bank is represented by the daily demand curves shown in Fig. 2. These curves present the hourly demands in kilovolt-amperes at typical seasons of the year and for loads on days when the factory is shut down. The rate paid for electricity, metered at the input terminals of the transformer bank, is:

Monthly Demand Charge

First 10 kva of maximum demand	$1.80 per kva
Next 40 kva of maximum demand	1.30 " "
Additional demand, in excess of 50 kva	0.90 " "

Monthly Energy Charge

First 1,000 kwhr	$0.03 per kwhr
Next 4,000 kwhr	0.02 " "
Additional energy, in excess of 5,000 kwhr	0.01 " "

Loss data for the two plans are given in the following table:

	Plan A (Used transformers)	Plan B (New transformer)
Core loss in watts		
Each transformer	230	
For the bank	690	480
Full-load copper loss in watts		
Each transformer	630	
For the bank	1,890	1,700

The installation costs for the two plans are as follows:

	Plan A (Used transformers)	Plan B (New transformer)
Delivered price of transformer bank	$375.00	$1,030.00
Contractor's installation price*	400.00	375.00
Overhead costs during installation, at 15%	116.00	211.00
Total installed cost	$891.00	$1,616.00
Salvage value at end of book life	50.00	150.00
Original installed cost less salvage value	$841.00	$1,466.00

* Contractor's price includes furnishing and installing the necessary cutouts, lightning arresters, grounds, and connections.

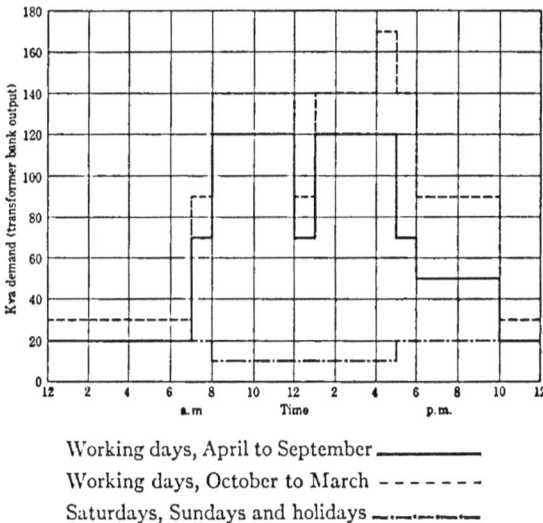

Working days, April to September ——————
Working days, October to March - - - - - -
Saturdays, Sundays and holidays —.—.—.—.—

FIG. 2. Demand curves for the industrial plant discussed in Art. 4.

The investment in the used transformers is to be written off in three years, but a ten-year depreciation period is to be allowed for the new unit. The straight-line method of computing depreciation is to be applied;

that is, the annual charge for depreciation is

$$\text{Annual depreciation} = \frac{\text{installed cost less salvage value}}{\text{book life}} \qquad [14]$$

The annual rates for fixed charges other than depreciation are:

Interest on investment	6 %
Taxes	2 %
Insurance	1.5%
Administration	3 %

Total annual fixed charge, excluding depreciation   12.5%

The used-transformer dealer agrees to clean and dry the windings and tank and to furnish new oil before delivering the transformers, including these operations in the quoted sale price. He also offers a one-year guarantee against defects, which is suitable to the prospective buyer.

Computation of Annual Load Losses: The annual energy loss due to transformer load losses is determined from Eq. 12, which is repeated here:

$$W_a = 8{,}760\lambda^2 P_{cu}, \qquad [15]$$

where

$W_a \equiv$ annual energy loss from load losses in kilowatthours,
$\lambda \equiv$ ratio of annual rms current I_a to rated full-load current I_{fl}.
$P_{cu} \equiv$ copper loss in kilowatts at rated full-load current.

If the voltage is assumed to be constant, the ratio λ is the same for kva demands as for currents. To determine λ, the annual rms demand must be computed. There are 117 days when the transformer installation is loaded in accordance with the dash-dot lines of Fig. 2. These 117 days include 52 Saturdays, 52 Sundays, 8 holidays, and 5 days shut down for inventory. Of the remaining 248 days in the year, 124 are taken to conform to the solid lines representing conditions from April to September, and 124 conform to the broken lines showing conditions from October to March. It is evident that such a clear-cut distinction does not exist in practice, but the average conditions represented in these load graphs are good enough for the present purpose.

The rms demand values for representative *single days* are calculated first:

Sundays, and so forth,

$$\text{Rms demand} = \sqrt{\tfrac{1}{24}[(10^2 \times 9) + (20^2 \times 15)]}$$

$$= 17 \text{ kva.} \qquad [16]$$

Working days, April to September,

$$\text{Rms demand} = \sqrt{\tfrac{1}{24}[(20^2 \times 9) + (50^2 \times 4) + (70^2 \times 3) + (120^2 \times 8)]}$$

$$\doteq 77.3 \text{ kva.} \qquad [17]$$

Working days, October to March,

$$\text{Rms demand} = \sqrt{\tfrac{1}{24}[(30^2 \times 9) + (90^2 \times 6) + (140^2 \times 8) + (170^2 \times 1)]}$$

$$= 100.5 \text{ kva} \qquad [18]$$

The values used in Eqs. 16–18 are taken from the appropriate load graphs, Fig. 2. For example, on Sundays, a load of 10 kva is present for 9 hours and a load of 20 kva is present for 15 hours.

The *annual* rms demand is next determined from the daily rms values given by Eqs. 16–18; thus

$$\text{Annual rms demand} = \sqrt{\tfrac{1}{365}[(17^2 \times 117) + (77.3^2 \times 124) + (100.5^2 \times 124)]}$$

$$= 74.5 \text{ kva.} \tag{19}$$

At constant voltage,

$$\lambda = \frac{\text{annual rms kva demand}}{\text{transformer rating}} = \frac{74.5}{150} = 0.496 \tag{20}$$

$$\lambda^2 = 0.246. \tag{21}$$

The annual energy consumed in load losses now can be determined from Eq. 15. For the three single-phase transformers, plan A,

$$W_a = 8,760(0.246)(1.89) = 4,070 \text{ kwhr per year.} \tag{22}$$

For the three-phase transformer, plan B,

$$W_a = 8,760(0.246)(1.70) = 3,660 \text{ kwhr per year.} \tag{23}$$

Computation of Annual Core Losses: The energy W_c consumed in core loss is, from Eq. 1:

For plan A,

$$W_c = 8,760(0.69) = 6,050 \text{ kwhr per year.} \tag{24}$$

For plan B,

$$W_c = 8,760(0.48) = 4,210 \text{ kwhr per year.} \tag{25}$$

Computation of Demand Resulting from Losses: The additional demand caused by transformer losses depends on the peak load, which is 120/150, or 80% of rating for six months, and 170/150, or 113% of rating for six months each year. For simplicity, assume that the peak-load power factor is 90% or better, so that the kw demand resulting from losses adds almost numerically to the kva demand of the load. The effect of the losses on the demand is now tabulated:

	Plan A (Used transformers)	Plan B (New transformer)
Added maximum demand due to copper loss, April–September	$(0.80)^2 \times 1.89 = 1.21$ kw	$(0.80)^2 \times 1.70 = 1.09$ kw
Added maximum demand due to copper loss, October–March	$(1.13)^2 \times 1.89 = 2.40$ kw	$(1.13)^2 \times 1.70 = 2.18$ kw
Average added maximum demand due to copper loss	1.80 kw	1.64 kw
Added maximum demand due to core loss	0.69 kw	0.48 kw
Average added maximum demand due to total transformer losses	2.49 kw	2.12 kw

Final Cost Comparison: In each month of the year, the maximum demand of the load is higher than 50 kva, so that, for purposes of comparison between the two plans, the effect of transformer losses on the demand is charged at the $0.90 step of the demand charge. From the April-to-September load graph, the kilovolt-ampere-hour consumption during this period is $[(20 \times 9) + (50 \times 4) + (70 \times 3) + (120 \times 8)]$, or 1,550 kvahr per working day It is evident that, with any practical value of power factor, the monthly energy consumption by the load is greatly in excess of 5,000 kwhr. Hence the cost of the transformer losses is all charged for at the energy rate of 1 cent per kwhr. The annual costs of the two plans can now be compared, as in the following tabulation:

	Plan A (Used transformers)	Plan B (New transformer)
Investment	$891.00	$1,616.00
Annual costs		
Depreciation (Eq. 14)	$280.30	$146.60
Other fixed charges at 12.5% of installed cost	111.40	202.00
Demand charge due to losses at $0.90 per kva (12 × 0.90 × kw demand due to losses)	26.90	22.90
Energy charge due to copper losses at $0.01 per kwhr	40.70	36.60
Energy charge due to core losses	60.50	42.10
Maintenance, estimated	50.00	30.00
Total annual comparative cost chargeable to transformers	$569.80	$480.20

The comparison shows that, for the first three years of ownership, the annual costs of the new three-phase transformer will be $89.60 less than the annual costs of the used transformers. After three years, the depreciation on the used transformers (plan A) will have been written off, and the annual charges against them will then be $569.80 less $280.30, or $289.50. Thus if the service life of the used transformers should extend beyond their book life of three years, the new transformer will then cost more to own and operate than the used transformers, because the depreciation charges against the new transformer continue for seven more years. The excess annual cost of plan B over plan A during these remaining seven years is $480.20 less $289.50, or $190.70. If the firm's capital position is sound, the extra cost of owning the new transformer, taken over a ten-year period, probably would be justified by the greater assurance of reliability it affords. However, if the industrial firm's capital position is insecure, so that the least possible immediate outlay is the guiding factor, the used transformers would probably be selected, in the hope that their service usefulness would extend beyond three years. It is interesting that the difference between the annual cost of losses of the two plans, resulting from better efficiency of the more modern unit, is only $26.50, an inconsiderable amount compared to the total cost of electrical energy handled by the transformers.

5. INDIRECT COSTS

Besides the fixed and operating costs directly chargeable to it, a transformer causes increases in both the fixed and operating costs in the circuit which energizes it.

The indirect fixed costs resulting from a transformer are less obvious and relatively less important than the direct fixed costs discussed in Art. 2. The energy losses in a transformer must be supplied by a prime mover and generator and transmitted through other transformers, feeder voltage regulators, line conductors, switches, and other circuit devices. The generating and transmitting equipment must therefore be of adequate size to handle not only the useful load, but also the load resulting from transformer losses. The generating and transmitting equipment must also supply the reactive component of the transformer exciting current, and at the usual power factors of commercial loads the exciting current increases the kilovolt-ampere load on the system. In individual cases, these considerations may seem trivial, but the magnitudes involved in the aggregate load of a large system are important. For example, in 1935 one public-utility company[2] operated distribution transformers having an aggregate rating of 70,415 kva on distribution circuits having an aggregate peak load of 47,224 kva. If it is assumed that during the system peak the transformers operated at an average efficiency of 97 per cent, the generating stations had to develop about 1,400 kw to supply the transformer losses alone, disregarding the effect of the losses and exciting currents on the circuits between the generators and the transformers. The plant investment reserved to generate this 1,400-kw load represents fixed charges of the type listed as indirect fixed costs, item A2a in the outline of Art. 1.

Additional indirect fixed costs are those resulting from the transformer voltage regulation. In public-utility systems, tap-changers or feeder voltage regulators vary the transmission-line or feeder input voltage to compensate approximately for the total voltage drop in the system between source and load.* If the feeder voltage regulator causes a 10 per cent boost in voltage at full load, and if the voltage drop through the distribution transformers is 1.5 per cent, then 15 per cent of the regulator fixed costs is chargeable to the transformers served by the feeder that is connected to the regulator. This type of indirect fixed cost is indicated as item A2b in the outline of Art. 1.

Besides causing additional fixed expenses, the added load produced by the internal losses and exciting current of a transformer causes extra losses in the energizing circuit. The cost of these extra losses is an indirect

[2] C. H. Lewis and E. H. Snyder, " Distribution Transformer Load Supervising Methods," *Edison Electric Institute Bulletin 5* (1937), 329–333.

* Tap-changing transformers and voltage regulators are discussed in Ch. XIX.

operating expense chargeable to the transformer, and is listed as item B2a in the outline of Art. 1. These external losses depend upon the effect which the exciting current and the load caused by losses exert on the total current supplied to the transformer. At unity power-factor rated load, the exciting current has a negligible effect on the external circuit. However, as the magnitude of the load current is reduced, or the power factor is made more lagging, the exciting current exerts an increasingly appreciable effect on the current in the external circuit, so that a charge must be made for the additional losses caused by the transformer exciting current. This indirect operating cost is small compared to the transformer direct costs.

A portion of the losses in the feeder voltage regulator located in the circuit supplying the transformer is also chargeable as a transformer operating cost. The portion of the voltage-regulator losses chargeable to the transformer usually is determined from the ratio of per cent transformer voltage regulation to per cent voltage boost furnished by the regulator. This type of indirect operating cost is listed as item B2b in the outline of Art. 1.

6. MAINTENANCE COSTS

Transformers require very little maintenance, but certain types of attention listed under item B3 in the outline, Art. 1, should be given at definite intervals. The insulating oil in large power transformers and in underground transformers should be tested annually for dielectric strength. Moisture or sludge in the oil tends to reduce its dielectric strength, and thus to promote faults in the windings. If the examination or test of the oil indicates a substandard condition, the oil must be filtered, or, in especially serious cases, replaced. In some large power transformers, an inert gas fills the space in the tank above the oil, to prevent oxidation of the oil. The inert gas may " breathe " through chemicals which remove any traces of oxygen from the gas. The supply of inert gas and of deoxidizing chemicals must be maintained, involving a small maintenance expense. Minor repairs are sometimes needed, such as replacement of a cracked bushing, replacement of blown protective fuses, adjustment of circuit-breaker contacts, replacement of ruptured lightning arresters, testing of ground resistance, and maintenance of good grounding conditions.

An important maintenance activity in connection with transformers is the supervision of transformer loading, which, in public-utility companies, is organized on a definite basis with special men assigned to it. Satisfactory methods of studying the loads on transformers include:

(a) Periodic "spot checks" of transformer loads by means of split-core ammeters;

(b) Calculation of transformer loads from maps of territory served, making use also of records of consumers' meter readings and well-established load factors and diversity factors;

(c) Observation of the readings of permanently installed thermometers which indicate the maximum temperature reached by the transformer oil;

(d) Load chart records made by recording instruments (applicable only to large power installations, on account of the high cost of purchasing and maintaining recording instruments).

The methods most commonly employed are those listed as (a) and (c) above, although excellent results are obtained by the calculation method (b) in the companies that employ it. All the methods involve some expense. The references given in the bibliography furnish more detailed information concerning transformer load-testing practices.

7. DETERMINATION OF MOST ECONOMICAL LOSS RATIO

The designer has considerable control over the price of a transformer, since within certain limits the production cost is inversely proportional to the product of the core loss and the full-load copper loss.[3] Usually the designer can lower the cost of a transformer, and thereby decrease the fixed charges against it, by increasing either of its losses, with a corresponding increase in the operating cost.

The ratio of core loss to full-load copper loss also can be varied by the designer. Since the core loss is present whenever the transformer is energized but the copper loss is important only when the transformer is loaded, the loss ratio is an important factor in determining the operating cost for any specified load curve.

A general relation can be derived which is useful for determining the most economical loss ratio for any specified loss product and annual rms demand ratio. Let

$L_s \equiv$ known loss product in kilowatts squared for a standard transformer having a desired rating,

$P_c \equiv$ unknown core loss, kilowatts,

$P_{cu} \equiv$ unknown copper loss at rated load, kilowatts,

$m \equiv$ ratio of maximum kva demand of load to transformer kva rating,

$D \equiv$ demand charge for transformer losses, dollars per year per kva,

$E \equiv$ energy charge for transformer losses, dollars per kilowatthour.

[3] W. A. Sumner and J. B. Hodtum, "Realigning Transformers with Distribution," *Elec. W.*, *105* (1935), 1586–1588.

If the core loss and copper loss are added, and if it is assumed that the kva demand due to losses adds numerically to the kva demand due to load (as is very nearly true for load power factors near unity), the annual costs of transformer losses are

$$\text{Annual cost of energy} = 8{,}760E(P_c + \lambda^2 P_{cu}) \text{ dollars}, \qquad [26]$$

$$\text{Annual cost of demand} = D(P_c + m^2 P_{cu}) \text{ dollars}, \qquad [27]$$

Total annual cost of losses

$$= (D + 8{,}760E)P_c + (m^2 D + 8{,}760\lambda^2 E)P_{cu}. \qquad [28]$$

Since the coefficients $(D + 8{,}760E)$ and $(m^2 D + 8{,}760\lambda^2 E)$ appear throughout the ensuing derivations, let

$$a \equiv D + 8{,}760E, \qquad [29]$$

$$b \equiv m^2 D + 8{,}760\lambda^2 E. \qquad [30]$$

In Eqs. 29 and 30, the symbol a represents the core-loss component of unit cost, expressed in dollars per year per kilowatt of core loss; while b represents the copper-loss component of unit cost, expressed in dollars per year per effective kva demand.* From the definitions of the symbols,

$$P_{cu} = \frac{L_s}{P_c}. \qquad [31]$$

Substituting Eqs. 29, 30, and 31 in Eq. 28 gives

$$\text{Total annual cost of losses} = aP_c + \frac{bL_s}{P_c}. \qquad [32]$$

For the lowest possible annual cost, the derivative with respect to P_c of the annual cost of losses is equated to zero:

$$\frac{d}{dP_c}(\text{annual cost}) = 0 = a - \frac{bL_s}{P_c^2}. \qquad [33]$$

from which

$$P_c = \sqrt{\frac{bL_s}{a}}, \qquad \blacktriangleright[34]$$

and

$$P_{cu} = \frac{L_s}{P_c} = \sqrt{\frac{aL_s}{b}}. \qquad \blacktriangleright[35]$$

The loss ratio for the lowest annual cost of losses is

$$\text{Loss ratio} = \frac{P_{cu}}{P_c} = \frac{L_s}{P_c^2} = \frac{a}{b}. \qquad \blacktriangleright[36]$$

* By effective kva demand is meant the demand during the entire year, corrected for its effect on cost as determined by the form of the electric rate.

Equation 36 states that, for maximum operating economy, the ratio of full-load copper loss to core loss should equal the ratio of the annual cost per kilowatt of core loss to the annual cost of copper loss per effective kva demand. The expression for the loss ratio, Eq. 36, is portrayed graphically in Fig. 3, which shows the effect of variations in any of the four quantities m, λ, D, and E. The outstanding fact is the marked negative slope of all the curves when the loss ratio for maximum economy is plotted against the annual rms demand ratio λ. The negative slope

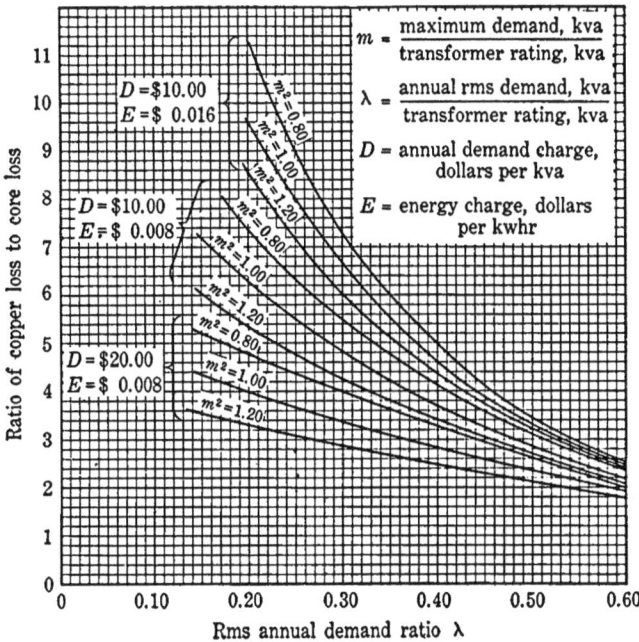

FIG. 3.   Loss ratio for lowest annual cost of losses as a function of the variables m, λ, D, and E.

indicates, as does Eq. 36, that as the copper loss increases because of increasing loads, the desirable ratio of copper loss to core loss decreases. It is thus apparent that in general a substation transformer, whose load is likely to have a relatively high rms demand ratio λ, should have a lower loss ratio than a distribution transformer.

The theoretical solution above gives no numerical indication of the seriousness of deviations from the optimum values. It is therefore of interest to compute the results of varying the loss ratio in a typical case. A standard 10-kva distribution transformer of 1940 design has a loss product of approximately 10,700 watts squared, or 0.0107 kw squared. Suppose it is desired to study the costs of losses of such a 10-kva transformer, with the loss product kept constant. Assume the following operating conditions:

$$m^2 = 1.15$$
$$\lambda^2 = 0.16$$

$D = \$10.00$ per year per kva demand

$E = \$0.009$ per kwhr.

From Eq. 36, the loss ratio for maximum economy is

$$\textbf{Loss ratio} = \frac{10 + 8,760(0.009)}{1.15(10) + 8,760(0.16)(0.009)} = 3.69 \qquad [37]$$

The annual cost of the losses is shown in Table I for several different loss ratios.

TABLE I

COMPARISON OF ANNUAL COSTS OF LOSSES IN A 10-KVA DISTRIBU-
TION TRANSFORMER, AS AFFECTED BY LOSS RATIO

1. Loss ratio (assumed)	6.00	4.50	3.69*	2.80	2.00
2. Product of losses, kw^2	0.0107	0.0107	0.0107	0.0107	0.0107
3. Core loss P_c, kw	0.0422	0.0488	0.0539	0.0619	0.0731
4. Copper loss P_{cu}, kw	0.254	0.219	0.199	0.173	0.146
5. Annual energy used in core losses, kwhr	370	428	472	542	640
6. Annual energy used in copper losses, kwhr	356	307	279	243	204
7. Total annual loss, kwhr	726	735	751	785	844
8. Maximum demand due to copper loss, kw	0.292	0.251	0.229	0.199	0.168
9. Maximum demand due to total loss, kva	0.334	0.300	0.283	0.261	0.241
10. Annual energy cost, dollars	6.54	6.62	6.76	7.07	7.59
11. Annual demand cost, dollars	3.34	3.00	2.83	2.61	2.41
12. Total annual cost of losses	9.88	9.62	9.59	9.68	10.00

* Theoretical ratio for lowest annual cost

$$\text{Line 3} = \sqrt{\frac{\text{line 2}}{\text{line 1}}}$$

$$\text{Line 4} = \frac{\text{line 2}}{\text{line 3}}$$

Line 5 = 8,760 × line 3
Line 6 = 8,760 λ^2 × line 4 = 1,400 × line 4
Line 7 = line 5 + line 6
Line 8 = m^2 × line 4 = 1.15 × line 4
Line 9 = line 3 + line 8
Line 10 = E × line 7 = \$0.009 × line 7
Line 11 = D × line 9 = \$10.00 × line 9
Line 12 = line 10 + line 11

Table I helps to preserve one's sense of proportion. The impression may have been gained from Eq. 36 that the optimum value of loss ratio gives annual costs markedly lower than those resulting from other loss ratios. The tabulation shows that a reasonable deviation from the theoretical optimum loss ratio is not serious, so that the loss ratio selected may be any value, reasonably near the correct one, which is most practical from the standpoint of the transformer manufacturer. It would be better to use transformers whose loss ratio P_{cu}/P_c is somewhat lower than the optimum value, rather than higher, in order to achieve maximum economy over a period of years during which the rms demand ratio λ will probably tend to increase. As distribution transformers are used in large numbers, a wide deviation from the theoretical loss ratio for maximum economy would be wasteful. For example, if in the case studied here, a loss ratio of 2.00 were used instead of 3.69, the annual cost of losses per thousand 10-kva transformers would be $410 higher than necessary, a possible saving that amply justifies the cost of a study to determine the best loss ratio. Standard distribution transformers in general have loss ratios appropriate for the load conditions prevailing in practice at the time of manufacture, and the transformer subcommittee of the American Institute of Electrical Engineers provides an opportunity for manufacturers and users to discuss trends in load conditions and to recommend future policies in the design of transformers for lowest cost of losses.

8. DETERMINATION OF MOST ECONOMICAL LOSS PRODUCT

The price of a transformer, within limits, is inversely proportional to the product of its core loss and its full-load copper loss. When a relatively high rate is paid for electricity, and under very heavy load conditions, the purchase of a transformer whose loss product is lower than standard and whose price is correspondingly higher may be justified. Conversely, if the loss product of a transformer is increased, with appropriate special provisions for cooling, the resulting lower price may be very desirable in some circumstances. The ideal combination of loss product and loss ratio can be determined analytically.

In addition to the symbols used in Art. 7, let

$C_s \equiv$ price in dollars of transformer having the standard loss product, L_s.

$l \equiv P_c P_{cu} \equiv$ loss product in kilowatts squared for maximum over-all economy, including fixed charges,

$c \equiv$ price in dollars of transformer having the loss product l,

$F \equiv$ rate* of annual fixed charges, decimal part of first cost.

---

* Includes interest, taxes, insurance, depreciation, overhead, and so forth.

Since the price is assumed to be inversely proportional to the loss product,

$$c = \frac{C_s L_s}{l} \qquad [38]$$

and, from the definitions of the symbols,

$$P_{cu} = \frac{l}{P_c}. \qquad [39]$$

The annual fixed charge is cF, or $C_s L_s F/l$, which is added to Eq. 32 (replacing L_s in Eq. 32 by l) to give the total annual cost. For convenience let

$$g = C_s L_s F. \qquad [40]$$

Then,

Total annual cost of losses and fixed charges $= aP_c + \dfrac{bl}{P_c} + \dfrac{g}{l}.$ [41]

To find the minimum annual cost, the two partial derivatives of the total annual cost are equated to zero, in turn.[4] Then

$$\frac{\partial}{\partial l}\left[aP_c + \frac{bl}{P_c} + \frac{g}{l}\right] = 0 \qquad [42]$$

$$\frac{b}{P_c} - \frac{g}{l^2} = 0 \qquad [43]$$

$$l = \sqrt{\frac{gP_c}{b}} \qquad [44]$$

$$\frac{\partial}{\partial P_c}\left[aP_c + \frac{b}{P_c}\sqrt{\frac{gP_c}{b}} + g\sqrt{\frac{b}{gP_c}}\right] = 0 \qquad [45]$$

$$a - \sqrt{\frac{gb}{P_c^3}} = 0 \qquad [46]$$

$$P_c = \sqrt[3]{\frac{bg}{a^2}} \qquad \blacktriangleright[47]$$

$$l = \sqrt{\frac{gP_c}{b}} = \sqrt[3]{\frac{g^2}{ab}} \qquad \blacktriangleright[48]$$

$$P_{cu} = \frac{l}{P_c} = \sqrt[3]{\frac{ag}{b^2}}. \qquad \blacktriangleright[49]$$

[4] See F. S. Woods, *Advanced Calculus* (new ed.; Boston: Ginn & Company, 1934), 117.

The loss ratio P_{cu}/P_c for maximum over-all economy is

$$\text{Loss ratio } \frac{P_{cu}}{P_c} = \frac{a}{b},$$

▶[50]

as in Eq. 36.

The loss product l as found from Eq. 48 may be less than, equal to, or greater than the known loss product L_s of a standard self-cooled transformer. If l is less than L_s, the purchase of a transformer at more than the standard price may be justified, as the saving in cost of losses may be greater than the additional fixed charges resulting from the higher price.

If l is greater than L_s, so that it is theoretically desirable to purchase a transformer whose loss product is higher than standard, it is possible to purchase a transformer

(a) of standard design, or
(b) of efficiency lower than standard, or
(c) of self-cooled rating lower than the standard self-cooled rating required for the load, but with an auxiliary cooling system for use during peak loads.

For a distribution application, (a) and (b) are the alternatives to be followed, in general. Certain lines of distribution transformers are manufactured in two efficiency classes called, respectively, high-efficiency and low-efficiency transformers. If a low-efficiency transformer which is suitable for the particular application in question is manufactured, its over-all costs are probably less than the over-all costs of the standard high-efficiency transformer, when l is greater than L_s. For a substation application, it is common practice to purchase a transformer having two ratings, one as a self-cooled transformer and one with fan cooling, as suggested in alternative (c).

9. SUMMARY OF TRANSFORMER COST DISCUSSION

Transformers are representative of all electric power equipment as far as the major elements of cost are concerned. Therefore, a study of transformer cost factors serves as a guide to methods of cost analysis for other types of power apparatus. However, transformers have such high efficiency that the numerical differences resulting from variations in the cost-controlling factors are, in general, not large; hence the use of absolute optimum values of the cost-controlling factors is not necessary.

The annual loss resulting from core loss is a constant, but the annual loss from copper loss is a function of the ratio of the annual rms demand to the full-load rating of the transformer, a low rms demand ratio indicating a high copper-to-core loss ratio and a high rms demand ratio indicating a low loss ratio, for economy. Fixed charges are usually the most expensive element in transformer costs, losses are somewhat less costly, while maintenance expenses are small compared to the costs of fixed charges and losses. It is good practice to overload transformers for short periods,* within the limits set by the American Standards Association (or the appropriate standards that were in vogue at the time the transformer was manufactured), in order to realize the lower fixed charges resulting from the saving in transformer size. Large allowances for future growth should not be made when a transformer rating is selected, except in localities where load increases are known to be occurring rapidly.

PROBLEMS

1. A manufacturer purchases power for his factory from a public-utility company at 2,400 v. The utility company's meter registers not only the actual demand and energy requirements of the factory load but also, in addition, the losses in the factory's transformer, which reduces the voltage to the value used in the factory. The meter is read to the nearest 0.5 kw of demand and the nearest 10 kwhr per month of energy usage. Power is purchased in accordance with the following rate:

Monthly demand charge

First 10 kw of demand	$2.00 per kw
Next 50 kw of demand	1.50 per kw
Additional demand	1.00 per kw

Energy charge

First 1,200 kwhr per month	$0.035 per kwhr
Next 1,300 kwhr per month	0.02 per kwhr
Additional energy	0.0125 per kwhr

Discount for purchasing power at 2,400 v, and owning transformers	10% of total bill each month

The manufacturer's substation consists of a 75-kva transformer, originally purchased 17 years ago, at which time the books of the firm were set up to write off the investment in the transformer in 15 years. Auxiliary equipment, including fusing protection and lightning arresters, have been maintained and replaced when necessary, and are still in satisfactory condition to continue in service. The power requirements of the factory are now expected to increase, and two possible solutions are available, as follows:

(1) A dealer in used transformers offers to supply a used 75-kva transformer of characteristics identical with the one already in service, at a delivered price of $120, and carrying the dealer's guarantee against defective operation for a period of one year. This transformer could be placed in parallel with the existing transformer at a cost of $60.00, according to an electrical contractor's bid. A reasonable life expectation on the two 75-kva transformers is 5 years.

* See Art. 8, Ch. XIV.

(2) A manufacturer bids to supply a new 150-kva transformer at a price of $753, f.o.b. factory. Freight on the transformer costs $18.00, and the contractor's estimate for delivery and installation is $60.00. The used-transformer dealer offers $60.00 for the old 75-kva transformer and will remove it at his expense. The losses of *each* of the old 75-kva transformers are:

No-load loss	400 w
Full-load copper loss	655 w

The losses of the new 150-kva transformer are:

No-load loss	540 w
Full-load copper loss	1,240 w

The manufacturer's accounting practice calls for the following charges:

Interest	7% per year on outstanding investment
Taxes	Not involved in this problem
Insurance	$1\frac{1}{2}\%$ of purchase price
Depreciation	Straight-line method — 15 years on new transformer, 5 years on used transformer
Maintenance	No difference in favor of either plan, but there is the probability that the old transformer will soon require an oil replacement at 25 cents per gallon. The 75-kva transformer contains 54 gallons of oil.

The estimated monthly average load curve for the new load is:

Kva in per cent of full load rating	Power factor in per cent	Hours per month at stated load
100	80	72
75	70	216
25	60	72
5	100	360

Determine which of the two alternatives the manufacturer should adopt, under each of the following circumstances:

(a) The manufacturer's business is very steady and enjoys an excellent credit rating. The firm has an established reputation for prompt shipments, and therefore places a very high value on the loss caused by shut-downs.

(b) The business is unsteady, credit rating marginal, and what business is carried on results from a low-priced product rather than from a high reputation for quality and service.

Note: An installation of this size would usually be three-phase, but the problem is entirely representative, even though solved as a single-phase problem.

2. The engineering department of a transformer manufacturing company is making a study to determine the best loss ratio for a new design of distribution transformers. Data for this study are furnished by several operating utility companies, representing considerable divergence of policy concerning load-building effort, rate structure, and engineering practice. The data received from each utility company are to be studied to determine the best transformer design for each company's requirements, after which a standard design will be made which appears to be the best compromise for general sale to the public-utility field.

Nine utility companies supply data concerning their practices. Some of these companies are willing to evaluate a portion of the transformer losses at the rate that would be paid for the energy by the company's customers, while other companies evaluate the losses at the costs of energy delivered to the transformers. The cost of losses in each of the nine companies is tabulated below:

Company Number	Demand charge D in dollars per year per kva maximum demand	Energy charge E in dollars per kilowatthour
1	0	0.030
2	0	0.020
3	0	0.010
4	10.00	0.030
5	10.00	0.020
6	10.00	0.010
7	24.00	0.020
8	24.00	0.010
9	24.00	0.005

This problem demonstrates the effect of load conditions and of the electric rate upon the desirable loss ratio. The students can be divided into groups, each group to complete the solution applying to the rate for losses in one of the utility companies. Each group of students is to determine the loss ratio P_{cu}/P_e which results in lowest annual operating cost for one of the companies listed above, using values of λ of 0.20, 0.30, 0.40, 0.50, and 0.60, and assuming that the maximum demand is rated load. Also, the saving in annual cost of losses resulting from using this optimum loss ratio rather than the standard loss ratio is to be determined for each value of λ for a group of one hundred 15-kva transformers. The standard loss ratio of a 15-kva distribution transformer can be taken as 3.40, and the standard loss product as 20,100 w^2.

The desired loss ratios as found by the various student groups can be plotted against the rms demand ratio λ for each electric rate, as a check on the accuracy of the results and to bring out the effect of changing the variables.

CHAPTER XVII

Self- and Mutual Inductances

As shown in Ch. XII, the fluxes linking the primary and secondary windings of a loaded transformer can be expressed in either of two ways. One method of combining the components represents the resultant fluxes linking each winding as the sum of a leakage flux and a resultant mutual flux;* thus

$$\varphi_1 = \varphi_{\ell 1} + \varphi \qquad [1]$$

$$\varphi_2 = \varphi_{\ell 2} + \varphi, \qquad [2]$$

where

φ_1 and φ_2 are the resultant fluxes linking the primary and secondary windings respectively,

$\varphi_{\ell 1}$ and $\varphi_{\ell 2}$ are the leakage fluxes,

φ is the resultant mutual flux produced by the simultaneous effects of the currents in both windings.

The other method expresses the resultant fluxes φ_1 and φ_2 as the sum of component fluxes produced by each current acting by itself;† thus

$$\varphi_1 = \varphi_{11} + \varphi_{12} \qquad [3]$$

$$\varphi_2 = \varphi_{22} + \varphi_{21}, \qquad [4]$$

where

φ_{11} is the flux linking winding 1 due to current i_1 acting by itself,

φ_{22} is the flux linking winding 2 due to current i_2 acting by itself,

φ_{21} and φ_{12} are the component mutual fluxes produced by the currents in windings 1 and 2 respectively.

Equations 1 and 2 are the basis for the method of analysis discussed in Arts. 3 and 4, Ch. XII, and developed in detail in Chs. XIII and XIV, while Eqs. 3 and 4 are the basis for the classical theory of electromagnetically coupled circuits, by which the voltage equations can be written as

$$v_1 = R_1 i_1 + L_1 \frac{di_1}{dt} + M \frac{di_2}{dt} \qquad \blacktriangleright[5]$$

$$v_2 = R_2 i_2 + L_2 \frac{di_2}{dt} + M \frac{di_1}{dt}, \qquad \blacktriangleright[6]$$

* See Eqs. 22 and 23, Ch. XII.
† See Eqs. 19 and 20, Ch. XII.

where

> v_1 and v_2 are the instantaneous terminal voltages of the windings,
> i_1 and i_2 are the instantaneous currents in them,
> R_1 and R_2 are their effective resistances,
> L_1 and L_2 are their self-inductances,
> M is the mutual inductance.

The purpose of this chapter is to develop transformer theory from the coupled-circuit point of view.[1]

Equations 5 and 6 neglect the distributed capacitances of the windings. They also neglect core loss and, because the equations are based on superposition of component fluxes produced by each current acting by itself, they imply linear relations between each component flux and the current producing it. The core loss in an iron-core transformer usually has little effect on any of the operating characteristics except the efficiency, and, in spite of the nonlinear magnetic characteristics of iron, Eqs. 5 and 6 are often applied to the analysis of such transformers. It is interesting to compare this method of analysis with the analytical theory based on the leakage and resultant mutual flux components of Eqs. 1 and 2, and to review some of the important points discussed in Ch. XII.

On the basis of Eqs. 1 and 2 it is shown in Art. 5, Ch. XII, that a transformer can be represented by the equivalent circuit of Fig. 6, Ch. XII, consisting of an ideal transformer whose ratio of transformation equals the ratio of the number of turns in the windings, in combination with a network representing the imperfections of the actual transformer. This equivalent circuit shows the factors that are important in determining the electrical characteristics of transformers, and its component parts have rather clear physical significances with relation to the magnetic conditions in a loaded transformer. Thus the leakage fluxes induce voltage drops which may be accounted for by leakage inductances in series with the ideal transformer, and since the leakage fluxes are partly in air these leakage inductances are approximately constant. In an iron-core transformer, the resultant mutual flux φ approximately equals the flux in the core and therefore the nonlinearity of the magnetic characteristics of the core is assumed to affect only the relation between this resultant mutual flux and the exciting current. If the exciting current is small compared with the load component of the primary current, or, if the amplitude of the variation of the resultant mutual flux is relatively small, the effects of magnetic nonlinearity are relatively unimportant, and an iron-core transformer under such conditions be-

---

[1] A brief discussion of transformers from the coupled-circuit equations is given in the volume on electric circuits (1940), Arts. 10 and 11, Ch. VI.

haves as an essentially linear circuit element. Hence the classical theory of linear coupled circuits, as expressed in Eqs. 5 and 6, is often applied to iron-core transformers, particularly in the analysis of communication circuits and of transformers having more than two windings.* Although the classical theory in terms of self- and mutual inductances does not lead to so clear a physical conception of the *internal* phenomena in an iron-core transformer as does the analysis in terms of leakage fluxes and a resultant mutual flux, it nevertheless is equally useful for determination of the *external* characteristics of a transformer as a circuit element when core loss and magnetic nonlinearity are neglected.

1. THE COUPLED–CIRCUIT EQUATIONS

Consider the transformer shown in Fig. 1. When the core loss and distributed capacitances are neglected and the inductances are assumed

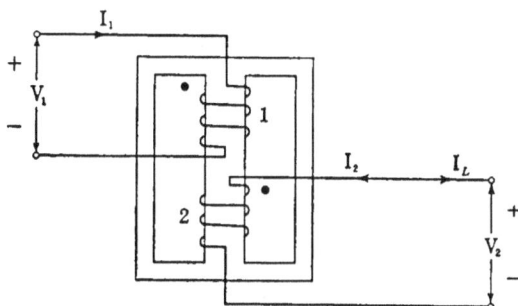

FIG. 1.   Positive directions of currents and voltages in Eqs. 7, 8, 9, and 10.

to be constant, the instantaneous voltage equations can be written as in Eqs. 5 and 6. If the currents and voltages are assumed to vary sinusoidally, these equations can be expressed in vector form as

$$V_1 = (R_1 + j\omega L_1)I_1 + j\omega M I_2 \qquad\qquad \blacktriangleright[7]$$

$$V_2 = (R_2 + j\omega L_2)I_2 + j\omega M I_1, \qquad\qquad \blacktriangleright[8]$$

where V_1 and V_2 are the vectors representing the terminal voltages of the two windings, and I_1 and I_2 are the vectors representing the currents in them. The algebraic signs in these equations depend on the assumed positive directions of the currents and voltages. In Eqs. 7 and 8, V_1 and V_2 represent the falls in potential through the windings in the same directions around the core, while I_1 and I_2 represent the currents in these same positive directions. Since positive currents in the directions of I_1 and I_2, Fig. 1, produce component fluxes in the same direc-

* Multicircuit transformers are discussed in Ch. XXVII.

tion, the self- and mutual-inductance terms in Eqs. 7 and 8 have the same signs. Because the secondary current usually produces a magneto-motive force which tends to oppose that of the primary current, the currents I_1 and I_2 usually are approximately opposite in phase. The direction of power flow is indicated by the phase relations between the currents and voltages. Thus if winding 1 acts as the primary, V_1 and I_1 differ in phase by less than 90 degrees but V_2 and I_2 differ in phase by more than 90 degrees.

Because of their symmetry, Eqs. 7 and 8 are convenient for the discussion of general problems in which the direction of the power flow is not specified. However, when winding 2 is known to be supplying power to a load, it may appear more natural to assume the positive direction of the current supplied to the load to be in the positive direc-

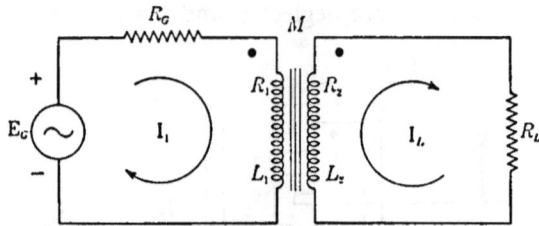

FIG. 2.   Circuit consisting of a generator, transformer, and resistance load.

tion of the fall in potential V_2 through the load, as shown by the arrow I_L in Fig. 1. Usually during most of the half-cycle when current is entering the dot-marked terminal of winding 1 current is simultaneously leaving the dot-marked terminal of winding 2, and therefore the currents I_1 and I_L usually are approximately in phase. In terms of the vector current I_L delivered to the load, the voltage equations can be written as

$$V_1 = (R_1 + j\omega L_1)I_1 - j\omega M I_L \qquad \blacktriangleright [9]$$

$$V_2 = j\omega M I_1 - (R_2 + j\omega L_2)I_L. \qquad \blacktriangleright [10]$$

For further discussion, consider the circuit of Fig. 2, consisting of a generator whose internal resistance is R_G, connected to the primary terminals of a transformer whose secondary supplies power to a resistance load R_L. This circuit is of considerable importance in communication networks, where the generator usually is a vacuum tube. In the volume on electronics it is shown that a vacuum tube can often be represented by a generator of constant internal resistance whose internal electromotive force is proportional to the voltage impressed on its grid. The output circuit of the tube often contains a battery (not shown in Fig. 2), which produces a direct current in the primary winding of the transformer. The fluxes produced by the alternating generator voltage are superposed on the magnetization produced by the direct current, and therefore, in so far as alternating currents are concerned, the effective inductances of the transformer, L_1, L_2, and M, are the incremental inductances, discussed in Arts. 14 and 15, Ch. VI. If the alternating generator voltage and the alternating components of the primary and secondary currents are assumed to vary sinusoidally, the vector equations for the circuit of Fig. 2 are

$$E_G = (R_G + R_1 + j\omega L_1)I_1 - j\omega M I_L \qquad \blacktriangleright[11]$$

$$0 = -j\omega M I_1 + (R_L + R_2 + j\omega L_2)I_L, \qquad \blacktriangleright[12]$$

where E_G is the vector representing the internal electromotive force of the generator and I_1, I_L are the vectors representing the alternating currents in the primary and secondary circuits. The positive directions are indicated in Fig. 2. These equations can be written in convenient and compact form in terms of the self- and mutual impedances of the circuits. Thus

$$E_G = Z_{11}I_1 + Z_{12}I_L \qquad [13]$$

$$0 = Z_{12}I_1 + Z_{22}I_L. \qquad [14]$$

For the circuit of Fig. 2, the self-impedances Z_{11} and Z_{22} are

$$Z_{11} = R_{11} + jX_{11} \qquad [15]$$

$$Z_{22} = R_{22} + jX_{22}, \qquad [16]$$

where

$R_{11} = R_G + R_1 =$ self-resistance of the loop comprising the generator and transformer primary,   [17]

$R_{22} = R_L + R_2 =$ self-resistance of the loop comprising the load and transformer secondary,   [18]

$X_{11} = \omega L_1 =$ self-reactance of the primary loop,   [19]

$X_{22} = \omega L_2 =$ self-reactance of the secondary loop.   [20]

For the polarities and positive directions shown in Fig. 2, the mutual impedance Z_{12} is

$$Z_{12} = -jX_m, \qquad [21]$$

where

$$X_m = \omega M = \text{mutual reactance.} \qquad [22]$$

From Eqs. 13 and 14, expressions can readily be obtained for the vector primary and secondary currents. From Eq. 14, the vector relation between the currents is

$$I_L = -\frac{Z_{12}}{Z_{22}}I_1. \qquad [23]$$

Substituting Eq. 23 in Eq. 13 gives

$$E_G = \left(Z_{11} - \frac{Z_{12}^2}{Z_{22}}\right)I_1. \qquad [24]$$

From Eq. 24,

$$I_1 = \frac{Z_{22}E_G}{Z_{11}Z_{22} - Z_{12}^2}, \qquad [25]$$

and from Eqs. 23 and 25,

$$I_L = \frac{-Z_{12}E_G}{Z_{11}Z_{22} - Z_{12}^2}. \qquad [26]$$

If the impedances in Eqs. 25 and 26 are expressed in terms of their resistance and reactance components, Eq. 25 becomes

$$I_1 = \frac{(R_{22} + jX_{22})E_G}{(R_{11} + jX_{11})(R_{22} + jX_{22}) + X_m^2} \qquad [27]$$

$$= \frac{(R_{22} + jX_{22})E_G}{R_{11}R_{22} - (X_{11}X_{22} - X_m^2) + j(R_{11}X_{22} + R_{22}X_{11})} \qquad \blacktriangleright[28]$$

Similarly Eq. 26 becomes

$$I_L = \frac{jX_mE_G}{R_{11}R_{22} - (X_{11}X_{22} - X_m^2) + j(R_{11}X_{22} + R_{22}X_{11})}. \qquad \blacktriangleright[29]$$

Equations 28 and 29 express the vector currents in the primary and secondary circuits of Fig. 2 in terms of the generator voltage E_G and the parameters of the transformer and its connected circuits.

2. COEFFICIENT OF COUPLING AND LEAKAGE COEFFICIENT

At this point it is convenient to introduce two new factors — the coefficient of coupling and the leakage coefficient. By definition,[2] the coefficient of coupling between two magnetically coupled circuits is

$$k \equiv \frac{M}{\sqrt{L_1L_2}}, \qquad \blacktriangleright[30]$$

or in terms of the self- and mutual reactances

$$k = \frac{X_m}{\sqrt{X_{11}X_{22}}} \qquad [31]$$

The coefficient of coupling is a measure of the closeness with which the magnetic fields of the two circuits are interlinked. If there were no

[2] See the volume on electric circuits (1940), p. 386.

magnetic leakage, the coefficient of coupling k would be unity, which is its greatest possible value. It is not unusual for the coefficient of coupling of an iron-core transformer to be as high as 0.998, but with air-core coils values of k greater than about 0.5 are attained only with special care in design. The leakage coefficient σ is defined as

$$\sigma \equiv 1 - k^2 \qquad\qquad \blacktriangleright[32]$$

$$= 1 - \frac{M^2}{L_1 L_2} = 1 - \frac{X_m^2}{X_{11} X_{22}}. \qquad [33]$$

Since the characteristics of magnetically coupled circuits depend to a considerable extent on magnetic leakage, the leakage coefficient is an important quantity in the theory of transformers and also in that of many other types of electromagnetic apparatus.

The significance of the coefficient of coupling in terms of the fluxes concerned in transformer theory can be shown by means of the relations among those fluxes and the self- and mutual inductances. By definition, the self-inductance of a winding is the flux linkage with the winding due to its own current, divided by the current; whence

$$L_1 \equiv \frac{\lambda_{11}}{i_1} = \frac{N_1 \varphi_{11}}{i_1} \qquad [34]$$

$$L_2 \equiv \frac{\lambda_{22}}{i_2} = \frac{N_2 \varphi_{22}}{i_2}, \qquad [35]$$

where

λ_{11} and λ_{22} are the flux linkages produced by the currents i_1 and i_2 respectively,

φ_{11} and φ_{22} are equivalent fluxes.*

Similarly, the mutual inductance of two windings is defined as the flux linkage with one winding per unit current in the other winding. In a conservative system, the mutual inductance of winding 1 with respect to winding 2 equals that of winding 2 with respect to winding 1; thus[3]

$$M \equiv \frac{\lambda_{21}}{i_1} = \frac{N_2 \varphi_{21}}{i_1} \qquad [36]$$

$$\equiv \frac{\lambda_{12}}{i_2} = \frac{N_1 \varphi_{12}}{i_2}, \qquad [37]$$

---

* The concept of equivalent fluxes is discussed in Art 1, Ch. XII.

[3] See R. R. Lawrence, *Principles of Alternating Currents* (2d ed.; New York: McGraw-Hill Book Company, Inc., 1935), 187.

where

λ_{21} is the flux linkage with winding 2 produced by current i_1,

λ_{12} is the flux linkage with winding 1 produced by current i_2,

φ_{21} and φ_{12} are equivalent mutual fluxes.

From these equations, the fluxes can be expressed in terms of the inductances; thus

$$\varphi_{11} = \frac{L_1 i_1}{N_1} \qquad [38]$$

$$\varphi_{22} = \frac{L_2 i_2}{N_2} \qquad [39]$$

$$\varphi_{21} = \frac{M i_1}{N_2} \qquad [40]$$

$$\varphi_{12} = \frac{M i_2}{N_1} \qquad [41]$$

Let k_1 be defined as the ratio of the mutual flux φ_{21} produced by the current i_1 to the total flux φ_{11} produced by i_1; thus

$$k_1 \equiv \frac{\varphi_{21}}{\varphi_{11}}. \qquad [42]$$

If the fluxes are expressed in terms of the self- and mutual inductances (as in Eqs. 38 and 40), Eq. 42 becomes

$$k_1 = \frac{\dfrac{M i_1}{N_2}}{\dfrac{L_1 i_1}{N_1}} = \frac{\dfrac{N_1}{N_2} M}{L_1}. \qquad [43]$$

Similarly, let k_2 be defined as the ratio of the mutual flux φ_{12} produced by the current i_2 to the total flux φ_{22} produced by i_2; then

$$k_2 \equiv \frac{\varphi_{12}}{\varphi_{22}} = \frac{\dfrac{N_2}{N_1} M}{L_2}. \qquad [44]$$

Multiplication of Eqs. 43 and 44 gives

$$k_1 k_2 = \frac{M^2}{L_1 L_2}. \qquad [45]$$

From Eq. 30, the right-hand side of Eq. 45 is the square of the coefficient of coupling, whence the coefficient of coupling k can be expressed as

$$k = \sqrt{k_1 k_2} = \sqrt{\frac{\varphi_{21}}{\varphi_{11}} \frac{\varphi_{12}}{\varphi_{22}}}. \qquad [46]$$

Therefore the coefficient of coupling k equals the geometric mean of the ratios of mutual to total flux produced by the separate effects of each current.

In Eqs. 28 and 29, the currents in the circuit of Fig. 2 are expressed in terms of the self- and mutual reactances of the transformer and the primary and secondary circuit resistances. The mutual reactance can be eliminated from these equations and the coefficient of coupling and leakage coefficient can be introduced in them. From Eq. 31,

$$X_m = k \sqrt{X_{11}X_{22}} \qquad\qquad [47]$$

and from Eq. 33

$$X_{11}X_{22} - X_m^2 = \sigma X_{11}X_{22}. \qquad\qquad [48]$$

On substitution of Eqs. 47 and 48 in Eqs. 28 and 29, the currents in the circuit of Fig. 2 can be expressed as

$$I_1 = \frac{(R_{22} + jX_{22})E_G}{R_{11}R_{22} - \sigma X_{11}X_{22} + j(R_{11}X_{22} + R_{22}X_{11})} \qquad \blacktriangleright[49]$$

$$I_L = \frac{jk\sqrt{X_{11}X_{22}}\, E_G}{R_{11}R_{22} - \sigma X_{11}X_{22} + j(R_{11}X_{22} + R_{22}X_{11})}. \qquad \blacktriangleright[50]$$

With an air-core transformer, the self- and mutual inductances are constant and the leakage coefficient is relatively large. The self- and mutual inductances can readily be determined from open-circuit tests and, if the effects of distributed capacitances are negligible, the performance of an air-core transformer can be computed satisfactorily through substitution of the measured values of the resistances and the self- and mutual reactances in Eqs. 28 and 29.

Quite a different situation arises, however, when the transformer has an iron core. Consider the quantity $X_{11}X_{22} - X_m^2$ in the denominators of Eqs. 28 and 29. Since the coefficient of coupling is nearly unity, X_m^2 is only slightly less than $X_{11}X_{22}$, and therefore very poor accuracy would result if the currents were computed by use of measured values of the self- and mutual reactances in Eqs. 28 and 29, since the important quantity $X_{11}X_{22} - X_m^2$ would then be determined from the difference between two large and nearly equal components. This difficulty is avoided in Eqs. 49 and 50, since the leakage coefficient can be determined accurately from open-circuit and short-circuit tests, as described in Art. 3.

3. MEASUREMENT OF THE PARAMETERS

When core loss, magnetic nonlinearity, and the effects of distributed capacitances are negligible, the electrical characteristics of a two-winding transformer can be expressed in terms of five independent parameters. In the coupled-circuit equations, Eqs. 5 and 6, these parameters are the effective resistances and self-inductances of the two windings and the mutual inductance between them. Since there are five independent parameters in the basic coupled-circuit equations, the number of independent parameters in all equations derived from them is five also, even when the five parameters are expressed in different forms which may be more convenient than the original coupled-circuit equations. For example, the currents in the circuit of Fig. 2 are expressed in Eqs. 49 and 50 in terms of the two winding resistances, the two self-reactances, and the leakage coefficient σ — five parameters of the transformer. The coefficient of coupling k which appears in Eq. 50 is not a sixth independent parameter, because it can be determined from the leakage coefficient σ by means of Eq. 32.

In order to predict the performance of a transformer, therefore, five independent characteristic constants must be known. The simple tests that may be performed to determine the five parameters are:

(1) Measurements of the direct-current resistances of the two windings,
(2) Measurements of the two self-inductances,
(3) Measurements of the mutual inductance,
(4) Measurements of the voltage ratio with one winding open-circuited.
(5) Measurements of the resistance and inductance with one winding short-circuited.

A sufficient number of these tests must be made to provide the values of the requisite five independent parameters. The selection of the best test procedure depends on numerous considerations such as the type and rating of the transformer to be tested, the accuracy desired, and the apparatus available.

The direct-current resistances of the windings can be measured with a Wheatstone bridge, or by measurement of the voltage drop through the winding while it is carrying a known direct current. With small transformers, the direct-current resistances of the windings are often assumed to equal their effective resistances. The self- and mutual inductances and the ratios of self- to mutual inductances may be determined from tests with one winding open-circuited, and short-circuit tests provide a means for determining the leakage coefficient and coefficient of coupling.

The five constants of a two-winding, air-core transformer can readily be determined through measuring the self-impedance of each of its windings with the other winding open-circuited, and through measuring the mutual inductance. Usually the measurements are made by means of a suitable alternating-current bridge. The only difficulties likely to be encountered are those arising from the effects of extraneous capacitances which, however, may be extremely important at high frequencies. Since there is no core loss, the apparent resistance of each winding with the other open-circuited is its effective resistance at the frequency employed in the measurement. If the effects of distributed capacitances and of skin effect are negligible, the resistances and inductances are substantially constant.

With the iron-core transformer, however, the measurements are considerably complicated by the nonlinear magnetic characteristics of the iron and by the core loss. Because of the latter, the apparent resistance of a winding when the other winding is open-circuited is not its effective resistance, but is a greater value; the effective resistances of the windings therefore cannot be obtained from open-circuit tests. Furthermore, the values of the self- and mutual inductances depend on the amplitude of the alternating flux density and, when there is direct current in one of the windings, upon the value of this direct current. The inductances must therefore be measured under conditions which approximate as closely as possible the conditions of use of the transformer. The values of the inductances also depend upon the previous magnetic history of the core, since residual magnetism has an appreciable effect on the apparent permeability of the core material. The apparent self- and mutual inductances, furthermore, depend on the waveform of the flux variation and the measured values also depend on the method of measurement.[4] For all these reasons, to attempt to obtain hair-splitting accuracy in the measurements of the self- and mutual inductances is useless, since it is practically certain that the values of the inductances under operating conditions will not exactly equal the measured values.

The characteristics of a transformer depend to a considerable extent on magnetic leakage, and fortunately this leakage can be determined accurately from the short-circuit tests described in part (c) of this article. Furthermore, the variability of the self- and mutual inductances is relatively unimportant when the amplitude of the alternating flux density is small, as it usually is in communication transformers.

3a. *Measurement of Self-inductances.* — Self-inductances can be measured by several methods. The choice of the best method is influenced by the magnitudes of the current, voltage, and power involved; by the

[4] E. Peterson, " Impedance of a Nonlinear Circuit Element," *A.I.E.E. Trans.*, *46* (1927), 528–534.

frequency at which the measurements are to be made; and by whether the incremental inductances must be measured in the presence of direct current.

When the measurements are made at power-system frequencies and when the no-load current, voltage, and power are sufficiently large, the open-circuit impedance of a transformer without direct current in its windings can be measured using standard alternating-current instruments, as in Fig. 3. This is the open-circuit test for determining the core loss and exciting current of a power-system transformer. The self-impedance Z_{oc} of the excited winding is

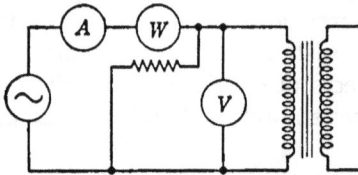

FIG. 3. Open-circuit test with standard alternating-current instruments.

$$Z_{oc} = \frac{V}{I_\varphi}, \qquad [51]$$

where V is the rms value of the applied voltage as read on the voltmeter, and I_φ is the rms value of the exciting current as determined from the reading of the ammeter A corrected if necessary for the currents in the voltmeter and in the potential coil of the wattmeter. The apparent resistance R_{oc} is

$$R_{oc} = \frac{P_{nl}}{I_\varphi^2}, \qquad [52]$$

where P_{nl} is the power absorbed by the transformer as determined from the reading of the wattmeter W corrected if necessary for the power losses in the voltmeter and in the potential coil of the wattmeter. Note that R_{oc} is *not* the effective resistance of the excited winding. The self-reactance of the winding equals its open-circuit reactance X_{oc} and is

$$X_{oc} = \sqrt{Z_{oc}^2 - R_{oc}^2}, \qquad [53]$$

from which the self-inductance may be computed if the frequency is known.

For higher frequencies and smaller currents, the self-inductances can be measured satisfactorily by use of a suitable alternating-current bridge. The measurements must be made at frequencies sufficiently low to avoid errors due to extraneous capacitances. If the incremental inductance is to be measured with direct current in the excited winding, the bridge circuit must be arranged to provide a path for the direct current, and the impedance of the direct-current source should not be included in the measurement. The modification of the Hay bridge[5]

[5] For a more complete description of this bridge, see F. E. Terman, *Measurements in Radio Engineering* (New York: McGraw-Hill Book Company, Inc., 1935), 53–58. For a description of another arrangement of the Hay bridge see B. Hague, *Alternating-Current Bridge Methods* (4th ed., London: Sir Isaac Pitman & Sons, Ltd., 1938), 391–392.

shown in Fig. 4 meets these requirements satisfactorily. In Fig. 4, *dc* is an adjustable source of direct current, *A* is a direct-current ammeter, *det* is a suitable detector, and *Tr* is the transformer to be tested. The condensers confine the direct current to the path through R_c and the transformer, and since the bridge can be balanced by adjustment of R_a and R_b, the direct current is not altered by the process of obtaining a balance. It can be shown that when the bridge is balanced, the incremental open-circuit inductance L_{oc} of the transformer is

$$L_{oc} = \frac{R_a R_c C_b}{1 + (R_b \omega C_b)^2},$$ [54]

and the apparent resistance R_{oc} is

$$R_{oc} = \frac{R_a R_b R_c (\omega C_b)^2}{1 + (R_b \omega C_b)^2}.$$ [55]

When the bridge is balanced for the fundamental-frequency components of the currents, it is not in balance for the harmonics introduced by the iron core of the transformer, and therefore a null reading cannot be obtained on a detector that is sensitive to the harmonics. This difficulty can be eliminated by use of a selective detector, such as a wave analyzer, tuned to the fundamental frequency. For frequencies from about 250 to 500 cycles per second, the detector may be a telephone receiver, used with a low-pass filter to reduce its sensitivity to the harmonics. The alternating flux density in the transformer is very nearly proportional to the alternating component of its terminal voltage divided

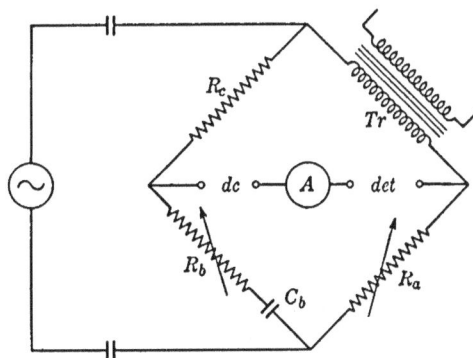

FIG. 4. Hay bridge for measurement of incremental inductance.

by the frequency. This alternating voltage can be measured by use of a vacuum-tube voltmeter of a type that measures only alternating voltages. If the alternating voltage is assumed to vary sinusoidally, the amplitude of the alternating flux can be computed by Eq. 19, Ch. VI.

When the current or voltage is so large that a bridge method becomes inconvenient, the incremental inductance can be measured by means of the series circuit shown in Fig. 5a, comprising a source of alternating voltage of sinusoidal waveform, a noninductive resistance *R* of known value, an adjustable direct-current source, and a direct-current ammeter

A in series with the transformer. The magnitudes of the alternating components of the voltage drops V_{12}, V_{23}, and V_{13} between the points indicated in the circuit diagram can be measured through connecting a suitable vacuum-tube voltmeter in turn between the pairs of points. The vacuum-tube voltmeter should be of a type that indicates only

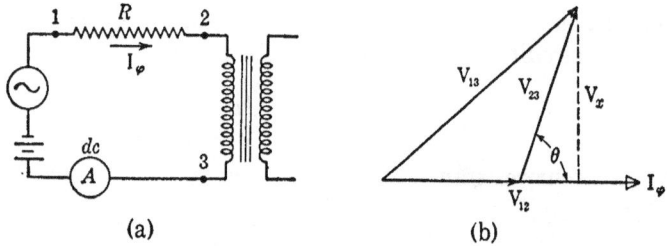

(a)   (b)

FIG. 5.   Series circuit for measurement of incremental inductance.

alternating voltages. If these alternating voltages are assumed to be sinusoidal in waveform, they can be represented by the vectors V_{12}, V_{23}, and V_{13} in Fig. 5b, forming a triangle. From the voltmeter readings V_{12}, V_{23}, and V_{13}, the open-circuit power factor of the transformer can be computed; it is given by

$$\cos\theta = \frac{V_{13}^2 - V_{12}^2 - V_{23}^2}{2V_{12}V_{23}}.\qquad [56]$$

The alternating exciting current I_ρ is

$$I_\varphi = \frac{V_{12}}{R};\qquad [57]$$

it can be computed from the voltmeter reading V_{12} and the known resistance R. The incremental inductance L_{oc} is then given by

$$L_{oc} = \frac{V_{23}\sin\theta}{\omega I_\varphi}.\qquad [58]$$

If the open-circuit power factor is so small that the reactance drop V_x in Fig. 5b substantially equals V_{23}, then

$$L_{oc} \approx \frac{V_{23}}{\omega I_\varphi} = \frac{1}{\omega}\frac{V_{23}}{V_{12}}R.\qquad [59]$$

3b. *Measurement of Mutual Inductances.* — The mutual inductance can also be determined from an open-circuit test, either by means of a suitable bridge or by measurement of the alternating current in the excited winding and the voltage induced in the open-circuited winding.

Since the current taken by the voltmeter should be small enough to have a negligible effect on the current in the excited winding and to produce negligible voltage drops in the transformer, a vacuum-tube voltmeter must be used with small transformers. If the primary is excited and the secondary is open-circuited, the mutual reactance X_m is

$$X_m = \frac{E_{oc2}}{I_{\varphi 1}}, \qquad [60]$$

where E_{oc2} is the voltage induced in the open-circuited secondary and $I_{\varphi 1}$ is the alternating exciting current in the primary. This equation assumes that the core-loss component of the exciting current is negligible and that the effective value of the exciting current therefore substantially equals the effective value of its reactive, or magnetizing, component.

The ratio of mutual to self-inductance is often more useful than the actual value of the mutual inductance, since this ratio is substantially constant in spite of the variable nature of the inductances. The ratio can be determined from measurement of the voltage E induced in the open-circuited winding and the alternating voltage V applied to the excited winding. Thus if the primary is excited and the secondary is open-circuited, the vector voltage equations are

$$V_1 = (R_1 + jX_{11})I_{\varphi 1} \qquad [61]$$

$$E_{oc2} = jX_m I_{\varphi 1}, \qquad [62]$$

where R_1 is the effective resistance of the primary winding and X_{11} is its self-reactance. From Eqs. 61 and 62, the ratio of the magnitudes of the voltages is

$$\frac{V_1}{E_{oc2}} = \frac{\sqrt{R_1^2 + X_{11}^2}}{X_m} = \frac{X_{11}\sqrt{1 + \left(\dfrac{R_1}{X_{11}}\right)^2}}{X_m}. \qquad [63]$$

Therefore

$$\frac{X_{11}}{X_m} = \frac{L_1}{M} = \frac{V_1}{E_{oc2}} \frac{1}{\sqrt{1 + \left(\dfrac{R_1}{X_{11}}\right)^2}}. \qquad [64]$$

Similarly, if a voltage V_2 is applied to the secondary and an induced voltage E_{oc1} is measured in the open-circuited primary,

$$\frac{X_m}{X_{22}} = \frac{M}{L_2} = \frac{E_{oc1}}{V_2} \sqrt{1 + \left(\frac{R_2}{X_{22}}\right)^2}, \qquad [65]$$

where R_2 is the effective resistance of the secondary winding.

If the frequency is so high that the winding resistances are small compared with their self-reactances, the inductance ratios very nearly equal the corresponding voltage ratios in Eqs. 64 and 65. However, the measurements must be made at frequencies sufficiently low to avoid errors caused by the effects of distributed capacitances.

In an iron-core transformer, the ratio of the mutual to the self-inductance approximately equals the turns ratio. Thus from Eq. 43

$$\frac{L_1}{M} = \frac{1}{k_1}\frac{N_1}{N_2},$$ ▶[66]

and from Eq. 44

$$\frac{M}{L_2} = k_2\frac{N_1}{N_2},$$ ▶[67]

where k_1 and k_2 are the ratios of mutual flux to total flux produced by each current acting by itself. Since most of the flux in an iron-core transformer links both windings, k_1 and k_2 are very nearly unity.

▶ Thus from Eqs. 66 and 67, L_1/M is slightly greater than the turns ratio N_1/N_2, and M/L_2 is slightly less than this turns ratio.◀

The geometric mean of the two inductance ratios given in Eqs. 66 and 67 is

$$\sqrt{\frac{L_1}{M} \times \frac{M}{L_2}} = \frac{N_1}{N_2}\sqrt{\frac{k_2}{k_1}}$$

or

$$\sqrt{\frac{L_1}{L_2}} = \frac{N_1}{N_2}\sqrt{\frac{k_2}{k_1}}.$$ ▶[68]

▶ The square root of the ratio of the self-inductances thus very nearly equals the turns ratio of an iron-core transformer.◀

Usually the inductance ratios in Eqs. 66, 67, and 68 differ from the turns ratio by only a fraction of a per cent and it is therefore commonly assumed that these inductance ratios equal the turns ratio. There are, however, no electrical measurements from which the turns ratio can be determined exactly.

3c. *Short-Circuit Tests.* — The short-circuit impedance of a transformer is the impedance measured at the terminals of one winding when the other winding is short-circuited. In Art. 8b, Ch. XIII, the short-circuit test is described as it is commonly applied to power-system transformers, and short-circuit impedance is shown to be very nearly equal to the equivalent impedance referred to the excited winding. Thus the short-circuit test is a simple means for determination of the combined effect of the magnetic leakage of both windings. In this article, some of the factors that must be considered in measuring the short-circuit impedance of a communication transformer are discussed, and it is shown that the leakage coefficient and coefficient of coupling can be determined from the results of a short-circuit and an open-circuit test.

With communication transformers, the short-circuit impedance can usually be measured satisfactorily by use of a suitable alternating-current bridge. Since the short-circuit reactance is almost entirely due to leakage fluxes which are not appreciably affected by the magnetic condition of the core, it is not necessary to measure the short-circuit impedance with superposed direct current in the transformer, even though the transformer may be operated in this way in service. Furthermore, except for the effects of distributed capacitances and skin effect, the short-circuit resistances and inductances are substantially constant, and therefore if these effects are of negligible importance the measurements may be made at any convenient voltage and frequency. With audio-frequency transformers, the measurements are usually made at a frequency of about 500 cycles per second, since this is a convenient frequency for bridge measurements and also is approximately the geometric mean of the frequency range over which these transformers are usually operated in service.

The measurements may be made with either winding short-circuited and the other winding excited, provided that for both arrangements the effects of distributed capacitances are negligible at the frequency employed in the test. They usually are negligible for the step-down output transformers used in communication circuits to couple the output of an amplifier to a low-impedance load, such as a loudspeaker. The effects of distributed capacitances, however, are usually more important in the step-up transformers used to couple a low-impedance generator to the grid of an amplifier tube, and therefore it is usually necessary to measure the short-circuit impedance of such an input transformer on its primary (low-voltage) side with the secondary (high-voltage) winding short-circuited. The effects of capacitances are then minimized since the winding with the greater capacitance is short-circuited.

If the secondary is short-circuited and a voltage V_1 is applied to the primary terminals, the vector voltage equations are

$$V_1 = Z_{11}I_1 + Z_{12}I_2 \qquad [69]$$

$$0 = Z_{12}I_1 + Z_{22}I_2, \qquad [70]$$

where Z_{11} and Z_{22} are the self-impedances of the windings. The mutual impedance Z_{12} is

$$Z_{12} = \pm jX_m = \pm j\omega M, \qquad [71]$$

where X_m is the mutual reactance. The algebraic sign in Eq. 71 is determined by the positive direction assumed for the secondary current. Eliminating the secondary current from Eqs. 69 and 70 gives

$$V_1 = \left(Z_{11} - \frac{Z_{12}^2}{Z_{22}}\right) I_1.$$ [72]

Compare with Eq. 24, p. 438. The only difference between Eqs. 24 and 72 is that the generator and load impedances are included in Eq. 24, whereas Eq. 72 applies to the transformer alone. The short-circuit impedance Z_{sc1} measured on the primary side with the secondary short-circuited is

$$Z_{sc1} = \frac{V_1}{I_1} = Z_{11} - \frac{Z_{12}^2}{Z_{22}}$$ ▶[73]

$$= Z_{11} - \frac{(\pm jX_m)^2}{Z_{22}} = Z_{11} + \frac{X_m^2}{Z_{22}}.$$ ▶[74]

Similarly, the short-circuit impedance measured on the secondary side with the primary short-circuited is

$$Z_{sc2} = Z_{22} - \frac{Z_{12}^2}{Z_{11}} = Z_{22} + \frac{X_m^2}{Z_{11}}.$$ ▶[75]

If the complex self-impedances Z_{11} and Z_{22} are expressed in terms of the resistances and self-reactances of the windings, the short-circuit impedance Z_{sc1} can be expressed as

$$Z_{sc1} = R_1 + jX_{11} + \frac{X_m^2}{R_2 + jX_{22}}$$

$$= R_1 + jX_{11} + \frac{X_m^2(R_2 - jX_{22})}{R_2^2 + X_{22}^2}$$ [76]

where R_1, R_2 are the effective resistances of the windings and X_{11}, X_{22} are their self-reactances. The short-circuit resistance R_{sc1} is

$$R_{sc1} = R_1 + \frac{X_m^2}{R_2^2 + X_{22}^2} R_2,$$ [77]

and the short-circuit reactance X_{sc1} is

$$X_{sc1} = X_{11} - \frac{X_m^2}{R_2^2 + X_{22}^2} X_{22}.$$ [78]

Note that the short-circuit resistance is greater than the effective resistance of the primary, because power is absorbed by the short-circuited secondary. On the other hand, the short-circuit reactance is less than the self-reactance of the primary, because of the demagnetizing effect of the secondary current. This modification of the apparent resistance and reactance is a general property of induced currents whether they flow in a discrete circuit, as in this analysis, or as eddy currents in masses of metal. It is interesting that the factor $X_m^2/(R_2^2 + X_{22}^2)$ in Eqs. 77 and 78 — which in a sense "refers" the secondary resistance and reactance to the primary — equals the square of the open-circuit voltage ratio. That is,

$$\frac{X_m^2}{R_2^2 + X_{22}^2} = \left(\frac{E_{oc1}}{V_2}\right)^2,$$ [79]

where E_{oc1} is the voltage that would be induced in the primary if the primary were open-circuited and a voltage V_2 were applied to the secondary.

Except at very low frequencies, the open-circuit voltage ratio of an iron-core transformer very nearly equals the turns ratio, and therefore, from Eq. 77,

$$R_{sc1} \approx R_1 + \left(\frac{N_1}{N_2}\right)^2 R_2.$$ [80]

By definition (see Art. 6, Ch. XIII), the right-hand side of Eq. 80 is the equivalent resistance referred to the primary, and therefore the short-circuit resistance very nearly equals the equivalent resistance. This fact is in agreement with the conclusion reached by a different line of reasoning in Art. 8b, Ch. XIII. The sum of the effective resistances of the two windings referred to a common side can be determined by measurement of the short-circuit resistance, but there is no way in which the effective resistance of each winding can be exactly determined from this measurement. Fortunately their combined value, the short-circuit or equivalent resistance, is all that is usually needed to determine the total resistive voltage drop in the transformer with sufficient accuracy for most purposes. With small audio-frequency transformers, the effective resistances of the windings are often assumed to equal their direct-current resistances.

If the short-circuit impedance is measured at a frequency so high that the secondary self-reactance is much greater than the secondary resistance, the following approximate expressions result from Eqs. 77 and 78:

$$R_{sc1} \approx R_1 + \left(\frac{X_m}{X_{22}}\right)^2 R_2$$ [81]

$$X_{sc1} \approx X_{11} - \left(\frac{X_m}{X_{22}}\right)^2 X_{22} = X_{11}\left(1 - \frac{X_m^2}{X_{11}X_{22}}\right),$$ [82]

and therefore, from Eq. 33, p. 439,

$$X_{sc1} \approx \sigma X_{11} = \sigma X_{oc1}, \qquad \blacktriangleright [83]$$

where σ is the leakage coefficient, and the self-reactance X_{11} equals the reactance X_{oc1} measured on the primary side with the secondary open-circuited. The reactance σX_{11} is sometimes called the *ideal short-circuit reactance* and is the hypothetical value of the short-circuit reactance for zero secondary resistance. The ideal short-circuit reactance σX_{11} is an important quantity not only in transformer theory but also in the theory of many types of rotating electrical machines. Often the secondary resistance is so small that the actual short-circuit reactance X_{sc1} approximately equals the ideal short-circuit reactance σX_{11}, as in Eq. 83.

The approximate value of the leakage coefficient can be determined from measured values of the short-circuit and open-circuit reactances. Thus, from Eq. 83,

$$\sigma \approx \frac{X_{sc1}}{X_{oc1}}. \qquad \blacktriangleright [84]$$

The coefficient of coupling k is related to the leakage coefficient σ by Eq. 32, p. 439; whence

$$k = \sqrt{1 - \sigma} \qquad [85]$$

$$\approx \sqrt{1 - \frac{X_{sc1}}{X_{oc1}}}. \qquad \blacktriangleright [86]$$

Although the approximate expressions, Eqs. 81–86, are usually sufficiently accurate, it should be noted that the neglect of secondary resistance causes a much greater error in Eq. 82 than in Eq. 81, since the right-hand side of Eq. 82 is the difference between two nearly equal quantities. The exact expression for the leakage coefficient can be derived from Eq. 78 and is

$$\sigma = \frac{X_{sc1}}{X_{oc1}} - \left(\frac{R_2}{X_{22}}\right)^2 \left(1 - \frac{X_{sc1}}{X_{oc1}}\right). \qquad [87]$$

Because the ratio of the short-circuit to the open-circuit reactance of an iron-core transformer is usually less than 0.01, the quantity $\left(1 - \frac{X_{sc1}}{X_{oc1}}\right)$ is very nearly unity, and therefore the approximate expression for σ, Eq. 84, can be used when $(R_2/X_{22})^2$ is negligible compared with X_{sc1}/X_{oc1}.

3d. *Conclusions Regarding Parameter Measurements.* — As pointed out at the beginning of this article, when the effects of capacitances, core loss, and magnetic nonlinearity are neglected, the performance of a transformer as a circuit element is expressible in terms of five independent parameters, three of which are concerned with the inductances and two with the resistances.

The self- and mutual inductances can be determined by means of the open-circuit tests described in parts (a) and (b) of this article, and the short-circuit inductances can be determined from the short-circuit test described in part (c). To determine the values of the three inductive parameters at least three independent inductance measurements must be made, and because of the importance of magnetic leakage one of these measurements should be a short-circuit test. For example, the three inductive parameters can be determined from measurements of the short-circuit inductance on one side, the open-circuit inductance of one winding, and a third test which may be a measurement of the open-circuit inductance of the other winding (for the same magnetic conditions in the core), or a measurement of the open-circuit voltage ratio (from which the ratio of self- to mutual inductance can be determined).

Simplifying assumptions can usually be made regarding the resistances. The combined resistance of both windings, as measured on short circuit, often is all that is needed to determine the total resistive voltage drop in the transformer. If, however, an approximate value of the effective resistance of each winding is required, the direct-current resistances can be measured and the ratio of effective to direct-current resistance can be determined by comparison of the measured short-circuit resistance with the short-circuit resistance computed by substitution of the direct-current resistances in Eq. 81. Often the effective resistances may be assumed to equal the direct-current resistances.

4. LEAKAGE INDUCTANCE

Since the factor k_1 introduced in Eq. 42, p. 440, is the ratio of the mutual flux φ_{21} to the total flux φ_{11} produced by i_1, then $(1 - k_1)$ is the ratio of the leakage flux $\varphi_{\ell 1}$ to the total flux φ_{11}; thus

$$1 - k_1 = 1 - \frac{\varphi_{21}}{\varphi_{11}} = \frac{\varphi_{11} - \varphi_{21}}{\varphi_{11}}. \qquad [88]$$

But the difference between the total flux φ_{11} and the mutual flux φ_{21} is the leakage flux $\varphi_{\ell 1}$ of winding 1 with respect to winding 2; whence

$$1 - k_1 = \frac{\varphi_{\ell 1}}{\varphi_{11}} = \frac{L_{\ell 1}}{L_1}, \qquad [89]$$

where $L_{\ell 1}$ is the leakage inductance of winding 1 with respect to winding 2. From Eq. 89,

$$L_{\ell 1} = (1 - k_1)L_1 = L_1 - k_1 L_1 \qquad [90]$$

and, on substitution of Eq. 43 in Eq. 90,

$$L_{\ell 1} = L_1 - \frac{N_1}{N_2} M. \qquad [91]$$

Similarly the leakage inductance $L_{\ell 2}$ of winding 2 with respect to winding 1 can be expressed as

$$L_{\ell 2} = L_2 - \frac{N_2}{N_1} M. \qquad [92]$$

In Eq. 91, $(N_1/N_2)M$ is the mutual inductance referred to the primary and is the component of the primary self-inductance due to the mutual flux. Similarly, in Eq. 92, $(N_2/N_1)M$ is the mutual inductance referred to the secondary. Note that if the windings have the same number of turns, the leakage inductances are the differences between the self- and mutual inductances. Although the self- and mutual inductances of an iron-core transformer are not constant, the difference between the self-inductance of a winding and the mutual inductance referred to this same winding equals the leakage inductance, as in Eqs. 91 and 92, and therefore is very nearly constant, since the leakage fluxes are partly in air.

Equations 91 and 92 suggest a theoretically possible method for determination of the leakage inductances. Thus if the turns ratio is known and if the self- and mutual inductances are measured or calculated from design data, these equations give the leakage inductances. However, this method is inherently inaccurate when used with *measured* values of the self- and mutual inductances of iron-core transformers. The leakage inductance of one winding of such a transformer often may be as small as 0.2 per cent of its self-inductance. For example, if the self-inductance of winding 1 is 10 henries, its leakage inductance may be about 0.02 henry. If the value of the leakage inductance is to be determined from Eq. 91 to the nearest millihenry — or within about 5 per cent of its true value — the value of the self-inductance must be measured to the nearest millihenry, or within 0.01 per cent of its true value, and the mutual inductance must be measured with the same per cent accuracy. Such precise measurements are impossible with iron-core transformers.

Although the leakage inductances of an iron-core transformer cannot be determined accurately through substituting *measured* values of the self- and mutual inductances in Eqs. 91 and 92, these equations are occasionally used in design calculations.[6] When they are, the self- and

---

[6] L. F. Blume, editor, *Transformer Engineering* (New York: John Wiley & Sons, 1938), 76–78.

mutual inductances are calculated from the geometry of the windings and the leakage inductances are then determined by the equations. Since the leakage inductances are not affected by flux which is wholly confined to the core, it is necessary only to calculate the self- and mutual inductances due to fluxes whose paths are partly outside the core as shown by the light unbroken flux lines in Figs. 2b and 3b, Ch. XII, p. 315. Simplifying approximations are usually made regarding the geometry of these core leakage fields, and empirical correction factors are applied to the calculated results. Since the leakage fields are not greatly affected by the core, the effect of the iron core is sometimes entirely neglected and the leakage inductances are computed as in Eqs. 91 and 92 from formulas giving the self- and mutual inductances of air-core coils.[7]

5. EQUIVALENT CIRCUITS FOR TWO-WINDING TRANSFORMERS

Representing a circuit element such as a transformer, electronic tube, or rotating machine by an equivalent circuit comprising resistance, inductance, and capacitance elements is often useful in analysis. Equivalent circuits often aid in visualizing the relations expressed in the fundamental theory. For example, the equivalent circuit of Fig. 6, Ch. XII, p. 327, clearly shows the factors of importance in the theory of transformers and therefore serves to suggest the simplifications that can often be made in the analytical treatment, as developed in Ch. XIII. This is not the only equivalent circuit which can be used to represent a two-winding transformer, however, for there is in fact an infinite number of equivalent circuits, some of which may have special uses.

Whenever a device is represented by an "equivalent" circuit, it is of the utmost importance to examine carefully in what respects the actual device and its "equivalent" circuit are equivalent, and how they differ. The "equivalent" circuit cannot be equivalent to the actual device in all respects or it would become identically the actual device. Failure to comprehend the limitations of an "equivalent" circuit may readily lead to erroneous conclusions.

The equivalent circuits derived in this article are based on the coupled-circuit theory of transformers, as expressed in Eqs. 5 and 6, p. 433. These equations, and therefore the equivalent circuits, neglect the effects of magnetic nonlinearity, of core loss, and of the distributed capacitances of the windings.

As given in Eqs. 5 and 6, the coupled-circuit voltage equations for a

[7] E. B. Rosa and F. W. Grover, "Formulas and Tables for the Calculation of Mutual and Self Inductance," *Sci. Paper Nat. Bur. Stand.*, *No. 169* (3d ed. [revised]; Washington: Government Printing Office, 1916).

transformer are

$$v_1 = R_1 i_1 + L_1 \frac{di_1}{dt} + M \frac{di_2}{dt} \qquad [93]$$

$$v_2 = R_2 i_2 + L_2 \frac{di_2}{dt} + M \frac{di_1}{dt}, \qquad [94]$$

where v_1, v_2 are the instantaneous terminal voltages of the two windings and i_1, i_2 are the instantaneous currents. The positive directions of these currents and voltages are indicated in Fig. 6a. Inspection of these equations shows that they apply not only to the transformer but also to the circuit of Fig. 6b, which therefore is equivalent to the transformer to the extent that its electrical characteristics are specified by Eqs. 93 and 94. Note that this equivalent circuit represents the equations for both steady-state and transient conditions. In the network of which the transformer is a part, the transformer can be replaced

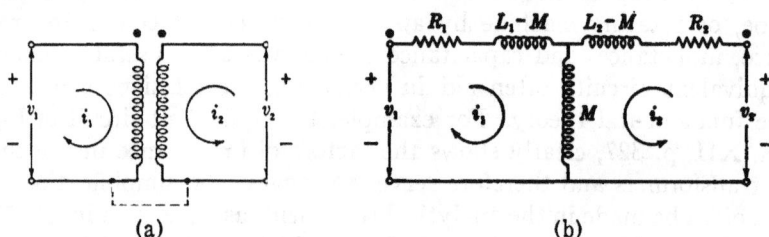

FIG. 6.   A transformer and the equivalent circuit derived from Eqs. 93 and 94.

by the equivalent circuit of Fig. 6b, provided that a direct connection between terminals of like polarity (shown by the broken line in Fig. 6a) either actually exists or can be made without alteration of the circuit conditions affecting the problem under consideration. Note that if the windings have the same number of turns, the branch inductances $L_1 - M$ and $L_2 - M$ are the primary and secondary leakage inductances. However, if the turns ratio is not unity, either $L_1 - M$ or $L_2 - M$ almost always is negative.   In general, a negative inductance is a fictitious quantity, but at a single frequency a negative inductance $-L$ has an impedance $-j\omega L$ which equals the impedance $1/j\omega C$ of a condenser whose capacitance C equals $1/\omega^2 L$.

For further discussion, assume that the currents and voltages vary sinusoidally so that the voltage equations can be written in vector form as

$$V_1 = (R_1 + jX_{11})I_1 + jX_m I_2 \qquad [95]$$

$$V_2 = (R_2 + jX_{22})I_2 + jX_m I_1. \qquad [96]$$

These equations, involving the five constants of the transformer

$(R_1, R_2, L_1, L_2,$ and $M)$, uniquely determine its electrical characteristics when any two of the vector currents and voltages are given either explicitly or by means of vector relations. In addition to the five fundamental constants, a sixth quantity may be introduced if special forms of the relations are desired for their convenience in analysis or in physical interpretation. Thus if Eq. 96 is multiplied by an arbitrarily chosen number A and the secondary current I_2 is divided by A, the voltage equations can be written as

$$V_1 = (R_1 + jX_{11})I_1 + jAX_m \frac{I_2}{A} \qquad [97]$$

$$AV_2 = A^2(R_2 + jX_{22})\frac{I_2}{A} + jAX_m I_1 \qquad [98]$$

If A in Eqs. 97 and 98 were the turns ratio N_1/N_2, then AV_2, I_2/A, $A^2(R_2 + jX_{22})$, and jAX_m would be the secondary voltage, secondary current, secondary self-impedance, and mutual impedance referred to the primary, as in Art. 4, Ch. XIII. Thus Eqs. 97 and 98 may be re-

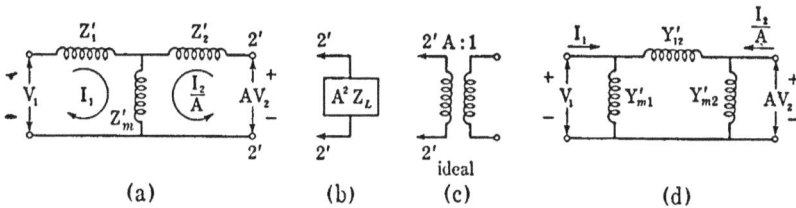

Fig. 7.　Generalized equivalent circuits for a transformer and its load.

garded as the coupled-circuit equations with the secondary voltage, secondary current, and the impedances " referred " to the primary in a more general sense than when A is the turns ratio.

Note that Eqs. 97 and 98 apply to the T circuit of Fig. 7a, in which the mutual impedance Z'_m " referred " to the primary is

$$Z'_m = jAX_m, \qquad [99]$$

and the branch impedances Z'_1 and Z'_2 " referred " to the primary are

$$Z'_1 = Z_{11} - Z'_m = R_1 + j(X_{11} - AX_m) \qquad [100]$$

$$Z'_2 = A^2 Z_{22} - Z'_m$$

$$= A^2 R_2 + j(A^2 X_{22} - AX_m) = A^2 R_2 + jA^2\left(X_{22} - \frac{X_m}{A}\right), \qquad [101]$$

where Z_{11} and Z_{22} are the vector self-impedances of the windings. If the secondary of the transformer supplies power to a load whose impedance is Z_L, the equivalent load in which a voltage AV_2 would produce a current I_2/A is one of impedance A^2Z_L, as indicated in Fig. 7b, whether A is a real or a complex quantity. If A is a real number, the voltage AV_2 and the current I_2/A are related to the actual secondary voltage V_2 and secondary current I_2 as are the primary and secondary voltages and currents in an ideal transformer whose ratio of transformation is A, as indicated in Fig. 7c. Thus a transformer is electrically equivalent to the T circuit of Fig. 7a with either the equivalent load A^2Z_L of Fig. 7b or the ideal transformer of Fig. 7c connected to the terminals 2' of the T circuit.

Since A may have any arbitrarily chosen value, there is an infinite number of these equivalent circuits, each corresponding to a particular value of A. Since any T circuit can be replaced by an equivalent π, there is also an infinite number of π equivalent circuits of the form shown in Fig. 7d, each corresponding to a particular set of values of the impedances in the T circuit of Fig. 7a. The vector relations[8] between the π admittances Y'_{12}, Y'_{m1}, and Y'_{m2} of Fig. 7d and the T impedances Z'_1, Z'_2, and Z'_m of Fig. 7a are

$$Y'_{12} = \frac{Z'_m}{D_z} \qquad [102]$$

$$Y'_{m1} = \frac{Z'_2}{D_z} \qquad [103]$$

$$Y'_{m2} = \frac{Z'_1}{D_z}, \qquad [104]$$

where

$$D_z = Z'_1Z'_2 + Z'_2Z'_m + Z'_mZ'_1. \qquad [105]$$

The following discussion treats some of the particular equivalent circuits which can be derived from the general equivalent circuit of Fig. 7 by the use of appropriate values of A.

Case I: $A = N_1/N_2$.

If A equals the turns ratio N_1/N_2, the impedance Z'_m is, from Eq. 99,

$$Z'_m = j\frac{N_1}{N_2}X_m \qquad [106]$$

and the branch impedances Z'_1 and Z'_2 are, from Eqs. 100 and 101,

$$Z'_1 = R_1 + j\left(X_{11} - \frac{N_1}{N_2}X_m\right) \qquad [107]$$

$$Z'_2 = \left(\frac{N_1}{N_2}\right)^2\left[R_2 + j\left(X_{22} - \frac{N_2}{N_1}X_m\right)\right]. \qquad [108]$$

---

[8] See the volume on electric circuits (1940), Eqs. 138, 139, and 140, p. 460.

From Eqs. 91 and 92, the reactances $X_{11} - (N_1/N_2)X_m$ and $X_{22} - (N_2/N_1)X_m$ are the leakage reactances X_{l1} of the primary and X_{l2} of the secondary, respectively. Hence when A equals N_1/N_2,

$$Z_1' = R_1 + jX_{l1} \qquad\qquad\qquad [109]$$

$$Z_2' = \left(\frac{N_1}{N_2}\right)^2 (R_2 + jX_{l2}). \qquad\qquad [110]$$

The impedances Z_1' and Z_2' are the leakage impedances of the primary and of the secondary referred to the primary, and the impedance Z_m' is the mutual or magnetizing reactance referred to the primary. Therefore when A equals the turns ratio, the equivalent circuit of Fig. 7 becomes the equivalent circuit of Fig. 6b, Ch. XIII, p. 344 (with core loss neglected). Thus the equivalent circuits involving leakage inductances may be regarded as special cases of the general equivalent circuit of Fig. 7.

Case II: A = 1.

Under these conditions, Eqs. 99, 100, and 101 reduce to

$$Z_m' = jX_m \qquad\qquad\qquad\qquad [111]$$

$$Z_1' = R_1 + j(X_{11} - X_m) \qquad\qquad [112]$$

$$Z_2' = R_2 + j(X_{22} - X_m). \qquad\qquad [113]$$

The equivalent circuit of Fig. 7 then reduces to that of Fig. 6b, p. 456.

Case III: $A = \sqrt{\dfrac{L_1}{L_2}}$.

For this value of A it can readily be shown that

$$Z_m' = jkX_{11} \qquad\qquad\qquad\qquad [114]$$

$$Z_1' = R_1 + j(1 - k)X_{11} \qquad\qquad [115]$$

$$Z_2' = \frac{L_1}{L_2}R_2 + j(1 - k)X_{11}, \qquad\qquad [116]$$

where k is the coefficient of coupling. This equivalent circuit is shown in Fig. 8a. Note that the reactances of the impedances Z_1' and Z_2' are equal in this equivalent circuit. If the core loss is assumed to vary as the square of the voltage across the shunt branch kL_1, the core loss can be accounted for by means of a resistance connected in parallel with kL_1, as indicated by r_c in Fig. 8a.

Case IV: $\quad A = \dfrac{L_1}{M} = \dfrac{1}{k}\sqrt{\dfrac{L_1}{L_2}}.$

Under these conditions, from Eqs. 99, 100, and 101,

$$Z'_m = jX_{11} \tag{117}$$

$$Z'_1 = R_1 + j0 \tag{118}$$

$$Z'_2 = \left(\frac{L_1}{M}\right)^2 R_2 + j\frac{\sigma X_{11}}{k^2}, \tag{119}$$

where σ is the leakage coefficient and equals $1 - k^2$. The equivalent circuit is shown in Fig. 8b. In this circuit, the series inductance accounting for magnetic leakage is entirely on the secondary side of the shunt inductance accounting for magnetizing current. The reactance σX_{11} is the ideal short-circuit reactance, which is slightly less than the short-circuit reactance X_{sc1} measured on the primary side (see Eq. 87), and therefore the reactance $\sigma X_{11}/k^2$ in Eq. 119 very nearly equals the short-circuit reactance X_{sc1}. A resistance r_c accounting for the core loss may be connected in parallel with L_1 if the core loss is assumed to vary as the square of the voltage across this shunt inductance. Since this voltage is the primary terminal voltage minus the primary resistance drop, it is the voltage induced by all the flux (both leakage and mutual) linking the primary winding.

Case V: $\quad A = \dfrac{M}{L_2} = k\sqrt{\dfrac{L_1}{L_2}}.$

For this value of A, the impedances become

$$Z'_m = jk^2 X_{11} \tag{120}$$

$$Z'_1 = R_1 + j\sigma X_{11} \tag{121}$$

$$Z'_2 = \left(\frac{M}{L_2}\right)^2 R_2 + j0. \tag{122}$$

See Fig. 8c. In this circuit, the series inductance is on the primary side of the shunt inductance, and the use of a shunt resistance r_c to account for core loss is based on the assumption that the core loss varies as the square of the voltage induced by the resultant flux linking the secondary.

Note that varying A between the values $(1/k)\sqrt{L_1/L_2}$ in Fig. 8b and $k\sqrt{L_1/L_2}$ in Fig. 8c moves the point of attachment of the shunt inductance from one end to the other of the series inductance, and slightly changes the ratio of the ideal transformer and the values of the shunt and series branches of the equivalent circuit. The coefficient of coupling of an iron-core transformer is usually so close to unity that the differences between the values of the transformation ratio and of the shunt and series inductances of Figs. 8b and 8c are ordinarily so small as to be entirely inappreciable.

Case VI: $A = \dfrac{Z_{11}}{jX_m} = \dfrac{V_1}{E_{oc2}}$.

One of the more useful forms of the equivalent circuit is obtained when the multiplying factor A is the complex ratio of self-impedance to

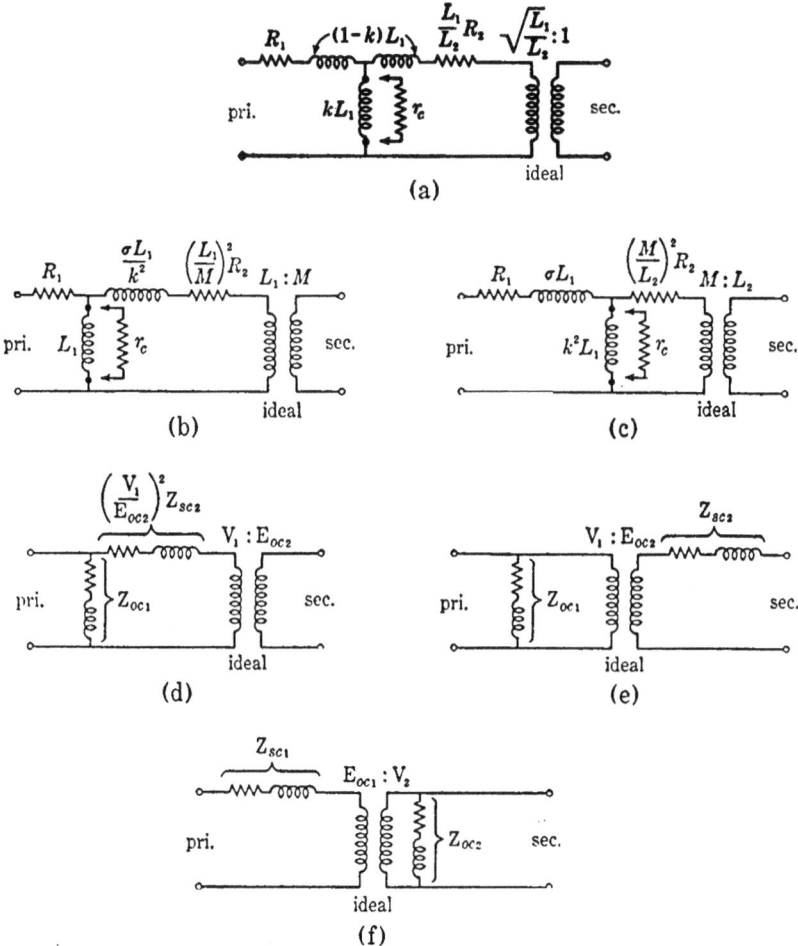

(a)

(b)   (c)

(d)   (e)

(f)

Fig. 8. A few of the equivalent circuits that can be derived from Fig. 7.

mutual impedance. Note that the ratio of self-impedance to mutual impedance equals the vector open-circuit voltage ratio; thus from the vector ratio of Eqs. 61 and 62, p. 447,

$$\frac{Z_{11}}{jX_m} = \frac{V_1}{E_{oc2}} = A, \qquad [123]$$

where E_{oc2} is the vector open-circuit voltage induced in the secondary when vector voltage V_1 is applied to the primary. When $A = Z_{11}/jX_m$ is substituted in Eqs. 99, 100, and 101,

$$Z_m' = Z_{11} = Z_{oc1} \tag{124}$$

$$Z_1' = 0 \tag{125}$$

$$Z_2' = \left(\frac{Z_{11}}{jX_m}\right)^2 Z_{22} - Z_{11} \tag{126}$$

$$= \left(\frac{Z_{11}}{jX_m}\right)^2 \left[Z_{22} - \frac{(jX_m)^2}{Z_{11}}\right] \tag{127}$$

$$= \left(\frac{V_1}{E_{oc2}}\right)^2 \left(Z_{22} + \frac{X_m^2}{Z_{11}}\right). \tag{128}$$

But by Eq. 75, p. 450,

$$Z_{22} + \frac{X_m^2}{Z_{11}} = Z_{sc2}, \tag{129}$$

where Z_{sc2} is the short-circuit impedance measured on the secondary side. Consequently,

$$Z_2' = \left(\frac{V_1}{E_{oc2}}\right)^2 Z_{sc2}. \tag{130}$$

The equivalent circuit therefore can be drawn as in Fig. 8d. If the short-circuit impedance is " referred " to the secondary side of the ideal transformer, the equivalent circuit becomes that shown in Fig. 8e, which should be compared with Fig. 9a, Ch. XIII, p. 356. In Fig. 8e, the transformer is represented, as in Thévenin's theorem, by its short-circuit impedance Z_{sc2} in series with its open-circuit secondary voltage E_{oc2}, the ratio of transformation is represented by an ideal transformer whose ratio is the vector open-circuit voltage ratio V_1/E_{oc2}, and the exciting current is taken into account by connection of the open-circuit impedance across the primary terminals.

Case VII: $A = \dfrac{jX_m}{Z_{22}} = \dfrac{E_{oc1}}{V_2}.$

It can readily be shown that the complex ratio of mutual impedance to secondary self-impedance equals the vector ratio of open-circuit voltage E_{oc1} measured on the primary side to vector voltage V_2 applied to the secondary, and that when A equals this complex ratio the equivalent circuit becomes that shown in Fig. 8f, which should be compared with Fig. 9b, Ch. XIII, p. 356.

The equivalent circuits of Figs. 8e and 8f are especially convenient for the representation of a transformer in terms of measured values of its constants, since the parameters in these equivalent circuits can readily be measured by means of the methods described in Art. 3. Usually the angle of the complex open-circuit voltage ratio is negligibly small, and therefore a circuit very nearly equivalent to the transformer can be determined from measurements of the short-circuit resistance and reactance on one side, the open-circuit resistance and reactance on the other side, and the *magnitude* of the open-circuit voltage ratio.

6. SUMMARY

It is of considerable interest to contrast the equivalent circuits involving leakage inductances with the general equivalent circuit of Fig. 7. If the multiplying factor A in Eqs. 97 and 98 is the ratio of turns, the series inductances in the equivalent circuit of Fig. 7 become the leakage inductances, which give a physical concept of the phenomena involved. On the other hand, if the multiplying factor A is any other number, the analysis reduces to a problem in pure mathematics in which certain transformations are made to aid in obtaining a desired solution. These mathematical transformations are chosen without regard to their physical meaning. It is sufficient that they are useful tools. Both methods of analysis are founded on the same physical laws which have been deduced from observing the results of certain events.

The physical concept has been created as an aid to understanding these events and, although it makes no pretense to being a complete representation, it has proved of inestimable value in strengthening the appreciation of the significance of the underlying physical laws. Possibly the most famous of these concepts was proposed by Faraday to aid in visualizing the phenomenon of electromagnetic induction. His conception of the magnetic field and its representation by lines of force has had an enormous influence on the development of electrical devices.

Notwithstanding the great value that the physical concept possesses, the insistence that it be applied to the solution of some problems may prove a handicap. This is illustrated by the theory of the transformer. If the physical concept of leakage inductance is used, only one basic equivalent circuit results. On the other hand, if the coupled-circuit equations are manipulated as though the problem were purely mathematical, the multiplying factor A in Eqs. 97 and 98 can assume any value, and an infinite number of equivalent circuits is obtained. In the analysis of some problems, it may be found that one of these equivalent circuits is to be preferred to any other. That is, it may be better to operate on the initial physical relations upon which the solution is to be built in a strictly mathematical sense without attempting to form a physical conception of the operations. Thus it is right to conclude that these two methods of analysis are supplementary and that it would be unwise to neglect developing the ability to apply the one or the other.

PROBLEMS

1. Prove that in any transformer the ratio of primary to secondary current with the secondary short-circuited equals the ratio of secondary to primary voltage with the primary open-circuited, if the effects of the nonlinear magnetic characteristics of the core are neglected.

2. When a voltage of 10 v at 600 $\sim$ is impressed on the primary winding of an air-core transformer with the secondary winding open-circuited, the primary current and power are 2 amp and 16 w, and the secondary voltage is 8 v. When a voltage of 20 v at 600 $\sim$ is impressed on the secondary with the primary open-circuited, the current and power are 1.0 amp and 16 w.

    (a) What are the values of the two self-inductances and of the mutual inductance?

    (b) What are the values of the coefficient of coupling k and of the leakage coefficient σ?

    (c) If a noninductive resistance of 14 ohms is connected across the secondary terminals, what current and power does the primary take when a voltage of 30 v at 600 $\sim$ is impressed across its terminals?

3. The following data have been obtained for an air-core transformer, by use of an alternating-current bridge operated at 1,000 $\sim$:

Primary winding resistance	= 3.0 ohms
Primary inductance with secondary open-circuited	= 217 mh
Secondary winding resistance	= 12.2 ohms
Secondary inductance with primary open-circuited	= 867 mh
Inductance of primary and secondary connected in series	= 1.84 h

In addition, the open-circuit voltage ratio and the short-circuit impedance referred to the primary have been measured at 1,000 $\sim$ as:

$$\frac{E_{oc2}}{V_1} = \frac{172 \text{ v}}{100 \text{ v}}$$

$$\frac{V_1}{I_{sc1}} = \frac{10 \text{ v}}{5.5 \text{ ma}} .$$

In using these data to compute the constants of the transformer, it is found that at least one of the measurements must be in error. If it may be assumed that only one of the measurements is incorrect,

    (a) Which measurement is in error?

    (b) What are the correct values of L_1, L_2, M, k, and σ?

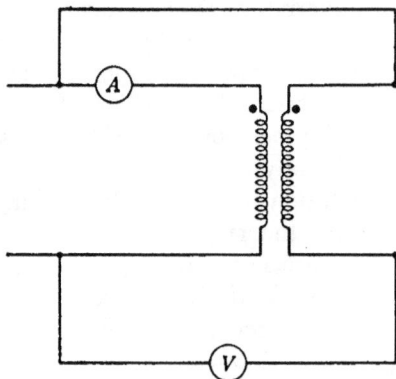

FIG. 9.   Air-core transformer, Prob. 4.

4. An air-core transformer having the same number of turns in the primary and secondary windings is tested as shown in Fig. 9. When a voltage of 110 v at 1,000 $\sim$ is applied to the primary, A reads 1.00 amp and V reads 10.0 v. When the windings are interchanged and the test repeated, A reads 0.95 amp and V reads 13.7 v. The

resistances of the primary and secondary windings are known to be 5.0 and 5.2 ohms, respectively. The resistance of the ammeter and the conductance of the voltmeter may be assumed to be negligibly small.

(a) Is there any inconsistency in the data?

(b) What are the values of L_1, L_2, M, k, and σ?

5. The following measurements have been made on an iron-core audio-frequency output transformer:

(1) With a direct-current bridge, the winding resistances have been found to be:

$$R_1 = 310 \text{ ohms}$$
$$R_2 = 2.22 \text{ ohms.}$$

(2) With an alternating-current bridge operated at $600 \sim$, the primary short-circuit resistance and inductance have been found to be:

$$R_{sc1} = 760 \text{ ohms}$$
$$L_{sc1} = 0.104 \text{ h.}$$

(3) With the secondary open-circuited and the primary connected as shown in Fig. 5, and supplied from a $600 \sim$ source, the following set of values has been obtained:

$$V_{12} = 49.7 \text{ v, a-c}$$
$$V_{23} = 40.0 \text{ v, a-c}$$
$$V_{13} = 65.3 \text{ v, a-c}$$
$$R = 50,000 \text{ ohms}$$
$$E_{oc2} = 3.98 \text{ v.}$$

From these data find:

(a) The apparent primary resistance,

(b) An estimate of the effective primary resistance,

(c) An estimate of the effective secondary resistance,

(d) The primary self-inductance,

(e) The secondary self-inductance,

(f) The mutual inductance,

(g) The leakage coefficient,

(h) The coupling coefficient.

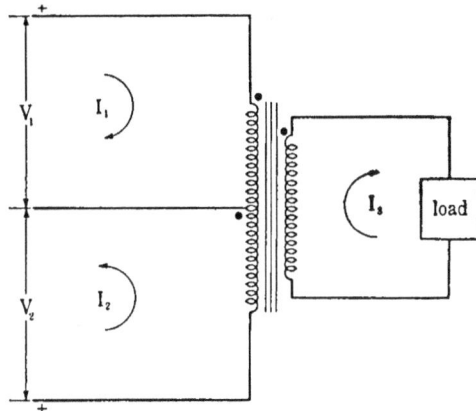

FIG. 10. Center-tapped transformer, Prob. 7.

6. For a representative audio-frequency communication transformer the ratio of ideal short-circuit inductance to open-circuit inductance is about 0.01. Compute the per cent errors in the following approximate relations for each of the three assumptions specified:

$$\sqrt{\frac{L_1}{L_2}} \approx \frac{N_1}{N_2} \quad ; \quad \frac{L_1}{M} \approx \frac{N_1}{N_2} \quad ; \quad \frac{M}{L_2} \approx \frac{N_1}{N_2}.$$

(a) Assume all the leakage flux to be associated with the primary.

(b) Assume all the leakage flux to be associated with the secondary.

(c) Assume the ratios of leakage to self-inductance to be the same for both windings.

7. A push-pull vacuum-tube power amplifier is usually coupled to its load by a transformer with a center-tapped primary, as shown in Fig. 10. In such an arrangement the alternating voltages applied between the outside terminals of the primary

and the center tap are 180 degrees out of phase and are approximately equal. Taking the transformer polarities and current directions to be as shown in Fig. 10, and assuming that the load may be represented by a pure resistance:

(a) Write the coupled-circuit equations for the network.

(b) Draw an equivalent circuit containing no ideal transformers.

8. A vacuum-tube amplifier operating in a restricted range and with sinusoidal input may often be represented by an equivalent sine-wave generator and series resistance. An amplifier for which these conditions are true is to supply power to a load which may be represented by a pure resistance of 8.0 ohms. The equivalent circuit of the vacuum-tube amplifier is a 100-v sine-wave generator in series with an 800-ohm resistance. To couple the amplifier to the load, a so-called matching transformer is to be used, of which there are two available. Data on these transformers are given below.

Which of these two transformers should be used to deliver maximum power to the load at a frequency of 1,000 $\sim$? Neglect core loss in the transformers and assume that the equivalent emf and internal resistance of the vacuum-tube amplifier are constant.

TRANSFORMER DATA AT 1,000 $\sim$

	Transformer A	Transformer B
L_{oc2}	57 mh	62 mh
$\dfrac{E_{oc1}}{V_2}$	10	12
R_{sc1}	400 ohms	300 ohms
L_{sc1}	36 mh	45 mh

DIRECT–CURRENT RESISTANCES

	Transformer A	Transformer B
R_1	182 ohms	140 ohms
R_2	1.90	1.01

9. For *Transformer A* of Prob. 8, specify the parameter values for equivalent circuits of the type shown in Figs. 6b, 8a, 8b, and 8d. State any necessary assumptions.

10. For a transformer in which a primary and a secondary terminal of unlike polarity are connected, find:

(a) An equivalent circuit containing no ideal transformers,

(b) An equivalent circuit containing an ideal transformer with a voltage ratio of $\dfrac{V_1}{E_{oc2}}$.

CHAPTER XVIII

Frequency Characteristics

A power-system transformer usually operates at substantially constant voltage and frequency, and its desired electrical characteristics usually are high efficiency, small voltage regulation, and small exciting current. In many important applications of transformers, however, voltage and frequency vary between wide limits. Probably the most important are in communication circuits for the transmission of speech and music, where transformers are required to operate over the entire audio band of frequencies.* Some of the important characteristics of transformers for variable-frequency service are discussed in this chapter. Although the emphasis here is on audio-frequency communication circuits, many important variable-frequency applications of transformers are also found in measurement and control circuits.

A transformer in a communication system usually is one element in a circuit connecting a source of electromotive force — for example, a microphone — to a load such as a loudspeaker. The internal electromotive force of the source is generated by the speech or music to be transmitted, and the function of the interconnecting circuit is to amplify the power output of the source and to apply to the load a voltage whose waveform matches that of the internal electromotive force of the source as closely as possible. Although the waveforms of the electromotive forces due to speech and music are extremely complicated and are, in fact, semitransient, much useful information can be obtained from a knowledge of the circuit behavior with sine-wave applied voltages. Any sustained periodic function of time can be expressed as a Fourier series of sine-wave harmonic components, each having a frequency which is an integer multiple of the fundamental frequency.† Fourier, the discoverer of this series, also discovered that any transient or nonperiodic function of time can be resolved into a continuous spectrum of sustained sine-wave harmonic components. This extension of the Fourier series is known as the Fourier integral method of analysis.[1] By superposition combined with Fourier's methods of analysis the response of any linear

---

* The limits of audibility are from about 16 to 20,000 cycles per second, but satisfactory reproduction of music is obtained if roughly the lower and upper octaves are omitted from this band of frequencies. For intelligible transmission of speech it is necessary to transmit only a band of frequencies from roughly 250 to 2,750 cycles per second. See D. G. Fink, *Engineering Electronics* (New York: McGraw-Hill Book Company, Inc., 1938), 272.

† See Art. 9, Ch. VI.

[1] See the reference volume, *The Mathematics of Circuit Analysis*, Ch. VII.

circuit to steady-state or transient source voltages of any waveform can be predicted if its response to sine-wave source voltage is known vectorially over the entire frequency range in question. The curve showing how the magnitude of the ratio of load voltage to sine-wave source voltage varies with frequency is often called the *frequency charac- teristic* or *frequency response* of the circuit, and the curve showing the phase angle between these voltages as a function of frequency is known as the *phase characteristic.** These characteristics depend not only on the constants of the transformer which may be in the circuit but also on the internal impedances of the source and load. For faithful reproduc- tion of speech and music, the waveforms of load voltage and source volt- age should be nearly the same, independent of voltage and frequency. That is, the response of the circuit to sine-wave source voltages should be essentially linear with respect to variations in the magnitude of the source voltage, the amplitude or frequency characteristic should be nearly " flat," and the time delay from source voltage to load voltage should be small and independent of frequency.

Among the most important applications of transformers in communi- cation circuits are their uses for changing the voltage of a source, for changing the apparent impedance of a load, for isolating direct current in one circuit from another, for permitting independent grounding of the primary and secondary circuits, and for reversing the phase of a voltage. For example, a source of signal voltage is often connected to the input of an amplifier by means of an *input transformer* whose purpose is to raise the signal voltage before it is applied to the grid of the first vacuum tube in the amplifier. The input transformer may also serve to provide a path for the direct current which may be present in the source — for example, when the source is a carbon-button microphone. In multi- stage amplifiers consisting of several vacuum tubes in cascade, *interstage transformers* are sometimes used to couple the output of one tube to the grid of the next stage of amplification. By a step-up turns ratio, an interstage transformer may be utilized to increase the voltage ampli- fication. An example of impedance transformation is the *output trans- former*, frequently used to couple a load, such as a loudspeaker, to the output of the last tube in an amplifier. The purpose of the output transformer is to change the apparent impedance of the load as viewed from the primary side of the transformer to a value which permits the

---

* These two curves give the amplitude and phase of the vector ratio of load voltage to source voltage as functions of the independent variable, frequency. A better name for the amplitude-frequency characteristic would be *amplitude characteristic*, but unfortunately inconsistent common terminology has named the amplitude-frequency characteristic after its independent variable — frequency — but the phase-frequency characteristic after its dependent variable — phase.

tube to supply the maximum amount of power to the load, with a specified amount of harmonic distortion. The output transformer also prevents the direct-current output of the tube from flowing in the load. In all these applications, the frequency and phase characteristics are important when the circuit must operate over a wide band of frequencies.

1. POWER RELATIONS

Transformers are often used in communication, measurement, and control circuits for their property of changing the apparent impedance of a load. If a load whose impedance is Z_L is connected to the secondary terminals of a transformer, the apparent impedance of this load as observed on the primary side of the transformer is $a^2 Z_L$, where a is the ratio of primary to secondary turns and the transformer is assumed to be ideal. An actual transformer has approximately the characteristics of an ideal transformer as long as the load impedance is large compared with the series equivalent impedance of the transformer and as long as the shunt exciting impedance of the transformer is large compared with the load impedance (when these impedances are all referred to the same side of the transformer).

Probably the most common application of transformers for changing impedances is the use of them to increase the power delivered to an impedance load by a generator of fixed internal impedance and fixed internal electromotive force. When both the resistance and reactance of the load can be adjusted independently, the maximum power is delivered to the load when the complex load impedance equals the conjugate of the complex generator impedance. If the load is purely resistive, the condition for maximum load power occurs when the load resistance equals the magnitude of the generator impedance. The simple case in which the impedances of both load and generator are essentially pure resistances arises frequently in communication circuits; the maximum power is then delivered to the load when the load and generator resistances are equal. The load and generator resistances then are said to be *matched*. Thus by the interposing of a suitable transformer between a resistive load and a resistive generator, each having fixed resistances, the apparent resistance of the load can be changed so as to bring about a large increase in the power delivered to the load.

Although the maximum power is delivered to a resistance load when its apparent resistance equals the internal impedance of the generator, the power delivered to the load is not greatly below the maximum possible value when the apparent load resistance differs quite appreciably from the generator impedance. Thus if E_G is the effective value of the generator internal electromotive force and R_G is its internal impedance (assumed to be purely resistive), the maximum power $P_{L\ max}$ is delivered to a resistive load when the apparent load resistance equals R_G. For these conditions, the current I in the series circuit is

$$I = \frac{E_G}{2R_G},$$ [1]

and the maximum load power is

$$P_{L\ max} = I^2 \times \text{(apparent load resistance)} = I^2 R_G$$ [2]

$$= \left(\frac{E_G}{2R_G}\right)^2 R_G = \frac{E_G^2}{4R_G}.$$ [3]

For any other value R_L' of the apparent load resistance, the current I' is

$$I' = \frac{E_G}{R_G + R_L'},$$ [4]

and the load power P_L is

$$P_L = (I')^2 R_L' = \left(\frac{E_G}{R_G + R_L'}\right)^2 R_L'.$$ [5]

The ratio of the load power P_L to the maximum possible load power $P_{L\ max}$ is

$$\frac{P_L}{P_{L\ max}} = \frac{\left(\dfrac{E_G}{R_G + R_L'}\right)^2 R_L'}{\dfrac{E_G^2}{4R_G}}$$ [6]

$$= \frac{4R_G R_L'}{(R_G + R_L')^2}$$ [7]

$$= \frac{4}{\dfrac{R_G}{R_L'} + 2 + \dfrac{R_L'}{R_G}}.$$ ▶[8]

Figure 1 shows the curve of the power ratio $P_L/P_{L\ max}$ as a function of the resistance ratio R_L'/R_G plotted on a logarithmic scale. From Eq. 8, if R_L'/R_G equals 3 (or $\frac{1}{3}$), the ratio $P_L/P_{L\ max}$ equals $\frac{3}{4}$. That is, if the apparent load resistance R_L' differs from the generator resistance R_G by a factor as great as 3 to 1, the power delivered to the load is still three-fourths of the maximum possible value. Thus the circuit conditions for satisfactory power transfer are not critically dependent on close equality between the generator and load resistances, and the use of a matching transformer would result in very little increase in load power if the load and generator resistances differ by a factor less than about 3 to 1.

Actually, many applications require that the apparent load resistance must not equal the generator resistance. One of the most important uses of impedance-changing transformers for increasing the power delivered to a load is illustrated by the output transformer used to couple the output of the last tube in an amplifier to the load, which may be a low-impedance loudspeaker. In these applications, distortion may be introduced by the tube characteristics unless the load impedance is properly adjusted to the tube. Thus for linear (Class A) operation of a vacuum tube, the variation in its plate current caused by the vary-

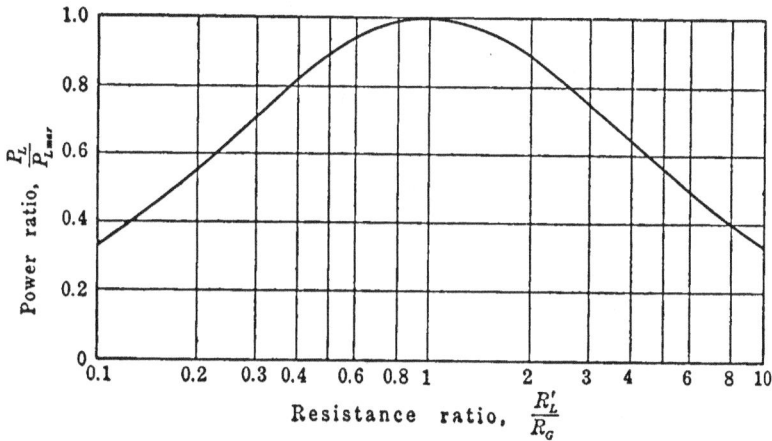

FIG. 1.   Curve of power ratio $P_L/P_{L\ max}$ as a function of the resistance ratio R'_L/R_G.

ing signal voltage impressed on its grid must be restricted to a limited range, and therefore the apparent impedance of the load must be such that the tube is not forced to operate beyond these limits of plate current.[2] That is, the apparent impedance of the load should be such that, for a specified amount of harmonic distortion, the maximum power is delivered to the load. For a triode tube operating as a linear amplifier, this load impedance usually is more than twice the dynamic plate resistance of the tube; for a pentode, the appropriate load impedance is but a fraction of the dynamic plate resistance. In other applications, as in push-pull amplifiers,[3] the maximum power output may be limited by the power that can be dissipated safely in the tubes; with various tubes the proper value of the apparent load impedance may then bear almost any relation to the dynamic plate resistances of the tubes.

[2] Class A amplifiers are discussed in the volume on electronics, Ch. VIII.
[3] Push-pull amplifiers are discussed in the volume on electronics, Ch. VIII.

2. FREQUENCY AND PHASE CHARACTERISTICS OF OUTPUT TRANS-
   FORMERS

Output transformers having turns ratios of about 10 or 20 to 1 are commonly used in radio receivers to connect the low-impedance voice coil of a loudspeaker to the output circuit of the power amplifier tube. The impedance of the voice coil of a dynamic-type loudspeaker may be of the order of magnitude of 10 ohms, whereas the proper load resistance for a power amplifier tube usually is of the order of magnitude of several thousand ohms. If the ratio a of primary to secondary turns is 20, the apparent impedance of a 10-ohm load is 4,000 ohms as observed at the primary terminals of the transformer, and thus, by the use of an output transformer having a suitable turns ratio, the load impedance can be properly adjusted to the tube.

In contrast with input and interstage transformers, discussed in

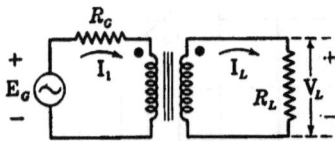

FIG. 2. Circuit containing an output transformer.

Art. 3, output transformers are called upon to deliver considerable amounts of power (as much as 20 watts in typical radio receivers). The load currents in output transformers thus are usually large compared with the audio-frequency charging currents in the distributed capacitances; the distributed capacitances therefore usually have negligible effects on the frequency and phase characteristics and usually are neglected in the analysis of output transformers.

The essential elements of a typical circuit containing an output transformer are shown in Fig. 2. The generator has an internal electromotive force E_G and an internal resistance R_G, and may represent the alternating-current output of a vacuum tube. The load is a resistance R_L. The input impedance of a loudspeaker is not purely resistive, nor is it constant; but it varies appreciably with the frequency and also, to some extent, with the acoustic characteristics of the room in which the loudspeaker is located. However, it would indeed be a difficult task to take into account variations in the load impedance, and therefore the frequency and phase characteristics of output transformers are specified on the basis of constant resistance loads. Although the characteristics with a constant resistance load may differ somewhat from the actual characteristics when the load is a loudspeaker, the following analysis, in which R_L is assumed constant, gives results that approximate fairly closely the characteristics of an output transformer in actual service.

In the theoretical analysis of the circuit of Fig. 2, the transformer can be treated in either one of two ways. The theory may be developed from the classical coupled-circuit equations, or the transformer may be represented by an equivalent circuit. The coupled-circuit equations readily yield a rigorous but somewhat complicated solution. On the other hand, the equivalent circuit readily lends itself to simplifications, resulting in a simple approximate solution sufficiently accurate for engineering use. Although the same approximate solution can be obtained through neglecting certain small terms in the mathematical expressions resulting from the coupled-circuit theory, the equivalent circuit probably gives a clearer understanding of the physical significances of these approximations. Since each method of attack has its advantages, both are presented below.

Experience shows that the effects of core loss on the frequency and phase characteristics are negligible over the normal operating range of output transformers. When core loss and distributed capacitances are neglected, the vector voltage equations for the circuit of Fig. 2 are, according to the coupled-circuit theory,

$$E_G = (R_{11} + jX_{11})I_1 - jX_m I_L \qquad [9]$$

$$0 = -jX_m I_1 + (R_{22} + jX_{22})I_L, \qquad [10]$$

where

> E_G is the vector representing the sine-wave internal electromotive force of the generator,
>
> I_1 and I_L are the vector primary and secondary currents,
>
> X_{11} and X_{22} are the primary and secondary incremental self-reactances,
>
> X_m is the incremental mutual reactance,
>
> R_{11} is the series resistance of the primary circuit and equals the sum of the generator resistance R_G and the primary winding resistance R_1,
>
> R_{22} is the series resistance of the secondary circuit and equals the sum of the secondary winding resistance R_2 and the load resistance R_L.

An expression for the load current I_L can be obtained from the simultaneous solution of these equations. This solution is given by Eq. 50, Ch. XVII, and can be expressed as

$$I_L = \frac{X_m E_G}{R_{11}X_{22} + R_{22}X_{11} + j(\sigma X_{11}X_{22} - R_{11}R_{22})}, \qquad [11]$$

where σ is the leakage coefficient, defined by Eq. 33, Ch. XVII. Division of the numerator and denominator of Eq. 11 by X_{22} gives

$$I_L = \frac{\dfrac{X_m}{X_{22}} E_G}{R_{11} + \dfrac{X_{11}}{X_{22}} R_{22} + j\left(\sigma X_{11} - \dfrac{R_{11}R_{22}}{X_{22}}\right)} \qquad [12]$$

$$= \frac{\dfrac{M}{L_2} E_G}{R_{11} + \dfrac{L_1}{L_2} R_{22} + j\left(\omega\sigma L_1 - \dfrac{R_{11}R_{22}}{\omega L_2}\right)}, \qquad [13]$$

where L_1, L_2, and M are the incremental self- and mutual inductances of the transformer. The term $R_{11}R_{22}/\omega L_2$ in the denominator of Eq. 13 can be expressed as

$$\frac{R_{11}R_{22}}{\omega L_2} = \frac{R_{11} \dfrac{L_1}{L_2} R_{22}}{\omega L_1}, \qquad [14]$$

and therefore, from Eq. 13,

$$I_L = \frac{\dfrac{M}{L_2} E_G}{R_{11} + \dfrac{L_1}{L_2} R_{22} + j\left(\omega\sigma L_1 - \dfrac{R_{11} \dfrac{L_1}{L_2} R_{22}}{\omega L_1}\right)}. \qquad [15]$$

Multiplication of Eq. 15 by L_2/M gives

$$\frac{L_2}{M} I_L = \frac{E_G}{R_{11} + \dfrac{L_1}{L_2} R_{22} + j\left(\omega\sigma L_1 - \dfrac{R_{11} \dfrac{L_1}{L_2} R_{22}}{\omega L_1}\right)}. \qquad [16]$$

The vector load voltage V_L is

$$V_L = I_L R_L = \frac{\dfrac{M}{L_2} E_G R_L}{R_{11} + \dfrac{L_1}{L_2} R_{22} + j\left(\omega\sigma L_1 - \dfrac{R_{11} \dfrac{L_1}{L_2} R_{22}}{\omega L_1}\right)}. \qquad [17]$$

Multiplication of Eq. 17 by L_1/M gives

$$\frac{L_1}{M}V_L = \frac{E_G\frac{L_1}{L_2}R_L}{R_{11} + \frac{L_1}{L_2}R_{22} + j\left(\omega\sigma L_1 - \frac{R_{11}\frac{L_1}{L_2}R_{22}}{\omega L_1}\right)}. \qquad [18]$$

Several interesting facts appear from a study of the above equations. The multiplying factor L_2/M on the left-hand side of Eq. 16 very nearly equals the ratio N_2/N_1 of secondary to primary turns, and therefore the left-hand side of Eq. 16 very nearly equals the load current referred to the primary. Similarly, L_1/M very nearly equals N_1/N_2 and therefore the left-hand side of Eq. 18 very nearly equals the load voltage referred to the primary. Also the multiplying factor L_1/L_2 which occurs with the load resistance R_L and with the secondary circuit resistance R_{22} very nearly equals $(N_1/N_2)^2$,

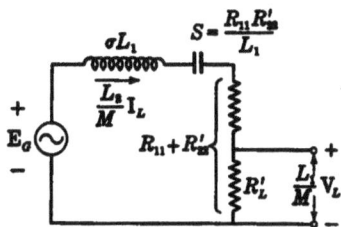

Fig. 3. Series *RLS* circuit which has the same frequency and phase characteristics as the circuit of Fig. 2.

and therefore $(L_1/L_2)R_L$ and $(L_1/L_2)R_{22}$ very nearly equal the load and secondary circuit resistances referred to the primary. To simplify the notation, let

$$\frac{L_1}{L_2}R_L \equiv R'_L \approx \text{load resistance referred to the primary} \qquad [19]$$

$$\frac{L_1}{L_2}R_{22} \equiv R'_{22} \approx \text{secondary circuit resistance referred to the} \qquad [20]$$
$$\text{primary.}$$

Also note that the inductance σL_1 very nearly equals the short-circuit inductance L_{sc1} measured on the primary side of the transformer, and the incremental self-inductance L_1 equals the open-circuit inductance L_{oc1} measured with the appropriate direct current in the primary.

It is interesting that Eq. 16 is of the same form as the expression for the current in a series *RLS* circuit whose resistance is $R_{11} + R'_{22}$, whose inductance is σL_1, and whose elastance is $R_{11}R'_{22}/L_1$. Equation 18 is the expression for the voltage across the portion R'_L of the resistance of this circuit. The circuit of Fig. 3 thus has frequency and phase characteristics identical with those of the output transformer circuit of Fig. 2.

From Eq. 18 with the simplified notation of Eqs. 19 and 20, the vector ratio of load voltage to generator electromotive force is

$$\frac{V_L}{E_G} = \frac{M}{L_1}\left[\frac{R_L'}{R_{11} + R_{22}' + j\left(\omega\sigma L_1 - \frac{R_{11}R_{22}'}{\omega L_1}\right)}\right] \qquad [21]$$

$$= \frac{M}{L_1}\frac{R_L'}{R_{11} + R_{22}'}\left[\frac{1}{1 + j\left(\frac{\omega\sigma L_1}{R_{11} + R_{22}'} - \frac{1}{\omega L_1}\frac{R_{11}R_{22}'}{R_{11} + R_{22}'}\right)}\right] \qquad [22]$$

In Eq. 22, $R_{11} + R_{22}'$ is the total series resistance of the generator, transformer windings, and load " referred " to the primary, while $R_{11}R_{22}'/(R_{11} + R_{22}')$ is the resistance of the parallel combination of the primary and secondary circuit resistances " referred " to the primary. In order further to simplify the notation, let

$$R_{11} + R_{22}' \equiv R_{se}', \qquad [23]$$

$$\frac{R_{11}R_{22}'}{R_{11} + R_{22}'} \equiv R_{par}'. \qquad [24]$$

Inserting this simplified notation in Eq. 22 and expressing the right-hand side in polar form give

$$\frac{V_L}{E_G} = \frac{M}{L_1}\frac{R_L'}{R_{se}'}\frac{1}{\sqrt{1 + \left(\frac{\omega\sigma L_1}{R_{se}'} - \frac{R_{par}'}{\omega L_1}\right)^2}}\left/\tan^{-1}\left(\frac{R_{par}'}{\omega L_1} - \frac{\omega\sigma L_1}{R_{se}'}\right)\right. . \qquad [25]$$

The frequency and phase characteristics are the magnitude and phase of this expression as functions of frequency.

The frequency $\omega_0/2\pi$ for which

$$\frac{\omega_0\sigma L_1}{R_{se}'} = \frac{R_{par}'}{\omega_0 L_1} \qquad [26]$$

is analogous to the resonant frequency of the series RLS circuit of Fig. 3. At this frequency, the voltage ratio has its greatest value and the phase shift is zero. From Eq. 26, this frequency f_0 is

$$f_0 = \frac{1}{2\pi}\sqrt{\frac{R_{par}'R_{se}'}{L_1\sigma L_1}} = \frac{1}{2\pi}\sqrt{\frac{R_{11}R_{22}'}{L_1\sigma L_1}}, \qquad \blacktriangleright[27]$$

and, since σL_1 very nearly equals the short-circuit inductance L_{sc1} measured on the primary side,

$$f_0 \approx \frac{1}{2\pi}\sqrt{\frac{R_{11}R_{22}'}{L_{oc1}L_{sc1}}}. \qquad \blacktriangleright[28]$$

At this frequency, the voltage ratio is

$$\left(\frac{V_L}{E_G}\right)_{max} = \frac{M}{L_1}\frac{R'_L}{R'_{se}}.$$ ▶[29]

At any other frequency, the relative magnitude of the voltage ratio as a fraction of its value at the frequency f_0 is

$$\text{Relative voltage ratio} = \frac{1}{\sqrt{1 + \left(\dfrac{\omega\sigma L_1}{R'_{se}} - \dfrac{R'_{par}}{\omega L_1}\right)^2}},$$ ▶[30]

and the phase angle is

$$\text{Phase angle} = \tan^{-1}\left(\frac{R'_{par}}{\omega L_1} - \frac{\omega\sigma L_1}{R'_{se}}\right).$$ ▶[31]

Typical frequency and phase characteristics are shown in Fig. 4, plotted on a logarithmic frequency scale which is convenient for showing

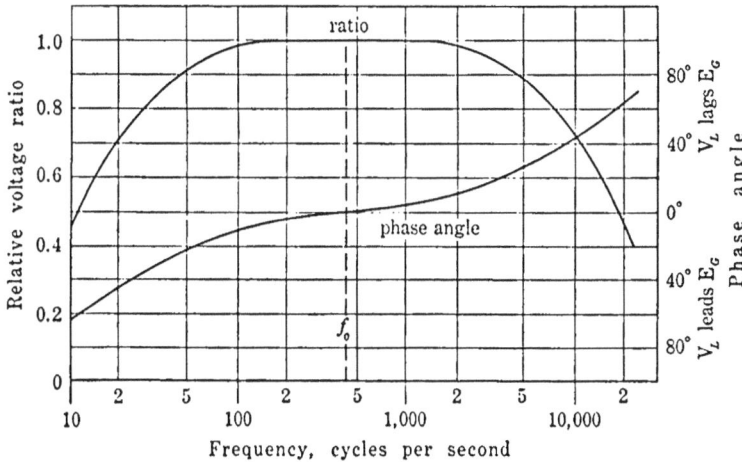

FIG. 4. Frequency and phase characteristics of a circuit containing an output transformer

a wide range of frequencies and which corresponds to the logarithmic nature of the musical scale.[4] At the "resonant" frequency f_0, Eq. 27, the relative voltage ratio equals unity, the actual voltage ratio is given by Eq. 29, and the phase angle between V_L and E_G is zero. If the frequency decreases, the ratio $R'_{par}/\omega L_1$ increases and becomes increasingly greater than the ratio $\omega\sigma L_1/R'_{se}$, as in the series RLS circuit of

---

[4] The voltage scale also is often expressed logarithmically — that is, in *decibels*. A discussion of the use of logarithmic scales for showing amplifier characteristics is given in the volume on electronics, Art. 2, Ch. IX.

Fig. 3, in which the capacitive reactance predominates over the inductive reactance at frequencies below resonance. Therefore, from Eqs. 30 and 31, as the frequency decreases the relative voltage ratio decreases and the load voltage advances in phase with respect to the generator electromotive force. On the other hand, if the frequency increases above f_0, the ratio $\omega\sigma L_1/R'_{se}$ increases and becomes increasingly greater than the ratio $R'_{par}/\omega L_1$, as in the series RLS circuit of Fig. 3 at frequencies above resonance, when the inductive reactance predominates. Therefore the relative voltage ratio also decreases at high frequencies, but the load voltage lags the generator electromotive force. However, if the ratios $R'_{par}/\omega_0 L_1$ and $\omega_0\sigma L_1/R'_{se}$ are sufficiently small, the frequency

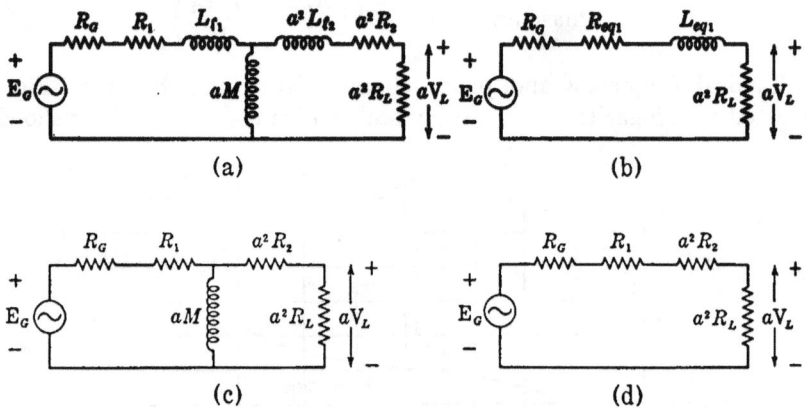

FIG. 5. Equivalent circuits for an output transformer. Circuit (a) applies over the whole audio range of frequencies. Circuit (b) is an approximate equivalent at high frequencies. Circuit (c) is an approximate equivalent at low frequencies. Circuit (d) is an approximate equivalent in the middle range of audio frequencies.

must be several octaves above or below the " resonant " frequency before one or the other of these terms becomes appreciable, and therefore the voltage ratio is very nearly constant over a wide range of frequencies, as in Fig. 4. The analogous RLS circuit of Fig. 3 is one in which the ratio of resonant-frequency reactance to resistance is small. On a logarithmic frequency scale, the frequency characteristic is symmetrical about f_0.

Before discussion of the simplifying approximations that can be made in this " exact " theoretical analysis, it is interesting to develop an approximate theory from the equivalent circuit of the transformer. The equivalent of the circuit of Fig. 2 when capacitances and core loss are neglected is shown in Fig. 5a, in which the transformer is replaced by its equivalent circuit comprising its winding resistances R_1 and $a^2 R_2$, its leakage inductances $L_{\ell 1}$ and $a^2 L_{\ell 2}$, and its magnetizing inductance

aM, all referred to the primary. The load resistance referred to the primary is a^2R_L, and the load voltage referred to the primary is aV_L, a being the ratio of primary to secondary turns.

Since the leakage inductance of a normal iron-core transformer is but a fraction of a per cent of the magnetizing inductance, these inductances are not simultaneously important. Thus when the frequency is sufficiently high to make the voltage drops in the leakage inductances important, the magnetizing reactance is so very high that its shunting effect is negligible and therefore at high audio frequencies the equivalent circuit can be simplified by omission of the magnetizing branch, as shown in Fig. 5b, in which R_{eq1} and L_{eq1} are the equivalent resistance and equivalent inductance referred to the primary. On the other hand, at low frequencies the magnetizing reactance ωaM is relatively small and its shunting effect is important, but the series voltage drops in the leakage inductances are then negligible. Thus at low frequencies the equivalent circuit simplifies to that shown in Fig. 5c. With a normal output transformer, there is usually an intermediate frequency range of several octaves in which the magnetizing reactance is so high that it produces negligible shunting effect but the leakage reactances are still so low that they cause negligible voltage drops. For this intermediate frequency range, the equivalent circuit can be approximately represented by the simple network of resistances shown in Fig. 5d.

From Fig. 5d, at intermediate frequencies,

$$aV_L = \frac{E_G \, a^2R_L}{R_{11} + a^2R_{22}}, \qquad [32]$$

where

   a^2R_L is the load resistance referred to the primary,
   R_{11} is the total resistance of the primary circuit,
   a^2R_{22} is the total resistance of the secondary circuit referred to the primary.

With an iron-core transformer, the discrepancy between the values of a^2 and of the inductance ratio L_1/L_2 is usually so small as to be undetectable, and therefore the resistances a^2R_L and a^2R_{22} very nearly equal the resistances R'_L and R'_{22} of Eqs. 19 and 20, and the denominator of Eq. 32 very nearly equals the series resistance R'_{se} of Eq. 23. Thus, from Eq. 32,

$$\text{Intermediate-frequency} \; \frac{V_L}{E_G} = \frac{1}{a} \frac{R'_L}{R'_{se}}. \qquad \blacktriangleright[33]$$

Since the discrepancy between the values of the turns ratio a and of the inductance ratio L_1/M is insignificant, Eq. 33 is almost identical with Eq. 29.

From the equivalent circuit of Fig. 5b, at high frequencies

$$aV_L = \frac{E_G\, a^2 R_L}{R_G + R_{eq1} + a^2 R_L + j\omega L_{eq1}} \qquad [34]$$

$$= \frac{E_G R_L'}{R_{se}'}\, \frac{1}{1 + j\dfrac{\omega L_{eq1}}{R_{se}'}}. \qquad [35]$$

Thus, from Eq. 35 expressed in polar form,

High-frequency $\dfrac{V_L}{E_G} =$

$$\frac{1}{a}\frac{R_L'}{R_{se}'}\, \frac{1}{\sqrt{1 + \left(\dfrac{\omega L_{eq1}}{R_{se}'}\right)^2}}\, \bigg/\tan^{-1}\left(-\frac{\omega L_{eq1}}{R_{se}'}\right). \qquad [36]$$

Note that R_L'/aR_{se}' is the intermediate-frequency voltage ratio and that the equivalent inductance L_{eq1} very nearly equals the short-circuit inductance L_{sc1} measured on the primary side. From Eq. 36, the magnitude of the relative voltage ratio at high frequencies is

$$\text{High-frequency relative voltage ratio} = \frac{1}{\sqrt{1 + \left(\dfrac{\omega L_{sc1}}{R_{se}'}\right)^2}}, \qquad \blacktriangleright[37]$$

and the high-frequency phase angle is

$$\text{High-frequency phase angle} = \tan^{-1}\left(-\frac{\omega L_{sc1}}{R_{se}'}\right). \qquad \blacktriangleright[38]$$

The minus sign in Eq. 38 indicates that the load voltage V_L lags the generator electromotive force E_G at high frequencies.

An analysis of the approximate equivalent circuit of Fig. 5c shows that at low frequencies

$$\text{Low-frequency } \frac{V_L}{E_G} = \frac{1}{a}\frac{R_L'}{R_{se}'}\, \frac{1}{\sqrt{1 + \left(\dfrac{R_{par}'}{\omega a M}\right)^2}}\, \bigg/\tan^{-1}\left(\frac{R_{par}'}{\omega a M}\right), \qquad [39]$$

where R_{par}' is the resistance formed by R_{11} and $a^2 R_{22}$ in parallel and very nearly equals the resistance R_{par}' of Eq. 24. Since the magnetizing inductance aM very nearly equals the open-circuit inductance L_{oc1} of the primary, the relative voltage ratio at low frequencies is

$$\text{Low-frequency relative voltage ratio} = \frac{1}{\sqrt{1 + \left(\dfrac{R_{par}'}{\omega L_{oc1}}\right)^2}}, \qquad \blacktriangleright[40]$$

and the phase angle at low frequencies is

$$\text{Low-frequency phase angle} = \tan^{-1}\left(\frac{R'_{par}}{\omega L_{oc1}}\right). \qquad \blacktriangleright[41]$$

The above approximate analysis assumes the leakage inductance to be so small compared with the magnetizing inductance that these inductances are never simultaneously important and that therefore the equivalent circuit can be simplified through considering different frequency ranges one at a time. It is interesting to compare the results of this analysis with the approximate results that can be obtained through simplifying the expressions, Eqs. 30 and 31, derived from the " exact " analysis. In Eqs. 30 and 31, if the self-inductance L_1 is sufficiently high compared with R'_{par}, the frequency must be relatively low when the ratio $R'_{par}/\omega L_1$ becomes important. But if the leakage coefficient is small, the ratio $\omega \sigma L_1/R'_{se}$ is then very small indeed at low frequencies. On the other hand, if the frequency is high enough so that $\omega \sigma L_1/R'_{se}$ is important, the ratio $R'_{par}/\omega L_1$ is then very small. Thus, with a small leakage coefficient, at high frequencies Eqs. 30 and 31 reduce to

$$\text{High-frequency relative voltage ratio} \approx \frac{1}{\sqrt{1 + \left(\dfrac{\omega \sigma L_1}{R'_{se}}\right)^2}} \qquad [42]$$

$$\text{High-frequency phase angle} \approx \tan^{-1}\left(-\frac{\omega \sigma L_1}{R'_{se}}\right), \qquad [43]$$

and at low frequencies

$$\text{Low-frequency relative voltage ratio} \approx \frac{1}{\sqrt{1 + \left(\dfrac{R'_{par}}{\omega L_1}\right)^2}} \qquad [44]$$

$$\text{Low-frequency phase angle} \approx \tan^{-1}\left(\frac{R'_{par}}{\omega L_1}\right). \qquad [45]$$

Since the resistances in Eqs. 30 and 31 differ from the corresponding resistances in Eqs. 37–41 only because of the insignificant difference between L_1/L_2 and a^2 and since the ideal short-circuit inductance σL_1 very nearly equals the actual short-circuit inductance L_{sc1}, the approximate results, Eqs. 42–45, obtained through simplifying the " exact " expressions, Eqs. 30 and 31, are practically identical with the results obtained from the approximate equivalent circuits.

According to Eqs. 40 and 41, at low frequencies the relative voltage ratio and phase angle are primarily determined by the value of the ratio $R'_{par}/\omega L_{oc1}$. The left-hand portion of Fig. 6 shows graphically the relations between these characteristics and the ratio $\omega L_{oc1}/R'_{par}$. At intermediate frequencies, the voltage ratio is approximately constant and the phase angle is small. At high frequencies, the relative voltage ratio and phase angle are primarily dependent on the value of the ratio $\omega L_{sc1}/R'_{se}$, as given by Eqs. 37 and 38 and shown graphically in the right-hand portion of Fig. 6.

The curves in Fig. 6 are applicable to any output transformer, since the variables involved are expressed as ratios. With the aid of these curves, it is a simple matter to compute the frequency and phase charac-

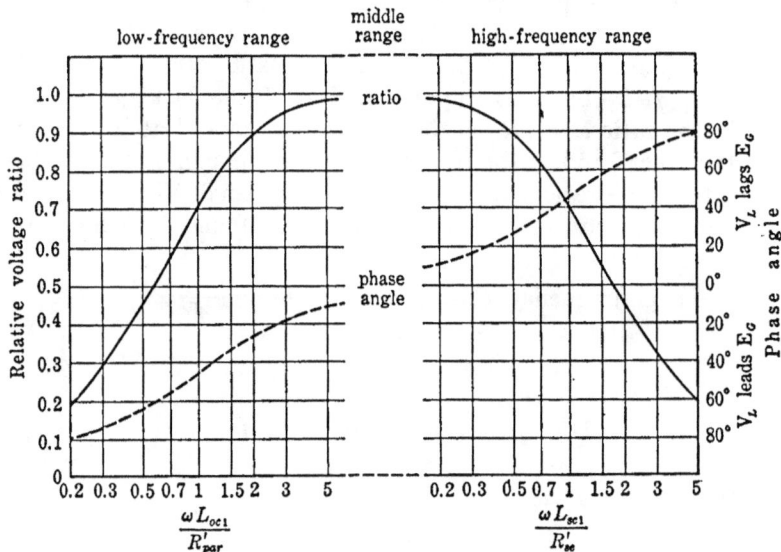

FIG. 6.   Universal frequency and phase characteristics of output transformers.[5]

teristics of any transformer if the load, generator, and winding resistances; the open-circuit and short-circuit inductances; and the turns ratio are known. For example, suppose it is desired to determine the frequency and phase characteristics of the circuit and transformer to which the following data apply:

$$R_G = 2{,}000 \text{ ohms} \qquad\qquad R_L = 10.0 \text{ ohms}$$

$$R_1 = 200 \text{ ohms} \qquad\qquad R_2 = 0.5 \text{ ohm}$$

$$L_{oc1} = 10.0 \text{ henries} \qquad\qquad L_{sc1} = 0.100 \text{ henry}$$

$$a = N_1/N_2 = 20.0$$

[5] After F. E. Terman and R. E. Ingebretsen, " Output Transformer Response," *Electronics,* *9* (January, 1936), 30–32.

From these data,

$$R_{11} = 2,200 \tag{46}$$

$$R_{22} = 10.5 \tag{47}$$

$$a^2 R_{22} = 4,200 \tag{48}$$

$$R'_{se} = 6,400 \tag{49}$$

$$R'_{par} = \frac{2,200 \times 4,200}{6,400} = 1,440. \tag{50}$$

From Eq. 40 or the curves of Fig. 6, the voltage ratio at low frequencies is $1/\sqrt{2}$ or 0.707 of its intermediate-frequency value when the frequency is f_l, the value for which

$$\frac{\omega_l L_{oc1}}{R'_{par}} = 1 \tag{51}$$

or

$$\omega_l = \frac{R'_{par}}{L_{oc1}} \qquad \blacktriangleright[52]$$

$$= \frac{1,440}{10} = 144. \tag{53}$$

The corresponding frequency f_l is $144/2\pi$, or 23.0 cycles per second. Also, from Eq. 37 or the curves of Fig. 6, at high frequencies the relative voltage ratio equals 0.707 when the frequency is f_h, the value for which

$$\frac{\omega_h L_{sc1}}{R'_{se}} = 1 \tag{54}$$

or

$$\omega_h = \frac{R'_{se}}{L_{sc1}} \qquad \blacktriangleright[55]$$

$$= \frac{6,400}{0.100} = 64,000. \tag{56}$$

The corresponding frequency f_h is $64,000/2\pi$, or 10,200 cycles per second. Other points on the frequency and phase characteristics can readily be determined in like manner.

When the relative voltage ratio equals 0.707, the power delivered to the load with a given generator electromotive force is $(0.707)^2$ or one-half the intermediate-frequency value, and therefore the frequencies at which the relative voltage ratio equals 0.707 are called the *half-power points*. The ratio between the frequency at the high-frequency half-power point and that at the low-frequency half-power point is a measure of the frequency range over which the circuit can be operated without causing the load power to fall below one-half its value for the best operating frequency. The relation between this frequency ratio ω_h/ω_l and the number of octaves n between the two half-power points is

$$\frac{\omega_h}{\omega_l} = 2^n, \qquad [57]$$

whence

$$n = \frac{\log \omega_h/\omega_l}{\log 2} = 3.32 \log \frac{\omega_h}{\omega_l}. \qquad [58]$$

From Eqs. 52 and 55,

$$\frac{\omega_h}{\omega_l} = \frac{R'_{ss}}{R'_{par}} \frac{L_{oc1}}{L_{sc1}}. \qquad \blacktriangleright[59]$$

Since the ratio of the short-circuit inductance to the open-circuit inductance very nearly equals the leakage coefficient σ,

$$\frac{\omega_h}{\omega_l} = \frac{1}{\sigma} \frac{R'_{ss}}{R'_{par}}. \qquad \blacktriangleright[60]$$

From Eqs. 37 and 40, the relative voltage ratio at a low frequency f_l equals the relative voltage ratio at a high frequency f_h if these frequencies have such values that

$$\frac{R'_{par}}{\omega_l L_{oc1}} = \frac{\omega_h L_{sc1}}{R'_{ss}} \qquad [61]$$

That is, the response is " down " by the same amount at a low and at a high frequency if these frequencies are such as to satisfy Eq. 61. On a logarithmic scale, the frequency f_0 midway between f_l and f_h is the geometric mean of f_l and f_h, or

$$f_0 = \sqrt{f_l f_h}$$

$$= \frac{1}{2\pi} \sqrt{\frac{R'_{par} R'_{ss}}{L_{oc1} L_{sc1}}} = \frac{1}{2\pi} \sqrt{\frac{R_{11} R'_{22}}{L_{oc1} L_{sc1}}}. \qquad [62]$$

Note that this geometric mean frequency equals the frequency for maximum voltage ratio as derived from the " exact " analysis and expressed in Eq. 28. Since the short-circuit inductance very nearly equals σL_{oc1},

$$f_0 = \frac{1}{2\pi L_{oc1}} \sqrt{\frac{R_{11} R'_{22}}{\sigma}}. \qquad \blacktriangleright[63]$$

The relations expressed in Eqs. 59, 60, and 63 are of the utmost importance in the design and application of output transformers.

▶ Equation 59 shows that the ratio of the highest to the lowest frequency at which the transformer can be used satisfactorily is proportional to the ratio of the open-circuit to the short-circuit inductance and is wholly independent of the magnitudes of these inductances. Thus the first requirement of a transformer to be used over a wide frequency range is that it must have a high ratio of self- to leakage inductance, or, in other words, a small leakage coefficient. ◀

The ratio of self- to leakage inductance is made large through using all possible means for increasing the self-inductance without also increasing the leakage inductance, and for reducing the leakage inductance without also reducing the self-inductance. Thus the self-inductance can be made large without producing a large leakage inductance by use of a core of high-permeability material with a suitable air gap to minimize the reduction in incremental permeability due to direct-current magnetization. The leakage inductance can be reduced without reducing the self-inductance through interleaving the primary and secondary windings and through properly proportioning the winding space. Note, however, that the frequency range ω_h/ω_l *cannot* be increased through increasing the numbers of turns in the windings. Increasing the numbers of turns in both windings (maintaining the same turns ratio) results in an increase in both the self- and leakage inductances, but not in an appreciable change in the leakage coefficient. The low-frequency response is improved by the increased self-inductance, but the high-frequency response is adversely affected by the increased leakage inductance. The geometric mean frequency is lowered, as shown by Eq. 63, but the ratio ω_h/ω_l is not changed appreciably. A transformer of a fixed shape and with a fixed space factor for its windings has approximately the same leakage coefficient regardless of the numbers of turns in its windings. Previous experience therefore enables a designer to anticipate approximately the frequency range to be expected for a particular core and winding arrangement, regardless of the numbers of turns in the windings. The numbers of turns determine the geometric mean frequency, however, and therefore in audio-frequency applications they are adjusted so that the geometric mean frequency, Eq. 63, is about 500 cycles per second when the transformer is used with the generator and load resistances for which it is designed.

Not only must an output transformer have satisfactory frequency and phase characteristics but also it must have the requisite power-handling capacity. The current-carrying capacity is limited by the temperature rise in the windings, as in a power-system transformer. The most important factor in determining the voltage rating is the amount of distortion that can be tolerated. The distortion caused by the peculiar waveform of the exciting current depends on the magnetic characteristics of the core material, the length of the air gap in the core, the direct-current magnetization, the amplitude of the alternating component of core flux density, and the internal impedance of the generator. The influence of all these factors on waveforms must be considered in the design of an output transformer. Since the amplitude of the alternating component of core flux density is proportional to the induced voltage and inversely proportional to the frequency, distortion with strong signals, rather than the falling-off in the frequency characteristic, may set the lower limit on the frequency range of an output transformer.

3. FREQUENCY AND PHASE CHARACTERISTICS OF INPUT AND INTERSTAGE TRANSFORMERS

Input and interstage transformers are used for similar purposes; namely, to connect a source of signal voltage to the grid of a vacuum tube. In many applications of vacuum tubes as linear amplifiers the power required by the grid of the tube is very small, and the grid cur-

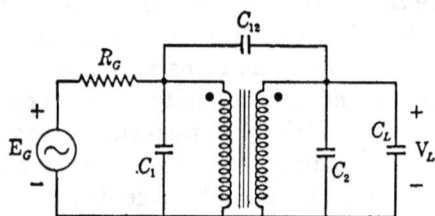

FIG. 7. Circuit containing an input or interstage transformer.

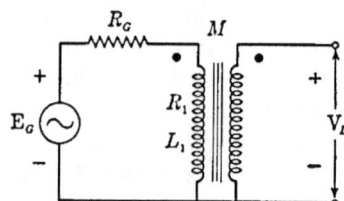

FIG. 8. Equivalent circuit at low and intermediate frequencies.

rent is very nearly only that taken by the small grid-to-cathode capacitance of the tube. This capacitance and the distributed capacitances of the transformer have important effects on the frequency and phase characteristics of the circuit.

Consider the circuit shown in Fig. 7, comprising a transformer, a generator whose internal resistance is R_G, and a load of capacitance C_L. When one primary and one secondary terminal of the transformer are joined, its distributed capacitances can be approximately represented by three lumped capacitances C_1, C_2, and C_{12}, as in Fig. 7. Let the internal electromotive force of the generator vary sinusoidally, and let E_G be the vector representing this source voltage. If the generator and transformer behave as essentially linear circuit elements, as they should where fidelity of waveform is important, the load voltage will also vary sinusoidally and can be represented by a vector V_L.

In a circuit designed to have a reasonably " flat " frequency response over the audio band of frequencies, the effects of capacitances of the tube and transformer usually are negligible when the frequency is below the middle of its range. Thus over the low and middle range of frequencies the circuit can be analyzed as if the secondary of the transformer were open-circuited, as shown in the approximate equivalent circuit of Fig. 8. Accordingly, the vector voltage equations for the primary and secondary circuits are

$$E_G = (R_{11} + j\omega L_1)I_1 \qquad [64]$$

$$V_L = j\omega M I_1, \qquad [65]$$

where

R_{11} is the total resistance of the primary circuit and equals the generator resistance R_G plus the transformer primary resistance R_1,

L_1 is the incremental self-inductance of the primary,

M is the incremental mutual inductance,

I_1 is the vector primary current, which is wholly exciting current since the secondary is open-circuited.

The vector ratio of load voltage to source electromotive force is

$$\frac{V_L}{E_G} = \frac{j\omega M}{R_{11} + j\omega L_1} = \frac{j\omega M}{j\omega L_1} \times \frac{1}{\dfrac{R_{11}}{j\omega L_1} + 1} \qquad [66]$$

$$= \frac{M}{L_1} \times \frac{1}{\sqrt{1 + \left(\dfrac{R_{11}}{\omega L_1}\right)^2}} \Big/ \tan^{-1}\frac{R_{11}}{\omega L_1}. \qquad \blacktriangleright[67]$$

At intermediate frequencies the primary circuit resistance R_{11} usually is negligible compared with the primary reactance ωL_1, and hence over this intermediate-frequency range the vector voltage ratio approximately equals M/L_1, which very nearly equals the ratio of secondary to primary turns. Thus the voltages are approximately in phase and the ratio of their magnitudes is very nearly independent of frequency and equal to the turns ratio, as in an ideal transformer. The frequency and phase characteristics of such a circuit are shown in Fig. 9. Note the flatness of the frequency characteristics in the middle range of frequencies.

As the frequency decreases, the primary reactance ωL_1 decreases and the ratio $R_{11}/\omega L_1$ increases. Thus, from Eq. 67, as the frequency decreases the voltage ratio V_L/E_G decreases and the load voltage V_L leads the source voltage E_G, as shown in the low-frequency portion of Fig. 9. Low frequencies therefore are not transmitted so readily as intermediate frequencies, and the circuit does not maintain the correct relations in magnitude and phase between the low- and intermediate-frequency components of a signal voltage having a complicated waveform.

The effects of the capacitances of the transformer and load become increasingly important as the frequency is increased above the middle

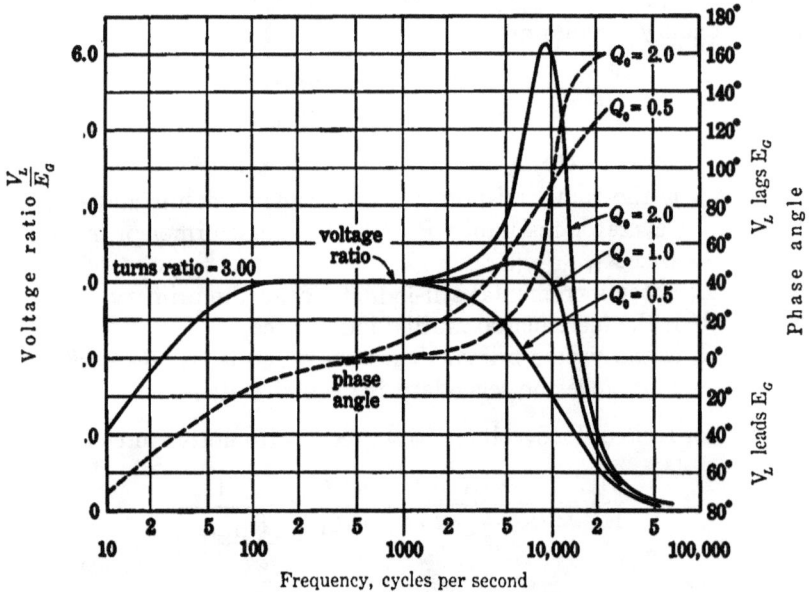

FG. 9.   Frequency and phase characteristics of a circuit containing an input or interstage transformer, showing the effects of varying the secondary resistance.

range. Figure 10a shows an approximate equivalent circuit including these capacitances when one primary and one secondary terminal are connected to each other. In Fig. 10a,

C_{10} is the primary shunt capacitance,
C_{12} is the interwinding capacitance,
C_{20} is the combined capacitance of the load and transformer secondary,
R_1 and R_2 are the effective resistances of the windings,
$L_{\ell 1}$ and $L_{\ell 2}$ are the leakage inductances,
M is the incremental mutual inductance,
a is the ratio of primary to secondary turns.

The values of the capacitances and their effects on the circuit performance depend not only on the internal arrangements of the windings but also on the relative polarity of the primary and secondary terminals that are joined. In the following analysis, terminals of like polarity are assumed to be connected. It is also assumed that the parameters of the equivalent circuit, Fig. 10a, are independent of frequency, although this assumption holds over only a limited range of frequency.

In analysis, it is convenient to " refer " the capacitances to the primary so that they may be transferred to the primary side of the ideal trans-

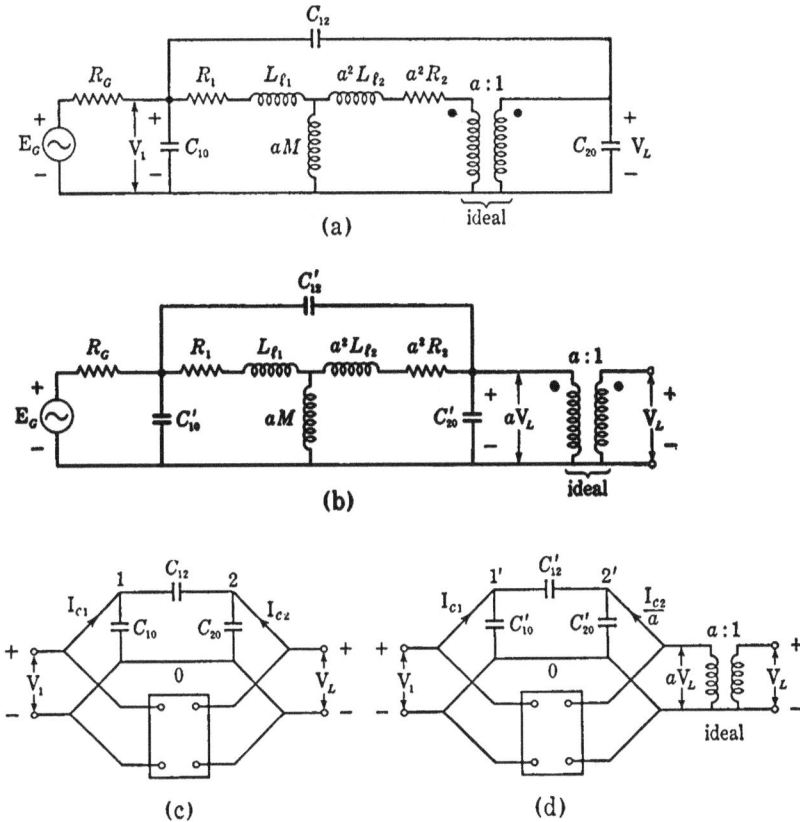

FIG. 10. Equivalent circuits at high audio frequencies.

former, as in Fig. 10b, in which C_{10}', C_{20}', and C_{12}' are capacitances that produce the same effect on the circuit as the capacitances C_{10}, C_{20}, and C_{12} in Fig. 10a. The capacitances in Fig. 10a form a π network, as shown in Fig. 10c, in which the box represents the inductive portion of the network of Fig. 10a. If the common terminal 0 is considered as the reference node, the currents I_{c1} and I_{c2} entering nodes 1 and 2, Fig. 10c, can be expressed as

$$I_{c1} = j\omega(C_{10} + C_{12})V_1 - j\omega C_{12}V_L \qquad [68]$$

$$I_{c2} = j\omega(C_{20} + C_{12})V_L - j\omega C_{12}V_1. \qquad [69]$$

These equations can be manipulated so that the variables are the secondary current and voltage referred to the primary — that is, I_{c2}/a and aV_L. Thus if Eq. 69 is divided by a, and V_L is multiplied by a, Eqs. 68 and 69 become

$$I_{c1} = j\omega(C_{10} + C_{12})V_1 - j\omega\frac{C_{12}}{a}aV_L \qquad [70]$$

$$\frac{I_{c2}}{a} = j\omega\left(\frac{C_{20} + C_{12}}{a^2}\right)aV_L - j\omega\frac{C_{12}}{a}V_1. \qquad [71]$$

When the capacitances are referred to the primary and are connected on the primary side of the ideal transformer, as in Figs. 10b and 10d, the current entering node 2', Fig. 10d, is I_{c2}/a, and the voltage of this node is aV_L. The current equations for nodes 1' and 2', Fig. 10d, are

$$I_{c1} = j\omega(C'_{10} + C'_{12})V_1 - j\omega C'_{12}aV_L \qquad [72]$$

$$\frac{I_{c2}}{a} = j\omega(C'_{20} + C'_{12})aV_L - j\omega C'_{12}V_1. \qquad [73]$$

Comparison of Eqs. 72 and 73 with Eqs. 70 and 71 shows that the relations among the referred values of the capacitances in Figs. 10b and 10d and their actual values in Figs. 10a and 10c must be

$$C'_{12} = \frac{C_{12}}{a} \qquad [74]$$

$$C'_{10} + C'_{12} = C_{10} + C_{12} \qquad [75]$$

$$C'_{20} + C'_{12} = \frac{C_{20} + C_{12}}{a^2}. \qquad [76]$$

From Eqs. 74 and 75,

$$C'_{10} = C_{10} + C_{12} - C'_{12} \qquad [77]$$

$$= C_{10} + C_{12}\left(1 - \frac{1}{a}\right), \qquad [78]$$

and, from Eqs. 74 and 76,

$$C'_{20} = \frac{C_{20} + C_{12}}{a^2} - C'_{12} \qquad [79]$$

$$= \frac{C_{20}}{a^2} + \frac{C_{12}}{a}\left(\frac{1}{a} - 1\right). \qquad [80]$$

Thus Eqs. 74, 78, and 80 give the referred values of the capacitances in terms of the actual values. With a step-up transformer, the ratio a of primary to secondary turns is less than unity and consequently the component $C_{12}\left(1 - \dfrac{1}{a}\right)$ of Eq. 78 is negative and the primary capacitance C'_{10} of Fig. 10b may be a hypothetical negative capacitance.

A complete analysis of the equivalent circuit of Fig. 10b is too complicated for engineering use and furthermore is probably not worth while since the representation of the distributed capacitances of the transformer by three constant capacitances is only approximately correct when the frequency varies. The equivalent circuit can be simplified, however, so as to show the nature of the phenomena involved.

At high audio frequencies, the reactance $\omega a M$ of the magnetizing branch usually is so high that its shunting effect is negligible and it is therefore usually omitted. Furthermore, the capacitance C'_{10} in Fig. 10b

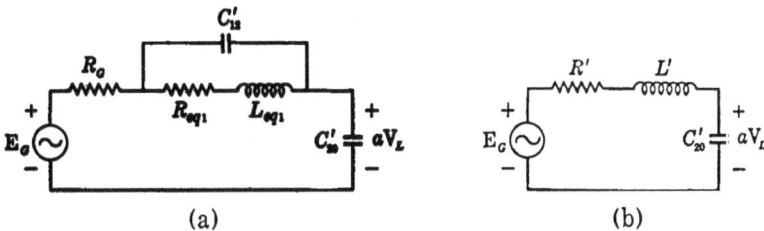

(a)   (b)

FIG. 11.   Approximate equivalent circuits at high audio frequencies.

frequently can be ignored, since its effects usually are relatively unimportant, because of its position in the circuit. The equivalent circuit hence can be simplified as shown in Fig. 11a, in which R_{eq1} and L_{eq1} are the equivalent resistance and equivalent reactance of the transformer referred to its primary side.

For frequencies below that at which the equivalent interwinding capacitance C'_{12} is in parallel resonance with the equivalent inductance L_{eq1}, the apparent reactance of this parallel combination is an inductive reactance, and the apparent resistance is larger than the equivalent resistance R_{eq1}. Although the values of apparent inductance and apparent resistance change with frequency, a fairly good approximation to the shape of the frequency characteristic can be obtained through assuming a constant value for the apparent inductance and ignoring the shunting effect of C'_{12} on the resistance. Accordingly, the equivalent circuit can be further simplified to that shown in Fig. 11b, in which R' is the generator resistance plus the equivalent resistance of the transformer and L' is the apparent inductance of the parallel combination of L_{eq1} and C'_{12}.

The resonant frequency f_0 of this series circuit is

$$f_0 = \frac{1}{2\pi \sqrt{L'C'_{20}}},$$ [81]

and at resonance the ratio Q_0 of apparent reactance to resistance is

$$Q_0 = \frac{\omega_0 L'}{R'}. \qquad [82]$$

For high values of Q_0, the voltage across the capacitance at resonance may exceed the generator voltage, and therefore when Q_0 is large the frequency characteristic has a sharp peak at the resonant frequency, as shown in the curve labeled $Q_0 = 2.0$ in Fig. 9. This peak may be objectionable because it exaggerates high-frequency components in the signal voltage. With lower values of Q_0, the resonant peak is suppressed; and, if Q_0 is between about 0.7 and 1.0, the frequency characteristic is substantially flat well into the range of frequencies where capacitances are significant, as shown in Fig. 9. As the frequency is increased above resonance, the reactances of the capacitances become smaller and eventually become practically zero, reducing the load voltage to a very small value, as shown in Fig. 9.

The circuit behavior at both low and high frequencies is markedly affected by the resistances and capacitances of the source and load, and therefore the source, transformer, and load in an amplifier must be properly suited to each other. Discussion of the complete amplifier characteristics is deferred to the volume on electronics, where the properties of electronic tubes are discussed. However, certain important conclusions with respect to the desirable qualities and design limitations of input and interstage transformers for broad-band amplifiers can be drawn from the present discussion.

The desirable features of such an amplifier are that the flat portion of its frequency characteristic should cover the range of frequencies to be transmitted, that the distortion introduced by the magnetization characteristics of the transformer should not be objectionable, and that the step-up turns ratio should be as high as possible in order to increase the voltage gain of the amplifier. The design of an audio-frequency transformer to meet these requirements is a balance among conflicting factors.

At low frequencies, the frequency characteristic is very nearly flat as long as the total primary circuit resistance R_{11} is small compared with the primary self-reactance ωL_1. (See Eq. 67.) Thus the region of essentially constant voltage gain is extended to lower frequencies if (1) the incremental self-inductance L_1 of the primary winding is made larger, (2) the primary winding resistance R_1 is made smaller, or (3) the generator resistance R_G is made smaller. Also, the region of essentially constant voltage gain is extended to higher frequencies if the resonant frequency f_0 (Eq. 81) is made higher by decreasing (1) the leakage inductances, (2) the interwinding capacitance, or (3) the secondary and load capacitances.

In order to obtain high primary incremental self-inductances, the cores are made of high-permeability low-loss magnetic materials. Silicon steel laminations are commonly used, and the cores of interstage transformers usually have a small air gap to reduce the direct-current magnetization due to the direct-current component of the plate current of the tube, and thereby to minimize the adverse effects of direct-current magnetization on the incremental permeability. Special alloys, such as permalloy, are sometimes used for the cores of high-fidelity audio-frequency transformers. Since the incremental permeability of permalloy is very adversely affected by direct-current magnetization, the circuit of the amplifier is arranged to eliminate the direct current from the primary winding when the transformer has a permalloy core. This circuit arrangement, known as "shunt feeding," is described in the volume on electronics, Art. 5, Ch. IX.

A high primary incremental self-inductance also is obtained if the cross-sectional area of the core and the number of turns in the primary winding are made as large as possible, consistent with other requirements. Increasing the core area and the number of primary turns reduces the amplitude of the alternating component of flux density and hence has the beneficial effect of reducing the distortion due to non-linear magnetic characteristics; at the same time, however, the high-frequency response is adversely affected. Thus an increase in the number of primary turns requires an increase in the number of secondary turns if the same turns ratio is maintained, and the leakage inductance increases with the number of turns. Likewise, increasing the core area with a fixed number of turns increases the capacitances and leakage inductances. Increasing these capacitances or the leakage inductances has the undesirable effect of lowering the resonant frequency. (See Eq. 81.) Interleaving the primary and secondary windings reduces the leakage inductances, but the advantage of lower leakage inductances is partly neutralized by the increased interwinding capacitance caused by interleaving, and hence interleaving must be done with care.

An increase in the step-up turns ratio for the purpose of increasing the voltage gain is accompanied by a decrease of the primary self-inductance, by increases in the secondary leakage inductance and distributed capacitance, or by both. Consequently, the frequency range over which the voltage ratio is essentially constant generally decreases as the turns ratio is increased, and the practical maximum for the step-up turns ratio has been found to be about 1 : 4 for audio-frequency interstage transformers.

A further deduction based on the curves of Fig. 9 is that to prevent an objectionable peak in the frequency response at the high frequencies near resonance, the Q_0 of the circuit (Eq. 82) must be not greater than about 1.0. The Q_0 may be reduced to this value through decreasing the leakage inductance (which also increases the resonant frequency) or through increasing the sum of the generator resistance and the transformer winding resistances. However, increasing the primary winding resistance or the generator resistance adversely affects the low-frequency response. The secondary winding resistance, on the other hand, does not affect the low-frequency response; hence the secondary windings of some interstage transformers are wound with high-resistivity wire to improve the high-frequency response. The effective Q_0 of the circuit may also be reduced and the voltage peak at resonance suppressed by connection of a high resistance across the secondary terminals of the transformer, although this results in some sacrifice of voltage gain at all frequencies.

PROBLEMS

1. An alternating-current generator with fixed electromotive force E and internal resistance R_G supplies power to a resistance load R_L. Assume that for any chosen value of R_L a transformer is available having a turns ratio which results in maximum load power. How large or how small must the ratio R_L/R_G be before the use of a transformer is warranted, if

    (a) the transformer efficiency is 85%?
    (b) the transformer efficiency is 50%?

2. The following data are known for an iron-core audio-frequency transformer:

Effective primary winding resistance	= 300 ohms
Effective secondary winding resistance	= 2 ohms
Primary open-circuit inductance	= 11 h
Primary short-circuit inductance	= 0.09 h
Turns ratio	= 20

If the primary circuit external to the transformer is a pure resistance of 2,000 ohms in series with an electromotive force of 125 v at 500 $\sim$:

    (a) What load resistance is required to make 500 $\sim$ be the frequency at which the voltage ratio is a maximum?
    (b) What is the power delivered to the load at 500 $\sim$ if the load resistance is the value found in (a)?

3. Starting from Eq. 25, p. 476, and making no further approximations, derive expressions for:

    (a) The upper and lower half-power angular frequencies ω_h and ω_l,
    (b) The band width $\omega_h - \omega_l$,
    (c) The selectivity factor $\dfrac{\omega_0}{\omega_h - \omega_l}$.
    (d) Show what terms or factors must be neglected in the expressions for ω_h and ω_l derived in part (a) to make these expressions identical with Eqs. 52 and 55.

4. A transformer is to be used to connect a dynamic microphone of 25 ohms resistance to a transmission line of 600 ohms resistance. The reactances of both the microphone and the line may be neglected. The transformer is to have the proper turns ratio to give maximum power transfer, and it is to have a drop in voltage ratio from the maximum of not more than 5% and a shift in phase of not more than ±20

degrees in the range from 50 $\sim$ to 8,000 $\sim$. Assuming that the resistances of the transformer are negligible, find its primary and secondary inductances and coefficient of coupling.

5. In a laboratory file, part of the information which is given concerning matching transformers is the values of external resistance in the primary and secondary circuits which will give a maximum power transfer at 500 $\sim$ for a given voltage in the primary circuit.

It is desired to use a telephone receiver of 10,000 ohms resistance and negligible inductance as a detector operating at 1,500 $\sim$ on an alternating-current bridge which presents an equivalent resistance of 100 ohms at its output terminal. If the resistances of the transformer windings are small enough to be neglected, what would be the external rated resistances for the transformer best suited for coupling the telephone receiver to the bridge?

6. A generator which has a generated voltage of

$$e_G = 15 \sin 11{,}500t + 5 \sin 34{,}500t \qquad [83]$$

and an internal resistance of 1,500 ohms is connected across the primary terminals of the input transformer for which the equivalent circuit is shown in Fig. 12.

(a) Plot the waveform of the input voltage.

(b) Plot the waveform of the output voltage, using the same voltage and time scales as in part (a). Any justifiable approximations may be made in determining this waveform.

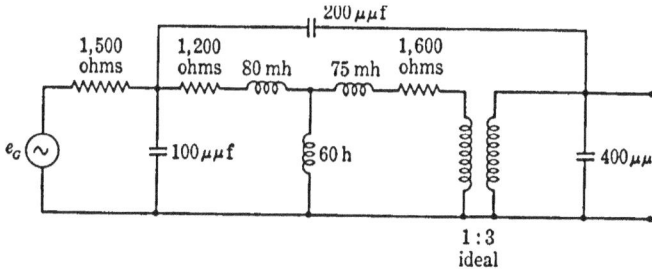

FIG. 12.   Equivalent circuit for an input transformer, Prob. 6.

7. The following data have been taken on an interstage transformer:

Primary short-circuit inductance   = 108 mh
Primary winding resistance   = 550 ohms
Secondary winding resistance   = 6,080 ohms
Open-circuit voltage ratio at 200 $\sim$ = 1 : 2.5
Primary open-circuit inductance   = 20 h

Furthermore, it is found that with the primary terminals of the transformer short-circuited and the secondary excited, parallel resonance occurs at 6,500 $\sim$.

(a) Determine the voltage ratio at the resonant frequency if the transformer is connected on its primary side to a source whose internal impedance is a resistance of 2,000 ohms. Neglect the interwinding capacitance and the capacitance of the load.

(b) Determine the frequency of the low-frequency half-power point when the transformer is used as in (a).

8. An input transformer with a turns ratio of 1 : 4 and a primary short-circuit inductance of 320 mh has as its resonant frequency 4,970 $\sim$. When supplied by a generator with an internal resistance of 20,000 ohms, the ratio of V_L/E_G at resonance is 3.2 : 1. If the resistance, mutual inductance, and equivalent primary shunt capacitance of the transformer may be neglected, what are the values of:

(a) The equivalent interwinding capacitance C_{12}?
(b) The equivalent secondary capacitance C_{20}?

CHAPTER XIX

Special Applications of Transformers in Power Systems

Many transformers are used in power systems to actuate measuring instruments and control relays, to obtain constant current for the operation of series-connected street-lighting circuits, and to regulate the voltage or to control the flow of reactive power in a feeder or transmission line. Some of these transformers are discussed briefly in this chapter.

1. INSTRUMENT TRANSFORMERS

Measuring instruments and relays for the operation of protective and control devices usually are connected to alternating-current power circuits through *instrument transformers* if the circuit voltage is higher than a few hundred volts. Direct connection between high-voltage circuits and instruments, which would be dangerous to operators and would require expensive and elaborately insulated instrument panels, is thus avoided. Even when the circuit voltage is not dangerously high, instrument transformers are commonly used for measurement of large currents, to avoid bringing heavy leads to the instrument panels.

Instrument transformers are classified as *potential transformers* or *current transformers* according to whether they are used in measurement of potential difference or current. The primary is suitably connected to the circuit whose voltage or current is to be measured, and the secondary is connected to the instruments or relays. When instrument transformers with the appropriate ratios of transformation are installed, 150-volt and 5-ampere instruments and relays can be used for circuits of any voltage and current ratings. With this standardization, which is the accepted American practice, the instrument and control-circuit costs can be minimized.

1a. *Potential Transformer.* — The primary of a potential transformer, wound for approximately the frequency and voltage to be measured, is connected across this voltage; the secondary, wound for approximately 115 volts, is connected to 150-volt measuring instruments. The secondary circuit should be grounded to safeguard the operator in the event of an insulation failure and also to prevent the accumulation of an electrostatic charge which might affect the readings of the instruments. The voltage V between the lines to which the primary of the potential transformer is connected is given by

$$V = \text{(transformer voltage ratio)} \times \text{(voltmeter reading)}. \qquad [1]$$

The scales of switchboard voltmeters may be graduated to indicate the line voltage directly.

The theory of a potential transformer is exactly the same as that of any other iron-core transformer. If it were not for the leakage-impedance voltage drops, the ratio of terminal voltages would be constant, independent of the voltage and of the current the secondary delivers to operate the instruments. This constant ratio would equal the ratio of turns in the two windings. Since the ratio of terminal voltages should be as nearly constant as possible, potential transformers are designed to have small resistances and small leakage reactances. The secondary burden* on a potential transformer usually consists of voltage-measuring instruments, and therefore usually is very nearly noninductive resistance.

Courtesy General Electric Co.

Fig. 1. An indoor 13,800 : 115-v potential transformer with current-limiting fuses. The output terminals are located within the metal cap in the foreground.

Because of the leakage-impedance voltage drops, there may be a small phase angle between the primary and secondary voltages of corresponding polarity. This phase angle may introduce an error, usually small, in the reading of a wattmeter whose potential coil is connected to the secondary of the transformer. Since the indication of a wattmeter is proportional to the cosine of the angle between the voltage across its potential coil and the current in its current coil, a given angular displacement in the voltage produces a greater error in the reading of the wattmeter when the angle between the voltage and current is near 90 degrees than when the angle is near zero degrees.

Potential transformers are readily designed whose ratios vary only a fraction of a per cent under varying conditions of voltage and burden and whose phase angles are less than 0.1 degree.

Although its output is but a few volt-amperes, the physical size and weight of a potential transformer for high voltage may be large, because high-voltage insulation is required.

1b. *Current Transformer.* — The primary of a current transformer, consisting of a few turns — sometimes even of a single turn — is con-

---

* The instrument load connected to the secondary of an instrument transformer is commonly called the *burden*.

nected in series with the circuit whose current is to be measured, and the secondary is connected to the current-measuring instruments. As in the use of potential transformers, the secondary circuits of current trans-

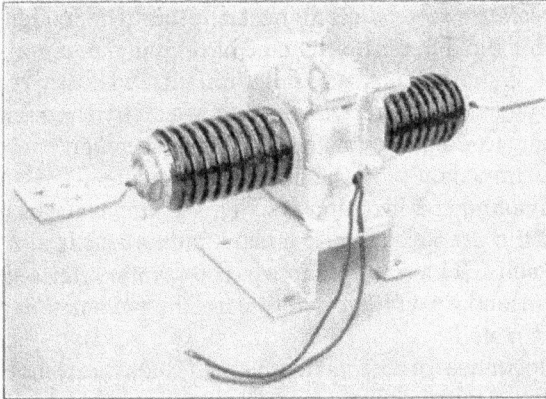

Courtesy Westinghouse Electric and Manufacturing Co.

FIG. 2. A current transformer for indoor service, typical of the design used for currents of 1,000 to 5,000 amp at about 25,000 v. The primary consists of a straight rod or bar, forming a single turn linked with the core. Current transformers for higher-voltage circuits are shown in Fig. 10, p. 279.

formers should be grounded. The current I in the primary circuit is given by

$$I = \text{(transformer current ratio)} \times \text{(ammeter reading)}. \qquad [2]$$

The scales of switchboard ammeters may be graduated to indicate the primary line current directly.

Current transformers are also used for comparison of two currents in different circuits or parts of the same circuit. For example, in Fig. 3, W represents the winding of one phase of a polyphase generator or motor, and CT', CT'' represent two identical current transformers. The relay coil R is connected across the secondaries of the two current transformers, whose secondary terminals are connected to each other with a relative polarity such that the current through the relay coil is the vector difference between the secondary currents in the two current transformers, as indicated in Fig. 3. If the insulation of the winding W is in good con-

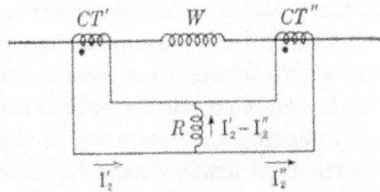

FIG. 3. Elementary circuit diagram showing the use of current transformers in differential-current relaying.

dition, the currents in the two transformers are equal and there is no current in the relay coil. In the event of a fault in the winding, however, the currents at its two ends differ by the current in the fault, and there is a proportional current in the relay coil, which can be arranged to operate a circuit breaker so as to disconnect the faulted machine. This circuit is the basis for the differential scheme of relaying, commonly used to protect machines, short feeders, parallel circuits, and other installations.

The theory of the current transformer is exactly the same as the theory of any other iron-core transformer. The secondary circuit is closed through the impedance of the instruments connected to it. Normally these are five-ampere instruments. The voltage across the secondary terminals is the drop through the instruments and leads, and usually is only a few volts. The voltage across the primary terminals is approximately the secondary voltage referred to the primary, and may be only a fraction of a volt.

If the impedance of the instruments is constant, both the induced voltage and the mutual flux are directly proportional to the secondary current. If it were not for the exciting current, the primary and secondary currents would be exactly inversely proportional to the number of turns in the two windings, and they would also be exactly in phase when considered in opposite directions about the core. If the magnetizing component of the exciting current were proportional to the mutual flux and if the core loss were proportional to the square of the mutual flux, the exciting current would be proportional to and have a fixed phase relation with the secondary current, and consequently the primary current would also be proportional to and have a fixed phase relation with the secondary current. Because of the nonlinearity of the magnetic circuit, however, the exciting current is not proportional to the secondary current, nor is its phase angle constant.

Since the exciting current alters the ratio and phase angle of the primary and secondary currents, it is made as small as possible through the use of high-permeability, low-loss steel in the construction of the magnetic circuit. Ingenious schemes have been developed which almost completely compensate for the effects of the exciting current.[1] The secondary leakage impedance and the impedances of the secondary leads and instruments also should be as low as possible, since any increase in these impedances increases the core flux and therefore the exciting current.

The exciting current depends on the degree of saturation of the core. For this reason, direct current should not be allowed to flow through a current transformer unless the transformer is afterwards thoroughly demagnetized to remove the residual magnetism. Also, the secondary winding should never be opened while the primary carries current. If

---

[1] For description of a number of compensating arrangements see B. Hague, *Instrument Transformers* (London: Sir Isaac Pitman & Sons, Ltd., 1936), 73–102.

the secondary is opened under these conditions, the primary current, being almost entirely determined by the circuit in which the primary is placed, is practically unchanged. Because of the removal of the demagnetizing effect of the secondary current, the flux then rises to a value determined by the primary current. Since a current transformer is normally operated at low flux density, the flux may increase considerably until it is limited by saturation, and the rms value of the voltage across the open secondary terminals may rise to a few hundred volts. Because of the distortion due to the then saturated condition of the core, the induced voltage is sharply peaked in waveform, and its peak value may be sufficiently high to be dangerous both to life and to the insulation of the transformer.

Another cause of serious overvoltages in current transformers is surges due to lightning or switching transients. To reduce the effects of these overvoltages *by-pass protectors*, similar to lightning arresters, often are connected in parallel with the primaries of current transformers.

1c. *Three-Phase Connections.** — Instrument transformers can be connected in a variety of ways for the measurement of voltage, current, and power in three-phase circuits. For example, if three potential trans-

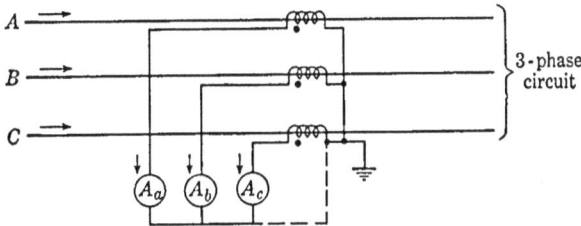

FIG. 4. Y-connected current transformers.

formers are connected in Δ–Δ, the three-phase line voltages can be measured by voltmeters connected between each pair of secondary terminals. If the primaries of three current transformers are connected one in series with each line, the three line currents can be measured by the circuit shown in Fig. 4, in which the secondaries are Y-connected and grounded, and the ammeters also are Y-connected. If the power circuit is a three-wire circuit without a neutral wire, the instantaneous sum of the three primary line currents flowing toward the load must be zero, and therefore the sum of the secondary currents must also be zero if the three current transformers are alike. Consequently the broken-line connection between the neutrals of the Y-connected secondaries and of the ammeters, Fig. 4, may be omitted. This connection, however, is necessary when the power circuit has a neutral wire.

* The general subject of three-phase connections of transformers is discussed in Chs. XXI–XXIV.

The voltages, currents, and power in a three-wire three-phase circuit can be measured by means of two identical potential transformers and two identical current transformers connected as shown in Fig. 5. The connection of the potential transformers in Fig. 5a is known as the

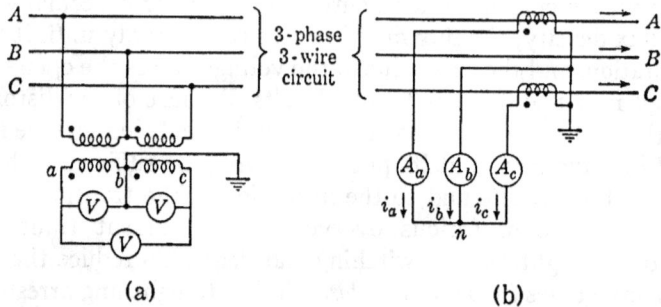

FIG. 5. Instrument-transformer connections for 3-wire 3-phase circuits; (a) open-delta connection of two potential transformers, (b) two current transformers.

open-delta connection. It is described in detail in Art. 5, Ch. XXI. In Fig. 5b, ammeters A_a and A_c are directly in series with the two current transformers, and therefore indicate the currents in lines A and C. From Kirchhoff's current law applied to point n, the relation among the secondary currents is

$$i_a + i_b + i_c = 0, \qquad [3]$$

and since i_a and i_c are proportional to the primary line currents i_A and i_C respectively, the current i_b in ammeter A_b is also proportional to the

FIG. 6. An outdoor oil-insulated 23,000-v 200-amp 3-wire 3-phase metering outfit. Each of the two foremost bushings encloses two leads conducting current to the primary terminals of a single-phase current transformer. The voltage across each current transformer primary winding is limited by the by-pass protector mounted on top of the bushing. The rear bushing houses one potential lead, the potential leads to the other two phases being connected inside the tank. The connections are given in Fig. 5. High-voltage single-phase outdoor potential transformers resemble this metering outfit, but have only two bushings and do not have by-pass protectors.

primary current i_B if the sum of the primary currents is zero, as it must be when the power circuit is a three-wire circuit. Often the potential and the current transformers are mounted in a single case, as shown in Fig. 6, in order to save space, weight, and cost. The combination is then known as a *metering outfit* The power in the three-phase three-wire circuit can be measured by the two-wattmeter method[2] if two wattmeters are connected with their current coils in series with ammeters A_a and A_c, Fig. 5b, and with their potential coils excited from voltages V_{ab} and V_{cb} respectively, Fig. 5a. Note that this arrangement gives the three-phase power and the current in phase B only when the three-phase circuit is a three-wire system without a neutral wire.

2. CONSTANT–CURRENT TRANSFORMER

In street lighting, arc, incandescent, or gaseous-discharge lamps are frequently used connected in series. In a series circuit, each lamp receives the same current, and the line current is much less than it would be if the lamps were supplied from a constant-voltage parallel circuit. For series connection, the lamps must all have the same current rating,* and the current in the circuit must be regulated so as to remain constant at the rated value. Common ratings for these series circuits are 6.6 and 7.5 amperes. *Constant-current transformers* are almost universally used for such circuits.

Figure 7 shows a photograph of a constant-current transformer, and Fig. 8 a simplified sketch of a slightly different arrangement. Both types, however, operate on the same principles, the essential features of which are shown in Fig. 8. The magnetic circuit is arranged like that of a shell-type transformer, except that the core is long enough to permit a considerable separation of the primary and secondary windings. The iron core CCC, Fig. 8, is operated at a flux density sufficiently high so that the core leakage flux is larger than in a constant-voltage power transformer. The primary and secondary windings 1 and 2 surround the central leg of the core. One winding is fixed; the other is movable and is supported from an arm pivoted at P. The leakage reactance of such a transformer is relatively large, and increases as the coils are separated. The weight of the movable winding is partly counterbalanced by a weight W hung from the sector S attached to the pivoted arm. The primary winding may be either the fixed or the movable winding. In the following discussion, assume that the primary is the fixed winding 1.

The core leakage flux due to the primary current is indicated by the

[2] See the volume on electric circuits (1940), Art. 10, Ch. X.

* Small auxiliary transformers connected in series with the constant-current circuit can be used to change the current supplied to a lamp whose rated current differs from of the series circuit. Sometimes these auxiliary transformers are autotransformers.

broken lines in Fig. 8. The instantaneous direction of this primary leakage flux is shown for the half-cycles when the primary current is in the direction shown in the figure by the dots and crosses (indicating the heads and tails of arrows pointing in the direction of the current). Since the primary and secondary currents are very nearly opposite in time phase, the direction of the secondary current is opposite to the direction of the primary

Fig. 7.   A station-type constant-current transformer used for power supply to a series street-lighting system. The protective wire enclosure which surrounds the transformer when it is in service has been removed. The rating is 2,300 v primary, 6.6 amp secondary, 60 ~, 60 kw output.

current except during a very small part of each cycle. By the left-hand rule for determining the direction of the force on a current-carrying conductor, it is seen that when the currents are in the directions shown in Fig. 8 an upward force is exerted on the secondary. A half-cycle later, both currents are reversed; the leakage flux therefore also is reversed, and the force still is upward. During all but the very small part of each half-cycle when the currents in the two coils are small and in the same direction, the force on the secondary is upward. The upward force, being proportional to the product of the alternating primary and secondary currents, is a double-line-frequency pulsating quantity having an average value which tends to separate the windings. On account of the mass of the secondary, the double-frequency variation in the force causes no appreciable motion of the secondary, though it may contribute appreciably to the noise radiated from the transformer.

Because of the electromagnetic force of repulsion between the two windings, the movable secondary adjusts its position with respect to the fixed primary so that the force just equals the unbalanced weight of the coil and the arm. If the impedance of the load is diminished, the current increases, the force of repulsion between the coils increases, and winding 2 moves farther away from 1, increasing the leakage reactance and diminish-

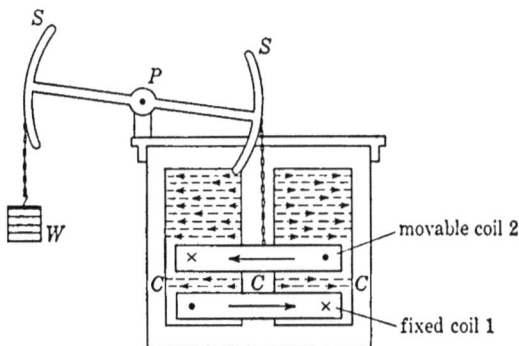

FIG. 8.   Sketch showing arrangement of core and windings in a constant-current transformer.

ing the current. In order to prevent coil 2 from oscillating about its balanced position, some means for damping the oscillations is provided, such as a dashpot connected to the movable system. By proper adjustment of the counterweight *W*, the shape of the sectors *S*, and the angle at which they are set, the transformer can be made to regulate for very nearly constant current over any desired range of load, and even when the secondary is short-circuited, provided the core is long enough to allow sufficient separation between the windings. When the secondary is short-circuited, the transformer delivers no load, because its output voltage is zero. As the resistance of the load increases, the windings move closer together in order to develop the voltage that is now necessary to maintain constant secondary current. Since the secondary current is constant, the power output increases with increase in load resistance until the windings finally come into contact. The power output then has its maximum value. After the windings come into contact, any further increase in the load resistance causes the secondary current and the power output to diminish, as in an ordinary constant-voltage transformer.

In a properly adjusted constant-current transformer, the secondary current is constant. If it were not for the variation in the magnitude and phase of the exciting current, the primary current also would be constant. Therefore the primary operates at a constant voltage and approximately constant current, and the change in power input caused by a change in the load is principally due to a change in the primary power factor. The secondary delivers power at constant current and variable voltage, and at a power factor determined by the load. For a load of incandescent lamps, the output power factor is very nearly unity.

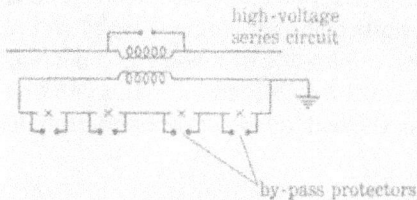

FIG. 9.   Circuit using a loop-insulating transformer.

A constant-current transformer is started with the secondary winding lifted to its highest position either manually or by automatic means. After the primary switch is closed, the secondary winding is released and allowed to take up the position corresponding to the load on the transformer.

The voltage to ground of parts of a series street-lighting circuit may reach dangerously high values — several thousand volts. To avoid the existence of such voltages between lighting fixtures and ground, street-lighting circuits often are broken up into a number of loops, each fed from a transformer whose primary is in series with the main circuit, as shown in Fig. 9. A transformer used in this way often is called a *loop-insulating transformer*, because it provides insulation between the main series circuit and the loop so that the loop may be grounded. By means of the appropriate ratio of current transformation, a loop-insulating transformer can also be used to supply a loop consisting of lamps whose rated current differs from that in the main series circuit.

To avoid the opening of the whole series circuit if one of the lamps should burn out, the lamps and primary windings of loop-insulating transformers usually are provided with by-pass protectors connected in parallel with them. If one of the lamps should burn out, the full voltage of the series circuit is then impressed on the by-pass protector. The excessive voltage breaks down the insulating film

FIG. 10. An insulating transformer used to supply low-voltage 20-amp current to one or two street lights from a high-voltage 6.6-amp constant-current circuit. This transformer is intended to be installed inside the base of an ornamental street-lighting pole.

Courtesy General Electric Co.

in the by-pass protector which then becomes conducting and permits the remainder of the circuit to continue in operation.

3. ADJUSTABLE–RATIO OR TAP–CHANGING TRANSFORMER

If a transformer is supplied with power through a transmission circuit whose impedance is relatively high, the primary terminal voltage may vary with changes in load over an undesirably large range, because of changes in the impedance drop in the transmission circuit. In such

Courtesy Westinghouse Electric and Manufacturing Co.

FIG. 11.   The tap-changing mechanism of a large power transformer. The contactors are operated in a definite sequence by cams mounted on a rotating shaft. The entire assembly is immersed in insulating oil.

instances, taps in the primary windings may be a desirable feature, in order to maintain the secondary voltage at its rated value under the varying conditions of load and power factor. The number of turns in use in the primary winding is selected by an arrangement of contactors so that the primary induced voltage per primary turn is maintained nearly constant. The flux is therefore nearly constant and the secondary induced voltage is nearly constant. Large tap-changing power transformers are also frequently used to control the flow of reactive power between two interconnected power systems or between component parts of the same system, at the same time permitting the voltages at specified points to be maintained at desired values.

Frequently the tap-changing apparatus must be designed so that the ratio of transformation can be adjusted under load without inter-

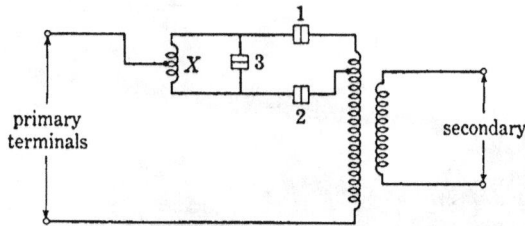

FIG. 12.   Elementary circuit for changing taps under load.

ruption of the load current. The essentials of one scheme for accomplishing this result are shown in Fig. 12. The coil X is a center-tapped iron-core reactor, and 1, 2, 3 are contactors. Suppose the transformer is operating on tap 1 when an increase in load reduces the primary voltage so that it is desired to shift to tap 2 in order to maintain constant secondary voltage. If the primary circuit is to be uninterrupted, contactor 2 must close before contactor 1 is opened. The purpose of the reactor is to prevent the coils between taps 1 and 2 from being short-circuited while contactors 1 and 2 are both closed. Although the voltage induced by the core flux in the coils between taps 1 and 2 is but a small fraction of the primary voltage, the leakage impedance of these coils also is small. Consequently, if taps 1 and 2 were short-circuited, the current in the coils between these taps might be 10 to 20 times rated primary current.

The sequence of events in changing from tap 1 to tap 2 is shown in Table I. Control of the contactors may be either manual or automatic. In the normal operating position, contactor 3 is closed, short-circuiting the reactor. The primary current then divides equally between the halves of the reactor, and, since the currents in the halves are in opposite directions, their resultant magnetomotive force is zero and the core flux in the reactor is zero. The impedance of the reactor is hence but one-half the leakage impedance of one section of its winding with respect to the other. In the second and fourth steps in the table, current flows in only one half of the reactor. There being no oppositely directed current in the other half, the magnetomotive force acting on its magnetic circuit equals the primary current times the turns in half the reactor. In order to prevent an undesirably large impedance drop in the reactor during the transition steps 2 and 4, the core of the reactor is sometimes designed to saturate when the current in either half of the reactor is near the full-

TABLE I

TAP CHANGING

Position	Contactor		
	1	2	3
1. Tap 1	0		0
2. Transition	0		
3. Tap 1½	0	0	
4. Transition		0	
5. Tap 2		0	0

0 = switch closed

load primary current. In the third step indicated in the table, the reactor is connected across the coils between taps 1 and 2. The current in the transformer coils between the taps is limited by the magnetizing reactance of the reactor. If this reactance is sufficiently great, the current in the transformer coils is not excessive. Thus the design of the reactor is a compromise which meets the various requirements in the most satisfactory manner.

4. VOLTAGE–REGULATING TRANSFORMERS

When several feeders are supplied from one bus, it is often necessary to regulate the voltages of the different feeders independently. Two types of voltage-regulating transformers commonly used for this purpose are the step regulator and the induction regulator.

4a. *Step Regulator.* — The *step regulator* usually consists of an auto-transformer whose series winding is provided with a number of taps

Courtesy Allis-Chalmers Manufacturing Co.

Fig. 13.   A 3-phase step feeder voltage regulator, typical of the design used for ratings below 750 kva at voltages below 15,000 v. This regulator has a voltage range of ±10 per cent in 32 steps of $\frac{5}{8}$ per cent each.

Courtesy Allis-Chalmers Manufacturing Co.

Fig. 14.   Tap-changing mechanism for the 3-phase regulator of Fig. 13. The contacts operate in an oil bath.

which may be changed under load and which are automatically selected by a voltage-sensitive control circuit. The series winding is connected so that its voltage adds to or subtracts from the voltage of the common winding. One of the simplest arrangements of a step-regulator circuit is shown in Fig. 15, in which the output is connected to the midpoint of the series winding so that, by selection of taps below or above this midpoint, the voltage of the series winding can be made to boost or buck the input voltage. The taps are selected by two movable fingers which slide over a set of fixed contacts. The reactor X limits the current in the turns that are short-circuited when the fingers are in the position bridging two taps. If the current is too great to be interrupted by the fingers without serious arcing, auxiliary contactors are connected in series with them, as shown in Fig. 15.

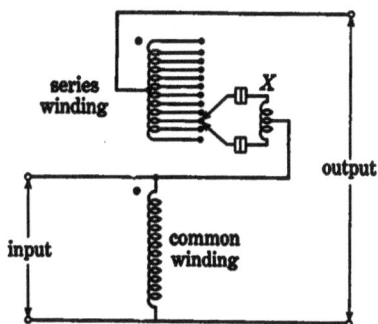

FIG. 15.   Circuit diagram for a step regulator.

4b. *Induction Regulator.* — The *single-phase induction regulator* is essentially a transformer with one of its windings mounted so that it can be turned into different positions with respect to the other.

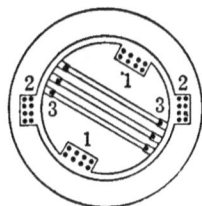

FIG. 16.   Schematic diagram of a single-phase induction regulator.

A schematic diagram of the regulator is shown in Fig. 16. In this figure, 1 and 2 are the coil sides of the primary and secondary windings respectively, and 3 indicates the coil sides of a short-circuited winding placed in quadrature with the primary winding. For simplicity, the primary and secondary windings are indicated as if each winding were concentrated in a single pair of slots. Actually, however, these windings are placed in a number of slots distributed around the peripheries of the stator and rotor. The arrangement of the rotor windings, 1 and 3, is shown in Fig. 17.

When the axes of the primary and secondary windings are in line, the mutual inductance is a maximum, and the voltage induced in the secondary winding is a maximum. When the axes are at right angles, the mutual inductance is zero, and the voltage induced in the secondary winding is zero. Any intermediate value of the secondary induced voltage can be obtained. The primary winding of the regulator is connected across the line whose voltage is to be regulated. The secondary usually is connected in series with the line as shown in Fig. 18. The voltage induced in the secondary then either adds to or subtracts from the primary line voltage according to the relative positions of the primary and secondary windings.

When the primary and secondary windings are in quadrature, there is no reaction between them, and the *self-impedance* of the secondary

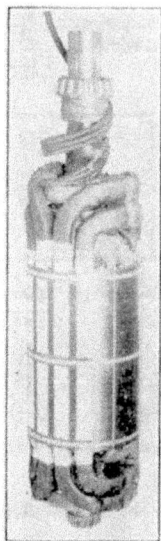

FIG. 17. The rotor of an oil-cooled single-phase induction voltage regulator. The coils at the left are active circuit coils; those at the right are the compensating coils. Connections to the external circuit are made through the flexible cables at the top.

Courtesy General Electric Co.

winding would be added in series with the transmission line were it not for the short-circuited winding 3, the purpose of which is to reduce the impedance of the secondary. When the primary and secondary windings are in quadrature, the short-circuited or compensating winding acts, with respect to the secondary winding, like the short-circuited secondary of a transformer, and reduces the apparent reactance of the secondary winding. The short-circuit impedance is much less than the self-impedance of the secondary.

The *polyphase* induction regulator consists of a stationary polyphase primary winding

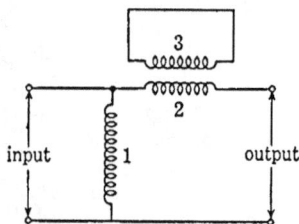

FIG. 18. Connections of a single-phase induction regulator.

and a movable polyphase secondary winding which can be turned into different positions with respect to the primary. The connections of a three-phase induction regulator are indicated in Fig. 19. The primary or stator windings, A, B, and C, are connected in Y to the primary line terminals A, B, and C. The secondary or rotor windings a, b, and c are connected in series with the secondary line terminals a, b, and c.

The theory of the polyphase induction regulator is like that of the polyphase wound-rotor induction motor.[3] According to polyphase induction-motor theory, the currents in the primary windings produce a rotating magnetic field of constant amplitude in the air gap, which induces in the secondary windings voltages of constant magnitude whose phases with respect to the primary voltages are varied when the rotor is turned into different positions with respect to the stator. If the small secondary leakage-impedance voltage drop is neglected, the secondary

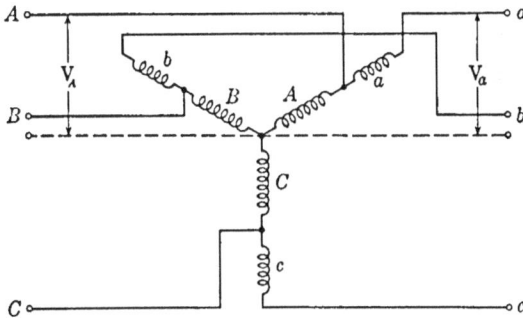

FIG. 19.   Connections of a 3-phase induction regulator.

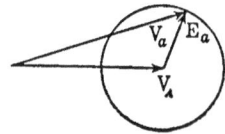

FIG. 20.   Vector diagram of voltages in a polyphase induction regulator.

phase-a terminal voltage equals the vector sum of the primary phase-A terminal voltage V_A and the secondary phase-a voltage E_a induced by the rotating magnetic field. For a fixed primary voltage, the locus of the vector secondary terminal voltage V_a is approximately a circle, as shown in Fig. 20. Thus any value of secondary terminal voltage V_a between the numerical sum and the numerical difference of V_A and E_a can be obtained through turning the rotor. For intermediate values of V_a, there is a phase displacement between V_A and V_a. In this respect, the polyphase induction regulator differs from the single-phase type. In the single-phase induction regulator, the secondary induced voltage is nearly in phase with the primary voltage, but varies in magnitude as the rotor is moved to different positions. Another difference between the two types is that the polyphase type requires no compensating winding.

The selection of the best type of feeder voltage regulator is typical of engineering problems in which economy is a major element. As a general guide, automatic feeder voltage regulators of the induction type are most logically applicable in areas having rather high load density, the regulator costs being spread over a large amount of business. Large step voltage

[3] See R. R. Lawrence, *Principles of Alternating-Current Machinery* (3d ed.; New York: McGraw-Hill Book Company, Inc., 1940), 480.

regulators are usually employed to control the loads on interconnecting circuits between substation buses, whereas small feeder step voltage regulators are used to maintain roughly the proper voltage on power lines of low load density, such as rural lines.[4] The high investment cost of

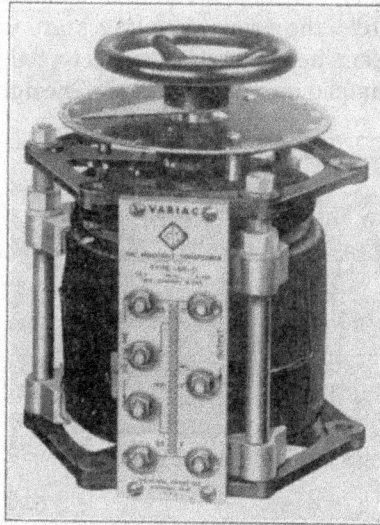

Courtesy General Radio Co.

FIG. 21. An adjustable autotransformer for laboratory use. The output voltage can be varied smoothly over a wide range by means of a sliding contact. The device is rated 115 v input, 60 ∼, 2 kva.

induction voltage regulators usually does not justify the application of them on lightly loaded lines, in comparison with the less expensive step voltage regulators.

In experimental work, a source of alternating-current power of a few kilovolt-amperes capacity whose voltage can readily be adjusted over a wide range is frequently required. Such a source can be obtained conveniently from a small transformer with a variable voltage ratio. Several types of variable voltage transformers are available commercially. One of them is shown in Fig. 21.

PROBLEMS

1. A current transformer which has a nominal ratio of transformation of 8 : 1 and a 5-amp secondary has the following constants:

$N_1 = 25$ turns	$R_2 = 0.51$ ohm
$N_2 = 196$ turns	$L_{p2} = 0.00105$ h

---

[4] G. H. Landis, " Voltage Regulation and Control in the Development of a Rural Distribution System," *A.I.E.E. Trans.*, 57 (Sept. 1938), 541–547.

The mean length of the magnetic circuit of the transformer is 20 in. and its net cross section is 4.0 sq in. The core material weighs 0.272 lb per cubic inch and has magnetic characteristics at low values of saturation as given in the following table:

$\mathcal{B}$	$\mathcal{H}$	*Core Loss per lb at 60* $\sim$
450 gausses	0.320 oersted	0.0098 w
400	0.315	0.0082
350	0.305	0.0067
300	0.290	0.0052
240	0.259	0.0036
180	0.224	0.0021
120	0.173	0.0008
60	0.105	0.00016

To simplify calculations, the magnetizing current may be assumed to be sinusoidal. If this transformer is used on a 60 $\sim$ circuit with a 5-amp ammeter which has a resistance of 0.178 ohm and an inductance of 0.000147 h:

(a) What is the actual ratio of transformation when the instrument reads 5.0 amp?

(b) What is the phase angle of the transformer when the instrument reads 5.0 amp?

2. A so-called two-stage current transformer with an auxiliary core and windings is sometimes used in place of an ordinary current transformer to reduce phase-angle and ratio errors. A diagram of such a transformer is shown in Fig. 22. The windings A' and B' have the same number of turns as the windings A and B, respectively, and

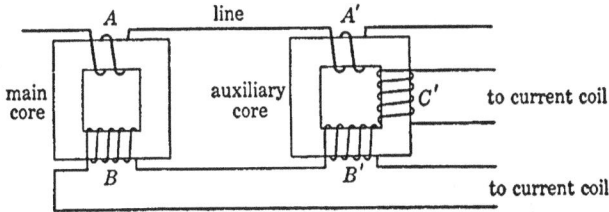

FIG. 22. Two-stage current transformer, Prob. 2.

are wound in the same sense; winding C' has the same number of turns as winding B'. The two-stage transformer is commonly used with instruments which have two separate current coils — one to carry the current from windings B and B' and the other to carry the current from winding C'. For a two-stage current transformer used in this way, explain, with the aid of a vector diagram, how the reduction in phase-angle and ratio errors is accomplished. If the auxiliary core were ideal, would the elimination of these errors be complete?

3. The three-phase power delivered by a three-wire transmission line to a balanced load is measured by a metering arrangement, as shown in Fig. 23. Wattmeter 1, which has its current coil in series with phase a and its potential coil across ab, reads

138 w; wattmeter 2, which has its current coil in series with phase c and its potential coil across cb, reads 406 w. The ratios and phase angles of the transformers for this load and instrument burden are: angle of lag of secondary voltage of potential transformers with respect to primary voltage, 0.167 degree; angle of lead of secondary

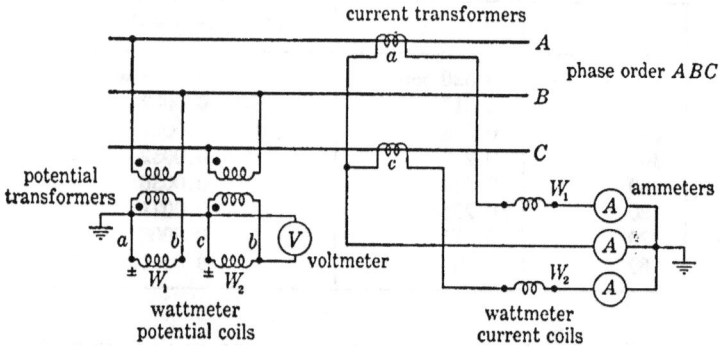

FIG. 23.   Three-wire metering connections, Prob. 3.

current of current transformers with respect to primary current, 0.833 degree; ratio of primary voltage of potential transformers to secondary voltage (20,800 : 100) $\times$ 1.004; ratio of primary current of current transformers to secondary current (300 : 5) $\times$ 1.006. Determine the total power delivered to the load and the power factor of the load.

4. A 4-kva 60 $\sim$ constant-current transformer with a primary-to-secondary turns ratio of 1 : 3.2 and a rated primary voltage of 230 v is adjusted to deliver a constant secondary current of 6.6 amp, for a range in load from short circuit to rated kva output. If the movable secondary is held down against the primary, a voltage of 63.5 v applied at the primary terminals causes a current of 6.6 amp to flow in the secondary when it is short-circuited through a low-impedance ammeter. The primary current and power for this test are 22.0 amp and 220 w.

(a) If the secondary is released and allowed to float, what is the leakage reactance at short circuit in per cent of the leakage reactance at full load?
(b) If exciting current and core loss are neglected, what is the input power factor when the transformer is supplying a resistive load of: (1) 50 ohms, (2) 125 ohms?

5. In the power distribution system shown in Fig. 24, A and B are generating

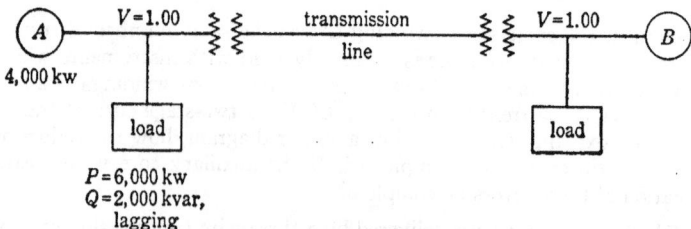

FIG. 24.   Elementary power system, Prob. 5.

stations connected by a high-voltage transmission line through transformers. Power loads are connected to the low-voltage buses at each end of the line. For normal operation, the power output of station A is held constant at the value indicated on the diagram, and fluctuations in load are handled by station B. For all loads, the voltages at both generating stations are held constant at the same value by control of the generator fields. The impedance of the line and transformers is $0.05 + j0.36$ per unit referred to rated voltage at either end of the line and to a 5,000-kva base.

(a) For the value of load specified on the diagram, find the reactive power delivered to A by the line.

(b) Because this type of operation causes an unwanted flow of reactive power in the transmission line for certain ranges of load, it is decided to replace the transformer at B by a tap-changing transformer to control the reactive power flow. Assume that the impedance of the transmission line and transformers is unchanged, and is not appreciably affected by a change in transformer taps. For the load specified in Fig. 24, find the per unit change in the high voltage at the B end of the line which the tap-changing transformer must provide in order to reduce the reactive power delivered to station A to zero.

6. Figure 25 represents the basic connections of an automatic feeder induction voltage regulator, and a control circuit known as a line-drop compensator. The relay coil K controls the regulator turning motor in such fashion as to maintain a constant voltage across coil K for all normal operating loads. The impedance of coil K is so

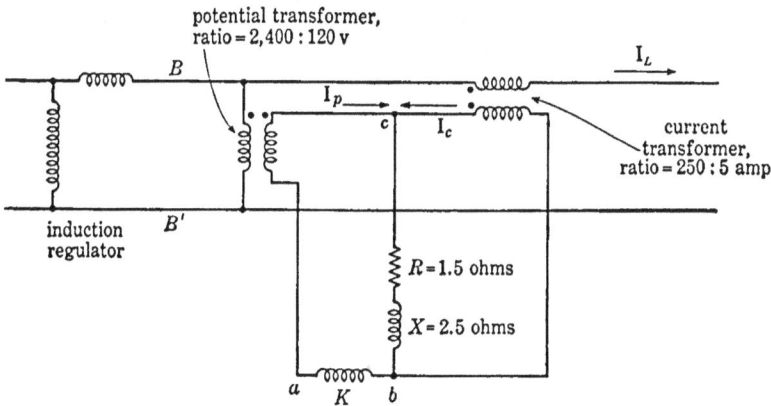

Fig. 25.   Induction regulator and line-drop compensator, Prob. 6.

high that $I_p \ll I_c$ except at extremely small loads. The line-drop-compensator impedance $R + jX$ is adjusted so that, for the values shown in Fig. 25, the voltage drop V_{cb} referred to the high-voltage side of the potential transformer equals the impedance voltage drop in the power line between the regulator output terminals BB' and the load.

The line delivers a normal full-load output to the load of 150 amp at 2,400 v. When a current I_L of 0.75 per unit at a power factor of 0.80, lagging, is delivered to the load at normal voltage, what is the per unit vector value of the output voltage $V_{BB'}$ referred to the load voltage as a base?

7. The connections for an induction regulator are shown in Fig. 26. Winding 1 is rated 10 amp and 220 v; winding 2 is rated 20 amp and 110 v.

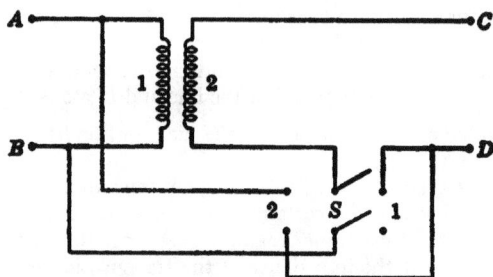

FIG. 26.   Induction regulator, Prob. 7.

(a) What ranges of voltage are available from the regulator if it is assumed to be ideal?
(b) What is the output current rating?
(c) With the switch S in position 2 the rotor is turned to the position which gives maximum output voltage when terminals CD are open and terminals AB are connected to a 220-v source. With the rotor held in this position the impedance measured at terminals CD with terminals AB short-circuited and switch S in position 2 is $0.160 + j0.510$ ohm. What is the maximum output voltage obtainable from the regulator when it is delivering rated current to an inductive load of 0.80 power factor, the input voltage being 220 v?

CHAPTER XX

Applications of Transformers in Telephone Systems

In Ch. XVIII reference is made to input, interstage, and output transformers as used in amplifiers. Certain other applications of transformers to communications are discussed in this chapter. These include uses in which the transformer is referred to as the induction coil, repeating coil, and hybrid coil or three-winding transformer.

1. INDUCTION COIL

The first commercial application of the transformer to communications was the use of the induction coil in telephony in 1877 to connect the carbon transmitter (microphone) to the line. For transmission over

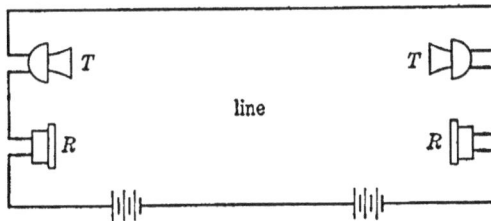

FIG. 1. Series telephone circuit.

short distances, the transmitters and receivers could be connected in series with the line, as in Fig. 1, but for transmission over any great distance this circuit was not feasible because of the battery voltage required. Consequently the battery and the primary winding of the induction coil were connected in series with the transmitter, and the receiver and the

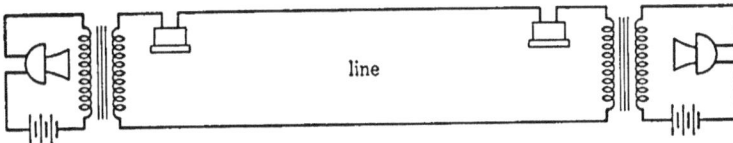

FIG. 2. Local-battery telephone circuit using induction coils.

secondary winding of the coil were connected in series with the line, as in Fig. 2. When used in this manner, the induction coil may be considered to be an elementary filter since it serves to separate the direct-

513

current carrier from the alternating-current components of the signal which carry the intelligence, and thereby reduces line losses. It also serves to match the low impedance of the transmitter (20 to 40 ohms) to the high impedance of the line (400 to 600 ohms) by use of a step-up turns ratio. Thus it results in extending the range of possible telephone transmission. Probably the induction coil is the most common type of transformer, since one is used in practically every telephone set.

Induction coils used to be made with a bundle of round iron wires as a core, square wooden or fiber spool ends for holding the windings, and a primary and secondary winding. The ends of the windings were brought out to terminals on the spool ends. Such coils had sufficiently high coefficients of coupling and satisfactory efficiencies, and the stray fields which they produced were of no serious disadvantage as long as two telephone sets were not placed near each other, or when the use of metal bell boxes prevented serious magnetic coupling between adjacent sets. Such a coil is illustrated in Fig. 3. In 1930, however, the use of Bakelite subscriber sets caused the problem of coupling to become im-

Courtesy Western Electric Co.

FIG. 3.   Local battery sidetone induction coil.

Courtesy Western Electric Co.

FIG. 4.   Modern common-battery anti-sidetone induction coil with closed magnetic circuit.

portant, and specifications were issued requiring telephone sets to be placed at least twelve inches apart. More recently induction coils have been made with a closed magnetic circuit, made of silicon-steel laminations, to avoid coupling with adjacent sets, and thus to permit bell boxes to be placed closer together. Such a coil is illustrated in Fig. 4.

The circuit of Fig. 2 shows that if the set on the left of the diagram is to transmit speech, a portion of the output of the induction coil is absorbed in the local receiver. The hearing of the speaker's voice by himself is called *sidetone*. As sidetone will be of greater volume than the received signal, the speaker is rendered temporarily less sensitive to the desired received signal, and the effective circuit efficiency is thereby reduced. In order to overcome this difficulty, anti-sidetone sets have been developed which reduce self-hearing.

The first anti-sidetone telephone set in common use was developed for switchboard operators. The circuit is shown in Fig. 5. It utilizes two

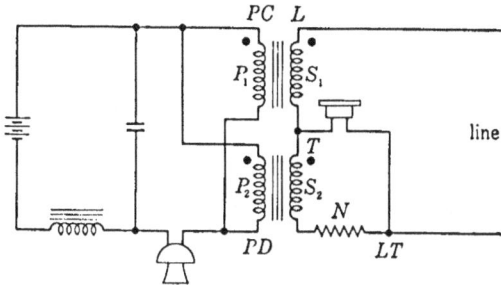

Fig. 5.　Operator's anti-sidetone telephone set.

induction coils with identical primary windings P_1 and P_2 connected in parallel. The secondary windings S_1 and S_2 are connected in series and each has the same number of turns, but S_2 is wound with smaller wire, which makes its resistance about 350 ohms greater than that of S_1. The secondaries, receiver, and line are connected in a bridge arrangement as shown in Fig. 5, where the 350-ohm increment in S_2 is indicated as a separate network N, which approximately balances the impedance of the average telephone line. When the user talks into the transmitter, equal voltages are induced in S_1 and S_2 from P_1 and P_2. If the network N balances the line exactly, no current flows in the receiver under this transmitting condition. For the usual condition of approximate balance, the receiver current, or sidetone, is small.

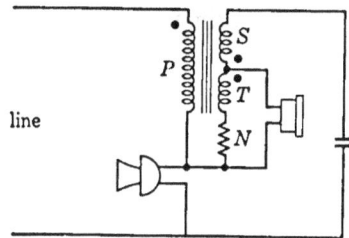

Fig. 6.　Circuit of the anti-sidetone common-battery induction coil.

The common-battery subscriber set is made anti-sidetone by means of a third winding placed on the induction coil as shown in Fig. 6. The third winding and network are so proportioned that when the set is transmitting, the voltage drop across the receiver is small.[1]

Anti-sidetone sets did not come into common subscriber use until

---

[1] For an explanation of the operation of this circuit, see C. O. Gibbon, " An Explanation of the Common Battery Anti-sidetone Subscriber Set," *B.S.T.J.*, *17* (April, 1938), 245–257.

about 1932. Because of the large number of telephone sets in the United States (about 25,000,000), the conversion has been expensive and therefore a slow process which is not yet complete.

2. REPEATING COIL

The next form of telephone transformer to come into common use was the repeating coil patented in 1892. The first repeating coils were made with four equal windings on one core to insure close coupling. These coils served to supply the common battery of a central office to the lines for signaling and later for talking purposes also.

The method of connection is illustrated in Fig. 7. Direct current flows over each loop to excite the transmitters and provide signaling current for the operation of the supervisory relays (not shown). Alternating current from either transmitter is passed by the repeating coil

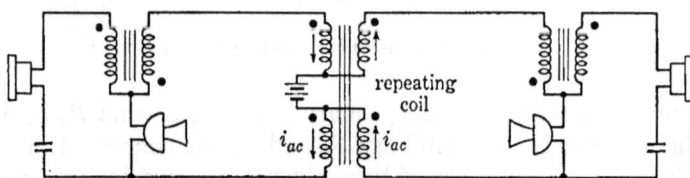

FIG. 7.   Repeating coil supplying battery current to two interconnected subscriber loops.

to the other loop. Because the coils are of unity turns ratio, the alternating primary and secondary currents shown by the arrows in Fig. 7 are very nearly equal, and the only alternating current that can flow in the battery is the very small exciting current of the repeating coil. The alternating voltage drop across the battery is the resistance drop in it due to the exciting current, and therefore is very small. As a result, any number of circuits can be connected through repeating coils and operated from the same battery without interference between the parties who are talking.

The early repeating coils were constructed like the early induction coils. They were given a different name, however, because they served to transfer or repeat alternating-current speech energy from one loop to another. Present-day repeating coils are usually wound on toroidal cores of magnetic material.

Repeating coils with ratios other than unity are used to match impedances when lines of different characteristics are connected. Many lines have characteristic impedances of the order of 600 ohms. Many loaded lines have higher impedances. It is the usual practice to use repeating coils to reduce the impedances of all circuits coming into any toll office to a common level, for example, 600 ohms. When this is done, all circuits terminating in the office may be interconnected without reflection effects.

Another very important use of the repeating coil is to provide a phantom or superposed circuit on two existing circuits. If two distant points are connected with two telephone circuits, a third, or phantom circuit may be provided if similar repeating coils are placed in each of the terminals of each side circuit, as shown in Fig. 8. These repeating coils are

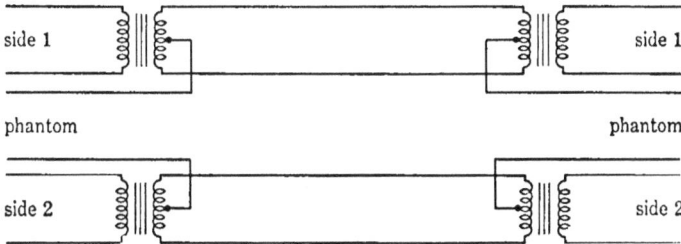

FIG. 8. Use of repeating coils to provide a phantom circuit.

usually of unity turns ratio. The line sides of the coils consist of two equal windings very carefully interwound so that the point of interconnection is the exact midpoint. The current from the upper wire of the phantom circuit flows to the midpoint and divides equally between the two windings and the two wires of the upper side circuit (if the resistances, inductances, and capacitances are equal). As a result this current produces

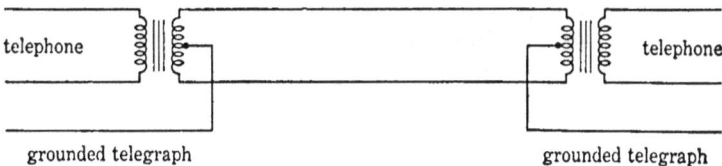

FIG. 9. Use of repeating coils to provide a simplex circuit.

no flux in the core and no coupling with this side circuit. The same condition exists in the other three transformers and consequently there is no interference between the phantom and the side circuits.

The repeating coils which are used to make up a phantom circuit must have sufficient magnetic material in their cores to pass the usual 20-cycle-per-second ringing current. Frequently an air gap is cut in the core so that it is readily demagnetized after direct current has passed through the windings.

Another application of the repeating coil is to provide what is known as a simplex circuit. Repeating coils are placed in the two ends of a telephone circuit, as shown in Fig. 9. The midpoint tap of the repeating coil then serves to make possible the utilization of the two telephone line wires in parallel for the line of a ground-return telegraph circuit. This ground-return circuit could of course be used for any other voice-frequency control.

Repeating coils with well-insulated windings are often used to connect telephone sets to a telephone line which parallels a high-voltage power line, for the purpose of reducing the shock hazard to station operators and others using such telephones.

3. HYBRID COIL OR THREE–WINDING TRANSFORMER

When a telephone circuit becomes so long that vacuum-tube amplifiers are required for satisfactory speech transmission, some means are required in the repeater to cause each amplifier to operate in the proper direction of transmission. A common way of introducing the two amplifiers in a two-wire line is the use of two hybrid coils or three-winding transformers, and two balancing networks.

Figure 10 shows a two-wire repeater with two amplifiers, two hybrid coils, and two balancing networks. Each hybrid coil consists of three windings wound on a common core. The windings 3–4 and 7–8 are

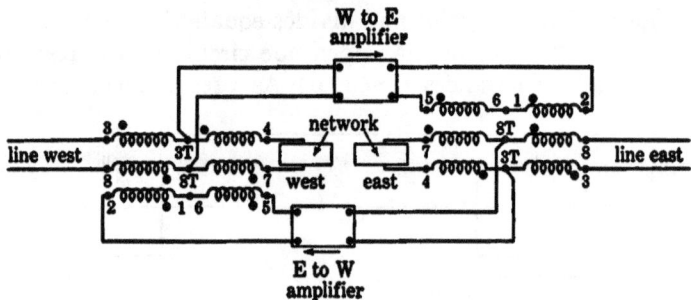

FIG. 10.   Two-wire repeater employing two amplifiers, two hybrid coils, and two balancing networks.

identical, with midtaps 3T and 8T accurately located. The balancing network is designed to have an impedance-frequency characteristic as nearly equal to that of the line as is economically feasible. The output of the E-to-W amplifier works into the winding 2–5, whose impedance matches the output impedance of the amplifier tube. If the network balances the line perfectly, half the amplifier output is dissipated in the network and is lost, while half is sent on the line to the west. No voltage difference results across the bridge points 3T–8T. Therefore none of the E-to-W amplifier output gets into the W-to-E amplifier. If the network on the right balances the line east, none of the output of the W-to-E amplifier gets into the E-to-W amplifier. If the balance between the line and network is not perfect, a portion of the E-to-W amplifier output gets into the W-to-E amplifier input. The same condition may exist on the east terminal. The sum of the two amplifications, expressed in decibels, must not equal the sum of the losses from the series windings 2–5 to the bridge points 3T–8T or oscillation will result and the circuit will become inoperative.

Input to the hybrid coil from the line west divides between the input to the W-to-E amplifier and the output of the E-to-W amplifier. The former is amplified and passed on to the line east and network east, while the latter is dissipated. If the impedance across 3T–8T is half the line and network impedances, no voltage exists across the network.

CHAPTER XXI

Three-Phase Connections; General Considerations

Nearly all electrical energy is generated in three-phase generators and transmitted over three-phase transmission lines.[1] Since it is often necessary to raise and lower the voltage several times between the generators and the loads, a great many transformers are used in three-phase systems. The transformations can be accomplished by banks of suitably connected single-phase transformers or by three-phase transformers in which the magnetic circuits of the three phases are interlinked. Three-phase transformers, certain of whose characteristics may differ from those of a similarly connected bank of single-phase transformers, are discussed in detail in Ch. XXVI.

Although nearly all electric power transmission is by means of three-phase systems, nearly one half of the energy is eventually used as single-phase power for domestic and small-power purposes. These single-phase loads are supplied from the low-voltage secondaries of distribution transformers whose primaries are connected to the three-phase transmission system, the single-phase loads being distributed among the phases of the three-phase system so as to result in an approximately balanced three-phase load.

A detailed analysis of three-phase transformer connections should answer various questions such as these: What are the relations among the currents and voltages in the transformers and those in the three-phase circuits? What should be the ratings of the transformers to supply a given load? Can the bank be used to supply single-phase and three-phase loads simultaneously? What are the effects of inequalities in the equivalent impedances, excitation characteristics, or turns ratios? What are the effects of harmonics in the exciting currents? How do the equivalent impedances affect the system voltages under normal conditions? How do the transformer connections, impedances, and methods of grounding influence the system voltages and currents during faults?

These and other questions can be answered analytically from the theory of a single transformer together with the relations among the currents and voltages in three-phase circuits. In its complete form, the theory of a transformer must take into account leakage impedances, exciting currents, magnetic nonlinearity, and, sometimes, distributed capacitances of the windings. Fortunately, it is rarely necessary that all these factors be given simultaneous detailed consideration. Many of the important

[1] Three-phase systems are discussed in this series in the volume on electric circuits (1940), Ch. X, pp. 514–577.

characteristics of the various connections can in fact be readily derived from an analysis in which the transformers are assumed to be ideal. Such simple analysis of a number of the common three-phase connections of single-phase transformers is given in this chapter, together with a qualitative discussion of some of the effects of the leakage impedances and exciting currents in actual transformers. Detailed analyses of impedance and excitation phenomena occupy subsequent chapters.

For use in three-phase circuits, transformers can be connected in various arrangements, some symmetrical and others unsymmetrical. If the connection is symmetrical, each primary phase is like the other two primary

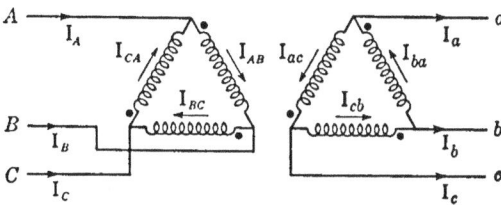

Fig. 1.   Δ–Δ connection.

phases, and each secondary phase is like the other two secondary phases. For example, if three identical transformers are connected either in Δ or in Y on the primary sides and also either in Δ or in Y on the secondary sides, the arrangement is symmetrical. There are four such symmetrical connections of two-winding transformers, namely:

(1) Δ–Δ
(2) Y–Y
(3) Δ–Y
(4) Y–Δ.

An important example of unsymmetrical arrangement is the open-delta or V connection which uses only two transformers. Since the symmetrical connections are not only the simpler to analyze but also are the more frequently used, they are discussed first.

1. Δ–Δ CONNECTION

The Δ–Δ arrangement can be used when neither the primary nor the secondary side requires a three-phase neutral connection. For reasons discussed subsequently, the Δ connection finds its chief applications in low- or moderate-voltage circuits, or when the current is large. The elementary connection diagram is shown in Fig. 1. The primary and secondary windings of the same transformer are parallel to each other. On both the primary and secondary sides, terminals of opposite polarity are connected together to form the Δ. The connection diagram may also be drawn as in Fig. 2, which shows a bank of Δ–Δ-connected distribution transformers, protected by fuse cutouts and lightning arresters on the primary sides. The type of fuse and arrester depends upon the current and voltage; the exposure of the high-voltage line to lightning; the location of the installation on a pole, on the ground, or in a manhole; and upon the experience and preferences of the operating company. When the trans-

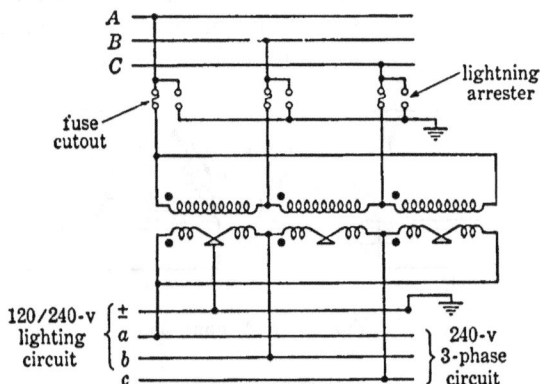

FIG. 2.   Δ–Δ connection of distribution transformers for simultaneous 240-v 3-phase and 120/240-v single-phase outputs.

formers have two 120-volt windings on their low-voltage sides, a single-phase, 120/240-volt three-wire lighting circuit can be connected to one of the transformers while the bank is simultaneously supplying 240-volt three-phase power.[2] The single-phase output unbalances the load on the bank, so that, if this output is an appreciable portion of the total load, a larger transformer may be required in the phase supplying it.

For satisfactory operation in a Δ–Δ connection, the transformers should have the same turns ratio; otherwise there is a circulating current in the transformers even when the bank delivers no load.* With a balanced three-phase load, the transformers should also have equal equivalent impedances if they are to share the load equally. If the equivalent impedances are not equal, the transformer having the smallest equivalent impedance has the largest current, the bank behaving in this respect in a manner somewhat similar to the behavior of impedances in parallel. Further discussion of this important property of the Δ–Δ connection is given in Art. 1b, Ch. XXIV. For the present it is sufficient to note that

---

[2] Three-wire systems are discussed in this series in the volume on electric circuits (1940), Ch. X, Art. 2, pp. 518–522. Also see p. 541 for further comments regarding systems of simultaneous single-phase and three-phase distribution.

* For further discussion of circulating currents see Art. 1a, Ch. XXIV.

there are parallel paths between each pair of line terminals in a Δ–Δ bank, and that the currents divide among these parallel paths in a manner determined by their impedances. Thus if three Δ–Δ-connected transformers of the same rating but different equivalent impedances supply power to a balanced load, the bank cannot deliver its full-load kilovolt-ampere rating without overloading the transformer with the smallest equivalent impedance. Therefore three identical transformers are commonly used when the load is balanced. This is the simplest condition to analyze.

1a. *Voltage and Current Relations in Balanced* Δ *Circuits.* — When the transformers are identical and the circuit is balanced, there is nothing to distinguish one phase from another except the phase displacement of 120 degrees between the currents and voltages of one phase and those of

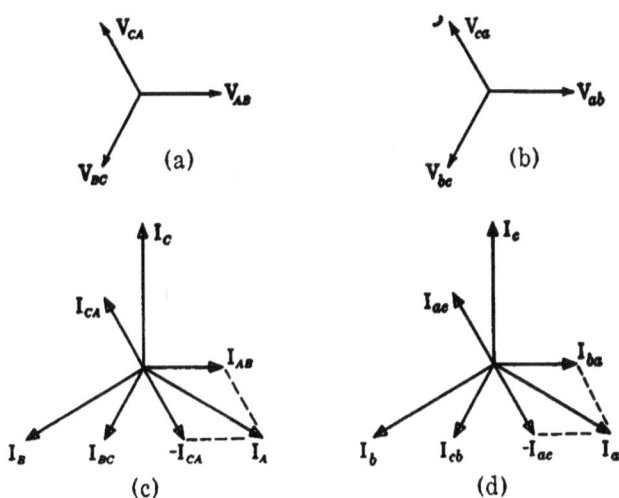

Fig. 3. Approximate vector diagrams for a Δ–Δ bank under balanced conditions.

another. Thus the currents and voltages in every phase can be determined by analysis of any one phase. Each transformer supplies one-third of the three-phase load. The primary and secondary terminal voltages of the transformers equal the corresponding three-phase line voltages and therefore, except for the voltage drops due to the leakage impedances, the ratio of the primary and secondary line voltages equals the turns ratio of the transformers. For the phase order ABC, these relations are shown vectorially in Figs. 3a and 3b, in which the voltages are assumed to vary sinusoidally and the vectors represent the falls in potential in the directions indicated by the order of the subscripts. For example, V_{AB} is the vector representing the fall in potential from line A to line B.

The capital letter subscripts indicate the primary phases and the lower-case subscripts the secondary phases. When the connections are made as in Fig. 1 and the positive directions of the terminal voltages are defined as above, the corresponding primary and secondary terminal voltages are very nearly in phase when the leakage-impedance voltage drops are small.

In Fig. 3c, the currents are assumed to vary sinusoidally and the vectors, I_{AB}, I_{BC}, I_{CA} represent the currents in the primaries in the directions indicated by the order of the subscripts. If the power factor is unity, these primary currents are in phase with the corresponding primary voltages, as shown in Figs. 3a and 3c. The currents in the transformers are related to the currents in the lines in the same manner that the currents in any balanced Δ are related to the line currents; that is, vectorially

$$I_A = I_{AB} - I_{CA} \qquad [1]$$

$$I_B = I_{BC} - I_{AB} \qquad [2]$$

$$I_C = I_{CA} - I_{BC}, \qquad [3]$$

where I_A, I_B, I_C are the vectors representing the primary line currents whose positive directions are shown by the arrows in Fig. 1. Equation 1 is represented vectorially by the vectors I_{AB}, $-I_{CA}$ and their sum I_A in Fig. 3c. When the currents are balanced and the phase order is ABC, I_{AB} and I_{CA} are equal in magnitude and I_{AB} lags I_{CA} by 120 degrees, whence

$$I_A = \sqrt{3}\, I_{AB} \underline{/-30^\circ}. \qquad \blacktriangleright[4]$$

▶ Thus when the currents are balanced and vary sinusoidally, the effective value of the line currents equals $\sqrt{3}$ times the effective value of the currents in the Δ-connected windings of the transformers. Furthermore there is a phase displacement* between the line currents and the Δ currents.◀

When the exciting currents are neglected, the primary and secondary currents produce equal and opposing magnetomotive forces, and therefore the primary and secondary currents are inversely proportional to the numbers of turns in the windings and are in phase when their positive directions are chosen as the directions of currents that would produce opposing magnetomotive forces, as shown by the arrow directions in

---

* It should be remembered that the power-factor angle is the angle of the *phase* current with respect to the *phase* voltage. In the Δ connection the voltage of any phase is the corresponding line-to-line voltage, but the Δ currents are not in phase with the line currents, and therefore the power-factor angle does *not* equal the phase angle between a line current and the corresponding line-to-line voltage.

Fig. 1. These secondary currents I_{ba}, I_{cb}, and I_{ac} are shown vectorially in Fig. 3d, in phase with the corresponding primary currents of Fig. 3c. The vector equations relating the secondary line currents and the secondary currents in the Δ-connected transformers are

$$I_a = I_{ba} - I_{ac} \qquad [5]$$

$$I_b = I_{cb} - I_{ba} \qquad [6]$$

$$I_c = I_{ac} - I_{cb}, \qquad [7]$$

where I_a, I_b, I_c are the vectors representing the secondary line currents in the arrow directions in Fig. 1. When the currents are balanced and the phase order is *abc*,

$$I_a = \sqrt{3}\, I_{ba} \underline{/-30°}, \qquad \blacktriangleright[8]$$

as shown vectorially in Fig. 3d. This relation is similar to Eq. 4 for the currents on the primary sides.

1b. *Summary; Δ Connection.* — The important facts regarding the Δ connection can be summarized as follows:

When transformer windings are connected in Δ, the three-phase line-to-line voltages equal the terminal voltages of the corresponding transformers.

In the Δ-Δ connection, corresponding primary and secondary line voltages are very nearly in phase.

When the circuit is balanced and the transformers are identical, each transformer of a Δ-Δ bank supplies one-third of the three-phase load and the effective values of the currents in Δ-connected windings are $\frac{1}{\sqrt{3}}$ times the effective values of the line currents.

Since there are parallel paths between every pair of line terminals on both the primary and secondary sides, the distribution of current among the transformers of a Δ-Δ bank depends on their equivalent impedances, the bank behaving in this respect in a manner analogous to the behavior of impedances in parallel.* Thus even if the line currents are balanced, the individual transformers of a Δ-Δ bank are loaded unequally unless their equivalent impedances are equal.

If the turns ratios of the three transformers in a Δ-Δ bank are unequal, there are circulating currents within the bank,† and therefore the transformers should have equal turns ratios.

---

* See Art. 1b, Ch. XXIV.
† See Art. 1a, Ch. XXIV.

2. Y–Y CONNECTION

Transformer windings are connected in Y when three terminals of like polarity are joined to form the neutral point of the Y, as shown in Fig. 4, in which N and n are the primary and secondary neutral points, the primary and secondary line terminals being A, B, C and a, b, c. The connection diagram may also be drawn as in Fig. 5, in which the primary and secondary windings of the same transformer are parallel to each other. Often the neutral points are grounded, as shown in Fig. 5.

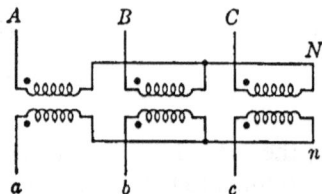

Although the characteristics of a Y–Y bank of single-phase transformers may be markedly influenced by the peculiar behavior of the harmonics in the exciting currents which in certain circumstances may bring about undesirable or even dangerous conditions, nevertheless some of the important properties of the Y–Y connection can be determined from a simple analysis in which the harmonics in the exciting currents are neglected.*

FIG. 4.   Y–Y connection.

2a. *Primary Neutral Connected to the Source Neutral.* — The simplest example is that of a Y–Y bank of transformers connected to the terminals

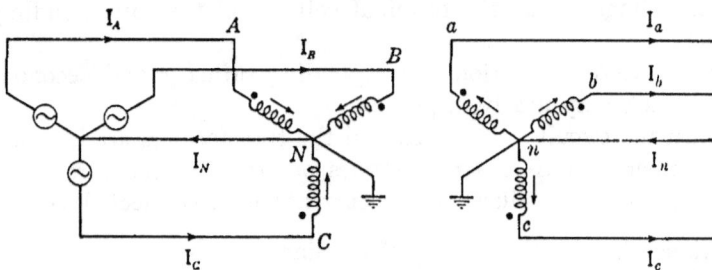

FIG. 5.   Y–Y bank of transformers with the primary neutral connected to the neutral of the source.

of a Y-connected source, with the neutral point N of the primary windings connected to the neutral of the source, as shown in Fig. 5. For the purposes of this discussion, the impedances of the four wires connecting the transformers to the source may be neglected. Under these conditions the primary voltages of the transformers are the corresponding line-to-neutral voltages of the source. If these voltages vary sinusoidally they can be represented by vectors V_{AN}, V_{BN}, and V_{CN}. As in any Y-connected circuit, the vector relations among these line-to-neutral voltages and the

* Harmonic phenomena are discussed in Ch. XXIII.

line-to-line voltages V_{AB}, V_{BC}, and V_{CA} are

$$V_{AB} = V_{AN} + V_{NB} = V_{AN} - V_{BN} \qquad [9]$$

$$V_{BC} = V_{BN} - V_{CN} \qquad [10]$$

$$V_{CA} = V_{CN} - V_{AN}. \qquad [11]$$

If the sinusoidal voltages to neutral are balanced and the phase order is

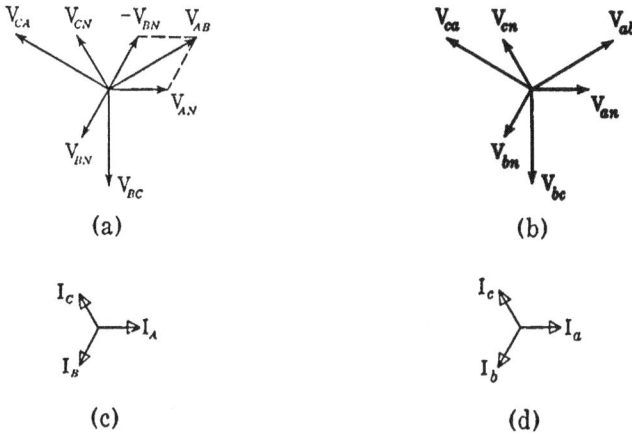

(a)   (b)

(c)   (d)

FIG. 6. Approximate vector diagrams for a Y–Y bank under balanced conditions.

ABC, the vector diagram of primary voltages is as shown in Fig. 6a, and Eq. 9 reduces to

$$V_{AB} = \sqrt{3}\, V_{AN}\,\underline{/30°}. \qquad \blacktriangleright[12]$$

The other line-to-line voltages are given by similar expressions and are also shown vectorially in Fig. 6a.

▶ Thus when the line-to-neutral voltages are balanced and vary sinusoidally, the effective value of the line-to-line voltages equals $\sqrt{3}$ times the effective value of the line-to-neutral voltages. It should also be noted that there is a phase displacement of 30 degrees between the line-to-neutral and the line-to-line voltages nearest in phase with each other.◀

If the leakage-impedance voltage drops are neglected, the secondary voltages to neutral equal the corresponding primary voltages to neutral divided by the ratio a of primary to secondary turns, and corresponding primary and secondary voltages are in phase when the connections are made as in Fig. 5. The vectors V_{an}, V_{bn}, and V_{cn}, which represent the secondary voltages to neutral, are shown in Fig. 6b. The vector secondary line-to-line voltages V_{ab}, V_{bc}, and V_{ca} are

$$V_{ab} = V_{an} - V_{bn} \qquad [13]$$

$$V_{bc} = V_{bn} - V_{cn} \qquad [14]$$

$$V_{ca} = V_{cn} - V_{an}, \qquad [15]$$

and, when the secondary voltages to neutral are balanced, sinusoidal in waveform, and of phase order abc, Eq. 13 reduces to

$$V_{ab} = \sqrt{3}\, V_{an} \underline{/30°}. \qquad \blacktriangleright[16]$$

The secondary line-to-line voltages are shown vectorially in Fig. 6b. Note that, except for the small leakage-impedance voltage drops, corresponding primary and secondary line-to-line voltages are in phase when the connections are made as in Fig. 5.

On both the primary and secondary sides, the currents in the transformers equal the corresponding line currents. If the circuit is balanced, the secondary currents delivered to the load are equal in magnitude and displaced in phase by 120 degrees, and if the power factor is unity the secondary currents are in phase with the corresponding secondary voltages to neutral,* as shown by the vectors, I_a, I_b, and I_c in Fig. 6d. Under balanced conditions, the vector sum of the secondary line currents equals zero and therefore there is no current in the secondary neutral wire, which may be omitted without change in the performance of the bank under balanced conditions.

If the exciting currents are neglected, the primary and secondary currents are inversely proportional to the ratio of primary to secondary turns, and these currents are in phase when their positive directions are chosen as the directions of currents that would produce opposing magnetomotive forces, as in Fig. 5. The vector diagram of the primary currents is shown in Fig. 6c. Except for the exciting currents, the vector sum of the balanced primary currents is zero, and therefore the current in the primary neutral wire is zero. However, it should not be concluded that the primary neutral wire may therefore be omitted, since the primary neutral connection has important effects on the exciting currents and on the behavior of the bank with certain types of unbalanced loads, as shown in part (b) of this article.

When the primary neutral is connected to the neutral of the source, as in Fig. 5, each transformer receives its power from one of the phases of the source and therefore each transformer can be loaded independently.

---

* Although the phase currents equal the corresponding line currents in a Y connection, the Y voltages to neutral are not in phase with the line-to-line voltages and therefore the power-factor angle does *not* equal the phase angle between a line current and the corresponding line-to-line voltage. The footnote on p. 530 should be re-read.

If the secondary voltage of the transformers is 120 volts, single-phase lighting loads can be connected from any one of the three phase wires a, b, or c, to the grounded neutral wire n, while three-phase motor loads can be operated from the three phase wires at a line-to-line voltage of $120\sqrt{3}$ or 208 volts.* Standard 220-volt three-phase motors usually can be operated at this reduced voltage without serious effects on their characteristics.†

Since the sum of the line currents delivered to the loads must equal the current returning to the transformers through the secondary neutral wire, the vector current I_n in the secondary neutral wire in the arrow direction, Fig. 5, is

$$I_n = I_a + I_b + I_c. \tag{17}$$

Similarly the vector current I_N in the primary neutral wire in the arrow direction is

$$I_N = I_A + I_B + I_C. \tag{18}$$

Except for the voltage drops caused by line and transformer impedances, the voltages at the loads are balanced if the source voltages to neutral are balanced and if the primary neutral is connected to the neutral of the source.

2b. *Isolated Primary Neutral.* — However, quite a different situation arises when the primary neutral is isolated from the neutral of the source. Under these conditions, the instantaneous sum of the primary currents supplied to the bank must be zero, or

$$i_A + i_B + i_C = 0, \qquad \blacktriangleright[19]$$

where i_A, i_B, i_C are the instantaneous values of the primary currents in the arrow directions, Fig. 7a. That is, the current supplied to any transformer must find its return path to the source through the primaries of the other transformers. This restriction on the primary currents has an important effect on the voltages to neutral and on the behavior of the bank with single-phase loads to neutral.

Consider the Y–Y bank, Fig. 7a, with the secondary lines open-circuited and with the neutrals isolated. Under these conditions, the primary currents are the exciting currents. If the line-to-line voltages applied to the primary terminals are balanced and sinusoidal in waveform, they can be represented by the vectors V_{AB}, V_{BC}, and V_{CA} of Fig. 7b, which form the three sides of an equilateral triangle. This result follows

---

* This three-phase four-wire system of distribution can also be supplied from transformers whose primaries are connected in Δ. See p. 540.

† Some manufacturers are now supplying standard three-phase induction motors, in sizes up to a few horsepower, with windings that can be connected for operation at circuit voltages of 220, 208, or 199 volts. (199 = 115 $\sqrt{3}$.)

from the fact that in any three-phase system the vector sum of the line-to-line voltages, taken in cyclic order, must be zero. Thus, on the primary side,

$$V_{AB} + V_{BC} + V_{CA} = 0. \qquad \blacktriangleright[20]$$

Similarly, on the secondary side

$$V_{ab} + V_{bc} + V_{ca} = 0. \qquad \blacktriangleright[21]$$

These equations demonstrate that the vectors representing the line-to-line voltages can always be drawn as the three sides of a triangle. If the line voltages are balanced, the voltage triangle is equilateral, as shown in Fig. 7b.

If the transformers have identical excitation characteristics, the exciting currents and voltages to neutral are also balanced. Although (as shown in Art. 2d, Ch. XXIII) the waveforms of the voltages to neutral

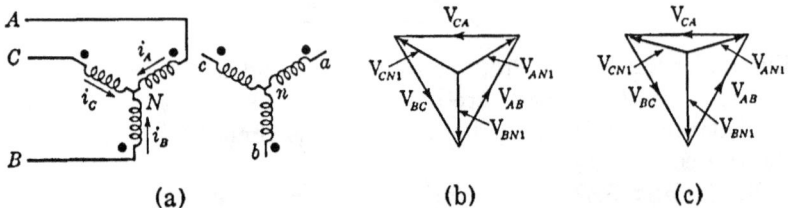

(a)   (b)   (c)

FIG. 7. Connections and voltage vector diagrams for a Y–Y bank of single-phase transformers with isolated neutrals.

may be far from sinusoidal, the fundamental components of these voltages can be represented by vectors V_{AN1}, V_{BN1}, and V_{CN1} connecting the apexes with the center of gravity of the line voltage triangle,[3] as shown in Fig. 7b.

Three transformers seldom have identical excitation characteristics, however, even when they are of the same design. Suppose one of the transformers — say the one in phase B — requires less exciting current than the other two. This transformer then has more than its normal exciting current forced upon it by the other two transformers, and therefore assumes more than its share of the voltage, as shown by the vector diagram of Fig. 7c. Thus in a Y–Y bank with isolated neutrals, the voltages to neutral are determined by the excitation characteristics of the transformers and are usually more or less unbalanced, even when the line-to-line voltages are balanced. This situation is undesirable.

When the primary neutral is isolated from the neutral of the source, the voltages to neutral may be unbalanced not only by inequalities in the excitation characteristics, but also by the connecting of unbalanced

---

[3] See the volume on electric circuits (1940), p. 531.

loads from line to neutral on the secondary side. The simplest case to consider is that of a single-phase load connected from one secondary line to the secondary neutral. Refer to Fig. 8a and note that, since the secondaries of transformers A and B are open-circuited, the only currents that can flow in their primaries are their exciting currents $i_{\varphi A}$ and $i_{\varphi B}$. Since the primary neutral is isolated, the instantaneous sum of the currents directed toward the primary neutral must be zero, or

$$i_{\varphi A} + i_{\varphi B} + i_C = 0; \qquad [22]$$

whence

$$i_C = -i_{\varphi A} - i_{\varphi B}, \qquad [23]$$

where i_C is the instantaneous value of the primary current in the loaded transformer. Thus the current that can flow in the primary of the loaded transformer is limited by the exciting currents of the other two. Any load current in transformer C disturbs the exciting currents of transformers

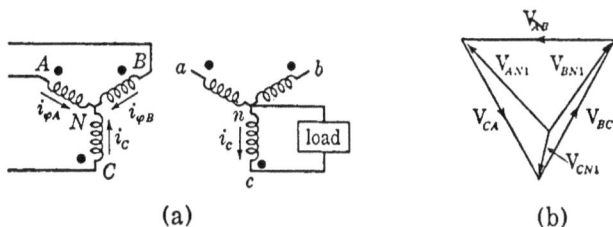

(a)   (b)

FIG. 8. Connections and voltage vector diagram of a single-phase load supplied from a Y–Y bank with isolated primary neutral.

A and B and as a result the voltages to neutral are greatly altered. Thus if one attempts to load the C transformer with a resistive or inductive load, its primary current must increase and force the exciting currents in transformers A and B to increase. Therefore the A and B transformers assume more than their shares of the voltages, and the voltage of the loaded C transformer may be greatly reduced by the attempt to load it, as indicated in the vector diagram,* Fig. 8b. A capacitive load to neutral on one of the transformers may actually reverse the phase of the voltage to neutral of the loaded transformer and produce voltages across the two unloaded transformers that are greater than the line-to-line voltages.

2c. *Summary; Y Connection.* — The important facts regarding the Y connection can be summarized as follows:

The currents in Y-connected windings equal the line currents.

---

* The voltages to neutral are nonsinusoidal in waveform and therefore the vectors V_{AN1}, V_{BN1}, V_{CN1} in Fig. 8b should be interpreted as representing the fundamental components of these voltages.

If the line-to-neutral voltages are balanced and sinusoidal in wave-form, the effective value of the voltages to neutral equals $\dfrac{1}{\sqrt{3}}$ **times the** effective value of the line-to-line voltages, and there is a phase displacement of 30 degrees between the line-to-neutral and the line-to-line voltages that are nearest in phase.

Corresponding primary and secondary line-to-line voltages are very nearly in phase in a Y–Y bank connected as in Fig. 5.

If the primary neutral is connected to the neutral of the source, each transformer can be loaded independently and therefore a Y–Y bank can be used to supply single-phase loads connected between any phase wire and the secondary neutral. The disadvantage of this arrangement is that it requires a fourth primary wire and that in certain circumstances objectionable voltages may be induced in neighboring communication circuits, caused by the unbalanced currents and by the third-harmonic components of the exciting currents.*

If the primary neutral is isolated from the neutral of the source, the primary current in each transformer must return to the source through the primaries of the other two transformers, and therefore the voltages to neutral may be greatly unbalanced by unbalanced loads connected to the secondary neutral, or by inequalities in the excitation characteristics of the transformers.

With isolated neutrals, the voltages to neutral of a Y–Y bank of single-phase transformers are nonsinusoidal in waveform.†

3. Comparison of Y and Δ connections

In many respects the Y and Δ connections have the properties of duality; that is, the voltage relations in one are similar to the current relations in the other. It is interesting to compare the Y and Δ connections from this point of view, since such a comparison shows some of the relative advantages of each.

The Y connection is rather similar to a series circuit, while the Δ connection has characteristics analogous to those of a parallel circuit. Each phase of the Y is connected in series with one of the lines, whereas each phase of the Δ is connected between a pair of lines; thus the Y phase currents equal the corresponding line currents, and the Δ phase voltages equal the corresponding line-to-line voltages. In the Y connection, there are two windings in series between each pair of line terminals, whereas in the Δ connection there are two parallel paths between each pair of line terminals. Thus each line-to-line voltage in a Y connection is the series combination of two Y phase voltages, and each line current

* See Art. 2b, Ch. XXIII.
† See Art. 2d, Ch. XXIII.

in a Δ connection is the parallel combination of two Δ phase currents; whence the voltage relations in a Y circuit are similar to the current relations in a Δ circuit.

If the voltages are balanced and harmonics are negligible, the effective value of the Y voltages to neutral equals $\dfrac{1}{\sqrt{3}}$ times the effective value of the line-to-line voltages. Therefore the Y connection is particularly suited for high-voltage windings, where insulation is a major problem, since for a specified line voltage the Y phase voltages are but $\dfrac{1}{\sqrt{3}}$ times the Δ voltages. Furthermore, with the Y connection the voltages between the coils and core can be reduced through grounding the neutral point. On the other hand, the Δ connection often results in a less expensive design when the currents are large, since for balanced conditions and specified line currents the effective value of the Δ currents is but $\dfrac{1}{\sqrt{3}}$ times the effective value of the Y currents.

When both the primary and secondary windings are connected in the same manner, the bank as a whole has the distinctive characteristics of the Y or Δ connection. Thus the Y–Y and Δ–Δ connections have contrasting characteristics. In the Y–Y connection with isolated neutrals, the line-to-neutral voltages are determined by the exciting admittances of the units, whereas in the Δ–Δ connection the Δ currents are determined by the equivalent impedances.* Hence, if the exciting impedances are unequal, the transformers in a Y–Y connection with isolated neutrals are subjected to unequal voltages, as in a series circuit whose component impedances are unequal; but if the equivalent impedances are unequal, the currents in Δ–Δ-connected transformers are unequal, as in a parallel circuit with unequal branch impedances.

Because of its sensitivity to voltage unbalance and because of undesirable harmonic phenomena, the Y–Y connection is seldom used without a neutral connection or other means of equalizing the voltages to neutral and eliminating the harmonics.†

With the Δ–Δ connection, the dependence of the transformer currents on the equivalent impedances may be an advantage or a disadvantage, depending on circumstances. For example, if the three transformers of a Δ–Δ bank have the same rating, it is desirable that they share the load equally, and if the load is balanced they should therefore have equal equivalent impedances. In these circumstances, the sensitivity of the Δ–Δ connection to inequalities in the equivalent impedances is a disadvantage. On the other hand, transformers of unequal kilovolt-ampere

---

* See Art. 1b, Ch. XXIV, for an analysis of this problem.
† See Arts. 2d and 3, Ch. XXIII.

ratings (but the same voltage ratings) may be used fairly successfully in a Δ–Δ bank, since the largest transformer usually has the smallest equivalent impedance and therefore automatically takes the greatest current. Although a disadvantage of this unsymmetrical arrangement is that the three-phase kilovolt-ampere rating of the bank is somewhat less than the sum of the ratings of the units, nevertheless there are circumstances in which this ability of the Δ–Δ connection to operate satisfactorily with units of unequal kilovolt-ampere ratings may be a distinct advantage.

The Δ–Y and Y–Δ connections, discussed below, partake of some of the characteristics of both the Y and Δ connections.

4. Δ–Y AND Y–Δ CONNECTIONS

Often it is desirable that transformer windings be connected in Y on one side; for example, when the circuit voltage is high, or when a symmetrical neutral connection is required for grounding or for supplying single-phase loads. The Δ–Y or Y–Δ connection is often well suited to these applications, since it does not have the often objectionable features of the Y–Y connection.

Because of the high circuit voltage, the Δ–Y connection is the usual one for the step-up power transformers at the sending end of a high-voltage transmission line, and, for the same reason, the step-down transformers at the receiving end usually are Y–Δ-connected. The neutral of the Y-connected high-voltage windings usually is grounded, in order to assure balanced voltage distribution between the lines and ground, and to reduce the voltages between the transformer coils and core. Sometimes the ground connection is made through a suitable impedance whose purpose is to limit the current resulting from a line-to-ground short circuit.

The Δ–Y connection is often used for step-down transformers when a low-voltage neutral connection is needed. For example, the step-down transformers installed in substations supplying intermediate-voltage primary distribution networks are often Δ–Y-connected. These primary distribution networks usually are three-phase four-wire systems with the neutral wire solidly grounded, the voltage to neutral being a few thousand volts.

The four-wire Y-connected 120/208-volt system of distribution described on p. 535 is often supplied from the Y-connected 120-volt secondaries of a Δ–Y bank of distribution transformers. The secondary neutral wire is grounded and 120-volt lighting loads can be supplied from the secondary lines to the neutral wire, while three-phase motor loads can be operated from the three line wires at a voltage of 120 $\sqrt{3}$ or 208 volts.

The Y–Δ connection may be used for step-down distribution transformers supplying three-phase and single-phase loads when a symmetrical three-phase neutral is not required on the low-voltage secondary side. If the Δ-connected secondaries are 120/240-volt split windings, they can be connected as are the secondaries of the Δ–Δ bank shown in Fig. 2, and power can be supplied simultaneously to 240-volt three-phase loads and to a three-wire 120/240-volt single-phase circuit with the neutral of the single-phase circuit grounded. The advantage of this Δ-connected system of distribution is that standard voltages of 120 and 240 volts are obtained, whereas the Y-connected system described in the pre-

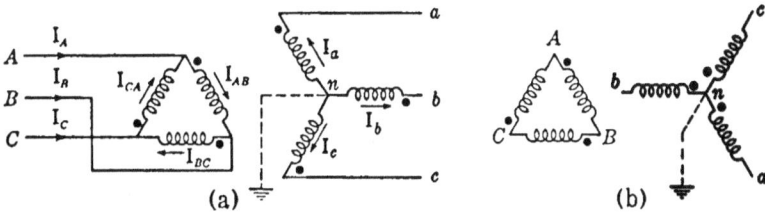

Fig. 9.   Δ–Y connection.

ceding paragraph results in the less common three-phase voltage of 208 volts. Furthermore, 240-volt power for electric ranges is available from the Δ-connected system, but not from the Y-connected system. The disadvantages of the Δ-connected system are its lack of symmetry with respect to currents and voltages. Thus, the single-phase loads on the Δ-connected system result in unbalanced currents, whereas, in the Y-connected system, the currents are approximately balanced if the single-phase loads are properly distributed among the phases. Furthermore, the ground connection of the Δ-connected system is unsymmetrical. In the Δ-connected system of Fig. 2, the voltage between line c and ground is 208 volts, but the voltages to ground are limited to 120 volts in the Y-connected system. The shock hazard is therefore slightly greater in the Δ-connected system, and sometimes more expensive wiring is required.

The analysis of the Δ–Y and Y–Δ connections is based on the fundamental relations for Δ- and Y-connected circuits, together with the relations among the primary and secondary voltages and currents in the transformers. The connections of a Δ–Y bank are shown in Fig. 9a, in which the primary and secondary windings of the same transformer are drawn parallel to each other.

On the Δ-connected side, the terminal voltages of the transformers equal the corresponding line-to-line voltages, and for balanced conditions and phase order ABC the vector diagram is shown in Fig. 10a. On the Y-connected side, the terminal voltages of the transformers equal the line-to-neutral voltages, and the vector diagram of Y voltages is shown in Fig. 10b, which should be compared with Fig. 6b, p. 533. For balanced conditions and phase order *abc*, the Y line-to-line voltage V_{ab} is

$$V_{ab} = \sqrt{3}\,V_{an}\,\underline{/30°},$$

[24]

as shown in Fig. 10b. When the leakage-impedance voltage drops are small, the primary and secondary terminal voltages of the transformers are very nearly in phase when their positive directions are taken with

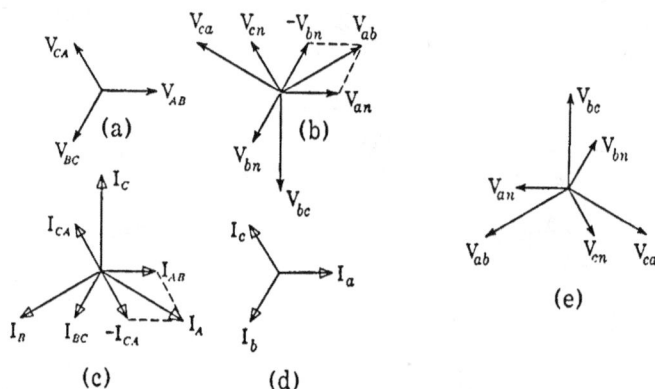

Fig. 10. Approximate vector diagrams for a Δ–Y bank.

the same relative polarity. Thus when the polarities of the connections are as in Fig. 9a, the Y line-to-neutral voltage V_{an} is very nearly in phase with the Δ line-to-line voltage V_{AB}, as in Fig. 10b, and therefore the line-to-line voltage V_{ab} on the Y-connected side leads the Δ voltage V_{AB} by approximately 30 degrees.

The Y connection may also be obtained as in Fig. 9b, in which the polarity of the neutral is the reverse of that for the connections of Fig. 9a. The corresponding Y voltages shown in Fig. 10e are therefore in phase opposition to the voltages of Fig. 10b; thus the Y-side line voltage V_{ab} lags the Δ line voltage V_{AB} by 150 degrees, and the Y-side line voltage V_{ca} lags the Δ line voltage V_{AB} by 30 degrees.

▶ Hence there is a phase angle of 30 degrees between the primary and secondary line voltages which have the minimum phase displacement, this displacement being either a lead or a lag, depending on the relative polarities of the primary and secondary connections. Contrast this condition with a Δ–Δ or a Y–Y bank, in which corresponding primary and secondary line voltages are very nearly in phase when the primary and secondary connections are made with the same relative polarities. On account of this phase difference, a Δ–Y or a Y–Δ bank should never be paralleled with a Δ–Δ or a Y–Y bank on both primary and secondary sides.*◀

* Parallel operation of polyphase banks is discussed in Art. 2, Ch. XXIX.

On the Y-connected side, the currents in the transformers equal the corresponding line currents. The vector diagram is shown in Fig. 10d, for balanced conditions and unity power factor. When the Y currents are balanced and vary sinusoidally, their vector sum is zero, and therefore there is no current in the neutral grounding connection (shown by the broken line in Fig. 9), which merely serves the purpose of maintaining the neutral at ground potential. The vector diagram of Δ currents is shown in Fig. 10c, which should be compared with Fig. 3c, p. 529. If the exciting currents are neglected, the primary and secondary currents produce equal and opposing magnetomotive forces, and therefore the Δ currents are in phase with the corresponding, oppositely directed Y currents.

When the three transformers have the same turns ratio and the Δ line voltages are balanced, the Y voltages to neutral must also be very nearly balanced when the leakage-impedance voltage drops are small. Contrast this situation with the behavior of a Y–Y bank whose voltages to neutral depend on the excitation characteristics of the units when the neutrals are isolated. Also, if the line currents are balanced on the Y-connected side, the Δ currents must be very nearly balanced, any unbalance in the Δ currents being only that due to differences in the relatively small exciting currents. Contrast this condition with the behavior of a Δ–Δ bank whose currents depend on the equivalent impedances of the units.

Thus for balanced line voltages and balanced line currents, the Δ connection balances the Y voltages to neutral, while the Y connection balances the currents in the Δ. This balancing action sometimes is an advantage of the Δ-Y and Y-Δ connections, since three transformers with the same voltage and current ratings automatically take nearly equal shares of a balanced three-phase load, even when their equivalent impedances and excitation characteristics are unequal.

5. OPEN–DELTA OR V CONNECTION

When three identical transformers have their windings connected either in Y or in Δ, the arrangement is symmetrical. Although the majority of three-phase transformations are by means of these symmetrical arrangements, an important example of an unsymmetrical arrangement is the open-delta or V connection which uses only two transformers. The elementary connections of an open-delta bank are shown in Fig. 11, and the connection diagram of an installation simultaneously supplying 240-volt three-phase and 120/240-volt three-wire single-phase loads is shown in Fig. 12.

Note that the open-delta connection is the equivalent of the Δ–Δ connection with one transformer removed. Thus if one of a group of Δ–Δ-connected transformers becomes defective so that it must be removed from the circuit, the two remaining transformers can still be used to transmit three-phase power, although with a reduced load capacity.

The open-delta connection not only is used as an emergency measure, but also often is installed in areas where the load is expected to grow, since, when the load has increased to a value exceeding the capacity of the open-delta bank, a third transformer can readily be added, con-

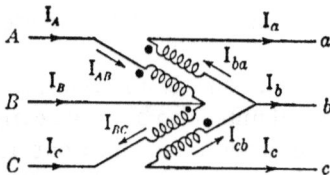

FIG. 11. Open-delta or V connection.

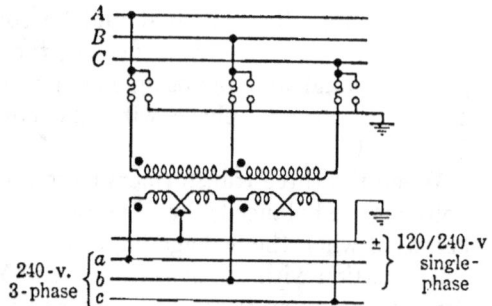

FIG. 12. Open-delta connection of distribution transformers for simultaneous 240-v 3-phase and 120/240-v single-phase outputs.

verting the open-delta to a Δ–Δ bank and thereby increasing the three-phase load capacity of the bank. Since only two transformers are involved, the open-delta connection is relatively simple to install and has a neat appearance on a pole-type installation.

Although the transformer that would be across phases CA to ca in the Δ–Δ connection is not present in the open-delta connection and therefore the bank is unsymmetrical with respect to the three lines, very nearly balanced secondary line voltages are obtained when balanced three-phase voltages are applied to the primary terminals. This result is due to the fact that in any three-phase system the vector sum of the line-to-line voltages, taken in cyclic order, must be zero,* and therefore the vectors representing the line-to-line voltages can be drawn as the three sides of a triangle. If the primary voltages are balanced, the primary voltage triangle is equilateral, as shown in Fig. 13a. The secondary line voltages V_{ab} and V_{bc} are obtained directly from the secondary terminals of the two transformers, and at no load these secondary voltages are very nearly in phase with and equal to the corresponding primary voltages divided by the ratio a of primary to secondary turns (assumed to be the same for both transformers). These secondary voltages are shown by the solid lines in Fig. 13b. The secondary line-to-line voltage V_{ca} is the closing side of the secondary line-voltage triangle and is shown by the broken line in Fig. 13b. From the similarity of the primary and

* See Eqs. 20 and 21, p. 536.

secondary voltage triangles, if the primary line voltages are balanced, the secondary line voltages also are very nearly balanced at no load. However, since the bank is unsymmetrical with respect to the three lines, the equivalent-impedance voltage drops in the transformers produce slight unbalance in the secondary voltages under load, but often this unbalance is of little consequence.

Inspection of Fig. 11 shows that the primary currents in the transformers equal the currents in lines A and C, and that the secondary currents equal the currents in lines a and c. Thus if V_2 and I_2 are the rated secondary voltage and current of each transformer in an open-delta bank, these are also the rated full-load values of the three-phase secondary line voltages and line currents for a balanced three-phase load, and therefore the three-phase rating of the bank is $\sqrt{3}\,V_2 I_2$ volt-amperes.*

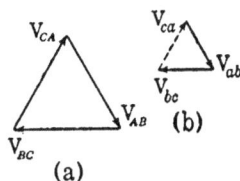

FIG. 13. Vector diagram of voltages of an open-delta bank.

---

*The expression for the volt-amperes delivered to a balanced three-phase load is $\sqrt{3}\,V_{line}\,I_{line}$, where V_{line} and I_{line} are the effective values of the line-to-line voltages and of the line currents. Though this relation is derived in most textbooks on three-phase circuits, it is so important that the derivation is repeated below.

Let V_{phase} and I_{phase} be the effective values of the voltage across each phase and of the current in each phase of a balanced three-phase load. The expression for the volt-amperes delivered to the three-phase load is $3V_{phase}I_{phase}$. If the load is Δ-connected,

$$V_{phase} = V_{line} \tag{25}$$

$$I_{phase} = \frac{I_{line}}{\sqrt{3}}, \tag{26}$$

and therefore

$$3V_{phase}I_{phase} = 3V_{line}\frac{I_{line}}{\sqrt{3}} = \sqrt{3}\,V_{line}I_{line}. \tag{▶27}$$

If the load is Y-connected,

$$V_{phase} = \frac{V_{line}}{\sqrt{3}} \tag{28}$$

$$I_{phase} = I_{line}, \tag{29}$$

and therefore

$$3V_{phase}I_{phase} = 3\frac{V_{line}}{\sqrt{3}}I_{line} = \sqrt{3}\,V_{line}I_{line}. \tag{▶30}$$

Thus, regardless of whether the load is connected in Δ or in Y, the expression for the volt-amperes delivered to the load is

$$3V_{phase}I_{phase} = \sqrt{3}\,V_{line}I_{line}. \tag{▶31}$$

On the secondary side of the open-delta connection, V_{line} equals the transformer voltage V_2 and I_{line} equals the transformer current I_2, whence the three-phase output of the bank is $\sqrt{3}\,V_2 I_2$ volt-amperes.

However, the bank consists of two transformers with a total single-phase rating of $2V_2I_2$. Thus the three-phase rating of the bank — $\sqrt{3}\, V_2I_2$ — is less than the combined single-phase ratings — $2V_2I_2$ — of the units installed in the bank. The ratio

$$\frac{\text{3-phase kva rating of the bank}}{\text{Sum of the kva ratings of the units installed}} \qquad \blacktriangleright[32]$$

is called the *apparatus economy* or *utilization factor* of the bank, and is a measure of the effectiveness with which the installed transformer capacity is utilized for supplying load. Thus for an open-delta bank of identical transformers supplying a balanced three-phase load

$$\text{Apparatus economy} = \frac{\sqrt{3}\, V_2I_2}{2V_2I_2} = \frac{\sqrt{3}}{2} = 0.866. \qquad [33]$$

That is, an open-delta bank of identical transformers is fully loaded when it is supplying a balanced three-phase load of 86.6 per cent of the total kilovolt-ampere rating of the installed transformers. It is therefore necessary to install a greater total transformer capacity in an open-delta than in a Δ–Δ bank capable of supplying the same balanced three-phase load. This is the principal disadvantage of the open-delta as compared with any of the symmetrical connections.

Because the apparatus economy of an open-delta bank is less than that of a Δ–Δ bank, the removal of a faulty transformer from a Δ–Δ bank reduces the three-phase load capacity of the bank by more than the kilovolt-ampere rating of the removed transformer. In a Δ–Δ bank with a balanced three-phase load, the effective value of the line currents equals $\sqrt{3}$ times the effective value of the currents in the transformers, whereas in the open-delta connection the line currents equal the transformer currents.

Therefore a bank of two transformers connected in open delta has a three-phase kilovolt-ampere rating of $\dfrac{1}{\sqrt{3}}$ or 0.577 of the rating of a Δ–Δ bank of three identical transformers.

6. Emergency Operation of Δ–Y and Y–Δ Banks

By means of an arrangement somewhat similar to the open-delta connection, a transmission line with a step-up bank of Δ–Y-connected transformers at its sending end and a step-down Y–Δ bank at its receiving end can be operated at a reduced maximum load with one phase faulted and removed from service. The neutrals at the sending and receiving ends must be connected, either through the ground or preferably with a neutral wire, as shown in Fig. 14a.

The system is unsymmetrical, and therefore the impedance drops are unbalanced; but, since these impedance drops often are small, the voltages at the secondary terminals of the step-down bank may not be seriously unbalanced. However, the currents in the transmission line are grossly unbalanced. If balanced currents are delivered to the load on the secondary side of the step-down bank, the vector diagram of secondary currents is as shown in Fig. 14b. Except for the exciting currents, the primary and secondary currents in the transformers produce equal and opposing magnetomotive forces. Thus for the positive directions of the currents shown by the arrows in Fig. 14a, the primary current I_A of transformer A is in phase with the secondary current I_a'', and the primary current I_C of transformer C is in phase opposition to the secondary current

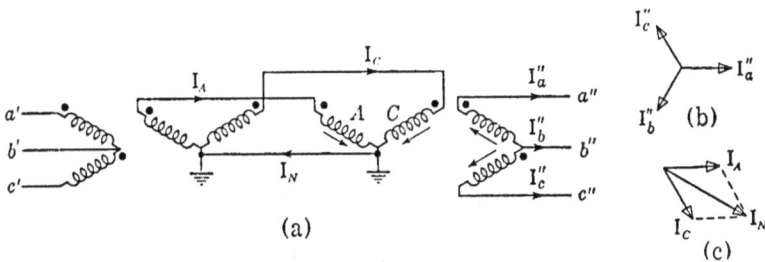

Fig. 14. Emergency operation of Δ-Y and Y-Δ banks, showing (a) connections, (b) vector diagram of balanced secondary load currents, and (c) vector diagram of currents in the high-voltage line.

I_c'', as shown in Fig. 14c. The neutral current I_N returning to the sending-end transformers is the vector sum of the primary line currents, as shown in Fig. 14c, and therefore, when the secondary currents are balanced, the effective value of the neutral current is $\sqrt{3}$ times the effective value of the phase currents in the transmission line. These unbalanced currents in the transmission line may produce objectionable induced voltages in adjacent communication circuits.

On the low-voltage sides the transformers are connected in open delta and therefore, as in Art. 5, the full-load capacity of the system with one phase faulted is 0.577 of the full-load capacity of the unfaulted system, and the apparatus economy is 0.866.

PROBLEMS

1. A bank of three single-phase transformers is to be used to step down the voltage of a 6,600-v distribution line to operate a group of 115-v incandescent lamps from a four-wire three-phase system. The transformers are normally connected Δ-Y with the neutral of the Y grounded; but, to provide a means for dimming the lamps to a small fraction of normal brilliancy, a change-over switch is arranged so that the sec-

ondaries may be connected in Δ ungrounded. The total power required by the lamps is 300 kw at unity power factor.

(a) What must be the primary current and voltage ratings of the transformers?

(b) What must be the secondary current and voltage ratings of the transformers?

(c) If the change-over switch is thrown and a short circuit occurs across one bank of lamps, what voltage will develop across the other two banks? Would this be likely to burn out any fuses?

(d) If the change-over switch is thrown and a fuse burns out in one line, what voltages will develop across the three banks of lamps?

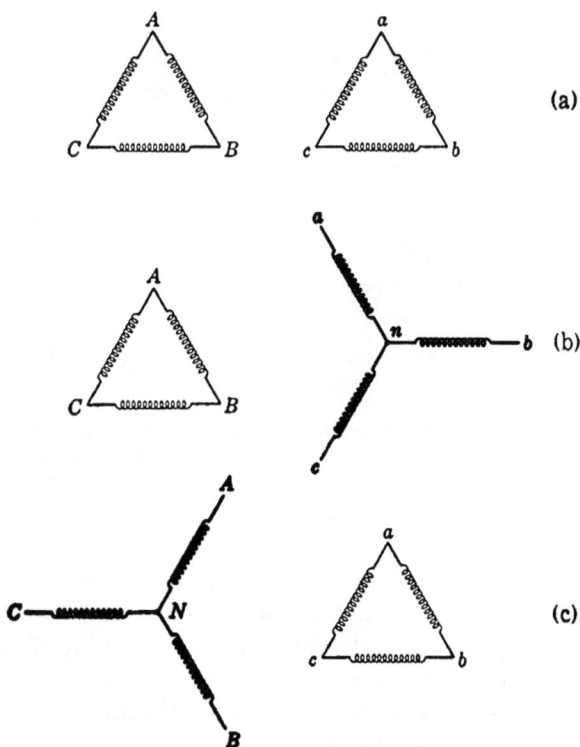

FIG. 15.   Transformer connections, Prob. 2.

2. Three 2,400 : 240-v single-phase transformers are to be connected in a symmetrical bank to a three-phase line of suitable voltage. There are no polarity markings on the transformers. For each of the connections shown in Fig. 15, describe a step-by-step test procedure for determining the correct connections with the use of a voltmeter only. In each instance tell what values the voltmeter might indicate.

3. It is necessary to step down the voltage of a three-phase three-wire 7,200-v line to obtain three-phase four-wire distribution at 120 v to neutral. The load consists of three-phase induction motors and of incandescent lamps. The induction-motor load is 100 kva at 0.70 power factor; the lighting load is 200 kva at unity power factor and is connected between the line wires and the neutral wire in such fashion that the total lighting load is balanced. If three single-phase transformers are to be used,

(a) How should they be connected, and why?

(b) What should be their primary and secondary voltage ratings?

(c) What are the full-load currents in their primary and secondary windings?

4. Each of the transformers of Prob. 3 has a short-circuit impedance of $5 + j20$ ohms referred to the high-voltage side. The transformers are connected as shown in Fig. 15c, with the primary neutral isolated, and balanced three-phase voltages of $7,200\sqrt{3}$, line-to-line, are applied to the Y-connected primaries.

(a) What will be the value of the current in the closed Δ of the secondaries if one transformer secondary is accidentally connected with its polarity reversed and if the exciting current is neglected?

(b) What will be the primary phase voltages V_{AN}, V_{BN}, and V_{CN} if coil ac is the one reversed? Is it permissible in this problem to neglect exciting currents?

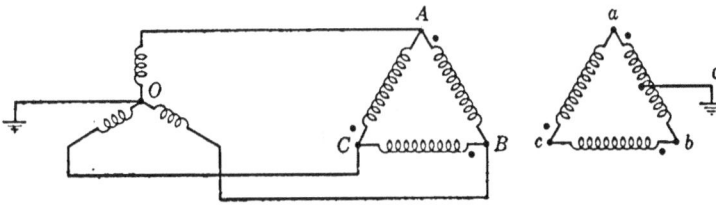

(a)

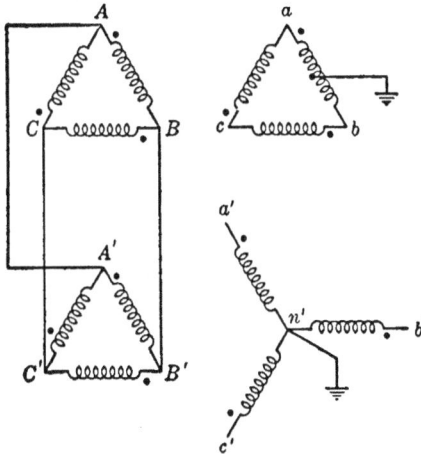

(b)

Fig. 16. Transformer connections, Prob. 5.

5. Figures 16a and 16b show portions of two distribution systems. Figure 16a represents a Δ-Δ three-phase connection of single-phase transformers supplied from a Y-connected source with the neutral grounded. Figure 16b represents a parallel three-phase connection of a Δ-Δ bank and a Δ-Y bank of single-phase transformers. All the transformers are identical and have one-to-one turns ratios. The primaries are supplied with balanced three-phase voltages of 440 v, line-to-line. Determine the voltages between the following sets of points.

(a) In Fig. 16a: co, Aa, and Cc.

(b) In Fig. 16b: aa' and cc'.

CHAPTER XXII

Effects of Transformer Impedances in Balanced Three-Phase Circuits

In Ch. XXI, the common three-phase arrangements of ideal transformers are discussed, and the major effects of the imperfections of actual transformers are mentioned qualitatively. This simple analysis, sufficiently accurate for many purposes, permits a comparison of the outstanding characteristics of the various three-phase arrangements. In many problems, however, the imperfections of transformers have important effects and must be taken into account quantitatively. For example, the leakage impedances of the transformers are important in problems involving voltage regulation, the division of current among parallel circuits, and the calculation of short-circuit currents, and in other problems embracing the impedances of transmission and distribution systems. It is the purpose of this chapter to develop a quantitative basis for the analysis of such questions, with emphasis on balanced conditions. Unbalanced conditions are discussed in Ch. XXIV.

1. EQUIVALENT CIRCUITS

In the following analysis it is convenient to represent each of the single-phase transformers composing the three-phase bank by an equiva-

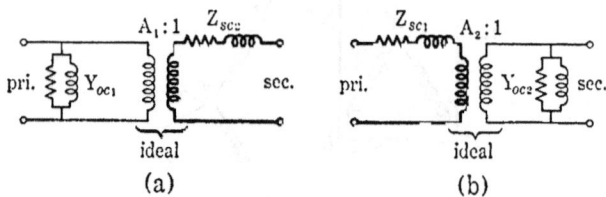

FIG. 1. Equivalent circuits for a single transformer.

lent circuit. Since the effects of the capacitances of the windings and of harmonics in the exciting currents usually are negligible, any one of the equivalent circuits derived in the preceding chapters may be used; those most useful for the present discussion are shown in Fig. 1.* In them, the transformer is represented, as in Thévenin's theorem, by its short-circuit impedance in series with its open-circuit voltage; the open-circuit voltage ratio is represented by an ideal transformer; and the excitation

---

* These equivalent circuits are described in Arts. 6 and 8c, Ch. XIII, and in Art. 5, Ch. XVII.

characteristics are represented by the open-circuit admittance; that is, in Fig. 1a,

> Z_{sc2} is the short-circuit impedance measured on the secondary side,
>
> Y_{oc1} is the open-circuit admittance measured on the primary side,
>
> A_1 is the vector ratio V_1/E_{oc2} of primary to secondary voltage measured with the secondary open;

and in Fig. 1b,

> Z_{sc1} is the short-circuit impedance measured on the primary side,
>
> Y_{oc2} is the open-circuit admittance measured on the secondary side,
>
> A_2 is the vector ratio E_{oc1}/V_2 of primary to secondary voltage measured with the primary open.

Except for the effects of capacitances and harmonics in the exciting current, these equivalent circuits are "exact" representations of the transformer. However, as shown in Art. 8, Ch. XIII, the open-circuit voltage ratios very nearly equal the turns ratio; that is,

$$A_1 \approx a \approx A_2, \qquad \blacktriangleright[1]$$

where a is the turns ratio N_1/N_2. The short-circuit impedances also very nearly equal the equivalent impedances referred to the same side; thus

$$Z_{sc1} \approx Z_{eq1} \qquad \blacktriangleright[2]$$

$$Z_{sc2} \approx Z_{eq2}, \qquad \blacktriangleright[3]$$

where Z_{eq1} and Z_{eq2} are the equivalent impedances. The open-circuit admittances are practically identical with the exciting admittances referred to the same side; that is,

$$Y_{oc1} \approx Y'_\varphi \qquad \blacktriangleright[4]$$

$$Y_{oc2} \approx Y''_\varphi, \qquad \blacktriangleright[5]$$

where Y'_φ and Y''_φ are the exciting admittances. The approximate relations, Eqs. 1–5, are so nearly correct that in the following analysis no distinction is made between the "exact" values of the parameters in Fig. 1 and their approximate values given by these equations. That is, the values of the parameters may be obtained from design data or from open-circuit and short-circuit tests taken on either side of the transformer, and these values may be used, without modification, in either the equivalent circuit of Fig. 1a (in which the exciting admittance is placed on the primary side) or in the equivalent circuit of Fig. 1b (in which the exciting admittance is placed on the secondary side). In many problems, the effects of the exciting current are so small that the exciting current may be neglected entirely and the transformer may be represented by its equivalent impedance in series with an ideal transformer. If desired, the equivalent impedances and exciting admittances in Fig. 1 may be referred to the other side of the transformer by multiplication or division, in the appropriate manner, by the square of the ratio of transformation.

The equivalent circuit of a three-phase bank of transformers can be drawn by connection of the equivalent circuits of the units in accordance with the connections of the bank. For example, Fig. 2a shows an equiva-

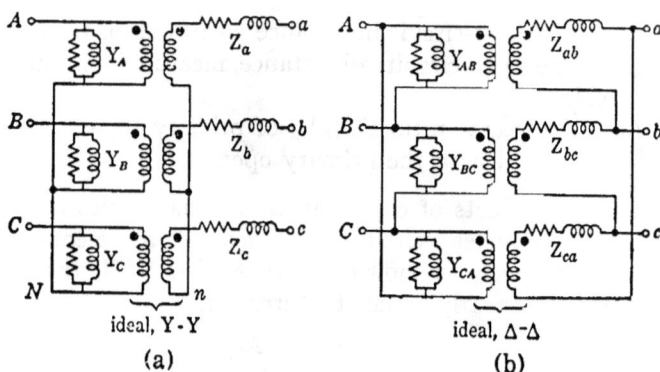

FIG. 2. Three-phase equivalent circuits; (a) Y–Y connection, and (b) Δ–Δ connection.

lent circuit for a Y–Y bank and Fig. 2b an equivalent circuit for a Δ–Δ bank. In Fig. 2, the Y's represent the open-circuit or exciting admittances and the Z's the short-circuit or equivalent impedances.

In the analysis of power systems, it is frequently necessary to combine the impedances of transformers with the impedances of transmission lines to which they are connected. Hence, it is often convenient to represent a Δ-connected group of windings by a Y-connected equivalent circuit, since in the Y connection the equivalent impedances representing the transformers are in series with the three line terminals, and therefore can be added directly to the phase impedances of the transmission circuits. It is well known that, as viewed from its three terminals, a Δ-connected group of circuit elements can be replaced by an equivalent Y connection.[1] Thus the Δ-connected exciting admittances Y_{AB}, Y_{BC}, Y_{CA} of Fig. 2b are equivalent to Y-connected admittances Y_A, Y_B, Y_C, whose values are given by the well-known relations

$$Y_A = \frac{Y_{AB}Y_{BC} + Y_{BC}Y_{CA} + Y_{CA}Y_{AB}}{Y_{BC}} \qquad [6]$$

[1] These Δ–Y transformations are discussed in this series in the volume on electric circuits (1940), pp. 146–150, 459–460, and 543.

$$Y_B = \frac{Y_{AB}Y_{BC} + Y_{BC}Y_{CA} + Y_{CA}Y_{AB}}{Y_{CA}} \qquad [7]$$

$$Y_C = \frac{Y_{AB}Y_{BC} + Y_{BC}Y_{CA} + Y_{CA}Y_{AB}}{Y_{AB}}. \qquad [8]$$

The impedances Z_{ab}, Z_{bc}, Z_{ca} of Fig. 2b, forming part of a Δ-connected system, furthermore, can be replaced by Y-connected impedances. Thus, by Thévenin's theorem, on its secondary side the Δ–Δ bank of Fig. 2b is equivalent to a Y-connected source, producing the same secondary line-to-line voltages on open circuit and connected in series with impedances whose values equal the Y equivalents of the short-circuit impedances of the transformers measured at their secondary terminals. Hence the Δ–Δ-connected ideal transformers in Fig. 2b can be replaced by a Y–Y bank that gives the same open-circuit voltages, and the short-circuit impedances of the transformers can be represented by Y-connected impedances in series with each line terminal. The relations between the short-circuit impedances Z_{ab}, Z_{bc}, Z_{ca} of the Δ-connected transformers and their Y-connected equivalents Z_a, Z_b, Z_c are given by the well-known relations

$$Z_a = \frac{Z_{ca}Z_{ab}}{Z_{ab} + Z_{bc} + Z_{ca}} \qquad [9]$$

$$Z_b = \frac{Z_{ab}Z_{bc}}{Z_{ab} + Z_{bc} + Z_{ca}} \qquad [10]$$

$$Z_c = \frac{Z_{bc}Z_{ca}}{Z_{ab} + Z_{bc} + Z_{ca}}. \qquad [11]$$

Therefore, in so far as its effects on the external circuits are concerned, a Δ-connected group of windings can be represented by a Y-connected equivalent circuit, as in Fig. 2a, provided the parameters of the equivalent Y are related to the actual parameters of the Δ-connected transformers as in Eqs. 6–11, and provided the open-circuit line-to-line voltages of the Y-connected equivalent circuit are the same as those of the Δ-connected winding.* With a Δ–Δ bank, that is, the ideal transformers may be replaced by a Y–Y bank giving the same open-circuit voltages. Similarly a Δ–Y (or a Y–Δ) bank often is represented by a Y–Y equivalent circuit, as in Fig. 2a, giving the same magnitudes of open-circuit line-to-line voltages. Because of the phase displacement introduced by the Δ–Y connection,† however, the Y–Y equivalent of a Δ–Y bank does not

---

* Note that the Y-connected equivalent circuit does not show the actual currents in the Δ-connected windings.

† See Art. 4, Ch. XXI.

show the correct phase relations between primary and secondary currents, or between primary and secondary voltages, although it does show correctly the relations among the currents and voltages on each side.

2. SINGLE-PHASE EQUIVALENT CIRCUITS FOR BALANCED CONDITIONS

When the transformers are identical and the currents and voltages are balanced, only the phase displacements of 120 degrees among their currents and among their voltages distinguish one phase from another. Therefore the currents and voltages in every phase can be determined by analysis of any one phase. Usually for these purposes it is convenient to consider all generators, transformer windings, and loads as if they were Y-connected. Thus the exciting admittances and equivalent impedances of Δ-connected transformer windings can be replaced by their Y-connected equivalents, as in Eqs. 6–11, which, for identical transformers, reduce to

$$Y_Y = 3Y_\Delta \qquad\qquad \blacktriangleright[12]$$

$$Z_Y = \tfrac{1}{3}Z_\Delta \qquad\qquad \blacktriangleright[13]$$

in which the subscript Δ indicates the Δ-phase admittance or impedance and the subscript Y the equivalent Y-phase value. For balanced conditions, the neutral points of all the equivalent Y-connected circuits can be considered directly connected together.

2a. *Illustrative Example of a Balanced Circuit.* — As an illustration of the analysis of a balanced system, consider the circuit shown in the schematic diagram of Fig. 3a, comprising a Y–Δ bank of three 1,000-kva 63,500 : 33,000-volt single-phase transformers connected on their primary sides to a balanced three-phase source (the 110,000-volt bus in a substation). The secondaries of this bank supply power to a Δ–Δ bank of three 1,000-kva 33,000 : 13,200-volt transformers through a three-phase transmission line.

The problem is to determine what voltage is required at the substation bus in order to maintain the rated line-to-line voltage of 13,200 volts at the secondary terminals of the Δ–Δ bank when this bank supplies a balanced three-phase load of 3,000 kva at unity power factor.

Data: The impedance of the 33,000-volt transmission line is

$$Z_{line} = 7.3 + j18.2 \text{ ohms per phase.} \qquad [14]$$

The equivalent impedance Z_R of each of the Δ–Δ transformers at the receiving end of the line is

$$Z_R = 1.71 + j9.33 \text{ ohms} \qquad [15]$$

referred to the low-voltage side. Their core loss is

$$P_c = 5.6 \text{ kw, each transformer,} \qquad [16]$$

and their magnetizing reactive kva is

$$(VI)_{mag} = 51 \text{ kvar, each transformer.} \qquad [17]$$

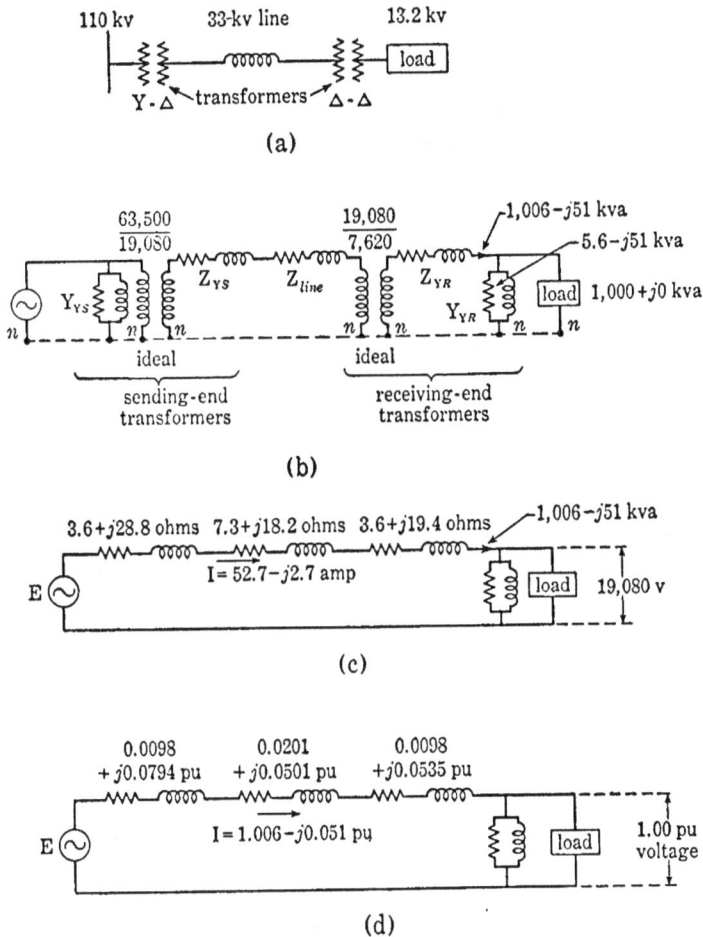

(a)

(b)

(c)

(d)

FIG. 3.   Single-phase diagram of the circuit discussed in Art. 2, and steps in the simplification of the circuit.

The average results of single-phase open-circuit and short-circuit tests taken on the sending-end transformers are:

Open-circuit test	Short-circuit test
$V = 33,000$ v	$V = 2,640$ v
$I = 1.24$ amp	$I = 30.3$ amp
$P = 5.30$ kw	$P = 9.81$ kw

In both these tests, the measurements were made on the low-voltage (33,000-volt) sides at rated frequency (60 cycles per second).

Solution: The first step is to reduce the circuit to a single phase of a Y-connected equivalent, as shown in Fig. 3b, in which the load and source are each considered as one phase of a Y-connected circuit, and each transformer bank is represented by one phase of a Y–Y equivalent comprising an ideal transformer in combination with a series impedance and a shunt admittance. The neutral points n of all the Y's can be considered connected, as shown by the broken line in Fig. 3b.

Although the exciting currents of the transformers ordinarily would be neglected in the solution of a problem of this kind, they will be included in the following solution, wherever they have any effect, for the sake of completeness.

The exciting currents may be accounted for by shunt admittances connected on either side of the transformers. It is convenient to connect the exciting admittances of the load-end transformers across the secondary terminals, since the excitation, given as

$$P_e - j(VI)_{mag} = 5.6 - j51 \text{ vector kva per phase,} \qquad [18]$$

then can be combined directly in parallel with the load, given as $1,000 + j0$ vector kva per phase. Thus the load and the excitation of the load-end transformers absorb a combined vector power of

$$1,006 - j51 \text{ vector kva per phase.} \qquad [19]$$

However, if the exciting admittances of the sending-end transformers are connected across their primary terminals, as in Fig. 3b, their exciting currents have no effect on the source voltage required to maintain rated voltage at the load, although they do affect the current taken from the source. When treated in this manner, the exciting admittances of the sending-end transformers thus do not enter into the solution of this problem, since the source current is not required.

The ratios of transformation of the Y–Y-connected ideal transformers in Fig. 3b are such that they produce the same open-circuit voltages as the bank they represent. The equivalent Y voltages to neutral for the 33,000 : 13,200-v, Δ–Δ bank at the load end are

$$\frac{33,000}{\sqrt{3}} = 19,080 \text{ v} \qquad [20]$$

$$\frac{13,200}{\sqrt{3}} = 7,620 \text{ v} \qquad [21]$$

and the equivalent Y voltages to neutral for the Y–Δ bank at the sending end are

$$\frac{110,000}{\sqrt{3}} \approx 63,500 \text{ v} \qquad [22]$$

and

$$19,080 \text{ v.} \qquad [23]$$

Thus the ratios of transformation of the ideal transformers are as marked in Fig. 3b.

The equivalent impedance of each of the Δ–Δ transformers at the receiving end is given as

$$Z_{\Delta R} = 1.71 + j9.33 \text{ ohms} \qquad [24]$$

referred to the low-voltage side. By Eq. 13, the equivalent Y-connected impedance is

$$Z_{YR} = \tfrac{1}{3}(1.71 + j9.33) = 0.57 + j3.11 \text{ ohms.} \tag{25}$$

Referred to the primary side of the receiving-end ideal transformer, this equivalent impedance is

$$a^2 Z_{YR} = \left(\frac{19{,}080}{7{,}620}\right)^2 (0.57 + j3.11)$$

$$= 3.6 + j19.4 \text{ ohms.} \tag{26}$$

From the short-circuit data, the short-circuit impedance of each sending-end transformer referred to the Δ-connected secondary side is

$$Z_{\Delta S} = \frac{V}{I} = \frac{2{,}640}{30.3} = 87.1 \text{ ohms} \tag{27}$$

$$R_{\Delta S} = \frac{P}{I^2} = \frac{9{,}810}{(30.3)^2} = 10.7 \text{ ohms} \tag{28}$$

$$X_{\Delta S} = \sqrt{Z_{\Delta S}^2 - R_{\Delta S}^2} = 86.4 \text{ ohms.} \tag{29}$$

By Eq. 13 the equivalent Y-connected impedance is

$$Z_{YS} = \tfrac{1}{3}(10.7 + j86.4) = 3.6 + j28.8 \text{ ohms.} \tag{30}$$

If the load and the equivalent impedance and exciting current of the receiving-end transformer are all referred to the primary side, and the source voltage is referred to the secondary side of the sending-end transformer, the ideal transformers may be omitted, and the circuit of Fig. 3b reduces to that shown in Fig. 3c, in which a load of $1{,}006 - j51$ vector kva per phase is supplied at a voltage of 19,080 v per phase through a series impedance whose value equals the sum of its three components, or

$$Z = Z_{YS} + Z_{line} + a^2 Z_{YR} = 14.5 + j66.4 \text{ ohms} \tag{31}$$

If the load voltage is taken as the reference vector, the vector current in the transmission line is

$$I = \frac{\text{kva per phase}}{\text{kv per phase Y}} = \frac{1{,}006 - j51}{19.08} \tag{32}$$

$$= 52.7 - j2.7 \text{ amp per phase Y.} \tag{33}$$

The required sending-end voltage now can be determined by adding the impedance drop in the system to the load voltage; thus

$$IZ = (52.7 - j2.7)(14.5 + j66.4)$$

$$= \quad 940 + j3{,}460 \text{ v Y} \tag{34}$$

$$V = 19{,}080 + j \quad 0 \tag{35}$$

$$E = \text{sum} = 20{,}020 + j3{,}460 \tag{36}$$

$$E = 20{,}300 \text{ v Y.} \tag{37}$$

The voltage E is the substation voltage to neutral, referred to the low-voltage secondary sides of the sending-end transformers. The actual substation voltage is

$$\frac{63{,}500}{19{,}080} \times 20{,}300 = 67{,}500 \text{ v to neutral} \tag{38}$$

or

$$\sqrt{3} \times 67,500 = 117,000 \text{ v line-to-line,} \qquad [39]$$

which is the sending-end voltage required to maintain rated voltage at the load end.

2b. *Illustrative Example Solved with Per Unit Quantities.* — The use of per cent or per unit quantities* often results in a considerable saving of time in power-circuit computations, and is particularly useful when three-phase transformer banks are involved, since the reduction factors for transformation ratios and Δ–Y conversion factors are automatically taken care of by the normal or base values in terms of which the currents. voltages, and circuit parameters are expressed. Furthermore, gross errors such as a misplaced decimal point are likely to be noticed, because of the evident absurdity of an incorrect result, and computations in the per unit system are therefore to some extent self-checking.

As an illustration of the use of per unit quantities in a three-phase problem, the illustrative example of part (a) of this article is solved below in these terms. The schematic circuit diagram is shown in Fig. 3a, and the data are given on pp. 554 and 555.

Solution: The first step in the solution is to choose suitable base values on which to express the loads, currents, voltages, and impedances. In this problem, the three-phase load and the three-phase ratings of both transformer banks are 3,000 kva, which is therefore a convenient base.

$$\text{Base kva} \equiv 3,000 \text{ kva, 3-phase} \qquad [40]$$

$$\equiv 1,000 \text{ kva per phase.} \qquad [41]$$

For the voltage bases, it is convenient to choose the rated values. Thus

$$\text{Base voltage} \equiv 13.2 \text{ kv } \Delta, \text{ for load end} \qquad [42]$$

$$\equiv 33.0 \text{ kv } \Delta, \text{ or } 19.08 \text{ kv Y,}$$
$$\text{for the transmission line} \qquad [43]$$

$$\equiv 110 \text{ kv } \Delta, \text{ or } 63.5 \text{ kv Y for the source.} \qquad [44]$$

The base currents are determined from the chosen base kva divided by the base voltages in kilovolts. Thus for the Δ-connected transformer windings,
Base current for load-end transformer secondaries

$$= \frac{1,000}{13.2} = 75.8 \text{ amp per phase } \Delta \qquad [45]$$

Base current for sending-end transformer secondaries

$$= \frac{1,000}{33.0} = 30.3 \text{ amp per phase } \Delta, \qquad [46]$$

and for the 33-kv transmission line,

$$\text{Base current} = \frac{1,000}{19.08} = 52.5 \text{ amp per phase Y.} \qquad [47]$$

---

* The per cent and per unit systems of expressing currents, voltages, and circuit parameters are discussed in Art. 6, Ch. XIV.

The base values of the impedances are determined from the base voltages divided by the base currents. Thus

Base impedance on the secondary sides of the load-end transformers

$$= \frac{13,200}{75.8} = 174 \text{ ohms per phase } \Delta \qquad [48]$$

Base impedance on the secondary sides of the sending-end transformers

$$= \frac{33,000}{30.3} = 1,090 \text{ ohms per phase } \Delta \qquad [49]$$

Base impedance for the 33-kv transmission line

$$= \frac{19,080}{52.5} = 363 \text{ ohms per phase Y.} \qquad [50]$$

The per unit values of the vector power absorbed by the load, and of the transformer and line impedances, now can be determined. Thus the three-phase load is given as 3,000 kva at unity power factor, and therefore equals the base kva, whence

$$\text{Load} = 1.00 + j0 \text{ per unit vector power.} \qquad [51]$$

The excitation of each load-end transformer at normal voltage is given in Eq. 18 as $5.6 - j51$ vector kva per phase (lagging current) and therefore, as a fraction of the per phase kva base, Eq. 41, the excitation is

$$\frac{5.6 - j51}{1,000} = 0.006 - j0.051 \text{ per unit vector power.} \qquad [52]$$

Thus the load and excitation of the receiving-end transformers together absorb

$$1.006 - j0.051 \text{ per unit vector power.} \qquad [53]$$

According to the conditions of the problem, the line-to-line load voltage is 13,200 v, or 1.00 per unit, and therefore, if the load voltage is taken as the reference vector, the vector current I in the transformers and transmission line is

$$I = \frac{\text{vector power}}{\text{voltage}} = 1.006 - j0.051 \text{ per unit current.} \qquad [54]$$

The equivalent impedance of the receiving-end transformers referred to their Δ-connected secondary sides is given in Eq. 15 as

$$Z_R = 1.71 + j9.33 \text{ ohms,}$$

whence, as a fraction of the base Δ-phase impedance (Eq. 48)

$$Z_R = \frac{1.71 + j9.33}{174}$$

$$= 0.0098 + j0.0535 \text{ per unit impedance.} \qquad [55]$$

Note that, although the receiving-end transformers are Δ-connected, the per unit value of the Δ impedance given in Eq. 55 is also the per unit value of the equivalent Y impedance, the factor of $\frac{1}{3}$ converting Δ to equivalent Y impedances in Eq. 13 being accounted for by the different base values for Δ- and Y-connected impedances. That is, in terms of ohmic values, by Eq. 13,

$$Z_Y(\text{ohms}) = \tfrac{1}{3} Z_\Delta(\text{ohms}).$$

But

$$\text{Base } Z_Y = \frac{\text{base } V_Y}{\text{base } I_Y} = \frac{(\text{base } V_\Delta)/\sqrt{3}}{\sqrt{3}(\text{base } I_\Delta)}$$

$$= \frac{\text{base } V_\Delta}{3(\text{base } I_\Delta)} = \tfrac{1}{3}(\text{base } Z_\Delta). \qquad [56]$$

Hence

$$Z_Y(\text{per unit}) = \frac{Z_Y(\text{ohms})}{\text{base } Z_Y} = \frac{\tfrac{1}{3}Z_\Delta(\text{ohms})}{\tfrac{1}{3}(\text{base } Z_\Delta)}$$

$$= \frac{Z_\Delta(\text{ohms})}{\text{base } Z_\Delta} = Z_\Delta(\text{per unit}). \qquad [57]$$

▶ Thus the per unit value of the Δ-phase impedance of the receiving-end transformers, Eq. 55, is also the per unit value of the equivalent Y-phase impedance. Furthermore, the conversion factor of a^2, which changes ohmic values of impedance on the secondary side to their ohmic values referred to the primary, can be omitted when the impedances are expressed as their per unit values, since the ratio of primary to secondary base impedances also equals a^2, and therefore the per unit values of impedance are the same when referred to either side.*◀

Equation 55 therefore also gives the per unit value of the equivalent Y impedance of the receiving-end transformers referred to the primary sides. The conversions of Eqs. 25 and 26, p. 557, which must be made when ohmic values are used, are thus unnecessary when the impedances are expressed as their per unit values.

The equivalent impedances of the sending-end transformers can be computed from the short-circuit data, p. 555, which can be converted to per unit values, as follows:

$$V = \frac{\text{s-c voltage}}{\text{base voltage } \Delta} = \frac{2,640}{33,000} = 0.0800 \text{ per unit voltage} \qquad [58]$$

$$I = \frac{\text{s-c current}}{\text{base current } \Delta} = \frac{30.3}{30.3} = 1.00 \text{ per unit current} \qquad [59]$$

$$P = \frac{\text{s-c kw}}{\text{base kva per phase}} = \frac{9.81}{1,000} = 0.0098 \text{ per unit power.} \qquad [60]$$

Whence the equivalent impedance Z_S of the sending-end transformers is

$$Z_S = \frac{V}{I} = 0.0800 \text{ per unit} \qquad [61]$$

$$R_S = \frac{P}{I^2} = 0.0098 \text{ per unit} \qquad [62]$$

$$X_S = \sqrt{Z_S^2 - R_S^2} = 0.0794 \text{ per unit} \qquad [63]$$

or

$$Z_S = 0.0098 + j0.0794 \text{ per unit impedance.} \qquad [64]$$

Although the sending-end transformers are Δ-connected on their secondary sides, the value of Z_S given in Eq. 64 also equals the per unit value of the equivalent Y impedance.

* See Art. 6, Ch. XIV.

The transmission-line impedance is given as

$$Z_{line} = 7.3 + j18.2 \text{ ohms}$$

or, as a fraction of base Y-phase impedance (Eq. 50),

$$Z_{line} = \frac{7.3 + j18.2}{363}$$

$$= 0.0201 + j0.0501 \text{ per unit impedance.} \qquad [65]$$

Therefore the system reduces to that shown in Fig. 3d, comprising a series impedance whose per unit value is

$$Z_S + Z_{line} + Z_R = 0.040 + j0.183 \text{ per unit impedance.} \qquad [66]$$

The per unit source voltage now can be determined by adding the per unit impedance drop in the system to the per unit load voltage; thus

$$IZ = (1.006 - j0.051)(0.040 + j0.183)$$

$$= 0.049 + j0.182 \text{ per unit voltage} \qquad [67]$$
$$V = 1.000 + j0 \qquad [68]$$
$$\overline{}$$
$$E = \text{sum} = 1.049 + j0.182 \qquad [69]$$
$$E = 1.065 \text{ per unit voltage.} \qquad [70]$$

Therefore the required substation bus voltage is 1.065 times its base value, or

$$1.065 \times 110,000 = 117,000 \text{ v, line-to-line.} \qquad [71]$$

This value agrees with the solution given in Eq. 39.

PROBLEMS

1. A generating station supplies power through step-up transformers over a transmission line and through step-down transformers to a substation. The nominal line voltage at the generating station is 6,600 v; at the transmission line, 110,000 v; and at the substation, 11,000 v. At the generating station are two paralleled banks of identical Δ–Y-connected transformers, each transformer having a rating of 5,000 kva. At the substation are three paralleled banks of identical Y–Δ-connected transformers, each transformer having a rating of 3,333 kva. Short-circuit data on one 5,000-kva transformer are:

$$\text{Power input} = 35.0 \text{ kw}$$
$$\text{Voltage} = 790 \text{ v}$$
$$\text{Current} = 758 \text{ amp}$$

Short-circuit data on one 3,333-kva transformer are:

$$\text{Power input} = 18.6 \text{ kw}$$
$$\text{Voltage} = 4,850 \text{ v}$$
$$\text{Current} = 52.4 \text{ amp}$$

The exciting currents of the transformers may be neglected. The transmission line may be represented by the equivalent circuit shown in Fig. 4.

Consider operation for a load of 30,000 kva at 0.92 power factor, lagging, and 11,000 v.

(a) Show the equivalent circuit of one phase of the whole system (line-to-neutral) referred to the 110,000-v line.
(b) What is the line voltage at the generating station?
(c) What is the line current at the generating station?
(d) What is the power factor at the generating station?

2. Eight Δ–Y-connected banks, each bank consisting of three 11,500 kva single-phase transformers, are connected in parallel between the 13.8-kv and 230-kv buses in a hydroelectric station. These transformers have 0.50% resistance, 10.0% reactance, 0.36% core loss, and 5.0% exciting current. Determine the full-load efficiency and regulation of these transformers for

(a) 0.80 power factor lagging load,
(b) 0.80 power factor leading load,
(c) Unity power factor load.

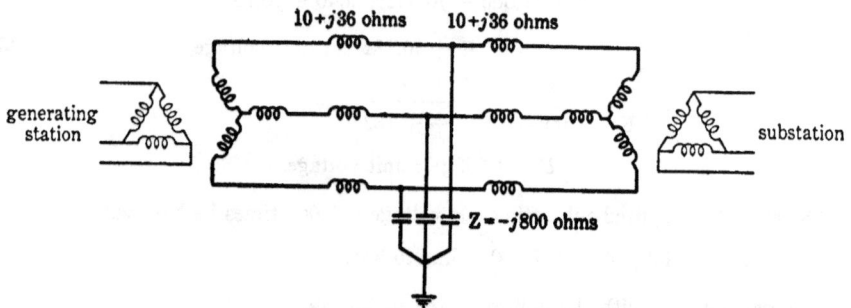

FIG. 4.   Power transmission circuit, Prob. 1.

3. A balanced three-phase load which operates at 2,400 v is to be supplied through transformers from a balanced three-phase system with a nominal line-to-line voltage of 13,800 v. The transformers are to be connected Y–Δ, the Y-connected side being the high-voltage side. The load has a daily demand schedule which is approximated by the following data:

Hours	Load	Power factor
5	1,200 kva	0.88 lagging
3	900 kva	0.91 lagging
2	300 kva	0.65 lagging
14	0	. . . .

(a) Determine the minimum kva rating of standard single-phase transformers which will be satisfactory for this application.
(b) Find the energy efficiency for the bank if the transformers have a per unit equivalent impedance of $0.0082 + j0.040$ and a core loss of 0.012. Assume that the load voltage is constant at 2,400 v.

4. A factory has three-phase power supplied from a substation some distance away as shown in Fig. 5. The three Y–Δ-connected single-phase transformers at the load are each rated 333 kva and have equivalent impedances of $0.145 + j0.780$ ohm, referred to the high-voltage side. The distribution line has an impedance per wire of $0.140 + j0.510$ ohm.

On the assumption that the impedances of the substation transformers are purely reactive, what would be the limiting value for the equivalent impedance of these transformers if the voltage regulation of the line and transformers is to be held to 10% for full load at 0.9 power factor?

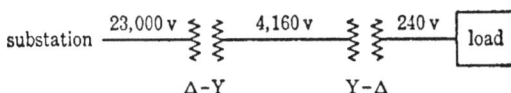

substation ——— 23,000 v ⧧⧧ 4,160 v ⧧⧧ 240 v [load]

Δ-Y   Y-Δ

Fig. 5.   Distribution system, Probs. 4 and 5.

5. In the distribution system of Prob. 4 let the substation transformers be rated 500 kva each and have an equivalent impedance of $0.081 + j0.710$ ohm referred to their low-voltage sides. Draw the equivalent single-phase circuit (line-to-neutral) and convert all values to per unit on a 1,500 kva three-phase base.

CHAPTER XXIII

Harmonic Phenomena in Three-Phase Circuits

Because of their small size, harmonics in the exciting currents of single-phase transformers usually are neglected. In many problems involving three-phase transformer banks they may likewise be neglected as in Ch. XXI, where the exciting currents are represented by equivalent sine waves. The peculiarities of the harmonic phenomena in three-phase systems may occasionally, however, have important effects on the system characteristics, particularly the behavior of Y–Y banks of single-phase transformers. Furthermore, in spite of their relatively small size, the exciting-current harmonics of a three-phase transformer bank may under certain conditions induce in neighboring communication circuits voltages that seriously interfere with the proper operation of the circuits. Both power-system and communication engineers should obviously be ready to avert situations of this sort. Discussion of these harmonic phenomena follows.

1. Δ–CONNECTED PRIMARIES

For the present, consider three transformers with their primaries connected in Δ, their secondaries being open-circuited and not connected to each other. If the transformers are identical and the line voltages are

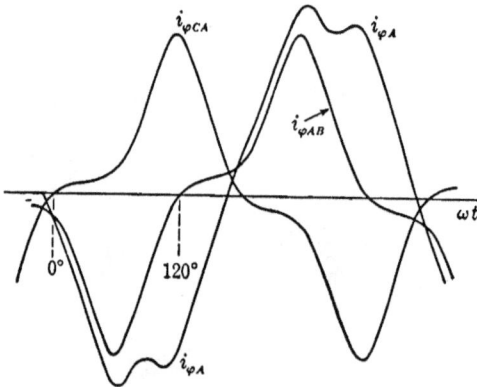

FIG. 1. Waveforms of the exciting currents in two phases of the Δ and in one of the lines supplying a group of Δ-connected primaries.

balanced, the waveforms of the exciting currents in the transformers are alike but differ in phase by 120 degrees. If the line voltages vary sinusoidally, the waveforms of the exciting currents have the general shape shown in Fig. 9, p. 170. Thus the exciting currents $i_{\varphi CA}$ and $i_{\varphi AB}$ in two of the transformers are of the form shown by the peaked waves in Fig. 1.

The exciting current supplied by line A is

$$i_{\varphi A} = i_{\varphi AB} - i_{\varphi CA} \quad [1]$$

and is shown by the double-topped wave. Note that the waveform of the exciting currents in the lines differs greatly from the waveform of the exciting currents in the Δ-connected transformers. The two reasons for the difference are discussed in the following two sections of this article.

1a. *Third Harmonics.* — First consider the third-harmonic components of the Δ currents. Figure 2a shows three currents i_{CA}, i_{AB}, i_{BC}, having identical waveforms and differing in phase by one-third of a cycle, or 120 degrees. These currents contain third-harmonic components which

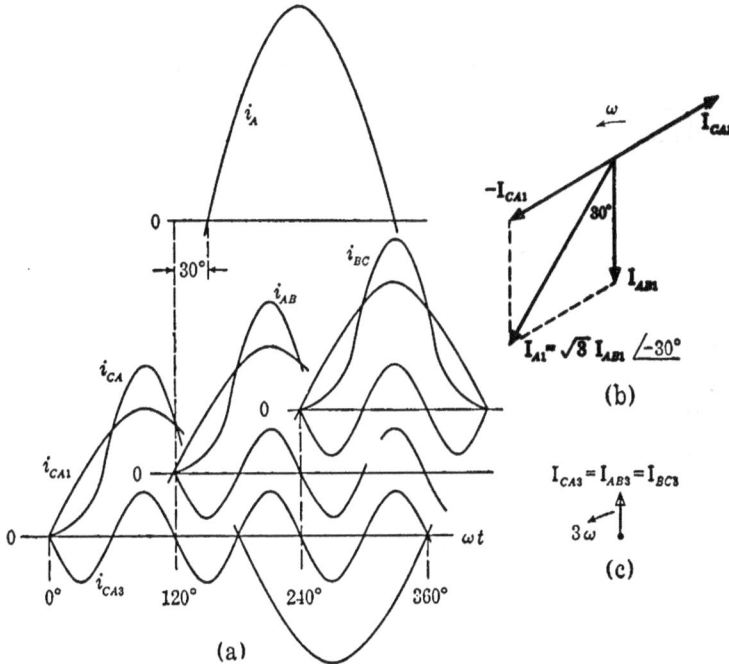

FIG. 2.   Balanced Δ currents with third-harmonic components, showing (a) waveforms, (b) fundamental vector diagram, and (c) third-harmonic vector diagram.

produce a peaked waveform, similar to the peaked waveform of the exciting currents. Since i_{AB} lags i_{CA} by one-third of a cycle, the third-harmonic component of i_{AB} lags the third-harmonic component of i_{CA} by three-thirds of one, or a whole, third-harmonic cycle, as shown in Fig. 2a. The third-harmonic components of the currents are thus in time phase,[*] as is also shown in the vector diagram of Fig. 2c. If i_{AB} and i_{CA} are the currents in phases AB and CA of a Δ, the line current i_A is their

---

[*] In the terminology of symmetrical components, the third-harmonic components are zero-sequence currents. The theory of symmetrical components is discussed in this series in the volume on electric circuits (1940), Ch. XI, pp. 578–605, and a brief discussion of its applications to transformers is given below in Art. 4, Ch. XXIV.

difference, as in Eq. 1. The difference between the fundamental components is shown vectorially in Fig. 2b. However, the difference between the instantaneously equal third-harmonic components is zero.

▶ Therefore there can be no third-harmonic currents in the lines when the transformers are identical and the voltages are balanced.◀

Nevertheless, if the line voltages vary sinusoidally, there must be third-harmonic components in the exciting currents of the Δ-connected transformers, because of the nonlinear magnetic characteristics of the cores.* These third-harmonic components, being instantaneously equal and in the same direction around the Δ, simply flow within the primary Δ, but do not appear on the lines. By the same reasoning it can be shown that all multiples of the third harmonic — the sixth, ninth, etc. — behave in a three-phase circuit in the same way as do the third harmonics. For example, the phase difference between the ninth-harmonic components of the current i_{CA} and i_{AB} is nine-thirds of one, or is three whole ninth-harmonic cycles, and hence these ninth-harmonic components are in phase and behave as do the third harmonics.

▶ Thus the third-, sixth-, ninth-, etc., harmonic currents may flow in a Δ but are excluded from the lines supplying the Δ when the circuit is perfectly balanced.◀

Although three transformers of identical design should have nearly identical characteristics, it is impossible to attain exact identity in materials and assemblies, and consequently there may be small third-harmonic components and multiples of the third harmonic in the line currents, caused by inequalities in the excitation characteristics of the three transformers. Usually only odd harmonics are present.

1b. *Fifth Harmonics.* — The waveforms of the Δ and of the line currents are different not only because third-harmonic components are present in the former but absent (for identical transformers) from the latter, but also for a second reason that concerns the effects of the fifth-harmonic components of the exciting currents. In the example of Art. 10, Ch. VI, the effective value of the fifth-harmonic component is about 10 per cent of the effective value of the exciting current and the fifth harmonic adds to the peak value. Figure 3a shows two identical peaked currents, i_{CA} and i_{AB}, containing fifth-harmonic components. Since i_{AB} lags i_{CA} by one-third of a cycle, the fifth-harmonic component of i_{AB} lags the fifth-harmonic component of i_{CA} by five-thirds of a fifth-harmonic cycle, as shown in Fig. 3a. This is equivalent to a lag of two-thirds, or a *lead* of one-third of a fifth-harmonic cycle.

---

* It is shown in Art. 10, Ch. VI, that the exciting current of a normal power-system transformer must have a third-harmonic component of about 40 per cent of the fundamental when the flux varies sinusoidally.

▶ Thus the phase order of the fifth harmonics is the *reverse* of the phase order of the fundamentals.* That is, when the fundamental component i_{AB1} *lags* the fundamental component i_{CA1} by one-third of its cycle, the fifth-harmonic component i_{AB5} *leads* the fifth-harmonic component i_{CA5} by one-third of a fifth-harmonic cycle. These phase relations are shown vectorially in Figs. 3b and 3c. ◀

If the currents i_{AB} and i_{CA} are the fundamental and fifth-harmonic components of the exciting currents in two of the Δ-connected trans-

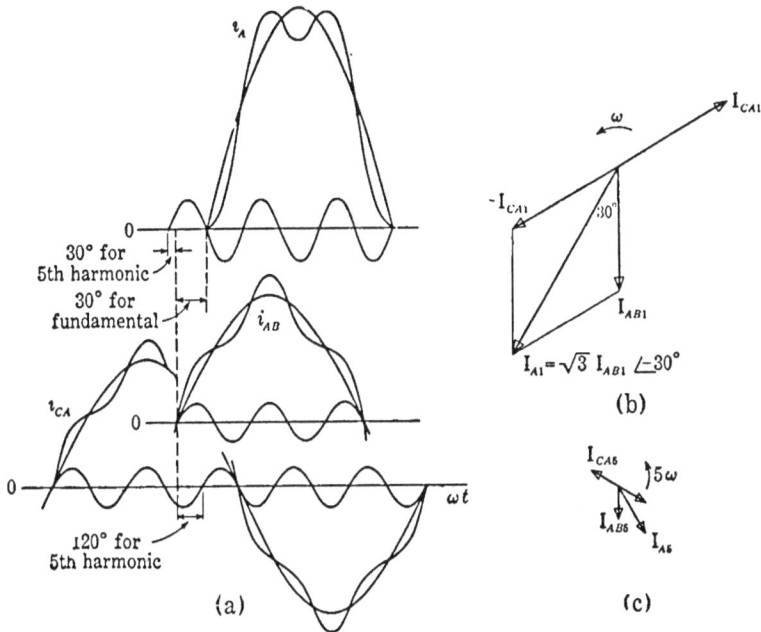

FIG. 3. Line currents and Δ currents with fifth-harmonic components, showing (a) wave-forms, (b) fundamental vector diagram, and (c) fifth-harmonic vector diagram.

formers, the line current i_A is their difference, as in Eq. 1. As shown in the vector diagrams, the fundamental component of the line current i_A equals $\sqrt{3}$ times the fundamental component of the Δ currents and *lags* the fundamental component of i_{AB} by 30 degrees or one-twelfth of a cycle, whereas the fifth-harmonic component of i_A equals $\sqrt{3}$ times the fifth-harmonic component of the Δ currents but *leads* the fifth-harmonic component of i_{AB} by 30 fifth-harmonic degrees or one-twelfth of a fifth-harmonic cycle. Thus the phase relation between the fundamental and

* In the terminology of symmetrical components, the fifth-harmonic components are negative-sequence currents.

fifth-harmonic components of the line currents differs from the phase relation between these components of the Δ currents and, although the line and Δ currents contain fundamental and fifth-harmonic components in the same proportions, their waveforms differ because of this phase shift. The Δ currents are peaked waves but the line currents are double-topped waves, as in Figs. 1 and 3a.

1c. *Summary of Phase Relations among the Harmonics.* — It can readily be shown that the phase order of the fundamental, fourth, seventh, etc., harmonics in a three-phase circuit is the same; that the phase order of the second, fifth, eighth, eleventh, etc., harmonics is the reverse of the phase order of the fundamentals; and that the third, sixth, ninth, etc., harmonics are in phase. If the phase order of the fundamentals is ABC, these facts are summarized in the table below, which applies not only to harmonics in the current waves but also to harmonics in the voltages. Usually only the odd harmonics are present.

PHASE RELATIONS AMONG THE HARMONICS IN THREE–PHASE CIRCUITS

Harmonic	*Phase Order*
1, 4, 7, 10, 13, etc.	ABC (positive sequence)
2, 5, 8, 11, 14, etc.	CBA (negative sequence)
3, 6, 9, 12, 15, etc.	in phase (zero sequence)

1d. *Effective Values of Δ and Line Currents.* — Because of the absence of third-harmonic components from the line currents, the effective value of the no-load line currents is not $\sqrt{3}$ times the effective value of the exciting currents in the Δ-connected transformers, but is less than this value. Thus if $I_{\varphi 1}$, $I_{\varphi 3}$, etc., are the effective values of the harmonic components in the three identical transformers, the effective value $I_{\varphi\Delta}$ of the exciting currents in the transformers is

$$I_{\varphi\Delta} = \sqrt{I_{\varphi 1}^2 + I_{\varphi 3}^2 + I_{\varphi 5}^2 + \cdots}. \qquad [2]$$

Since the effective values of the harmonic components in the line currents are $\sqrt{3}$ times the effective values of these harmonic components in the Δ currents, but the third harmonic and its multiples are absent from the line currents, the effective value $I_{\varphi\ line}$ of the line currents is

$$I_{\varphi\ line} = \sqrt{3}\ \sqrt{I_{\varphi 1}^2 + 0 + I_{\varphi 5}^2 + \cdots}. \qquad [3]$$

If the harmonics that are multiples of the third harmonic are neglected,

$$I_{\varphi\ line} = \sqrt{3}\ \sqrt{I_{\varphi\Delta}^2 - I_{\varphi 3}^2}. \qquad [4]$$

In the example of Art. 10, Ch. VI, the effective value of the third harmonic is 41.5 per cent of the exciting current of the transformer. Thus if three such transformers were connected in Δ, the line current would be

$$I_{\varphi\ line} = \sqrt{3}\ \sqrt{I_{\varphi\Delta}^2 - (0.415 I_{\varphi\Delta})^2}$$

$$= I_{\varphi\Delta} \sqrt{3} \sqrt{1 - 0.415^2}$$

$$= 0.91 \sqrt{3}\ I_{\varphi\Delta}. \qquad [5]$$

That is, since the third-harmonic currents are present in the Δ but absent from the lines, the ratio of the line currents to the Δ-phase currents is less than the value $\sqrt{3}$ applying to sine-wave currents. Although this effect of the third-harmonic currents is seldom of much practical importance and is often neglected, it explains certain phenomena which superficially might appear puzzling. For example, when three identical transformers are connected in Δ to a balanced three-phase line and the line currents and voltages are measured, the three-phase exciting volt-amperes $\sqrt{3}\ V_{line} I_{\varphi\ line}$ is less than 3 times the exciting volt-amperes of a single transformer as measured in a single-phase test at the same voltage. The total three-phase power input must, however, be the sum of the core losses in the three transformers. Thus the three-phase no-load power factor of a Δ-connected bank is a few per cent higher than the single-phase no-load power factor of the units.

1e. *Δ–Y and Δ–Δ Connections.* — The third-harmonic components of the exciting currents in the Δ-connected primaries cause third-harmonic leakage-impedance voltage drops in the primary leakage impedance of each transformer. Since the sine-wave applied voltage equals the voltage drop due to the primary leakage impedance plus the counter electromotive force generated by the mutual flux, this electromotive force must contain a third-harmonic component equal and opposite to the third-harmonic primary leakage-impedance voltage drop, and the mutual flux accordingly must adjust itself to generate this small third-harmonic voltage, which usually is only about 0.1 per cent of rated voltage. A small third-harmonic electromotive force is therefore generated in the secondaries also. Since the waveforms of the electromotive forces generated in the three identical transformers are alike but displaced in phase by one-third of a cycle, the third-harmonic components of the secondary electromotive forces in the three transformers are alike but displaced in phase by three-thirds of a third-harmonic cycle, and thus are in time phase.

If the secondaries are connected in Y with isolated neutral, the third-harmonic exciting currents required to permit the very nearly sinusoidal variations of the mutual fluxes are confined to the primary Δ, and the primary leakage-impedance voltage drops appear as small components of the secondary voltages to neutral, but, as shown in Art. 2d, these third harmonics are not present in the line-to-line voltages.

However, if the secondaries are connected in Δ, the third-harmonic components of the three secondary voltages are in phase in the same direction around the Δ, and therefore they produce a small third-harmonic current in the secondary Δ. Third-harmonic exciting currents are thus present in both the primary and secondary Δ's, and the third-harmonic magnetomotive forces necessary to permit the very nearly sinusoidal variations of the mutual fluxes are produced by the combined effects of these primary and secondary exciting currents. It can be shown that the third-harmonic exciting currents in the primary and secondary Δ's are inversely proportional to the leakage impedances of the primary and secondary at third-harmonic frequency, the currents and impedances being referred to the same side.

2. Y–CONNECTED PRIMARIES

Since, as shown in Art. 2, Ch. XXI, the characteristics of the Y connection are greatly influenced by isolation of the primary neutral, the following discussion is subdivided according to whether the primary neutral is isolated or connected to the neutral of the source, and whether the secondaries are Y- or Δ-connected.

In certain circumstances, the currents in a three-phase power line may induce objectionable voltages in parallel communication circuits. A brief discussion of this inductive interference is included in this article, because the offending power-line currents often are the third-harmonic components of the exciting currents resulting from Y-connected transformer windings whose neutral point is connected to some other system neutral.

2a. Y–Y *Connection: Primary Neutral Connected to the Source Neutral.* — First, consider the simple example of a Y–Y bank of three identical single-phase transformers whose secondaries are open-circuited. The primary neutral is connected to the neutral of a three-phase Y-connected source of balanced sine-wave voltages, as shown in Fig. 4a. Figure 4b shows oscillograms of the voltage v_{AN} applied to one of the transformers, of the exciting current $i_{\varphi A}$ in this transformer, and of the current i_N in the neutral wire. When the applied voltage is a sine wave, the exciting current contains only odd harmonic components, of which the third harmonic is the most prominent. If the three transformers are identical and the voltages are balanced, the exciting currents $i_{\varphi A}$, $i_{\varphi B}$, and $i_{\varphi C}$ are identical except for a relative phase displacement of 120 degrees.

The current in the neutral wire N is the sum of the currents $i_{\varphi A}$, $i_{\varphi B}$, and $i_{\varphi C}$. When the transformers are identical and the voltages are balanced, the fundamental components of the three exciting currents are three sine waves of the same amplitude displaced in phase by 120 degrees;

their sum is therefore zero and hence there is no fundamental current in the neutral wire. The third-harmonic components of the three exciting currents, however, are in time phase (as shown in Art. 1a), and hence the third-harmonic component of the neutral current equals 3 times the third-harmonic component of the exciting currents in the transformers.

From the table on p. 568, it can be seen that only the third-harmonic components and multiples of the third harmonic are in phase, all other harmonics of the three exciting currents being displaced in phase by 120 degrees.

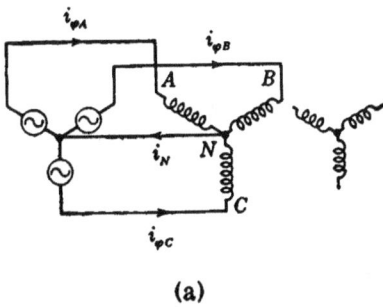

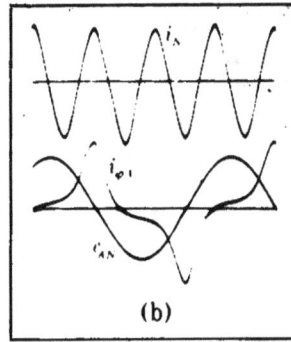

Fig. 4. Y–Y bank of transformers; (a) connections, and (b) oscillograms of the primary voltage to neutral v_{AN}, the exciting current $i_{\varphi A}$, and the current i_N in the primary neutral wire.

▶ Therefore, when the transformers are identical and the voltages are balanced, the neutral current comprises only the odd harmonics that are multiples of the third harmonic. The ninth and higher harmonics usually are small, and therefore the neutral current is approximately a triple-frequency sine wave whose effective value approximately equals 3 times the effective value of the third-harmonic component of the exciting currents in the transformers. ◀

This fact is shown in the oscillogram of Fig. 4b. The small fundamental-frequency component of the neutral current i_N is due to small inequalities in the excitation characteristics of the three transformers.

The third-harmonic components of the exciting currents produce small third-harmonic voltage drops in the primary leakage impedances of the transformers, and therefore, when the primary applied voltages to neutral vary sinusoidally, the electromotive forces induced by the mutual flux contain small third-harmonic components. These appear as small components of the secondary voltages to neutral but are not present in the secondary line-to-line voltages, as shown in part (d) of this article.

In order to maintain a proper sense of proportion, comment on the order of magnitudes of the exciting currents is justified. The exciting current of a typical power-system transformer is about 5 per cent of rated current, the third-harmonic component being about 40 per cent of the exciting current, or about 0.4 × 5 or 2 per cent of rated current. The neutral current is 3 times the third-harmonic current, or about 6 per cent of rated current.   The neutral current thus is of the same order of magnitude as the exciting currents in the three line wires, but all these currents are only a few per cent of rated current.

Although the characteristics of a Y–Y bank usually are satisfactory when the primary neutral is connected to the neutral of the source, the third-harmonic line currents resulting from this connection may cause unpleasant inductive interference in parallel communication circuits, as shown in the following discussion.

2b. *Inductive Interference Caused by Third-Harmonic Exciting Currents.* — Although the effective value of the third-harmonic exciting currents usually is only about 2 per cent of the rated full-load value of the line currents, the inductive effect of the exciting currents on a communication circuit parallel to the power circuit may be greater than the effect of balanced three-phase full-load currents in the power lines. For example, consider Fig. 5 which shows a three-phase power line *ABC* with flat spacing, and a parallel open-wire telephone circuit *ab*. For simplicity, the two circuits are assumed to be in the same plane. Let the positive directions of the currents in the power line be away from the observer, as indicated by the crosses representing the tails of arrows pointing in these positive directions. The power-line currents produce a magnetic field linking the communication circuit, as shown by the space vector $\mathfrak{B}$ representing the resultant flux density, which is the sum of the components produced by each current acting separately. According to the Biot-Savart law,[1] the component flux densities are proportional to the instantaneous values i_A, i_B, i_C of the currents and inversely proportional to the distances x_A, x_B, x_C from each power-line conductor; thus

$$\mathfrak{B} = \mu_0 \left(\frac{2i_A}{x_A} + \frac{2i_B}{x_B} + \frac{2i_C}{x_C} \right),$$ [6]

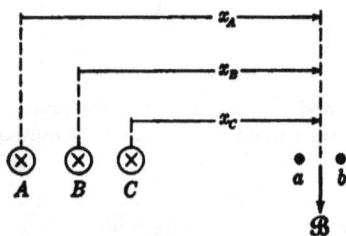

FIG. 5.   Parallel power and communication circuits.

[1] The Biot-Savart law is discussed in this series in the volume on electric circuits (1940), pp. 45, 52.

where μ_0 is the permeability of free space. Any consistent unrationalized system of units may be used in this equation. The flux linkage with the communication circuit is the integral of the flux density $\mathcal{B}$ over the area enclosed between the two communication wires.

If the currents in the power line are balanced, sine-wave currents differing in time phase by 120 degrees (as are the fundamental, fifth-, seventh-, etc., harmonic components of balanced three-phase currents), their instantaneous sum is zero, and, if the distances x_A, x_B, x_C from the three phase wires to the communication circuit were the same, the sum of the component flux densities produced by the currents would also be zero. That is, the component flux densities produced by the phase currents tend to cancel one another, the resultant flux density depending on the differences among the distances x_A, x_B, and x_C. The resultant flux density produced by balanced fundamental, fifth-, seventh-, etc., harmonic components of the power-line currents is therefore relatively small when the distances x_A, x_B, x_C are large compared with the spacings between the power-line conductors.

However, if the power line supplies excitation to a bank of Y–Y-connected transformers whose primary neutral is connected to the neutral of the source, the power-line currents contain third-harmonic components. For simplicity, assume that the neutral connection is through the ground, and that the inductive effect of the third-harmonic ground current is negligible, owing to the height of the circuits above ground. The third-harmonic components of the three line currents are in time phase and therefore the instantaneous values of the component third-harmonic flux densities are directly additive.* Hence relatively small third-harmonic components of the three-phase currents may produce a greater flux linkage with the communication circuit than is produced by much larger balanced, fundamental currents. Furthermore, the time rate of change of the third-harmonic flux linkage is 3 times the time rate of change of the fundamental flux linkage, and therefore the triple-frequency voltage induced in the communication circuit by a given third-harmonic flux is 3 times as great as the fundamental-frequency voltage that would be induced by an equal fundamental flux. Also, the response of a telephone receiver and of the ear to a signal whose frequency is 3 times the fundamental of ordinary power-system frequencies is greater than the response to a fundamental-frequency signal of the same strength. All these factors tend to make the inductive interference due to relatively small

---

* This situation is also true for the component flux densities produced under certain unbalanced conditions, such as that illustrated to the right of the fault in Fig. 3, Ch. XXIV, where the line currents are equal and in time phase. In the analysis of three-phase circuits by the method of symmetrical components, such currents are called the zero-sequence components. See Art. 4, Ch. XXIV.

third-harmonic currents in the power line more objectionable than the interference due to much larger balanced fundamental currents.

When it is necessary to expose an open-wire communication circuit to the inductive effects of a power line, the interference may be reduced through properly transposing the communication line. It is also advisable to use connections of the power-system transformers that do not permit third-harmonic exciting currents to flow in the power circuit From these considerations, Y-connected primaries with their neutral point connected to the neutral of the source may be objectionable.

2c. Y–Δ *Connection.* — Consider a bank of three identical transformers whose primaries are connected in Y and whose secondaries are connected in series preparatory to being connected in Δ, as in Fig. 6; that is, one corner of the Δ is open. The primary neutral is connected to the neutral of a three-phase Y-connected source of balanced sine-wave voltages. On the primary side, the bank behaves in the same manner as the Y–Y bank discussed in part (a) of this article; that is, the primary voltages vary sinusoidally, each transformer receives its exciting current from the primary lines, and the third-harmonic components of the exciting currents return to the source through the neutral wire, as shown in the oscillograms of Fig. 4b.

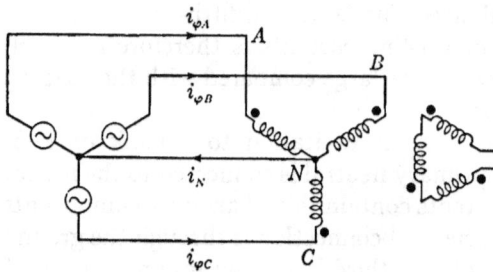

Fig. 6. Y–Δ bank of transformers with one corner of the Δ open.

The third-harmonic components of the exciting currents produce small third-harmonic voltage drops in the primary leakage impedances of the transformers, and therefore the electromotive forces induced by the mutual flux contain small third-harmonic components, which appear as small components of the secondary voltages of the transformers. Since the third-harmonic electromotive forces in the three transformers are in time phase, the voltage across the open corner of the secondary Δ contains a third-harmonic component equaling 3 times the third-harmonic electromotive force induced in each secondary. Since the fundamental, fifth-, seventh-, etc., harmonic voltages in the three secondaries are equal in magnitude and displaced in phase by 120 degrees, their sums are zero, and therefore there are no voltages of these frequencies across the open corner of the secondary Δ. Hence, if harmonics of higher order than the seventh are neglected, the voltage across the open corner of the secondary Δ equals 3 times the third-harmonic electromotive force generated in each secondary, when the transformers are identical and the applied voltages are balanced. Although this third-harmonic voltage usually is very small compared with the secondary voltages across the terminals of each transformer, the third-harmonic secondary electromotive forces have an important effect on the excitation phenomena when the Δ is closed. This is discussed below.

If the secondary Δ is now closed, the third-harmonic secondary electromotive forces produce a third-harmonic secondary current which circulates in the secondary Δ. The third-harmonic magnetomotive forces necessary to permit the very nearly sinusoidal variations of the fluxes in the three transformers are now produced by the combined action of third-

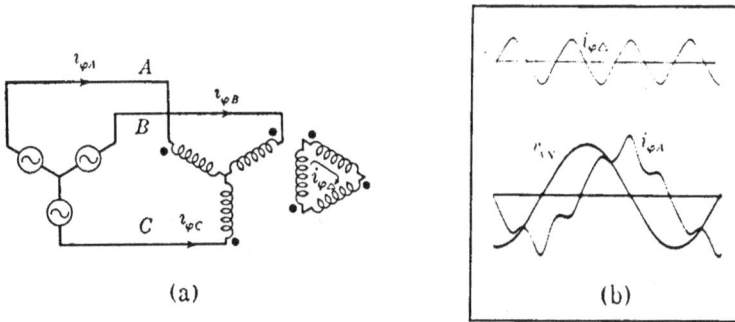

Fig. 7. Y–Δ bank with isolated neutral; (a) connections, and (b) oscillograms of the voltage to neutral v_{AN}, the exciting current $i_{\varphi A}$, and the exciting current $i_{\varphi\Delta}$ in the secondary Δ.

harmonic exciting currents in both the primary and secondary windings, as in the Δ–Δ connection discussed in Art. 1e, and therefore the third-harmonic components of the primary currents are reduced when the secondary Δ is closed.

If the secondary Δ is closed and the neutral wire is disconnected, as in Fig. 7a, the waveforms of the voltage to neutral v_{AN}, of the exciting current $i_{\varphi A}$ in one of the primaries, and of the exciting current $i_{\varphi\Delta}$ in the secondary Δ are those shown in the oscillograms of Fig. 7b. The third-harmonic components of the primary currents can no longer exist, since their return path in the neutral wire is open-circuited. Hence the waveform of the exciting current $i_{\varphi A}$ in Fig. 7b differs from the waveform of $i_{\varphi A}$ in Fig. 4b, the principal difference being the absence of third-harmonic components when the neutral of the Y is isolated. The closed secondary Δ provides a path for third-harmonic currents, however, and the third-harmonic components of the magnetomotive forces necessary to permit the very nearly sinusoidal variations of the mutual fluxes are provided by the third-harmonic exciting current $i_{\varphi\Delta}$ in the secondary Δ. Since a third-harmonic electromotive force is necessary to produce this third-harmonic Δ current, the mutual flux adjusts itself to contain the third-harmonic component required to generate this small third-harmonic secondary electromotive force. Therefore the mutual flux induces a third-harmonic component in the voltage to neutral on the primary side, but usually this third-harmonic voltage is very small and the waveform of the primary voltage to neutral remains essentially sinusoidal, as shown in the oscillogram of v_{AN}, Fig. 7b.

From this discussion and the discussion of Δ–Y and Δ–Δ connections given in Art. 1e, it is possible to draw an important conclusion with regard to the effects of Δ-connected windings on excitation phenomena.

▶ The third-harmonic exciting currents necessary for sinusoidally varying flux flow within the Δ-connected windings of a Δ–Δ, Δ–Y, or Y–Δ bank, but (for identical transformers) are not present in the three-phase lines connected to the transformers when the neutrals of Y-connected windings are isolated from other system neutrals. ◀

On the other hand, the third-harmonic exciting currents are present in the primary lines supplying a Y–Y bank whose primary neutral is connected to the source neutral, and therefore this arrangement may cause inductive-interference troubles. If the third-harmonic exciting currents are eliminated from the lines by isolation of the neutrals of a Y–Y bank, the voltages to neutral may become badly distorted, as is shown below.

2d. Y–Y *Connection with Isolated Neutrals.* — The following discussion applies to a Y–Y-connected bank of three single-phase transformers. It is shown in Art. 1, Ch. XXVI, that a three-phase shell-type transformer has essentially the same characteristics as a bank of three single-phase transformers, and therefore the discussion also applies to the Y–Y connection of three-phase shell-type units. However, this discussion does *not* apply to a Y–Y-connected, three-phase, core-type transformer; its excitation characteristics, discussed in Art. 2a, Ch. XXVI, differ markedly from the characteristics of a Y–Y-connected bank of three single-phase transformers.

The Y–Y connection of single-phase transformers should be used only after a careful study of the conditions in which the bank operates, since the connection has characteristics that under certain circumstances may be undesirable or even dangerous. These characteristics are discussed below.

Consider a Y–Y bank of single-phase transformers with isolated neutrals excited from a balanced three-phase source whose voltages are sinusoidal in waveform. Assume the bank is delivering no load. The connections are shown in Fig. 8a.

Since the primary neutral is isolated from the neutral of the source, the instantaneous sum of the exciting currents supplied to the bank must be zero; that is, the exciting current supplied to any transformer must find its return path to the source through the primaries of the other transformers. As shown in Art. 2b, Ch. XXI, the result of this interdependence of the exciting currents on one another is that the voltages across the transformers are determined by their excitation characteristics. Since three transformers seldom have identical excitation characteristics, even when they are of the same design, the voltages to neutral usually are more or less unbalanced, even though the line-to-line voltages are balanced. This is an undesirable situation.

The isolated neutral also has an important effect on the harmonics in the exciting current. For this discussion, assume the transformers to

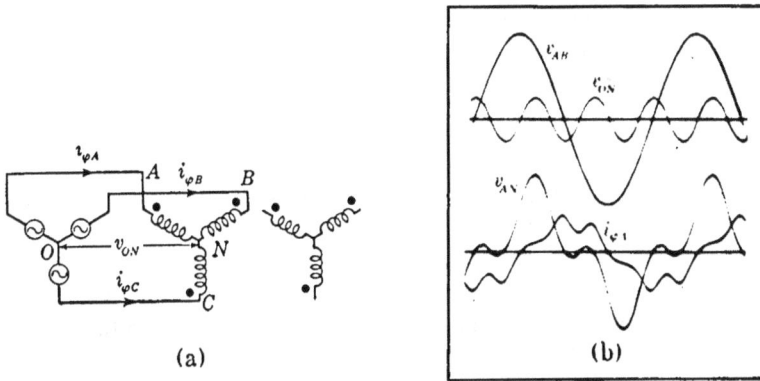

FIG. 8. Y–Y bank of single-phase transformers; (a) connections, and (b) oscillograms of the line-to-line voltage v_{AB}, the voltage to neutral v_{AN}, the exciting current $i_{\varphi A}$, and the voltage v_{ON} between the neutral O of the source and the primary neutral N.

have identical excitation characteristics. Then the individual exciting currents supplied to the transformers are equal in magnitude and waveform but differ in phase by one-third of a cycle. Therefore the third-harmonic components of the three exciting currents, if they existed, would be in time phase and their sum would *not* be zero.

▶ However, since the instantaneous sum of the exciting currents *must* be zero, because of the isolated neutral, there can be no third-harmonic or multiple of the third-harmonic component in the exciting currents, and therefore the waveform of the exciting current in each transformer differs from the waveform required to produce a sinusoidal variation of its core flux.◀

The oscillogram of $i_{\varphi A}$, Fig. 8b, is a typical example of the waveform of the exciting currents of a Y–Y bank with isolated neutral. It is interesting to compare this waveform with the waveform of the exciting current when the flux varies sinusoidally, the latter being shown in Fig. 9, p. 170. When the flux variation is sinusoidal, the waveform of the exciting current is sharply peaked, the peak in the current corresponding to the bending over of the hysteresis loop at its tips. This sharp peak in the exciting-current wave is the cause of the relatively large third-harmonic component of the exciting current for sine-wave flux. When this third-harmonic component is suppressed by the isolated neutral of the Y connection, the peak value of the exciting current is reduced and the waveform often becomes double topped, as in Fig. 8b. The double top is principally due to the presence of fifth-harmonic components. It is of interest that the double top on the exciting-current wave indicates a re-entrance in the hysteresis loop near each of its tips. Note that only the third-harmonic components and harmonic components whose frequencies are multiples of the third-harmonic frequency are suppressed by the isolated primary neutral. All other harmonic components of the three exciting currents can flow, since they differ in phase by 120 degrees and their sums are zero.

▶ Since a third-harmonic component in the exciting current is necessary if the flux is to vary sinusoidally, the suppression of these third-harmonic exciting currents prevents the flux from varying sinusoidally and introduces in the core fluxes third-harmonic components which induce third-harmonic components in the primary and secondary voltages in each transformer. At ordinary core flux densities, these third-harmonic voltages usually are 30 to 70 per cent of the fundamental component of the voltage to neutral Thus even when the voltages applied to the line terminals vary sinusoidally, the waveform of the voltages to neutral is nonsinusoidal when the neutral of the bank is isolated from the neutral of the source.◀

Since the relations between the instantaneous values of the line-to-line voltages and line-to-neutral voltages are

$$v_{AB} = v_{AN} - v_{BN} \tag{7}$$

$$v_{BC} = v_{BN} - v_{CN} \tag{8}$$

$$v_{CA} = v_{CN} - v_{AN}, \tag{9}$$

and, since the third-harmonic components of the three voltages to neutral, and the multiples of the third harmonic, are in phase and equal in magnitude, the differences between the third-harmonic components of any two voltages to neutral, as on the right-hand sides of Eqs. 7, 8, and 9, are zero under balanced conditions.

▶ Hence there may be third-harmonic components and multiples of the third harmonic in the voltages to neutral without the presence of these components in the line-to-line voltages.◀

No other harmonic components can appear, however, in the line-to-neutral voltages unless they are also present in the line-to-line voltages. Hence if the line-to-line voltages are balanced and vary sinusoidally, the line-to-neutral voltages comprise fundamental components whose effective value equals $1/\sqrt{3}$ or 0.577 of the effective value of the line-to-line voltages, and third-harmonic components, and its multiples, whose effective values are determined by the nonlinear magnetic characteristics of the core.

If the ninth and higher harmonics are neglected and sine-wave line-to-line voltages are assumed, the effective value V_Y of the voltages to neutral is

$$V_Y = \sqrt{V_{Y1}^2 + V_{Y3}^2}, \qquad [10]$$

where V_{Y1} is the effective value of the fundamental component and V_{Y3} is the effective value of the third-harmonic component. The third-harmonic component is usually 30 to 70 per cent of the fundamental. If, as a representative value,

$$V_{Y3} = 0.50 \, V_{Y1}, \qquad [11]$$

then

$$V_Y = V_{Y1} \sqrt{1 + 0.25} = 1.12 V_{Y1} \qquad [12]$$

$$= 1.12 \frac{V_{line}}{\sqrt{3}}, \qquad [13]$$

where V_{line} is the effective value of the balanced sinusoidally varying line-to-line voltages, and the transformers are assumed to have identical excitation characteristics. Thus the effective value of the voltages to neutral is no longer the value $V_{line}/\sqrt{3}$ that would be obtained if the voltages to neutral varied sinusoidally, but is greater than this value.

Furthermore, the maximum values of the fundamental and third-harmonic components of the voltages to neutral occur at approximately the same time in the cycle. Therefore the waveform of the voltages to neutral is sharply peaked, as shown by the oscillogram of v_{AN} in Fig. 8b. The peak value of the voltage across each transformer is 30 to 70 per cent greater than the peak value of the fundamental component and thus may be nearly as great as the peak value of the line-to-line voltages. This is an undesirable condition, since the voltage stress on the insulation is increased by the third-harmonic voltages.

When the voltages of the Y-connected source are balanced and vary sinusoidally and the transformers have identical excitation characteristics, the third-harmonic components of the voltages across the transformers appear as a triple-frequency voltage between the neutral of the source and the primary neutral of the transformers, as shown in the oscillogram of v_{ON} in Fig. 8b. If the neutral of the source is grounded, a triple-frequency voltage, whose effective value usually is 30 to 70 per cent of the line-to-neutral voltage, exists between the neutral point of the primary windings and ground. If neither the neutral of the source nor the primary neutral is grounded, the voltages between the lines and ground, and between the primary neutral and ground, are determined by the capacitances of the lines to ground and of the transformer windings to ground.

▶ In certain circumstances the third-harmonic vo_tages may be greatly increased by resonance phenomena. This dangerous condition may arise when the bank is connected to a long transmission line or cable and when the neutral point of the transformer windings is grounded, as shown in Fig. 9.◀

With this connection, the neutral of the transformer bank is at ground potential and therefore, although the third-harmonic voltages induced in each transformer are not present in the line-to-line voltages, they appear as third-harmonic components of the voltages between the line wires and ground. These third-harmonic voltages produce third-harmonic

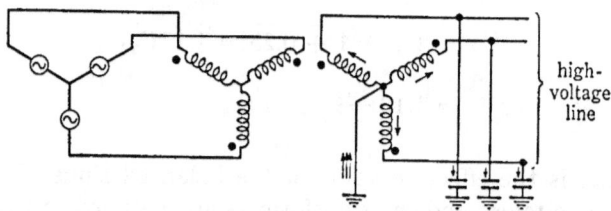

FIG. 9. Paths of third-harmonic currents produced when a Y–Y bank with grounded neutral is connected to a long transmission line. This connection should be avoided, since excessive third-harmonic voltages may be developed by resonance.

exciting currents in the series circuits consisting of the capacitances of the line wires to ground and the grounded windings of the transformers, as shown in Fig. 9. If the triple-frequency capacitive reactances of the line wires to ground approximately equal the triple-frequency magnetizing reactances of the transformers, a condition approaching series resonance exists, and the third-harmonic voltages from the lines to neutral may become dangerously high. Voltages as high as three times normal line-to-neutral voltage have been measured in these circumstances.

3. EQUALIZATION OF NEUTRAL VOLTAGES IN Y–Y BANKS

In spite of its peculiarities, there are occasions when the Y–Y connec- tion is desirable. Sometimes Y–Y banks with isolated neutrals are used, the voltages to neutral then being more or less unbalanced and their waveforms being nonsinusoidal. The voltages to neutral may be equalized and the third-harmonic voltages eliminated by connection of the primary neutral to the neutral of the source, but there may be objections to this arrangement on the grounds of inductive interference.* If it is not practicable to connect the primary neutral to the source neutral, there are several other ways of accomplishing the same results. Two of these ways are described below.

3a. *Y–Δ Grounding Transformers.* — The third-harmonic voltages to neutral of a Y–Y bank can be practically eliminated and the neutral voltages equalized by arrangement of the circuit so that the exciting currents necessary for these purposes can flow in the secondaries. One method of accomplishing this result is shown in Fig. 10, in which a Y–Δ bank with grounded primary neutral is connected to the secondary terminals of the

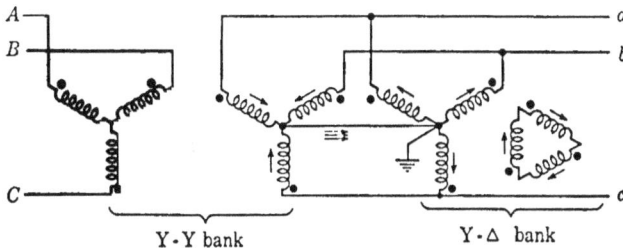

FIG. 10.   Use of a Y–Δ bank of grounding transformers to equalize the neutral voltages and eliminate third-harmonic voltages in a Y–Y bank. The arrows show the circuits for the third-harmonic exciting currents of the Y–Y bank.

Y–Y bank whose neutral voltages are to be equalized. The secondary neutral of the Y–Y bank is connected to the primary neutral of the Y–Δ bank, either through the grounding of both neutrals or preferably with a neutral wire as in Fig. 10. In this circuit, the third-harmonic components of the exciting currents of the Y–Y bank flow in its secondary windings, their paths being indicated in Fig. 10. When these triple-frequency currents flow in the primary windings of the Y–Δ bank, they induce triple-frequency currents in its Δ-connected secondaries whose magnetomotive forces oppose and nearly equal the magnetomotive forces of the triple-frequency primary currents. Therefore the impedance introduced by the Y–Δ bank and opposing the circulation of the third-harmonic exciting currents of the Y–Y bank is only the relatively small triple-frequency equivalent impedance of the Y–Δ bank referred to its primary. Since these triple-frequency currents are equal and in time phase (assuming identical transformers in the Y–Y bank), they simply circulate within the secondary Δ of the Y–Δ bank. In addition to the third-harmonic exciting currents

* See Art. 2b.

of the Y–Y bank shown in Fig. 10, the third-harmonic exciting currents of the Y–Δ bank flow in its Δ-connected secondaries in the normal manner described in Art. 2c, and the fundamental, fifth-, seventh-, etc., harmonic components of the exciting currents of both banks are supplied by the lines.

Sometimes the Y–Δ bank supplies a secondary load, but at other times no load is connected to its secondary Δ, the sole purposes of the Y–Δ bank being to eliminate third-harmonic voltages, to equalize the voltages to neutral of one or more Y–Y banks, and to provide a system ground free from the dangers of third-harmonic resonance. Sometimes autotransformers connected in zigzag are used as grounding transformers, as described in Art. 4a, Ch. XXV. Usually the grounding transformers are located in the same substation as the Y–Y bank.

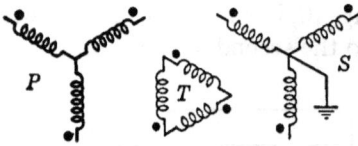

FIG. 11.   Y–Y–Δ connection.

Although under balanced conditions the currents in the grounding transformers are only their own exciting currents plus the third-harmonic components of the exciting currents of the Y–Y bank, large currents may flow in the grounding transformers in the event of a line-to-ground short circuit. Under these conditions the currents in a Y–Δ bank are as shown in Fig. 3, p. 597. The grounding transformers therefore must have sufficient current-carrying capacity to withstand the effects of line-to-ground faults.

3b.   Y–Y–Δ *Connection.* — The third-harmonic voltages to neutral of a Y–Y bank of single-phase transformers can be greatly reduced and the neutral voltages can be equalized if each transformer is provided with a third winding, or tertiary, and these windings are connected in Δ, as in Fig. 11. With this arrangement of circuits, the third-harmonic exciting currents necessary to maintain sinusoidal variations of the core fluxes can flow in the Δ, as in the Y–Δ connection described in Art. 2c. The advantage of this arrangement is that the third-harmonic exciting currents need not flow in the lines where they might cause inductive interference with telephone circuits. If the excitation characteristics of the transformers are not identical, the tertiary Δ also provides a circuit in which a single-phase, or zero-sequence, exciting current may flow to compensate for the inequalities in the excitation characteristics and prevent the unbalance in the voltages to neutral that would otherwise occur in a Y–Y bank with isolated neutrals.

The design of the tertiary windings is determined by the system connections and the results that the tertiary Δ is meant to accomplish. For example, if both the primary and secondary neutrals are isolated and the tertiary Δ supplies no load, the only currents that can flow in the tertiary windings are the third-harmonic and zero-sequence exciting currents, and consequently the windings can be relatively light. However, the neutral of the high-voltage Y is usually grounded, as in Fig. 11, and sometimes both neutrals are grounded. Under these conditions relatively large currents may be induced in the tertiary windings by line-to-ground faults on the high-voltage system, and the tertiary windings must be able to withstand the heating and mechanical forces occasioned by them. Often the tertiary Δ supplies a load; for example, auxiliary circuits in a substation, or static or synchronous condensers for power-factor and voltage control. Under these conditions, the tertiary Δ must be able to withstand the effects of short circuits at its own terminals. The analysis of three-winding transformers under load is taken up in Ch. XXVII.

4. SUMMARY OF HARMONIC PHENOMENA

The important points brought up in the preceding discussion of harmonic phenomena in balanced three-phase circuits can be summarized as follows:

The third-harmonic components of currents or voltages in balanced three-phase circuits are equal and in time phase. (That is, they are zero-sequence quantities.)

The phase order of the fifth-harmonic components is opposite to the phase order of the fundamentals. (That is, they are negative-sequence quantities.)

Because they are in phase, third-harmonic currents cannot flow in the lines of a balanced three-phase system, unless a return path is provided for them through a neutral connection. However, third-harmonic currents can flow in Δ-connected circuits without being present in the lines connected to the Δ.

Because their inductive effects are directly additive, third-harmonic (or other zero-sequence) currents in the lines of a three-phase system may cause serious inductive interference in communication circuits parallel to the power lines.

Because the instantaneous sum of the line-to-line voltages of a three-phase system (taken in cyclic order) must be zero, third-harmonic components cannot be present in the line-to-line voltages of a balanced three-phase system, since these components would be in phase and therefore their sum would not be zero. However, third-harmonic components may exist in the Y voltages to neutral without being present in the line-to-line voltages.

The magnetic characteristics of iron require that the exciting current of a transformer for sinusoidally varying flux must contain a third-harmonic component whose value normally is about 40 per cent of the fundamental component of the exciting current.

The application of these general principles to three-phase connections of single-phase transformers results in the following conclusions:

If the transformer voltages are to vary sinusoidally, the third-harmonic exciting currents must be permitted to flow, either in Δ-connected windings, or through a neutral connection.*

Although the characteristics of the Y–Y connection with a neutral wire are satisfactory in so far as the behavior of the bank is concerned, the necessity for a fourth wire is a disadvantage, and the presence of third-harmonic exciting currents in the three-phase lines may result in inductive-interference difficulties.

If the third-harmonic exciting currents are suppressed, as in the Y–Y connection with isolated neutrals, the voltages to neutral may become unbalanced,† and may contain relatively large third-harmonic components. Under certain conditions, these third-harmonic voltages to neutral may be greatly exaggerated by resonance phenomena. Thus the Y–Y connection of single-phase transformers should be used with caution.

PROBLEMS

1. A bank of identical Δ–Δ-connected single-phase transformers is excited from a three-phase system supplying balanced sinusoidal voltages of 6,600 v, line-to-line. The transformers are rated 6,600 : 440 v, 100 kva, and 60 ∿. The leakage impedances of the two windings of the transformers may be assumed to be vectorially equal when referred to the same side.

While the transformers are operating at no load, low-impedance ammeters are connected in both the primary and secondary deltas. The primary ammeter reads 0.820 amp and the secondary ammeter reads 2.65 amp.

What would be the current per phase in the primary windings at no load if the transformers were reconnected Y–Y, the primary neutral was connected to the source neutral, and balanced sinusoidal voltages of 6,600 v to neutral were applied?

2. A 66,000-v three-phase transmission line is supplied from a Y-connected 6,600-v three-phase alternating-current generator through a Y–Δ-connected bank of single-phase transformers. The neutrals of the generator and of the primary side of the transformer bank are grounded. Each of the transformers is rated 1,000 kva, 3,810 : 66,000 v, and 60 ∿. On short circuit with a 60 ∿ voltage of 0.06 per unit impressed on its high-voltage winding, each of the transformers draws rated current with a power input of 0.008 per unit.

When they are delivering rated balanced load, the transformers become very hot. Upon investigation the alternating-current generator is found to have a terminal voltage which contains a 15% third harmonic. On the assumption that exciting currents are negligible, determine the full-load copper loss in the transformers as a percentage of what it would be if the primary terminal voltages were exactly sinusoidal.

---

* Note that this is not true of three-phase core-type transformers, discussed in Art. 2, Ch. XXVI.

† See Art. 2b, Ch. XXI.

3. A 30-kva three-phase transformer bank consists of three identical 10-kva 2,400 : 240-v 60 $\sim$ distribution transformers connected in Y–Y with the neutral points ungrounded. The line terminals of the high-voltage primaries are directly connected to the secondary line terminals of a balanced three-phase 1,500-kva Δ–Y-connected substation transformer bank whose secondary line-to-line voltages are sinusoidal in waveform and have an effective value of 4,160 v. The secondary neutral of the 1,500-kva substation transformer bank is solidly grounded. A voltmeter connected between ground and the primary neutral of the 30-kva distribution transformer bank reads 1,000 v. What is the effective value of the line-to-neutral voltage across the secondaries of the distribution transformers?

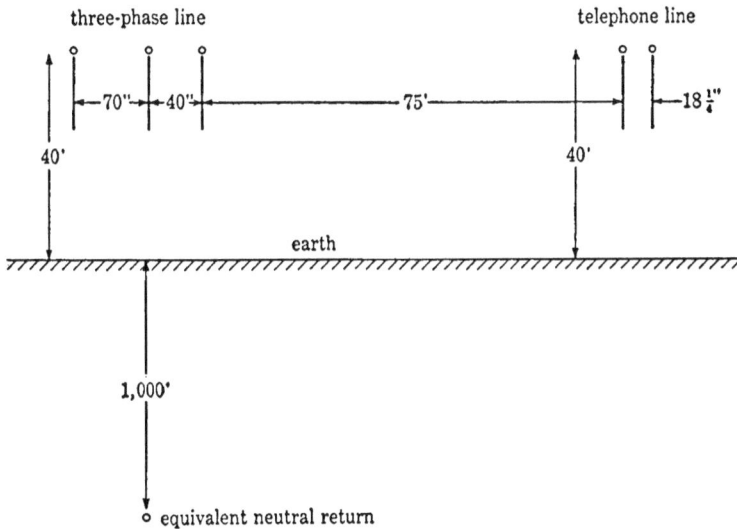

Fig. 12. Separation and spacings of power line
and telephone line, Prob. 4.

4. A three-wire three-phase transmission line runs parallel to a telephone line for several miles. The spacings and separation between the two lines are given in Fig. 12. The conductivity of the earth in the region over which these lines are strung is such that a ground return for the three-phase system, in case one exists, may be represented by a single-wire image 1,000 ft below the surface of the earth.

(a) If the transmission line ends in a Δ-connected bank of identical transformers, if the load is balanced, and if balanced sinusoidal voltages are impressed at the sending end of the line, what is the induced voltage in the telephone line per mile per ampere of three-phase current?

(b) If the transmission line ends in a Y-connected bank of identical transformers and the neutral is grounded, what frequency currents flowing in the line probably give particular trouble from inductive interference with the telephone line? What is the induced voltage in the telephone line per mile per ampere from each of these currents which has a frequency less than 1,000 $\sim$ per second?

5. Three identical 60 $\sim$ single-phase transformers are rated 1,328 : 230 v. When rated sinusoidal voltage at rated frequency is impressed on the high-voltage side of

any transformer with the low-voltage side open, the current is

$$i = 3 \sin \omega t + 1.1 \sin (3\omega t + \alpha_3) + 0.1 \sin (5\omega t + \alpha_5).$$

The third-harmonic leakage impedances of the two windings of each transformer are equal when referred to the same side.

If these transformers are connected Y–Δ to a three-phase 2,300-v 60 ~ circuit with balanced sinusoidal voltages, what effective current would flow in an ammeter placed in the closed circuit formed by the Δ-connected secondaries when the secondary load is zero:

(a) When there is no connection between the neutral of the Y-connected source of power and the neutral of the Y-connected primaries?

(b) When the neutral of the transformer primary windings is connected through a conductor of negligible impedance to the neutral of a generator whose line-to-neutral voltages are sinusoidal and balanced?

CHAPTER XXIV

Unbalanced Conditions in Three-Phase Transformer Banks

The three preceding chapters are devoted primarily to analysis of the operation of symmetrical transformer banks in balanced three-phase circuits. Practical problems involving unbalanced conditions, which may result from dissymmetry in the bank or from unbalanced or single-phase loads or short circuits, are next to be discussed.

For analysis of unbalanced conditions in which the impedances of rotating machines play an important part, the method of symmetrical components is almost indispensable.[1] Most problems in which the impedances of transformer banks are the principal controlling factors can, however, be solved satisfactorily through combining the theory of a single transformer with the current and voltage relations for three-phase circuits. As a review, and for convenient reference, the simplified transformer theory and the three-phase circuit equations are summarized below.

Transformer Equations: Usually the exciting currents of the transformer may be neglected, and the primary and secondary currents assumed to produce equal and opposing magnetomotive forces. Thus the vector relation between the primary current I_1 and the oppositely directed secondary current I_L is

$$I_L = aI_1, \qquad [1]$$

where a is the turns ratio N_1/N_2.

The vector relation between the primary and secondary terminal voltages is

$$\frac{V_1}{a} = V_2 + I_L Z_{eq2}, \qquad [2]$$

where Z_{eq2} is the equivalent impedance referred to the secondary. Alter-

[1] The elementary theory of symmetrical components is discussed in this series in the volume on electric circuits (1940), Ch. XI, pp. 578–605. For more extended treatments of the subject see W. V. Lyon, *Applications of the Method of Symmetrical Components* (New York: McGraw-Hill Book Company, Inc., 1937); C. F. Wagner and R. D. Evans, *Symmetrical Components* (New York: McGraw-Hill Book Company, Inc., 1933). A summary of the applications of symmetrical components to transformer problems is given in L. F. Blume, editor, *Transformer Engineering* (New York: John Wiley & Sons, 1938), Ch. vi.

natively, the voltage equation referred to the primary is

$$V_1 = aV_2 + I_1 Z_{eq1},$$ [3]

where Z_{eq1} is the equivalent impedance referred to the primary.

The transformer hence is characterized by Eq. 1 and either Eq. 2 or Eq. 3.

Line-Voltage Equations: The vector sum of the line-to-line voltages, taken in cyclic order, is zero:

$$V_{AB} + V_{BC} + V_{CA} = 0$$ [4]

$$V_{ab} + V_{bc} + V_{ca} = 0,$$ [5]

where the capital letter subscripts indicate the primary phases and the lower-case subscripts the secondary phases.

Y-Voltage Equations: The vector relations among the line-to-line voltages and the line-to-neutral voltages are

$$V_{AB} = V_{AN} - V_{BN}$$ [6]

$$V_{BC} = V_{BN} - V_{CN}$$ [7]

$$V_{CA} = V_{CN} - V_{AN}$$ [8]

$$V_{ab} = V_{an} - V_{bn}$$ [9]

$$V_{bc} = V_{bn} - V_{cn}$$ [10]

$$V_{ca} = V_{cn} - V_{an}.$$ [11]

Note that only three of the four relations, Eqs. 4, 6, 7, and 8 (or Eqs. 5, 9, 10, and 11), are independent equations, since any one of them can be derived from the other three.

Line-Current Equations: For Y-connected circuits with neutral wires, the primary current equation is

$$I_A + I_B + I_C = I_N,$$ [12]

where I_A, I_B, I_C are the vector line currents flowing into the primaries, and I_N is the vector neutral current returning to the source. For the secondary currents,

$$I_a + I_b + I_c = I_n,$$ [13]

where I_a, I_b, I_c are the vector secondary line currents flowing toward the load, and I_n is the vector neutral current returning from the load.

For Δ-connected circuits, or Y-connected circuits without neutral wires,

$$I_A + I_B + I_C = 0$$ [14]

$$I_a + I_b + I_c = 0.$$ [15]

Relations among Line and Δ *Currents:* The vector relations among the line currents and the Δ phase currents are

$$I_A = I_{AB} - I_{CA} \qquad [16]$$

$$I_B = I_{BC} - I_{AB} \qquad [17]$$

$$I_C = I_{CA} - I_{BC} \qquad [18]$$

$$I_a = I_{ba} - I_{ac} \qquad [19]$$

$$I_b = I_{cb} - I_{ba} \qquad [20]$$

$$I_c = I_{ac} - I_{cb}. \qquad [21]$$

Note that only three of the four relations, Eqs. 14, 16, 17, and 18 (or Eqs. 15, 19, 20, and 21), are independent equations, since any one of them can be derived from the other three.

As examples of the applications of these equations, several problems arising in connection with the use of Δ–Δ banks are discussed in the following article.

1. Unbalanced conditions in Δ–Δ banks

Since the Δ–Δ connection provides two parallel paths between each pair of line terminals on both the primary and secondary sides, the currents in the transformers depend not only on the currents taken by the load but also on the characteristics of the transformers. This important property of the Δ–Δ connection is discussed qualitatively in Arts. 1 and 3, Ch. XXI. As a consequence of it, there are numerous problems involving the operation of Δ–Δ banks under unbalanced conditions due either to unbalanced loads, or to dissymmetry in the bank caused by unequal ratios of transformation or unequal equivalent impedances. A few of these problems are discussed below.[2]

1a. *Circulating Currents in* Δ–Δ *Banks, Due to Unequal Ratios.* — As mentioned in Art. 1, Ch. XXI, inequalities in the ratios of transformation of the three transformers cause circulating currents in Δ–Δ banks. These currents can be readily computed by the application of Thévenin's theorem.[3]

Consider the bank of transformers shown in Fig. 1a, in which the primaries are connected in Δ and the secondaries are connected in series preparatory to being connected in Δ. The secondary Δ may be completed by the closing of switch *K*.

---

[2] For a more detailed discussion of Δ–Δ banks, see L. F. Blume, editor, *Transformer Engineering* (New York: John Wiley & Sons, 1938), 172–180.

[3] Thévenin's theorem is discussed in this series in the volume on electric circuits (1940), pp. 144-146 and 469–470.

If the ratios of transformation of the three transformers are all the same, no voltage exists across the open switch K (neglecting a small third-harmonic component), and therefore no current (except a small third-harmonic exciting current) results if switch K is closed. However, if the ratios of transformation are not equal, a voltage E_{20} exists across the open switch K, this voltage being equal to the vector sum of the open-circuit secondary voltages; that is,

$$E_{20} = \frac{V_{AB}}{a_{AB}} + \frac{V_{BC}}{a_{BC}} + \frac{V_{CA}}{a_{CA}}, \qquad [22]$$

where a_{AB}, a_{BC}, a_{CA} are the ratios of transformation N_1/N_2 of the three transformers, and very nearly equal the open-circuit voltage ratios. Now, when switch K is closed a secondary current results. By Thévenin's theorem, this secondary current equals the open-circuit voltage E_{20}

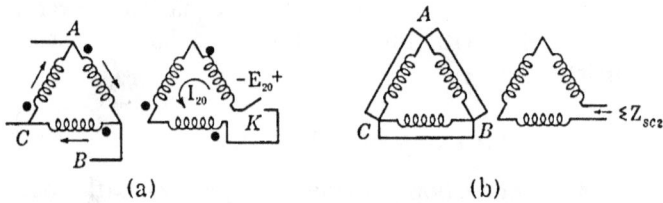

Fig. 1. Illustrating the discussion of circulating currents in Δ–Δ banks.

divided by the impedance measured at the open corner of the Δ with the applied voltages V_{AB}, V_{BC}, V_{CA} short-circuited. From Fig. 1b it is evident that the impedance equals the vector sum of the transformer short-circuit impedances Z_{sc2} measured at their secondary terminals with their primary terminals short-circuited. Thus the circulating current I_{20} in the secondary Δ is

$$I_{20} = \frac{E_{20}}{\Sigma Z_{sc2}}. \qquad [23]$$

The currents in the primary Δ then can be determined from Eq. 1; for example,

$$I_{AB} = \frac{I_{20}}{a_{AB}}. \qquad [24]$$

The primary line currents are then given by Eqs. 16, 17, and 18; for example,

$$I_A = I_{AB} - I_{CA} = \frac{I_{20}}{a_{AB}} - \frac{I_{20}}{a_{CA}} = \left(\frac{a_{CA} - a_{AB}}{a_{AB}a_{CA}}\right) I_{20}. \qquad [25]$$

Note that the circulating current I_{20} is limited by the relatively small short-circuit impedances, and consequently fairly small inequalities in the ratios of transformation may result in relatively large circulating currents in the bank. Thus the Δ-Δ connection of transformers with unequal ratios should be avoided, but this situation may arise accidentally in a Δ-Δ bank of transformers with tap-changing equipment, in the event that the tap changers do not operate simultaneously.

Also note that, even though the circulating currents within the bank may be relatively large, the primary line currents may be small, since they depend on the differences between two ratios of transformation, as shown by Eq. 25. Therefore relatively large circulating currents may exist in the bank without their presence being detectable from measurements of the line currents.

1b. *General Equations for* Δ-Δ *Banks; Equal Ratios of Transformation.* — If the exciting currents are neglected, the primary current equals the oppositely directed secondary current in the same transformer when these currents are both referred to the same side. Thus if the three transformers have the same ratio a of primary to secondary turns,

$$I_{ba} = aI_{AB} \tag{26}$$

$$I_{cb} = aI_{BC} \tag{27}$$

$$I_{ac} = aI_{CA}, \tag{28}$$

where I_{AB}, I_{BC}, I_{CA} are the vectors representing the primary currents in the right-hand-screw direction about positive flux and I_{ba}, I_{cb}, I_{ac} are the vectors representing the oppositely directed secondary currents.

When the exciting currents are neglected, the vector voltage equations are

$$V_{AB} = aV_{ab} + I_{AB}Z_{AB} \tag{29}$$

$$V_{BC} = aV_{bc} + I_{BC}Z_{BC} \tag{30}$$

$$V_{CA} = aV_{ca} + I_{CA}Z_{CA}, \tag{31}$$

where Z_{AB}, Z_{BC}, Z_{CA} are the equivalent impedances of the transformers referred to their primary sides. These voltage equations also may be referred to the secondary sides. Since the sum of the line voltages equals zero (Eqs. 4 and 5), the sum of Eqs. 29, 30, and 31 shows that

$$I_{AB}Z_{AB} + I_{BC}Z_{BC} + I_{CA}Z_{CA} = 0. \qquad ▶[32]$$

If the currents and impedances are referred to the secondary sides, a similar relation is true; that is,

$$I_{ba}Z_{ab} + I_{cb}Z_{bc} + I_{ac}Z_{ca} = 0, \qquad ▶[33]$$

where the equivalent impedances Z_{ab}, Z_{bc}, Z_{ca} are referred to the secondaries.

The currents and voltages now can be determined for any specified operating conditions. For example, assume that two of the secondary line voltages and two of the secondary line currents are given vectorially. The third secondary line voltage and the third secondary line current then can be determined, since the vector sum of the line voltages equals zero (Eq. 5), and the vector sum of the line currents also equals zero (Eq. 15). The currents in the transformers can then be determined through substituting in Eq. 33 the values

$$I_{ac} = I_{ba} - I_a \qquad\qquad [34]$$

$$I_{cb} = I_b + I_{ba} \qquad\qquad [35]$$

obtained from Eqs. 19 and 20. The result is

$$I_{ba}Z_{ab} + (I_b + I_{ba})Z_{bc} + (I_{ba} - I_a)Z_{ca} = 0 \qquad\qquad [36]$$

or

$$I_{ba} = \frac{I_a Z_{ca} - I_b Z_{bc}}{Z_{ab} + Z_{bc} + Z_{ca}} \cdot \qquad\qquad [37]$$

The other secondary currents can be determined in like manner. The primary currents are then known from Eqs. 26, 27, and 28 and the primary voltages can be determined from Eqs. 29, 30, and 31.

Examination of Eq. 37 shows that the currents in the transformers depend on their equivalent impedances. Thus, if the line currents are balanced, the currents in the transformers are not balanced unless the complex equivalent impedances are equal. In general, the transformer having the smallest equivalent impedance has the greatest current, the bank behaving in this respect in a manner somewhat similar to the behavior of impedances in parallel.

Thus if three Δ–Δ-connected transformers of the same rating but unequal equivalent impedances supply power to a balanced load, the bank cannot deliver its full kilovolt-ampere rating without the current exceeding the rated value in the transformer with the smallest equivalent impedance.

For this reason it is preferable to use identical transformers for Δ–Δ connection when the load is balanced, although departure from this practice may be desirable if the load is unbalanced. For further discussion of this point, see Art 3, Ch. XXI.

2. SINGLE–PHASE CURRENTS IN THREE–PHASE BANKS

Since single-phase loads almost always are supplied from three-phase systems, and furthermore since single-phase short circuits may occur on these systems, it is often necessary to determine the distribution of single-phase currents in three-phase transformer banks.

Figure 2 shows a number of three-phase arrangements of transformers

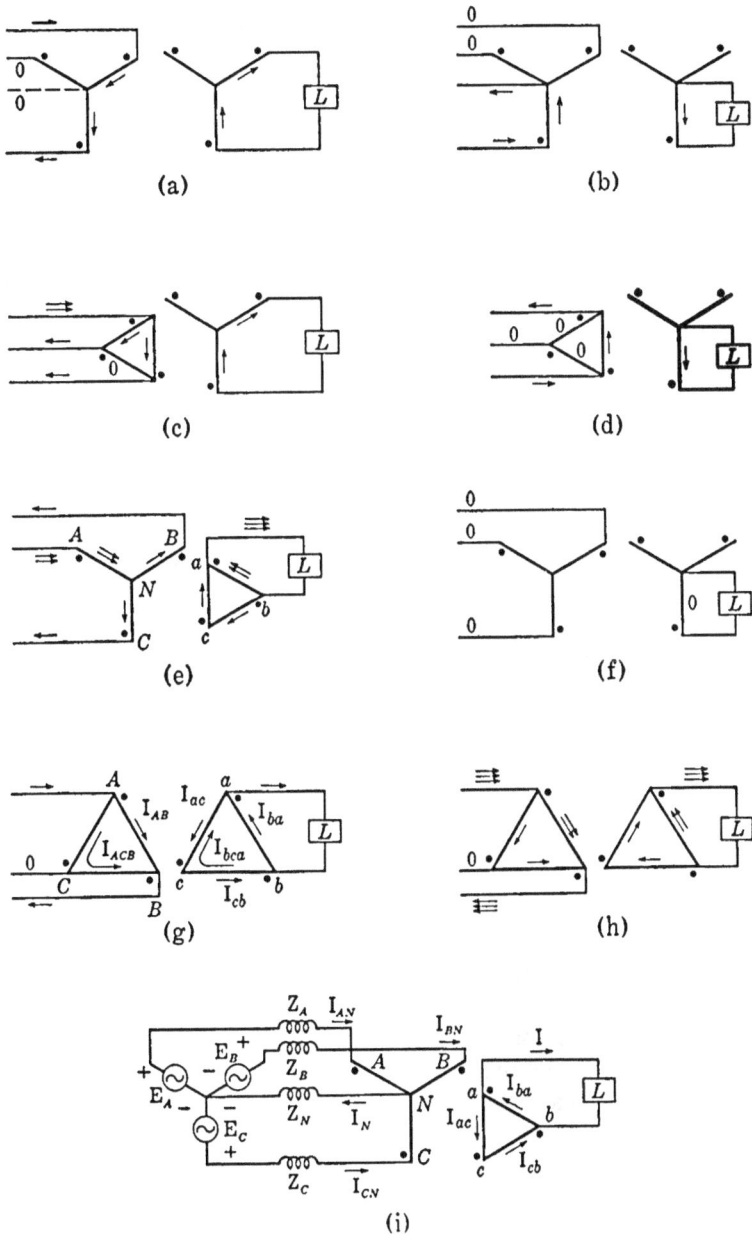

FIG. 2.　Single-phase currents in 3-phase banks.

supplying single-phase loads. The transformer windings are indicated by heavy lines, the primary and secondary windings of the same transformer being drawn parallel to each other, and primary and secondary terminals of the same polarity being indicated by dots. The resulting currents (neglecting exciting currents) are shown by arrows, each arrow representing one unit of current on the basis of one-to-one turns ratio. In (a), (b), (c), (d), and (e) the current distributions are fixed by the transformer connections alone and are determined from the fact that, if there is current in the secondary of any transformer, there must be an equal and opposite current in its primary (on the basis of one-to-one turns ratio and negligible exciting current). In (a), (b), (c), and (d), the secondaries are Y-connected, and therefore the single-phase secondary current can flow in only one series path, but in (e) the secondaries are Δ-connected, and the single-phase secondary current divides between the two parallel paths ba and bca. Since in (e) the currents in the secondaries of transformers bc and ca are equal, their primary currents must also be equal. The currents I_{NB} and I_{NC} returning to the source through transformers B and C therefore each equal one-half the current I_{AN} in transformer A, as shown by the arrows in (e). Thus the secondary currents I_{bc} and I_{ca} in transformers B and C are also each one-half the current I_{ba} in the secondary of transformer A. Transformer A hence supplies two-thirds of the single-phase load current I and transformers B and C each supply the remaining one-third, as indicated by the arrows in (e).

In Fig. 2f — which shows the Y–Y connection of single-phase transformers with isolated primary neutral — if current flows in the primary of one transformer it must return to the source through the primaries of the other two, and therefore the single-phase current that can be supplied from line to neutral on the secondary side is limited to a small value determined by the excitation characteristics of the two unloaded transformers.* This characteristic of a Y–Y bank of single-phase transformers is discussed in Art. 2b, Ch. XXI.

In the Δ–Δ bank shown in Fig. 2g there are parallel paths in both the primary and secondary circuits, and the distribution of single-phase current among the transformers is not determined solely by the connections, but depends on the equivalent impedances of the transformers. The current thus is supplied partly by transformer ba and partly by the path bca consisting of the series combination of transformers bc and ca in parallel with transformer ba. The current in the secondary line c is zero and, from inspection of Fig. 2g,

$$I_{ac} = I_{cb} = -I_{bca}, \qquad [38]$$

---

* However, a Y–Y-connected, three-phase, core-type transformer can supply a fairly large single-phase load from a secondary line to neutral, even when the primary neutral is isolated. See Art. 2b, Ch. XXVI.

where I_{bca} is the current from b to a through the path bca. Substituting Eq. 38 in Eq. 33 gives

$$I_{ba}Z_{ab} - I_{bca}(Z_{bc} + Z_{ca}) = 0, \qquad [39]$$

whence

$$\frac{I_{ba}}{I_{bca}} = \frac{Z_{bc} + Z_{ca}}{Z_{ab}}. \qquad \blacktriangleright[40]$$

That is, the currents are inversely proportional to the equivalent impedances of the parallel paths ba and bca through the transformer bank. If the transformers are identical, two-thirds of the load current is supplied by transformer ba, and one-third by transformers bc and ca in series, as shown in (h).

Another circuit in which there are parallel paths on both the primary and secondary sides is the Y–Δ connection with the primary neutral connected to the neutral of the source, as shown in Fig. 2i. In this circuit, the current distribution depends on the impedances, not only of the transformers, but also of the source. The primary current equation is

$$I_{AN} + I_{BN} + I_{CN} = I_N. \qquad [41]$$

If the exciting currents are neglected the relations among the oppositely directed primary and secondary currents are

$$I_{ba} = aI_{AN} \qquad [42]$$

$$I_{cb} = aI_{BN} \qquad [43]$$

$$I_{ac} = aI_{CN}, \qquad [44]$$

where a is the ratio of primary to secondary turns. Since the secondary line c is open,

$$I_{cb} = I_{ac}. \qquad [45]$$

The relation between the current I delivered to the load and the secondary currents in the transformers is

$$I = I_{ba} - I_{ac}. \qquad [46]$$

Let E_A, E_B, E_C be the vector electromotive forces of the source and let Z_N be the complex impedance in the neutral wire. Also let Z_A, Z_B, Z_C be the complex impedances in each primary phase, these impedances each being the vector sum of the source impedance, the line impedance, and the equivalent impedance of the transformer referred to its primary side. The voltage equations for the three phases are

$$E_A = I_{AN}Z_A + I_NZ_N + aV_{ab} \qquad [47]$$

$$E_B = I_{BN}Z_B + I_NZ_N + aV_{bc} \qquad [48]$$

$$E_C = I_{CN}Z_C + I_NZ_N + aV_{ca}, \qquad [49]$$

where V_{ab}, V_{bc}, V_{ca} are the vector secondary terminal voltages. **Note that, since these are the** line-to-line voltages, their vector **sum is zero, or**

$$V_{ab} + V_{bc} + V_{ca} = 0. \qquad [50]$$

Equations 41 to 50 inclusive are the general relations for a Y–Δ bank with a single-phase load. When the impedances are unequal or the source voltages E_A, E_B, and E_C are unbalanced, these ten equations involving the circuit constants and 14 vector currents and voltages can be solved simultaneously if the circuit constants and four independent vector currents or voltages are known. The general solution is rather complicated.

However, the relations among the currents are greatly simplified if the source voltages E_A, E_B, E_C are balanced and the impedances Z_A, Z_B, Z_C are equal. If the source voltages are balanced, their vector sum is zero, and since the vector sum of the secondary terminal voltages also is zero (Eq. 50), the sum of Eqs. 47, 48, and 49 is

$$0 = (I_{AN} + I_{BN} + I_{CN})Z + 3I_N Z_N, \qquad [51]$$

where Z is the impedance of each primary phase. But the vector sum of the primary line currents equals the neutral current I_N (Eq. 41). Thus, from Eq. 51,

$$0 = I_N(Z + 3Z_N) \qquad [52]$$

or

$$I_N = 0. \qquad [53]$$

Hence with balanced source voltages and equal primary phase impedances, there is no current in the neutral wire, and therefore the current distribution is the same as it would be if the primary neutral wire were disconnected. This distribution is shown in Fig. 2e.

3. LINE–TO–GROUND FAULT ON THE PRIMARY SIDE OF A Y–Δ BANK WITH GROUNDED NEUTRAL

Another situation in which there are single-phase currents in a Y–Δ bank is illustrated in Fig. 3, which shows a Y–Δ bank with grounded neutral at the receiving end of a transmission line. There is a line-to-ground fault F on the phase C conductor. For the present, assume that the neutral of the Y is the only system ground other than the fault. The fault current flows from phase C to ground and returns to the transmission system through the ground connection at the neutral of the Y. Since some of this current flows from the neutral of the Y through the primary of transformer C, as shown by the current I_{NC} in Fig. 3, there must be an oppositely directed current I_{ca} in the secondary of transformer

C, which must also flow in the secondaries of the other two transformers, as indicated by the arrows in Fig. 3. There must also be, therefore, oppositely directed currents in the primaries of transformers B and C. Since the same current flows in the three secondaries, the three primary currents are all equal and in phase, and hence must each be one-third

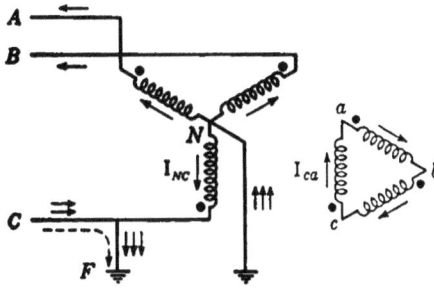

Fig. 3.　Line-to-ground fault on the primary side of a Y–Δ bank with grounded primary neutral.

of the fault current. The current distribution is thus as shown by the full-line arrows in Fig. 3, where each arrow represents one-third of the fault current. In the terminology of symmetrical components,[4] these equal and in-phase currents are zero-sequence currents.

If the neutral of the source is also grounded, the source supplies an additional component to the fault current, as indicated by the broken-line arrow in Fig. 3.

4. SYMMETRICAL–COMPONENT ANALYSIS

Simple problems in which the impedances of the transformers are the principal controlling factors can usually be solved satisfactorily by the simple methods discussed in the preceding articles. For more complicated problems, however, such as those involving the impedances of transmission lines and rotating machines, the method of symmetrical components is usually more expeditious.[*] If a network analyzer is available, and if the complexity of the system warrants the use of it, the behavior of the system can be determined experimentally through properly setting up and interconnecting the zero-, positive-, and negative-sequence equivalent networks of the complete system.[5] In problems of this kind, the first objective is the determination of the behavior of each part of the system. The transformers here play an important rôle.

[4] See the volume on electric circuits (1940), Ch. XI, pp. 578–605.
[*] References to texts on the subject of symmetrical components are given in footnote 1, p. 587.
[5] See the volume on electric circuits (1940), Art. 11, Ch. XI, pp. 598–603.

The following discussion is confined to the treatment of unbalanced conditions resulting from unbalanced loads or short circuits at one or more points on an otherwise symmetrical system. In such a system, nothing distinguishes one phase from another, except at the points of unbalance; that is, the impedances of the three phases of the system are equal. Consequently, if the unbalanced voltages and currents are resolved into three balanced sets of components — the zero-, positive-, and negative-sequence sets — the system can be analyzed as a balanced-circuit problem on a per phase basis for each set.

If the voltages and currents existing before the unbalance is applied have the phase order *abc*, the positive-sequence components of the voltages and currents in the three phases for the unbalanced conditions form balanced sets whose phase order is *abc*. The impedances of rotating machines, transmission lines, and transformer banks are the same for positive-sequence currents and voltages as for balanced conditions, and the positive-sequence equivalent network of the system on a per phase basis is the same as for normal balanced conditions.

The negative-sequence components of the voltages and currents in the three phases form balanced sets whose phase order is *acb*. The only difference between the positive- and negative-sequence sets is their phase order. The impedances of static apparatus, such as transmission lines and transformers, are independent of phase order, and the portions of the negative-sequence network that represent them are the same as the corresponding portions of the positive-sequence network. Rotating machines, however, present different values of impedance to positive- and negative-sequence currents, and ordinarily generate no negative-sequence internal electromotive forces. Consequently they are represented in the negative-sequence network by values of impedance different from those in the positive-sequence network, and their internal electromotive forces are short-circuited in the negative-sequence network.

The zero-sequence components of the voltages and currents in the three phases also form symmetrical sets but with a form of symmetry different from that existing for the positive- and negative-sequence components.[6] By definition, the vector zero-sequence component I_0 of the vector currents I_a, I_b, I_c in the phases of a three-phase system is

$$I_0 = \tfrac{1}{3}(I_a + I_b + I_c). \tag{54}$$

The zero-sequence components of the three currents are equal and in phase with each other, in contrast with the positive- and negative-sequence components, which are equal in magnitude but differ in phase by 120 degrees; that is, for the zero-sequence components of I_a, I_b, and I_c,

$$I_{a0} = I_{b0} = I_{c0} = I_0. \tag{55}$$

[6] See the volume on electric circuits (1940), Arts. 4 and 9, Ch. XI, pp. 584–587 and 595.

▶ From Eq. 54, zero-sequence currents can exist only when the circuit is arranged so that the vector sum of the currents in the three phases is not forced to be zero.◀

This fact means that zero-sequence currents cannot exist in symmetrical Y-connected rotating machines, transformer banks, or transmission lines unless one or more neutral points are grounded or interconnected. For example, zero-sequence currents could not exist in the primary windings of the transformers in Fig. 3 if the neutral point N were ungrounded. Since the paths of the zero-sequence currents differ from those of the positive- and negative-sequence currents, the impedances to zero-sequence currents in rotating machines and transmission lines differ from the impedances to other sequence currents.

Zero-sequence currents can exist, however, in the phases of Δ-connected circuits. In this arrangement, the zero-sequence components of the three Δ currents, being equal and in phase with each other, simply circulate around the Δ but do not flow in the lines connected to the Δ, as in the secondary windings of the transformer bank shown in Fig. 3. Since the vector sum of the three line-to-line voltages of a three-phase system taken in cyclic order must always be zero, the existence of zero-sequence components in the line-to-line voltages is impossible. Thus, even though zero-sequence currents exist in the Δ-connected secondaries of Fig. 3, they produce no zero-sequence components in the line-to-line voltages.

From the foregoing discussion, it should be evident that the connections of transformer banks have important influences on the zero-sequence currents. The general principles can be summed up rather simply.[7]

Zero-sequence currents can exist in the lines terminating in a Y-connected group of windings only when the neutral point is grounded or connected to a neutral wire. If the neutral point is isolated, the circuit is open-circuited in so far as zero-sequence currents are concerned.

The lines terminating in a Δ-connected group of windings are open-circuited in so far as zero-sequence currents are concerned, since no neutral connection exists to provide a return path for them. Zero-sequence circulating currents may be induced in the Δ, however, if zero-sequence currents exist in other windings with which the Δ-connected group are coupled inductively.

If the circuits are arranged so that zero-sequence currents can exist in both the primary and secondary windings, then zero-sequence currents on one side induce on the other side zero-sequence currents which produce equal and opposing magnetomotive forces (exciting currents

[7] For further discussion, see the references given in footnote 1, p. 587. A number of arrangements of transformers are discussed in detail in L. F. Blume, editor, *Transformer Engineering* (New York: John Wiley & Sons, 1938), 142–155.

being neglected). The impedance to zero-sequence currents introduced
by such a transformer bank is therefore the equivalent or short-circuit
impedance per phase.   For example, a Y–Y-connected group of identical
transformers with *both* neutral points grounded is equivalent in the zero-
sequence network to the short-circuit impedance of one of the transformers in series with the primary and secondary circuits (it being understood, of course, that all currents, voltages, and impedances are referred
to a common base). The equivalent circuit for zero sequence is shown
in Fig. 4a. If zero-sequence currents exist in the Y-connected windings of

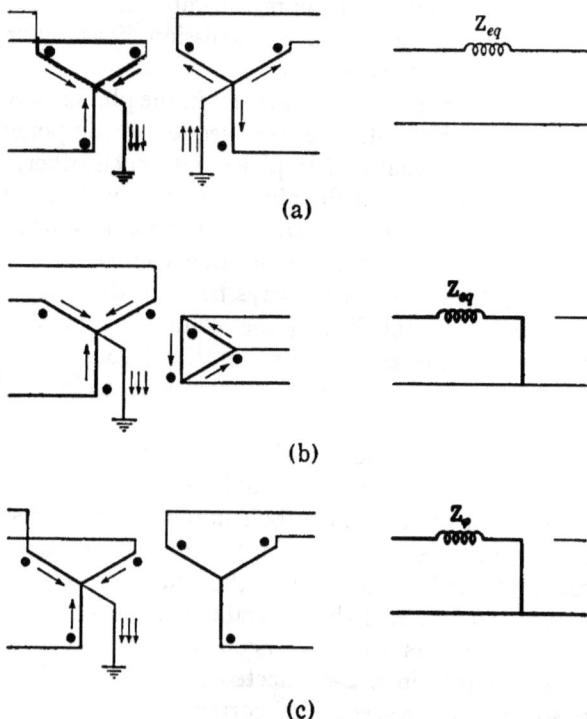

(a)

(b)

(c)

Fig. 4.   Zero-sequence equivalent circuits for several arrangements of transformers.

a Y–Δ bank whose Y neutral is grounded, the zero-sequence currents on
the Y-connected side induce in the Δ zero-sequence currents which
simply circulate within it, as shown in Fig. 3. Thus the zero-sequence
impedance of the transformer bank per phase as viewed from its Y-
connected side equals the equivalent impedance of one of the transformers. No zero-sequence currents can exist in the external circuits connected to the Δ, however, and the bank therefore acts as an open circuit
to zero-sequence currents in the external circuit on the Δ-connected side,
as shown in Fig. 4b.

If, however, the transformer connections are arranged so that zero-sequence currents can exist on one side but not on the other, then the impedance to zero-sequence currents on the side where they can exist is the open-circuit or exciting impedance of one phase of the bank. On the other side, the bank acts as an open circuit for zero-sequence currents. Such is the situation in the arrangement shown in Fig. 4c.

By application of the general principles illustrated in the foregoing discussion, the distribution of zero-sequence currents in any transformer bank that involves any combination of Y- and Δ-connected windings can be determined. If the bank consists of multicircuit transformers, the impedance phenomena (discussed in Ch. XXVII) are somewhat more complicated, but the same general principles can be applied for determination of the distribution of zero-sequence currents. These same principles also apply to three-phase transformers, as well as to banks of three single-phase units, the only difference being that the zero-sequence exciting impedance of a three-phase core-type transformer (discussed in Art. 2b, Ch. XXVI) is much smaller than that of a similar bank of three single-phase units. Sometimes transformers are used in which the phases are interconnected in a zigzag arrangement. The behavior of such transformers with zero-sequence currents in them is discussed in Art. 4a, Ch. XXV.

PROBLEMS

1. Three 2,400 : 240-v 10 kva 60 ∿ transformers are connected in Δ on both sides and are used to step down a balanced three-phase voltage of 2,400 v to supply a balanced Δ-connected inductive load of impedance $6.0 + j2.0$ ohms per phase. One of the regular transformers burns out and is replaced by an old transformer rated 2,400 : 230 v 10 kva and 60 ∿. The equivalent impedance of each of the two original transformers is $0.06 + j0.20$ ohm, referred to the secondary, and the equivalent impedance of the substituted transformer is $0.11 + j0.36$ ohm, referred to the low-voltage side.

Determine whether any of the transformers are overloaded.

2. Three single-phase transformers, each rated 100 kva 11,000 : 2,200 v and 60 ∿, are connected to an 11,000-v three-phase line with both their primaries and secondaries in Δ. Conventional short-circuit tests on the individual transformers yield the following data:

SHORT–CIRCUIT DATA

Transformer	Current	Voltage	Power
A	9.10 amp	312 v	1,000 w
B	9.10	402	1,130
C	9.10	450	1,200

For the condition that the line currents are balanced, determine the greatest kva load these transformers can supply without overload on any one of them.

3. Three identical 100-kva transformers are connected to a balanced three-phase circuit. For the following connections, what is the greatest single-phase load that can be applied between any pair of line terminals on the secondary side without overheating any one of the transformers?

(a) Δ–Δ,

(b) Y–Δ,

(c) Δ–Y.

Assume that the transformers are operating at rated frequency and voltage.

4. Three nearly identical single-phase transformers rated 10 kva, 2,400 : 120 v and 60 $\sim$ are connected in Δ on their high-voltage sides to a balanced 2,400-v 60 $\sim$ three-phase circuit. The low-voltage windings are connected in Y to a three-phase four-wire distribution system, so that 120-v lighting loads may be connected from any line to the neutral wire, and small power loads may be operated at 208 v between lines.

Each of the three transformers when operated with the low-voltage winding short-circuited draws 4.6 amp and 170 w at a voltage of 110 v.

What will be the voltages between the pairs of secondary line terminals if a single-phase lighting load of 83 amp at unity power factor is connected between one secondary terminal and the secondary neutral?

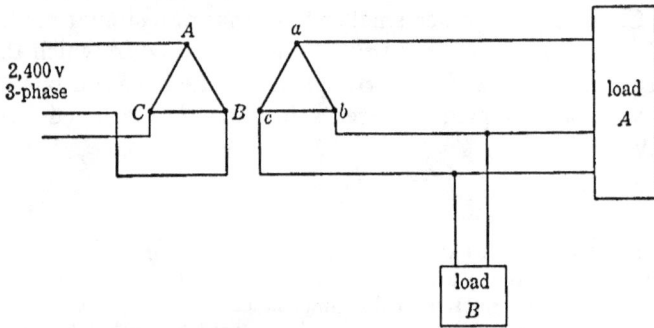

Fig. 5.   Connection of transformers supplying a 3-phase and a single-phase load, Prob. 8.

5. On short circuit, three similar single-phase transformers, rated 11,000 : 2,200 v, 100 kva, and 60 $\sim$, each take 9.1 amp and 1,000 w when a voltage of 310 v is impressed on their high-voltage windings. These three transformers are connected in Δ on both sides, and their primaries are connected to a three-phase 60 $\sim$ 11,000-v circuit. A single-phase load of 0.090 h inductance and 36 ohms resistance is connected between one pair of secondary terminals. Find the values of the secondary terminal voltages.

6. Two transformers, each with voltage ratings of 2,400 : 12,000 v, are connected in open delta, and balanced three-phase voltages of rated value are impressed on the low-voltage sides. A single-phase load current of 90 amp at 0.80 power factor, lagging, is taken from the open side of the secondary. If the equivalent impedances of the two transformers referred to the secondary are $1.0 + j8.0$ and $0.7 + j5.0$ ohms, what is the load voltage, and what is the secondary voltage of each of the transformers?

7. Two 100-kva 6,600 : 230-v 60 $\sim$ single-phase transformers are connected in open delta to a three-phase system with voltages of 6,600 v between each pair of line conductors. With the low-voltage windings short-circuited and 390 v at 60 $\sim$ applied to the high-voltage windings, each transformer draws 17 amp and 980 w.

If a single-phase load of impedance $0.36 + j0.30$ ohm at $60 \sim$ is placed across the secondary terminals of each of the transformers, what is the magnitude of the secondary voltage across the open side of the V?

8. Three transformers are to be connected Δ–Δ as shown in Fig. 5 to supply power to a substantially balanced three-phase load A and to a single-phase load B. The operating conditions at the loads are:

	Continuous load	*Two-hour peak load following continuous load*	*During two-hour peak load*	
			Load voltage	*Power factor*
Load A	56 kva	75 kva	220 v	0.70
Load B	30	40	230	0.90

The per unit equivalent complex impedance of a standard distribution transformer may be assumed to have the same value, referred to the rating of the transformer, for all ratings from 15 to 100 kva.

Specify the minimum permissible kva rating of each transformer and compute the ratio of its peak load to its continuous rating.

Justify any approximations or assumptions made in the course of the solution.

CHAPTER XXV

Three-Phase Connections of Autotransformers

Autotransformers may be connected in many ways for the transformation of polyphase voltages.[1] The common three-phase connections discussed in this chapter may be used either with three single-phase autotransformers or with a three-phase autotransformer with its windings for the three phases on a three-phase magnetic core.[*] The present discussion is chiefly concerned with the first possibility.

▶ Autotransformers are seldom used when the voltage ratio is greater than about two-to-one, since their advantages in comparison with two-circuit transformers are insignificant for higher voltage ratios, and since with higher voltage ratios the danger is greater that overvoltages will be produced on the low-voltage system by transients and faults. ◀

1. Y CONNECTION OF AUTOTRANSFORMERS

Three single-phase autotransformers can be connected in Y, as indicated in Fig. 1. Under these conditions, the behavior of the bank is in many respects similar to that of a bank of three Y–Y-connected two-circuit transformers. If the neutral is isolated, as in Fig. 1, the voltages to neutral are unbalanced unless the transformers have identical excitation characteristics. Furthermore, the line-to-neutral voltages include relatively large third-harmonic components caused by the suppression of the third-harmonic components of the exciting currents.[†] If the neutral is grounded but all other system neutrals are isolated, these third-harmonic voltages to neutral may be greatly

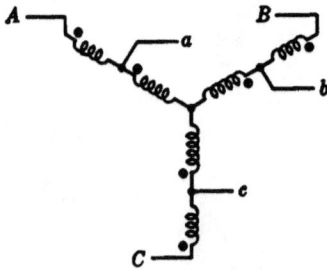

FIG. 1. Y-connected autotransformers.

intensified by resonance of the triple-frequency magnetizing reactances of the transformers and the triple-frequency capacitive reactances of the transmission lines to ground.[‡] If the transmission lines have appreciable capacitance to ground, the neutral of the autotransformers

---

[1] For a more detailed discussion of autotransformer connections, see L. F. Blume, editor, *Transformer Engineering* (New York: John Wiley & Sons, 1938), Chs. viii and xvi.

[*] See Ch. XXVI.

[†] This is not true of a three-phase, three-legged, core-type autotransformer. See Art. 2a, Ch. XXVI.

[‡] See Art. 2d, Ch. XXIII.

should therefore not be grounded unless the system neutral is also solidly grounded elsewhere, or unless the autotransformers are provided with a tertiary Δ, as described below.

If the neutral of the autotransformers is isolated and the system neutrals are not grounded elsewhere, a line-to-ground fault on the high-voltage system impresses abnormally high voltages between two of the low-voltage terminals and ground. Consequently, autotransformers with isolated neutrals are not often used on ungrounded systems, unless the voltage ratio is close to unity.

The third-harmonic voltages can be practically eliminated by the same methods used for this purpose with Y–Y banks of two-circuit transformers. The most satisfactory method is to provide each autotransformer with a tertiary winding and to connect these tertiaries in Δ, thus providing a circuit for the third-harmonic components of the exciting currents. Under these conditions, it is safe to ground the neutral of the autotransformers, and the dangers of abnormal voltage stresses are greatly reduced.

2. Δ CONNECTION OF AUTOTRANSFORMERS

Three autotransformers can be connected in Δ as indicated in Fig. 2. A possible disadvantage of this connection is that the secondary line voltages are not in phase with the primary line voltages. The greatest

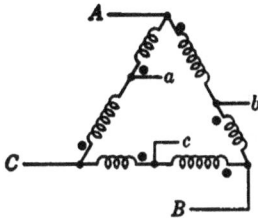

FIG. 2. Δ–connected autotransformers.

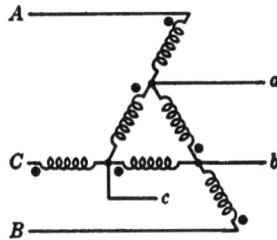

FIG. 3. Extended-Δ connection of autotransformers.

ratio of transformation that can be obtained, moreover, is two to one. As in the Δ–Δ connection of two-circuit transformers, the third-harmonic components of the exciting currents flow around the Δ but do not appear in the line currents.

Autotransformers can also be connected in Δ as indicated in Fig. 3, in which the series windings are connected in series with the high-voltage lines and the common windings are connected in Δ. As in the Δ connection of Fig. 2, the primary and secondary line voltages are not in phase.

3. Open–delta connection of autotransformers

Unlike the Δ connection, the open-delta connection of autotransformers, shown in Fig. 4, is not restricted to ratios of transformation less than two to one. Furthermore, except for the voltage drops due to the leakage impedances, the primary and secondary line voltages are in phase.

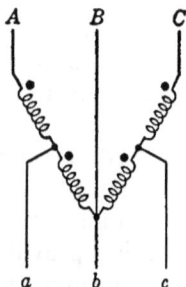

The connection has some of the characteristics of the open-delta connection of two-circuit transformers.* Its apparatus economy is 86.6 per cent; that is, the three-phase rating of the bank is 86.6 per cent of the combined single-phase ratings of the two autotransformers. Because the connection is unsymmetrical, the impedance drops in the transformers introduce a slight and ordinarily negligible unbalance in the circuit.

Fig. 4. Open-delta connection of auto-transformers.

If the primary terminals are supplied from a Y-connected source whose neutral is grounded, the secondary voltages to ground are unbalanced, since the same voltage to ground is impressed on the secondary conductor b as exists between the primary conductor B and ground, but the voltages between the secondary lines ab and bc differ from the voltages between the primary lines AB and BC.

▶ Hence, if the primary and secondary sides of an open-delta bank of autotransformers are both connected to Y-connected circuits, the neutral on only one side of the bank should be grounded, since there is a voltage difference between the neutrals of the primary and secondary circuits.◀

If both the primary and the secondary Y were grounded, the voltage difference would be short-circuited, and a short-circuit current would flow in both the primary and secondary circuits. This displacement between the primary and secondary neutrals is a feature that may be a disadvantage of the open-delta connection of autotransformers.

3a. *Starting Compensators.* — The most common application of the open-delta connection of autotransformers is in the starting compensators used to apply reduced voltages to the armature terminals of three-phase motors in order to limit their starting currents to values that are safe for the motors and do not produce objectionable voltage drops in the supply lines. The starting compensator consists of two single-phase autotransformers, or a three-phase unit, connected in open delta, together with suitable switching arrangements. The elementary connections are shown in Fig. 5, in which M, S, and R are manually — or magnetically — operated switches, or oil circuit breakers. To start the motor, switches M and S

_____
* See Art. 5, Ch. XXI.

are closed and reduced voltages are applied to the motor through the autotransformers. When the motor has attained its maximum speed with the reduced voltages, switches M and S are opened and switch R is closed, the autotransformers being thus disconnected and full line voltages being applied to the motor. The switches are usually interlocked to assure the proper sequence of operation, and often the compensator is provided with suitable undervoltage and overcurrent protection. Since the normal service required of the compensator is intermittent and of short duration, the autotransformers are designed to operate at high current densities in the windings and high flux densities in the cores, for economy and compactness. They would overheat rapidly if kept in operation continuously.

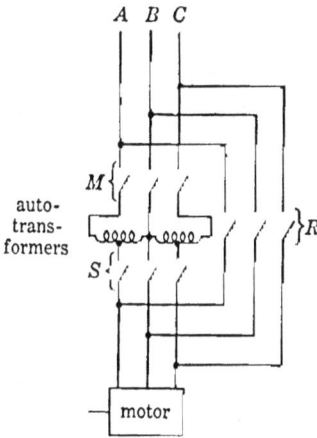

FIG. 5. One-step starting compensator connections.

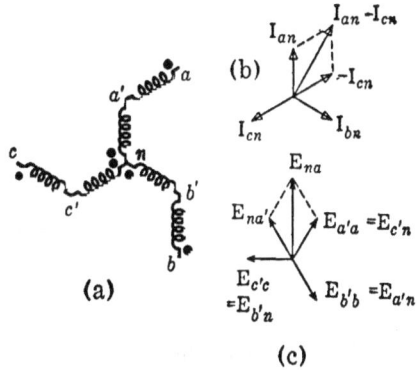

FIG. 6. Interconnected-Y or zigzag connection, and vector diagrams of the currents and voltages.

4. INTERCONNECTED–Y OR ZIGZAG CONNECTION

Transformers of polyphase groups are sometimes arranged with the windings on different transformers connected in series. One type of this interconnection is illustrated in Fig. 6a, which may represent either autotransformers or the primary windings of three-winding transformers whose secondaries are not shown in the figure. The windings shown parallel to each other are on the same transformer; for example, windings aa' and nc'. All the windings have the same number of turns. Note that terminals of the same relative polarity are connected, and that the circuit is symmetrical.

If balanced currents I_{an}, I_{bn}, I_{cn} flow into the terminals a, b, c, in the transformer whose windings are aa' and nc' the net ampere-turns due to these currents are vectorially

$$N(\mathbf{I}_{aa'} + \mathbf{I}_{nc'}) = N(\mathbf{I}_{an} - \mathbf{I}_{cn}), \tag{1}$$

where N is the number of turns in one winding. For the phase order abc, these vector relations are shown in Fig. 6b, whence

$$N(\mathbf{I}_{an} - \mathbf{I}_{cn}) = \sqrt{3}\, N\mathbf{I}_{an}\, \underline{/-30°}. \tag{2}$$

If the windings on the same transformer were connected in series, as in the Y connection of Fig. 1, the ampere-turns of the transformer in phase a would be $2N\mathbf{I}_{an}$. That is, in the zigzag connection the net magnetomotive force is $\sqrt{3}/2$ or 0.866 of what it would be if the windings on the same transformer were connected in series.

Also, if the applied voltages are balanced, the voltages induced in the windings by the resultant core fluxes are balanced, as shown in Fig. 6c, in which the vectors represent the electromotive forces or rises in potential in the directions indicated by the order of the subscripts. Hence

$$\mathbf{E}_{na} = \mathbf{E}_{na'} + \mathbf{E}_{a'a} \tag{3}$$

$$= -\mathbf{E}_{a'n} + \mathbf{E}_{a'a} \tag{4}$$

$$= \sqrt{3}\, \mathbf{E}_{a'a}\, \underline{/30°}. \tag{5}$$

If the windings on the same transformer were connected in series, as in the Y connection of Fig. 1, the line-to-neutral induced voltage $\mathbf{E}_{na}$ would be $2\mathbf{E}_{a'a}$. Thus, in the zigzag connection, the induced line-to-neutral voltage for a given core flux is $\sqrt{3}/2$ or 0.866 of the value it would be for a Y connection. Consequently, for a given line-to-neutral voltage $\mathbf{E}_{na}$ 1/0.866 or 1.15 times as many turns are required when the windings are connected in zigzag, Fig. 6a, as when the windings are connected in Y, if the magnitude of the flux in the magnetic core is to be the same. There is a phase displacement of 30 degrees between the line-to-neutral voltages and the voltages induced in the corresponding windings in the zigzag connection.

If the applied voltages are balanced and the neutral n is connected to the neutral of the source, there may be equal third-harmonic components in the currents flowing into the transformers at the terminals a, b, c. However, these third-harmonic currents produce no net magnetizing effect, since they flow in opposite directions in the two windings on each core; that is, the third-harmonic component of the phase-a current flows in the direction aa', while an equal and in-phase third-harmonic component of the phase-c current flows in the opposite direction $c'n$, and, since windings aa' and nc' are on the same core, the net third-harmonic magnetomotive force acting on the core is zero.* Hence, whether the neutral point n is isolated or connected to the neutral of the source, there is no net third-harmonic magnetomotive force acting on each core, and consequently the absence of the third-harmonic magnetomotive forces (which are necessary to permit sinusoidal variations of the fluxes) results in a third-harmonic component in the voltage induced in each of the six windings. Since these third-harmonic voltages in the individual windings are equal and in time phase, and since the line-to-neutral voltage E_{na} equals the difference of the voltages generated in two windings (Eq. 4), there is no third-harmonic component in the line-to-neutral voltage as there would be in a Y-connected bank with isolated neutral.

4a. *Interconnected-Y Grounding Transformers.* — If the neutral point n, Fig. 6a, is grounded, interconnected autotransformers are described as grounding transformers, the sole purpose of the bank being to ground the neutral of an otherwise isolated-neutral system. If the entire system to which the transformers are connected is symmetrical, the line-to-neutral voltages at the transformers are balanced, and the currents flowing into the transformers are just sufficient to excite them. If in the system there is unsymmetry due, for example, to a line-to-ground fault, so that the line-to-neutral voltages at the transformers are no longer balanced, there may be component currents in the phases an, bn, and cn of the grounding transformers that are equal to each other, and in time phase. If the analysis of the unsymmetrical system is made by the method of symmetrical components, the component currents that are equal and

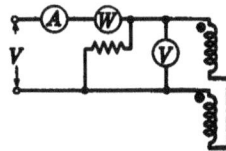

FIG. 7. Circuit for measuring the zero-sequence impedance of a bank of interconnected-Y autotransformers.

in time phase are the zero-sequence components. Zero-sequence currents flowing from a to a' and from c' to n in the same transformer produce no net magnetization of the core, since these currents flow in opposite directions about the core. Consequently the line-to-neutral voltage necessary to maintain this zero-sequence current equals the product of the current and the sum of the leakage impedances of the windings aa' and $a'n$. When the transformers are identical, this impedance can be measured through connecting the two windings of one transformer in series opposition — as indicated in Fig. 7 — applying a low voltage, and measuring the voltage, current, and power. The combined leakage impedance of the two windings being in the neighborhood of 0.002 of the

*Since the zero-sequence currents of the method of symmetrical components are equal and in phase, they behave in the same way as do third-harmonic components of balanced exciting currents. This characteristic of zero-sequence currents is discussed in part (a) of this article.

exciting impedance, the transformers offer a very much smaller impedance to zero-sequence currents than to the balanced exciting currents flowing when the entire system is balanced. Therefore they require very little exciting current under normal conditions, but provide a relatively low impedance path to ground, thereby maintaining the system neutral at ground potential. In the event of a line-to-ground fault on the system, the grounding transformers permit a fault current to flow that is sufficiently large to cause proper operation of protective relays.

PROBLEMS

1. Three equal 2 : 1-ratio autotransformers are arranged in Δ, as in Fig. 2, p. 605. With balanced applied voltages and a balanced load of I amperes per phase on the secondary circuit abc, what are the magnitudes of the primary line current I_A, and of the winding currents I_{Aa} and I_{Ab}? What are the phase relations of these currents?

2. The 2 : 1-ratio autotransformers in Fig. 2, p. 605, supply a single-phase load of I amperes connected to the terminals ab. What are the line currents at each of the primary terminals A, B, and C?

3. Three 2 : 1-ratio autotransformers are arranged in Y, as in Fig. 1, p. 604. A lamp load requiring I amperes is connected across the secondary terminals ab, and another load requiring the same current at 0.50 power factor, lagging current, is connected across the terminals bc. What are the currents in each of the windings of each transformer?

4. Three autotransformers are connected in extended Δ, as in Fig. 3, p. 605. The ratio of primary to secondary line voltages is 2 : 1. What must be the ratio of turns in the two windings of each transformer? What is the phase relation of the primary voltage V_{AB} to the secondary voltage V_{ab}? If a balanced lamp load is connected to the secondary terminals abc, what is the phase relation of the primary line current I_A to the primary line voltage V_{AB}?

5. A balanced resistance load is connected to the secondary terminals of each of three groups of autotransformers arranged as shown in Figs. 1, 2, and 3. Compare the effects produced by opening the circuit breaker in line A in these three arrangements.

CHAPTER XXVI

Three-Phase Transformers

In the transformation of three-phase power, a three-phase transformer is frequently preferred to a bank of three single-phase units. The three-phase transformer usually is less expensive than a bank of single-phase transformers of the same total rating because windings in the three-phase unit are placed on a common magnetic core rather than on three independent ones, the consolidation resulting in an appreciable saving in materials. As with single-phase transformers, arrangement of the magnetic circuit divides three-phase transformers into the *shell type*, in which the magnetic circuit is a shell encircling the windings, and the *core type*, in which the magnetic circuit is a core surrounded by the windings.

1. SHELL TYPE

Consider three single-phase shell-type transformers with their cores pushed together so that they touch, as indicated in Fig. 1a. The only difference between this arrangement of three individual single-phase

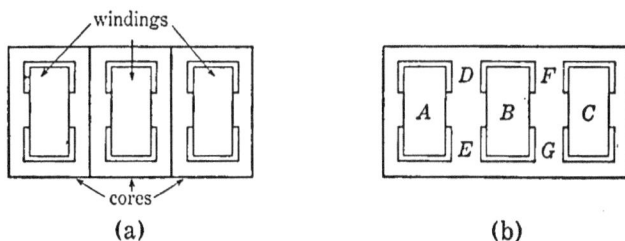

FIG. 1. The shell-type transformer; (a) three single-phase transformers, and (b) the corresponding 3-phase unit.

transformers and a three-phase shell-type transformer, shown in Fig. 1b, is that the core laminations of the latter are interwoven; the three parts of the core are not separated. This blending of the magnetic cores causes the core fluxes of adjacent phases to merge in the parts of the core indicated by D, E, F, G in Fig. 1b, and, as shown below, makes possible an appreciable saving of core material.

Assume that the transformer is operating with balanced sine-wave induced voltages. Then the core fluxes linking the three primary windings are balanced and sinusoidal in waveform, and can be represented by the vectors Φ_A, Φ_B, Φ_C in Fig. 2a. The three primary windings can be con-

611

nected symmetrically in the circuit so that the positive directions of the core fluxes Φ_A, Φ_B, Φ_C are the same, as in Fig. 2b; or the connections of the middle phase can be reversed so that the positive directions are as indicated in Fig. 2c. In Fig. 2b, the fluxes in the portions D and E of the core, Fig. 1b, are each $\frac{1}{2}(\Phi_A - \Phi_B)$, and from the vector diagram of Fig. 2a the magnitude of each of these fluxes is $(\sqrt{3}/2)\phi$, where ϕ is the magnitude of each of the core fluxes Φ_A, Φ_B, Φ_C. However, in Fig. 2c (in which the connections of the middle phase are reversed), the fluxes in D and E are each $\frac{1}{2}(\Phi_A + \Phi_B)$, and from the vector diagram the magnitude of each of these fluxes is $\frac{1}{2}\phi$, or $1/\sqrt{3}$ of their magnitude in Fig. 2b. Consequently, if all parts of the core are to operate at the same

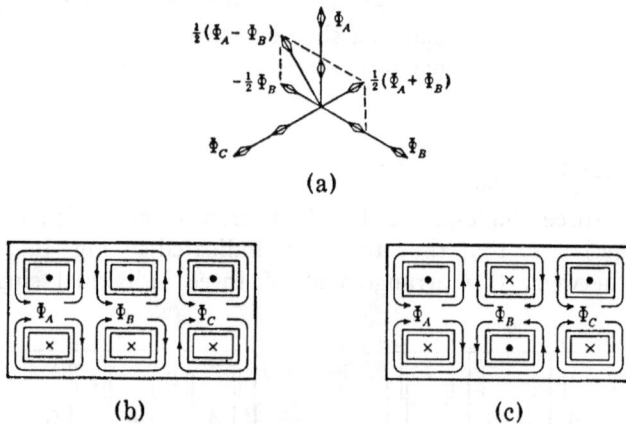

(a)

(b)   (c)

FIG. 2.   Flux relations in a 3-phase shell-type transformer; (a) vector diagram for balanced phase voltages, (b) positive directions of the fluxes for symmetrically connected windings, and (c) positive directions of the fluxes when the connections of the middle phase are reversed. The standard method of connection is that shown in (c).

maximum flux density, the cross-sectional area of the core in the spaces D, E, F, G between the phases need be but $1/\sqrt{3}$ or 0.58 as great when the connections of the middle phase are inverted, Fig. 2c, as when the connections are made so that the positive directions of the core fluxes are the same, Fig. 2b.

▶ Therefore the connections of the middle phase of a three-phase shell-type transformer are always inverted so that the positive directions of the fluxes are as shown in Fig. 2c.◀

When the connections are made as in Fig. 2c, the magnitude of the fluxes in the parts D, E, F, G of the core is the same as that of the fluxes in the corresponding parts of the core in a single-phase shell-type transformer. Consequently the core of a three-phase shell-type transformer contains less material than do the cores of three single-phase shell-type transformers designed for the same core flux and the same maximum flux density. The shaded portion of Fig. 3 shows the core material that can be saved in the three-phase design.

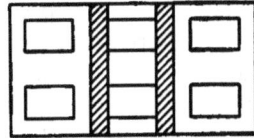

FIG. 3. Comparison of single-phase and 3-phase shell-type designs. The shaded portion shows the core material saved by the 3-phase design.

In all parts of the core, except as noted above, the fluxes are the same as they would be in single-phase transformers. The merging of the fluxes in no important way affects the operation of a three-phase shell-type transformer except to modify the waveforms and slightly to unbalance the magnitude of the exciting currents.

Courtesy Westinghouse Electric and Manufacturing Co.

Courtesy Pennsylvania Transformer Co.

FIG. 4. The operating parts of a shell-type 3-phase autotransformer rated 60 ∼ 36,000 kva 34,500 : 13,800 v.

FIG. 5. The core, coils, and lead-in assembly of a heavy-duty 3-phase transformer used for power supply to an electric furnace having a capacity of 75 tons for each charge. The rating is 25 ∼ 15,000 kva 6,600 : 242-v, Δ–Δ, with an assortment of taps. This transformer has a 5-legged core. Its electrical characteristics are similar to those of a 3-phase shell-type unit.

▶ Thus there is substantially no difference between the operating characteristics of a three-phase shell-type transformer and a similarly connected group of three single-phase transformers.◀

2. Three–legged core type

Consider a group of three single-phase core-type transformers whose windings are placed on but one leg of each core, as illustrated diagrammatically in Fig. 6a. If the induced voltages are balanced and vary sinusoidally, the core fluxes are also balanced and sinusoidal in waveform, and consequently their instantaneous sum is zero. If the transformers are arranged so that the positive directions of their core fluxes are the same in the adjoining legs of the cores, as in Fig. 6a, the instantaneous sum of the fluxes in these legs is zero. Hence, if the cores are joined at the top and bottom, forming two Y-shaped yokes as in Fig. 6b, the core legs at the center, shown by the broken line, can be omitted without dis-

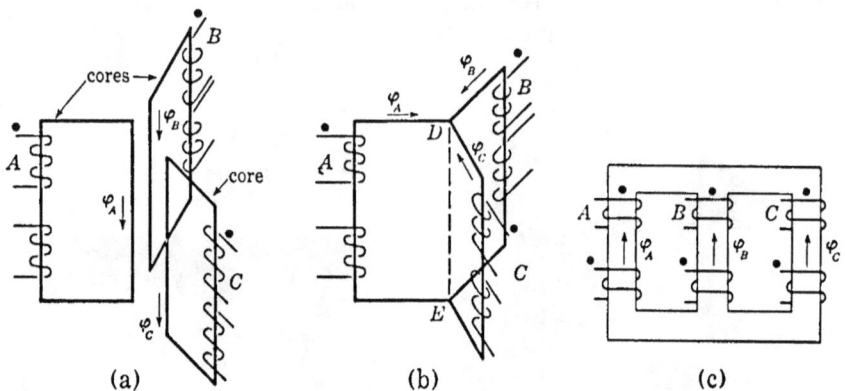

FIG. 6.   Synthesis of a 3-phase, 3-legged, core-type transformer from three single-phase units.

turbance to the fluxes or the exciting currents. Conditions in the magnetic circuit are similar to those in an electric circuit comprising a Y-connected source and a Y-connected load, the center core legs being the magnetic equivalent of a low-impedance electrical connection between the source and load neutrals. In the electric-circuit analogy, if the instantaneous sum of the line currents delivered to the load is zero, there is no current in the neutral conductor, which therefore may be omitted without disturbance to the circuit. However, in the electric circuit, if the instantaneous sum of the line currents is not zero when the source and load neutrals are connected, the omission of the neutral conductor would disturb the circuit. Similarly, in the magnetic circuit of Fig. 6b, if for any reason the sum of the core fluxes in the broken-line central legs is not zero, the removal of these core legs would derange the distribution of the fluxes. The unbalanced flux would have to close its circuit through the high-reluctance air path from the upper yoke D to the lower yoke E, Fig. 6b. The electric-circuit analogy is the insertion of a high impedance in the neutral conductor.

In the actual three-legged core-type transformer, the dotted central core legs of Fig. 6b do not appear, the horizontal portions of the core of the B transformer are reduced to zero, and the cores are arranged in the same plane, as indicated in Figs. 6c and 7. The result is an appreciable saving in core material. This construction slightly unbalances the magnetic circuits so that the exciting currents are somewhat unbalanced.

FIG. 7.   The core and coils of an air-cooled, 3-phase, 3-legged, core-type power transformer.

In this respect, the core- and shell-type transformers are alike. However, in the shell type there is a closed iron path for the flux of each phase, whereas in the three-legged core type unbalanced flux must close its circuit through an air path.

▶ Thus the characteristics of three-phase core- and shell-type transformers differ when the instantaneous sum of the phase fluxes is not zero. With this exception — which, however, may be important under the circumstances discussed below — there is substantially no difference between the electrical characteristics of a three-phase core-type and a similarly connected three-phase shell-type transformer or group of single-phase transformers.◀

The instantaneous sum of the phase fluxes may not be zero either because of third-harmonic fluxes resulting from excitation phenomena in the Y–Y connection, or because of unbalanced currents with zero-sequence components. The characteristics of three-phase, three-legged, core-type transformers under these two conditions are discussed in the following two sections of this article.

2a. *Excitation Phenomena in* Y–Y-*Connected, Three-Legged, Core-Type Transformers.* — If the primary windings are connected in Y with their neutral isolated, the sum of the exciting currents must be zero. If the magnetic circuit were perfectly symmetrical, and balanced sine-wave line voltages were applied to the three primary line terminals, the exciting currents would be balanced; that is, they would be equal in magnitude and waveform, and displaced in phase by one-third of a cycle. The unbalance in the magnetic circuit due to the three-phase core-type construction is so small that for the purposes of this discussion the exciting currents may be considered as substantially balanced.

If the sum of three balanced currents is zero (because of the isolated neutral), the currents can contain no third-harmonic components, or multiples of the third harmonic. This is like the condition existing when three single-phase transformers are connected in Y with isolated neutral.

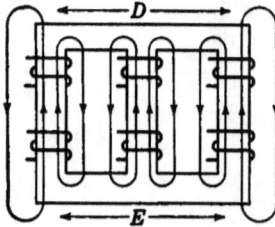

Fig. 8. Paths of third-harmonic and other zero-sequence fluxes in a 3-phase, 3-legged, core-type transformer.

With single-phase transformers, the suppression of the third-harmonic components of the exciting currents introduces third-harmonic components in the fluxes and hence there are relatively large third-harmonic components in the line-to-neutral voltages of the bank. The objectionable features of these third-harmonic voltages are discussed in Art. 2d, Ch. XXIII.

The behavior of a three-phase, three-legged, core-type transformer, however, is quite different. The third-harmonic components in the core fluxes are in time phase, and consequently the magnetic circuit for third-harmonic flux is completed through air from the upper yoke *D* to the lower yoke *E*, as indicated in Fig. 8. The high reluctance of this air path limits the third-harmonic fluxes to small values, and therefore the third-harmonic components of the line-to-neutral voltages are relatively small.

▶ Thus a three-legged core-type transformer may be Y–Y connected with isolated neutrals without the production of the large third-harmonic voltages to neutral that would be present in a bank of single-phase transformers or in a three-phase shell-type transformer connected in the same manner. ◀

The magnetic characteristics of iron demand a third-harmonic component in the magnetizing force when the flux density varies sinusoidally; yet in the three-legged core-type transformer a substantially sinusoidal variation of the fluxes results even when the third-harmonic components of the exciting currents are suppressed by the Y–Y connection with isolated neutrals. Superficially, these two statements appear contradictory, but further inquiry shows that they are not. Actually, there are relatively small third-harmonic components in the fluxes, whose paths are in air from the upper yoke D to the lower yoke E of Fig. 8. To create these third-harmonic fluxes, there must be a third-harmonic fall in magnetic potential through the air path from D to E, and therefore an equal rise in magnetic potential acting on the vertical legs of the core from E to D. Consequently, acting on each vertical leg of the core is a resultant magnetic potential difference, due in part to the magnetomotive force of the exciting current and in part to the effect of the third-harmonic flux and its multiples. The third-harmonic flux adjusts itself so that the third-harmonic rise in magnetic potential acting on each vertical leg of the core from E to D — as required for the existence of the third-harmonic flux in the air path from D to E — equals the third-harmonic magnetomotive force required by the iron. Since the reluctance of the air path from D to E is relatively high, a relatively small third-harmonic flux results in the relatively much larger third-harmonic magnetomotive force required for the iron, and therefore the flux variation is very nearly sinusoidal.

2b. *Zero-Sequence Currents in Three-Legged Core-Type Transformers.* — Consider a Y–Y-connected, three-legged, core-type transformer supplying power to a Y-connected three-phase load. The secondary neutral is connected to the neutral of the load. If the load is unbalanced, the instantaneous sum of the line currents delivered to the load may not be zero and therefore there may be current in the secondary neutral wire. If the circuit is analyzed by the method of symmetrical components, the current in each phase contains a zero-sequence component that equals one-third of the neutral current.

If the primary neutral is connected to the neutral of the source, each primary phase receives its power independently of the other phases. Under these conditions and except for exciting currents, the primary and secondary currents produce equal and opposing magnetomotive forces, and therefore the primary and secondary currents are inversely proportional to the numbers of turns in the primary and secondary windings, and each phase of the transformer is equivalent to a single-phase ideal transformer in series with the equivalent or short-circuit impedance of the two windings. Thus, when the primary neutral is connected to the neutral of the source, there is no essential difference between the behavior of a Y–Y-connected, three-legged, core-type transformer and a similarly connected bank of single-phase transformers.

However, if the primary neutral is isolated, the instantaneous sum of the primary line currents must be zero; that is, there can be no zero-sequence primary currents. Except for the effects of exciting currents, the instantaneous sum of the secondary currents must therefore be zero also; that is, the only zero-sequence secondary currents that can exist are exciting currents, since they are present in only one winding of each phase without inducing an equal and opposite load component of current in the other winding.

With a Y–Y bank of single-phase transformers (or a similarly connected three-phase shell-type transformer), when the primary neutral is isolated from the neutral of the source, any attempt to supply current to a single-phase load connected to the secondary neutral will force more than normal exciting currents on the two unloaded transformers, and thereby will greatly unbalance the voltages to neutral.* When the voltages to neutral become unbalanced, the fluxes also become unbalanced; that is, their instantaneous sum is not zero, and they contain zero-sequence components.

In a three-legged core-type transformer, however, the unbalanced zero-sequence flux must complete its circuit through the high-reluctance air path from the upper yoke D to the lower yoke E of Fig. 8. In this respect, the flux due to unbalanced zero-sequence currents is like the third-harmonic flux described in part (a) of this article.

▶ Thus in a three-legged core-type transformer the unbalanced zero-sequence flux is limited to a relatively small value, and therefore the voltages to neutral tend to remain balanced, in spite of the unbalanced magnetomotive forces resulting from the single-phase load to neutral. A single-phase load equal to several per cent of the three-phase rating can be delivered from line to neutral of a Y–Y-connected, three-legged, core-type transformer without seriously unbalancing the phase voltages, even when the primary neutral is isolated.[1]◀

3. OPEN–DELTA CONNECTION OF THREE–PHASE TRANSFORMERS

If one phase of a Δ–Δ-connected three-phase transformer becomes damaged, the transformer can be reconnected in open delta and used to supply loads up to $1/\sqrt{3}$ or 0.58 of its normal rating.† Both the primary and secondary windings of the damaged phase should be disconnected from the circuit.

With a shell-type transformer, these windings may be either open-

---

* See Art. 2b, Ch. XXI.

[1] For further discussion of this point, see L. F. Blume, editor, *Transformer Engineering* (New York: John Wiley & Sons, 1938), 188–190.

† See Art. 5, Ch. XXI.

circuited or short-circuited. If they are open-circuited, voltages are induced in them whose magnitudes may be about one-quarter of rated voltage. If they are short-circuited, the induced voltages produce short-circuit currents, but the magnetic effect of these induced currents forces most of the flux to follow paths that do not link the short-circuited windings. Therefore the currents in the short-circuited windings are relatively small.

▶ With a three-legged core-type transformer, however, both windings of the damaged phase must be *open-circuited*, since the return paths for the fluxes of the other two phases must be through the core of the damaged phase. Thus, if the fault in the damaged phase is an internal short circuit, it is impossible to operate a three-legged core-type transformer in open delta without first opening the tank and removing the short circuit.◀

4. COMPARISON OF THREE–PHASE AND SINGLE–PHASE TRANSFORMERS

As shown in the preceding articles, the merging of the magnetic circuits results in an appreciable saving of core material in a three-phase transformer as compared with a bank of three single-phase units having similar characteristics. Furthermore, while three single-phase units may require six high-voltage and six low-voltage bushings, a three-phase transformer requires only three of each, the connections between phases being made at an internal terminal board. The single larger tank for a three-phase transformer may cost less than the three smaller tanks for a bank of single-phase units. More expensive radiators or other means of cooling may, however, be required for the three-phase unit.

The result of these savings in materials is that in power-transformer sizes a three-phase transformer usually costs and weighs less than a bank of single-phase transformers having similar characteristics. Certain exceptions to this statement should be noted, however. In small sizes (that is, for bank ratings below roughly 300 kva, or 100 kva per phase) and for standard distribution-circuit voltages, there is a much greater demand for single-phase transformers than for three-phase units; consequently, the lower manufacturing cost resulting from quantity production of single-phase units offsets the possibly greater cost of materials. In these ratings, therefore, there is no substantial difference in the costs of a three-phase transformer and a bank of single-phase transformers. Furthermore, in small sizes the three-phase unit may actually contain more material and weigh more than three single-phase units of the same bank rating, particularly when the wound-core design* is used for the single-phase units.

* See Art. 1, Ch. XI.

Courtesy Westinghouse Electric and Manufacturing Co.

FIG. 9. A 33,000 : 5,000-v 3-phase portable substation for emergency use. The substation includes high-voltage disconnecting switch, lightning arresters, 3-phase transformer, and low-voltage circuit breaker.

FIG. 10. A 750-kva 13,800 : 4,800-v 3-phase unit substation. In addition to the transformer, the equipment includes high-voltage disconnecting switches and fuses, low-voltage circuit breaker, and a 13,800-v automatic tap changer.

The principal advantages and disadvantages of a three-phase transformer as compared with a bank of single-phase units can be summarized as follows:

Advantages of Three-Phase Transformers
1. Usually cost less.
2. Usually weigh less.
3. Occupy less space.
4. Only one unit to handle and connect.

Disadvantages
1. Greater weight per unit.
2. Greater cost of spare units.
3. Greater cost of repairs.

PROBLEMS

1. A single-phase voltage of 110 v at rated frequency is impressed on the low-voltage winding on the middle leg of a three-phase three-legged core-type transformer. The current is 1.0 amp when all other windings are open-circuited.

(a) What are the open-circuit voltages of each of the other low-voltage windings?

(b) If one of the low-voltage windings is short-circuited, what is the open-circuit voltage of the other low-voltage winding? Estimate the current in the short-circuited winding.

(c) If one of the low-voltage windings is short-circuited and a resistance of 10 ohms is connected across the other low-voltage winding, estimate the current in each low-voltage winding.

Assume that the resistances of the windings are negligibly small and that there is no magnetic field outside the core.

2. The low-voltage windings of a three-phase three-legged core-type transformer are connected in Δ and balanced three-phase voltages of 440 v at rated frequency are impressed at the terminals of the Δ.

If the low-voltage winding on the middle leg is disconnected from the others and from the three-phase supply, what voltage is generated in it? Does the disconnection cause the no-load currents in the other low-voltage windings to increase or diminish?

3. Rated voltage of 2,400 v at rated frequency is impressed on the middle low-voltage winding of a three-phase shell-type transformer. Estimate the open-circuit voltage generated in each of the other two low-voltage windings.

CHAPTER XXVII

Multicircuit Transformers

On many occasions single-phase or three-phase transformers are used to interconnect three or more independent circuits. Although two-winding transformers can be employed, the same results can be obtained at lower cost and with smaller losses by use of transformers having three or more independent windings.

An example of a single-phase application is a distribution transformer supplying a 120/240-volt three-wire circuit from its two secondaries, each of which can be loaded independently. The analysis of such a transformer when it is supplying unequal 120-volt loads must be based on three-circuit transformer theory. A rather similar application in communication circuits is found in transformers with center-tapped secondaries, as used in push-pull amplifiers.[1]

A common three-phase application of three-circuit transformers is in the Y–Y–Δ connection described in Art. 3b, Ch. XXIII. The Δ-connected tertiary windings provide a circuit for the third-harmonic components of the exciting currents, the lack of which may introduce undesirable third-harmonic components in the voltages to neutral of the Y-connected windings. The tertiary windings also may supply power to a local load, such as auxiliary apparatus or synchronous condensers used for power-factor control or voltage regulation.[2]

Often it is desirable to supply an important distribution system from two separate transmission systems — which may be of different voltages — so that in the event of a fault on either transmission system the distribution system still can be supplied from the other source. Under these conditions, three-circuit transformers generally are used.

Sometimes a large distribution system is divided into two portions, each one supplied from an independent group of windings on a three-phase bank of three-circuit transformers. When the load is subdivided in this manner, the currents resulting from accidental short circuits can be reduced without adversely affecting the voltage regulation. Thus, with a subdivided load, each secondary winding supplies only part of the load current, and therefore it may be designed to have a larger leakage reactance than would be permissible for the same voltage regulation with

---

[1] Push-pull amplifiers are discussed in this series in the volume on electronics, Art. 14, Ch. VIII, and Arts 1 and 2, Ch. X.

[2] For a discussion of the use of multicircuit transformers with synchronous condensers, see H. P. St.Clair, " The Use of Multiwinding Transformers with Synchronous Condensers for System Voltage Regulation," *A.I.E.E. Trans.*, 59 (1940), 212–217.

an undivided load. The larger leakage reactance results in smaller short-circuit currents.

Another common use of multicircuit transformers is in transformer banks for phase transformation. By the use of a suitably connected bank of multicircuit transformers, six-phase power, for example, can be obtained from a three-phase supply. These transformer connections are described in Ch. XXVIII.

Many types of multicircuit transformers are used. The external circuits may all be connected to separate windings that are insulated from each other, or portions of the windings may be common to two or more circuits, as in autotransformers.* The transformers used in three-phase arrangements may be single-phase transformers, or they may be three-phase core- or shell-type transformers with several windings for each phase. The three-phase windings may be connected in a variety of ways, the commonest arrangements being the Y or Δ connection.

Since the performance of balanced three-phase circuits can be determined by analysis of a single phase,† and since (except for excitation phenomena) the per phase characteristics of a three-phase transformer do not differ from those of a single-phase transformer, the theory of a multicircuit single-phase transformer developed in this chapter applies on a per phase basis to three-phase problems involving either three-phase transformers or banks of single-phase transformers.‡ Furthermore, the analysis applies to the external characteristics of autotransformers as well as to transformers having separate windings, since the characteristics of an autotransformer as viewed from its terminals can be computed from its open-circuit and short-circuit impedances by exactly the same methods of analysis that apply to a transformer having separate windings.§ Although the currents in some of the windings of an autotransformer differ from those in the external circuits, they can readily be computed from the circuit currents.

Important problems arising in connection with the use of multicircuit transformers are those of voltage regulation, short-circuit currents, load

---

* See Ch. XV.

† The analysis of unbalanced circuits by the method of symmetrical components also involves the same per phase analysis.

‡ In the analysis of polyphase circuits in which there are cross connections between the windings on different magnetic cores — as in the zigzag connection discussed in Art. 4, Ch. XXV, and in many circuits for phase transformation — the following analysis applies on a single-phase basis to all the windings on any magnetic core, but it may *not* be permissible to consider each phase of the electric circuit as a single-phase problem, because each phase current flows in two or more cross-connected windings and affects two or more magnetic cores. For an analysis of circuits involving cross-connected windings, see A. Boyajian, " Progress in Three-Circuit Theory," *A.I.E.E. Trans.*, *52* (September, 1933), 914–917.

§ See Ch. XV.

division between circuits, and the behavior of multicircuit transformers in parallel with other transformers. In order to solve these problems, it is necessary to know the characteristics of a multicircuit transformer as an impedance element in the circuits to which it is connected. The following discussion aims to develop methods of analysis on which the solution of these and similar problems can be based.

1. General theory of multicircuit transformers

Figure 1 shows diagrammatically a transformer having n windings, all placed on a common magnetic core. This circuit may represent either a single-phase transformer or one phase of a three-phase transformer. Except at very high frequencies or during rapidly varying transient conditions, the currents taken by the capacitances of the transformer are negligibly small and therefore the currents in the circuits are unchanged if all terminals of the same relative polarity are connected and grounded, as shown by the broken-line connection in Fig. 1.

If the circuits connected to the transformer are ungrounded, the voltages between each terminal and ground are determined by the capacitances of the external circuits and of the transformer windings with respect to each other and with respect to ground. Thus the circuit of Fig. 1, in

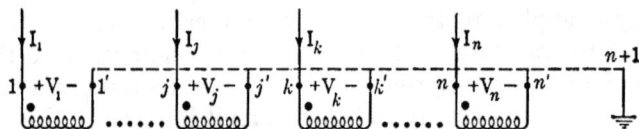

FIG. 1.   Circuit diagram for an n-circuit transformer.

which all terminals of the same relative polarity are arbitrarily grounded, may not give the correct voltages between one circuit and another, and from the various circuits to ground. In the analysis of problems concerning the characteristics of transformers at high frequencies or during rapidly varying transient conditions — such as transients caused by lightning — these voltages may be important and the arbitrary grounding of terminals of like polarity may not be permissible. However, so long as the currents taken by the capacitances are negligibly small, the relations among the terminal voltages of the various windings and the currents in them are unaltered by the broken-line ground connection in Fig. 1. The transformer then may be analyzed as an n-loop circuit with $n + 1$ terminals, or as a circuit with $n + 1$ nodes, n of which are independent.

In spite of the nonlinear magnetic characteristics of iron, it is shown in preceding chapters that a two-winding transformer often may be analyzed as a linear circuit element,* and the same is true of a multicircuit transformer. The nonlinear magnetic characteristics of the core affect only the excitation phenomena, but, since the exciting current usually is only a few per cent of the rated circuit currents, it is sufficiently accurate usually to assume that the exciting current for sine-wave flux is sinusoidal in waveform. Therefore the theory of linear coupled circuits, discussed in the volume on electric circuits, Ch. VIII, may be used as a basis for the analysis of multicircuit transformers.

1a. *The Loop-Voltage Equations.* — An n-circuit transformer can be analyzed as an n-loop circuit, each loop having electromagnetic coupling with all other loops. Thus the following steady-state vector voltage equations can be written:[3]

$$V_1 = Z_{11}I_1 + Z_{12}I_2 + \cdots + Z_{1j}I_j + Z_{1k}I_k + \cdots + Z_{1n}I_n \quad [1]$$

$$V_2 = Z_{21}I_1 + Z_{22}I_2 + \cdots + Z_{2j}I_j + Z_{2k}I_k + \cdots + Z_{2n}I_n \quad [2]$$

$$\cdots\cdots\cdots\cdots\cdots\cdots\cdots\cdots\cdots\cdots\cdots\cdots\cdots\cdots\cdots\cdots\cdots$$

$$V_j = Z_{j1}I_1 + Z_{j2}I_2 + \cdots + Z_{jj}I_j + Z_{jk}I_k + \cdots + Z_{jn}I_n \quad [3]$$

$$\cdots\cdots\cdots\cdots\cdots\cdots\cdots\cdots\cdots\cdots\cdots\cdots\cdots\cdots\cdots\cdots\cdots$$

$$V_n = Z_{n1}I_1 + Z_{n2}I_2 + \cdots + Z_{nj}I_j + Z_{nk}I_k + \cdots + Z_{nn}I_n, \quad [4]$$

in which the V's are the vector terminal voltages of the windings, the I's are the vector currents in them and the Z's are their complex self- and mutual impedances. The positive directions of the currents and voltages are shown in Fig. 1, in which terminals of the same relative polarity are indicated by dots. The meanings of the impedances can be explained by means of a typical equation. For example, in Eq. 3, Z_{jj} is the complex self-impedance of circuit j, and Z_{jk} is the complex mutual impedance between circuits j and k. Thus

$$Z_{jj} = R_j + jX_{jj}, \quad [5]$$

where R_j is the effective resistance and X_{jj} is the self-inductive reactance of the winding. If core loss is neglected, the resistive components of the mutual impedances are zero, and, for the assumed positive directions of the currents and voltages in Fig. 1, the mutual reactances are positive; that is

$$Z_{jk} = jX_{jk}, \quad [6]$$

where X_{jk} is the mutual reactance — a positive quantity. It is well known that the mutual reactance of circuit j with respect to circuit k

---

* For further discussion of the application of linear-circuit theory to iron-core transformers, see Art. 1, Ch. XVII.

[3] See the volume on electric circuits (1940), Art. 5, Ch. VIII, p. 439.

equals that of circuit k with respect to circuit j, or

$$Z_{jk} = Z_{kj}.$$ [7]

Theoretically, then, the behavior of an n-circuit transformer for any specified operating condition can be computed if the complex values of the impedance coefficients in Eqs. 1–4 are known from measurements or from computations based on dimensions. Note the open-circuit character of the impedance coefficients in these equations. Thus the self-impedance of any circuit equals its impedance with all other circuits open, and the mutual impedance between any two circuits, say Z_{jk}, equals the vector ratio of the open-circuit voltage produced in circuit j to the current in circuit k, all circuits except circuit k being open. Therefore the impedances on Eqs. 1–4 can be determined experimentally by means of a number of open-circuit tests.

Just as in the analysis of a two-winding transformer,* however, inaccurate results would be obtained if the performance of a multicircuit transformer were computed by substitution of measured values of the self- and mutual impedances in Eqs. 1–4. The voltage drops in the transformer depend primarily on magnetic leakage — that is, on relatively small differences among the various self- and mutual impedances — and in order to determine these small differences with reasonable accuracy, the self- and mutual impedances would have to be measured with very great accuracy. Such measurement would be very difficult indeed, owing to the nonlinear magnetic characteristics of the core. Therefore, although Eqs. 1–4 are useful as a starting point for the analysis of multicircuit transformers, it is necessary to manipulate them into other forms before they can be used in numerical computations.

1b. *Short-Circuit Driving-Point and Transfer Admittances.* — Equations 1–4 can be solved for the currents by Cramer's rule, yielding the following vector equations:[4]

$$I_1 = \frac{M_{11}}{D_z}V_1 + \frac{M_{21}}{D_z}V_2 + \cdots + \frac{M_{j1}}{D_z}V_j + \frac{M_{k1}}{D_z}V_k + \cdots + \frac{M_{n1}}{D_z}V_n$$ [8]

$$I_2 = \frac{M_{12}}{D_z}V_1 + \frac{M_{22}}{D_z}V_2 + \cdots + \frac{M_{j2}}{D_z}V_j + \frac{M_{k2}}{D_z}V_k + \cdots + \frac{M_{n2}}{D_z}V_n$$ [9]

$$\cdots\cdots\cdots\cdots\cdots\cdots\cdots\cdots\cdots\cdots\cdots\cdots\cdots\cdots\cdots\cdots$$

$$I_j = \frac{M_{1j}}{D_z}V_1 + \frac{M_{2j}}{D_z}V_2 + \cdots + \frac{M_{jj}}{D_z}V_j + \frac{M_{kj}}{D_z}V_k + \cdots + \frac{M_{nj}}{D_z}V_n$$ [10]

$$\cdots\cdots\cdots\cdots\cdots\cdots\cdots\cdots\cdots\cdots\cdots\cdots\cdots\cdots\cdots\cdots$$

$$I_n = \frac{M_{1n}}{D_z}V_1 + \frac{M_{2n}}{D_z}V_2 + \cdots + \frac{M_{jn}}{D_z}V_j + \frac{M_{kn}}{D_z}V_k + \cdots + \frac{M_{nn}}{D_z}V_n$$ [11]

---

* See Art. 2, Ch. XVII.

[4] See the volume on electric circuits (1940), Art. 5, Ch. VIII, p. 440.

in which D_z is the determinant of the Z's in Eqs. 1–4, and M_{jk} is the cofactor of the jth row and kth column including the sign factor $(-1)^{j+k}$. Because of the symmetry of Eqs. 1–4,

$$M_{jk} = M_{kj}. \qquad [12]$$

Each term on the right-hand sides of Eqs. 8–11 is a component of one of the currents produced by one of the terminal voltages acting alone, the other terminal voltages all being zero. Thus the coefficient of each V term is representable as an admittance. In general,

$$y_{jk} = \frac{M_{kj}}{D_z} \qquad [13]$$

is the vector current in circuit j resulting from unit vector voltage V_k applied to circuit k when all other terminal voltages are zero — that is, short-circuited. The admittance y_{jk} is called the *short-circuit transfer admittance* between circuits j and k, and from Eq. 12 it is evident that

$$y_{jk} = y_{kj}. \qquad [14]$$

When the voltage and current are in the same circuit so that the admittance has the form

$$y_{jj} = \frac{M_{jj}}{D_z}, \qquad [15]$$

the admittance is called the *short-circuit driving-point admittance*, since it is the vector current in circuit j per unit vector voltage applied to the same circuit when all other circuits are short-circuited. When rewritten in terms of the short-circuit driving-point and transfer admittances, Eqs. 8–11 become

$$I_1 = y_{11}V_1 + y_{12}V_2 + \cdots + y_{1j}V_j + y_{1k}V_k + \cdots + y_{1n}V_n \qquad \blacktriangleright[16]$$

$$I_2 = y_{21}V_1 + y_{22}V_2 + \cdots + y_{2j}V_j + y_{2k}V_k + \cdots + y_{2n}V_n \qquad \blacktriangleright[17]$$

$$\cdots\cdots\cdots\cdots\cdots\cdots\cdots\cdots\cdots\cdots\cdots\cdots\cdots\cdots$$

$$I_j = y_{j1}V_1 + y_{j2}V_2 + \cdots + y_{jj}V_j + y_{jk}V_k + \cdots + y_{jn}V_n \qquad \blacktriangleright[18]$$

$$\cdots\cdots\cdots\cdots\cdots\cdots\cdots\cdots\cdots\cdots\cdots\cdots\cdots\cdots$$

$$I_n = y_{n1}V_1 + y_{n2}V_2 + \cdots + y_{nj}V_j + y_{nk}V_k + \cdots + y_{nn}V_n. \qquad \blacktriangleright[19]$$

As shown subsequently, these equations are suitable for computational purposes, since the short-circuit admittances in Eqs. 16–19 readily can be measured or computed from dimensions and are directly related to the leakage-impedance characteristics of the transformer.

1c. *The Node-Current Equations.* — In the preceding two sections of this article, the loop-voltage equations, Eqs. 1–4, are taken as the starting point for the analysis of the n-circuit transformer. The solution of these equations yields the more useful loop-current equations, Eqs. 16–19.

Alternatively, the current equations can be obtained directly if the node-current equations are written for the circuit of Fig. 1, the grounded terminals $1'$, $2'$, $\cdots n'$ being taken as the reference node.[5] Thus

$$I_1 = Y_{11}V_1 + Y_{12}V_2 + \cdots + Y_{1j}V_j + Y_{1k}V_k + \cdots + Y_{1n}V_n \quad [20]$$

$$I_2 = Y_{21}V_1 + Y_{22}V_2 + \cdots + Y_{2j}V_j + Y_{2k}V_k + \cdots + Y_{2n}V_n \quad [21]$$

$$\cdots\cdots\cdots\cdots\cdots\cdots\cdots\cdots\cdots\cdots\cdots\cdots\cdots\cdots\cdots$$

$$I_j = Y_{j1}V_1 + Y_{j2}V_2 + \cdots + Y_{jj}V_j + Y_{jk}V_k + \cdots + Y_{jn}V_n \quad [22]$$

$$\cdots\cdots\cdots\cdots\cdots\cdots\cdots\cdots\cdots\cdots\cdots\cdots\cdots\cdots\cdots$$

$$I_n = Y_{n1}V_1 + Y_{n2}V_2 + \cdots + Y_{nj}V_j + Y_{nk}V_k + \cdots + Y_{nn}V_n. \quad [23]$$

In these equations the I's are the vector currents entering the free terminals, the V's are the vector potentials of these terminals with respect to the grounded reference node, and the Y's are the complex self- and mutual admittances of the nodes. For example, Y_{jj} is the self-admittance of node j and equals the vector current entering terminal j per unit vector voltage applied to terminal j, all other terminals being connected to ground — that is, short-circuited. Similarly Y_{jk} is the mutual admittance of nodes j and k, and equals the vector current entering terminal j per unit vector voltage applied to terminal k, all terminals but terminal k being grounded, or short-circuited. As in linear networks,

$$Y_{jk} = Y_{kj}. \quad [24]$$

Note that, in spite of the similarity in notation, the admittances in Eqs. 20–23 are *not* the reciprocals of the corresponding impedances in Eqs. 1–4; that is,

$$Y_{jk} \text{ (Eq. 22)} \neq \frac{1}{Z_{jk}} \text{ (Eq. 3)}.$$

The impedances in Eqs. 1–4 are *open-circuit* impedances, but the admittances in Eqs. 20–23 are *short-circuit* admittances.

Since the short-circuit admittances depend principally on magnetic leakage, and since magnetic leakage is the major cause of voltage drops in a transformer, the coefficients in Eqs. 20–23 are directly related to the important factors influencing the effects of the transformer as an impedance element in the circuits to which it is connected. Furthermore, the short-circuit admittances readily can be computed from the geometrical arrangements of the windings, or measured by the tests de-

---

[5] See the volume on electric circuits (1940), Art. 13, Ch. VIII, p. 463.

scribed in Art. 3 of this chapter. Therefore Eqs. 20–23 are in a convenient form for the analysis of multicircuit transformers.

1d. *Comparison of Loop and Node Equations.* — It is interesting to compare the loop equations, Eqs. 16–19, with the node equations, Eqs. 20–23. The V's, which are considered as the loop voltages in Eqs. 1–4 and 16–19, become the node voltages in Eqs. 20–23. The I's are considered as loop currents in Eqs. 1–4 and 16–19, and as currents entering the nodes in Eqs. 20–23. The y's in Eqs. 16–19 are the short-circuit driving-point and transfer admittances, whereas the Y's in Eqs. 20–23 are the self- and mutual admittances of the nodes. In spite of these differences in terminology, the V's, I's, and admittances of Eqs. 16–19 represent exactly the same physical quantities as the corresponding quantities in Eqs. 20–23, and therefore the loop and node methods of analysis lead to exactly the same current equations.

In the general theory of linear networks, the geometry of the network may be such that the electromotive forces acting in the loops, in terms of which the loop-voltage equations are written, do not coincide with the node potentials, in terms of which the node currents are expressed. It may not, therefore, be possible to connect one terminal of every driving point, as in Fig. 1, this common terminal then being considered the reference node. In such general cases, the self- and mutual admittances of the node method are not the same as the short-circuit driving-point and transfer admittances of the loop method; but, in Fig. 1, the loop voltages and the node potentials are the same, the loop currents and the node currents are the same, and therefore the admittances of the loop and node equations are the same also.

1e. *Currents, Voltages, and Circuit Parameters Referred to a Common Base.* — In all the preceding analysis of this article, the V's, I's, Z's, and Y's may be the actual voltages, currents, impedances, and admittances, but in the analysis of transformers usually it is more convenient to express the voltages, currents, impedances, and admittances on a common base. Thus they may all be expressed in per unit* on a common kilovolt-ampere base, or they may all be referred to one winding in a manner similar to that employed in the analysis of two-winding transformers.† When they are so referred, the transformer is analyzed as an equivalent transformer with the same number of turns in all its windings.

For example, the current I_k in circuit k can be referred to circuit j through division by the ratio

$$a_{jk} = \frac{N_j}{N_k} \qquad [25]$$

---

* See Art. 6, Ch. XIV.
† See Art. 4, Ch. XIII.

of turns N_j in winding j to turns N_k in winding k; hence, the current in circuit k referred to circuit j is

$$\frac{I_k}{a_{jk}} = I_k \frac{N_k}{N_j}. \qquad [26]$$

Note that this is the current, flowing in a winding of N_j turns, that would produce the same magnetic effect as the current I_k in the actual winding k. Also, as in two-circuit transformer theory, the voltage of circuit l referred to circuit j is

$$a_{jl}V_l = \frac{N_j}{N_l} V_l, \qquad [27]$$

where V_l is the actual voltage of circuit l and a_{jl} is the turns ratio N_j/N_l. Just as in two-circuit transformer theory, the referred voltage is the voltage of an equivalent winding having the same geometrical arrangement as the actual winding l, but with the same number of turns as the winding j to which the voltage is referred.

From the expressions for the referred values of current and voltage (Eqs. 26 and 27), the referred values of impedance and admittance readily can be obtained. For example, if $I_{k(l)}$ is the current produced in circuit k by voltage V_l applied to circuit l — all circuits but circuit l being short-circuited — then the short-circuit transfer admittance Y_{kl} between circuits k and l is

$$Y_{kl} = \frac{I_{k(l)}}{V_l}. \qquad [28]$$

If both the current and the voltage are referred to a third circuit — say circuit j — the quotient of the referred current and the referred voltage is the short-circuit transfer admittance between circuits k and l referred to circuit j and is, by Eqs. 26 and 27,

$$\frac{I_{k(l)}/a_{jk}}{a_{jl}V_l} = \frac{1}{a_{jl}a_{jk}} \frac{I_{k(l)}}{V_l} = \frac{Y_{kl}}{a_{jl}a_{jk}}. \qquad [29]$$

From a line of reasoning similar to that explained in the derivation of Eq. 29, the referred value of any other circuit parameter can be obtained when needed.

▶ In the following analysis, it is always assumed that the currents, voltages, impedances, and admittances either are all expressed in the per unit system on a common kilovolt-ampere base or are all referred to the same circuit.◀

2. EQUIVALENT CIRCUITS

Although the performance of an n-circuit transformer can be computed analytically from Eqs. 16–19 or 20–23, it is often more convenient to represent the transformer by an equivalent circuit and to determine the currents and voltages from the equivalent circuit.[6]

If an n-circuit transformer is to be represented by an equivalent circuit, it is evident that the equivalent circuit must have the same number of free terminals as the transformer. Consequently, in the most general case, the equivalent circuit for an n-circuit transformer must have $2n$ free terminals. However, as shown in Art. 1, when the currents taken by the capacitances of the windings are negligibly small the relations among the terminal voltages of the various windings and the currents in them are unaltered if terminals of the same relative polarity are considered connected, as shown by the broken-line connection in Fig. 1. Thus, except in the analysis of problems in which the capacitances of the windings are important (as in the determination of voltage stresses between ungrounded windings or during rapidly varying transient conditions), an n-circuit transformer can be analyzed as a circuit with $n + 1$ terminals, and therefore with these same restrictions it can be represented by an equivalent circuit with $n + 1$ terminals.

It is also evident that the equivalent circuit must have as many independent branches as there are independent coefficients in the equations expressing the current-voltage relations, Eqs. 16–19 or 20–23. The right-hand side of each of these equations comprises n terms, and contains one self-admittance and $n - 1$ mutual admittances. Thus the complete set of n current equations contains n self-admittances and $n(n - 1)$ mutual admittances, but each of the mutual admittances appears twice, whence there are $\dfrac{n(n - 1)}{2}$ independent mutual admittances. Therefore the n-circuit transformer is characterized by n self-admittances and $\dfrac{n(n - 1)}{2}$ mutual admittances, or a total of

$$n + \frac{n(n - 1)}{2} = \frac{n(n + 1)}{2} \tag{30}$$

independent admittance coefficients. Consequently the equivalent circuit must have this same number of independent branches. The right-hand side of Eq. 30 is the general expression for the number of possible combinations of $n + 1$ things taken two at a time.

[6] For a general discussion of the theory of equivalent networks for electromagnetically coupled circuits, see F. M. Starr, " Equivalent Circuits," *A.I.E.E. Trans.*, *51* (June, 1932), 287–298.

▶ Therefore, when it is permissible to consider one terminal of each circuit connected to the corresponding terminals of all the other circuits (as by the broken-line connection in Fig. 1), an n-circuit transformer can be represented by an equivalent circuit having $n + 1$ terminals and an admittance link connecting each terminal to all other terminals.◀

For example, consider the four-circuit transformer indicated by the windings 11′, 22′, 33′, 44′ in Fig. 2a. This circuit may represent either a single-phase four-circuit transformer in a single-phase circuit, or one phase of a three-phase bank of four-circuit transformers. In Fig. 2a, windings 1 and 2 are connected to generators, and windings 3 and 4 supply power to loads. Terminals 1′, 2′, 3′, 4′ are assumed to be connected, as shown by the broken line. An equivalent circuit is shown in Fig. 2b, and consists of a five-terminal network with admittance links connecting each terminal to all other terminals. There are ten independent admittance links, a number which agrees with the value given by substitution of $n = 4$ in Eq. 30.

In Fig. 2b, the admittances $Y_{\overline{10}}, Y_{\overline{20}}, Y_{\overline{30}}, Y_{\overline{40}}$ terminating at the common terminal 0 are in shunt with the generator and load circuits and represent the excitation characteristics of the transformer. Theoretically, the values of these admittances can be determined from the results of a number of open-circuit tests,* but, since the exciting current usually is only a few per cent of the

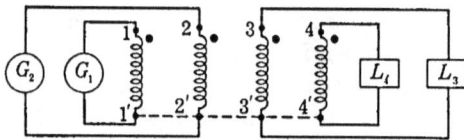

(a)

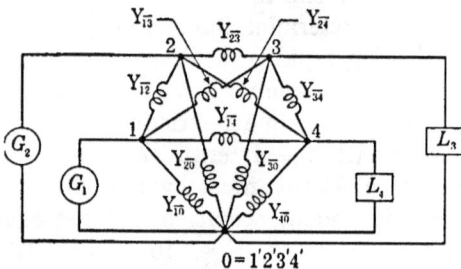

(b)

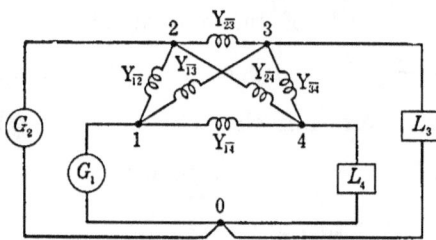

(c)

Fig. 2. Circuit diagram and equivalent circuits for a four-circuit transformer.

---

* The practical difficulties encountered in an attempt to determine the exact values of these admittances are discussed in Art. 5d of this chapter.

full-load currents, it can be treated by approximate methods. In fact, it usually may be neglected. However, if it is necessary to include the exciting current, either of two approximations usually is permissible. Thus it may be assumed that the n shunt admittances of an n-circuit transformer all have the same value and equal $1/n$th of the admittance of any circuit measured with all other circuits open. Or all but one of the shunt admittances may be neglected, and it may be assumed that the exciting current flows in this one shunt admittance, whose value equals the open-circuit admittance. When the open-circuit admittances are referred to a common base, very nearly the same value of admittance is obtained regardless of the circuit used for the measurement of the admittance.

The admittances $Y_{\overline{12}}$, $Y_{\overline{13}}$, $Y_{\overline{14}}$, $Y_{\overline{23}}$, $Y_{\overline{24}}$, $Y_{\overline{34}}$ form a network connecting the generators and loads through which the load currents must flow. These branches of the equivalent circuit represent the effects of winding resistances and magnetic leakage. The voltage drops in them are of major importance in determining the effects of the transformer as an impedance element in the circuits to which it is connected. The values of these admittances can be computed from design data or can be determined from the results of short-circuit tests, as described in Art. 3.

2a. *Equivalent Circuit Neglecting Exciting Current.* — Since the exciting current usually is small, the shunt admittances representing the excitation characteristics usually are so small (or their impedances are so large) that the effects of these large impedances in parallel with the generator and load circuits are negligible and therefore the shunt branches may be considered as open circuits. The equivalent circuit then reduces to a simpler form.

For example, if the excitation admittances $Y_{\overline{10}}$, $Y_{\overline{20}}$, $Y_{\overline{30}}$, $Y_{\overline{40}}$ are omitted, the equivalent circuit of the four-circuit transformer of Fig. 2b reduces to the network shown in Fig. 2c, comprising four free terminals 1, 2, 3, 4 with an admittance link connecting each terminal to every other terminal — a total of six links, as compared with the ten links in the "exact" equivalent circuit of Fig. 2b. The generator and load circuits are connected to the terminals 1, 2, 3, 4 of the equivalent circuit and to a common point or reference node 0, but the terminal 0 has no connection with the equivalent circuit.

In the general case of an n-circuit transformer, neglecting the exciting current reduces the number of terminals of the equivalent circuit from $n + 1$ to n and eliminates the n shunt links representing the excitation characteristics.

▶ Therefore, if exciting current is neglected, an n-circuit transformer can be represented by an n-terminal equivalent circuit with a link connecting each terminal to every other terminal.◀

2b. *Relations between the Equivalent Circuit and the Current-Voltage Equations.* — The reader should be careful to note the distinction between the admittance links in the equivalent circuit and the short-circuit transfer admittances in Eqs. 20–23. Thus the symbol $Y_{\overline{12}}$ (with a bar above the subscripts) represents an admittance link in the equivalent circuit, the bar indicating that the symbol represents a link connecting terminals 1 and 2. The symbol Y_{12} (without the bar) is the short-circuit transfer admittance in Eq. 20. It is shown below that for the positive directions chosen in this discussion,

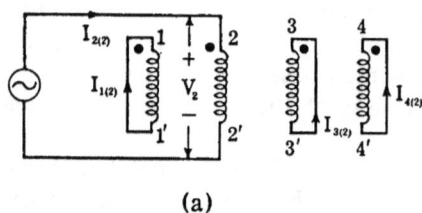

$$Y_{12} = -Y_{\overline{12}}. \qquad [31]$$

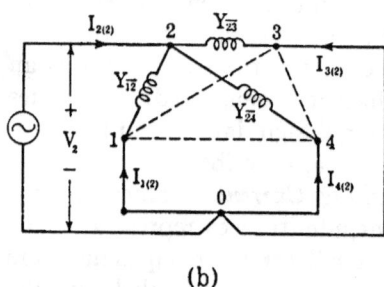

(a)

(b)

FIG. 3. Circuit diagram and equivalent circuit for a four-circuit transformer with windings 1, 3, and 4 short-circuited. The positive directions of the voltage and currents are the same as in Fig. 1.

In order to prove the validity of Eq. 31 consider the four-circuit transformer shown in Fig. 3a with circuit 2 excited at a reduced voltage and circuits 1, 3, and 4 short-circuited. Let $I_{1(2)}, I_{3(2)}, I_{4(2)}$ be the vector currents produced in the short-circuited windings by the vector voltage V_2 applied to circuit 2. The positive directions of the voltage and currents are indicated in Fig. 3a and conform with the positive directions adhered to throughout this analysis, as shown in Fig. 1. From Eq. 20, the short-circuit transfer admittance Y_{12} is the vector current entering terminal 1 per unit vector voltage applied to circuit 2, all other circuits but circuit 2 being short-circuited. Therefore

$$Y_{12} = \frac{I_{1(2)}}{V_2}. \qquad [32]$$

For the short-circuit conditions of Fig. 3a, the equivalent circuit reduces to that shown in Fig. 3b, in which all terminals but terminal 2 are short-circuited to the common terminal 0. The branches shown by the broken lines are short-circuited and do not carry current, and therefore only the branches connected to terminal 2 need be considered. Since on short circuit the exciting current is very small indeed, the equivalent circuit of Fig. 3b — from which the exciting admittances are omitted — is very nearly an "exact" equivalent of the short-circuited transformer. The voltage V_2 and the currents $I_{1(2)}$, $I_{3(2)}$, $I_{4(2)}$ are shown in Fig. 3b with the same positive directions as in Fig. 3a. In Fig. 3b, the current in the admittance link $Y_{\overline{12}}$ in the direction of the voltage drop V_2 across $Y_{\overline{12}}$ is $-I_{1(2)}$, whence

$$-I_{1(2)} = Y_{\overline{12}}V_2 \qquad [33]$$

or

$$Y_{\overline{12}} = -\frac{I_{1(2)}}{V_2}. \qquad [34]$$

The negative signs in Eqs. 33 and 34 are necessary for conformity with the assumed positive direction of $I_{1(2)}$. By comparison of Eqs. 32 and 34 it is seen that for the positive directions chosen in this analysis

$$Y_{12} = -Y_{\overline{12}},$$

as given in Eq. 31. In general, then, the relation between the short-circuit transfer admittance Y_{jk} and the admittance link $Y_{\overline{jk}}$ is

$$Y_{jk} = -Y_{\overline{jk}}. \qquad \blacktriangleright[35]$$

The short-circuit driving-point admittances of Eqs. 20–23 also are expressible, by simple relations, in terms of the admittance links in the equivalent circuit. For example, the short-circuit driving-point admittance Y_{22} of Eq. 21 is the admittance of circuit 2 with all other circuits short-circuited, as in Fig. 3a. Inspection of the equivalent circuit of Fig. 3b shows that for these short-circuit conditions the admittance links $Y_{\overline{12}}$, $Y_{\overline{23}}$, $Y_{\overline{24}}$ are in parallel, and therefore the short-circuit driving-point admittance Y_{22} of circuit 2 is

$$Y_{22} = Y_{\overline{12}} + Y_{\overline{23}} + Y_{\overline{24}}. \qquad \blacktriangleright[36]$$

Hence, the very small exciting current on short circuit being neglected, the short-circuit driving-point admittance of any circuit equals the vector sum of the admittance links connecting the free terminal of this circuit to the free terminals of the other circuits in the equivalent network of Fig. 2c.

3. Measurement of the Admittances

The values of the admittance coefficients in Eqs. 20–23 and of the branch admittances in the equivalent circuit can be determined experimentally by means of a number of tests in which a low voltage is applied to each winding in turn with all other windings short-circuited. The applied voltage and the currents in all windings are measured vectorially — that is, their effective values are measured with ordinary alternating-current instruments, and the phase relations are determined by the methods discussed below.

Since the short-circuit admittances are determined principally by magnetic leakage, their values are very little affected by the nonlinear magnetic characteristics of the core, and therefore the values of these admittances as determined from short-circuit tests at reduced voltage are very nearly the same as their values for normal voltage conditions.

If circuit k is excited and all other circuits are short-circuited, the complex short-circuit driving-point admittance Y_{kk} of circuit k is

$$\mathbf{Y}_{kk} = Y_{kk}\underline{/\theta_{kk}} = G_{kk} + jB_{kk} = \frac{\mathbf{I}_{k(k)}}{\mathbf{V}_k}, \qquad \blacktriangleright [37]$$

where

> $\mathbf{V}_k$ is the vector voltage applied to circuit k,
> $\mathbf{I}_{k(k)}$ is the vector current in circuit k,
> G_{kk} is the short-circuit driving-point conductance,
> B_{kk} is the short-circuit driving point susceptance,
> θ_{kk} is the phase angle of $\mathbf{I}_{k(k)}$ with respect to $\mathbf{V}_k$.

▶ In Eq 37 and in all the following equations, it is assumed that the voltages and currents are all referred to a common base before their values are substituted in the equations. The admittances then are their referred values.◀

The short-circuit driving-point conductances of all circuits are positive quantities, since power is absorbed by the excited winding. Thus if P_{kk} is the reading of a wattmeter connected to read the power supplied to the excited winding k,

$$G_{kk} = +\frac{P_{kk}}{V_k^2}. \qquad [38]$$

The short-circuit driving-point susceptances of all circuits are *negative* quantities, since the current in the excited winding lags the applied voltage, as in any inductive circuit. Therefore

$$B_{kk} = -\sqrt{\left(\frac{I_{k(k)}}{V_k}\right)^2 - G_{kk}^2}. \qquad [39]$$

The short-circuit transfer admittances and the admittances of the links in the equivalent circuit of Fig. 2c can be determined from measurement of the vector currents in the short-circuited windings. Thus if winding k is excited and all other windings are short-circuited, the complex value of the short-circuit transfer admittance Y_{jk} between circuits j and k is

$$\mathbf{Y}_{jk} = Y_{jk}\underline{/\theta_{jk}} = G_{jk} + jB_{jk} = \frac{\mathbf{I}_{j(k)}}{\mathbf{V}_k}, \qquad \blacktriangleright [40]$$

where

    V_k is the vector voltage applied to circuit k,

    $I_{j(k)}$ is the vector current in circuit j,

    G_{jk} is the short-circuit transfer conductance,

    B_{jk} is the short-circuit transfer susceptance,

    θ_{jk} is the phase angle of $I_{j(k)}$ with respect to V_k.

The positive directions of the voltage and current are as indicated in Fig. 1.

▶ As shown in Eq. 35, the admittance of the link $Y_{\overline{jk}}$ in the equivalent circuit is the negative of the short-circuit transfer admittance Y_{jk}. ◀

Note that the value of the short-circuit transfer admittance Y_{jk} also can be determined if winding j is excited, with all other windings short-circuited, and the voltage applied to circuit j and the current in circuit k is measured. Thus the value of each admittance can be determined from two independent sets of measurements.

Great care must be taken in determining the algebraic signs of the mutual conductances and susceptances. Before the n-circuit transformer is discussed, it may be helpful to consider the two-circuit transformer shown in Fig. 4a with its secondary short-circuited. The positive directions of the voltage and currents are indicated in Fig. 4a and correspond to the positive directions in the n-circuit transformer of Fig. 1. Since the primary and secondary currents produce very nearly equal and opposing magnetomotive forces, the secondary current I_2 is very nearly in phase opposition to the primary current when the positive directions of both currents are chosen in the same direction with respect to the core flux, as indicated by the arrows and polarity dots in Fig. 4a. Thus the vector diagram of the short-circuited two-winding transformer is as shown in Fig. 4b, in which the applied primary voltage V_1 is the reference vector. The short-circuit transfer admittance between primary and secondary is, vectorially,

$$Y_{12} = G_{12} + jB_{12} = \frac{I_2}{V_1}. \qquad [41]$$

Inspection of Fig. 4b shows that G_{12} is a *negative* quantity, since the real component of I_2 is negative, but B_{12} is a *positive* quantity, since the

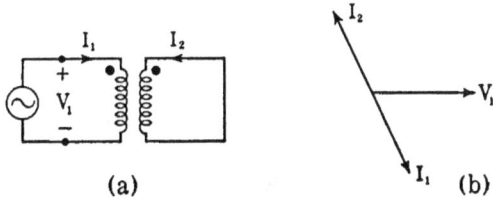

FIG. 4.   Circuit diagram and vector diagram for a short-circuited two-winding transformer.

imaginary component of I_2 is positive. However, the admittance link in the equivalent circuit is the negative of the corresponding short-circuit transfer admittance; hence the admittance link $Y_{\overline{12}}$ has a *positive* conductance and a *negative* susceptance, and therefore is representable physically by an impedance element comprising a resistor and an inductance coil.

With a multicircuit transformer, if all the windings but the excited winding are short-circuited, the currents in the majority of the short-circuited windings are in phase opposition to the current in the excited winding when the positive directions of all currents are chosen in the same direction with respect to the core flux, as in Fig. 1. However, because of the relations that may exist among the leakage fluxes with some arrangements of the windings, the current in one or more of the short-circuited windings may be reversed, with corresponding reversals in the signs of the conductances and susceptances associated with this circuit. In these circumstances, the equivalent circuit would contain links having *negative* conductances and *positive* susceptances. At a constant frequency, the positive susceptance can be represented by a capacitance; but, although a negative conductance introduces no difficulty in the mathematical treatment, there is no simple way of representing it by physical elements. Thus if a multicircuit transformer is to be represented on a network analyzer* by means of its equivalent circuit, either the negative conductances must be omitted — as often is permissible, since they usually are small — or the equivalent circuit must be changed to some form permitting combination of the negative resistance elements in series with positive resistances of external circuits to leave a net positive resistance that can be represented on the network analyzer.[7]

The magnitude of each mutual conductance and susceptance can be determined by the use of suitably connected wattmeters. Thus if P_{jk} is the reading of a wattmeter whose current coil is connected in series with the short-circuited winding j and whose potential coil is in parallel

---

* A network analyzer is a device consisting of a large number of resistors, inductance coils, capacitors, and alternating voltage sources which can be readily connected in a great variety of ways, and with arrangements for conveniently measuring the current, voltage, and power at any point in the network. In the study of complicated power systems, where laborious computations would be necessary if analytical methods were used, it is often more economical to do the work experimentally by representing the power system on a network analyzer on a per phase basis at a small convenient scale. The use of a network analyzer in power-system studies is mentioned briefly in this series in the volume on electric circuits (1940), p. 442.

[7] For further discussion of negative conductances and of other forms of the equivalent circuit, see F. M. Starr, "An Equivalent Circuit for the Four-winding Transformer," *G. E. Rev.*, *36* (March, 1933), 150–152; L. F. Blume, editor, *Transformer Engineering* (New York: John Wiley & Sons, 1938), 103–107, 122.

with the voltage V_k applied to the excited winding k,

$$G_{jk} = \pm \frac{P_{jk}}{V_k^2}. \tag{42}$$

The mutual susceptance is then given by

$$B_{jk} = \pm \sqrt{\left(\frac{I_{j(k)}}{V_k}\right)^2 - G_{jk}^2}. \tag{43}$$

The sign of the mutual conductance can be determined if the polarities of the wattmeter connections for up-scale deflection are compared with their polarities for a known direction of power flow. It should be remembered that the sign of the conductance $G_{\overline{jk}}$ in the equivalent circuit is the negative of the sign of the short-circuit transfer conductance G_{jk}. However, the wattmeter gives no indication of the sign of the mutual susceptance.

The algebraic sign of the mutual susceptance can be determined by use of a properly connected phase-angle meter, power-factor meter, or reactive-power meter. However, there are many opportunities for experimental error, such as the accidental reversal of a connection, especially when instrument transformers are used. As an alternative method of determining the signs of the susceptances — which should always be used as a check on the results of other measurements — use can be made of the fact that the exciting current on short circuit is very small indeed. That is, the resultant magnetomotive force of all the currents is very nearly zero, whence, for the positive directions of the currents indicated in Fig. 1, the vector sum of the currents referred to the same circuit is very nearly zero. Since the currents are either approximately in phase or in phase opposition, the *algebraic* sum of the currents also is approximately zero. Thus, if circuit k is excited,

$$I_{k(k)} \approx -[\pm I_{1(k)} \pm \cdots \pm I_{j(k)} \pm I_{l(k)}, \pm \cdots \pm I_{n(k)}], \tag{44}$$

where the I's are the absolute values of the referred currents. The signs can be determined by trial so as to satisfy Eq. 44. Most of the currents in the short-circuited windings must be in phase opposition to the current in the excited winding, as indicated by the minus sign preceding the right-hand side of Eq. 44, and therefore most of the terms within the brackets are positive. For this normal phase relation, the susceptance of the associated admittance link in the equivalent circuit is *negative* — that is, it is an inductive link — and the sign of the corresponding transfer susceptance in Eqs. 20–23 is *positive*. These signs are the same as for the two-circuit transformer of Fig. 4. If any of the currents are reversed from this normal phase relation — usually the small currents — the fact that they must be given negative signs within the brackets of Eq. 44 discloses the reversal. The signs of the mutual susceptances associated with these reversed currents are also reversed.

As a final check on the results, use can be made of the generalized form of Eq. 36, which states that the short-circuit driving-point admittance of any circuit very nearly equals the vector sum of the admittance links associated with that circuit. Thus the following algebraic sums must be very nearly true for the conductances and susceptances:

$$G_{kk} = G_{\overline{1k}} + G_{\overline{2k}} + \cdots + G_{\overline{jk}} + G_{\overline{lk}} + \cdots + G_{\overline{nk}} \qquad [45]$$

$$B_{kk} = B_{\overline{1k}} + B_{\overline{2k}} + \cdots + B_{\overline{jk}} + B_{\overline{lk}} + \cdots + B_{\overline{nk}}. \qquad [46]$$

The short-circuit driving-point conductance G_{kk} is always a positive quantity, and the short-circuit driving-point susceptance B_{kk} is always a negative quantity.

3a. Relations of the Admittances to Two-Circuit Transformer Theory. — The preceding discussion shows that the values of the admittance links in the equivalent circuit of an n-circuit transformer can be determined from the results of tests in which each winding in turn is excited at a reduced voltage with *all* other windings short-circuited. Alternatively, the admittance links can be evaluated from the results of a number of tests in which each winding is excited in turn with *one* other winding short-circuited. For example, if winding j is excited and winding k is short-circuited, all other windings being open-circuited, the presence of these open-circuited windings has no effect on the currents. Thus the impedance $Z_{(jk)}$ observed at the terminals of the excited winding is the short-circuit or equivalent impedance of windings j and k acting as a two-circuit transformer, and very nearly equals the sum of the leakage impedances of the two windings with respect to each other.* This approach to the n-circuit transformer from two-circuit theory is useful when the equivalent-circuit admittances are to be determined from design formulas, since the equivalent impedance of any pair of windings can be computed from well-known design formulas for two-circuit transformers.†

The number of independent equivalent impedances of an n-circuit transformer equals the number of possible combinations of n windings taken two at a time. The equivalent circuit (exciting current being neglected) comprises n terminals with an admittance link connecting each pair of terminals, and therefore the number of independent admittance links also equals the number of possible combinations of n things taken

---

* The notation $Z_{(jk)}$, with the subscripts enclosed in parentheses, is used to designate the short-circuit or equivalent impedance of windings j and k acting as a two-circuit transformer, the parentheses being intended to convey the idea that the two windings are considered as a two-circuit pair.

† For example, see Art. 9, Ch. XIII.

two at a time. Thus, exciting current being neglected, the number of independent equivalent impedances equals the number of independent parameters required to specify completely the characteristics of the transformer as a circuit element.

The expressions for the admittance links in terms of the equivalent impedances are rather complicated and are not derived in this text.[8] An explanation of the method of derivation is given below, however, to emphasize the fact that the admittance links in the equivalent circuit are intimately related to the leakage impedances of the transformer. The brief discussion which follows should also serve as a basis for further study.

The relations among the equivalent impedances and the admittance links of the equivalent circuit can be visualized from the equivalent circuit of the n-circuit transformer as it appears when acting as a short-circuited two-circuit transformer. For example, consider a four-circuit transformer with winding 2 excited and winding 1 short-circuited, the other windings being open-circuited. Under these conditions, the equivalent circuit of Fig. 2c reduces to that shown in Fig. 5. Examination of Fig. 5 shows that the equivalent impedance $Z_{(12)}$ of windings 1 and 2 equals the impedance of the network 1234 between terminals 1 and 2. When the impedances are all referred to a common base and the exciting current is negligible, the same result would be obtained if circuit 1 were excited and circuit 2 short-circuited. Thus it is possible to write an equation expressing each of the six two-circuit equivalent impedances $Z_{(12)}$, $Z_{(13)}$, $Z_{(14)}$, $Z_{(23)}$, $Z_{(24)}$, $Z_{(34)}$ in terms of the six admittance links of the

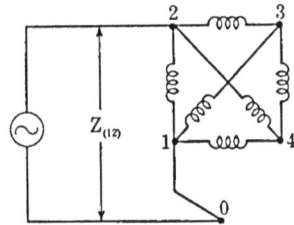

FIG. 5. The equivalent impedance $Z_{(12)}$ equals the impedance of the network 1234 between terminals 1 and 2.

equivalent circuit. Then these equations can be solved for the six admittance links in terms of the six equivalent impedances. The expressions are rather long, and therefore the procedure of Art. 3 is a simpler way to determine the admittance links by tests. The designer may however be forced to use the more complicated procedure outlined above.

4. SUMMARY OF MULTICIRCUIT TRANSFORMER THEORY

The important points brought up in the preceding discussion of multicircuit transformers can be summarized briefly as follows:

[8] For a complete derivation of these relations see L. F. Blume, editor, *Transformer Engineering* (New York: John Wiley & Sons, 1938), 118–120.

Some of the chief problems in the use of multicircuit transformers are concerned with voltage regulation, short-circuit currents, load division between circuits, and the behavior of multicircuit transformers in parallel with other transformers.

Except when there are cross connections between the phases,* problems involving polyphase arrangements of multicircuit transformers or autotransformers in balanced polyphase circuits can be solved on a per phase basis by means of the theory of a single-phase, n-circuit transformer.

When the effects of winding capacitances are neglected, an n-circuit transformer can be analyzed as an n-loop or as an n-node circuit.

The current-voltage equations can be expressed in convenient form in terms of the short-circuit driving-point and transfer admittances, and, if the values of these short-circuit admittances are known, the performance of an n-circuit transformer as a circuit element can be computed by means of Eqs. 16–19 or 20–23.

An alternative method of expressing the characteristics of an n-circuit transformer is by means of an equivalent circuit comprising a number of simple impedance links. The principal advantages of the equivalent circuit are that often the equivalent circuit aids in visualizing the problem, and that for the solution of complicated networks the transformer can be represented on a network analyzer by means of its equivalent circuit.

When it is permissible to consider one terminal of each winding connected to the corresponding terminal of every other winding, an n-circuit transformer can be represented by an equivalent circuit with $n + 1$ terminals and a link connecting each terminal to every other terminal.

The exciting current usually is small, and when it is neglected the equivalent circuit reduces to a network having n terminals and a link connecting each terminal to all other terminals. This network contains the minimum number of accessible terminals and branches required to represent an n-circuit transformer with negligible exciting current, and it is therefore the simplest equivalent circuit that can be found. However, it is not unique. Other equivalent circuits can be derived which may be more useful in certain circumstances. (See p. 638.)

The values of the short-circuit driving-point and transfer admittances and of the admittance links in the equivalent circuit can be determined from the results of a number of tests in which each winding in turn is excited at a reduced voltage with *all* other windings short-circuited.

Alternatively, the admittances can be evaluated in terms of the equivalent impedances of pairs of windings.

* See the footnote‡ on p. 623.

5. THREE-CIRCUIT TRANSFORMERS

The commonest multicircuit transformers are those having three independent circuits — a primary, a secondary, and a tertiary. Some of the applications of these three-circuit transformers in single-phase and three-phase circuits are mentioned at the beginning of this chapter.

A single-phase three-circuit transformer can be analyzed by means of the general theory of single-phase n-circuit transformers developed in the preceding articles of this chapter, and this theory can be applied on a per phase basis to the analysis of three-phase problems.* The three-circuit transformer can also be treated by methods that do not apply when more than three independent circuits are involved. Both methods of analysis are presented in the following discussion.

5a. From Multicircuit Theory. — According to the general theory of n-circuit transformers, a three-circuit transformer can be represented by a four-terminal equivalent circuit with a link connecting each terminal to every other terminal. This statement assumes that three terminals

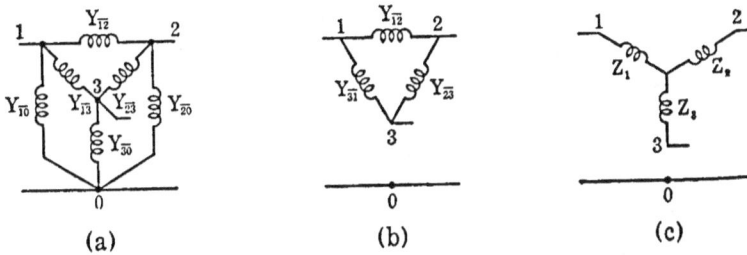

FIG. 6. Equivalent circuits for a three-circuit transformer. The " exact " equivalent circuit (a) reduces to the approximate equivalent circuits (b) and (c) when exciting current is neglected.

of the same relative polarity can be considered as connected. Thus an " exact " equivalent circuit of a three-circuit transformer is the four-terminal six-branch circuit of Fig. 6a. The external circuits are connected between terminals 1, 2, 3, and the common terminal 0. This circuit may represent either a single-phase three-circuit transformer in a single-phase circuit, or one phase of a three-phase bank of three-circuit transformers.

As shown in the discussion of n-circuit transformers, the links of the equivalent circuit terminating at the common terminal 0 represent the excitation characteristics of the transformer, and, since the exciting current usually is small, the admittances of these links usually are small.

* For comments regarding situations in which there are interconnections between the phases, see the footnote‡ on p. 623.

They are therefore almost always omitted, and the equivalent circuit then reduces to the simple Δ arrangement shown in Fig. 6b.

Since a Δ arrangement of linear circuit elements always can be replaced by an equivalent Y arrangement, the Δ equivalent circuit of Fig. 6b can be converted to the Y equivalent circuit shown in Fig. 6c. According to the well-known Δ–Y transformation formulas,[9] the relations among the impedances Z_1, Z_2, Z_3 of the Y-connected equivalent circuit (Fig. 6c) and the admittances $Y_{\overline{12}}$, $Y_{\overline{23}}$, $Y_{\overline{31}}$ of the links in the Δ equivalent circuit (Fig. 6b) are

$$Z_1 = \frac{Y_{\overline{23}}}{Y_{\overline{12}}Y_{\overline{23}} + Y_{\overline{23}}Y_{\overline{31}} + Y_{\overline{31}}Y_{\overline{12}}} \qquad [47]$$

$$Z_2 = \frac{Y_{\overline{31}}}{Y_{\overline{12}}Y_{\overline{23}} + Y_{\overline{23}}Y_{\overline{31}} + Y_{\overline{31}}Y_{\overline{12}}} \qquad [48]$$

$$Z_3 = \frac{Y_{\overline{12}}}{Y_{\overline{12}}Y_{\overline{23}} + Y_{\overline{23}}Y_{\overline{31}} + Y_{\overline{31}}Y_{\overline{12}}}. \qquad [49]$$

Thus a three-circuit transformer can be represented by a simple equivalent circuit comprising three impedances in series with the three external circuits, as in the Y or three-pointed star arrangement of Fig. 6c. In this respect, a three-circuit transformer differs from a transformer with four or more independent circuits, since an n-circuit transformer is characterized by more than n independent admittances when n is greater than three, and consequently such a transformer cannot be represented by an n-pointed star.

5b. *The Three-Circuit Transformer as a Three-Terminal Network.* — The simple three-terminal equivalent circuits representing a three-circuit transformer can also be derived without the use of n-circuit theory. Consider the three-circuit transformer shown in Fig. 7a. The external circuits are connected to terminals 11′, 22′, and 33′. At power-system frequencies, the currents taken by the capacitances of the windings are negligible and therefore conditions in the circuit are unaltered if terminals 1′, 2′, 3′ are considered to be connected, forming a common terminal 0, as shown by the broken-line connections in Fig. 7a.* The external circuits then may be considered as connected between the free terminals 1, 2, 3, and the common terminal 0. Thus the three-circuit transformer is equivalent to a circuit element having four accessible terminals or points of entry, as indicated by the rectangular box in Fig. 7b.

Let I_1, I_2, I_3 be the vector currents entering the three terminals of corresponding polarity, as indicated by the polarity dots and arrows in

---

[9] See the volume on electric circuits (1940), Eqs. 135, 136, and 137, p. 460.

* For further discussion of this point, see p. 624.

Fig. 7a. These also are the currents entering the terminals 1, 2, 3 of Fig. 7b, and therefore the current leaving the rectangular box at the common terminal $1'2'3'$ must equal the vector sum $I_1 + I_2 + I_3$ of the currents entering the three free terminals, as indicated in Fig. 7b.

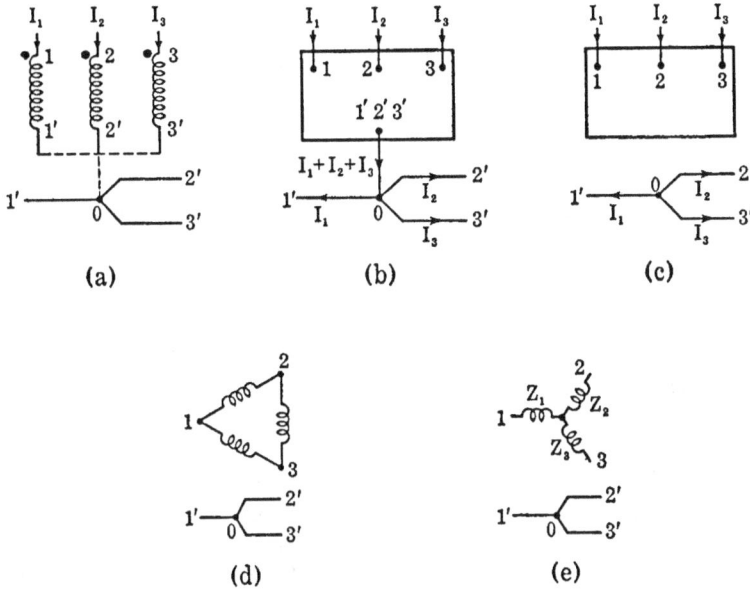

Fig. 7. Schematic diagram of a three-circuit transformer (a), and successive steps in the derivation of the approximate equivalent circuits (d) and (e).

Since the exciting current usually is small, it is almost always neglected; that is, the resultant magnetomotive force of all the currents is almost always assumed to be zero. Thus, if the positive directions of the currents are all chosen in the same direction about the core, as indicated by the arrows and polarity dots in Fig. 7a, and if the currents are all referred to a common base, their vector sum is zero, or

$$I_1 + I_2 + I_3 = 0. \tag{50}$$

On this assumption, then, there is no current in the connection between the common terminal $1'2'3'$ and the junction point 0 of the external circuits. Therefore, when the currents are all referred to a common base, terminal $1'2'3'$ may be omitted, and the three-circuit transformer is equivalent to a three-terminal circuit element, as shown by the box in Fig. 7c.

Except for excitation phenomena, an iron-core transformer behaves as an essentially linear circuit element, and therefore the three-terminal network within the box, Fig. 7c, is a network of essentially linear impedances. As far as its external characteristics are concerned, any linear network having three accessible terminals or points of entry is equivalent to either a Δ or a Y arrangement of impedances.

▶ Therefore, when exciting current is neglected and all currents and voltages are referred to a common base, a three-circuit transformer can be represented by either a Δ equivalent circuit (Fig. 7d), or a Y (Fig. 7e). Usually the Y equivalent circuit, Fig. 7e, is more useful, since the impedances representing the transformer can then be combined in series with the impedances of external circuits.◀

▶ Note that these Δ- and Y-connected equivalent circuits represent the relations among the voltages and currents in the three circuits of a single-phase three-circuit transformer. They do *not* represent equivalent Δ- and Y-connected three-phase circuits.◀

5c. Determination of the Equivalent-Circuit Parameters. — As shown in part (a) of this article, the mesh equivalent circuit of an n-circuit transformer reduces, for a three-circuit transformer, to the Δ of Fig. 6b; that is, the Δ of Fig. 6b is simply a special case of n-circuit theory. The admittances of the links in the Δ equivalent circuit thus can be determined by the same methods that apply in general to n-circuit transformers. Their values therefore can be determined experimentally by means of short-circuit tests in which a low voltage is applied to each winding in turn with *both* other windings short-circuited. These tests are described in detail in Art. 3. The impedances of the equivalent Y, Fig. 6c, can then be computed from the Δ admittances by means of Eqs. 47, 48, and 49.

However, a simpler method of determining the impedances of the Y equivalent circuit is to evaluate them in terms of the short-circuit or equivalent impedances of each pair of windings acting as a two-circuit transformer. The equivalent impedance of each pair of windings can be computed from well-known two-circuit design formulas,* or can be determined experimentally by means of three simple short-circuit tests, described below.

For example, suppose a low voltage is applied to winding 1 with winding 2 short-circuited and winding 3 open-circuited, as shown in Fig. 8a. Under these conditions, the equivalent circuit of Fig. 7e reduces to that shown in Fig. 8b. Let V_1, I_1, and P_1 be the measured values of the voltage, current, and power supplied to winding 1. Then the magnitude of the equivalent or short-circuit impedance $Z_{(12)}$ of windings 1 and 2 is

$$Z_{(12)} = \frac{V_1}{I_1}, \qquad [51]$$

---

* For example, see Art. 9, Ch. XIII.

and its resistance and reactance components are

$$R_{(12)} = \frac{P_1}{I_1^2} \tag{52}$$

$$X_{(12)} = \sqrt{Z_{(12)}^2 - R_{(12)}^2}. \tag{53}$$

The subscripts associated with the equivalent impedance and its resistance and reactance components are enclosed in parentheses to indicate that the two designated windings are acting as a two-circuit pair. The order of the subscripts has no significance. Thus, if the currents and voltages

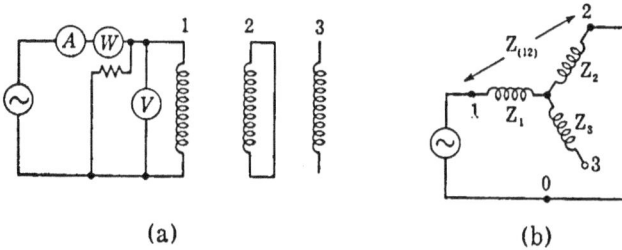

(a)   (b)

FIG. 8.   Short-circuit test for measuring the equivalent impedances; (a) circuit diagram and (b) the corresponding equivalent circuit.

are referred to a common base, except for the effects of the very small exciting current on short circuit, the same value of equivalent impedance would be obtained if winding 2 were excited and winding 1 short-circuited.

Inspection of Fig. 8b shows that with winding 1 excited and winding 2 short-circuited, the short-circuit impedance $Z_{(12)}$ is the series combination of the two branch impedances Z_1 and Z_2 of the Y equivalent circuit. Therefore

$$Z_{(12)} = Z_1 + Z_2. \tag{54}$$

Similarly, the relations among the short-circuit impedances $Z_{(23)}$ and $Z_{(31)}$ and the equivalent-circuit impedances Z_1, Z_2, Z_3 are

$$Z_{(23)} = Z_2 + Z_3 \tag{55}$$

$$Z_{(31)} = Z_3 + Z_1. \tag{56}$$

Equations 54, 55, and 56 can be solved for the branch impedances Z_1, Z_2, Z_3 of the Y equivalent circuit in terms of the short-circuit impedances $Z_{(12)}$, $Z_{(23)}$, $Z_{(31)}$ yielding the following expressions:

$$Z_1 = \tfrac{1}{2}[Z_{(12)} + Z_{(13)} - Z_{(23)}] \qquad \blacktriangleright[57]$$

$$Z_2 = \tfrac{1}{2}[Z_{(21)} + Z_{(23)} - Z_{(13)}] \qquad \blacktriangleright[58]$$

$$Z_3 = \tfrac{1}{2}[Z_{(31)} + Z_{(32)} - Z_{(12)}]. \qquad \blacktriangleright[59]$$

Note the form of these expressions; for example, the impedance representing circuit 1 in the Y equivalent circuit of Fig. 7e equals one-half of the quantity: the vector sum of the two equivalent impedances of circuit 1 taken with each of the other two circuits, minus the vector equivalent impedance of circuits 2 and 3.

▶ Thus the three impedances of the Y equivalent circuit, Fig. 7e, can be determined from the measured or computed values of the three short-circuit or equivalent impedances of each pair of windings acting as a two-circuit transformer.◀

Although the equivalent impedances $Z_{(12)}$, $Z_{(23)}$, $Z_{(31)}$ must all have positive resistances and positive (that is, inductive) reactances, it is quite possible that, for some arrangements of the windings, one of the impedances Z_1, Z_2, or Z_3 in the equivalent circuit may have a negative reactance.[10] For example, if the winding arrangement is such that the equivalent reactance $X_{(13)}$ of windings 1 and 3 is greater than the sum of the equivalent reactances $X_{(21)}$ and $X_{(23)}$, then, by Eq. 58, the reactance X_2 of impedance Z_2 is negative. One of the branches of the equivalent circuit may, in fact, have a negative resistance, but it should be borne in mind that the equivalent circuit shows only the *external* behavior of the transformer, and a negative resistance in one branch of the equivalent circuit does *not* signify a negative load loss in one of the windings. The equivalent circuit gives the correct total load loss for any load conditions, but it does *not* show the way in which the load losses are distributed among the windings.

Although negative impedances present no difficulties in the analytical treatment, their presence makes it difficult to represent a three-circuit transformer on a network analyzer* by means of its equivalent circuit. A negative reactance can be represented (at a constant frequency) by a capacitance, but the reactance usually is small, and therefore an inconveniently large capacitance usually is required. A negative resistance presents still greater difficulties, since it cannot be represented by any simple physical elements. The difficulties usually are avoided through combining the negative impedance of the branch of the equivalent circuit with the positive impedances of the external circuits connected to the transformer. Since the negative impedance in the equivalent circuit usually is small, the series combination usually is a positive impedance.

5d. *Equivalent Circuits Including Exciting Current.* — Except for the peculiarities of excitation phenomena caused by the nonlinear magnetic characteristics of the core, the four-terminal six-branch equivalent cir-

---

[10] For a discussion of these negative impedances, see L. F. Blume, editor, *Transformer Engineering* (New York: John Wiley & Sons, 1938), 103–107.

* See the footnote on p. 638.

cuit of Fig. 6a, p. 643, is an exact equivalent of a three-circuit transformer. It is very seldom indeed, however, that the exciting current is important enough to require the use of this relatively complicated equivalent circuit.

Even if it were desired to use this equivalent circuit, considerable difficulty would be encountered in the attempt to obtain accurate values of the three excitation admittances $Y_{\overline{10}}$, $Y_{\overline{20}}$, and $Y_{\overline{30}}$ of Fig. 6a. Theoretically, the values of these admittances can be determined from the results of three open-circuit tests — that is, by measurement of the exciting current first on the primary side, then on the secondary, and finally on the tertiary. If the admittances of the links $Y_{\overline{12}}$, $Y_{\overline{23}}$, and $Y_{\overline{13}}$ of Fig. 6a are known from the results of short-circuit tests, the three open-circuit tests in theory give enough information for the three excitation admittances to be evaluated. Ordinarily, however, when the voltages and exciting currents are referred to a common base, the three values of the exciting current for the same applied voltage are found to be the same to within the accuracy of the measurements. That is, instead of giving three independent items of information, the three open-circuit tests actually give only one. Or it can be said that, because the short-circuit admittances $Y_{\overline{12}}$, $Y_{\overline{13}}$, $Y_{\overline{23}}$ of Fig. 6a are very much larger than the excitation admittances, all the open-circuit tests measure essentially the same admittance, namely, that of $Y_{\overline{10}}$, $Y_{\overline{20}}$, and $Y_{\overline{30}}$ in parallel, and it is impossible from this single measured value to determine the separate values of the individual components. About the best that can be done is to assume that the admittances $Y_{\overline{10}}$, $Y_{\overline{20}}$, $Y_{\overline{30}}$ are equal and that each equals one-third of the open-circuit admittance.

A much simpler method of approximately including the effects of exciting current is shown in Fig. 9, in which the approximate equivalent circuit of Fig. 7e is modified through connection of an admittance equal to the open-circuit admittance Y_{oc} from the junction point of the Y impedances Z_1, Z_2, Z_3 to the common terminal 0 of the external circuits. As a matter of fact, the equivalent circuit of Fig. 9 is probably as accurate a representation of the transformer as is the "exact" equivalent circuit of Fig. 6a, because the exact values of the excitation admittances in Fig. 6a usually cannot be determined.

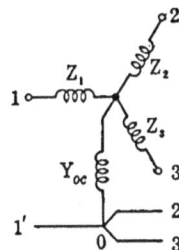

FIG. 9. An equivalent circuit that approximately takes into account the exciting current.

5e. *Summary of Three-Circuit Transformer Theory.* — The preceding discussion of three-circuit transformer theory can be summarized as follows:

Problems involving the effects of a three-circuit transformer as an impedance element — such those concerned with voltage regulation, short-circuit currents, and load division among circuits or among parallel transformers — can be solved with the aid of an equivalent circuit. In problems involving complicated networks, a three-circuit transformer can be represented on a network analyzer* by means of its equivalent circuit.

When one terminal of each winding can be considered as connected to the corresponding terminals of the other two windings, a three-circuit transformer can be represented by a four-terminal six-branch equivalent circuit, as in Fig. 6a.

When exciting current is neglected, the equivalent circuit of a three-circuit transformer reduces to a three-terminal network, which can be represented by either a Δ or a Y arrangement of circuit elements. The Y arrangement generally is the more useful, since the impedances representing the transformer are then in series with the impedances of external circuits, and can be combined readily with them.

Unlike the three-circuit transformer, a transformer with four or more circuits cannot be represented by an impedance in series with each external circuit, but requires a more complicated network to represent it.

The impedances of the Y-connected equivalent circuit of a three-circuit transformer can be determined readily from the equivalent or short-circuit impedances of each pair of windings acting as a two-circuit transformer, as in Eqs. 57, 58, and 59.

The effects of exciting current can be included approximately by connection of the open-circuit admittance from the junction point of the Y to the common point of the external circuits, as in Fig. 9.

5f. *Illustrative Example of Three-Circuit Transformer Calculations.* — A 5,000-kva 3-phase 60 ∼ 3-winding 14,400 : 2,400 : 575-volt transformer is to be used as a station-service transformer to supply turbine and boiler auxiliaries in a generating station. The 14,400-volt winding is rated 5,000 kva; the 2,400- and 575-volt windings are each rated 2,500 kva. All three windings are Δ-connected.

To avoid excessive interrupting duty on the 575-volt switchgear, the transformer is designed so that a solid, symmetrical, three-phase short circuit directly at the terminals of the 575-volt winding, with rated voltage sustained on the 14,400-volt winding, will cause steady-state currents limited to 25,000 amperes in the lines emanating from the terminals of the 575-volt winding. When the transformer is designed to meet these requirements, the short-circuit reactance of the 14,400 : 2,400-volt windings is found to be 6.0 per cent on a 5,000-kva base, and the short-circuit reactance of the 2,400 : 575-volt windings is 10.0 per cent on a 2,500-kva base. Winding resistances are small enough to be neglected in this problem.

Because of these special conditions, it is advisable to determine whether

* See the footnote on p. 638.

the full-load voltages of the 575- and 2,400-volt buses will be unduly low. These voltages are to be computed for 2,500-kva loads at 0.85 power factor, lagging, on each bus and with rated voltage impressed on the 14,400-volt winding.

Solution: The constants of the equivalent circuit will first be found. Let the 14,400-, 2,400-, and 575-v windings be numbered 1, 2, and 3, respectively. Then, in per unit on a 2,500-kva base.

$$X_{(12)} = 0.06 \times \frac{2,500}{5,000} = 0.03 \tag{60}$$

$$X_{(23)} = 0.10. \tag{61}$$

On a 2,500-kva base, unit current in the 575-v lines is

$$\frac{2,500 \times 1,000}{\sqrt{3} \times 575} \text{ , or 2,510 amp.}$$

Hence the 25,000-amp short-circuit current is

$$\frac{25,000}{2,510} \text{ , or 9.96 per unit,}$$

and

$$X_{(31)} = \frac{1}{9.96} = 0.1003. \tag{62}$$

From Eqs. 57, 58, and 59, since winding resistances are negligible,

$$X_1 = \tfrac{1}{2}(0.03 + 0.1003 - 0.10) = 0.0151 \tag{63}$$

$$X_2 = \tfrac{1}{2}(0.03 + 0.10 - 0.1003) = 0.0149 \tag{64}$$

$$X_3 = \tfrac{1}{2}(0.1003 + 0.10 - 0.03) = 0.0851. \tag{65}$$

The equivalent circuit of the transformer is given in Fig. 10, where the conditions of the problem are also presented. Inspection of Fig. 10 shows that in a straightforward solution of the problem the transformer currents would have to be taken as unknowns

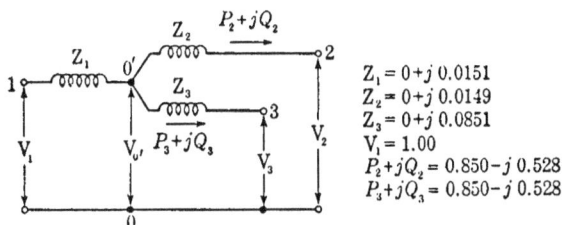

FIG. 10. Equivalent circuit for the three-winding transformer of Art. 5f, showing conditions of the problem. All quantities are expressed in per unit on a 2,500-kva base.

and an unwieldy set of simultaneous equations (some quadratics) would have to be established. To avoid the excessive labor and possible inaccuracy of such a solution, a method of successive approximations will be adopted.

Recognizing that all windings carry somewhere near their rated currents, one might roughly estimate a voltage drop of 1.5% in Z_2, and 8.5% in Z_3. The current in Z_1 is approximately twice the unit current (on a 2,500-kva base), and therefore the voltage drop in Z_1 is approximately 3%. The voltage drops subtract vectorially from the rated voltage applied to winding 1, and consequently the terminal voltages of windings 2 and 3 are somewhat greater than the arithmetic differences between the applied voltage V_1 and the voltage drops. Approximate values of the terminal voltages V_2 and V_3 can be obtained, however, if the vector relations are ignored; thus

$$V_2 \approx 1.00 - 0.03 - 0.01 = 0.96 \qquad [66]$$

$$V_3 \approx 1.00 - 0.03 - 0.08 = 0.89. \qquad [67]$$

On this basis, the winding currents are

$$I_2 \approx \frac{1.00}{0.96} = 1.04, \qquad [68]$$

$$I_3 \approx \frac{1.00}{0.89} = 1.12, \qquad [69]$$

and
$$I_1 \approx 1.04 + 1.12 = 2.16. \qquad [70]$$

The vector power input to winding 1 can now be estimated if the inductive reactive power (I^2X) absorbed by the transformer leakage reactances is added to the vector power outputs of windings 2 and 3. The reactive power absorbed by the leakage reactances is

$$\Sigma I^2X = I_1^2X_1 + I_2^2X_2 + I_3^2X_3 \qquad [71]$$

$$\approx (2.16)^2 (0.0151) + (1.04)^2 (0.0149) + (1.12)^2 (0.0851) \qquad [72]$$

$$\approx 0.19. \qquad [73]$$

By convention, the algebraic sign associated with inductive reactive power is negative,[11] and therefore the vector power input $P_1 + jQ_1$ to winding 1 is

$$P_1 + jQ_1 = P_2 + jQ_2 + P_3 + jQ_3 - j\Sigma I^2X \qquad [74]$$

$$\approx 2(0.850 - j0.528) - j0.19 \qquad [75]$$

$$\approx 1.70 - j1.25. \qquad [76]$$

A solution for V_2 and V_3 can be formulated from this input, giving values which, compared with those estimated above, will serve as a guide for a second approximation if one is necessary. Since V_1 equals 1.00 per unit, the per unit vector current I_1 in winding 1 equals the per unit vector power $P_1 + jQ_1$ when V_1 is the reference vector. Thus

$$I_1 = 1.70 - j1.25. \qquad [77]$$

The vector voltage $V_{0'}$ of the common point $0'$ in Fig. 10 is

$$V_{0'} = V_1 - jI_1X_1 \qquad [78]$$

$$= 1.00 + j0 - j0.0151(1.70 - j1.25) \qquad [79]$$

$$= 0.981 - j0.0256. \qquad [80]$$

The vector terminal voltages of windings 2 and 3 now can be determined.

$$V_2 = V_{0'} - jI_2X_2 \qquad [81]$$

$$= 0.981 - j0.0256 - j0.0149\left(\frac{1.70 - j1.25}{2}\right) \qquad [82]$$

$$= 0.972 - j0.0383. \qquad [83]$$

[11] See the volume on electric circuits (1940), Art. 21, Ch. IV.

The magnitude of V_2 is

$$V_2 = 0.972 \text{ per unit, or 2,330 v.} \qquad [84]$$

The vector voltage V_3 is

$$V_3 = V_{0'} - jI_3X_3 \qquad [85]$$

$$= 0.981 - j0.0256 - j0.0851 \left(\frac{1.70 - j1.25}{2} \right) \qquad [86]$$

$$= 0.928 - j0.0979. \qquad [87]$$

The magnitude of V_3 is

$$V_3 = 0.930 \text{ per unit, or 535 v.} \qquad [88]$$

Carrying through this process again with these voltages as second approximations yields practically identical results. The full-load voltages of the transformer are, then, 2,330 and 535 v.

PROBLEMS

1. A four-circuit transformer has windings with ratings as follows:

Winding	Voltage	Capacity
1	76,800 v	30,000 kva
2	38,500	15,000
3	6,350	20,000
4	6,350	20,000

A series of tests performed with one winding excited and all the others short-circuited yields the data tabulated below. The data are all per unit on a 30,000-kva rated-voltage base.

Winding excited	Voltage applied	Current coil in winding	Current	Wattmeter reading
1	0.0478	1	1.00	0.00733
1	0.0478	2	0.980	0.00710
1	0.0478	3	0.041	0.000458
1	0.0478	4	0.019	0.000137
2	0.0314	1	0.644	0.00306
2	0.0314	2	1.00	0.00483
2	0.0314	3	0.448	0.00217
2	0.0314	4	0.095	0.000434
3	0.0366	1	0.031	0.000268
3	0.0366	2	0.522	0.00295
3	0.0366	3	1.00	0.00495
3	0.0366	4	0.445	0.00174

Determine the values of the parameters for an equivalent circuit of the type shown in Fig. 2c.

2. Find the short-circuit impedance between the two high-voltage windings of the transformer of Prob. 1.

3. If the primary voltage of the transformer of Prob. 1 is adjusted so that winding 2 is supplying full load at rated voltage and unity power factor when windings 3 and 4 are supplying no load, what per cent of rated voltage will appear across the terminals of winding 2 when full-load current at unity power factor is taken from winding 3? Neglect the transformer equivalent resistances.

4. One equivalent circuit for a four-winding transformer is shown in Fig. 11a, where the Z's include both resistance and reactance.

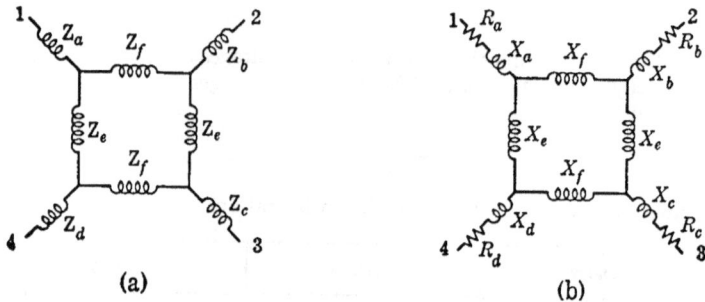

(a)   (b)

FIG. 11.   Equivalent circuits for 4-circuit transformer, Prob. 4.

(a) Determine the values of the impedances of the circuit of Fig. 11a in terms of the short-circuit impedances of each pair of windings.

(b) A modification of the circuit of Fig. 11a which is convenient for use on a network analyzer is shown in Fig. 11b. Determine the values of the parameters of the circuit of Fig. 11b in terms of the short-circuit impedances of each pair of windings. What are the advantages of the circuit of Fig. 11b?

5. The following data apply to a three-circuit power transformer:

RATINGS

Winding	Capacity	Voltage
1	10,000 kva	63,500 v
2	6,667	11,000
3	5,000	7,580

SHORT–CIRCUIT DATA

(Rated current in short-circuited winding)

Winding short-circuited	Voltage applied to winding	Voltage	Power
2	1	4,490 v	46,100 w
3	2	588	35,000
3	1	5,800	35,800

OPEN–CIRCUIT DATA

Winding excited	Voltage applied	Current	Power
1	63,500 v	1.12 amp	17,900 w

(a) Determine the values of the parameters for an equivalent circuit of the type shown in Fig. 9. Express all values in per cent on a 10,000-kva rated-voltage base.

(b) Determine the values of the parameters for an equivalent circuit of the type shown in Fig. 6a. Express all values in per cent on a 10,000-kva rated-voltage base, and assume that the exciting admittances $Y_{1\sigma}$, $Y_{2\sigma}$, and $Y_{3\sigma}$ are equal.

6. Three transformers identical with the one of Prob. 5 are connected in a symmetrical three-phase bank to step down the voltage of a balanced three-phase 110,000-v transmission line. Windings 2, connected in Δ, supply a balanced 18,000-kva inductive load at 0.80 power factor. Windings 3, connected in Y, supply a balanced 10,000-kva inductive load at 0.90 power factor.

(a) What must be the value of the primary line-to-line voltages to maintain rated voltage at the 18,000-kva load?

(b) For the condition of part (a), what would be the voltage at the 10,000-kva load?

(c) For the condition of part (a), what would be the efficiency of the transformer?

7. A 5-kva single-phase distribution transformer has a 2,400-v winding and two identical symmetrically arranged 240-v windings which for normal operation are connected in parallel. The equivalent impedance with the two low-voltage windings in parallel is $16.0 + j23.2$ ohms referred to the high-voltage side.

The transformer is used to supply two small single-phase motors, one low-voltage winding being connected directly to each motor. If one motor requires 2 kva at 0.75 power factor and the other requires 2 kva at 0.85 power factor, what are the terminal voltages at the motors when the primary terminal voltage is 2,400 v? State all assumptions made in the solution.

8. Three identical three-circuit transformers with ratings and short-circuit characteristics specified in Prob. 5 are connected with their two sets of low-voltage windings in Δ and their high-voltage windings in Y.

If balanced three-phase line-to-line voltages of 110,000 volts are applied to the high-voltage windings and a short circuit appears across the 11,000-v winding of one of the transformers, what will be the value of the short-circuit current?

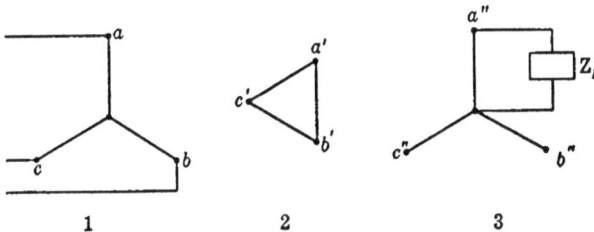

FIG. 12.   Three-phase connection of 3-circuit transformer, Prob. 9.

9. Three identical three-circuit transformers are connected in Y–Δ–Y as shown in Fig. 12. Balanced three-phase line voltages are applied to one set of Y-connected windings and the neutral of this set is left ungrounded. A single-phase load is taken from line to neutral of the other Y-connected set. Find the value of the load current in terms of the applied voltage, the equivalent-circuit transformer impedances, and the impedance of the load referred to a common base.

CHAPTER XXVIII

Transformer Connections for Phase Transformation

On many occasions it is desired to transform power from one polyphase system to another with a different number of phases. For example, although three-phase generation, transmission, and distribution are standard practice for modern power systems, many of the two-phase distribution networks[1] used in early developments are still in existence, since the expense of modernization and conversion to three-phase distribution often is prohibitively high. Many power companies with three-phase generating stations and transmission systems therefore still must distribute a considerable amount of two-phase power. These two-phase networks usually are supplied from the three-phase system through a bank of suitably connected transformers.

In other applications, moreover, six-phase or twelve-phase power may be desired, particularly in substations for converting alternating-current to direct-current power by means of synchronous converters or electronic rectifiers.[2] Increasing the number of phases of the alternating-current supply to an electronic rectifier decreases the objectionable harmonics in the alternating-current input and results in a smoother voltage wave-form on the direct-current output side.[3] Many six-phase rectifiers are in operation, and some of the larger installations are supplied with twelve-phase power. With the synchronous, or rotary, converter — a rotating machine that combines in one armature winding the functions of an alternating-current synchronous motor and a direct-current generator[4] — physical size decreases and efficiency increases as the number of phases of the alternating-current supply is increased. A large number of phases, however, requires a complicated armature winding and has certain other undesirable features, so that most synchronous converters are supplied with six-phase power, although some large twelve-phase machines are in operation.

It would rarely be economically possible to supply rectifiers or syn-

---

[1] Two-phase systems are discussed in this series in the volume on electric circuits (1940), Ch. X, Art. 3, p. 522.

[2] For illustrations showing some of the features of such a substation, see the volume on electric circuits (1940), pp. 129–133.

[3] See the volume on electronics, Art. 1, Ch. VII.

[4] Synchronous converters are discussed in most textbooks on alternating-current machinery; for example, see R. R. Lawrence, *Principles of Alternating-Current Machinery* (3d ed.; New York: McGraw-Hill Book Company, Inc., 1940), 417–467.

chronous converters from six- or twelve-phase transmission lines, but to convert power received from a three-phase system to six- or twelve-phase power by means of a transformer bank is a simple matter.

The number of transformer arrangements for obtaining phase transformation for these and allied purposes is unlimited. Only some of the common arrangements are considered in this chapter.

1. SYMMETRICAL THREE–PHASE TO SIX–PHASE CONNECTIONS

A balanced six-phase system of voltages consists of six voltages of equal magnitude with a phase displacement of 360/6 or 60 degrees between the voltages of consecutive phases taken in cyclic order.

Several arrangements are commonly used to transform three-phase into six-phase power. The simplest consist of a three-phase transformer — or a bank of three single-phase transformers — having a primary winding and two independent secondary windings for each phase. The three-phase primaries, which may be connected either in Y or in Δ, sometimes have important effects on harmonic phenomena, as shown in part (e) of this article. The secondaries can be connected in a variety of ways.

1a. *Six-Phase Star Connection.* — Consider a transformer bank comprising three transformers, each with a single primary winding and two identical secondaries. The actual arrangement of the transformers is shown in Fig. 1a, and the connections also are shown in simplified form in Fig. 1b, in which all the windings of each transformer are drawn parallel to each other. The primaries are connected to a balanced three-phase source. The secondaries are Y-connected in two groups, S' and S'', *of opposite polarity;* that is, the neutral point n' of the S' group is formed by junction of the unmarked terminals of three secondaries, while the neutral point n'' of the S'' group is formed by the dot-marked terminals connected together.

The vector diagrams of terminal voltages (neglecting leakage-impedance voltage drops) are shown in Fig. 1c, in which the V's represent falls in potential in the directions indicated by the order of the subscripts. The secondary voltages of the S' group are in phase with the corresponding primary voltages, but those of the S'' group are in phase opposition. A relative phase displacement of 60 degrees thus exists between the three-phase systems of voltages obtained from the two groups of secondaries. Therefore, if the two neutral points n' and n'' are connected together, as indicated by the broken-line connections n in Figs. 1a and 1b, six-phase voltages can be obtained from the secondaries. This interconnection of the two neutral points is equivalent to superposing the two secondary Y's, as indicated in Fig. 2a, so that they form a six-pointed star. In Fig. 2a, the secondary terminals are designated a, b, c, d, e, f to correspond with the cyclic order of the six-phase voltages. The vector diagrams of the six-phase voltages to neutral and of the line currents delivered to a balanced, unity-power-factor, six-phase load are shown in Fig. 2b. Since

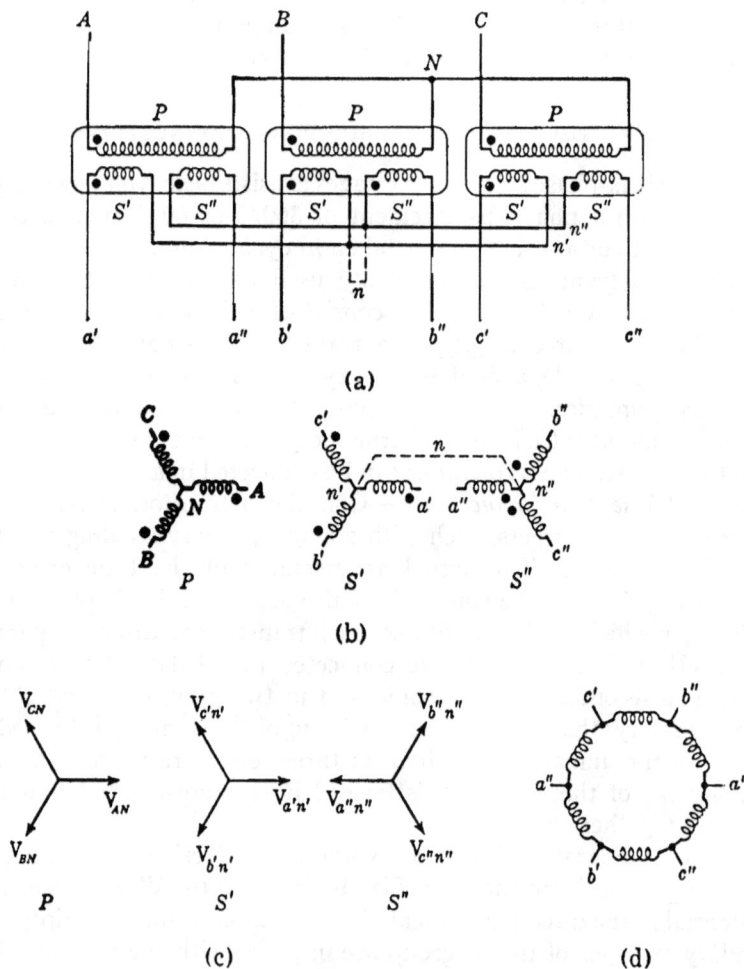

(a)

(b)

(c)   (d)

FIG. 1. Three-phase-Y to six-phase-double-Y connection, and vector diagrams of the voltages. When the neutral points n' and n'' are connected together by the broken-line connection n, the double-Y connection becomes the six-phase star shown in Fig. 2. A balanced, six-phase load is shown in (d).

in the six-phase star connection all the unmarked terminals of the S' group of secondaries are joined to all the dot-marked terminals of the S'' group, the actual secondary connections can be made as in Fig. 2c, in which the broken-line connection n forms the neutral point of the six-phase star. Not all four secondary terminals of each transformer need therefore be accessible, a single secondary winding with a center tap being sufficient for the six-phase star connection.

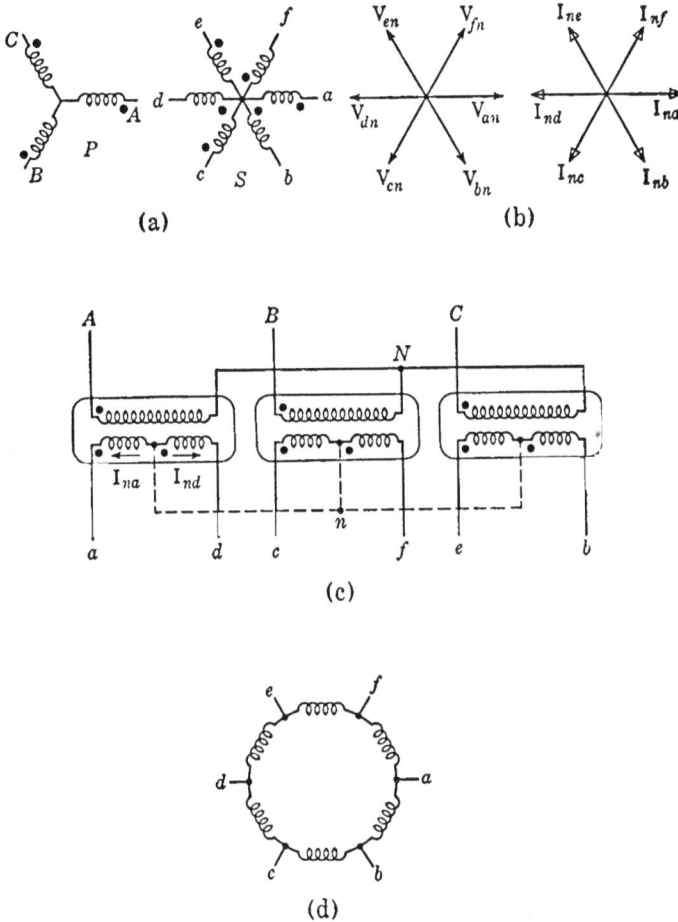

(a)

(b)

(c)

(d)

FIG. 2.   Three-phase-Y to six-phase-star connection, and vector diagrams of balanced six-phase voltages and currents. When the broken-line neutral connection is omitted from (c), the six-phase star becomes the diametrical connection. A balanced, six-phase load is shown in (d).

If the transformer bank is connected to a synchronous converter which supplies power to a three-wire system on its direct-current side, the neutral point n of the six-phase star-connected secondaries can be used to obtain the neutral of the direct-current system.

The secondary voltages of the six-phase star connection form a true six-phase system; that is, definite voltages exist between each line terminal and all the other line terminals. Several other arrangements which give balanced six-phase voltages only when the transformer windings are connected to a balanced six-phase load are described below.

1b. *Diametrical Connection.* — If the secondary terminals of the six-phase star arrangement, Fig. 2c, are connected to a balanced six-phase load, such as the synchronous converter armature indicated diagrammatically in Fig. 2d, the six-phase currents delivered to the load are balanced, as shown in the vector diagram of Fig. 2b. Consider the secondaries of one of the transformers — say the one whose secondaries supply phases a and d of the six-phase system. The currents I_{na} and I_{nd} are indicated by the arrows beside the secondary windings of Fig. 2c. From the vector diagram, Fig. 2b, the currents I_{na} and I_{nd} are equal in magnitude but are in phase opposition, and therefore their vector sum, which equals the current entering the center tap through the broken-line neutral connection of Fig. 2c, is zero. That is, the same current flows in the same direction in both secondaries. Consequently, the broken-line neutral connection in Fig. 2c can be omitted without disturbing the circuit. When the neutral connection of the six-phase star is omitted, the arrangement is called the diametrical connection, because each secondary is connected across diametrically opposite points on the six-phase load, as indicated by the lettering of the terminals in Figs. 2c and 2d.

Note that, if the secondary terminals are disconnected from the load, no definite voltages exist between one set of secondaries and another, and the arrangement then breaks down to three independent single-phase systems that are not in phase. It is only when the phases are interconnected through a balanced six-phase load that the diametrical connection gives balanced six-phase voltages.

1c. *Double-Y Connection.* — Consider the double-Y arrangement shown in Figs. 1a and 1b. If the six-phase load is balanced, each Y-connected group of secondaries delivers balanced three-phase currents. As far as fundamental currents are concerned, the vector sum of three balanced three-phase currents is zero, and therefore there is no current in the broken-line connection n, Figs. 1a and 1b, joining the neutral points of the two Y's. This connection hence can be omitted, leaving the two Y's independent of each other except for their interconnection through the load. The isolation of the two neutral points may, however, have important effects on the harmonic phenomena described in part (e) of this article.

Note that, like the diametrical connection, the double-Y connection gives balanced six-phase voltages only when the two Y-connected groups of secondaries are properly connected to a balanced six-phase load, such as the armature windings of a synchronous converter, indicated diagrammatically in Fig. 1d. The manner in which the connections should be made is indicated by the lettering of the terminals in Figs. 1a and 1d.

The double-Y connection is often used in electronic rectifier circuits, the neutrals of the Y's then being connected through an autotransformer, or center-tapped reactor, known as an interphase transformer. This circuit, whose behavior is markedly affected by the characteristics of the rectifier, is discussed in detail in the volume on electronics.

1d. *Double-Δ Connection.* — Two three-phase systems of voltages differing in phase by 60 degrees are obtained when each of the two groups of secondaries is connected in Δ, the polarity of one group being opposite to the polarity of the other, as indicated by the connections and vector diagrams shown in Fig. 3. This arrangement is called the double-Δ connection. As with the double-Y connection, balanced six-phase voltages result if the two three-phase systems are properly connected to a balanced six-phase load, as indicated in Fig. 3d; however, if the load is disconnected, the secondaries do not give six-phase voltages, but merely produce two independent three-phase systems. The double-Δ connection sometimes is used to supply six-phase power to synchronous converters when a six-phase neutral connection is not required.

1e. *Effects of the Primary Connections.* — Under certain conditions the primary connections may have important effects on the performance of the circuit, because of the peculiar behavior of third-harmonic currents and voltages in three-phase systems.* Third-harmonic currents or voltages may be present either because of the excitation characteristics of the transformers, or because they are produced by the load. The basic principle in these situations where harmonic phenomena are important can be stated as follows.

If the primary windings are connected in Δ, third-harmonic currents can flow in them, but, if the primaries are connected in Y with isolated neutral, balanced third-harmonic primary currents cannot exist. Therefore the Y connection with isolated neutral should not be used when the secondary circuits do not provide a path for the third-harmonic components of the exciting current required to produce a sine-wave flux, or when the load demands the presence of third-harmonic currents in the primaries. For example, in some circuit arrangements, the secondary currents in a transformer bank supplying power to a six-phase electronic rectifier may contain large third-harmonic components, and, if the primaries are not connected in Δ so as to permit the existence of third-harmonic primary currents, the performance of the rectifier may be adversely affected. These rectifier circuits are discussed in detail in the volume on electronics.

In considering the third-harmonic phenomena, one should remember

* See Ch. XXIII.

that the behavior of a three-phase, three-legged, core-type transformer is quite different from the behavior of a three-phase shell-type unit or a bank of three single-phase transformers.* For example, the waveform of the flux in a three-phase core-type transformer is very nearly sinusoidal

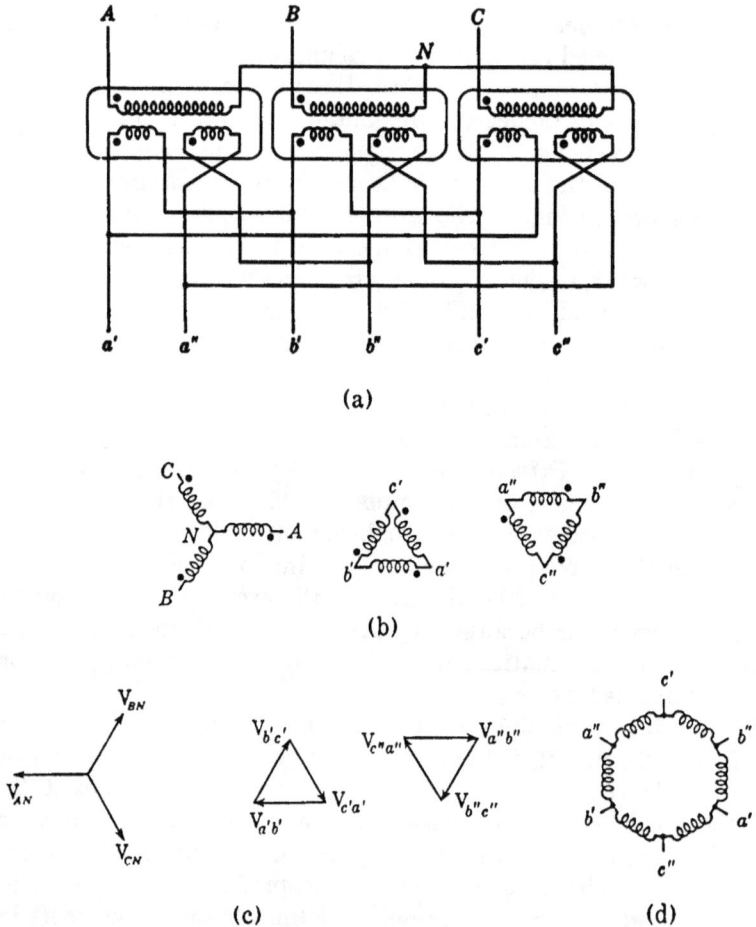

(a)

(b)

(c)   (d)

FIG. 3. Three-phase-Y to six-phase-double-Δ connection, and vector diagrams of the voltages. A balanced, six-phase load is shown in (d).

even if the third-harmonic components of the exciting currents are suppressed. Therefore the Y connection with isolated neutrals often can be used for the primary windings of a three-phase core-type transformer in circumstances where such a connection would be undesirable with a three-phase shell-type transformer or a bank of three single-phase units.

* See Art. 2a, Ch. XXVI, p. 616.

2. PHASE TRANSFORMATION WITH CROSS–CONNECTED SECONDARIES

An unlimited number of transformer arrangements can be devised to change power from any number of phases to any other number of phases. The general principle on which these phase transformations are based is discussed below.

Consider two transformers A and B, shown in Fig. 4a. Each has a primary winding and several independent secondaries, two of which are shown in Fig. 4a. Let the primaries form part of a polyphase arrangement,

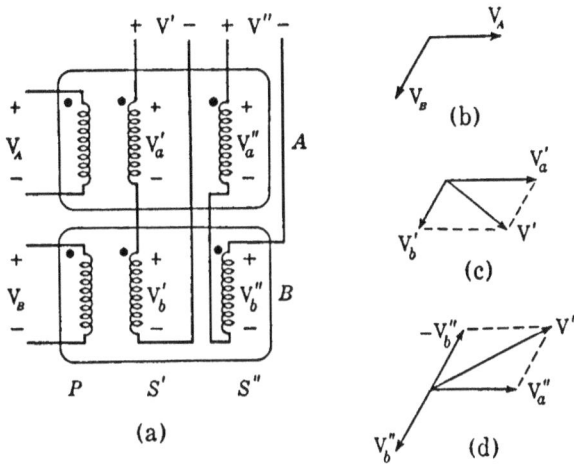

FIG. 4. Illustrating the general principles of phase transformation with interconnected secondaries.

and let V_A and V_B be the vector voltages applied to them from the polyphase primary circuit. These are shown vectorially in Fig. 4b, in which the phase angle between the primary voltages depends on the number of phases of the primary circuit and on the manner in which the primaries are connected to this circuit. The voltages of the S' group of secondaries are V'_a and V'_b, and those of the S'' group are V''_a and V''_b, the positive directions of the voltages being indicated in Fig. 4a. If the leakage-impedance voltage drops are neglected, the secondary voltages V'_a and V''_a are in phase with the primary voltage V_A, and the secondary voltages V'_b and V''_b are in phase with the primary voltage V_B, as shown in the vector diagrams.

Let the two secondaries of the S' group be connected in series, as shown in Fig. 4a. Then the resultant voltage V' is the vector sum of V'_a and V'_b, as in Fig. 4c. It is evident that adjusting the magnitudes of the two secondary voltages V'_a and V'_b (by selection of the appropriate numbers of turns in the two secondary windings) can make the voltage V' have any desired value and assume any phase position within the angle included between the vectors V'_a and V'_b. If the connections of the secondary of the B transformer are reversed, as in the connections of the S'' coils in Fig. 4a, the resultant voltage V'' is as shown in the vector diagram of Fig. 4d. The numbers of turns in the secondary windings can be adjusted so that the voltages V' and V'' are equal in magnitude and differ in phase by any chosen angle, thus forming part of a polyphase system.

▶ Therefore, with a group of transformers having a sufficient number of independent secondary windings and with proper arrangement of the interconnections among windings and their numbers of turns, it is possible to obtain a polyphase system of voltages having any desired number of phases from any other polyphase system.◀

2a. *Forked-Y or Double-Zigzag Connection.* — As an example of the use of interconnected windings, consider the arrangement shown in Fig. 5a, comprising a three-phase transformer — or a bank of three single-phase transformers — having a primary winding and three independent secondary windings for each phase. All windings on the same transformer are drawn parallel to one another. The primary windings may be connected either in Y or in Δ; they are shown Δ-connected in Fig. 5a. The manner of making the secondary connections is indicated in the figure. When the three-phase voltages applied to the primaries are balanced, the secondaries deliver balanced six-phase voltages, as shown in the vector diagram, Fig. 5b. Note that the six-phase line-to-neutral voltage equals $\sqrt{3}$ times the voltage of one secondary winding.

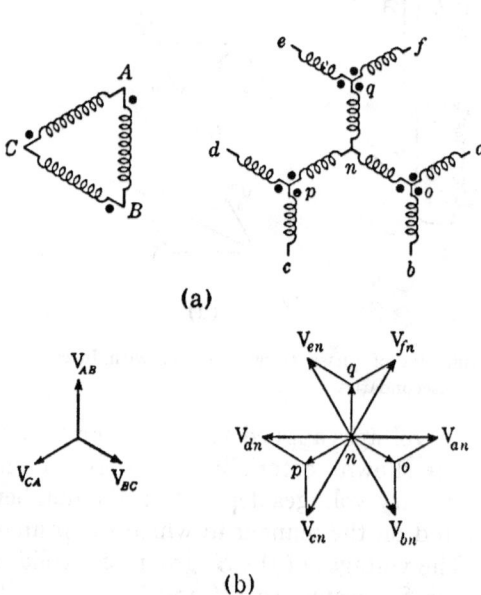

(a)

(b)

FIG. 5. Six-phase forked-Y or double-zigzag connection.

The forked-Y connection often is used to supply six-phase power to an electronic rectifier.

2b. *Quadruple-Zigzag Connection.* — A twelve-phase system of voltages can be obtained from a three-phase system if each of three transformers has six independent secondaries which are interconnected as shown in Fig. 6. The windings drawn parallel to one another are on the same magnetic core. The polarities of the windings are indicated. The neutral of the twelve-phase system is n and the line terminals are $a, b, c \cdots l$. The primaries of the three transformers (not shown in Fig. 6) may be connected either in Δ or in Y. It can be shown that for balanced, twelve-phase output, there should be 0.366 times as many turns in each of the twelve small secondary windings as in the six star-connected windings.

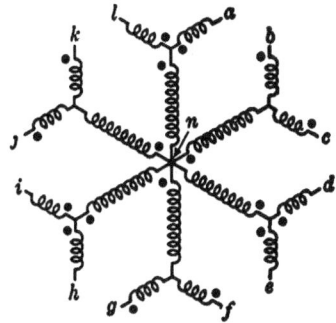

An arrangement similar to this connection is often used to supply twelve-phase power to an electronic rectifier.

Fig. 6. Arrangement of the secondaries for 12-phase quadruple zigzag connection.

3. THREE-PHASE TO TWO-PHASE TRANSFORMATION

A balanced two-phase system of voltages comprises two voltages of equal magnitude differing in time phase by 90 degrees.[5] Three common

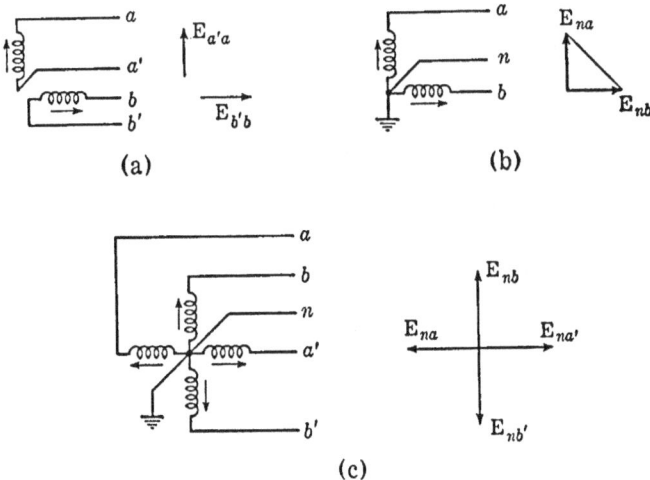

(a)   (b)

(c)

Fig. 7. Three common arrangements of 2-phase systems; (a) 4-wire, (b) 3-wire, (c) 5-wire. The 5-wire arrangement (c) can also be regarded as a star-connected 4-phase system.

arrangements of two-phase systems are shown in Fig. 7, in which the coils represent generator or transformer windings whose induced electromotive forces (rises in potential) in the arrow directions differ in time phase by 90 degrees. In the two-phase four-wire system of Fig. 7a, the circuits of

[5]See the volume on electric circuits (1940), p. 523.

the two phases are independent of each other. If the windings are connected at one end, the two-phase three-wire system of Fig. 7b is obtained, and if the windings are joined at their midpoints, as in Fig. 7c, a five-wire system results. The arrangement of Fig. 7c can be described more accurately as a four-phase system, but, according to accepted usage, all the systems shown in Fig. 7 generally are regarded as two-phase.

For changing three-phase to two-phase power, a great variety of transformer arrangements are available, some of which are adaptable to autotransformers.[6] A number of these circuits can also be arranged to deliver

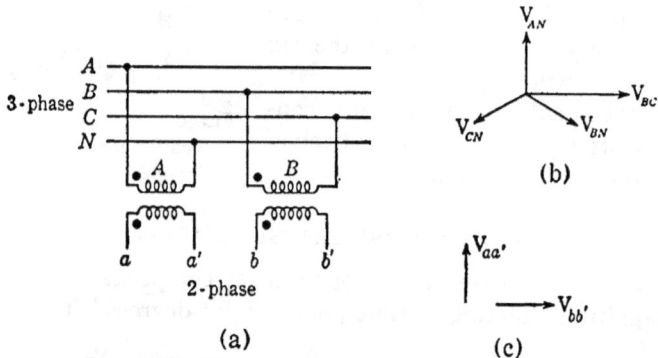

FIG. 8.   Transformer connections for obtaining 2-phase power from a 4-wire 3-phase system.

two-phase, three-phase, and three-wire single-phase power simultaneously. A few of the common arrangements are described below.

3a. *Four-Wire–Three-Phase to Two-Phase.* — Since the line-to-line voltage V_{BC} is in quadrature with the line-to-neutral voltage V_{AN} of a balanced Y-connected three-phase system, balanced two-phase voltages can be obtained from the secondaries of two transformers whose primaries are connected as shown in Fig. 8a. In this diagram, the secondaries are independent of each other, but if desired they can be connected for three-wire or five-wire two-phase output. The vector diagrams of primary and secondary voltages are shown in Figs. 8b and 8c, from which it can be seen that the primary voltage of the B transformer is $\sqrt{3}$ times that of the A transformer, and therefore the turns ratio N_1/N_2 of the B transformer must be $\sqrt{3}$ times that of the A transformer for balanced two-phase output. The principal disadvantages of this arrangement are that a four-wire three-phase circuit is required and that the three-phase currents are grossly unbalanced.

[6] For a description of a large number of these arrangements see L. F. Blume, editor, *Transformer Engineering* (New York: John Wiley & Sons, 1938), 219–230 and 252–256; also J. B. Gibbs, *Transformer Principles and Practice* (New York: McGraw-Hill Book Company, Inc., 1937), 82–86 and 92–98.

Several variations of this arrangement are in use. For example, if either the line-to-line or the line-to-neutral voltage of the three-phase system is the same as the desired two-phase voltage, one of the transformers can be omitted and the other replaced by an autotransformer.

3b. *Scott Connection.* — The circuit shown in Fig. 9a is probably the most widely used arrangement of transformers for three-phase to two-phase transformation. It was invented by C. F. Scott and is known as the Scott connection. It consists of two transformers, as shown in Fig. 9a.

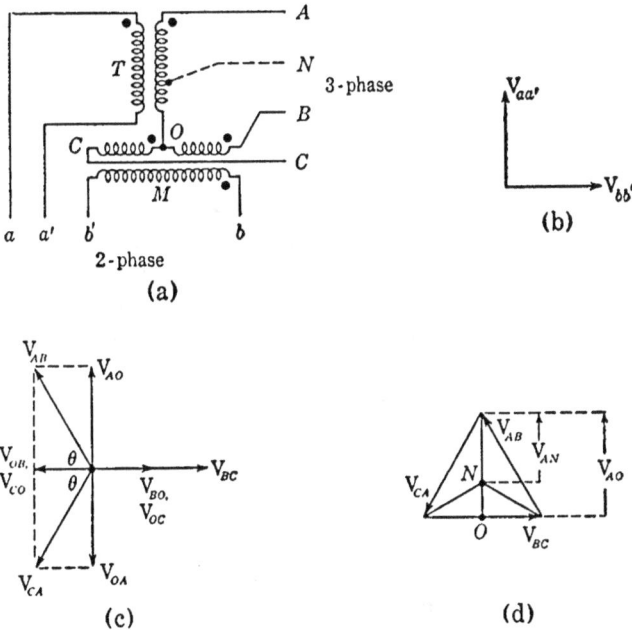

Fig. 9. Scott connection for 3-phase to 2-phase transformation, and vector diagrams of the voltages.

Transformer M — known as the " main " transformer — has a single winding on its two-phase side bb', and a center-tapped winding BOC on its three-phase side, while transformer T — known as the " teaser " — has a single winding on each side. The two transformers have different ratios of transformation, as shown in the following discussion.

Although the usual direction of power flow is from the three-phase to the two-phase system, it is simpler to explain the operation of the bank on the assumption that the two-phase windings aa' and bb' are the primaries. If balanced two-phase voltages $V_{aa'}$ and $V_{bb'}$, Fig. 9b, are applied to the two-phase windings and leakage-impedance voltage drops are neglected, the vector voltages across the three windings on the three-phase side are as indicated by the vectors V_{AO}, V_{BO}, and V_{OC} in Fig. 9c, where the vectors represent falls in potential in the directions indicated by the order of the subscripts. According to the polarity markings, V_{AO} is in phase with $V_{aa'}$, and V_{BO}, V_{OC} are in phase with $V_{bb'}$, as shown in Fig. 9a. In Fig. 9a, note that the line-to-line voltages on the three-phase side are

$$V_{AB} = V_{AO} + V_{OB} \tag{1}$$

$$V_{BC} = V_{BO} + V_{OC} \tag{2}$$

$$V_{CA} = V_{CO} + V_{OA}. \tag{3}$$

Therefore these vector voltages are as shown in Fig. 9c. If they are to form a balanced three-phase system, the angles θ, Fig. 9c, must be 60 degrees, and hence

$$\frac{V_{AO}}{V_{AB}} = \sin 60° = 0.866. \tag{4}$$

▶ On its three-phase side, the " teaser " transformer T operates at 0.866 of the three-phase line-to-line voltage. Therefore, if the " main " and " teaser " transformers have the same number of turns in their two-phase windings, there must be 0.866 as many turns in the three-phase winding AO of the " teaser " as there are in the complete winding BC of the " main " transformer. ◀

From the vector diagram, which can be drawn as an equilateral triangle, as shown in Fig. 9d, it can be seen that the voltage V_{AO} of the " teaser " transformer is in phase with the voltage of phase A to the neutral of a symmetrical Y system of voltages, and therefore a symmetrical three-phase neutral can be obtained if a tap point N is located on the three-phase winding of transformer T so that the voltages V_{AN}, V_{BN}, V_{CN} are equal. Since, from Eq. 4,

$$V_{AO} = \frac{\sqrt{3}}{2} V_{line}, \tag{5}$$

and, since for a balanced three-phase system,

$$V_{AN} = \frac{1}{\sqrt{3}} V_{line}, \tag{6}$$

then

$$\frac{V_{AN}}{V_{AO}} = \frac{1}{\sqrt{3}} \times \frac{2}{\sqrt{3}} = \frac{2}{3}. \tag{7}$$

Consequently, if the three-phase winding of the " teaser " transformer T is tapped at a point N so that two-thirds of the turns are between A and N, Fig. 9a, the point N is the neutral of a balanced three-phase system.

3c. T *Connection.* — If both primary and secondary sides are arranged like the three-phase side in the Scott connection, power is transformed from three-phase to three-phase. This arrangement, called the T connection, is a means of transforming three-phase power with two transformers. It can be used either with autotransformers or with transformers having separate primary and secondary windings. As with the open-delta connection,* its principal disadvantages are that it is unsymmetrical and that its apparatus economy is less than that of a symmetrical arrangement. Although the T-to-T connection does not change the number of phases, as do all the other arrangements described in this chapter, it should be discussed at this time, because of its similarity to the Scott connection.

When both primary and secondary windings are connected in T, the primary and secondary voltages of the " main " transformer equal the three-phase primary and secondary line-to-line voltages, but the primary and secondary voltages of the " teaser " transformer equal 0.866 of the primary and secondary line voltages. (See Eq. 4.) This arrangement differs from the Scott connection, in that the ratios of transformation of both transformers should be the same. The currents in both transformers are the three-phase line currents. The " teaser " transformer may have the same current and voltage ratings as the " main " transformer, but it would then operate at 0.866 of its rated voltage. It is better to use a " teaser " transformer designed to operate at 0.866 of the rated three-phase line voltage, but then the two transformers are not interchangeable.

The expression for the three-phase volt-amperes of a balanced three-phase load is $\sqrt{3}\ V_{line}I_{line}$, and therefore, if V_2 and I_2 are the rated secondary voltage and current of the " main " transformer, the three-phase volt-ampere rating of the bank is $\sqrt{3}\ V_2I_2$. The apparatus economy is the ratio of the three-phase rating of the bank to the sum of the ratings of the transformers.† If the " teaser " transformer has the same current and voltage ratings as the " main " transformer, the sum of the ratings of the transformers is $2V_2I_2$ and the apparatus economy is

$$\frac{\sqrt{3}\ V_2I_2}{2V_2I_2} = 0.866. \qquad [8]$$

This is the same as the apparatus economy of an open-delta bank. However, if the " teaser " transformer is designed to operate at 0.866 of the three-phase line voltage, the apparatus economy is somewhat improved,

* See Art. 5, Ch. XXI.
† See Eq. 32, Ch. XXI, p. 546.

and is

$$\frac{\sqrt{3}\,V_2I_2}{V_2I_2 + 0.866\,V_2I_2} = \frac{\sqrt{3}}{1 + \dfrac{\sqrt{3}}{2}} = 0.928. \tag{9}$$

The T-to-T connection is not often used because, if the transformers are interchangeable, it has no advantages over the open-delta arrangement.

4. SUMMARY

Although only a few of the large number of transformer arrangements for phase transformation are described in this chapter, enough examples have been cited to show the general principles. The major points brought up in the preceding discussion can be summarized as follows:

A six-phase system of voltages can be obtained through combining two three-phase systems of opposite polarity.

A voltage having any desired magnitude and phase position can be obtained through combining the voltages of the secondaries of two transformers whose primary voltages differ in phase. Thus it is possible to transform polyphase power at any number of phases to polyphase power at any other number of phases by the use of a bank of transformers having a sufficient number of interconnected secondary windings. The forked-Y and quadruple-zigzag connections, Figs. 5 and 6, are examples of such arrangements.

Three-phase to two-phase transformation can be accomplished by use of the fact that the line-to-neutral voltage V_{AN} of a balanced three-phase system is in quadature with the line-to-line voltage V_{BC}. (See Fig. 8.)

Two voltages in time quadrature can be obtained from a three-phase system by means of the T connection, as in Fig. 9a. Thus the T connection can be used for three-phase to two-phase transformation, as in the Scott connection, Fig. 9.

If both primary and secondary windings of a bank of two transformers are connected in T, three-phase voltages can be transformed without a change in the number of phases.

PROBLEMS

1. A six-phase synchronous converter requires 1,000 kva at 212 v. The converter is to be supplied through a bank of single-phase transformers from a three-phase 6,600-v transmission line. For each of the following connections, what must be the transformer voltage and current ratings?

(a) Δ to double-Δ,
(b) Δ to double-Y,
(c) Δ to diametrical.

2. Three identical single-phase transformers are connected in Δ to a three-phase line with balanced voltages. The secondaries are arranged in quadruple zigzag to give a twelve-phase system. If the voltage to neutral of the twelve-phase system is to be 2,400 v. what must be the voltage rating of each secondary winding?

3. Three identical transformers, each rated 2,300 : 424 v, 60 ∼, and 100 kva, are connected in Δ on their primary sides to a three-phase line with balanced line-to-line voltages of 2,300 v. The secondary of each transformer is center-tapped, and the secondaries are connected to give a six-phase system with 212 v between adjacent line terminals. Short-circuit data on one of these transformers with the high-voltage winding excited and one of the secondary terminals short-circuited to the center tap are: 104 v, 21.5 amp, 375 w, 60 ∼.

If a single-phase inductive load of 470 amp and 0.80 power factor is supplied from one pair of adjacent six-phase terminals, what will be the voltage across the loaded phase?

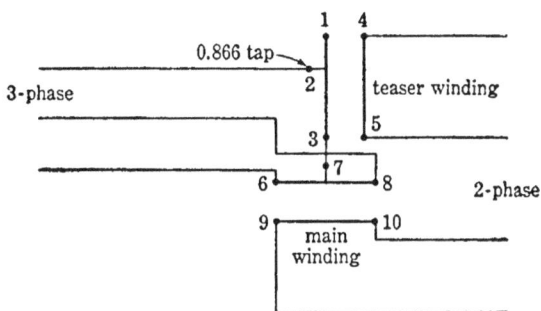

Fig. 10.   Scott connection, Prob. 6.

4. Three identical four-winding single-phase 60 ∼ transformers are built for Δ connection on their primary sides and forked-Y connection on their secondary sides. When the six-phase output terminals are short-circuited, a balanced three-phase voltage of 5.2% is necessary to cause rated current to flow. Under this short-circuit condition the power factor is 0.29.

The transformer bank is to supply a balanced six-phase load of 300 kva with a voltage between adjacent lines of 230 v. The line-to-line primary voltage is 2,300 v.

(a) On the assumption that the transformers are ideal, find the voltage and current ratings of each winding.

(b) For a load power factor of 0.90, lagging, find what balanced primary voltages must be impressed to maintain the output voltage at 230 v.

5. Show that, when two transformers are used in the Scott connection (Fig. 9) for transforming from three-phase to two-phase power, a balanced two-phase load will cause balanced currents to flow in the three-phase line. Assume that the transformers are ideal.

6. Two 10-kva 2,400 : 240-v 60 ∼ single-phase transformers are to be used in the Scott connection, as in Fig. 10. The following data have been obtained for these transformers:

TRANSFORMER A

Connection: Terminals 4 and 5 short-circuited; 2 and 3 excited.
Data: Power = 76.0 w; voltage = 57.8 v; current = 4.16 amp.

TRANSFORMER B
Connection: Terminals 9 and 10 short-circuited; 6 and 8 excited.
Data: Power = 80.2 w; voltage = 60.3 v; current = 4.16 amp.

TRANSFORMER B
Connection: Terminals 6 and 8 connected together; 6 and 7 excited.
Data: Power = 4.52 w; voltage = 4.80 v; current = 4.16 amp.

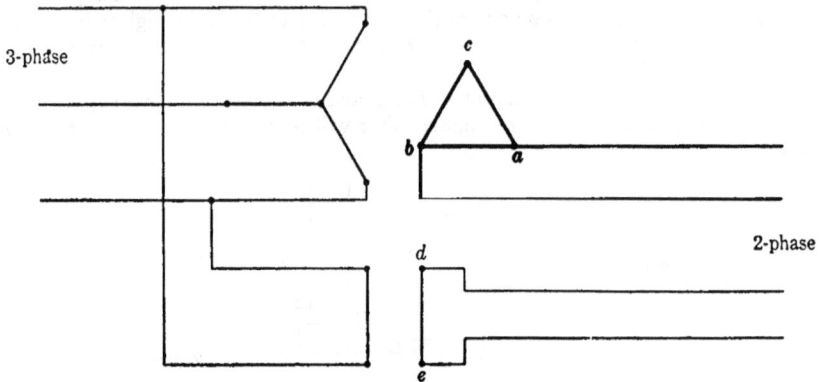

FIG. 11.   Three-phase to 2-phase connection, Prob. 7.

If the three-phase line-to-line voltage is held at 2,400 v, determine the per cent voltage regulation for each of the following load conditions:

(a) At the terminals 9 and 10, for a single-phase inductive load of 6 kva at 0.75 power factor across these terminals, the secondary winding of the teaser transformer being open-circuited.

(b) At the terminals 4 and 5, for a single-phase inductive load of 6 kva at 0.75 power factor across these terminals, the secondary winding of the main transformer being open-circuited.

(c) At both pairs of secondary terminals if the loads of parts (a) and (b) are applied simultaneously

7. One method of transforming from three-phase to two-phase power is shown in Fig. 11. For this connection:

(a) Specify the transformer ratios which will give two-phase line-to-line voltages equal to the line-to-line voltages of the three-phase system. Show that V_{ab} and V_{de} are in quadrature.

(b) Prove that a balanced two-phase load will produce balanced currents in the three-phase line. Assume that the transformers are ideal.

CHAPTER XXIX

Parallel Operation of Transformers

Transformers are said to be connected in parallel when both their primary windings and their secondary windings are connected in parallel. They often are operated in parallel in single-phase combinations and also in three-phase groups, because a growing load may make it necessary to increase the kilovolt-ampere capacity of an existing bank by the addition of new transformers in parallel with it, or because it may be desirable to supply an important load from several groups of transformers in order to maintain continuity of service despite failure of one of the transformers or its associated circuits.

The principal problem arising in connection with the parallel operation of transformers is the determination of the manner in which the load is shared among the units. Best results are obtained when the load is divided among the units in proportion to their kilovolt-ampere ratings, and when the secondary current in each transformer is in phase with the load current. If the load is divided among the units in any other manner, the full-load capacity of the bank is less than the sum of the kilovolt-ampere ratings of the units.

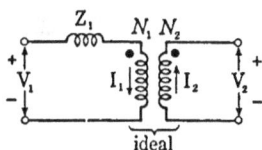

FIG. 1. Equivalent circuit of a single transformer.

1. PARALLEL OPERATION IN SINGLE–PHASE CIRCUITS

In the following analysis, the exciting currents are neglected and therefore the equivalent circuit for any one of the transformers is as shown in Fig. 1. Thus the vector relation between the primary and secondary terminal voltages is

$$V_1 \cdot I_1 Z_1 = a V_2, \qquad [1]$$

where

V_1 and V_2 are the vector terminal voltages,
I_1 is the vector primary current in the transformer,
Z_1 is its complex equivalent impedance referred to the primary,
a is its turns ratio N_1/N_2.

Alternatively, the voltage equation referred to the secondary is

$$\frac{V_1}{a} - I_2 Z_2 = V_2, \qquad [2]$$

673

where

I_2 is the vector secondary current,

Z_2 is the complex equivalent impedance referred to the secondary.

The positive directions of the voltages and currents are indicated in Fig. 1. For these positive directions

$$I_2 = aI_1 \qquad [3]$$

when the exciting current is neglected.

When the transformers are operated in parallel, the terminal voltages of all primary windings are the same, and likewise the terminal voltages of all secondary windings are the same. The total primary current taken by the group is the vector sum of the individual primary currents, and the total current delivered to the load is the vector sum of the individual secondary currents. These facts together with Eqs. 1, 2, and 3 are the basis for the analysis of all problems concerning parallel operation of transformers.

If the turns ratios of two transformers are not equal, the secondary induced voltages of the transformers are unequal when their primaries are connected to the same source, and therefore, if their secondaries are connected in parallel, a circulating current must flow, even at no load. Since this is an undesirable situation, transformers operated in parallel usually have equal turns ratios.

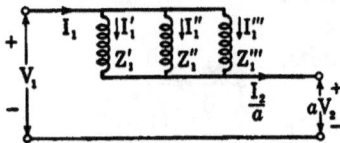

FIG. 2. Equivalent circuit for three transformers with equal turns ratios operating in parallel.

1a. *Equal Ratios of Transformation.* — When the turns ratios of the paralleled transformers are equal, their primary terminal voltages are equal, their secondary terminal voltages referred to the primary also are equal, and therefore their vector equivalent-impedance voltage drops likewise must be equal. That is, from Eq. 1,

$$V_1 - aV_2 = I_1'Z_1' = I_1''Z_1'' = I_1'''Z_1''' \cdots , \qquad \blacktriangleright[4]$$

where the primes, double primes, triple primes indicate the individual transformers. The equivalent circuit representing these conditions is shown in Fig. 2.

▶ Therefore, when the turns ratios are equal, the currents in the paralleled transformers are related to one another as are the currents in impedances connected in parallel.◀

▶ It follows from Eq. 4 that, if transformers operating in parallel are to share the load in proportion to their kilovolt-ampere ratings, their equivalent-impedance voltage drops at full load must be equal. That is, the transformers must all have the same *per unit* equivalent impedance. Or it can be said that the *ohmic* values of their equivalent impedances must be inversely proportional to their kilovolt-ampere ratings.◀

Furthermore, for best results the several transformers should all have the same ratio of equivalent reactance to equivalent resistance. The effects of inequalities in the reactance-to-resistance ratios can be seen by study of Fig. 3, which shows the vector diagram of two transformers operating in parallel. The equivalent impedances of the two transformers are equal in magnitude, and therefore the currents I_1' and I_1'' in them are equal in magnitude. However, the currents are not in phase unless the ratios of equivalent reactance to equivalent resistance are the same for both transformers. The angle between the currents is

$$\angle \frac{I_1''}{I_1'} = \tan^{-1}\frac{X_1'}{R_1'} - \tan^{-1}\frac{X_1''}{R_1''}. \qquad [5]$$

Since the currents are not in phase, the current in each transformer is greater than half their combined current, and therefore the kilovolt-ampere output of the pair is less than the sum of the kilovolt-ampere outputs of the individual transformers. Thus the full-load capacity of the combination is less than the sum of the ratings of the units.

However, the requirement that the reactance-to-resistance ratio be the same for paralleled transformers is of minor importance compared with the requirement that their per unit equivalent impedances be the same. For example, consider two transformers having equivalent impedances that are equal in magnitude but with the widely different reactance-to-resistance ratios of 10 and 3. As in Eq. 5, the difference between the angles of these impedances equals the angle between the vector currents in the two transformers, and is

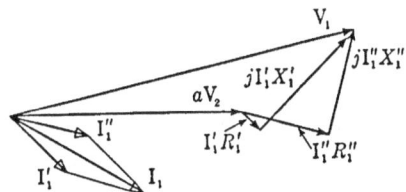

FIG. 3. Vector diagram for two transformers operating in parallel, showing the effects of unequal equivalent reactance-to-resistance ratios. The voltage drops are greatly exaggerated.

$$\tan^{-1}10 - \tan^{-1}3 = 12.7°. \qquad [6]$$

The ratio of the magnitude of the vector sum to the numerical sum of the currents is

$$\cos\frac{12.7°}{2} = 0.994. \qquad [7]$$

Thus the currents in the transformers usually are so nearly in phase that their vector sum substantially equals their numerical sum.

▶ Therefore, when transformers having equal turns ratios are connected in parallel, the *magnitudes* of the equivalent impedances are generally important in determining the division of current among the transformers. The angles of these equivalent impedances usually are of relatively little importance, unless it is necessary to know the actual vector values of the transformer currents.◀

The division of current among a number of parallel branches is best determined in terms of the admittances, the current in each branch being directly proportional to the admittance of the branch. If the total current supplied to the transformers on the primary side is I_1, the currents in the individual transformers are

$$I_1' = \frac{Y_1'}{Y_1} I_1 \qquad\qquad ▶[8]$$

$$I_1'' = \frac{Y_1''}{Y_1} I_1 \qquad\qquad ▶[9]$$

$$I_1''' = \frac{Y_1'''}{Y_1} I_1 \qquad\qquad ▶[10]$$

. . . .

where

I_1 is the total current,
I_1', I_1'', I_1''' $\cdots$ are the currents in the individual transformers,
Y_1', Y_1'', Y_1''' $\cdots$ are the reciprocals of their equivalent impedances,
Y_1 is the admittance of the parallel combination; that is

$$Y_1 = Y_1' + Y_1'' + Y_1''' \cdots. \qquad\qquad [11]$$

In an exact analysis, Eqs. 8–11 should be interpreted as vector relations, but if only the magnitudes of the currents are required it usually is sufficiently accurate to use these relations as if they were algebraic equations.

If the full-load equivalent-impedance voltage drops of the transformers are not the same, the transformers do not share the load in proportion to their kilovolt-ampere ratings. For example, consider a number of transformers connected in parallel. Let $(IZ)_{fl}'$, $(IZ)_{fl}''$, $(IZ)_{fl}'''$ be the magnitudes of the equivalent-impedance voltage drops in each transformer at its rated current. The full-load impedance drops may be expressed in volts, in per cent, or in per unit. Let the impedance drops be arranged in ascending order of magnitude, $(IZ)_{fl}'$ being the smallest. Let $(kva)'$, $(kva)''$, $(kva)'''$ be the kilovolt-ampere ratings of the several transformers.

When the transformers are connected in parallel, their equivalent-impedance voltage drops must be the same. As the output of the parallel combination is increased, the equivalent-impedance voltage drop in each transformer also increases proportionally and finally becomes equal to $(IZ)'_{fl}$, the full-load equivalent-impedance voltage of the first transformer. The first transformer is then delivering its rated kilovolt-ampere output $(kva)'$. Since the output of a transformer and its equivalent-impedance voltage drop are directly proportional, the output of the second transformer is

$$\frac{(IZ)'_{fl}}{(IZ)''_{fl}} \, (kva)''.$$

The total kilovolt-ampere output $(kva)_L$ of the group substantially equals the numerical sum of the individual outputs and is

$$(kva)_L = (kva)' + \frac{(IZ)'_{fl}}{(IZ)''_{fl}} \, (kva)'' + \frac{(IZ)'_{fl}}{(IZ)'''_{fl}} \, (kva)''' + \cdots. \quad [12]$$

If the total output is increased beyond this value, the common equivalent-impedance voltage increases, and the output of the first transformer exceeds its rated output. Hence $(kva)_L$, Eq. 12, is the greatest load that can be delivered by the transformers without overloading the transformer whose full-load equivalent-impedance voltage is the smallest. It may happen that a greater output can be delivered if this transformer is removed from the circuit.

Transformers can be adjusted so that they share the load properly by the addition of impedances in series with the transformers having the smaller values of per unit impedance. Theoretically it is necessary to adjust both the added resistance and reactance so that the voltage drop through the equivalent impedance of each transformer, its leads, and its added impedance is vectorially the same when the current in each transformer has its rated value. Usually, however, the angles of the impedances are relatively unimportant, and a properly adjusted reactor is all that is necessary. With low-voltage high-current transformers, sufficient reactance may often be secured if one of the low-voltage leads simply is surrounded with a suitably proportioned laminated iron core.

Proper load division between two transformers can also be obtained by the use of a small autotransformer connected as in Fig. 4. The load current I_L is forced to divide between the two windings of the autotransformer so that the net magnetomotive force acting on its core is zero. Therefore the components I'_L and I''_L of the load current are inversely proportional to the numbers of turns in the two windings of the autotransformer, and adjustment of these turns will divide the load current properly. In addition to the load components I'_L and I''_L, a small exciting current flows in the autotransformer and magnetizes its core in such a manner that the voltages of the autotransformer add to the output voltage of the main transformer whose secondary voltage tends to be the lower, and subtract from the output voltage of the other main transformer.

1b. *Unequal Ratios of Transformation.* — Although transformers connected in parallel should ideally have the same turns ratio, sometimes

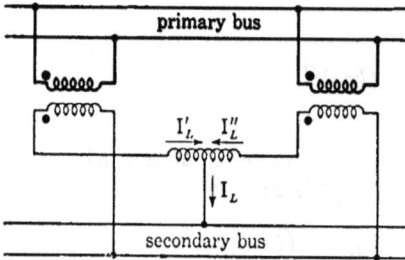

Fig. 4. Circuit using an autotransformer to obtain proper load division between two transformers operating in parallel.

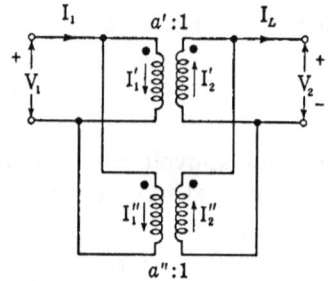

Fig. 5. Two transformers with unequal ratios of transformation operating in parallel.

transformers whose ratios of transformation are different must be operated in parallel, as for example during an emergency, or when transformers having tap-changing equipment are operating on different taps.

Consider two transformers having unequal ratios of transformation connected in parallel, as shown in Fig. 5. When the currents, voltages, and impedances are referred to the secondary sides, the vector voltage equations for the two transformers are

$$\frac{V_1}{a'} - I'_2 Z'_2 = V_2 \tag{13}$$

$$\frac{V_1}{a''} - I''_2 Z''_2 = V_2; \tag{14}$$

whence

$$\frac{V_1}{a'} - I'_2 Z'_2 = \frac{V_1}{a''} - I''_2 Z''_2, \tag{15}$$

where

I'_2, I''_2 are the vector secondary currents,

Z'_2, Z''_2 are the complex equivalent impedances referred to the secondary sides,

a', a'' are the turns ratios, N_1/N_2.

▶ Note that the equivalent-impedance voltage drops are not equal, as they would be if the turns ratios were equal, and therefore the total current does not divide between the transformers as it would between paralleled impedances.◀

The division of current between the transformers can be determined as follows; from Eq. 15:

$$I_2'Z_2' = I_2''Z_2'' + \frac{V_1}{a'} - \frac{V_1}{a''}. \qquad [16]$$

If $I_2'Z_2''$ is added to both sides of Eq. 16,

$$I_2'(Z_2' + Z_2'') = (I_2' + I_2'')Z_2'' + \frac{V_1}{a'} - \frac{V_1}{a''}. \qquad [17]$$

In this equation, $I_2' + I_2''$ is the total current delivered to the load, and hence the expression for the secondary current I_2' in the prime transformer in terms of the total current I_L delivered to the load is

$$I_2' = \frac{I_L Z_2''}{Z_2' + Z_2''} + \frac{\dfrac{V_1}{a'} - \dfrac{V_1}{a''}}{Z_2' + Z_2''}. \qquad [18]$$

The expression for the secondary current I_2'' in the double-prime transformer can be written by interchanging the primes and double primes; thus

$$I_2'' = \frac{I_L Z_2'}{Z_2' + Z_2''} + \frac{\dfrac{V_1}{a''} - \dfrac{V_1}{a'}}{Z_2' + Z_2''}. \qquad [19]$$

From Eqs. 18 and 19, the secondary currents in the transformers can be expressed as

$$I_2' = I_L' + I_{nl}' \qquad [20]$$
$$I_2'' = I_L'' + I_{nl}'', \qquad [21]$$

where

$$I_L' = \frac{I_L Z_2''}{Z_2' + Z_2''} \qquad [22]$$

$$I_L'' = \frac{I_L Z_2'}{Z_2' + Z_2''} \qquad [23]$$

$$I_{nl}' = \frac{\dfrac{V_1}{a'} - \dfrac{V_1}{a''}}{Z_2' + Z_2''} \qquad [24]$$

$$I_{nl}'' = \frac{\dfrac{V_1}{a''} - \dfrac{V_1}{a'}}{Z_2' + Z_2''}. \qquad [25]$$

In these expressions, each secondary current is the sum of two components. The first components I'_L and I''_L are inversely proportional to the equivalent impedances of the transformers and are the secondary currents that would flow if the turns ratios of the two transformers were the same. The second components I'_{nl} and I''_{nl} are independent of the load current I_L and depend on the inequality of the turns ratios. These are the currents that would flow in the secondaries even if the bank delivered no load. Since

$$I'_{nl} = -I''_{nl} , \qquad [26]$$

this no-load current simply circulates through the two secondaries in series. Because the impedance opposing the flow of this circulating current is the relatively small series combination of the equivalent impedances of the two transformers, a relatively small inequality in the two turns ratios may produce a rather large circulating current. It is generally

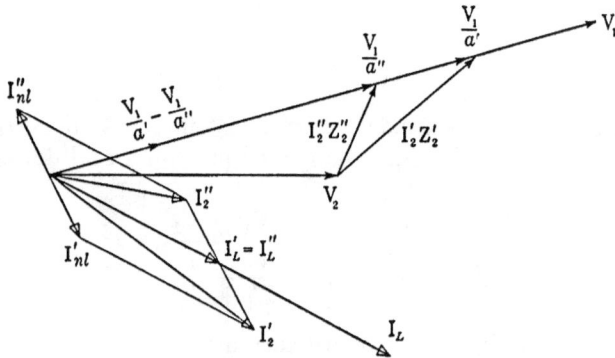

FIG. 6. Vector diagram for two paralleled transformers whose equivalent impedances are vectorially equal, but whose turns ratios are unequal.

considered poor practice to operate transformers in parallel when the circulating current in any transformer exceeds ten per cent of its rated current; consequently the turns ratios generally must be very nearly equal.

Figure 6 shows the vector diagram for two paralleled transformers whose ohmic equivalent impedances are vectorially equal, but whose turns ratios are unequal. When the equivalent impedances are vectorially equal, the load components I'_L and I''_L each equal one-half of the total load current I_L and are in phase with the load current, as shown in Fig. 6. If the turns ratios were equal, I'_L and I''_L would be the secondary currents in the transformers. The circulating currents due to unequal ratios are I'_{nl} and I''_{nl}, and their phase positions depend on the angle of the series combination of the equivalent impedances; thus I'_{nl} lags the voltage difference $(V_1/a') - (V_1/a'')$ by the impedance angle, as shown in Fig. 6. In accordance with Eqs. 20 and 21, the actual secondary currents are I_2' and I_2''. Examination of Fig. 6 shows that the effects of the circulating currents depend on the power factor of the load, and are greater for lagging power factors than for power factors near unity.

2. PARALLEL OPERATION OF THREE-PHASE GROUPS

Three-phase groups of transformers can be operated in parallel on both their primary and secondary sides provided they have the same ratio of primary to secondary line-to-line voltages, and provided the corresponding secondary line voltages of the groups are in phase when the groups are connected in parallel on their primary sides alone. Thus Δ–Δ groups can be operated in parallel with other Δ–Δ groups or with Y–Y groups.

▶ However, a Δ–Δ group or a Y–Y group should *not* be connected in parallel with a Δ–Y or Y–Δ group, since these give a minimum phase displacement of 30 degrees between primary and secondary line voltages.*◀

2a. Symmetrical Arrangements in Balanced Circuits. — In the following discussion, it is assumed that the transformers of any three-phase group are alike, but the transformers of one group may be different from those of another paralleled group. It is also assumed that the three-phase currents and voltages in the circuits connected to the transformers are balanced.

Consider two groups of transformers connected in parallel on both their primary and secondary sides. Assume that the connections of the two groups are the same. If the primary windings of both groups are connected in Δ, each transformer is connected between a pair of line terminals, and therefore the primary windings of the two transformers connected to corresponding phases are in parallel. Similarly, if the secondaries are connected in Δ, the secondaries of the two transformers connected to corresponding phases are in parallel. If the primary windings are connected in Y, the voltages to neutral for corresponding phases of the two groups are vectorially equal, since balanced conditions are assumed, and the neutral points of the two Y's are therefore at the same potential. Consequently the neutral points of the two Y's can be considered as if they were interconnected, even though they may not be actually connected to each other. When the neutral points are connected, the primary windings of the two transformers connected to corresponding phases are in parallel. Similarly, if the secondary windings are connected in Y, the secondary neutral points can be considered as if they were connected when the

* See Art. 4, Ch. XXI.

circuit is balanced, and the secondary windings of the two transformers connected to corresponding phases hence are in parallel.

Therefore, when similarly connected symmetrical three-phase groups of transformers are operated in parallel on both their primary and secondary sides, each transformer is in parallel with the corresponding transformers of the other groups. The total current of any phase consequently divides among the transformers of that phase in the same manner as it would in a single-phase circuit.

▶ Thus for balanced conditions, the behavior of paralleled symmetrical three-phase groups can be determined by analysis of any one phase as a single-phase problem. For best operating results, the transformers should all have equal ratios of transformation, and their per unit equivalent impedances should be equal vectorially.◀

2b. Unsymmetrical Δ–Δ Groups. — In the event of its being necessary to remove one or more transformers from a number of paralleled Δ–Δ groups, the division of current among the remaining transformers can be determined as in the following illustration. Assume that one transformer is removed from three paralleled groups of equal transformers, each of which has an equivalent impedance Z referred to the secondary side. The resulting equivalent impedance on one side of the Δ is $Z/2$, and on each of the other two sides is $Z/3$. The load currents divide among the three Δ branches as they would among three Δ–Δ-connected transformers having equivalent impedances $Z/2$, $Z/3$, $Z/3$.

In Eq. 37, Ch. XXIV, p. 592, it is shown that for an unbalanced Δ–Δ group, the vector relation between the current I_{ba} in one branch of the Δ and the currents I_b and I_a in the lines connected to this branch is

$$I_{ba} = \frac{I_a Z_{ca} - I_b Z_{bc}}{Z_{ab} + Z_{bc} + Z_{ca}}, \qquad [27]$$

where Z_{ab}, Z_{bc}, Z_{ca} are the equivalent impedances. If ab is the branch containing two transformers

$$Z_{ab} = \frac{Z}{2} \qquad [28]$$

$$Z_{bc} = Z_{ca} = \frac{Z}{3}; \qquad [29]$$

whence

$$I_{ba} = \frac{(I_a - I_b)\dfrac{Z}{3}}{\dfrac{Z}{2} + \dfrac{Z}{3} + \dfrac{Z}{3}}$$

$$= \frac{2}{7}(I_a - I_b). \qquad [30]$$

If the three-phase line currents I_a and I_b are balanced and of phase order *abc*, it can be seen from a vector diagram of the line currents that

$$I_a - I_b = \sqrt{3}\, I_a \underline{/30°}. \tag{31}$$

Therefore, from Eq. 30,

$$I_{ba} = \frac{2\sqrt{3}}{7} I_a \underline{/30°}. \tag{32}$$

Thus the relation between the effective value I_{line} of the line currents and the effective value of the current in phase *ba* of the Δ is

$$I_{line} = \frac{7}{2\sqrt{3}} I_{ba}. \tag{33}$$

If the rated full-load secondary current of each transformer is I_2, the full-load value of I_{ba} is $2I_2$, since this phase comprises two transformers in parallel. Consequently the greatest line current that can be delivered without overloading the two transformers of phase *ba* is

$$I_{line} = \frac{7}{2\sqrt{3}} \times 2I_2 = \frac{7}{\sqrt{3}} I_2. \tag{34}$$

The volt-amperes of this greatest balanced load are

$$\sqrt{3}\, V_{line} I_{line} = 7 V_2 I_2, \tag{35}$$

where V_{line} is the line voltage and equals the secondary voltage V_2 of the transformers. Therefore, for a balanced load, the maximum safe output of the eight transformers is seven times the rating of one transformer, or the apparatus economy of this particular unsymmetrical arrangement[1] is $\frac{7}{8}$.

PROBLEMS

1. The following data are given for two single-phase 11,000 : 2,300-v 60 $\sim$ transformers:

Transformer	*Rating*	*Short-circuit data*		
		Voltage	*Current*	*Power*
A	100 kva	265 v	9.1 amp	1,000 w
B	500	345	45.5	3,370

[1] For further discussion of unsymmetrical Δ-Δ arrangements, see L. F. Blume, editor, *Transformer Engineering* (New York: John Wiley & Sons, 1938), 172–177.

If these transformers are operated in parallel to supply an inductive load at 0.80 power factor from an 11,000-v 60 ~ circuit:

(a) What is the maximum kilowatt load they can supply without causing their secondary voltages to drop below 2,250 v?

(b) At this total load, what kva output will each transformer deliver?

2. Two 500-kva 11,000 : 2,200-v 25 ~ transformers yield the following data when operated at rated current with their secondaries short-circuited:

Transformer	Voltage	Power
A	265 v	3,300 w
B	945	3,980

If these two transformers are operated in parallel and the primary voltage is adjusted so that the secondary voltage is maintained at 2,200 v,

(a) What is the greatest load at unity power factor that can be applied to the two transformers in parallel without overloading either of them?

(b) Under the condition of (a) what is the primary voltage?

3. Ratings and short-circuit data for a group of single-phase transformers are:

Transformer	Rating	Voltage	Short-circuit data		
			Voltage	Current	Power
A	100 kva	2,300 : 230 v	119 v	45.0 amp	1,000 w
B	100	2,300 : 230	154	40.0	1,300
C	200	2,300 : 230	106	80.0	1,580
D	300	2,300 : 230	132	125.0	3,100

(a) Which two of these transformers would operate most successfully in parallel?

(b) If the load to be supplied at 230 v were 400 kva at unity power factor and all the above transformers were available to be used in a parallel bank to supply the load, which transformers would you recommend using? Why? What additional data would be needed to determine the choice completely?

4. Two identical banks of transformers, each connected in V, are paralleled on both their low- and high-voltage sides, but with the open sides of the V's between different phases of a balanced three-phase supply. What is the greatest balanced three-phase output that can be taken from the two banks of transformers as compared with the output either can give alone?

5. Show that it is possible to parallel a Δ–Y-connected bank of transformers with a Y–Δ-connected bank of transformers and achieve satisfactory operation. State how the transformer ratios and constants must be related.

6. Two banks of Δ–Δ-connected single-phase transformers are connected in parallel to supply a balanced load consisting of three noninductive resistors connected in Y to the low-voltage terminals of the system. Each resistor has a resistance of 0.088 ohm. The rating of each of the six single-phase transformers is 100 kva, 11,500 : 230 v, and 60 ~.

The transformers in each bank may be considered identical, with short-circuit impedances given by the following data:

Bank	Voltage	Current	Power
A	300 v	8.7 amp	1.2 kw
B	450	8.7	1.1

If balanced three-phase voltages of rated value are applied to the transformer primaries:

(a) What are the magnitudes of the currents taken by the load?
(b) What are the magnitudes of the currents in the primaries of each bank of transformers?

7. Two balanced three-phase loads are supplied through a Y–Δ–Δ-connected bank of three-winding single-phase transformers from a high-voltage transmission line. The following data apply to each of the three-winding transformers:

RATINGS

Winding	Capacity	Voltage
1	10,000 kva	63,500 v
2	5,000	6,600
3	7,500	13,200

SHORT–CIRCUIT DATA

Windings	Impedance on 10,000-kva rated-voltage base
1–2	11%
2–3	10
3–1	13

The peak load supplied by the 6,600-v secondaries is 4,000 kva per phase at 0.80 power factor, lagging, and the peak load supplied by the 13,200-v tertiaries is 7,000 kva at unity power factor.

It is expected that the load demand on the 6,600-v circuit will increase to 6,000 kva per phase at 0.80 power factor, lagging, and it is decided to install a Y–Δ-connected bank of two-winding single-phase transformers in parallel with the primary and secondary windings of the three-winding transformer.

What should be the ratings of the transformers of the auxiliary bank, and what should be their per cent equivalent impedances?

Preferred Ratings of Power and Distribution Transformers

Tables I, II, and III give schedules of preferred ratings of power transformers. They represent standard American practice as compiled by the American Institute of Electrical Engineers, National Electrical Manufacturers Association, Electric Light and Power Group, and the Association of American Railroads, under the sponsorship of the American Standards Association.[1]

The term *rated circuit voltage* appears in Tables II and III. For the purpose of fixing a value to be used in designing and testing electrical apparatus, the rated circuit voltage of a circuit is defined as the highest rated voltage of the apparatus supplying it. It is expected that the operating voltage of the apparatus connected to a circuit of a given class will not normally exceed the rated circuit voltage recommended for that class.

In Tables II and III, Y ratings appear in the voltage columns, as, for example, 2,400/4,160 Y. This shows that the winding so rated is suitable for use in one phase of a Y-connected transformer or bank in a circuit having the line-to-line voltage indicated by the Y rating; in such a transformer, the winding is insulated for the Y voltage. Thus, a 2,400/4,160-Y rating implies a normal winding voltage of 2,400 volts but with the winding and bushings insulated for 4,160 volts.

TABLE I

PREFERRED KVA RATINGS OF SINGLE- AND THREE-PHASE TRANSFORMERS

Single-Phase				Three-Phase			
$1\frac{1}{2}$	75	833	6,667	10	200	2,000	12,000
3	100	1,000	8,333	15	300	2,500	15,000
5	150	1,250	10,000	25	450	3,000	20,000
$7\frac{1}{2}$	200	1,667	12,500	$37\frac{1}{2}$	600	3,750	25,000
10	250	2,000	16,667	50	750	5,000	30,000
15	333	2,500	20,000	75	1,000	6,000	37,500
25	500	3,333	25,000	100	1,200	7,500	50,000
$37\frac{1}{2}$	667	4,000	33,333	150	1,500	10,000	.....
50	...	5,000	.....	...	....	.....	.....

[1] "American Engineering and Industrial Standards," No. C57: *Proposed American Standards; Transformers, Regulators, and Reactors* (New York: American Standards Association, 1940).

TABLE II
PREFERRED RATINGS OF TRANSFORMERS
SINGLE PHASE — 500 KVA AND BELOW

Used primarily to step down from a transmission or distribution voltage to a utilization voltage.

Rated Circuit Voltage, Volts	Maximum and Minimum Rated Kva	High Voltage				Low Voltage			
		Rated Voltage	Rated-Kva Taps		Reduced-Kva Tap	Rated Voltage			
480	1.5-100	480	456	432	.....	120/240		600	
600	1.5-100	600	570	540	.....	120/240		600	
2,400 and 4,160	1.5- 50	2,400/4,160 Y	.....	.....	.....	120/240	240/480	600	
	75-500	2,400/4,160 Y	2,280	2,160	.....	120/240	240/480	600	
4,800	1.5-100	2,400/4,800	.....	.....	.....	120/240	240/480	600	
	150-500	2,400/4,800	2,280	2,160	.....	120/240	240/480	600	
7,200 and 12,000	1.5-500	7,200/12,470 Y	6,875	6,545	6,220	120/240	240/480	600	
	5-500	7,200/12,470 Y	6,840	6,480	.....	.....	240/480	600	2,400
	3-500	12,000	11,400	10,800	.....	120/240	240/480	600	
	5-500	12,000	11,400	10,800	.....	.....	240/480	600	2,400/4,160 Y
13,800	1.5-500	7,620/13,200 Y	7,240	6,860	.....	120/240	240/480	600	
	5-500	7,620/13,200 Y	7,240	6,860	.....	.....	240/480	600	2,400
	3-500	13,200	12,540	11,880	.....	120/240	240/480	600	
	5-500	13,200	12,540	11,880	.....	.....	240/480	600	2,400/4,160 Y
23,000	10-500	22,000	20,900	19,800	.....	120/240	240/480	600	2,400/4,160 Y
	10-500	12,700/22,000 Y	12,070	11,430	.....	120/240	240/480	600	2,400
34,500	15-500	33,000	31,350	29,700	.....	120/240	240/480	600	2,400/4,160 Y
	15-500	19,050/33,000 Y	18,100	17,150	.....	120/240	240/480	600	2,400
46,000	25-500	44,000	41,800	39,600	.....	120/240	240/480	600	2,400/4,160 Y
	25-500	25,500/44,000 Y	24,140	22,860	.....	120/240	240/480	600	2,400
69,000	50-500	66,000	62,700	59,400	.....	120/240	240/480	600	2,400/4,160 Y
	50-500	38,100/66,000 Y	36,200	34,290	.....	120/240	240/480	600	2,400

TABLE III

PREFERRED RATINGS OF TRANSFORMERS

THREE PHASE — 450 KVA AND BELOW

Used primarily to step down from a transmission or distribution voltage to a utilization voltage.

Rated Circuit Voltage, Volts	Maximum and Minimum Rated Kva	High Voltage			Low Voltage		
		Rated Voltage	Rated-Kva Taps		Rated Voltage		
2,400 and 4,160	10–150	2,400/4,160 Y	......	......	240/480	600	
	200–450	2,400/4,160 Y	2,280	2,160	240/480	600	
4,800	10–450	4,800 Y	4,560	4,320	240/480	600	
7,200	10–450	7,200 Y	6,840	6,480	240/480	600	
	10–450	7,200 Y	6,840	6,480	......	:::	2,400 / 2,400/4,160 Y
	10–450	7,200 Δ	6,840	6,480	......	:::	
12,000	10–450	12,000	11,400	10,800	240/480	600	
	10–450	12,000	11,400	10,800	......	:::	2,400 / 2,400/4,160 Y
	10–450	12,000	11,400	10,800	......	:::	
13,800	15–450	13,200 Y	12,540	11,880	240/480	600	
	15–450	13,200 Δ	12,540	11,880	......	:::	2,400 / 2,400/4,160 Y
23,000	25–450	22,000 Y	20,900	19,800	240/480	600	
	25–450	22,000 Δ	20,900	19,800	......	:::	2,400 / 2,400/4,160 Y
34,500	50–450	33,000 Y	31,350	29,700	240/480	600	
	50–450	33,000 Δ	31,350	29,700	......	:::	2,400 / 2,400/4,160 Y
46,000	50–450	44,000 Y	41,800	39,600	240/480	600	
	50–450	44,000 Δ	41,800	39,600	......	:::	2,400 / 2,400/4,160 Y
69,000	150–450	66,000 Y	62,700	59,400	240/480	600	
	150–450	66,000 Δ	62,700	59,400	......	:::	2,400 / 2,400/4,160 Y

Standard Terminal Markings for Power and Distribution Transformers

When the windings of transformers are connected in parallel or in polyphase groups, the connections must be made with the correct relative polarities. In order to simplify making these connections, the American Standards Association has adopted certain standard terminal markings[1] with which any one who uses power and distribution transformers should be familiar.

Standard markings for two-winding power and distribution transformers are shown in Fig. 1. The high-voltage terminals are labeled H_1, H_2, and the low-voltage terminals X_1, X_2, where H_1 and X_1 are terminals for which the polarities of the voltages induced by the resultant core flux are the same; that is, if leakage-impedance voltage drops are neglected, the terminal H_1 is at a positive potential with respect to H_2 during the time when the terminal X_1 is positive with respect to X_2. For example, tracing the windings of Fig. 1 from their left-hand terminals to their right-hand terminals shows the core to be encircled in the same direction by both windings; hence the voltages induced by the resultant core flux

FIG. 1. Standard terminal markings.

are in phase in the directions shown by the plus and minus signs in Fig. 1. The left-hand terminals are therefore of the same relative polarity, as indicated by the dots, and should be labeled H_1 and X_1, as in Fig. 1.

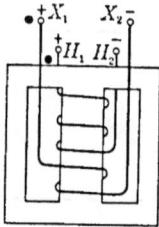

The leads usually are brought out on opposite sides of the case, either at the top of the tank or through the cover, as indicated in Fig. 2. In Fig. 2a, high-voltage and low-voltage terminals of the same relative polarity are adjacent, so that if adjacent primary and secondary terminals — say H_2 and X_2 — are connected, with either winding excited from a suitable alternating voltage, the voltage between terminals H_1 and X_1 is very nearly the difference between the effective values of the voltages V_H and V_X on the high- and low-voltage sides. The arrangement of external terminals in Fig. 2a is therefore called *subtractive external polarity*. In Fig. 2b, when adjacent high-voltage and low-voltage terminals H_2 and X_1 are connected, the voltage between terminals H_1 and X_2 is very nearly the sum of the effective values of the voltages V_H and V_X,

[1] " American Engineering and Industrial Standards," No. C6: *American Standard Rotation, Connections, and Terminal Markings for Electric Power Apparatus* (New York: National Electric Manufacturers Association, 1936).

and the external arrangement of terminals in Fig. 2b is therefore called *additive external polarity*. If the external polarity is unknown, it can be determined through testing the transformer as indicated in Fig. 2. The voltage should be applied to the high-voltage winding and should be relatively low; that is, about 110 volts, to safeguard the tester and to permit the use of ordinary voltmeters.

Whether the external polarity is additive or subtractive is determined solely by the manner in which the leads from the windings are connected to the external terminals, and is entirely independent of the internal

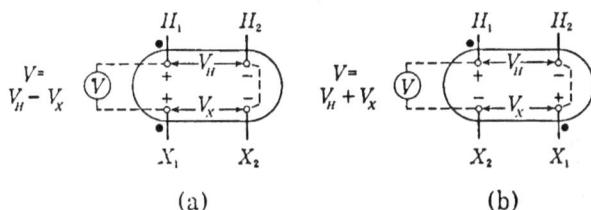

FIG. 2.   Plan showing location of terminals on cover, for (a) subtractive polarity, and (b) additive polarity.

arrangement of the windings. The external terminal markings contain all the information that a lineman needs in order correctly to connect a transformer in its circuit. Designers of transformers, however, are also interested in the *internal polarity* of the windings;[2] that is, whether the voltage between adjacent portions of the high- and low-voltage windings is the sum or difference of the voltages in the windings. The internal polarity is determined by the arrangement and method of winding the coils and can be changed only through rewinding the transformer. The external polarity can be changed through reversing the connections inside the tank between either the primary or the secondary leads and their external terminals.

The distribution transformers used to supply domestic loads usually have two 120-volt secondary windings which may be connected either in series or in parallel. The standard terminal markings for these split secondaries are shown in Fig. 3. The terminals of one secondary winding are X_1, X_2, and those of the other are X_3, X_4, where X_1 and X_3 are terminals of corresponding polarity. Sometimes all four secondary terminals are brought out through the case, and the connections between the secondary windings are made outside the transformer. Many modern designs, however, locate inside the tank a terminal board at which the appropriate connections between the windings are made, so that only

[2] J. B. Gibbs, *Transformer Principles and Practice* (New York: McGraw-Hill Book Company, Inc., 1937), Ch. vii, 48–54.

three secondary bushings are needed to supply a three-wire service. Note that the leads to the secondary terminals X_2 and X_3 are crossed inside the tank. With this arrangement of secondary terminals, the two secondary windings can be connected either in series or in parallel by

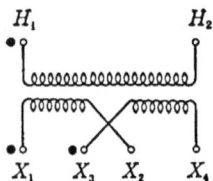

FIG. 3. Standard terminal markings for split secondaries.

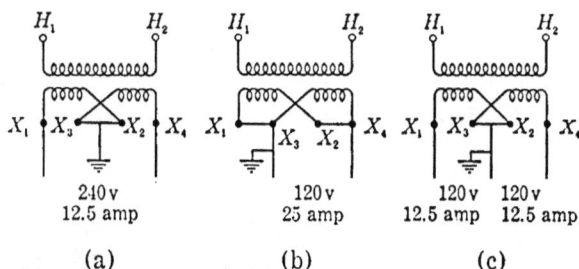

240 v
12.5 amp

(a)

120 v
25 amp

(b)

120 v 120 v
12.5 amp 12.5 amp

(c)

FIG. 4. Connections of 120-volt secondaries; (a) series, (b) parallel, (c) three-wire. The current ratings are for a 3-kva transformer.

connection of the appropriate pairs of adjacent secondary terminals, as shown in Fig. 4. With an internal terminal board, these connections can be made conveniently with suitable connecting links.

For information regarding the standard terminal markings of transformers having more complicated winding arrangements, the reference given in footnote 1 should be consulted. Such transformers usually have an explanatory connection diagram given with their nameplate.

Bibliography

This list of references supplements the series of footnote references inserted in the text. Although this bibliography is in no way to be taken as an exhaustive and final collection, it is sufficiently comprehensive and varied to allow the acquisition of an extensive knowledge of the subject. Many of the works cited contain further references, so that an exhaustive bibliography can readily be compiled with the listed material as a basis.

The following abbreviations for periodicals and circulars are used in footnotes and in this bibliography:

A.I.E.E. Trans.	American Institute of Electrical Engineers *Transactions*
B.S.T.J.	Bell System Technical *Journal*
Brit. I.R.E.J.	British Institution of Radio Engineers *Journal*
Bul. Nat. Bur. Stand.	National Bureau of Standards *Bulletin*
Circ. Nat. Bur. Stand.	National Bureau of Standards *Circular*
E.E.	*Electrical Engineering*
Elec. J.	*Electric Journal*
Elec. W.	*Electrical World*
E. u. M.	*Elektrotechnik und Maschinenbau*
E.T.Z.	*Elektrotechnische Zeitschrift*
G.E. Rev.	*General Electric Review*
Gen. Rad. Exp.	*General Radio Experimenter*
Instrs.	*Instruments*
I.E.E.J.	Institution of Electrical Engineers *Journal*
I.R.E. Proc.	Institute of Radio Engineers *Proceedings*
J.O.S.A. and R.S.I.	*Journal* of Optical Society of America and *Review of Scientific Instruments*
J. Res. Nat. Bur. Stand.	National Bureau of Standards *Journal of Research*
M.E.	*Mechanical Engineering*
Philips Tech. Rev.	*Philips Technical Review*
Phil. Trans.	*Philosophical Transactions*
Phys. Rev.	*Physical Review*
R.G.E.	*Revue Générale de l' Électricité*
R.S.I.	*Review of Scientific Instruments*
Sci. Paper Nat. Bur. Stand.	National Bureau of Standards *Scientific Paper*

1. THEORY OF MAGNETISM (CHAPTER I)

R. M. Bozorth, "Present Status of Ferromagnetic Theory," *A.I.E.E. Trans.*, *54* (1935), 1251–1261. A discussion of the structure of atoms, molecules, and crystals of the transition metals; experimental evidence supporting the domain theory.

R. M. Bozorth, "The Physical Basis of Ferromagnetism," *B.S.T.J.*, *19* (1940), 1–39. A comparison of the Ewing and the Weiss theories of magnetism and a description of some basic experiments in ferromagnetism.

L. B. Loeb, *Atomic Structure* (New York: John Wiley & Sons, 1938). An excellent introductory treatise showing the historical development of the Bohr atom and the later wave-mechanical or mathematical concepts of the atom.

K. K. Darrow, " Spinning Atoms and Spinning Electrons," *B.S.T.J.*, *16* (1937),

319–336. A word picture showing how angular momentum and magnetic moment determine magnetic properties of atoms and molecules.

F. Bitter, *Introduction to Ferromagnetism* (New York: McGraw-Hill Book Company, Inc., 1927). This book presents a discussion of the theories of magnetism and the methods of preparing magnetic materials which are or may become commercially useful.

W. Shockley, " The Quantum Physics of Solids — I," *B.S.T.J.*, *18* (1939), 645–723. Beginning with a lucid and pictorial description of the energy levels of electrons in atoms, the author leads the reader through a discussion of diatomic molecules and finally into the concept of energy bands in crystals.

2. PROPERTIES OF FERROMAGNETIC MATERIALS (CHAPTER I)

Allegheny Steel Company, Brackenridge, Pa., *Magnetic Core Materials Practice* (1937). Representative normal magnetization and core-loss curves for the various grades of sheet steel, with a discussion of applications and standard testing methods.

Carnegie-Illinois Steel Corporation, Pittsburgh, Pa., *Electrical Steel Sheets, Technical Bulletin No. 2* (1941). A compilation of normal magnetization curves, hysteresis loops, and loss curves for many grades of sheet steels. Applications and standard methods of testing are included.

V. E. Legg, " Survey of Magnetic Materials and Applications in the Telephone System," *B.S.T.J.*, *18* (1939), 438–464. Magnetic materials are evaluated with respect to apparatus applications, especially in communication systems.

G. W. Elmen, " Magnetic Alloys of Iron, Nickel, and Cobalt," *A.I.E.E. Trans*, *54* (1935) 1292–1299.

V. E. Legg and F. J. Given, " Compressed Powdered Molybdenum Permalloy for High-Quality Inductance Coils," *B.S.T.J.*, *19* (1940), 385–406.

W. F. Randall, " Nickel-Iron Alloys of High Permeability with Special Reference to Mumetal," *I.E.E.J.*, *80* (1937), 647–667.

L. R. Jackson and H. W. Russell, " Temperature-Sensitive Magnetic Alloys and Their Uses," *Instr.*, *11* (1938), 280–282.

General Electric Company, *The Story of the G. E. Spirakore Transformer*, GES–2038 (October, 1938); *Chapter Two in the Story of the G. E. Spirakore Transformer*, GES–2447 (January, 1941). A description of an ingenious method of assembling the core to effect a saving of space and to utilize the preferred grain orientation.

Westinghouse Electric & Manufacturing Company, *Distribution Transformers with Hipersil Cores*, R–992 (April, 1941); *Hipersil Transformers*, B–2287 (April, 1941). These special pamphlets show the application of materials having particularly desirable magnetic properties in the direction of rolling.

J. H. Goss, " Permanent Magnets," *Metals and Alloys*, *15* (1942), 576–582. The new designs and manufacturing techniques for the production of Alnico magnets are discussed.

F. G. Bailey, " The Hysteresis of Iron and Steel in a Rotating Magnetic Field," *Phil. Trans.*, *187* (1896), 715–746.

F. Brailsford, " Rotational Hysteresis Loss in Electrical Sheet Steels," *I.E.E.J.*, *83* (1938), 566–575. This article describes a method of measurement, gives data, and contains a good bibliography.

J. Q. Adams, " Alnico — Its Properties and Possibilities," *G.E. Rev.*, *41* (1938), 518–523.

B. Jonas and H. J. Meerkamp van Embden, " New Kinds of Steel of High Magnetic Power," *Philips Tech. Rev.*, *6* (1941), 8–11. A summary of the progress in permanent-magnet alloys since cobalt steel. A curve of $(\mathfrak{BH})_{max}$ from 1890 to 1940 is included.

W. E. Ruder, "New Magnetic Materials," *I.R.E. Proc.*, *30* (1942), 437–440. A brief summary of both soft and hard magnetic materials, including Alnico V.

3. DESIGN OF MAGNETIC DEVICES (CHAPTERS III, IV, V)

Herbert C. Roters, *Electromagnetic Devices* (New York: John Wiley & Sons, 1942). An outstanding book on the design of magnetic devices.

S. Evershed, " Permanent Magnets in Theory and Practice," *I.E.E.J.*, *58* (1920), 780–837, and *63* (1925), 725–821. The classic treatment of permanent-magnet circuit design.

A. J. Corson, " Magnetic Materials," *G.E. Rev.*, *45* (1942), 573–575. This article describes the development of new d-c and a-c indicating instruments using the new magnetic alloys.

A. S. Langsdorf, *Principles of Direct-Current Machines* (5th ed., New York: McGraw-Hill Book Company, Inc., 1940), 401–419. A chapter on the computation of the magnetization curve of a dynamo from design data.

R. G. Kloeffler, J. L. Brenneman, and R. M. Kerchner, *Direct-Current Machinery* (New York: The Macmillan Company, 1934), 54–65. A discussion of the magnetization curve of a dynamo.

F. Massa, *Acoustic Design Charts* (Philadelphia: The Blakiston Company, 1942). Charts with numerical examples showing method of use in design of acoustic devices.

4. MAGNETIC TESTING (CHAPTERS I, V)

Thomas Spooner, *Properties and Testing of Magnetic Materials* (New York: McGraw-Hill Book Company, Inc., 1927). Basic considerations for a large variety of magnetic tests with circuit diagrams and instructions for performing the tests.

R. L. Sanford, " Magnetic Testing," *Circ. Nat. Bur. Stand.*, *C415* (1937). The testing of ring and bar samples of magnetic materials is described. Circuit diagrams are given.

" Standard Methods of Test for Magnetic Properties of Iron and Steel," *Standards of the American Society for Testing Materials*, *A-34* (1940). This pamphlet contains the standard definitions and tests for magnetic materials.

B. J. Babbitt, " An Improved Permeameter for Testing Magnet Steel," *J.O.S.A. and R.S.I.*, *17* (1928), 47–58.

R. L. Sanford and E. G. Bennett, " An Apparatus for Magnetic Testing at High Magnetizing Forces," *J. Res. Nat. Bur. Stand.*, *10* (1933), 567–573.

R. L. Sanford and E. G. Bennett, " An Apparatus for Magnetic Testing at Magnetizing Forces up to 5,000 Oersteds," *J. Res. Nat. Bur. Stand.*, *23* (1939), 415–425.

D. E. Foster and A. E. Newlon, " Measurement of Iron Cores at Radio Frequencies," *I.R.E. Proc.*, *29* (1941), 266–276. A method of measuring the loss factor and effective permeability of iron at radio frequencies is presented and sample calculations are given.

A. E. Kettner, " Magnetic Measurements," *G.E. Rev.*, *45* (1942), 633–636. A résumé of modern methods of commercial testing with pictures of apparatus.

B. M. Smith, " Alnico, Properties and Equipment for Magnetization and Test," *G.E. Rev.*, *45* (1942), 210–213. Electromagnets for magnetizing Alnico, a saturation permeameter, and a photoelectric hysteresigraph are described and pictured. Curves are given for Alnico V.

B. M. Smith and C. Concordia, " Measuring Core Loss at High Densities," *E.E.*, *51* (January, 1932), 36–38.

G. Camilli, " A Flux Voltmeter for Magnetic Tests," *A.I.E.E. Trans.*, *45* (1926) 721–728.

5. Iron–Core Reactors; Transformer Excitation Characteristics; Model Theory (Chapters VI, VII)

The references listed in this section discuss the general properties of iron-core reactors and transformers, nonsinusoidal waves, design problems, dimensional analysis, and model theory. Several references to articles on the analysis and applications of circuits containing nonlinear iron-core reactors are given in the bibliography of the volume on electric circuits (1940), 765.

N. Partridge, " An Introduction to the Study of Harmonic Distortion in Audio Frequency Transformers," *Brit. I.R.E.J., 3* (June–July, 1942), 6–21.

E. Peterson, " Harmonic Production in Ferromagnetic Materials at Low Frequencies and Low Flux Densities," *B.S.T.J., 7* (1928), 762–796. This paper deals primarily with effects in multichannel communication circuits.

Philip Franklin, *Differential Equations for Electrical Engineers* (New York: John Wiley & Sons, 1933). Fourier series are discussed in Chs. ii and viii, and a short bibliography is given at the end of the book.

R. E. Doherty and E. G. Keller, *Mathematics of Modern Engineering*, Vol. I (New York: John Wiley & Sons, 1936). Part III, Ch. ii, gives a treatment of Fourier series and examples of engineering applications. A discussion of harmonic analysis of a function given in graphical form is included.

R. R. Lawrence, *Principles of Alternating Currents* (2d ed.; New York: McGraw-Hill Book Company, Inc., 1935). Nonsinusoidal currents and voltages, their representation by Fourier series, effective values, power, and harmonic analysis are discussed in Ch. iv.

J. M. Bryant, J. A. Correll, and E. W. Johnson, *Alternating-Current Circuits* (3d ed.; New York: McGraw-Hill Book Company, Inc., 1939). Nonsinusoidal waveforms are discussed in Ch. xvi.

J. W. Butler and E. B. Pope, " The Effect of Overexciting Transformers on System Voltage Wave Shapes and Power Factor," *A.I.E.E. Trans., 60* (1941), 49–53.

L. B. Arguimbau, " Losses in Audio-Frequency Coils," *Gen. Rad. Exp., 11, No. 6* (November, 1936), 1–4. This paper gives a simple equivalent circuit for an iron-core coil, and discusses the effects of air gaps on the quality factor.

P. K. McElroy and R. F. Field, " How Good Is an Iron-Cored Coil?," *Gen. Rad. Exp., 16, No. 10* (March, 1942), 1–12. A continuation and extension of the subject discussed in the Arguimbau paper (see preceding reference).

H. A. Wheeler, " Formulas for the Skin Effect," *I.R.E. Proc., 30* (September, 1942), 412–424. A part of this paper is concerned with equivalent circuits for iron-core coils.

L. W. Chubb and T. Spooner, " The Effect of Displaced Magnetic Pulsations on the Hysteresis Loss of Sheet Steel," *A.I.E.E. Trans., 34, Part 2* (1915), 2671–2692.

J. D. Ball, " The Unsymmetrical Hysteresis Loop," *A.I.E.E. Trans., 34, Part 2* (1915), 2693–2715.

T. Spooner, " Effect of a Superposed Alternating Field on Apparent Magnetic Permeability and Hysteresis Loss," *Phys. Rev., 25* (1925), 527–540.

R. F. Edgar, " Loss Characteristics of Silicon Steel at 60 Cycles with D-C Excitation," *A.I.E.E. Trans., 52* (September, 1933), 721–724.

J. Minton and I. G. Maloff, " Design Methods for Soft Magnetic Materials in Radio." *I.R.E. Proc., 17* (June, 1929), 1021–1033. Discusses the design of audio-frequency transformers whose windings carry direct current.

D. C. Prince, " Direct-Current Reactor Design," *G.E. Rev., 27* (June, 1924), 380–383. Discusses the design of smoothing inductances for use in the output circuits of electronic rectifiers.

A. Boyajian, " Theory of D-C Excited Iron-Core Reactors," *A.I.E.E. Trans., 43*

(1924), 919–936. A comprehensive treatment of saturable-core reactors and a discussion of some applications as control devices. A good discussion of waveforms is included.

C. R. Hanna, "Design of Reactances and Transformers Which Carry Direct Current," *A.I.E.E. Trans.*, *46* (1927), 155–158. A direct method of determining the optimum air-gap length for reactors or transformers in which the alternating component of flux density is small.

C. V. Aggers and W. E. Pakala, "Direct-Current Controlled Reactors," *Elec. J.*, *34* (February, 1937), 55–59. Discusses the design of three-legged, saturable-core reactors.

P. W. Bridgman, *Dimensional Analysis* (revised ed., New Haven: Yale University Press, 1931). A general discussion of the principles of similarity between physical systems, with illustrations of the applications of dimensional analysis to model experiments.

E. Buckingham, "On Physically Similar Systems; Illustrations of the Use of Dimensional Equations," *Phys. Rev.*, *4* (1914), 345–376. A derivation of certain important general theorems regarding dimensional analysis, with illustrative applications.

Sir Richard Glazebrook, editor, *Dictionary of Applied Physics*, Vol. I: *Mechanics, Engineering, Heat* (London: Macmillan and Company, Ltd., 1922), 81–96. A discussion of the principles of dynamical similarity. Model theory is discussed in Part III of this article.

H. B. Brooks, "Design of Standards of Inductance, and the Proposed Use of Model Reactors in the Design of Air-Core and Iron-Core Reactors," *J. Res. Nat. Bur. Stand.*, *7* (August, 1931), 289–328. A comprehensive paper, with illustrations of model theory as applied to the design of inductance standards, current-limiting reactors, and iron-core reactors with air gaps.

F. Emde, "Die Berechnung von Eisendrosseln mit grosser Zeitkonstante," *E. u. M.*, *48* (1930), 521–530. A comprehensive treatment of reactor design.

H. A. W. Klinkhamer, "Equivalent Networks with Highly Saturated Iron Cores with Special Reference to Their Use in the Design of Stabilisers," *Philips Tech. Rev.*, *2* (September, 1937), 276–281. Discusses general principles of model theory applied to networks containing nonlinear elements, and gives an illustration of the design of such a network.

6. Thermal Characteristics of Electric Apparatus (Chapter VIII)

References pertaining to thermal characteristics of transformers are listed in Section 11 of this bibliography.

F. M. Clark, "Factors Affecting the Mechanical Deterioration of Cellulose Insulation," *A.I.E.E. Trans.*, *61* (October, 1942), 742–749.

J. J. Smith and J. A. Scott, "Temperature Aging of Class A Insulation," *A.I.E.E. Trans.*, *58* (September, 1939), 435–444.

C. F. Hill, "Temperature Limits Set by Oil and Cellulose Insulation," *A.I.E.E. Trans.*, *58* (September, 1939), 484–491.

J. A. Scott and B. H. Thompson, "Temperature-Aging Tests on Class-A-Insulated Fractional-Horsepower Motor Stators," *A.I.E.E. Trans.*, *61* (July, 1942), 499–501.

A. D. Moore, "Dissipation of Heat by Radiation," *A.I.E.E. Trans.*, *49* (January, 1930), 359–365.

H. C. Hottel, "Radiant Heat Transmission," *M.E.*, *52* (July, 1930), 699–704.

E. Griffiths and A. H. Davis, "The Transmission of Heat by Radiation and Convection," *Special Report No. 9* (London: H. M. Stationery Office, 1931).

A. D. Moore, *Fundamentals of Electrical Design* (New York: McGraw-Hill Book Company, Inc., 1927). Thermal problems are discussed in Chs. xiv, xv, and xvi.

" American Institute of Electrical Engineers Standards," No. 1: *General Principles upon Which Temperature Limits are Based in the Rating of Electrical Machinery and Apparatus* (New York: American Institute of Electrical Engineers, 1940).

" American Engineering and Industrial Standards," No. C50: *American Standards for Rotating Electrical Machinery* (New York: American Standards Association, 1936).

" American Institute of Electrical Engineers Standards," No. 11: *American Tentative Standards for Railway Motors and Other Rotating Electrical Machinery* (New York: American Institute of Electrical Engineers, 1937).

R. E. Hellmund, " Classification and Co-ordination of Short-Time and Intermittent Ratings and Applications," *A.I.E.E. Trans., 60* (July, 1941), 792–798.

P. L. Alger and T. C. Johnson, " Rating of General-Purpose Induction Motors," *A.I.E.E. Trans., 58* (September, 1939), 445–459.

L. F. Hildebrand, " Duty Cycles and Motor Rating," *A.I.E.E. Trans., 58* (Sentemper, 1939), 478–483.

R. C. Freeman and A. U. Welch, " The Service Factor Rating of Arc-Welding Generators and Transformers," *A.I.E.E. Trans., 60* (April, 1941), 137–141.

7. TEXTBOOKS ON TRANSFORMERS

In this section are listed textbooks dealing with transformer theory, and discussing the construction, design, operation, and applications of transformers in power systems.

R. R. Lawrence, *Principles of Alternating-Current Machinery* (3d ed.; New York: McGraw-Hill Book Company, Inc., 1940), Chs. ix to xix.

J. M. Bryant and E. W. Johnson, *Alternating-Current Machinery* (New York: McGraw-Hill Book Company, Inc., 1935), Chs. i to xi. A bibliography is included at the end of each chapter.

A. S. Langsdorf, *Theory of Alternating-Current Machinery* (New York: McGraw-Hill Book Company, Inc., 1937), Ch. iv.

A. F. Puchstein and T. C. Lloyd, *Alternating-Current Machines* (2d ed.; New York: John Wiley & Sons, 1942), Chs. xi to xx.

L. F. Blume, editor, *Transformer Engineering* (New York: John Wiley & Sons, 1938). An excellent book on technical problems of an advanced nature incidental to the application and operation of transformers in power systems.

S. A. Stigant and H. M. Lacey, *The J. and P. Transformer Book* (6th ed.; London: Johnson and Phillips Ltd., 1935). A practical technology of the power transformer.

J. L. la Cour and K. Faye-Hansen, *Die Transformatoren* (3d ed., Berlin: J. Springer, 1936).

R. Richter, *Elektrische Maschinen.* Bd III: *Die Transformatoren* (Berlin: J. Springer, 1932). This book includes a bibliography of 167 items.

E. G. Reed, *Essentials of Transformer Practice* (2d ed.; New York: McGraw-Hill Book Company, Inc., 1927). Discusses the theory, design, and operation of power-system transformers from a practical engineering point of view.

J. B. Gibbs, *Transformer Principles and Practice* (New York: McGraw-Hill Book Company, Inc., 1937). A nonmathematical description of some of the types of power-system transformers and the principles underlying their operation.

E. G. Reed, *Transformer Construction and Operation* (New York: McGraw-Hill Book Company, Inc., 1928). Nonmathematical and descriptive.

J. H. Kuhlmann, *Design of Electrical Apparatus* (2d ed.; New York: John Wiley & Sons, 1940). Chapters xxii to xxv deal with the design of power-system transformers.

8. Physical Features of Transformers (Chapter XI)

K. K. Paluev, " Power Transformers with Concentric Windings," *A.I.E.E. Trans.*, *55* (June, 1936), 649–659. Traces the development of large high-voltage power transformers, and describes the sending-end transformers of the Boulder Dam–Los Angeles line.

W. G. James and F. J. Vogel, " Power Transformers for 287.5 Kv Service," *A.I.E.E. Trans.*, *55* (May, 1936), 438–444. Describes the receiving-end transformers of the Boulder Dam–Los Angeles line.

E. D. Treanor, " The Wound-Core Distribution Transformer," *A.I.E.E. Trans.*, *57* (November, 1938), 622–625. Describes an ingenious design and manufacturing process for producing a distribution transformer with preformed coils encircled by two cores wound from continuous steel strips.

J. O. Fenwick and D. E. Wiegand, " New Transformer Lowers Copper Loss, Improves Regulation," *Elec. W.*, *115* (1941), 972–973. Describes a type of distribution transformer made from a preformed core of continuous steel strip, and a method of winding cylindrical coils on the preformed core.

J. K. Hodnette and C. C. Horstman, " Hipersil, a New Magnetic Steel and Its Use in Transformers," *The Westinghouse Engineer*, *1* (August, 1941), 52–56.

P. Sporn and H. V. Putman, " A New Transformer for Base-Load Stations," *A.I.E.E. Trans.*, *60* (October, 1941), 916–918. Describes improvements in large power transformers resulting from the use of Hipersil core material.

J. F. Peters, " High Power Audio Transformers," *A.I.E.E. Trans.*, *55* (January, 1936), 34–36. Describes a 7.5-kw and a 180-kw audio transformer built for radio broadcast station WLW.

A. G. Ganz and A. G. Laird, " Improvements in Communication Transformers," *A.I.E.E. Trans.*, *54* (December, 1935), 1367–1373.

E. M. Hunter and J. C. Page, " Standardized Load-Center Unit Substations for Low-Voltage A-C Systems," *A.I.E.E. Trans.*, *61* (July, 1942), 519–525.

F. Meyer, " Fluid Filling-Media for Electrical Apparatus," *I.E.E.J.*, *86* (April, 1940), 313–320.

L. H. Burnham and S. T. Maunder, " Pyranol Power Transformers," *G.E. Rev.*, *42* (June, 1939), 236–239.

9. Insulation and Surge Phenomena

A.I.E.E. Committee on Electrical Machinery (Transformer Subcommittee), " Protection of Power Transformers against Lightning Surges," *A.I.E.E. Trans.*, *60* (June, 1941), 568–577. A bibliography of 25 items is included with this article.

Surge Phenomena — Seven Years' Research for the Central Electricity Board (London: The British Electrical and Allied Industries Research Association, 1941). Part III of this comprehensive report deals with surge voltage distribution in transformers.

D. F. Miner, *Insulation of Electrical Apparatus* (New York: McGraw-Hill Book Company, Inc., 1941). Chapter viii deals with transformers and reactors.

R. Rüdenberg, " Surge Characteristics of Two-Winding Transformers," *A.I.E.E. Trans.*, *60* (December, 1941), 1136–1144.

L. F. Blume, editor, *Transformer Engineering* (New York: John Wiley & Sons, 1938). Characteristics of insulating materials, transient voltage characteristics of transformers, and insulation co-ordination are discussed in Chs. xv, xvii, and xviii.

K. K. Palueff, " Effect of Transient Voltages on Power Transformer Design," *A.I.E.E. Trans.*, *48* (July, 1929), 681–701; " The Behavior of Transformers with Neutral Isolated or Grounded through an Impedance," *A.I.E.E. Trans.*, *49* (July,

1930), 1179–1190; " Non-Resonating Autotransformer," *A.I.E.E. Trans.*, *50* (June, 1931), 803–809; K. K. Palueff and J. II. Hagenguth, " Transition of Lightning Waves from One Circuit to Another through Transformers," *A.I.E.E. Trans.*, *51* (September, 1932), 601–615.

L. V. Bewley, " Transient Oscillations in Distributed Circuits with Special Reference to Transformer Windings," *A.I.E.E. Trans.*, *50* (December, 1931), 1215–1233.

II. L. Thomas, " Insulation Stresses in Transformers, with Special Reference to Surges and Electrostatic Shielding," *I.E.E.J.*, *87* (October, 1940), 427–443.

H. O. Stephens, " Shielded Concentric Cylindrical Windings in High-Voltage Power Transformers," *G.E. Rev.*, *45* (December, 1942), 705–709.

II. V. Putman, " Surge-Proof Transformers," *A.I.E.E. Trans.*, *51* (September, 1932), 579–584.

10. MAGNETIC FLUX DISTRIBUTION AND LEAKAGE IMPEDANCE
(CHAPTERS XII, XIII, XIV)

K. B. McEachron, " Magnetic Flux Distribution in Transformers," *A.I.E.E. Trans.*, *41* (1922). 247–261.

O. G. C. Dahl, " Separate Leakage Reactance of Transformer Windings," *A.I.E.E. Trans.*, *44* (1925), 785–791.

A. Boyajian, " Resolution of Transformer Reactance into Primary and Secondary Reactances," *A.I.E.E. Trans.*, *44* (1925), 805–810.

L. F. Blume. editor. *Transformer Engineering* (New York: John Wiley & Sons, 1938). Transformer impedance calculations are discussed in Ch. iv.

B. Hague, *Electromagnetic Problems in Electrical Engineering* (London: Oxford University Press, 1929). Magnetic leakage in transformers is discussed in Ch. xii, and interesting flux plots are shown.

A. L. Morris, " The Influence of Various Factors upon the Leakage Reactance of Transformers," *I.E.E.J.*, *86* (May, 1940), 485–495.

II. B. Dwight and L. S. Dzung, " A Formula for the Reactance of the Interleaved Component of Transformers," *A.I.E.E. Trans.*, *56* (November, 1937), 1368–1371.

H. O. Stephens, " Transformer Reactance and Losses with Nonuniform Windings," *A.I.E.E. Trans.*, *53* (February, 1934), 346–349.

H. L. Cole, " Reactance and Stray Losses in Power Transformers," *A.I.E.E. Trans.*, *53* (February, 1934), 338–342.

G. Kapp, " Ein Beitrag zur Vorausberechnung der Streuung in Transformatoren," *E.T.Z.*, *19* (1898), 244–246. This article and the articles by Rogowski, listed below, are classics on the subject of transformer leakage reactance.

W. Rogowski and K. Simons, " Die Streuung bei Wechselstromtransformatoren und Kommutatormotoren," *E.T.Z.*, *29* (1908), 535–538 and 564–567.

W. Rogowski, " Ueber das Streufeld und den Streuinduktionskoeffizienten eines Transformators mit Scheibenwicklung und geteilten Endspulen," *Mitteilungen über Forschungsarbeiten auf dem Gebiete des Ingenieurwesens*, *71* (Berlin: J. Springer, 1909).

W. Rogowski, " Ueber die Streuung des Transformators," *E.T.Z.*, *31* (1910), 1033–1036 and 1069–1071.

W. Knaak, " Beitrag zur Berechnung der Streuung bei symmetrischen Scheibenwicklungen," *E.T.Z.*, *60* (1939), 47–48.

W. Knaak, " Zusätzliche Verluste durch Streufelder in den Wicklungen von Transformatoren," *E. u. M.*, *57* (1939), 89–93.

E. Roth, " Étude analytique du Champ de Fuites des Transformateurs et des Efforts mécaniques exercés sur les Enroulements," *R.G.E.*, *23* (1928), 773–787.

11. Voltage Regulation, Losses, Rating, and Thermal Characteristics (Chapter XIV)

" American Engineering and Industrial Standards," No. C57: *Proposed American Standards; Transformers, Regulators, and Reactors* (New York: American Standards Association, 1940). This publication represents standardized practices in the United States relating to rating and other characteristics of power and distribution transformers, instrument transformers, and other induction apparatus. Although these proposed standards have not been approved by the American Standards Association, they are the most up-to-date source of information on the subject of transformer standards, and contain data from the standards of the American Institute of Electrical Engineers and the National Electrical Manufacturers Association. A proposed Test Code and Guides for Operation of Transformers are included.

A.I.E.E. Committee on Electrical Machinery (Transformer Subcommittee), " Interim Report on Guides for Overloading Transformers and Voltage Regulators," *A.I.E.E. Trans., 61* (September, 1942), 692–694.

F. J. Vogel and T. K. Sloat, " Emergency Overloads for Oil-Insulated Transformers," *A.I.E.E. Trans., 61* (September, 1942), 669–673.

F. J. Vogel and Paul Narbutovskih, " Hot-Spot Winding Temperatures in Self-Cooled Oil-Insulated Transformers," *A.I.E.E. Trans., 61* (March, 1942). 133–136.

V. M. Montsinger, " Effect of Load Factor on Operation of Power Transformers by Temperature," *A.I.E.E. Trans., 59* (November, 1940), 632–636.

H. V. Putman and W. M. Dann, " Loading Transformers by Copper Temperature," *A.I.E.E. Trans., 58* (October 1939), 504–509.

L. F. Blume, editor, *Transformer Engineering* (New York: John Wiley & Sons, 1938), Ch. ix.

V. M. Montsinger, " Temperature Limits for Short-Time Overloads for Oil-Insulated Neutral Grounding Reactors and Transformers," *A.I.E.E. Trans., 57* (January, 1938), 39–44.

V. M. Montsinger, " Loading Transformers by Temperature," *A.I.E.E. Trans., 49* (April, 1930), 776–790.

L. C. Nichols, " Effects of Overloads on Transformer Life," *A.I.E.E. Trans., 53* (December, 1934), 1616–1621.

12. Economy in Power–System Transformer Applications (Chapter XVI)

D. J. Bolton, *Electrical Engineering Economics* (London: Chapman & Hall, Ltd., 1928). Factors influencing the economic choice of transformers are discussed in Ch. ix.

E. G. Reed, *Essentials of Transformer Practice* (2d ed.; New York: D. Van Nostrand Company, Inc., 1927). Transformer loading and cost problems are discussed in Chs. xxxiii to xxxvi.

J. B. Gibbs, *Transformer Principles and Practice* (New York: McGraw-Hill Book Company, Inc., 1937). The economies that may result from replacement of old transformers by new ones are discussed in Ch. xxii.

R. L. Brown, " When Transformers Get Old," *Factory Management and Maintenance, 96* (1938), 97–99. A discussion of economies resulting from replacement or rebuilding of transformers having high core losses.

S. Lenard, " L'Évaluation et la Capitalisation des Pertes dans les Transformateurs," *Conférence Internationale des Grands Reseaux Électriques, 1* (1937), Art. 101, 1–23. A mathematical treatment of transformer loss costs on a capitalized-value basis, with a comprehensive bibliography of European literature.

R. Winfrey, " Statistical Analyses of Industrial Property Retirements," *Iowa State College Engineering Experiment Station Bulletin 125* (December, 1935)

W. A. Sumner, " Modern Load Cycles Warrant New Transformer Characteristics," *Elec. W., 115* (1941), 1304–1306. Describes improved distribution-transformer characteristics resulting from changed loss-ratio design based on use of core material with high saturation flux density.

M. De Merit, " Distribution Transformer Electrical Performance Characteristics," *Electric Light and Power, 17* (November, 1939), 37–41. Discusses the effects of demand and energy costs and load cycles on the most economical loss product and loss ratio for small distribution transformers.

W. A. Sumner and J. B. Hodtum, " Realigning Transformers with Distribution," *Elec. W., 105* (1935), 1586–1588. Discusses the interrelations among first cost, loss product, loss ratio, and other transformer characteristics.

M. F. Beavers, " Selecting Distribution-Transformer Size," *E.E., 59* (October, 1940), 407–412. Discusses a method of converting a fluctuating daily load curve into a thermally equivalent form, from which permissible overloads can be determined. A bibliography of 17 items is included with this article.

C. H. Lewis and E. H. Snyder, " Distribution Transformer Load Supervising Methods," *Edison Electric Institute Bulletin 5* (1937), 329–333.

R. Rader, " Choosing Transformer Sizes for Distribution Circuits," *Elec. W., 107* (1937), 272–273.

F. H. Ferguson, " Distribution Transformer Loading," *Electrical West, 78* (April, 1937), 33–35.

13.   COMMUNICATION TRANSFORMERS (CHAPTERS XVII, XVIII, XX)

E. A. Guillemin, *Communication Networks*. Vol. I: *The Classical Theory of Lumped Constant Networks* (New York: John Wiley & Sons, 1931). Chapter viii, Secs. 1 and 2, deal with the steady-state solution to the transformer problem and transformer equivalent circuits.

K. S. Johnson, *Transmission Circuits for Telephonic Communication* (6th ed.; New York: D. Van Nostrand Company, Inc., 1939). Chapters vi and vii deal with transformers.

F. E. Terman, *Measurements in Radio Engineering* (New York: McGraw-Hill Book Company, Inc., 1935). The measurement of incremental inductance is discussed on pp. 53–58, and measurements of audio-frequency transformer constants on pp. 195–202.

F. E. Terman, *Radio Engineering* (2d ed.; New York: McGraw-Hill Book Company, Inc., 1937). Input and interstage transformers are discussed in Sec. 45, pp. 188–202, and output transformers in Sec. 57, pp. 296–301.

F. E. Terman and R. E. Ingebretsen, " Output Transformer Response," *Electronics, 9* (January, 1936), 30–32. A discussion of a universal frequency-response chart for output transformers.

F. E. Terman, " Universal Amplification Charts," *Electronics, 10* (June, 1937), 34–35. A discussion of universal frequency-response charts for resistance-capacitance-coupled amplifiers and for input and interstage transformers.

P. W. Klipsch, " Design of Audio-Frequency Amplifier Circuits Using Transformers," *I.R.E. Proc., 24* (February, 1936), 219–232. An article dealing with design of interstage transformers, with a discussion of high-frequency resonance effects and an analysis of the smoothing effects of shunt resistance connected across the secondary.

Glenn Koehler, " The Design of Transformers for Audio-Frequency Amplifiers with Preassigned Characteristics," *I.R.E. Proc., 16* (December, 1928), 1742–1770. This article treats design aspects of input and interstage transformers. A discussion of the effects of interleaving the windings on the leakage inductances and distributed capacitances is included.

L. A. Kelley, " Transformer Design," *Radio Engineering, 14* (December, 1934), 7–11, and *15* (February, 1935), 16–19. The first part of this article deals with transformers having resistance terminations, and the second part with transformers whose function is voltage transformation.

R. C. Hitchcock and W. O. Osbon, " Design of the Output Transformer," *Electronics, 1* (1930), 381–383 and 427–429.

J. G. Story, " Design of Audio-Frequency Input and Intervalve Transformers," *Wireless Engineer, 15* (February, 1938), 69–80.

P. W. Willans, " Low-Frequency Intervalve Transformers," *I.E.E.J., 64* (1926), 1065–1083.

B. S. Cohen, A. J. Aldridge, and W. West, " The Frequency Characteristics of Telephone Systems and Audio-Frequency Apparatus, and Their Measurement," *I.E.E.J., 64* (1926), 1023–1050.

W. L. Casper, " Telephone Transformers," *A.I.E.E. Trans., 43* (1924), 443–455.

J. F. Peters, " High Power Audio Transformers," *A.I.E.E. Trans., 55* (January, 1936), 34–36. Describes a 7.5-kw and a 180-kw audio transformer built for radio broadcast station WLW.

C. H. Crawford and E. J. Thomas, " Silicon Steel in Communication Equipment," *A.I.E.E. Trans., 54* (December, 1935), 1348–1353. Discusses, among other things, the selection of core material for audio-frequency transformers and the effects of core material on frequency characteristics.

A. G. Ganz and A. G. Laird, " Improvements in Communication Transformers," *A.I.E.E. Trans., 54* (December, 1935), 1367–1373. Discusses effects of improved design and materials on frequency characteristics.

C. O. Gibbon, " An Explanation of the Common Battery Anti-Sidetone Subscriber Set," *B.S.T.J., 17* (April, 1938), 245–257.

14. INSTRUMENT TRANSFORMERS (CHAPTER XIX)

B. Hague, *Instrument Transformers* (London: Sir Isaac Pitman & Sons Ltd., 1936).

A.I.E.E. Committee on Protective Devices (Current-Transformer Subcommittee), " Current- and Potential-Transformer Standardization," *A.I.E.E. Trans., 61* (September, 1942), 698–706.

A. C. Schwager, " Current-Transformer Performance Based on Admittance-Vector Locus," *A.I.E.E. Trans., 61* (January, 1942), 26–30.

E. C. Wentz, " A Simple Method for Determination of Ratio Error and Phase Angle in Current Transformers," *A.I.E.E. Trans., 60* (October, 1941), 949–954.

G. Camilli, " New Developments in Current-Transformer Design," *A.I.E.E. Trans., 59* (1940), 835–842.

A. T. Sinks, " Computation of Accuracy of Current Transformers," *A.I.E.E. Trans., 59* (December, 1940), 663–668.

G. Camilli and R. L. Ten Broeck, " A Proposed Method for the Determination of Current-Transformer Errors," *A.I.E.E. Trans., 59* (September, 1940), 547–550.

A. C. Schwager and V. A. Treat, " Shaping of Magnetization Curves and the Zero Error Current Transformer," *A.I.E.E. Trans., 52* (March, 1933), 45–52.

G. Camilli and L. V. Bewley, " Surge Protectors for Current Transformers," *A.I.E.E. Trans., 55* (March, 1936), 254–260.

O. A. Knopp, " Some Applications of Instrument Transformers," *A.I.E.E. Trans., 55* (May, 1936), 480–489. Describes instrument transformers of special design for use in calibration of meters and standard instruments.

J. H. Buchanan, " Design, Construction, and Testing of Voltage Transformers, *I.E.E.J.*, *78* (March, 1936), 292–309.

A. M. Wiggins, " Parallel Operation of Current Transformers for Totalizing Two or More Circuits," *Elec. J.*, *26* (August, 1929), 379–381.

E. J. Boland, " Autotransformer Connections for Varhour Metering," *G.E. Rev.*, *43* (July, 1940), 298–301.

G. Camilli, " Cascade-Type Potential Transformers," *G.E. Rev.*, *39* (February, 1936), 95–99.

15. VOLTAGE–REGULATING TRANSFORMERS (CHAPTER XIX)

L. F. Blume, editor, *Transformer Engineering* (New York: John Wiley & Sons, 1938). The characteristics of tap-changing transformers and associated apparatus are discussed in Chs. x, xi, and xii; applications for voltage-control equipment in Ch. xiii; and phase-angle control in Ch. xiv.

G. H. Landis, " Voltage Regulation and Control in the Development of a Rural Distribution System," *A.I.E.E. Trans.*, *57* (September, 1938), 541–547. A discussion of the fields of application of step voltage regulators, boosters, and capacitors for voltage regulation on rural distribution lines.

J. E. Clem, " Equivalent Circuit Impedance of Regulating Transformers," *A.I.E.E. Trans.*, *58* (1939), 871–873.

J. E. Hobson and W. A. Lewis, " Regulating Transformers in Power-System Analysis," *A.I.E.E. Trans.*, *58* (1939), 874–883.

W. J. Lyman and J. R. North, " Application of Large Phase-Shifting Transformer on an Interconnected System Loop," *A.I.E.E. Trans.*, *57* (October, 1938), 579–587.

L. F. Blume, " Characteristics of Load Ratio Control Circuits for Changing Transformer Ratio under Load," *A.I.E.E. Trans.*, *51* (December, 1932), 952–956.

A. Palme, " Transformers with Load Ratio Control," *A.I.E.E. Trans.*, *50* (March, 1931), 172–177.

H. B. West, " Tap Changing under Load for Voltage and Phase-Angle Control," *A.I.E.E. Trans.*, *49* (July, 1930), 839–845.

L. H. Hill, " Transformer Tap Changing under Load," *A.I.E.E. Trans.*, *46* (1927), 582–589.

L. F. Blume, " Characteristics of Interconnected Power Systems as Affected by Transformer Ratio Control," *A.I.E.E. Trans.*, *46* (1927), 590–598.

H. C. Albrecht, " Transformer Tap Changing under Load," *A.I.E.E. Trans.*, *44* (1925), 581–585.

16. HARMONIC PHENOMENA IN THREE–PHASE CIRCUITS (CHAPTER XXIII)

R. R. Lawrence, *Principles of Alternating Currents* (2d ed.; New York: McGraw-Hill Book Company, Inc., 1935). The fundamental theory of harmonic phenomena in polyphase circuits is discussed in Ch. xi.

O. G. C. Dahl, *Electric Circuits, Theory and Applications.* Vol. I: *Short-Circuit Calculations and Steady-State Theory* (New York: McGraw-Hill Book Company, Inc., 1928). Harmonic phenomena in power systems are discussed in Ch. viii.

L. F. Blume, editor, *Transformer Engineering* (New York: John Wiley & Sons, 1938). Harmonic phenomena in three-phase transformer banks are discussed in Chs. vii and xvi.

O. G. C. Dahl, " Transformer Harmonics and Their Distribution," *A.I.E.E. Trans.*, *44* (1925), 792–805.

T. H. Morgan, C. A. Bairos, and G. S. Kimball, " The Triple-Harmonic Equivalent Circuit in Three-Phase Power Transformer Banks," *A.I.E.E. Trans.*, *52* (March, 1933), 64–71.

T. C. Lennox, "Circulation of Harmonics in Transformer Circuits," *A.I.E.E. Trans.*, *45* (1926), 708–710.

C. T. Weller, "Experiences with Grounded-Neutral, Y-Connected Potential Transformers on Ungrounded Systems," *A.I.E.E. Trans.*, *50* (March, 1931), 299–316.

A. Boyajian and O. P. McCarty, "Physical Nature of Neutral Instability," *A.I.E.E. Trans.*, *50* (March, 1931), 299–316.

C. W. La Pierre, "Theory of Abnormal Line-to-Neutral Transformer Voltages," *A.I.E.E. Trans.*, *50* (March, 1931), 328–342.

K. E. Gould, "Instability in Transformer Banks," *A.I.E.E. Trans.*, *46* (1927), 676–682.

H. S. Shott and H. A. Peterson, "Criteria for Neutral Stability of Transformer Circuits," *A.I.E.E. Trans.*, *60* (November, 1941), 997–1002.

B. G. Gates, "Neutral Inversion in Power Systems," *I.E.E.J.*, *78* (1936), 317–325.

A. Boyajian, "Inversion Currents and Voltages in Auto-Transformers," *A.I.E.E. Trans.*, *49* (April, 1930) 810–818.

J. L. Cantwell, "Frequency Tripling Transformers," *A.I.E.E. Trans.*, *55* (July, 1936), 784–790.

17. UNBALANCED CONDITIONS IN THREE-PHASE TRANSFORMER BANKS
(CHAPTER XXIV)

L. F. Blume, editor, *Transformer Engineering* (New York: John Wiley & Sons, 1938), Chs. vi and vii.

W. V. Lyon, *Applications of the Method of Symmetrical Components* (New York: McGraw-Hill Book Company, Inc., 1937), Ch. v.

C. F. Wagner and R. D. Evans, *Symmetrical Components* (New York: McGraw-Hill Book Company, Inc., 1933), Ch. vi.

O. G. C. Dahl, *Electric Circuits, Theory and Applications*. Vol. I: *Short-Circuit Calculations and Steady-State Theory* (New York: McGraw-Hill Book Company, Inc., 1928), Ch. iv.

A. N. Garin, "Zero-phase-sequence Characteristics of Transformers." Part I: "Sequence Impedances of a Static Symmetrical Three-Phase Circuit and of Transformers," *G.E. Rev.*, *43* (March, 1940), 131–136; Part II: "Equivalent Circuits for Transformers," *G.E. Rev.*, *43* (April, 1940), 174–179.

18. MULTICIRCUIT TRANSFORMERS (CHAPTER XXVII)

L. F. Blume, editor, *Transformer Engineering* (New York: John Wiley & Sons, 1938), Ch. v.

O. G. C. Dahl, *Electric Circuits, Theory and Applications*. Vol. I: *Short-Circuit Calculations and Steady-State Theory* (New York: McGraw-Hill Book Company, Inc., 1928). Multicircuit transformers are discussed in Ch. ii.

F. M. Starr, "Equivalent Circuits," *A.I.E.E. Trans.*, *51* (June, 1932), 287–298. A comprehensive paper dealing with equivalent circuits for inductively coupled circuits.

F. M. Starr, "An Equivalent Circuit for the Four-Winding Transformer," *G.E. Rev.*, *36* (March, 1933), 150–152.

A. Boyajian, "Theory of Three-Circuit Transformers," *A.I.E.E. Trans.*, *43* (1924), 508–528.

A. Boyajian, "Progress in Three-Circuit Theory," *A.I.E.E. Trans.*, *52* (September, 1933), 914–917. A treatment of three-phase banks with interconnections among the phases.

A. N. Garin and K. K. Paluev, "Transformer Circuit Impedance Calculations,"

A.I.E.E. Trans., *55* (June, 1936), 717–730. Describes a general method for determination of the short-circuit impedances of any arrangement of transformer windings.

J. E. Clem, " An Exact Formula for Transformer Regulation," *A.I.E.E. Trans.*, *55* (May, 1936), 466–471. Gives voltage-regulation formulas for two- and three-winding transformers.

R. D. Evans, " Simplified Computation of Voltage Regulation with Four-Winding Transformers," *E.E.*, *58* (October, 1939), 420–421.

G. W. Bills and C. A. MacArthur, " Three-Winding Transformer Ring-Bus Characteristics," *A.I.E.E. Trans.*, *61* (December, 1942), 848–849. A study of a bus arrangement for reducing fault currents and improving the transient stability of a large power system.

D. D. Chase and A. N. Garin, " Split Winding Transformers," *A.I.E.E. Trans.*, *53* (June, 1934), 914–922. A discussion of the application of and advantages derived from the use of split-winding transformers for separating bus sections in large generating stations.

H. P. St. Clair, " The Use of Multiwinding Transformers with Synchronous Condensers for System Voltage Regulation," *A.I.E.E. Trans.*, *59* (1940), 212–217.

19. PHASE TRANSFORMATION; PARALLEL OPERATION OF TRANSFORMERS
(CHAPTERS XXVIII, XXIX)

L. F. Blume, editor, *Transformer Engineering* (New York: John Wiley & Sons, 1938). Transformer arrangements for three-phase to two-phase transformation are discussed on pp. 219–230. Autotransformer connections for phase transformation are discussed on pp. 252–257.

J. B. Gibbs, *Transformer Principles and Practice* (New York: McGraw-Hill Book Company, Inc., 1937). Parallel operation of transformers is discussed in Ch. x, and phase transformation in Chs. xi and xii.

M. Macferran, " Parallel Operation of Transformers Whose Ratios of Transformation Are Unequal," *A.I.E.E. Trans.*, *49* (January, 1930), 125–131.

F. M. Starr, " Operation of Load-Ratio Control Transformers Connected in Parallel and in Networks," *A.I.E.E. Trans.*, *60* (1941), 1274–1280.

A. C. Seletzky, " Load Loci for Transformers in Parallel," *A.I.E.E. Trans.*, *56* (November, 1937), 1379–1384.

F. M. Starr and D. L. Beeman, " Load-Balancing Transformers for A-C Secondary Network Systems," *G.E. Rev.*, *43* (November, 1940), 452–458.

R. C. Ghen and C. F. Avila, " Balancing Parallel Transformers," *Elec. W.*, *116* (1941), 1406–1407.

Index

www.ingramcontent.com/pod-product-compliance
Lightning Source LLC
Chambersburg PA
CBHW070708220326
41598CB00026B/3674